Hyperthermia In Cancer Therapy

Hyperthermia In Cancer Therapy

Edited by
F. Kristian Storm, M.D.
Associate Professor of Surgery
Division of Oncology
UCLA School of Medicine
Los Angeles, California

RC271
T5
H97
1983

G. K. Hall Medical Publishers
Boston, Massachusetts

Copyright © 1983 by G. K. Hall & Co.

G. K. Hall Medical Publishers
70 Lincoln Street
Boston, MA 02111

All rights, including of translation into other languages, reserved. Photomechanical reproduction (photocopy, microcopy) of this book or parts thereof without special permission of the publisher is prohibited.

83 84 85 86 / 4 3 2 1

Main entry under title:

Hyperthermia in cancer therapy.

Bibliography.
Includes index.
1. Cancer—Treatment. 2. Thermotherapy. I. Storm, F. Kristian. [DNLM: 1. Neoplasms—Therapy. 2. Fever Therapy. QZ 266 H9985]
RC271.T5H97 1983   616.99'40632   82–11710
ISBN 0-8161-2170-2

The authors and publisher have worked to ensure that all information in this book concerning drug dosages, schedules, and routes of administration is accurate at the time of publication. As medical research and practice advance, however, therapeutic standards may change. For this reason, and because human and mechanical errors will sometimes occur, we recommend that our readers consult the *PDR* or a manufacturer's product information sheet prior to prescribing or administering any drug discussed in this volume.

Designed by Jack Schwartz. Copyedited by Susan Glick under the direction of Lucie Ferranti. Produced by Carole Rollins and Sandra McLean. Composed in 10 pt Baskerville by The Saybrook Press.

Contributors

Luigi Aloe, M.D.
Regina Elena Institute for Cancer Research
Rome, Italy

E. Ronald Atkinson, Ph.D.
American Hospital Supply Corporation
Evanston, Illinois

Haim I. Bicher, M.D., Ph.D.
Department of Therapeutic Biology
Henry Ford Hospital
Detroit, Michigan

Joan M. Bull, M.D.
Division of Cancer Treatment, Medical Branch
National Cancer Institute
Bethesda, Maryland

Stuart K. Calderwood, Ph.D.
Cancer Research Unit
University Department of Clinical Biochemistry
Royal Victoria Infirmary
Newcastle-Upon-Tyne, England

Renato Cavaliere, M.D.
Regina Elena Institute for Cancer Research
Rome, Italy

Thomas C. Cetas, Ph.D.
Department of Radiology, Radiation Oncology Division
Department of Electrical Engineering
University of Arizona Health Sciences Center
Tucson, Arizona

Chung-Kwang Chou, Ph.D.
Bioelectromagnetics Research Laboratory
School of Medicine
University of Washington
Seattle, Washington

Douglas A. Christensen, Ph.D.
Department of Electrical Engineering and Bioengineering
University of Utah
Salt Lake City, Utah

Stephen F. Cleary, Ph.D.
Department of Biophysics
Medical College of Virginia
Virginia Commonwealth University
Richmond, Virginia

John A. Dickson, M.D., Ph.D.
Cancer Research Unit
University Department of Clinical Biochemistry
Royal Victoria Infirmary
Newcastle-Upon-Tyne, England

Franco Di Filippo, M.D.
Regina Elena Institute for Cancer Research
Rome, Italy

Robert S. Elliott, Ph.D.
Department of Electrical Sciences and Engineering
University of California at Los Angeles
Los Angeles, California

Eugene W. Gerner, Ph.D.
Department of Radiology, Radiation Oncology Division
University of Arizona Health Sciences Center
Tucson, Arizona

Leo E. Gerweck, Ph.D.
Department of Radiation Medicine
Massachusetts General Hospital
Harvard Medical School
Boston, Massachusetts

Beppino C. Giovanella, Ph.D.
The Stehlin Foundation for Cancer Research
Houston, Texas

Arthur W. Guy, Ph.D.
Bioelectromagnetics Research Laboratory
School of Medicine
University of Washington
Seattle, Washington

William H. Harrison, B.A.
Department of Electrical Sciences and Engineering
University of California at Los Angeles
Los Angeles, California

Kurt J. Henle, Ph.D.
Medical Research Service
Veterans Administration Medical Center
Little Rock, Arkansas

Fred W. Hetzel, Ph.D.
Department of Therapeutic Radiology
Henry Ford Hospital
Detroit, Michigan

Rakesh K. Jain, Ph.D.
Department of Chemical Engineering
Carnegie-Mellon University
Pittsburgh, Pennsylvania

Contributors

Padmakar P. Lele, M.D., Ph.D.
Department of Mechanical Engineering
Massachusetts Institute of Technology
Cambridge, Massachusetts

Michael R. Manning, M.D.
Department of Radiology, Radiation Oncology Division
University of Arizona Health Sciences Center
Tucson, Arizona

Bruno Mondovi, M.D.
Institute of Applied Biochemistry
University of Rome
Rome, Italy

Giorgio Monticelli, M.D.
1st Orthopedic Clinic
University of Rome
Rome, Italy

Guido Moricca, M.D.
Regina Elena Institute for Cancer Research
Rome, Italy

Donald L. Morton, M.D.
Department of Surgery, Division of Oncology
University of California at Los Angeles
Los Angeles, California

Pier Giorgio Natali, M.D.
Regina Elena Institute for Cancer Research
Rome, Italy

Jens Overgaard, M.D.
The Institute of Cancer Research, Radiumstationen
Aarhus, Denmark

Leon C. Parks, M.D.
Department of Surgery
University of Mississippi Medical Center
Jackson, Mississippi

Alessandro Rossi-Fanelli, M.D.
Institute of Biological Chemistry
University of Rome
Rome, Italy

Taljit S. Sandhu, Ph.D.
Department of Therapeutic Radiology
Henry Ford Hospital
Detroit, Michigan

Francesco Saverio Santori, M.D.
1st Orthopedic Clinic
University of Rome
Rome, Italy

Sudhir A. Shah, M.D.
Cancer Research Unit
University Department of Clinical Biochemistry
Royal Victoria Infirmary
Newcastle-Upon-Tyne, England

George V. Smith, M.D.
Department of Surgery
University of Mississippi Medical Center
Jackson, Mississippi

Chang W. Song, Ph.D.
Department of Therapeutic Radiology,
    Section of Radiobiology
University of Minnesota
Minneapolis, Minnesota

F. Kristian Storm, M.D.
Department of Surgery, Division of Oncology
University of California at Los Angeles
Los Angeles, California

Antonio Varanese, M.D.
Regina Elena Institute for Cancer Research
Rome, Italy

Contents

FOREWORD ix
PREFACE xi

CHAPTER 1
Background, Principles, and Practice 1
   F. Kristian Storm

CHAPTER 2
Bioheat Transfer: Mathematical Models of Thermal Systems 9
   Rakesh K. Jain

CHAPTER 3
Arrhenius Analysis of Thermal Responses 47
   Kurt J. Henle

CHAPTER 4
Thermosensitivity of Neoplastic Cells In Vitro 55
   Beppino C. Giovanella

CHAPTER 5
Thermosensitivity of Neoplastic Tissues In Vivo 63
   John A. Dickson
   Stuart K. Calderwood

CHAPTER 6
Thermotolerance 141
   Eugene W. Gerner

CHAPTER 7
Histopathologic Effects of Hyperthermia 163
   Jens Overgaard

CHAPTER 8
Blood Flow in Tumors and Normal Tissues in Hyperthermia 187
   Chang W. Song

CHAPTER 9
Physiology and Morphology of Tumor Microcirculation in Hyperthermia 207
   Haim I. Bicher
   Fred W. Hetzel
   Taljit S. Sandhu

CHAPTER 10
Thermometry and Thermography 223
   Douglas A. Christensen

CHAPTER 11
Hyperthermia Techniques and Instrumentation 233
   E. Ronald Atkinson

CHAPTER 12
Physical Models (Phantoms) in Thermal Dosimetry 257
   Thomas C. Cetas

CHAPTER 13
Physical Aspects of Localized Heating by Radiowaves and Microwaves 279
   Arthur W. Guy
   Chung-Kwang Chou

CHAPTER 14
Physical Aspects of Localized Heating by Magnetic-Loop Induction 305
   F. Kristian Storm
   William H. Harrison
   Robert S. Elliott
   Donald L. Morton

CHAPTER 15
Animal and Clinical Studies with Microwave and Radiowave Hyperthermia 315
   F. Kristian Storm
   Donald L. Morton

CHAPTER 16
Physical Aspects and Clinical Studies with Ultrasonic Hyperthermia 333
   Padmakar P. Lele

CHAPTER 17
Regional Perfusion Hyperthermia 369
   Renato Cavaliere
   Bruno Mondovi
   Guido Moricca
   Giorgio Monticelli
   Pier Giorgio Natali
   Francesco Saverio Santori
   Franco Di Filippo
   Antonio Varanese
   Luigi Aloe
   Alessandro Rossi-Fanelli

Contents

CHAPTER 18
Systemic Hyperthermia: Background and Principles — 401
Joan M. Bull

CHAPTER 19
Systemic Hyperthermia by Extracorporeal Induction: Techniques and Results — 407
Leon C. Parks
George V. Smith

CHAPTER 20
Thermoradiotherapy: Molecular and Cellular Kinetics — 447
Leo E. Gerweck

CHAPTER 21
Interstitial Thermoradiotherapy — 467
Michael R. Manning
Eugene W. Gerner

CHAPTER 22
Clinical Thermoradiotherapy — 479
Haim I. Bicher
Taljit S. Sandhu
Fred W. Hetzel

CHAPTER 23
Immunologic Aspects of Hyperthermia — 487
John A. Dickson
Sudhir A. Shah

CHAPTER 24
Bioeffects of Microwave and Radiofrequency Radiation — 545
Stephen F. Cleary

INDEX — 567

Foreword

Hyperthermia at temperatures above 41°C has been used sporadically as an agent for cancer therapy since the early 1900s. Although there were encouraging results, there was not enough consistency to encourage continued and sustained efforts in applying hyperthermia either by itself or combined with radiation or other agents. During the last few years, however, results from studies of cell cultures and animals, as well as from a few preliminary clinical trials, have provided some enthusiasm for systematically investigating the potential use of hyperthermia in cancer therapy.

Results in vitro have provided a rationale for considering the use of hyperthermia in cancer therapy. First, hyperthermia at temperatures above 41°C kills mammalian cells and sensitizes them to ionizing radiation, with the degree of killing and radiosensitization varying greatly with only a 0.5°C change in temperature. Therefore, with inadequate temperature control and measurement in clinical studies in the past, consistent results could not have been expected. Second, hyperthermia selectively kills and radiosensitizes cells that are relatively resistant to ionizing radiation; these populations are those in the process of synthesizing DNA and those existing in the hypoxic compartment of tumors. These hypoxic cells are probably also at a low pH and under nutrient or energy deprivation, all of which sensitize cells to hyperthermia. Therefore, the action of hyperthermia by itself or interacting with ionizing radiation should selectively kill those tumor cells surviving a dose of radiation. This same rationale has been offered for considering the use of high linear energy transfer (LET) radiation and electronic affinic compounds in radiation therapy. Third, hyperthermia has been shown to eliminate or reduce recovery from sublethal and potentially lethal radiation damage; this may have a selective effect on tumor cells, especially those existing in the G_0 compartment. Finally, the toxicity of electron affinic compounds for hypoxic cells and the toxicity of several chemotherapeutic agents can be enhanced greatly by hyperthermia. Thus, basic studies of cell cultures indicate that there are reasons to believe that hyperthermia may enhance the therapeutic efficacy of radiation and chemotherapeutic agents used in clinical practice. Furthermore, since the toxic effects of radiation, radiosensitizers, or chemotherapeutic agents are greatly enhanced by hyperthermia, and since heat and radiation act in a complementary way (e.g., killing of hypoxic and S-phase cells), the combined modality approach appears to offer the greatest potential for use of hyperthermia.

Studies in animals and humans during the last few years have indicated that there may be some merit to the rationale proposed above. For example, hyperthermia has been shown to have a selective effect on the radiation response of chronically hypoxic tumor cells and, under certain conditions, can improve the therapeutic ratio (i.e., the radiation response of the tumor relative to that of normal tissue). As predicted from studies of cell cultures, however, variations in the sequence between administration of hyperthermia and radiation may alter greatly the effectiveness of hyperthermia for improving the therapeutic ratio. Beneficial clinical results are still largely anecdotal but do suggest that careful clinical studies need to be carried out in which variations in sequencing are considered as the temperature profiles in tumors and normal tissues are monitored. Also, physiologic changes, such as changes in oxygen tension and pH, occur from hyperthermic treatment and should be considered in relation to the effectiveness of hyperthermic treatments. Preliminary data from animal and human studies indicate that hyperthermia may indeed enhance the effectiveness of radiation and chemotherapy, but these data also indicate that various parameters must be monitored and carefully evaluated to determine the reasons for both successes and failures.

Issues to be resolved or confirmed are: the mechanisms of action for heat killing and heat radiosensitization; the importance of the particular temperature used and its duration; the importance of sequence (heat before radiation or radiation before heat); whether heat and the second agent of radiation or chemotherapy should be separated by a long interval such that the two agents act independently, or by a short interval such that the two agents interact additively or synergistically; the best interval between combined heat and x-ray fractions; how to heat the tumor with precise temperature control; the recognition and possible use of thermal tolerance for heat alone and possibly for heat combined with radiation, especially if tolerance is reduced for tumor cells at low pH; heat effects on the immunologic response and on destroying or spreading metastases; and physiologic changes that are especially important for heat before irradiation and that may involve changes in pH, nutrients, oxygen consumption, blood flow, and oxygen concentration. For example, when heat is delivered before radiation, the frequently observed decrease in oxygen concentration might be expected to increase the hypoxic radioresistant fraction in the tumor. The most important observation to be made, however, is

Foreword

whether there indeed will be a differential effect between tumor and normal tissue when both are at the same temperature. If there is no differential effect, then the tumor will have to be heated selectively to achieve any therapeutic efficacy. These studies of therapeutic gain or efficacy are most important for fractionated doses and when tumor cure is compared with late effect in normal tissue.

The immediate challenge is to develop and make available heating and thermometry equipment that is essential for conducting definitive experiments in small and large animals and in humans. Then, as basic mechanisms involved in heat inactivation, heat radiosensitization, and heat-drug interactions are understood, results obtained in vitro and in small and large animals, as well as in phase I and II clinical trials, should lead to meaningful phase III trials. It is hoped that phase III clinical trials will demonstrate that hyperthermia, when applied properly, is a beneficial agent for cancer therapy: that it will provide an improvement in the therapeutic index defined as a percentage of a group of treated cancer patients who remain both free of recurrence and free of severe complications over a given period of time.

This volume reviews and addresses the questions and issues mentioned above. The authors hope that its readers will gain an insight into both the problems and the potential benefits associated with hyperthermia in cancer therapy.

William C. Dewey, Ph.D.
Department of Radiology and
Radiation Biology
Colorado State University
Fort Collins, Colorado

Preface

Hyperthermia in Cancer Therapy is a succinct state-of-the-art text on thermal therapy, its history, its current status, and its potential use. No other medical field has required the interaction of such a multitude of experts for its development—mathematicians, engineers, cell biologists, biophysicists, pharmaco-kineticists, radiation therapists, medical oncologists, computer programmers, and oncologic surgeons. To this end, the text contains over 2000 references from many of these specialists.

There will be 750,000 new cases of cancer diagnosed this year in the United States, and 450,000 patients will die of this disease. Of the one in four Americans who will be afflicted by cancer in a lifetime, only one in three will be cured by present methods of therapy that include surgery, radiation therapy, chemotherapy, and immunotherapy. This book was written for all those investigators who have given and who will give a significant portion of their life to the development of hyperthermia as a fifth treatment of cancer. All contributors are recognized national and international authorities. Each author reviews the principles of hyperthermia in his or her own way, and in some instances redundancy has been allowed for comprehension. Indeed, the writing of this book has served to define more clearly those areas of agreement and disagreement as the strategies for treatment for cancer patients improve and evolve.

This book is dedicated to Donald L. Morton, M.D., my teacher, colleague, and friend, who has helped me to understand the many rewards of scientific investigation. This book is dedicated also to my wife, Patty, for her unselfish and continued encouragement, to my parents, Dorothy and Fred Storm, whose love and sacrifices have been my inspiration, and to my young daughters, Sandy, Lori, and Darlene, for their gift of time to complete the necessary editing.

F. Kristian Storm, M.D.

In Memory

Curtis C. Johnson, Ph.D.
1932–1978

Professor Curtis C. Johnson was a dedicated scholar whose research efforts had a direct impact on advancing the state of knowledge of hyperthermia. Educated at the California Institute of Technology and Stanford University, he held industrial positions before joining the University of Utah faculty in electrical engineering in 1961. After an intervening faculty appointment at the University of Washington, he rejoined the University of Utah in 1971, and was professor and chairman of the department of bioengineering there until his untimely death in March 1978. Dr. Johnson's research interests were in ultrasonics, optics, semiconductor devices, and medical instrumentation, but his greatest scientific and leadership contributions were in the field of biological effects of electromagnetic waves. Here he defined and elucidated many aspects of this important subject, including the application and control of microwave energy to generate tissue hyperthermia. His observations will stimulate investigators for generations to come. The medical community will be forever grateful.

Douglas Christensen, Ph.D.

William Lyman Caldwell, M.D.
1924–1979

The untimely death of William Lyman Caldwell was a great loss to the radiotherapy community and an even greater loss to the hyperthermia community. He keenly recognized the current needs and the potential of this modality. He first launched his efforts in this field in 1977, charged with an unfaltering zeal to explore all scientific, clinical, and fiscal avenues to allow the field of hyperthermia a fair chance to make a significant impact on cancer treatments.

He was born on November 12, 1924, in Honolulu, Hawaii. After receiving his medical degree from Stanford University in 1955, and following his internship at San Francisco General Hospital, he became a member of the United States Army Medical Corps, serving as a resident in general practice at Fort Knox, Kentucky and in radiology at Stanford University Hospitals. He was certified by the American Board of Radiology in 1969.

Dr. Caldwell spent a decade serving in various capacities at major facilities, such as the United States Army Hospital in San Juan, Puerto Rico; the Royal Marsden Hospital in Sutton, Surrey; the Institute of Cancer Research in London; and the Vanderbilt University Hospital and School of Medicine in Nashville, Tennessee. In 1971 he began his affiliation with the University of Wisconsin Hospitals as Director of the Division of Radiation Oncology. In 1973 he became an Associate Director of the Wisconsin Clinical Cancer Center and contributed significantly to the process of integrating the various cancer treatment modalities with new ways to improve cancer treatment.

Dr. Caldwell served on a number of national scientific committees related to cancer and to radiation therapy. In recognition of his contributions as a member of the National Cancer Institute clinical cancer program, project review committee, he was appointed chairman of that committee in May 1979. He was recognized as an authority on radiation therapy of the bladder, and was author of a textbook, *Cancer of the Urinary Bladder*.

Dr. Caldwell's death saddened us indeed. The loss of his constructive impact on the field of radiation oncology, as well as on the hyperthermia community, will be deeply felt. His dynamic participation in current and future efforts to vitalize multidisciplinary cancer management will certainly be missed. He was dedicated to teaching, to scientific research, and to cancer patients, and was an extremely sensitive and kind person. He treated his fellow workers with respect and valued their contributions and ideas. He made life happy, harmonious, and enjoyable, not only for himself, but for those around him.

Bhudatt Paliwal, Ph.D.

Fire will succeed when all other methods fail.
                                Hippocrates, 400 B.C.

CHAPTER 1

Background, Principles, and Practice

F. Kristian Storm

Early History
Selective Thermosensitivity
Systemic Hyperthermia
Thermoradiotherapy
Thermochemotherapy
Immune Correlates
Selective Tumor Heating
Electromagnetic Hyperthermia
Temperature Measurement
Hyperthermia Treatment Schedules
Unresolved Questions

EARLY HISTORY

Hyperthermia, from the Greek *hyper*, meaning beyond, above, over, or excessive, and *therme*, heat, has been applied to various ailments, including cancer, since ancient times. The early translations of Ramajama (2000 B.C.), Hippocrates (400 B.C.), and Galen (200 A.D.) record the use of *ferrum candens* (red-hot irons) and chemical caustics in the treatment of small, nonulcerating cancers. After the Renaissance there were numerous case reports of spontaneous tumor regressions in patients with erysipelas, smallpox, influenza, tuberculosis, and malaria. A factor common to all these illnesses was an infectious fever of about 40°C, which lasted for several days.

The first documented evidence that elevated temperatures might have a selective effect on tumors usually is ascribed to Busch, who in 1866 reported the disappearance of a histologically verified sarcoma of the face after two attacks of erysipelas. Many reports from the late nineteenth and early twentieth centuries describe the regression of primary and secondary tumors after infection with pyrogenic bacteria (Bruns 1887; Coley 1893, 1896; Rohdenburg 1918). In 1896, Coley introduced febrile therapy with "mixed toxins of erysipelas and *B. prodigensus*" for the treatment of malignant tumors and reported a disease-free survival of one to seven years in 3 of 17 inoperable carcinomas and 7 of 17 inoperable sarcomas. Interestingly, these original works stated that for the toxin to be effective a fever of 39°C to 40°C had to be maintained for several days. Nauts, Fowler, and Bogatko (1953) later repeated Coley's work and found 25 of 30 selected patients with soft tissue sarcoma, lymphosarcoma, and carcinoma of the cervix and breast alive and disease-free at 10 years. The earlier results achieved with the original highly pyrogenic agents were never equaled with lesser agents, however, and it has been suggested that the fever itself, rather than the toxin, was the tumoricidal agent. F. Westermark in 1898 placed water-circulating cisterns at 42°C to 44°C into inoperable carcinomas of the uterus for 48 hours and observed palliative shedding of many tumors, although subsequent healing was rare. Also in the 1890s, d'Arsonval, Telsa, and others simultaneously reported upon the healing effects of high frequency currents on deep tissues. Using this information, Nagelschmidt in 1926 coined the term *diathermy*, to deep heat, and the use of electromagnetic heating of tumors rapidly followed.

SELECTIVE THERMOSENSITIVITY

The modern era of hyperthermia investigation has since provided mounting evidence to support a hypothesis that cancer cells are more selectively sensitive to heat than are normal cells. In 1927, N. Westermark heated Flexner-Jobling carcinoma and Jensen's sarcoma in rats using diathermy and found total tumor regression after 180 minutes at 44°C or 90 minutes at 45°C, while normal tissues were not damaged under similar conditions. He was the first to introduce the concepts of dose-time thermal effect and histopathologic examination of treated tumors for evidence of heat-induced necrosis. Stevenson (1919) and Rohdenburg and Prime (1921) also were among the first to investigate the relationship between treatment times and temperature of animal tumors. Crile (1961, 1962) extended the work of Westermark and confirmed a dose-response relationship between heat and tumor cure. Above 42°C, he found that for each 1°C increase in tumor temperature, the exposure time needed for tumor cure in mice could be halved. Crile also discovered the phenomenon of delayed cell killing after hyperthermia; tumors excised immediately after heating were able to grow in recipient animals on transplantation, but tumors removed four hours after heating did not grow when transplanted. In 1961, Crile heated mice feet implanted with sarcoma 180 in a water-bath for various times and temperatures and found that 70 minutes at 43°C "cured" most of the tumors, while only one of six normal feet was lost after 384 minutes at the same temperature. In 1963, Crile found that preheating tumors could make them less sensitive to a further heating dose (thermal tolerance), and that clamping the tumor blood supply increased heat sensitivity. In 1966, Bender and Schramm studied the effects of 30 minutes of heat on human cells in vitro and found that tumor cell lines were killed at 45°C to 46°C, whereas normal cells were killed at one degree higher temperature. In 1967, Cavaliere caused irreversible damage to Novikoff hepatoma cells by incubation at 42°C to 44°C, which did not occur in either normal or regenerating rat liver cells, or in minimal deviation hepatoma 5123. Two years later (1969), co-worker Mondovi reported that the selective inhibitory effect of high temperature on tumor cells was paralleled by a higher sensitivity of these cells to the polyene antibiotic filipin as well as ethanol, and that ethanol inhibition was also enhanced by higher temperature. The effects were tentatively ascribed to an alteration in cell and/or lysosomal membrane permeability. In clinical trials, these investigators performed regional limb perfusions with blood prewarmed to 41.5°C to 43.5°C in 22 patients with large, recurrent, or single metastatic cancers localized to the extremity. All gross tumor disappeared in 10 patients, although three had disease recurrence, five had regression, three failed to respond, and four were not evaluable. The complication rate was high, with six deaths and three immediate amputations; however, 8 of 22 patients had complete massive tumor necrosis. In 1971, Muckel and Dickson treated highly malignant squamous carcinoma VX-2 in rabbit extremities with water-bath hyperthermia at 42°C. After three one-hour applications there was widespread tumor necrosis, subsequently replaced by connective tissue, and survival was prolonged by 18 months in 50% of the treated animals. All control animals died within 10 weeks. In the next year, Overgaard and Overgaard (1972) reported permanent cures of transplantable mouse mammary carcinoma using shortwave-induced hyperthermia in the range of 41.5°C to 43.5°C. A definite relationship could be established between temperature and exposure time. They found that the heat induced distinct histologic changes in the tumor cells but not in stromal or vascular cells within the tumor, or in normal surrounding tissue. They saw rapid, autolytic disintegration of heat-damaged tumor cells, followed by a marked increase in connective tissue stroma and scar formation. Giovanella, Lohman, and Heidelberger in 1970, and Giovanella and associates in 1973, studied the effects of hyperthermia on normal (embryonic) and neoplastic (methyl-cholanthrene sarcoma) mesenchymal cells derived from C57BL/6 mice. They found that 95% of all cultures of tumor-derived and tumor-producing cells died after two hours at 42.5°C, whereas only 43% of all cultured normal and non-tumor-producing cells died under similar conditions. When a cell subline derived from a non-tumor-producing line acquired high tumor-producing ability, it also acquired greater thermosensitivity. These results suggested that the acquisition of malignant potential, both in vivo and in vitro, is accompanied by decreased thermotolerance. In 1976, Mendecki, Friedenthal, and Botstein applied microwave-induced hyperthermia of 43°C for 45 minutes to superficially implanted mammary carcinoma in C3H mice and found that the neoplasm completely disappeared after two treatments. All mice in the treated group survived four months, while all nontreated controls died within 4 weeks. In 1977, Dickson and Shah and Dickson with others reported shortwave heating of VX-2 carcinomas in rabbit extremities at 47°C for 30 minutes. Skin and normal muscle remained 3°C to 4°C below minimal tumor temperature. After one application, 7 of 10 tumors regressed completely. In a similar experiment, Marmor, Hahn, and Hahn (1977) found that shortwave heating of EMT-6 sarcomas in mice for five minutes at 44°C resulted in cure of nearly 50% of the tumors.

SYSTEMIC HYPERTHERMIA

Clinical trials also support the potential usefulness of systemic (total-body) hyperthermia. In 1935, Warren produced systemic hyperthermia in the range of 41.4°C to 43°C in a heating cabinet using a combination of carbon filament lamps and radiofrequency diathermy. He reported an immediate improvement in constitutional symptoms, as well as tumor regression of varying degrees, with remissions from one to six months. Pettigrew and associates in 1974 described treatment of 38 terminal cancer patients who were immersed in molten wax for total-body hyperthermia at 41.8°C for an average of four hours. An objective response—weight gain or pain relief plus measured tumor regression or histologic evidence of necrosis—was seen in 18 of 38 cases. Four patients died from disseminated intravascular coagulation. Larkin, Edwards, and Smith in 1976 reported their experience with total-body hyperthermia applied by a water-circulating suit. Nineteen patients were maintained at 41.5°C to 42°C for two to five hours, with an objective tumor response of about 70%. Complications included one death, transient cardiac arrhythmias in 15%, superficial burns in 15%, and transient respiratory distress in 11% of patients. These complications were attributed to the many hours of anesthesia time initially required to raise and maintain body temperature in these critically ill patients. Since these pioneering efforts, heat therapy techniques have been refined. Bull recently defined the physiologic effects of total-body hyperthermia as a single agent and in combination with chemotherapy, and Parks has achieved a practical reduction in total treatment time by the use of extracorporeal circulation of heated blood.

THERMORADIOTHERAPY

As attention has focused increasingly on local cancer therapy, hyperthermia has been combined with radiation therapy, both external beam and interstitial, in an effort to produce a synergistic and augmented response. Several investigators have concluded that hypoxic cells may be at least as sensitive to hyperthermia as oxygenated cells, forming the rationale for combined therapy, since hypoxic cells seem to be more radioresistant. Ben-Hur, Elkin, and Bronk (1974), however, suggested that the primary effect of hyperthermia was to inhibit cellular recovery from sublethal radiation damage. Connor and co-workers (1977) found that if tumor cells were exposed to hyperthermia followed by 600 rad, there was a greater than 3-log increase in cell kill at 43°C compared to 37°C. He suggested that clinical doses for local and regional treatment using radiation plus hyperthermia would lie in the range of 200 to 600 rad per fraction. Manning and colleagues (1982) have performed phase I studies that suggest that effective thermal doses may be in the range of 43°C to 45°C, combined with high doses of radiation, or an equivalent of 4000 rad in four weeks. Kim, Kim, and Hahn (1975) and Kim and colleagues in 1977 reported their experience with hyperthermia and radiation for cutaneous cancers in man. With fractionated doses to 800 to 2400 rad followed by 43.5°C surface heating by water bath or microwaves, 7 of 10 patients showed significant prolonged benefits by combination therapy when compared to radiation alone. Hornback and colleagues (1979) have treated 70 patients with advanced malignancies with a combination of microwaves (434 MHz, extrapolated to produce 41°C at 7 to 8 cm from phantom models) and standard radiation therapy fractions and total doses. There were no complications to combined therapy, and no patient developed symptomatic or unusually sensitive skin reactions in or around the treatment area. Of 21 patients who received a full course of therapy, 16 (80%) had complete regression of all local tumor, and nine of these remained free of disease for 9 to 14 months.

THERMOCHEMOTHERAPY

The combination of hyperthermia and chemotherapy has been under investigation ever since the realization that heat may alter tumor cell membrane permeability and enhance uptake of chemotherapeutic agents. In 1970, Giovanella, Lohman, and Heidelberger found that temperatures from 37°C to 40°C had little effect on L1210 leukemia cells, but a lethal effect could be achieved between 41°C and 42°C, such that a 4-log kill was observed at 42°C in three hours. Moreover, a 100-fold kill enhancement was observed with the addition of dihydroxybutylaldehyde, with no increase in toxicity. DL-glyceraldehyde, melphalan, and sodium oxamate were also more active in combination with heat. In vitro data from Hahn and co-workers (Hahn 1974; Hahn, Braun, and Har-Kedar 1975; Hahn and Pounds 1976) suggested benefit using hyperthermia with Adriamycin, bleomycin, the nitrosoureas, cisplatin, and possibly other drugs. In 1976, Goss and Parsons reported on four human fibroblast strains and seven melanoma cell lines exposed to various concentrations of melphalan alone and in combination with heat at 42°C for four hours. They found that sensitivity to melphalan was usually accompanied by sensitivity to heat, and that combined treatment was not only synergistic but increased the differential between fibroblast and melanoma lines. Reporting clinical trials in limb perfusion, Stehlin and colleagues (1975) found an increased response from 35% to 80% by the addition of heat of 40.5°C to 41.5°C to melphalan perfusion for regionally metastatic melanoma.

IMMUNE CORRELATES

Several investigators have suggested that tumor regression after hyperthermia may be, in part, due to some augmentation of the immune system. Few studies are available and much data is needed. Crile (1963) showed that tumor cell killing in vivo occurred after development of a host inflammatory reaction and that heat could be potentiated by serotonin, a chemical mediator of inflammation. Heat has been shown also to have increased effectiveness against the more immunogenic tumors (Dickson 1978; Dickson and Shah 1980), and the effects of heat may be enhanced by immunostimulation (Szmigielski and Janiak 1978; Dickson and Shah 1980) or reduced by immunosuppression (Dickson and Shah 1980). Goldenberg and Langner (1971) found growth inhibition of GW-77 human colonic tumors growing in hamster cheek pouches after shortwave diathermy heating, as well as growth inhibition of contralateral, presumably normothermic cheek pouch tumors. Marmor, Hahn, and Hahn (1977) found that EMT-6 sarcomas implanted in mice were highly sensitive to cure by radiofrequency heating. Cell kill as assessed by cloning efficiency of treated and immediately excised tumors, however, was insufficient to account for the in vivo cure rate. This led the authors to suggest that delayed killing might be the result of destruction of tumor blood vessels and possibly of stimulation of a tumor-directed immune response.

SELECTIVE TUMOR HEATING

Most studies to date have dealt with moderate hyperthermia in the range of 42°C to 43°C alone, or in combination with radiation or chemotherapy, based upon the evidence of selective thermal sensitivity of tumor cells. Lethal temperature/exposure time relationships have been established for many cell lines, and for each degree above 42°C the time for tumor destruction seems to be approximately halved (Dickson and Shah 1977; Dickson et al. 1977; Overgaard and Overgaard 1972). Storm and colleagues (1980a) have found similar dose/time relationships in human tumors. Several investigators, however, have shown that at temperatures approaching 45°C the differential susceptibility between malignant cells and normal cells decreases, and host tolerance becomes the prime consideration (Hardy et al. 1965). In 1927, Nils Westermark succeeded in achieving marked regression of rodent tumors at 45°C to 50°C using radiofrequency without the attendant destruction of surrounding tissues, and said that such experimental tumors could not derive any great influence through their circulation, "their thermoregulation being apparently bad." His was the first suggestion that reduced blood flow in tumors might allow selective tumor heating to very high temperatures without normal tissue injury. Hahn, Braun, and Har-Kedar reached a similar conclusion in 1975. Therapeutic hyperthermia in higher temperature ranges seemed feasible with the realization that some solid tumors might act as heat reservoirs to retain heat because of their abnormal vascularity and relatively poor blood flow. Natadze (1959) found that adrenalin, histamine, and actylcholine had no effect on tumor blood flow, and Gullino and Grantham (1961) found that experimental hepatomas had a 20-fold smaller blood supply than host liver. Shibata and MacLean (1965) evaluated cancers in man and found the blood supply to be poorer in all tumors studied. LeVeen and co-workers (1976) subsequently found that tumor blood flow was only 2% to 15% that of surrounding tissue using isotope dilution techniques and reaffirmed the theory that tumors retain more heat than normal tissue because of differential blood flow. Storm and colleagues (1979a, 1980a) recently suggested that while ambient blood flow may differ in tumor and normal tissue, the inability to regulate and augment flow in response to hyperthermia may be the determining factor in achieving selective tumor heating.

ELECTROMAGNETIC HYPERTHERMIA

Hyperthermia has been applied by various means, including fluid immersion, irrigation, regional perfusion, and electromagnetic waves. It is the latter, in the form of microwaves or radiofrequency waves, that appears to be the most practical and efficient for producing localized hyperthermia. All these forms of electromagnetic energy seem to cause tissue heating by a similar mechanism. Energy is transferred into tissue by a field interaction that causes oscillation of ions in the tissue, or changes in the magnetic orientation of molecules, which is locally converted into heat. The energy of a microwave or shortwave quantum is only about 10^{-5} eV and therefore is insufficient to produce ionization or excitation (Milroy and Michaelson 1971). The biological effects of electromagnetic waves seem to be primarily, and possibly solely, due to heat production. The absorption and penetration characteristics of electromagnetic waves are dependent, however, on tissue composition and interfaces (viz., skin/muscle/fat/bone). Moreover, the depth of penetration often is limited and is frequency-dependent (Guy 1975). Satisfactory heating is limited presently to depths of 2 to 3 cm with commercially available diathermy apparatus. In an attempt to overcome limited penetration, several investigators have designed specialized equipment in the 915-MHz and 2450-MHz microwave bands; however, even with surface cooling, documented temperatures of only 42°C to 44°C have been possible at only 2- to 3-cm depths, with a continuously decreasing thermal gradient with increasing depths. Microwave phase array is being investigated in several centers to obviate this problem. For these reasons, clinical trials using microwaves have so far been limited to superficial cancers. LeVeen and co-workers (1976) applied standard radiofrequency diathermy techniques at 13.56 MHz to 21 patients and achieved tumor temperatures over 46°C in three cases, 8°C to 10°C higher than adjacent normal tissue. Tumor necrosis or substantial regression of cancer was reported in all cases. They also found, however, that energy was best transmitted to surgically exposed tumors to avoid the undesirable heating of skin and subcutaneous tissue, which occasionally resulted in burns.

More recently, Storm and colleagues (1979a, 1979b, 1980a, 1980b, in press) and Storm and Morton (1979) developed a noninvasive circumferential electrode that creates a magnetic field whose field lines are coaxial with the body portion being treated, thereby producing deep internal hyperthermia without attendant surface tissue injury. Using this method, the field is not focused, and selective tumor heating is dependent upon the inability of tumors to dissipate heat. Alternately, Lele and Hahn have indicated independently that it may be possible to produce safe and effective deep hyperthermia with focused narrow-beam ultrasound (acoustic waves), although the exact location of the tumor must be known.

A major unknown factor in hyperthermia is dosimetry, the ability to quantitate absorbed heat. Without it, standardization and comparison of treatments is impossible. The measurement has been particularly elusive in electromagnetic hyperthermia. For microwave frequencies, only 40% of incident energy is absorbed, and even this value is extremely variable, depending on relative water concentrations and the presence of interfaces. Because of these many variables, microwave energy absorption can in no way be assessed on the basis of power output of the microwave generator. Several investigators have attempted to overcome this problem using reflectometers and field-strength meters with feedback loops; however, no reliable system has yet been reported in this frequency range. Ultrasonic energy absorption has been readily measured. Storm suggests that radiofrequency absorption may be monitored with greater than 95% accuracy employing specialized equipment in certain circumstances (magnetic-loop induction) (see chapter 14).

TEMPERATURE MEASUREMENT

Temperature measurement, an essential prerequisite to hyperthermic treatments in humans, presents another obstacle to heat therapy. The most reliable and accurate measuring devices seem to be thermocouples and thermistors. Unfortunately, the metal in these instruments distorts the electromagnetic field from radiofrequency and microwave applicators, which in turn causes independent heating of the thermometer and surrounding tissue, resulting in erroneous readings. Christensen, Cetas, and Bowman have been experimenting to allow temperature recording with relative or completely nonconducting materials, and research is currently underway at several centers with liquid crystal, solid crystal, and viscometric thermometers.

HYPERTHERMIA TREATMENT SCHEDULES

Another unresolved area in hyperthermic therapy involves treatment scheduling. Palzer and Heidelberger (1973) studied quantitative killing of HeLa cells in vitro by a cloning assay and found that the cells could not only recover from hyperthermic damage, but that this phenomenon appeared to be cell cycle phase dependent. Cells seemed to be more sensitive to heat during late-S or early G_2 phases by sevenfold, compared to other phases of the cell cycle. Bhuyan and colleagues (1977) found that CHO cells were most sensitive to hyperthermia in the mid- and late-S phase and that G_1 and G_2 cells were the least sensitive. Gerner and co-workers (1976) studied HeLa cells at temperatures from 41°C to 45°C and again found that cell killing increased exponentially with time at elevated temperatures. Surviving cells, however, seemed to develop a transient state of thermotolerance that only began to abate as the cells divided, and was completely lost to progeny of previously heated cells. This study suggested that temporally spaced thermal doses may allow transiently induced thermal resistance to subside for more effective therapy. The import of their studies bears upon scheduling in all clinical trials.

UNRESOLVED QUESTIONS

This brief historical introduction is intended to highlight some of the areas of hyperthermia that have come under intense investigation. It should be evident that hyperthermia is a rapidly evolving field with a multitude of unresolved questions. Hyperthermia clearly is an effective tumoricidal agent; but how should it best be used, and when, and with what other therapies? What techniques will prove most efficacious for a particular tumor, and what is the optimum treatment schedule? How does hyperthermia work, and will it work in all tumors and in all patients?

Do metastases selectively grow in specific organs because of the ambient hospitable temperature (e.g., skin 33°C, lung 35°C, liver 38°C), and if so, is the hyperthermia treatment result more dependent upon the *relative* change in temperature rather than on the absolute temperature achieved? (Storm, in press).

Does in situ thermal tumor destruction augment the immune response? Can the understanding of thermal cell kill kinetics eventually provide a guide to treatment with a prediction of response? And finally, do as yet unrecognized hazards exist, and if so, how might they be remedied? Pragmatically speaking, the answers to these questions will not come from individual achievement, but rather from the cumulative effort of all who are committed to the cure of cancer, including physicians and biological and physical scientists.

The resolve to develop hyperthermia as a potentially fifth form of cancer therapy must be tempered by judgment and logic. It has been said repeatedly that "if cancer were easy to cure, it would have been cured a long time ago." The development of hyperthermia as a clinical tool will depend on rigorously controlled scientific investigation. At this time, human hyperthermic therapy must be considered experimental and should not be used in lieu of proved methods of cancer treatment.

References

Bender, E., and Schramm, T. Untersuchungen zur thermosensibilitat von tumor und normalzellen in vitro. *Acta Biol. Med. Ger.* 17:527–543, 1966.

Ben-Hur, E.; Elkin, M. M.; and Bronk, B.V. Thermally enhanced radioresponse of cultured Chinese hamster cells—inhibition of repair of sublethal damage and enhancement of lethal damage. *Radiat. Res.* 58:38–51, 1974.

Bhuyan, B. K. et al. Sensitivity of different cell lines and of different phases in the cell cycle to hyperthermia. *Cancer Res.* 37:3780–3784, 1977.

Bruns, P. Die heilwirkung des erysipels auf geschwulste. *Beitr. Klin. Chir.* 3:443–466, 1887.

Busch, W. Über den einfluss welchen heftigere erysipeln zuweilen auf organisierte neubildungen ausüben. *Verhandl. Naturh. Preuss. Rhein. Westphal.* 23:28–30, 1866.

Cavaliere, R. et al. Selective heat sensitivity of cancer cells—biochemical and clinical studies. *Cancer* 20:1351–1381, 1967.

Cetas, T. C. Temperature measurement in microwave diathermy fields: principles and probes. In *Cancer therapy by hyperthermia and radiation. Proceedings international symposium on hyperthermia, April 1975.* Washington, D.C.: American College of Radiology Publishers, 1975, pp. 193–203.

Coley, W. B. The treatment of malignant tumors by repeated inoculations of erysipelas—with a report of ten original cases. *Am. J. Med. Sci.* 105:487–511, 1893.

Coley, W. B. The therapeutic value of the mixed toxins of erysipelas and *Bacillus prodigeosus* in the treatment of inoperable malignant tumors. *Am. J. Med. Sci.* 112:251–281, 1896.

Connor, W. G. et al. Prospects for hyperthermia in human cancer therapy, II. Implications of biological and physical data for applications of hyperthermia in man. *Radiology* 123:497–503, 1977.

Crile, G., Jr. Heat as an adjunct to the treatment of cancer—experimental studies. *Cleve. Clin. Q.* 28:75–89, 1961.

Crile, G., Jr. Selective destruction of cancers after exposure to heat. *Ann. Surg.* 156:404–407, 1962.

Crile, G., Jr. The effects of heat and radiation on cancers implanted into the feet of mice. *Cancer Res.* 23:372–380, 1963.

De Lateur, B. J. et al. Muscle heating in human subjects with 915 MHz microwave contact applicator. *Arch. Phys. Med. Rehabil.* 51:147–151, 1970.

Dickson, J. A. The sensitivity of human cancer to hyperthermia. In *Proceedings of conference on clinical prospects for hypoxic cell sensitizers and hyperthermia*, eds. W. L. Caldwell and R. E. Durand. Madison, Wisc.: University of Wisconsin Press, 1978, pp. 174–193.

Dickson, J. A. et al. Tumor eradication in the rabbit by radio frequency heating. *Cancer Res.* 37:2162–2169, 1977.

Dickson, J. A., and Shah, S. A. Technology for the hyperthermic treatment of large solid tumors at 50°C. *Clin. Oncol.* 3:301–318, 1977.

Dickson, J. A., and Shah, S. A. Hyperthermia and the immune response in cancer therapy: a review. *Cancer Immunol. Immunother.* 9:1–10, 1980.

Gerner, E. W. et al. A transient thermotolerant survival response produced by single thermal dose in HeLa Cells. *Cancer Res.* 36:1035–1040, 1976.

Gerweck, L. E.; Gillette, E. L.; and Dewey, W.C. Killing of Chinese hamster cells in vitro by heating under hypoxic or aerobic conditions. *Eur. J. Cancer* 10:691–693, 1974.

Giovanella, B. C. et al. Selective lethal effect of supranormal temperatures on mouse sarcoma cells. *Cancer Res.* 33:2568–2578, 1973.

Giovanella, B. C.; Lohman, W. A.; and Heidelberger, C. Effects of elevated temperatures and drugs on the viability of L-1210 leukemia cells. *Cancer Res.* 30:1623–1631, 1970.

Goldenberg, D. M., and Langner, M. Direct and abscopal antitumor action of local hyperthermia. *Z. Naturforsch.* 266:359–361, 1971.

Goss, P., and Parsons, P. G. The effect of hyperthermia and melphalan on survival of human fibroblasts strains and melanoma cell lines. *Cancer Res.* 37:152–156, 1977.

Gullino, P. M., and Grantham, F. H. Studies on the exchange of fluids between host and tumor. II. The blood flow of hepatomas and other tumors in rats and mice. *J. Natl. Cancer Inst.* 27:1465–1491, 1961.

Guy, A. W. Physical aspects of the electromagnetic heating of tissue volume. In *Cancer therapy by hyperthermia and radiation. Proceedings of an international symposium on hyperthermia, April 1975*. Washington, D.C.: American College of Radiology Publishers, 1975, pp. 179–192.

Hahn, G. M. Metabolic aspects of the role of hyperthermia in mammalian cell inactivation and their possible relevance to cancer treatment. *Cancer Res.* 34:3117–3123, 1974.

Hahn, G. M.; Braun, J.; and Har-Kedar, I. Thermochemotherapy: synergy between hyperthermia (42–43) and adriamycin (or bleomycin) in mammalian cell inactivation. *Proc. Natl. Acad. Sci. USA* 72:937–940, 1975.

Hahn, G. M., and Pounds, D. Heat treatment of solid tumors: why and how. *Appl. Radiol.* 5:131–134, 1976.

Hardy, J. D. et al. Skin temperature and cutaneous pain during warm water immersion. *J. Appl. Physiol.* 20:1014–1021, 1965.

Harisiadis, L. et al. Hyperthermia: biological studies at the cellular level. *Radiology* 117:447–452, 1975.

Hornback, N. B. et al. Preliminary clinical results of 433 megahertz microwave therapy and radiation therapy on patients with advanced cancer. *Cancer* 40:2854–2863, 1977.

Kim, J. H. et al. Local tumor hyperthermia in combination with radiation therapy. *Cancer* 40:161–169, 1977.

Kim, S. H.; Kim, J. H.; and Hahn, E. W. Enhanced killing of hypoxic tumor cells by hyperthermia. *Br. J. Radiol.* 48:872–874, 1975.

Larkin, J. M.; Edwards, W. S.; and Smith, D.E. Total body hyperthermia and preliminary results in human neoplasms. *Surg. Forum* 27:121–122, 1976.

LeVeen, H. H. et al. Tumor eradication by radio frequency therapy. *JAMA* 235:2198–2200, 1976.

Manning, M. R. et al. Clinical hyperthermia: results of a phase I trial employing hyperthermia alone or in combination with external beam or interstitial radiotherapy. *Cancer* 49:205–216, 1982.

Marmor, J. B.; Hahn, N.; and Hahn, G. M. Tumor cure and cell survival after localized radio frequency heating. *Cancer Res.* 37:879–883, 1977.

Mendecki, J.; Friedenthal, E.; and Botstein, C. Effects of microwave-induced local hyperthermia on mammary adenocarcinoma in C3H mice. *Cancer Res.* 36:2113–2114, 1976.

Milroy, W. C., and Michaelson, S. M. Biological effects of microwave radiation. *Health Phys.* 20:567–575, 1971.

Mondovi, B. et al. The biochemical mechanism of selective heat sensitivity of cancer cells—studies on cellular respiration. *Eur. J. Cancer* 5:129–136, 1969.

Muckle, D. S., and Dickson, J. A. The selective inhibitory effect of hyperthermia on the metabolism and growth of malignant cells. *Br. J. Cancer* 15:771–778, 1971.

Nagelschmidt, F. *Lehrbuch der diathermie*, Aufe III, 1926.

Natadze, T. G. Regulation of blood circulation in malignant tumors. *Vopr. Onkol.* 5:14–23, 1959.

Nauts, H. C.; Fowler, G. A.; and Bogatko, F. A. A review of the influence of bacterial infection and of bacterial products (Coley's toxins) on malignant tumors in man. *Acta Med. Scand.* 276:1–103, 1953.

Overgaard, K., and Overgaard, J. Investigations on the possibility of a thermic tumor therapy-shortwave treatment of a transplanted isologous mouse mammary carcinoma. *Eur. J. Cancer* 8:65–78, 1972.

Palzer, R. J., and Heidelberger, C. Influence of drugs and synchrony on the hyperthermic killing of Le Ha cells. *Cancer Res.* 33:422–427, 1973.

Pettigrew, R. T. et al. Clinical effects of whole body hyperthermia in advanced malignancy. *Br. Med. J.* 4:679–682, 1974.

Rohdenburg, G. L. Fluctuations in the growth of malignant tumors in man, with special reference to spontaneous recession. *J. Cancer Res.* 3:193–225, 1918.

Rohdenburg, G. L., and Prime, F. The effect of combined radiation and heat in neoplasms. *Arch. Surg.* 2:116, 1921.

Shibata, H. R., and MacLean, L. D. Blood flow to tumors. *Prog. Clin. Cancer* 2:33–47, 1965.

Stehlin, J. S. et al. Results of hyperthermic perfusion for melanoma of the extremities. *Surg. Gynecol. Obstet.* 140:339–348, 1975.

Stevenson, H. N. The effect of heat upon tumor tissues. *J. Cancer Res.* 4:54, 1919.

Storm, F. K. et al. Normal tissue and solid tumor effects of hyperthermia in animal models and clinical trials. *Cancer Res.* 39:2245–2251, 1979a.

Storm, F. K. et al. Human hyperthermic therapy: relationship between tumor type and capacity to induce hyperthermia by radio frequency. *Am. J. Surg.* 138:170–174, 1979b.

Storm, F. K. et al. Hyperthermia therapy for human neoplasms: thermal death time. *Cancer* 46:1849–1854, 1980a.

Storm, F. K. et al. Hyperthermia in cancer treatment: potential and progress. In *Practical oncology for the primary care physician*, vol. 1, ed. G. Sarna. Boston: Houghton Mifflin, 1980b, pp. 42–52.

Storm, F. K. et al. Clinical radiofrequency hyperthermia: a review. *J. Natl. Cancer Inst.*, in press.

Storm, F. K., and Morton, D. L. Treatment of metastatic disease. In *Advances in surgery*, vol. 13, eds. G. L. Jordan et al. Chicago: Year Book, 1979, pp. 33–68.

Szmigielski, S., and Janiak, M. Reaction of cell-mediated immunity to local hyperthermia of tumors and its potentiation by immunostimulation. In *Proceedings of the second international symposium on cancer therapy by hyperthermia and radiation*, eds. C. Streffer et al. Baltimore and Munich: Urban & Schwarzenberg, 1978, pp. 80–88.

Warren, S. L. Preliminary study of the effect of artificial fever upon hopeless tumor cases. *Am. J. Roentgenol.* 33:75–85, 1935.

Westermark, F. *Zentralbl. f. Gynak.* 49, 1898.

Westermark, N. The effect of heat upon rat tumors. *Skand. Arch. Physiol.* 52:257–322, 1927.

CHAPTER 2

Bioheat Transfer: Mathematical Models of Thermal Systems

Rakesh K. Jain

Introduction
Distributed Parameter Approach
Lumped Parameter Approach
Thermal Energy Absorbed During Ultrasound, Microwave, and Radiofrequency Heating
Temperature Distributions During Normothermia
Temperature Distributions During Hyperthermia
Summary and Recommendations

INTRODUCTION

Despite recent and extensive efforts to treat cancer by hyperthermia (Dietzel 1975; Robinson and Wizenberg 1976; Lett and Adler 1976; Rossi-Fanelli et al. 1977; Streffer et al. 1978; Caldwell and Durand 1978; Milder 1979; Jain and Gullino 1980a; Dewey and Dethlefsen 1980), the long-term survival results in patients taken collectively from different treatment centers have been less than impressive. The full potential of hyperthermia, used alone or in conjunction with other currently available methods for human cancer treatment (table 2.1), has not been realized. Some of the causes for its failures are (1) the lack of data on the susceptibility of tumors to various thermal doses, as determined by the temperature and duration of heating; (2) the technical difficulties of monitoring the temperature of internal tumors and measuring and controlling heat transfer from the energy source to the tumor; (3) the lack of precise knowledge and control of temperature distributions within tumors and the surrounding normal tissues during local or whole-body hyperthermia; (4) poor understanding of the biochemical, physiologic, and immunologic responses of normal and neoplastic tissues at elevated temperatures; and (5) the paucity of data on optimal sequencing of hyperthermia with other modalities of cancer treatment to minimize the damage to normal tissues while maximizing it to neoplastic tissue.

Table 2.1
Current Methods of Cancer Treatment

Methods*	Major Problems
Surgery	Only the primary or localized tumor mass and the surrounding tissue can be removed.
Radiotherapy	Hypoxic cells, which represent a major fraction of tumor, are resistant to radiotherapy.
Chemotherapy	Most anticancer agents are also toxic to normal tissues.
Immunotherapy	Only the residual tumor can be treated.
Hyperthermia	Thermal dosimetry is poorly understood.

*In most cases, a combination of two or more methods is used for cancer treatment.

This work was supported in part by National Science Foundation grants ENG-78-22814 and ENG-78-25432, by American Cancer Society grant PDT-150, and by a Career Development Award CA-00643 from the U.S. Public Health Service.

The author also acknowledges Dr. Sudhir Shah for his many helpful comments, Daniel Solomon and Robert Peloso for their help in the literature search, and Maria Strati for typing the manuscript.

The principal objective of this review is to present various theoretical frameworks that can be used to estimate heat transfer from an external or internal source to a tissue, and to predict resulting temperature distributions in the normal and neoplastic tissues of various mammals during normothermia and hyperthermia. This information is important for improving tumor detection by thermography and in designing heating protocols for hyperthermic treatment. Whereas the response of the normal and neoplastic tissues to thermal stress depends upon the absolute temperature obtained, the duration of that temperature, and the treatment history, there are many physical, physiologic, biochemical, immunologic, and structural factors that must be considered in evaluating the effectiveness of hyperthermia in cancer treatment. Since these factors have been discussed in depth by this author elsewhere (Jain, 1982b), this presentation will focus on heat transfer and temperature distribution.

There are four major problems in analyzing heat transfer in normal and neoplastic tissues during hyperthermia:

1. The exact description of convective heat transfer in tissues is mathematically intractable. In most cases, therefore, a simplified scaler term is used to describe the heat dissipation by blood.
2. Actual geometries of tumors and normal tissues are complex. While finite element techniques can be used to solve the system equations for irregular geometries, most investigators have obtained numerical and analytical solutions for "simple" geometries.
3. The physiologic parameters (i.e., blood flow and metabolic heat generation) and biophysical parameters (i.e., thermal, electrical, and acoustic properties) are not available for most neoplastic tissues. In addition, these parameters vary during the course of the treatment.
4. Analytic expressions for the thermal energy absorbed in a tissue from microwave, radiofrequency, and ultrasonic fields are not available for realistic geometries. Therefore, prediction of temperatures in the presence of these fields involves numerical solution of two problems: energy absorption and dissipation. Each is complex by itself.

In light of these problems, two approaches have been used by investigators in this area of research: distributed and lumped parameter. While the former approach provides a detailed picture of the temperature field, it may require large computational time. The latter approach, at the cost of detailed spatial information, provides adequate information on the average temperature distribution with little computational effort. This chapter reviews both approaches and compares the numerical results with the data available in the literature. Wherever possible, outstanding problems in the prediction of temperature distribution are identified. Finally, some directions for future research in heat transfer related areas are indicated.

DISTRIBUTED PARAMETER APPROACH

The temperature field in a tissue is determined by heat conduction and convection, metabolic heat generation, thermal energy transferred to the tissue from an external source or the surrounding tissue, and the tissue geometry. Thermal conduction is characterized by a thermal conductivity, k, at steady state and by a thermal diffusivity, α, in transient state. Thermal convection is characterized by the topology of the vascular bed and the blood flow rate, which is subject to the thermal regulation.

The Bio-Heat Transfer Equation

The most common representation of the spatial and temporal distribution of temperature in living systems is the *bio-heat transfer equation* (BHTE). It was first suggested by Pennes (1948) in the following form:

$$\rho C \frac{\partial T}{\partial t} = \nabla \cdot (k \nabla T) + Q_b + Q_m. \qquad (1)$$

Here, T is the tissue temperature, C is the tissue heat capacity, ρ is the tissue density, k is the tissue thermal conductivity, Q_m is the rate of metabolic heat generation, and Q_b is the rate of heat exchange with blood. The evaluation of Q_m from the oxygen consumption of the tissue is discussed in the next section. The estimation of Q_b is discussed below.

Within the vasculature of a tissue, blood flows in all directions, and the local direction of convection depends on the vascular morphology of the tissue. The situation is even more complex in tumors where the direction and magnitude of blood flow are not fixed. In tumors, a venous capillary may behave as an arterial capillary at a different time. Therefore, the local description of the convective heat transfer term, Q_b, in tissues would be a time-dependent vector. This is an enormously complex problem that thus far has proved mathematically intractable. In order to circumvent a mathematical description of the details and complexities of the microcirculation in a capillary bed, two approaches have been taken by investigators in this area of research.

In the first approach, the convection term is replaced by a "diffusion" type of term, and the heat transfer in tissues is described in terms of an effective thermal conductivity (k_{eff}):

$$\rho C \frac{\partial T}{\partial t} = \nabla \cdot (k_{eff} \nabla T) + Q_m. \quad (2)$$

Implicit in this equation is the assumption that because of a large vascular surface area the blood temperature equilibrates with the tissue temperature. The concept of effective thermal conductivity has been used by several investigators in thermal physiology (Trezak and Jewett, 1970). Jain and Wei (1977) also have used this concept to describe the drug distribution in tumors.

In the second approach, the convective heat transfer term, Q_b, is replaced by the thermal energy brought in by the arterial blood minus the thermal energy carried away with the venous blood:

$$Q_b = \eta P \rho_b C_b (T_a - T). \quad (3)$$

Here, ρ_b is the blood density, C_b is the blood heat capacity, T_a is the arterial blood temperature, and η is a measure of the effectiveness of heat transfer between the tissue and the venous blood ($0 \leq \eta \leq 1$); η is equal to 1 when venous blood is in complete equilibrium with tissue. It is reasonable to expect that because of slow blood flow, η will be close to 1. Although it is possible to introduce a value of η different from 1 and carry it through, it introduces no new insight and changes the numerical value of the blood flow rate slightly.

The BHTE with both of these assumptions has been solved for various tissue geometries, initial and boundary conditions.

Owing to scalar treatment of the convective heat transport by blood, the BHTE recently has come under serious criticism. In a New York Academy of Sciences conference organized by Jain and Gullino (1980a), the limitations of the BHTE were discussed at length, and various alternatives were presented. Considering tissue as porous media, Wulff (1974) has introduced the blood velocity vector, $\vec{u}_b$, in the BHTE. Unfortunately, the complex nature of the system defies any attempt to specify the circulation vector at the microscopic level.

A second criticism of the BHTE is due to its failure to account for the countercurrent heat exchange in the capillary bed. Assuming that the velocity vector is one-dimensional, Mitchell and Myers (1968), and later Keller and Seiler (1971), have analyzed the countercurrent heat transfer in tissues. While the model of Keller and Seiler attempts to account for the continuous changes in temperature of the arterial and venous networks as one moves from the body core to the skin, the model does not account for the heterogeneous structure of the vascular network or the continuous variation between arterial and venous temperatures in the tissue. Recently, Weinbaum (1980) proposed a two-phase model that provides the temperature distribution along the arterial and venous networks, as well as the temperature variation in the extravascular space, as a function of the distribution of the collateral circulation and a simple vessel-branching law. While this model is elegant conceptually, it is impractical for most tissues where the vascular morphology is much more complex and where the velocity vector changes its magnitude and direction randomly.

Klinger (1974, 1980) has also provided recently the use of the Green's function to obtain an exact analytic solution of the diffusion equation with convection terms without making any special assumptions concerning the velocity field. The absence of a detailed knowledge of the convection field limits the use of this approach.

One of the most significant improvements in the BHTE has been made by Chen and Holmes (1980). These authors point out that in addition to the blood perfusion term, similar to equation (3), the blood flow in the microvasculature may have at least two contributions to heat transfer: a contribution proportional to the local blood velocity vector, $\vec{u}_b$, and a contribution proportional to the temperature gradient, similar to the effective thermal conduction term in equation (2). They also suggest that in some circumstances, these two additional contributions may be negligible compared to the perfusion term.

While these various improvements in the BHTE provide new insight into the heat transfer process in a capillary bed, mathematical complexity in describing the microcirculation and vascular topology makes their application to normal and neoplastic tissues nearly impossible. Therefore, in this article, we use the bio-heat transfer equation of Pennes to describe heat transfer and temperature distribution in tissues. The next section presents values of the parameter incorporated in Pennes's model.

Parameter Values

Thermal Properties

A comprehensive tabulation of the thermal properties of tissues that were reported up to 1980 can be found in Eberhart and Shitzer (1982). This compilation is an extension of previous efforts (Chato, 1969; Bowman, Cravalho, and Woods 1975) and contains limited data on neoplastic tissues.

Using a noninvasive probe technique, Jain, Grantham, and Gullino (1979) recently have measured the thermal conductivity (3 mW/cm − K) and thermal diffusivity (10^{-3} cm^2/sec) of a tumor of mammary origin, Walker-256 carcinoma. In this study, tumors weighing from 2 to 11 g and with blood perfusion rates between 1 and 6 hr^{-1} were used. While the effective thermal conductivity of these tumors decreased as they grew larger, no definite correlation was found between the true thermal conductivity of tumor and its weight. When the blood flow rate of a tumor was modified by inducing hypo- or hypervolemia into the animal, its effective thermal conductivity (as measured by the temperature rise in the heating probe embedded in tumor) varied proportionally to the square root of the perfusion rate (Peclet number). Figure 2.1 suggests that in order to obtain a biologically significant increase in the effective thermal conductivity, a substantial increase in the blood flow rate is needed. In these experiments, thermal conductivity measured in vitro was found to be within 10% of the in vivo value (Jain and Gullino 1980b). Shah and co-workers (1981) and Shah and Jain (1981) recently have measured thermal conductivity, diffusivity, and perfusion rates of various animal tumors by noninvasively implanting a probe. Holmes and Chen (1979) and Bowman (1980) also have measured thermal properties of tumors using invasive-probe techniques (table 2.2).

Until more data on the thermal properties of tumors are collected, an order of magnitude estimate of the thermal properties of tumors can be obtained using the correlation of Cooper and Trezak (1971). This correlation is a modification of the one developed by Poppendiek and colleagues (1967) and relates the thermal properties of the tissue to its composition (water, protein, and fat). The need for more accurate measurements and predictions of thermal properties of tumors over a wide range of temperatures is urgent.

Table 2.2
Thermal Conductivity of Various Animal and Human Normal Tissues and Tumors

Species	Tumor	k (mW/cm−°K)
Rat*		
	Walker-256 carcinoma	3.2 ± 0.9
Human†		
Breast		
	Atrophic normal tissue	4.99 ± 0.04
	Scirrhous carcinoma	3.97 ± 0.04
	Mucinous (colloid) carcinoma	5.27 ± 0.41
Colon		
	Normal	5.56 ± 0.09
	Metastatic colon carcinoma	5.56 ± 0.12
Liver		
	Normal	5.72 ± 0.09
	Metastatic colon carcinoma	5.20 ± 0.08
	Normal	5.08 ± 0.11
	Metastatic pancreatic carcinoma	5.62 ± 0.21
Lung		
	Normal	5.18 ± 0.21
	Squamous cell carcinoma	6.66 ± 0.18
Pancreas		
	Normal	3.45 ± 0.05
	Metastatic carcinoma	4.78 ± 0.39
	Normal	4.68 ± 0.06
	Metastatic gastric carcinoma	4.92 ± 0.54
Nerve		
	Acoustic schwannoma	5.81 ± 0.17

SOURCE: Jain, Grantham, and Gullino 1979; Bowman 1980.
*In vivo.
†In vitro.

Blood Flow Rate

Blood flow rates and volumes of various organs and normal tissues of mammals (mouse, rat, hamster, dog, swine, rabbit, monkey, and human) were compiled recently by Bischoff (1975), Jain, Weissbrod, and Wei (1980), and Townsend, Jain, and Eddy (1981). Blood flow rates of various animal and human tumors during normothermia are given in table 2.3. In general, perfusion rates of tumors are less than those of normal tissues, with the exception of a canine lymphosarcoma. In addition, the average blood flow rate of an animal tumor decreases as it grows larger, with the exception

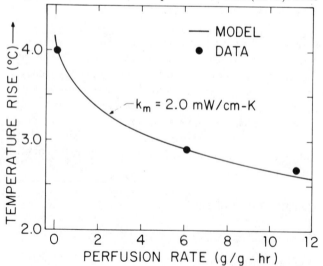

Figure 2.1. The effect of blood flow on the effective thermal conductivity of tumors. Temperature rise is inversely proportional to the effective thermal conductivity. The *points* represent data and the *solid line* represents the solution of the bio-heat transfer equation. (Jain, Grantham, and Gullino 1979. Reprinted by permission.)

Table 2.3
Blood Flow Rates of Various Animal and Human Tumors

Reference	Tumor	Species	Blood Flow (ml/g/min)	Method
Gullino and Grantham 1961a, 1961b	Hepatoma-5123	Rat	0.1–0.17	Direct collection of efferent blood
	Novikoff's hepatoma	Rat	0.02–0.05	Direct collection of efferent blood
	Walker-256 carcinoma	Rat	0.03–0.1	Direct collection of efferent blood
Song, Payne, and Levitt 1972	Walker-256 carcinoma	Rat	0.16–0.48	Radioactive microspheres
Kjartansson 1976	Sarcoma	Rat	0.04–0.21	^{133}Xe clearance
	Sarcoma	Rat	0.22–0.58	Plethysmography
Vaupel 1975	DS-carcinosarcoma	Rat	0.07–0.32	Direct collection of efferent blood
Dickson and Calderwood 1980	Yoshida sarcoma	Rat	0.07	Uptake of ^{86}Rb
Allen et al. 1975	Nerve and brain tumors	Rat	0.44–0.79	Uptake of ^{14}C-antipyrine
Takacs, Debreczeni, and Farsang 1975	Gnerin carcinoma	Rat	0.20–0.21	Uptake of ^{86}Rb
Moller and Bojsen 1975	DMBA-induced adenocarcinoma	Rat	0.025	^{133}Xe clearance
Endrich et al. 1979a	BA-1112 rhabdo-myosarcoma	Rat	$10-10^{-5} \frac{ml}{min}$	RBC velocity measurement
Rogers 1968	Melanoma	Hamster	0.60	^{131}I-antipyrine uptake
Townsend, Jain, and Eddy 1981	Cervical carcinoma	Hamster	0.22	^{133}Xe clearance
Robert, Martin, and Burg 1967	Sarcoma	Mouse	0.01–0.22	^{133}Xe clearance
Peterson et al. 1969	Sarcoma	Mouse	0.04–0.19	^{133}Xe clearance
Kallman, de Nardo, and Stasch 1972	Sarcoma	Mouse	0.07–0.14	^{133}Xe clearance
Peterson et al. 1969	Mammary carcinoma	Mouse	0.01–0.17	^{133}Xe clearance
Gump and White 1969	VX-2 carcinoma	Rabbit		^{85}Kr clearance
Slotman et al. 1980	VX-2 carcinoma	Rabbit	0.24–1.13	Radioactive microspheres
Straw et al. 1974	Lymphosarcoma	Dog	0.63–43.4	Thermal dilution technique
Mäntylä, Kuikka, and Rekonen 1976; Mäntylä 1979	Lymphoma	Human	0.34 ± 0.21	^{131}Xe clearance
	Anaplastic carcinoma	Human	0.15 ± 0.11	^{131}Xe clearance
	Differentiated tumors	Human	0.23 ± 0.15	^{131}Xe clearance
Plengvanit et al. 1972	Liver carcinoma	Human	0.12	^{131}Xe clearance

of the data of Slotman and colleagues (1980) on VX-2 carcinoma in rabbits. This relationship has not been found to be valid in the human tumors (Mäntylä, 1979).

After analyzing the perfusion rate, P, of various experimental tumors, Gullino (1968) proposed the following simple relationship for a tumor with weight, W, in grams:

$$\log P = -0.667 (\log W) - a \quad (P \text{ in ml/g/min}), \tag{4a}$$

where, a, a constant, is about 0.37 for Walker-256 carcinoma.

Song and associates (1980) obtained the following best-fitting curve for their data on blood flow rate of Walker-256 as a function of W(g) in the range 0.05 to 5.0 g:

$$\log P = -0.1721(\log W)^2 - 0.5382(\log W) - 0.5255 \quad (P \text{ in ml/g/min}). \quad \textbf{(4b)}$$

Vaupel (1975) found an exponential decay in blood flow rate of DS-carcinosarcoma in the range 3 to 13 g:

$$\log P = -0.100 W - 0.315 \quad (P \text{ in ml/g/min}). \quad \textbf{(4c)}$$

While it is well established that as a tumor grows larger its center becomes necrotic and its average perfusion rate decreases, at least in animal tumors, approximately according to the various fits possible in equation **(4)**, limited data and constitutive relationships are available for the blood flow distribution in tumors.

Goldacre and Sylvén (1962) have comprehensively reviewed the work done on the distribution of blood-borne dyes in the tumors. By changing the color of the systemic blood with lisammine green, they have isolated the poorly perfused region of several transplanted tumors in mice (mammary carcinoma, sarcoma-37, Ehrlich-Landschutz ascites) and Walker-256 carcinosarcoma in Wistar rats. These investigators did not propose a quantitative relationship of the spatial heterogeneities in blood flow. Using available data, Jain and Wei (1977) proposed the following two-zone model for Walker-256 carcinosarcoma in rats:

$$P = \begin{cases} P_N & \text{(in necrotic zone)} \\ 10 P_N & \text{(in viable zone)} \end{cases}, \quad \textbf{(5a)}$$

where P_N is the perfusion rate in the necrotic zone. Recently, using a thermal dilution technique (Gullino, Jain, and Grantham 1982), we have found that bloodflow rate can be nearly zero in some sections of tumors and around three times the average perfusion rate in other sections.

Shibata and MacLean (1966) have used radioactive microspheres to study the relative perfusion rates of several human and animal tumors. Their results on distribution ratio of radioactivity between the periphery and the center of Walker-256 carcinoma (grown in rat thigh muscle) show that the ratio of radioactive counts between periphery and center ranges from 0.9 to 9.8, and it seems to have no correlation with tumor size. A definite decrease in perfusion rate is evident as the size of tumor increases, however.

Employing the thermal dilution technique of Tuttle and Saddler (1964), Straw and co-workers (1974) measured the regional perfusion rates of a canine lymphosarcoma. As shown in figure 2.2, the perfusion rate is as high as 43.4 ml/gm/min in the periphery and as low as 0.63 ml/gm/min in the center. We have been able to fit their data using the following parabolic relationship:

$$P = 1.2 + 15.1 \, (r/R)^2, \quad \textbf{(5b)}$$

where r and R are the radial positions in the tumor and tumor radius, respectively. In our previous works, we have used equations **(5a)** and **(5b)** in estimating temperature distributions in tumors, after normalizing these equations so that both lead to the same average perfusion rate, as given by equation **(4a)**. (Jain 1978, 1980a).

Using a sandwich tumor preparation developed by Reinhold and van den Berg-Blok (1980), Endrich and associates (1979a) have recently studied the blood flow distribution in BA-1112 rhabdomyosarcoma in rats. They have divided the two-dimensional tumor tissue into five-regions on the basis of perfusion rates (fig. 2.3). Our current efforts are directed toward developing detailed constitutive relationships to describe regional perfusion rates of VX-2 carcinoma grown in a transparent ear chamber placed in a rabbit ear (Dudar and Jain 1982; Nugent and Jain 1982; Zawicki and Jain 1982[1]).

[1]D. Zawicki and R. K. Jain, unpublished results.

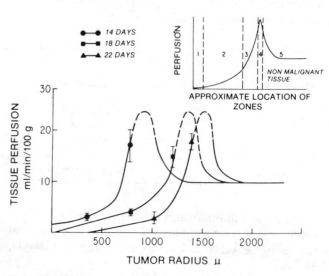

Figure 2.2. Regional distribution of blood flow in a canine lymphosarcoma.

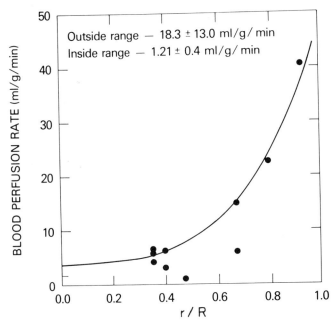

Figure 2.3. Regional distribution of blood flow in a two-dimensional (sandwich) tumor. (Endrich et al. 1979a. Reprinted by permission.)

Metabolic Heat Generation

This is, perhaps, the most elusive term in the BHTE and is, therefore, neglected by most investigators. The most widely used approach in estimating this term has been to set Q_m equal to oxygen consumption multiplied by the caloric value of oxygen (Gemmill and Brobeck 1968). Implicit in this approach is the assumption that oxygen, and not glucose, determines the heat generation in a tissue. Oxygen consumption of various mammalian tissues can be found in the literature (Altman and Ditmer 1964; Huckaba and Tam 1980). In vivo oxygen and glucose consumption and lactate production of various animal tumors are given in table 2.4 (Aisenberg 1961; Gullino 1976; Vaupel 1974).

Gullino, Grantham, and Courtney (1967a, 1967b), and Gullino with others (1967), have shown that oxygen consumption of tumors is linearly related to its perfusion rate:

$$Q_{O_2} = -0.16 + 2.03\, P, \qquad (6)$$

where Q_{O_2} is in mM/hr/100 g, and P is in l/hr/100 g.

Similarly, Vaupel (1975, 1980) has shown that oxygen consumption by DS-carcinosarcoma decreases exponentially with tumor weight, in a fashion analogous to the perfusion rate:

$$\log Q_{O_2} = -0.107\, W - 1.469, \qquad (7)$$

where Q_{O_2} is in ml/g/min. Vaupel (1980) has also found maldistribution of oxygen concentration and consumption in tumors similar to their local perfusion rates.

Using equation (6), Sien and Jain (1979) have developed the following constitutive relationship between the metabolic heat generation and the local blood supply:

$$Q_m = (6 \times 10^4\, P - 0.22)\ \text{mW/cm}^3. \qquad (8)$$

This relationship includes the effect of the respiratory coefficient in heat generation. In addition, this equation allows for incorporation of inhomogeneities in metabolic heat generation, analogous to the nonuniform perfusion rate.

Figure 2.4 shows the relationship among the tumor blood-flow rate (equation **4a**), oxygen consumption rate (equation **6**), and metabolic heat-generation rate (equation **8**) of tumors as a function of their weight.

Table 2.4
In Vivo Oxygen and Glucose Consumption and Lactate Production of Rat Tumors

Tumor	Weight Range (g)	Weight Doubling Time (days)	moles/g/hr			Reference
			Oxygen	Glucose	Lactate	
Walker-256 carcinoma	2.0–9.8	33	0.24	0.54	0.38	Gullino, Grantham, and Courtney 1967a, 1967b; Gullino et al. 1967; Gullino 1976
Hepatocarcinoma-5123	3.8–7.0	196	0.42	0.31	0.20	
Fibrosarcoma-4956	5.0–12.7	61	0.08	0.25	0.17	
DS-carcinoma	3.2–13	?	0.13	—	—	Vaupel 1975; Vaupel et al. 1980

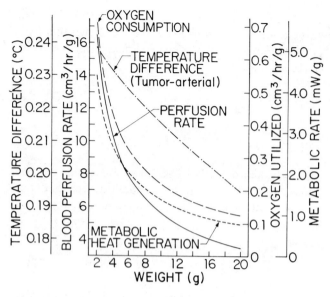

Figure 2.4. The effect of tumor weight on its oxygen consumption, blood flow rate, metabolic heat generation rate, and temperature rise owing to metabolic heat generation (Sien and Jain 1979. Reprinted by permission.)

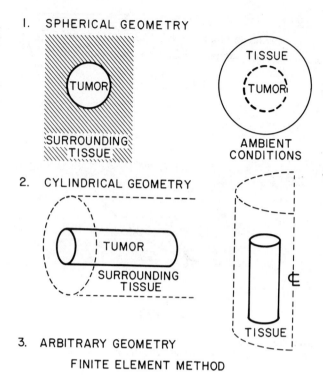

Figure 2.5. Typical model systems that approximate the physical and anatomic situation. While a tumor may have a more complex geometry, simple geometries (such as cylindrical or spherical) make it possible to obtain analytical solutions in most cases. (Jain 1980a. Reprinted by permission.)

Tissue Geometry

Once all the model parameter values are specified, the geometry of the model system must be defined. Depending upon the information desired, either a particular organ (or tissue region) or the whole mammalian body may be considered as the region in which the BHTE must be solved. Both of these approaches have been discussed in depth in Eberhart and Shitzer (1982) for application in the normal tissues; therefore, we will focus our attention on tumors.

The situation for tumor is more complex. Figure 2.5 shows some of the model geometries, which have been used to approximate thermal interaction between a tumor and the surrounding tissues. While it is known that a tumor may infiltrate the surrounding tissue in a complex geometric fashion, because of simple geometries (i.e., cylinder, sphere) it is possible to obtain analytic solutions in most cases. For more complex geometries, finite difference or finite element methods are necessary to calculate the temperature field in the tumor and the surrounding normal tissues.

Boundary Conditions

In order to solve the BHTE for a given geometry, the following boundary conditions must be specified. Within the tissue region or an organ, heat flux and temperature at various interfaces must be continuous. If the tissue or organ containing the tumor is exposed to the ambient temperature, heat exchange with the external environment by conduction, convection, radiation, and evaporation must be accounted for.

Heat source terms must be added either in the bio-heat equation or in the boundary conditions, depending on the method used for inducing hyperthermia. (See later sections of this chapter for details of hyperthermia technology.) During surface heating by hot air or water or molten wax bath, heat transfer by conduction and convection to the overlying skin must be added to the boundary condition. During infrared- or visible radiation-induced heating, a radiation term must be added to the skin boundary condition. During hyperthermic perfusion, the arterial temperature, T_a, must be set equal to the experimentally set afferent blood temperature in the BHTE. During volume heating by ultrasound, microwaves, or radiofrequency currents, an additional term describing temporal and spatial distribution of absorbed energy must be added to the right hand side of the BHTE. The techniques for estimating the volume heating term are discussed in a later section.

LUMPED PARAMETER APPROACH

While the distributed parameter approach discussed above describes the detailed temporal and spatial distribution of temperature in a tissue, the solution of system equations is often tedious and requires precise knowledge of the tissue geometry, anisotropy, and orientation with respect to the surrounding tissues or heat source. Lumped parameter models overcome these problems at the cost of detailed information, which may not be needed in some cases of interest. This section offers a brief discussion of various lumped parameter models, with descriptions of various heat transfer mechanisms that should be incorporated in such models.

Compartmental Approach

In general, lumped parameter models describing the mammalian thermal system have focused either on specific organs or the whole body (for review, see Hardy, Gagge, and Stolwijk 1970; Fan, Hsu, and Hwang 1971; Hwang and Konz 1977; Stolwijk 1980; Huckaba and Tam 1980); analyses of the thermal interaction between a tumor and the host have been limited (Sien and Jain 1979, 1980; Chrysanthopoulos and Jain 1980; Jain 1980a, 1980b). The whole-body model developed by Huckaba and co-workers is discussed first, followed by our model for a tumor-bearing host.

The whole-body lumped parameter model of Huckaba and Tam is based on the distributed parameter models developed by Stolwijk, Hardy, and Gagge. In this model, the body is divided into one spherical and ten cylindrical segments: head, neck, upper trunk, lower trunk, upper arms, forearms, hands, fingers, thighs, legs, and feet. In the case of extremities (e.g., forearms and legs) single cylinders are used to represent corresponding pairs of segments together (fig. 2.6). Note that this symmetry assumption will not work for the tumor-bearing organs.

Each segment is further divided into four subsections: skin, fat, muscle, and core (fig. 2.7). Unlike the distributed parameter model, each subsection is assumed to have spatially uniform temperature, which is given by an unsteady-state energy-balance equation as shown below:

$$\begin{bmatrix} \text{Net accumulation of} \\ \text{thermal energy in} \\ \text{each subsection} \end{bmatrix} = \begin{bmatrix} \text{metabolic} \\ \text{heat} \\ \text{production} \end{bmatrix}$$

$$+ \begin{bmatrix} \text{Heat gained} \\ \text{by conduction} \\ \text{from interacting} \\ \text{subsection(s)} \end{bmatrix} - \begin{bmatrix} \text{Heat lost to the} \\ \text{perfusing blood} \end{bmatrix}$$

$$- \begin{bmatrix} \text{Heat lost to the} \\ \text{environment} \end{bmatrix}. \quad (9a)$$

The detailed expression for each of the above terms is discussed in the next section.

In addition to the transient energy balance equation for each of the 44 subsections, the following balance is written for the central blood pool:

$$\begin{bmatrix} \text{Net accumulation} \\ \text{of thermal energy} \\ \text{in blood pool} \end{bmatrix} = \begin{bmatrix} \text{Energy brought} \\ \text{in with the} \\ \text{venous blood} \end{bmatrix}$$

$$- \begin{bmatrix} \text{Energy carried} \\ \text{away with the} \\ \text{arterial blood} \end{bmatrix}. \quad (9b)$$

These 45 equations, after substituting appropriate parameter values, are solved numerically to obtain transient temperature distributions in humans during hypothermia and hyperthermia (Huckaba and Tam 1980).

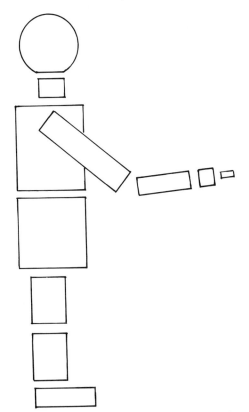

Figure 2.6. A schematic diagram showing the geometric arrangement of the various segments of a human body for thermal modeling purposes. Note that the presence of a tumor will require additional compartments.

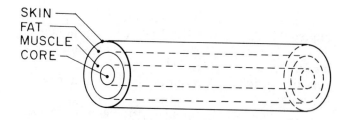

Figure 2.7. Each segment of the mammalian body is divided into four concentric layers: core, muscle, fat, and skin.

In our analyses, a mammalian species is considered to be comprised of a tumor and of normal tissues, represented by compartments interconnected in an anatomic fashion. Because of our focus on interaction between the tumor and the surrounding tissues, we lump all the normal tissues, except those next to the tumor, into a single compartment. This approach is illustrated in figure 2.8 for a rat carrying a subcutaneous tumor, in which the model consists of the following seven compartments: the tumor, the surrounding normal tissue, the body (which represents the remaining normal tissues), the skin directly above tumor, the skin above surrounding normal tissue, the rest of the skin, and the central blood pool. For predictions of intratumor temperature distributions, it is necessary to subdivide the neoplastic tissue and the skin above it into N equal compartments, where N is determined by the spatial resolution and precision desired in the computed temperature field (fig. 2.9). While dividing compartments further in this model increases the spatial resolution, its advantages are offset by the larger number of parameters necessary to formulate the system equations. In the limits of an infinite number of subcompartments, this model is equivalent to a distributed parameter model. The detailed schematic diagrams of various versions of our model are given elsewhere for the rat, rabbit, swine, dog, baboon, and humans (Sien 1978; Chrysanthopoulos 1979; Volpe 1981).

Once the number of compartments has been specified, the analysis consists of applying unsteady-state energy balance equations (cf equation 9) to each compartment. Upon substitution of suitable numerical values for the various parameters and heat flux terms, the set of coupled, nonlinear ordinary differential equations is solved numerically.

Heat Transfer Mechanisms

Under normal physiologic conditions, the following heat transfer terms should be incorporated in a lumped or distributed parameter model: metabolic heat generation; conduction and convection within the body; heat exchange with the environment by radiation, conduction, and convection; heat loss from the skin by evaporation of sweat and water diffused across the skin; and respiratory heat loss. Expressions for each of these terms, along with the parameter values, are given by Cooney (1976) and will not be discussed here.

During hyperthermia, terms representing the heat input to a specific tissue or whole body must be

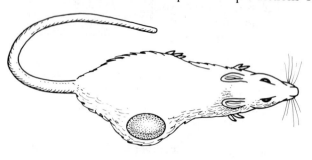

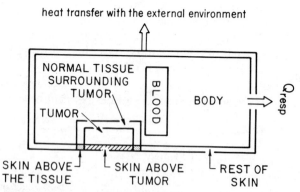

Figure 2.8. Schematic of a compartmental model for the analysis of temperature distribution in a tumor-bearing mammal during hyperthermia. In this case, a rat is represented by seven compartments interconnected in anatomic fashion: tumor, skin above tumor, normal tissue surrounding the tumor, skin above the normal tissue, rest of body, rest of skin, and central blood pool.

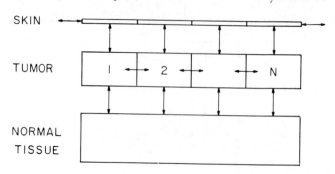

Figure 2.9. For simulations of intratumor temperature gradients, it is necessary to subdivide the tumor and skin above it into N subcompartments. N is determined by the spatial resolution and precision desired in the calculated temperature distribution.

added to the proper system equation(s). For example, during whole-body or local hyperthermia induced by a hot air/water/wax bath or infrared or visible radiation, a heat flux term is added to the skin surface area interacting with the source. During hyperthermia produced by radiofrequency currents, microwaves, or ultrasound, a heat flux term is added to the section of body being heated. During hyperthermia with blood perfusion, the afferent blood temperature is set at a desired value, and the efferent blood is circulated to the central blood pool, or to the extracorporeal device used for heating the blood. Suitable numerical values for the various parameters involved in these heat flux terms are given elsewhere (Sien and Jain 1979, 1980; Chrysanthopoulos and Jain 1980; Volpe and Jain 1982; Volpe 1981).

Finally, the incorporation of thermoregulation of normal and neoplastic tissues in our model warrants discussion. The first reaction of an alive animal exposed to high ambient temperature is to increase the blood flow to the skin, which in turn increases the heat flux from the skin to the surrounding environment (Cunningham 1970). This type of physical thermoregulation is effective only when the skin temperature is higher than the ambient temperature; when the ambient temperature is higher than the body temperature, more drastic means of cooling are needed. Animals like the rat, rabbit, dog, and swine increase their evaporative cooling by panting. Shallow breathing occurs, which increases the ventilation rate in the upper parts of the respiratory system only, which provides the desirable increase in evaporation. Perspiration, which is a major heat loss modality in human systems, does not constitute a significant cooling mechanism in these animals. Table 2.5 summarizes the thermoregulatory mechanisms used by various homotherms.

There are two ways in which thermoregulation can be included in such a model: feedforward and feedback controls (Hardy, Gagge, and Stolwijk 1970; Stolwijk 1980; Huckaba and Tam 1980). While the

Table 2.5
Thermoregulatory Mechanisms in Homotherms

Animal	Rectal Temperature			Critical Air Temperature		Temperature Regulating Mechanisms			Thermoneutrality Zone °C
	Normal °C	Minimum °C	Maximum °C	Low °C	High °C	Sweating	Shivering	Panting	
Man	37	21	44	1	32	+	+	−	24−31
Camel	34−40	—	—	—	—	+	+	−	—
Cat	39	19	44	—	36	−	+	+	24−27
Cattle, Brahman	38−39	—	—	1	32	−	—	+	10−27
Cattle, dairy	38−39	—	43	—	24	−	+	+	5−16
Dog	38−39	24	42	−80	42−58	−	+	+	18−25
Donkey	36−38	—	—	—	—	+	+	—	—
Goat	38−39	—	—	—	—	−	+	+	20−26
Guinea pig	39	17	43	−15	32	−	—	+	30−31
Horse	38	—	—	—	—	+	—	—	—
Monkey	37−39	19	43	—	40	+	+	−	27−30
Mouse, white	37	10	—	10	37	−	—	−	30−33
Rabbit	39	20	42	−29	32	−	+	+	28−32
Rat, white	37.5	15	44	−10	32	−	+	−	28−30
Seal	37	—	—	−30	—	—	—	—	−10−+30
Sheep	39	—	42	—	32	+	—	+	13−31
Swine	37−38	—	42−45	—	30	−	—	+	0−20
Chicken	41−42	15	47	−35	32	−	+	+	19−29
Pigeon	43	—	47	−85	42	−	+	+	20−30

SOURCE: Altman and Ditmer 1964.
NOTES: Critical air temperature: Air temperature at which the normal animal first begins to show a change in deep-body temperature.
Temperature regulating mechanisms: + = present; − = absent.
Thermoneutrality zone: The range of air temperature at which the normal animal has the lowest metabolic rate.

Table 2.6
Effect of Hyperthermia on Tumor Blood Flow Rate

Reference	Tumor (Host)	Measurement Techniques	°C × Time	Observation
Gullino, Yi, and Grantham 1978; Gullino 1980	W-256 carcinoma (1.7–8.5 g) (Sprague Dawley rats)	Direct collection of tumor efferent blood	40–43 × 1 hr	No significant change during hyperthermia
Song 1978; Song et al. 1980; Song 1980	W-256 carcinoma (0.3–5 g) (Sprague Dawley rats)	Radioactive microsphere method	43 × ½ hr 43 × 1 hr 45 × 1 hr	No significant change 18, 48, and 72 hrs after heating No significant change immediately after heating Significant decrease 3 hrs after hyperthermia
Dickson and Calderwood 1980	Yoshida sarcoma (1–1.5 g) (Wistar rats)	The fractional distribution of ^{86}Rb	42 × 1 hr	No change during heating, decrease to zero in the next hour, and recovered to control level 12 hrs after heating
Vaupel et al. 1980	DS-carcinosarcoma (3–5 g) (Sprague Dawley rats)	Direct collection of tumor efferent blood	35–44 × 20 min	Increases blood flow up to 39.5°C and decreases after that below control value at 44°C (no change in large tumors of 5.4–8.5 g)
Bicher et al. 1980	Mammary adenocarcinoma (0.3–0.6 g) (C3H mouse)	Hydrogen clearance method	Continuous increase of temp. from 32 to 45 in 30–40 min	Blood flow increases up to 41°C, decreases after 42°C
Robinson, McCulloch, and McCready 1980	Transplants of spontaneous tumors (C3H mice)	Fractional distribution of ^{86}Rb	37.5, 42, 44	No change as the temperature was raised from ambient to 44°C
Sutton 1980	Ependymoblastoma (C57BL/6 mice)	^{133}Xe clearance method	40, 42, 45 × 15–75 min	Increased up to 30–45 min at 40–42°C, and rapidly decreased after that; decreased to ½ control value in 15 min at 45°C, and 60 min at 42°C
Emami et al. 1981	BA-1112 Rhabdomyosarcoma (WAG-Rij rats)	^{15}O clearance method	41 × 40 min 42 × 30 min	Decreased by 50%
Reinhold, Blachiewicz, and van den Berg-Blok 1978; Reinhold and van den Berg-Blok 1980	BA-1112 Rhabdomyosarcoma (transparent chambers in WAG-Rij rats)	Microphotography and histopathology	42.5 × 3 hrs	Permanent damage to microvasculature in 140 ± 60 min. The same effect was observed at 42°C with either glucose, misonidazole, or 5-TG
Endrich et al. 1979b	BA-1112 Rhabdomyosarcoma (transparent chambers in WAG-Rij rats)	Red blood cell velocity measured using a photodiode method	27–42	Continuous heating from 26° to 35°C increased RBC velocity, but the functional capillaries remained the same in number. Heating at 40°C caused a significant decrease in both RBC velocity and the number of functional capillaries. Heating at 40°C to 42°C for 60 min caused permanent vascular damage
Eddy 1980; Eddy, Sutherland, and Chmielewski 1980	Squamous carcinoma (hamster cheek pouch)	Microphotography and histopathology	41–45 × 30 min	Hemorrhage and stasis of vessels applied at 43°C or fractionated heat applied at 42°C with 1-hr interval

NOTE: A marked dilation and congestion of vessels followed by massive hemorrhage and necrosis has been observed in rat tumors by Emami and co-workers (1981), in human lung tumors by Sugaar and LeVeen (1979), in human intraabdominal sarcomas by Storm and co-workers (1979), and in various animal and human tumors by Overgaard and Nielson (1980), using histopathology. In addition, inhibition of tumor blood flow has been measured by Algire and Legallais (1951), Dickson and Calderwood (1980), and Shah and co-workers (1981) after glucose injection into the host.

Table 2.7
Effect of Hyperthermia on Oxygen and Glucose Consumption of Transplanted Tumors in Sprague Dawley Rats

Tumor	Method	Temperature	Oxygen Consumption	Glucose Consumption	Reference
Walker-256: 2.3–13.2 g	Differences in A-V concentration in a tissue isolated preparation	40–41.8°C × 1 or 3 hrs	No significant change	No significant change	Gullino, Yi, and Grantham 1978; Gullino 1980
DS-carcinoma: 3–5 g		Stepwise increase from physiologic level (33–36°C) to 44°C with 20 min at each temp.	Increased from 6.1 l/g/min to 9.3 at 39.5°C, and then decreased to initial value at 42–44°C. No change in large tumors	Increased from 0.28 l/g/min to 0.41 at 39.5°C and then decreased to 0.30 at 42°C, and to 0.22 at 44°C. No change in large tumors	Vaupel et al. 1980

NOTE: Several investigators have measured these consumption rates in tumor slices in vitro (Westermark 1927; Dickson 1977).

latter model is more sophisticated and realistic, we have used the former approach in our analyses because of its simplicity and because of our lack of understanding of the physiologic feedback control system. In panting animals, we introduce thermoregulation by an increase in the respiration rate as a function of temperature and time. Similar increases are incorporated in the blood flow rates and the metabolic heat generation rates of various organs, as reported in the literature (Thauer 1965; Esmay 1969; Hardy, Gagge, and Stolwijk 1970; Hales 1974; Stolwijk 1980; Huckaba and Tam 1980).

Unlike normal tissues, the physiologic response of tumors is poorly understood. As shown in table 2.6, most tumors exhibit an increase in their blood supply up to 40°C to 41°C, and at higher temperatures blood flow begins to decrease. Walker-256 carcinoma is an exception to this rule, and does not show any increase in its blood flow up to 45°C to 46°C, and then shows a marked impairment in its blood supply at higher temperatures. As shown in table 2.7, the oxygen consumption of tumors at elevated temperatures follows essentially the same pattern as their blood supply.

With our current understanding of tumor thermoregulation, the following points must be borne in mind while modeling tumor thermal behavior at elevated temperatures:

1. In the absence of any tumor data, the following assumption may be used in developing a mathematical model. Most tumors show a moderate increase in their blood supply up to 40°C to 41°C, and then their supply may be impaired. In normal tissues, the blood flow may increase up to 44°C, and then may decrease at higher temperatures and/or longer duration of heating. This differential physiologic response may account, in part, for the selective lethal response to hyperthermia.
2. Each tumor is different, and therefore mathematical generalizations about tumors are hard to make. In addition to the intertumor differences, heterogeneities within a tumor make the situation more complex for modeling purposes.

THERMAL ENERGY ABSORBED DURING ULTRASOUND, MICROWAVE, AND RADIOFREQUENCY HEATING

In the past decade, three methods for heating a deep-seated tumor have received considerable attention: ultrasound (US), microwaves (MW), and radiofrequency (RF) (fig. 2.10). Various aspects of applying these techniques, including the advantages and disadvantages of each, have been reviewed recently by many investigators (Schwan 1965; Har-Kedar and Bleehen 1976; Dobson 1980; Hunt 1980; Jain 1982b). This section discusses the quantitative aspects of thermal energy generated[2] in tissues while applying these tech-

---

[2]Heat generated per unit volume (W/cm^3) is referred to as SAR (specific absorption rate) (see also chapter 13).

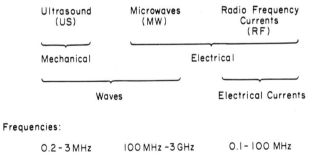

Figure 2.10. Classification of deep heating methods on the basis of heat-producing mechanisms. (Schwan 1965; Hunt 1980. Reprinted by permission.)

Table 2.8
Velocity of Ultrasound in Tissues

Tissue	Velocity (m/sec)
Muscle	1585
Liver	1590
Spleen	1555
Kidney	1560
Brain	1540
Fat	1440
Bone, skull	3360

SOURCE: Schwan 1965.
NOTE: Standard deviation about 20 m/sec. No noticeable dependence of velocity data on temperature and frequency exists. The data apply to human and animal tissues.

niques. This quantity is essential to compute the resulting temperature distributions.

Ultrasound

Ultrasonic heating is based on the absorption of high energy waves by the tissue. The mechanical energy carried by the longitudinal waves (the particles oscillate in the direction of wave propagation) is converted into thermal energy by frictional losses. Depending on the need and the instrumentation, the beam may be focused or unfocused, stationary or translocating. In addition, more than one beam may be used simultaneously or in a predetermined sequence to increase heat deposition (Marmor et al. 1978; Lele 1980; Pounds 1980[3]) (see chapter 16).

Wavelength, λ, and frequency, f, of the beam are related to the speed of propagation, C, in the medium by the following equation:

$$\lambda f = C. \qquad (10)$$

The velocity of sound in most tissues (except bone) is approximately equal to that in water, about 1500 m/sec (table 2.8). Air, on the other hand, does not allow the transmission of US from the source to the tissue. Therefore, US cannot be used to heat tissues that contain even a minute amount of air, such as the chest cavity.

The physical property that determines the absorption of US energy in a tissue is the attenuation (or absorption) coefficient, α, of the tissue. As a plane parallel beam penetrates homogeneous tissue, its intensity, I, decreases exponentially with distance, x, as a result of absorption:

$$I = I_o \exp(-\alpha x). \qquad (11)$$

Values of α as a function of the wave frequency for various normal tissues are shown in figure 2.11. Note that the value of α is largest for the bone, smallest for the fat, and in between these values for the muscle, suggesting a similar trend in the energy absorption by these tissues.

While equation (11) has been used by many investigators to compute temperature distributions in tissues, one must be aware of the following constraints while using it:

1. As a beam traverses away from the transducer, its cross-section increases linearly with the distance from the source. The divergence of beam is inversely proportional to its frequency. Note that the depth of penetration ($D = 1/\alpha$) is proportional to the frequency. These two opposite requirements restrict the operational range between 0.2 and 3 MHz, with an optimum around 1 MHz.

2. Owing to constructive and destructive interference of waves arriving at a point near the source, the intensity of a beam goes through maximum and minimum in space. These interference patterns are quite complex in the near field of the transducer.

3. Heterogeneities in the transducer may lead to cross-sectional variation in the intensity, even in the far field of the transducer.

4. Near bones or metal implants, the longitudinal waves turn into transverse (or shear) waves, which are absorbed more rapidly, leading to "hot spots." In addition, reflections by the bone may create complex standing wave patterns. Recently, Chan, Sigelmann, and Guy (1974) have computed the energy distribution resulting from the change of longitudinal waves into shear waves.

5. In the case of a focused beam, the intensity distribution in the focal region is Gaussian, the dimensions of which are difficult to predict theoretically.

In the light of these constraints, investigators have either arbitrarily assumed equation (11) to be valid (Guttner 1954; Schwan, Carstensen, and Li 1954; Chan et al. 1973; ter Haar 1980) or fitted the measured intensity distribution with a set of empirical relationships (Robinson and Lele 1969; ter Haar 1980; Parker and Lele 1980).

---

[3]D. Pounds, personal communication.

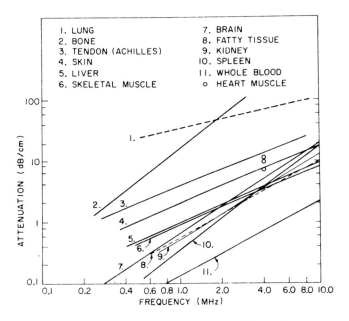

Figure 2.11. Ultrasound absorption coefficients for human tissue. (The *dashed line* represents canine data.) The data were compiled by Goss, Johnston, and Dunn (1978) and plotted by Hahn and co-workers (1980). (Reprinted by permission.)

Microwaves

Microwave diathermy, similar to ultrasound, is based on the absorption of high energy electromagnetic waves by the tissue. The energy carried by these waves is converted into thermal energy by dielectric and resistive losses. Similar to ultrasonic waves, the product of the frequency, f, and wavelength, λ, of microwaves is equal to the speed of these waves, C, in the medium. (In air and vacuum, $C = 3.0 \times 10^8$ m/sec.) While frequencies between 10 MHz and 100 GHz can be used in principle, the Federal Communications Commission (FCC) allows the use of 13.56, 27.12, 40.68, and 915 MHz, and 2.45, 5.8, and 22.125 GHz for the industrial, medical, and scientific purposes (IMS) in the United States. Outside the United States, a frequency of 433 MHz is also allowed. Use of an unassigned frequency requires careful metal screening (viz., a Faraday cage) so that the radiated energy is less than 15 μV/m at a distance of 1000 feet from the applicator (Dobson 1980). In addition, there is a limit on occupational exposure of 10 mW/cm power averaged over any 0.1 hour period for 10-MHz to 100-GHz range.

Similar to US, the strength of electrical field, E, resulting from the absorption of plane parallel microwaves in a homogeneous tissue decreases exponentially according to equation **(11)** (Bladel 1964). The absorption coefficient, α, is related to the dielectric constant, ε, the electrical conductivity, σ (mho/cm), and wavelength in air, λ(cm), as follows:

$$\alpha = \left(\frac{2\pi}{\lambda}\right)\sqrt{2\varepsilon}\left[\sqrt{1 + \left(\frac{60\lambda\sigma}{\varepsilon}\right)} - 1\right]^{1/2}. \quad (12)$$

The absorbed power density, Q, resulting from the electrical field, E, is given by:

$$Q = \frac{\sigma}{2} E^2. \quad (13)$$

Since both σ and ε are functions of the wave frequency and temperature, Q will depend upon the frequency and temperature.

A number of investigators have measured ε and σ of various normal tissues. (Durney et al. 1978; Burdette, Cain, and Seals 1980; Peloso 1981). The data on neoplastic tissues are scant, however. Recently, Schepps and Foster (1980) have measured σ and ε for hemangiopericytoma, intestinal leiomyosarcoma, and splenic hematoma, and Hahn and co-workers (1980) have measured for a canine fibrosarcoma.

We have measured σ and ε of various mammary carcinomas and 9L-glioma at 37°C and 43°C and have compared the results with normal tissues (Peloso 1981). Shown in figures 2.12 and 2.13 are our data on σ and ε, respectively, for 9L-glioma. These data have been fitted by the following equations:

$$\log \varepsilon = a + b \log f + \frac{C}{\log f} \quad (14a)$$

$$\log \frac{\sigma}{\omega \varepsilon_0} = d + e \log f. \quad (14b)$$

The parameters incorporated in two equations were estimated for various tumors at 37°C and 43°C, and are compared in table 2.9. Here, $\omega = 2\pi f$ and $\varepsilon_0 = 1/(36\pi \times 10^{11})$ in farads/cm.

For normal tissues, both ε and σ vary with the temperature according to the following relationships (Schwan 1965):

$$\frac{\Delta \varepsilon}{\varepsilon} \approx -.05\%/°C \text{ and } \frac{\Delta \sigma}{\sigma} \approx 2\%/°C. \quad (14c)$$

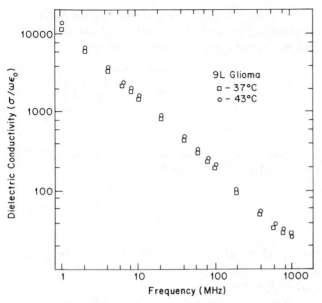

Figure 2.12. Electrical conductivity ($\sigma/\omega\epsilon_0$) of 9L-glioma at 37°C and 43°C as a function of frequency. (Peloso 1981.)

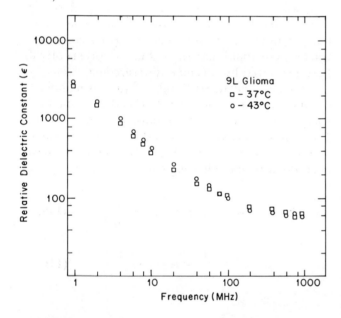

Figure 2.13. Relative dielectric constant (ϵ) of 9L-glioma at 37°C and 43°C as a function of frequency. (Peloso 1981.)

For neoplastic tissues, we have found that the temperature coefficients are of the same order of magnitude (Peloso 1981).

While evaluation of heating patterns is relatively straightforward for a homogeneous tissue exposed to far-field radiation, the following factors make it diffi-

Table 2.9
Parameters for σ (Electrical Conductivity) and ϵ (Dielectric Constant) for Various Tumors in Rats

Tumor (°C)	Parameters				
	a	b	c	d	e
9L Glioma (37)	−25.4	0.62	344	21.9	−0.902
9L Glioma (43)	−21.2	0.48	313	22.0	−0.901
Walker-256 (37)	−28.7	0.70	377	21.3	−0.872
Walker-256 (43)	−29.3	0.70	387	21.2	−0.856
MTW9A (37)	−27.8	0.72	346	21.1	−0.859
MTW9A (43)	−28.4	0.74	353	21.7	−0.895
MNU (37)	−27.8	0.75	337	22.1	−0.910
MNU (43)	−33.7	0.89	353	22.7	−0.941
MTW9 (37)	−22.9	0.60	301	21.6	−0.890
MTW9 (43)	−27.5	0.72	342	21.7	−0.890
13762 (37)	−28.0	0.70	361	21.1	−0.862
13762 (43)	−24.1	0.60	324	21.3	−0.867

SOURCE: Peloso 1981.

cult to compute heating patterns for heterogeneous media with realistic geometries:

1. The microwaves are transmitted through, absorbed by, and reflected at biological interfaces as a function of the tissue size and geometry, tissue composition and properties, wave frequency, and source design. Absorption is high, and depth of penetration ($D = 1/\alpha$) is low in tissues of high water content (e.g., muscle, brain, internal organs, skin), and the opposite is true in tissues of low water content (e.g., fat, bone). In addition, interference between the transmitting and reflecting waves coming from an interface separating tissues of different complex dielectric constant, $\epsilon^* (= \epsilon - i\sigma/\omega\epsilon_o)$, may create standing waves, resulting in hot spots. Figures 2.14 and 2.15 show the heating patterns generated in two-layer (fat-muscle) and four-layer (fat-muscle-bone-muscle) models of tissue exposed to plane waves. Analytic and numerical results for plane wave absorption by tissues of various geometries have been obtained by many investigators, as summarized in table 2.10.

2. Although noncontact-type applicators are used widely to induce hyperthermia, the direct contact applicators are being developed because they reduce unwarranted reflection from the skin, making their use safer to both the patient and the operator. The complex near field heating patterns generated by the contact applicators, however, have hindered the progress in their design. Investigators have, therefore, taken two approaches to optimize an applicator design: measure experimentally the heating patterns in a test material (Kantor 1977) or evaluate numerically the heat deposition using various novel solution techniques. Table 2.11 summarizes the attempts of the latter type for various types of contact apertures.

3. In order to heat large volumes of the body (e.g., chest cavity) or a deep-seated tumor, many investigators have used multiple applicators placed in a specified geometry (Holt 1975; Turner 1980). Recently, Turner and Kumar (1980) have obtained approximate numerical solutions for such applicators by replacing the original aperture field with an array of point-source dipole radiators. Shown in figure 2.16 is the result of such a computation for a homogeneous tissue-cylinder phantom heated by an annular array applicator. Note that the analysis is in qualitative agreement with the data.

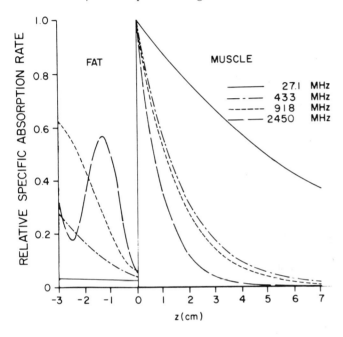

Figure 2.14. Relative absorbed power density patterns in a two-layer model (fat and muscle) exposed to a plane wave microwave source. (Johnson and Guy 1972. Reprinted by permission. © 1972 by the IEEE.)

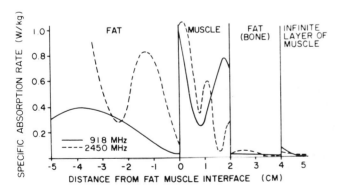

Figure 2.15. Relative absorbed power density patterns in a four-layer model (fat, muscle, bone, and muscle) exposed to a plane wave microwave source. (Johnson and Guy 1972. Reprinted by permission. © 1972 by the IEEE.)

4. In order to overcome the poor penetration problem associated with the microwaves in the GHz range, several investigators place the needlelike microwave applicators inside the tumor or in an adjacent cavity. Insertion of radioactive needles is a common practice in radiotherapy. While the heating pattern around these probes are being studied experimentally by many investigators (Arcangeli et al. 1982; Trembly et al. 1980) theoretical attempts to predict the patterns have not been reported.

Radiofrequency Heating

Heating by radiofrequency[4] currents can be achieved by coupling the electromagnetic energy to the tissue either capacitively (also referred to as dielectrically) or inductively. These techniques can be applied either invasively or noninvasively. In principle, electrical currents with frequency greater than 10 kHz (to avoid muscle contractions associated with electrical shock) may be used for RF heating; however, owing to FCC regulations, 13.56- and 27.12-MHz frequencies are used most commonly.

In capacitive heating mode, high frequency currents are passed through the tissue volume via appropriately placed electrodes. The current density in the tissue can be controlled by varying either the interelectrode distance or the electrode geometry. Conduction currents lead to ohmic (or resistive) heating, and the displacement currents lead to the dielectric heating. Therefore, from the knowledge of the electromagnetic properties (σ and ε), wave frequency, and geometry, it is possible to calculate heat generation in the tissue. When the ohmic heating dominates, (i.e., when $\sigma > \omega\varepsilon$), the heat generated, Q, is given by:

$$Q = \frac{I^2}{\sigma} \frac{\text{watt}}{\text{cm}^3}, \qquad (15)$$

where, I is the current density (ampere/cm³).

In an inductive heating mode, RF currents are passed through a well insulated, flexible heavy wire arranged in regular coils. Alternating current passing through the coils generates an oscillating magnetic field, which in turn produces eddy currents in the tissue volume. The current density, I, is proportional to the rate of change of the magnetic field, $\frac{dB}{dt}$, and to

---

[4]Also referred to as high frequency (HF) or ultrashort wave (USW) diathermy. These names incorrectly suggest that waves are involved in heating the tissue.

Table 2.10
Theoretical Analysis of Power Absorbed by Tissues Exposed to Plane Microwaves

Tissue Model	Reference
Plane tissue layers (semi-infinite slab)	
Isotropic	
2-Layer model (fat-muscle)	Schwan 1953; Schwan and Piersol 1954, 1955
3-Layer model (skin-fat-muscle)	Schwan and Li 1956; Schwan 1965
4-Layer model (fat-muscle-bone-muscle)	Johnson and Guy 1972
Anisotropic	
3-Layer model (skin-fat-muscle)	Johnson et al. 1975
Cylindrical tissue layers	
Homogeneous tissue	Massoudi, Durney, and Johnson 1979*
3-Layer model (fat-muscle-bone)	Ho et al. 1969; Ho 1975b
Spherical tissue layers	
Homogeneous sphere	Anne 1963; Kritikos and Schwan 1972, 1975; Lin et al. 1973; Lin, Guy, and Johnson 1973,[†] Ho 1975a; Johnson and Guy 1972
2-Layer model (fat-muscle)	Anne 1963
3-Layer model (skin-fat-muscle)	Hand 1977
4-Layer model (brain-skull-fat-skin)	Kritikos and Schwan 1976
5-Layer model (brain-CSF-dura-bone/fat-skin)	Ho 1975a
6-Layer model (brain-CSF-dura-bone-fat-skin)	Shapiro, Lutomirski, and Yura 1971; Joines and Spiegel 1974; Weil 1975; Neuder et al. 1976
Prolate spheroid and ellipsoid models	
Homogeneous spheroid	Johnson, Durney, and Massoudi 1975; Durney, Massoudi, and Johnson 1975; Wu and Lin 1977; Barber 1977[‡]; Rowlinson and Barber 1979*
Homogeneous ellipsoid	Massoudi, Durney, and Johnson 1977
Long cylinders of arbitrary cross-section	
Homogeneous tissue	Wu and Tsai 1977[§]
3-Layer model (skin-fat-muscle)	Neuder and Meijer 1976[‖]
Finite biological bodies	
Finite planar model	Livesay and Chen 1974
Isolated section of body	Guru and Chen 1976; Rukspollmuang and Chen 1979
Part-body and multibody analysis	Gandhi, Hagerman, and D'Andrea 1979
Block model of human	Chen and Guru 1977a, 1977b; Hagerman and Gandhi 1979

*Geometric optics technique.
[†]Radiofrequency (1–20 MHz); solution obtained by combination of quasi-static electric and magnetic solutions.
[‡]Extended-boundary condition method.
[§]Integral equations and moment method technique.
[‖]Finite element method.

Table 2.11
Sources Other Than a Plane Wave

Tissue Model	Source	Reference
Plane tissue layers (semi-infinite slab)		
2-Layer model (fat-muscle)	Dielectric loaded dipole-corner reflector applicator	Guy and Lehman 1966
	Direct contact rectangular aperture	Guy 1971
Cylindrical tissue layers		
3-Layer model (fat-muscle-bone)	Direct contact aperture	Ho et al. 1971
Spherical tissue layer		
1- and 5-layer model (brain-CSF-dura-bone/fat-skin)	Direct contact applicator	Ho 1975a
6-Layer model (brain-CSF-dura-bone-fat-skin)	Loop or dipole antennas	Hizal and Baykal 1978

the tissue conductivity, σ. Therefore, the heat generated, Q, is given by:

$$Q = \frac{\sigma}{2} \left(\frac{dB}{dt}\right)^2 \frac{\text{watt}}{\text{cm}^3}. \qquad (16)$$

Since σ for fat and skin is an order of magnitude less than for the muscle, inductive heating, unlike the capacitive heating, permits heating of deeper tissues more easily. In addition, skin cooling by air or water may be used in both modes of heating to help protect surface tissues during deep-heat application.

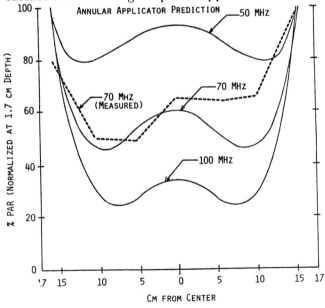

Figure 2.16. Relative power absorption rates in a homogeneous thorax model exposed to an annular array applicator at three frequencies: 50, 70, and 100 MHz. The *broken line* shows a typical measured pattern at 70 MHz. (Turner and Kumar 1980. Reprinted by permission.)

Similar to microwaves, RF heating may be produced by one or many applicators surrounding the tissue volume. In addition, the probes may be implanted inside the tumor or in a cavity adjacent to it. Unlike microwaves, however, not much effort has been put into computing heat distribution in tissues.

Guy, Lehmann, and Stonebridge (1974) have recently computed specific heat generation rates in a two-layer (fat-muscle) model exposed to a flat "pancake" induction coil. These investigators found heat generated to be nonuniform radially (toroidal in shape) and exhibiting a maximum temperature at the muscle-fat interface (see chapter 13).

Sustained effort is now needed to predict power distribution in normal and neoplastic tissues to exploit optimally these novel methods of heating.

TEMPERATURE DISTRIBUTIONS DURING NORMOTHERMIA

A proper understanding of temperature distribution during normothermia is needed for improving thermography for cancer detection (Amalric et al. 1975; Gershon-Cohen et al. 1964).

If the tumor is considered infinite and system parameters are constant, the temperature of the tumor, T, is equal to the arterial blood temperature, T_a, plus the temperature rise owing to metabolic heat generation, T_m: $T = T_a + T_m$ where $T_m = Q_m/\rho_b C_p P$. Shown in figure 2.4 is the value of T_m as a function of tumor weight (Sien and Jain 1979). It is of interest to note that T_m is less than 0.25°C and does not change significantly with tumor weight.

For tumors of finite radii, which interact thermally with the surroundings at temperature T_s, the calculated temperature profiles are shown in figure 2.17. The results indicate that the temperature profile throughout a small tumor is lower than that in large tumors for a fixed, uniform blood-flow rate. This result is not surprising if one realizes that the temperature profile is determined by both the total metabolic heat generation rate (which is proportional to the tumor volume) and the total rate of heat loss to the surroundings (which is proportional to the outer surface area). Consequently, for a fixed P, Q_m, and T_s, a higher total-heat generation rate per unit surface area in large tumors leads to a higher temperature profile in large tumors when compared to small ones (Jain 1978).

Figure 2.18 shows the temperature distribution in human mammary epitheliomas, measured by Gautherie and co-workers (1972). These data suggest that temperatures are maximum at the centers of these tumors, and are in qualitative agreement with our analysis (fig. 2.17) (Jain 1979). For quantitative comparison, some additional assumptions must be made about the thermal symmetry of the system. In reality, the tumor is located between the skin exposed to the ambient air at 25°C, and the chest at 36°C to 37°C. Hence, the temperature of surroundings, T_s, is not uniform. It is possible, however, to set bounds on the temperature distributions in the tumor by considering

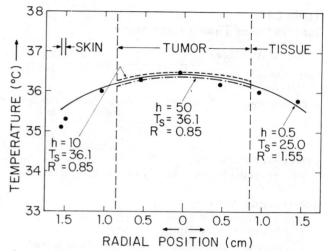

Figure 2.18. Comparison of model predictions with the data on temperature distributions in a human mammary epithelioma and the surrounding tissue. *Solid line* (———) refers to the first approach, and *dotted lines* (- - - and -----) refer to the second approach discussed in the text. (Jain 1980a. Reprinted by permission.)

two limiting cases: (1) the tumor and the surrounding tissue can be considered as a sphere of radius, R = 1.55 cm, exposed to ambient air temperature (25°C); or (2) the tumor alone (R = 0.85 cm) is surrounded by a tissue at 36.1°C. The results of computation using these two approaches are compared with data in figure 2.18.

The effect of nonuniformities in the perfusion rate on the temperature profile is shown in figure 2.19 for a tumor of radius R = 1.5 cm. The results are similar for tumors of different radii. Because of its large blood-flow rate and higher rate of metabolism, the temperature in the periphery of a necrotic tumor is higher than that in a tumor with a uniform blood-flow rate, although both tumors have the same average perfusion rate. Inhomogeneities in blood-flow rate, therefore, tend to make the temperature profile flat. These results are in qualitative agreement with the data of Gautherie (1980). Our current efforts are directed toward improving this analysis by considering thermal asymmetry of the tumors.

TEMPERATURE DISTRIBUTIONS DURING HYPERTHERMIA

Currently available "physical" methods of producing hyperthermia can be divided into two categories: surface heating and volume heating.

In surface heating, the temperature is increased using one or more of the following methods: immersion in hot-water bath or hot-air incubator, exposure

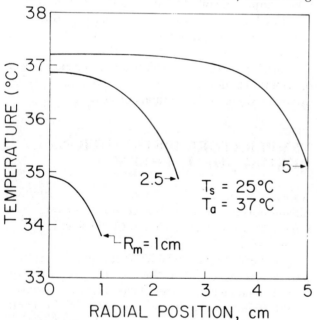

Figure 2.17. Effect of tumor radius on the temperature distribution in uniformly perfused tumors of various radii. For an infinite tumor, initial temperature would be uniform at 37.226°C (T = $T_a + T_m$). Note that blood perfusion rate is assumed independent of tumor weight in these simulations. (Jain 1980a. Reprinted by permission.)

Bioheat Transfer: Mathematical Models of Thermal Systems

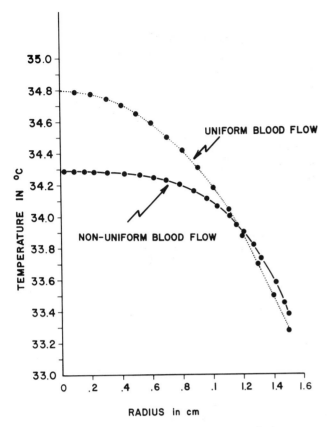

Figure 2.19. Effect of nonuniformity in perfusion rate on temperature distribution in a 1.5 cm (radius) tumor. Note that nonuniformity in these calculations is considered parabolic. (Jain 1980a. Reprinted by permission.)

to visible or infrared radiation, or direct contact with an interface at a higher temperature. Heat transfer, then, takes place from the skin to the underlying tissue by conduction and convection. In volume heating, the temperature of underlying tissue is brought to an elevated temperature either by perfusing the tissue with preheated blood or by exposing it to US, MW, or RF currents. The only "nonphysical" method of inducing hyperthermia is the injection of mixed bacterial toxins (MBT), or some other pyrogenic agents, into the host (Coley 1893).

Depending on the surface area or volume heated, hyperthermia can be divided into three categories: (1) local, (2) regional, and (3) whole body. In local hyperthermia, tumor mass and minimal surrounding tissue are heated; in regional hyperthermia, usually the limb containing the tumor mass is heated; in whole-body hyperthermia (WBH), the temperature of the entire host is elevated.

No matter which method is used to induce hyperthermia, it is essential to monitor the resulting temperature distributions in the normal and neoplastic tissues serially or continuously and to control the power input and position of heating source so that the temperatures are maintained in the desired range for an optimal time. This type of control can be achieved either manually or automatically and is incorporated in most commercial or in-house-built hyperthermia systems currently in use.

The objective of this section is to discuss the quantitative aspects of heat transfer and temperature distributions during hyperthermia.

Hot-Water Bath and Moist Hot-Air Incubator

The two simplest methods for surface heating in vitro and in vivo employ a hot-water bath and a moist hot-air incubator. They have been used to induce localized, regional, and whole-body hyperthermia in animals and humans. The principal mode of heat transfer in these methods is conduction.

For the past 10 years, several research groups in Europe have been using a hyperthermia cabin designed by Siemens, in which hot air (50°C to 60°C) is circulated to maintain the temperature of the patient around 41°C to 42°C. In order to increase the initial rate of heating, this unit has provisions to install a 27-MHz RF heating unit and a 433-MHz MW applicator. The cabin is made from a transparent polymeric material that allows observation of the patient (Reinhold et al. 1980).

Recently, we have used the lumped parameter model to simulate the temperature distribution data in a tumor-bearing rat during whole-body and local hyperthermia induced by a water bath. Figure 2.20 shows the temperature data obtained by Gullino, Yi, and Grantham (1978) in Walker-256 carcinoma (2.5 g) during whole-body hyperthermia. In these experiments, the tumor-bearing Sprague-Dawley female rat (200 g), anesthetized with urethan (1 mg/kg), was immersed vertically, tail down, into a well stirred, heated waterbath (10 cm^3, 41°C) until the water reached its mandible. As shown in figure 2.20 the computed values adequately reproduce those measured experimentally: a rapid elevation of 4°C to 5°C during the first five minutes and a slower increment of 4°C to 5°C for the remaining 20 minutes; the time required to reach 41°C was within the experimental range of 20 to 30 minutes. The normal tissues followed the same time course (Sien and Jain 1979).

Figure 2.21 shows the temperature data of Dickson and Suzanger (1974) in Yoshida sarcoma transplanted in the foot of Wistar rats (200 g) during localized hyperthermia. In these experiments, the rat, anesthetized with pentobarbital sodium (Nembutal™) (2 ml/kg), was placed over a water bath (15 cm^3, 42.7 ± 0.05°C),

29

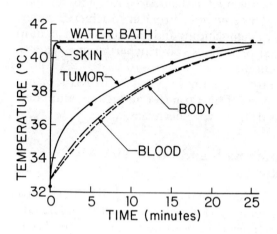

Figure 2.20. Temperature distributions in Walker-256 carcinoma (2.5 g) and normal tissues of a Sprague Dawley female rat (200 g) during whole-body hyperthermia induced by immersing the rat in a constant temperature water bath (41°C) (Gullino, Yi, and Grantham 1978). *Points* represent the data points, and *lines* represent the model predictions. (Sien and Jain 1979. Reprinted by permission.)

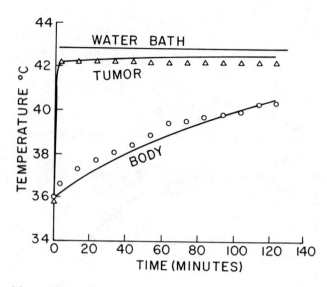

Figure 2.21. Temperature distributions in Yoshida sarcoma and normal tissues of a Wistar rat (200 g) during localized hyperthermia induced by immersing the tumor-bearing foot of the rat in a constant temperature water bath (42.7°C) (Dickson and Suzanger 1974.) *Points* represent the data points and *lines* represent the model predictions. (Chrysanthopoulos and Jain 1980. Reprinted by permission.)

and the tumor-bearing foot was inserted into the water bath through a 10-cm diameter padded opening to a depth that permitted complete submersion of the tumor. As shown in figure 2.21, the computed values agree closely with those measured experimentally: a rapid elevation of 6°C during the first two to three minutes, followed by a slow asymptotic convergence to the waterbath temperature. The temperature of the normal tissues followed a smooth time course to reach 40°C in about 120 minutes (Chrysanthopoulos and Jain 1980).

Stolwijk (1980) has used the whole-body model to simulate the temperature distributions in normal humans, when heat is deposited into the trunk core via a water bath. In these computations, heat loss by sweat evaporation is set equal to zero (fig. 2.22). Huckaba and Tam (1980) also have analyzed the temperature rise in various organs of a human placed in an incubator. We are improving the models developed by Stolwijk and Huckaba and Tam to predict temperature distributions in a cancer patient. As seen in figure 2.23, we are able to describe the transient data of Huckaba and Tam adequately (Volpe and Jain 1982). We have also used this model to compare various modes of inducing whole-body and local hyperthermia in patients (Volpe and Jain 1982).

In order to exploit whole-body hyperthermia effectively, it is essential to measure as many thermal, physiologic, and biochemical parameters as possible without harming a patient, and to incorporate this information in future analyses.

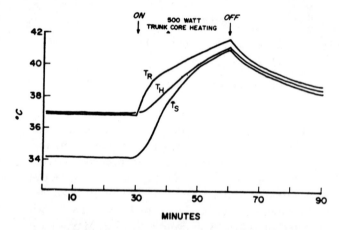

Figure 2.22. Simulation of the effect of depositing 500 watts of heat into the trunk core of a man for a period of 30 minutes. Sweat is not allowed to evaporate. Shown are the trunk core temperature (T_R), brain temperature (T_H), and mean skin temperature ($\overline{T}_s$). (Stolwijk 1980. Reprinted by permission.)

Space-Suit and Blanket Techniques

These two techniques also are based on conduction of heat from a hot fluid (usually air or water) to the skin, except that the patient is not in direct contact with the fluid. Containment of fluid in a suit or thermal blanket makes the procedure more manageable and convenient (Herman, Zukoski, and Anderson, in press). The models discussed can easily be modified to predict temperature distributions during the use of these methods (Volpe and Jain 1982).

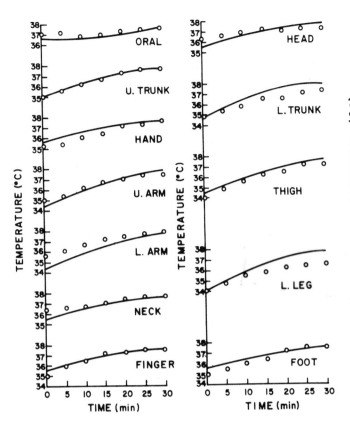

Figure 2.23. Comparison of computed *(lines)* and measured *(dots)* temperatures of the various segments of a human subject undergoing whole-body heating. (Volpe and Jain 1982. Reprinted by permission.)

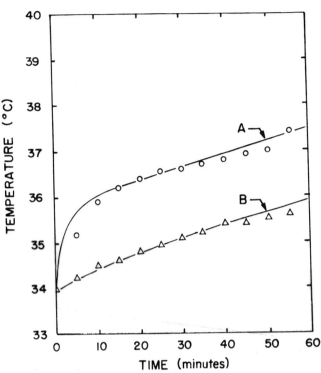

Figure 2.24. Comparison between model simulations and experimental data on intratumor temperature distributions during localized hyperthermia. *Points* represent the experimental data measured by thermistors *A* and *B* placed 1 cm apart in the tumor, and *solid lines* represent model simulations. Note the intratumor temperature gradient of 2.2°C/cm after 20 minutes. (Sien and Jain 1980. Reprinted by permission.)

Infrared and Visible Radiation

Every object with temperature above absolute zero emits infrared radiation. The rate of emission is proportional to the fourth power of absolute temperature. Similarly, very hot bodies (> 2000°K) and ionized gases emit visible radiation. By exposing the whole body or part of it to the radiation source, the temperature of the skin is brought up, and heat absorbed is transferred to the underlying tissues by convection and conduction. See Stoner (1965) for the details of source design and applications.

We have recently measured the temperature distributions inside a tumor during localized hyperthermia induced by infrared radiation (Sien and Jain 1980). Figure 2.24 shows our data on intratumor temperature distributions measured by two thermistors placed 1 cm apart in Walker-256 during localized hyperthermia. In these experiments, a specified portion of skin above the tumor was heated by a diathermic lamp. As shown in figure 2.24, the computed values agreed adequately with the data: a rapid elevation of 2°C during the initial 10 minutes and a slower increment for the remainder of the experiment in the thermistor close to the heated skin area; the initial temperature rise, measured further away by the second thermistor, was less than 1°C in 20 minutes. A temperature difference of 2.2°C between two points in the tumor persisted after 20 minutes. This kind of information should be kept in mind while heating patients for cancer treatment, especially when the precise location of a tumor is not known.

Pettigrew Technique (Hot-Wax Bath)

An elegant, effective, and simple method for producing whole-body hyperthermia was introduced by Henderson and Pettigrew (1971) in the Western General Hospital, Edinburgh, Scotland. In this method, a patient is enclosed in a sealed covering of polyethylene sheet to eliminate evaporative heat losses, and then immersed in a bath of molten paraffin wax having a melting point of 46°C. Heat is transferred from the wax to the patient's skin mostly by conduction and, consequently, wax begins to solidify at the polyethylene envelope surrounding the patient. As a result, an increasingly thick layer of solid wax begins to form and is in equilibrium with the skin temperature on one side of the envelope and at its melting point (46°C) on the other side. The skin temperature rises very rapidly in the beginning and then approaches steady state slowly as the thickness of solid wax increases. The wax film acts as an insulator to minimize heat loss by conduction and convection from the patient.

Decubitus ulcer (pressure sore) is a common problem in patients lying on their back for a long time. Since the density of the molten wax is slightly higher than the patient's, the buoyancy force provides an ideal support to the patient.

In order to increase the initial rate of heating, some investigators provide a vent for their patients by a mixture of hot air (usually, 50% mixture of helium and oxygen). Others use a microwave diathermy apparatus (433 MHz) to increase the tumor temperature (Levin, Wasserman, and Blair 1980).

Recently, Law and Pettigrew (1980) have developed a simple mathematical model to predict the core and skin temperature during hyperthermia using Pettigrew technique (fig. 2.25). While this is in qualitative agreement with their data, it underpredicts the thickness of solid wax and, as a result, leads to discrepancies in heat transfer to the patient. Nevertheless, this model can be improved easily to provide more detailed and accurate information on the temperature distribution in a patient (Volpe and Jain 1982).

Regional and Systemic Perfusion with Heated Blood

While the technology for isolated perfusion of an organ or a limb has existed for a long time, heating of limbs containing tumors started in 1967 (Cavaliere et al. 1967, 1980; Stehlin 1969, 1980) and led to significant improvements in the long-term survival of patients. In this technique, blood is heated to about 43°C in a heat exchanger, then circulated through the limb to heat the muscle to 40°C to 42°C by convection. There are two major problems in this technique: (1) loss of heat through the skin, which is compensated for by the use of infrared lamps or warm rubber blankets; and (2) exchange of heat with the systemic circulation. Surrounding tissues make the selective heating difficult with regional perfusion (Volpe and Jain 1982) (see chapter 17).

Implantation of a Heat Source and Electrocoagulation

Direct implantation of a heat source into the tumor has been used by Sutton (1971) in the treatment of malignant gliomas of the brain. The procedure showed promising results for early gliomas prior to widespread invasion of normal brain.

Strauss (1969) also electrocoagulated ($\geq 70°C$) a large number of tumors by bringing a resistive heat source in contact with the tumors. Strauss obtained permanent cures in advanced patients, which he attributed to the stimulation of the host immune response.

We have recently obtained analytic and numerical solutions for transient and steady-state temperature distributions in and around the heating probe for both uniformly and nonuniformly perfused tumors (fig. 2.26). These solutions are valid for several types of heating functions, such as a step input, a single or a series of pulses, a transient input of the form $a + b/\sqrt{t}$ (Jain 1978, 1979, 1980a, 1980b).

Shown in figure 2.27 is the temperature profile in and around a tumor with a 1.5-cm radius before the onset of heating. Note that, as expected, the temperature profile within the probe is flat. Temperature in the center of a necrotic tumor is lower when compared to a viable tumor, although both tumors have the same average perfusion rate. These results are analogous to the case when the tumor is not grown around a probe (fig. 2.19).

Shown in figure 2.28 is the steady-state temperature profile for the same probe and tumor resulting from a step heat input. Since the temperature response is linear with respect to the magnitude of step input, $Q(W/cm^3)$, the temperature is nondimensionalized in the following manner:

$$\Theta = \frac{T - T_o}{QR^2/3k}. \qquad (18)$$

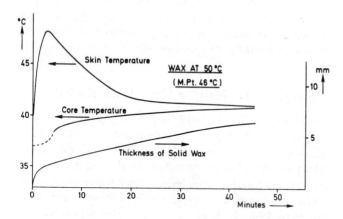

Figure 2.25. Computed skin and core temperatures in a patient undergoing whole-body hyperthermia via the hot-wax bath technique. Note that the molten wax solidifies with time. (Law and Pettigrew 1980. Reprinted by permission.)

Bioheat Transfer: Mathematical Models of Thermal Systems

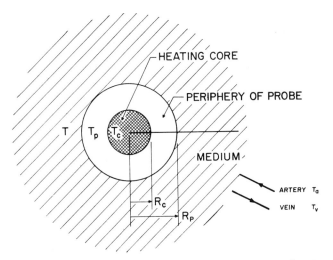

Figure 2.26. Schematic diagram of the tumor (medium) containing a spherical heating probe in its center. This model is applicable in estimating temperature distributions in tumors during localized hyperthermia. T is the temperature of the tumor (medium). T_p, T_c, T_a, and T_v represent the respective temperatures at the periphery of the probe, heating core, artery, and vein, at a heating core radius of R_c (core) and R_p (periphery). (Jain 1978. Reprinted by permission.)

Pyrogenic Agents and Bacterial Toxins

Coley (1893) obtained remarkable success in controlling advanced disease in many patients by injecting mixed bacterial toxins (MBT) that induced fever. Owing to poor biological characterization, however, many investigators have been unable to reproduce Coley's work. To this author's knowledge, no attempt has been made to model temperature distributions during induced fever.

Ultrasound

Although ultrasound has received special attention in inducing hyperthermia for cancer treatment, there is a paucity of data and analyses of the resulting temperature distributions in tissues. The first attempts to compute temperature distributions during ultrasonic therapy are attributed to Guttner (1954) and Schwan, Carstensen, and Li (1954). Schwan, Carstensen, and Li calculated temperature distributions in a two-layer

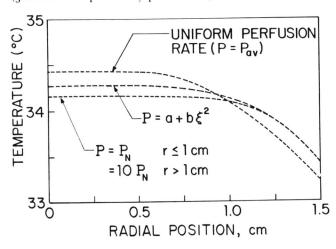

Figure 2.27. Effect of nonuniformities in perfusion rate on the temperature distribution in a 1.5-cm (radius) tumor grown around a 0.5-cm (radius) heating probe, as shown in figure 2.26. Note that the heater is off. (Jain 1978. Reprinted by permission.)

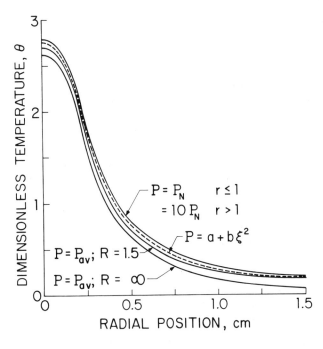

Figure 2.28. Effect of nonuniformities in perfusion rate on steady-state temperature rise in the probe and tumor as a result of step heat input. The probe radius is 0.5 cm, and the tumor radius is 1.5 cm. (Jain 1978. Reprinted by permission.)

Here, T_o is the steady-state temperature before the onset of heating (fig. 2.27), T is the steady-state temperature during heating, and R_p is the radius of the heating probe. As expected, because of a poor convective heat transfer, the temperature rise in a necrotic tumor is higher than that in a viable tumor. Temperature rise in an infinite and uniformly perfused tumor is the lowest (fig. 2.28).

Table 2.12
Temperature Probes for Measurement in Electromagnetic Fields

Sensor	Leads	Temperature Range, °C	Resolution °C	Diameter mm	Advantages
Thermistor	High-resistance leads (carbon impregnated plastic)	Adequate	0.1	1–2	Relatively accurate, reproducible, and stable
Liquid crystal	Fiber optic	35–50	0.1–0.5	1–2	Uses commercially available equipment*
Birefringent crystals (lithium tantalate)	"	18–49	0.1	1	Available commercially.[†] These devices have the greatest inherent sensitivity in the required temperature range.
Semiconductor (gallium arsenide)	"	20–50	0.1–0.2	0.25	Available commercially.[†] Currently available optical fibers exhibit minimum transmission loss in the wavelength range of gallium arsenide emission; relatively small in size.
Phosphors	"	9–250	0.025	0.4	Available commercially.[‡] Best prospects for accurate, low-cost, interchangeable, small probes.

*Ramal Inc., Sandy, Utah.
[†]Yellow Springs Instrument Co., Yellow Springs, Ohio.
[‡]Luxtron Corp., Santa Clara, California.

(fat-muscle) model exposed to plane parallel (unfocused) waves, whereas Guttner carried out the analysis for a three-layer (fat-muscle-bone) model. Both groups of investigators neglected convection and diffusion in their analyses and calculated temperature rise on the basis of the local absorption of ultrasonic energy given by equation (**11**). Chan and colleagues (1973) modified the analysis of Guttner by including conduction and convection in their layer model. In 1974, these authors presented a detailed analysis for the reflection and transmission of ultrasonic waves, as well as conversion of longitudinal waves into shear waves at the tissue boundaries. These authors also studied the role of the angle of incidence on the relative heating pattern; however, no attempt was made to compute the resulting temperatures (Chan, Sigelmann, and Guy 1974). Recently, ter Haar (1980) has repeated these calculations for an unfocused beam impinging on a three-layer model to study the sensitivity of temperature distributions to the wave frequency, incident energy (I_o, in equation **11**), skin temperature (which can be controlled by a water bath), tissue composition, and blood flow rate. In one set of computations, ter Haar has also used the measured intensity distribution from an ultrasonic source to compute temperature distributions in a layered model. Her results are in qualitative agreement with her data obtained in the hind leg of an anesthetized pig.

One of the first attempts to simulate temperature elevations caused by a stationary focused beam was by Robinson and Lele (1969). Although the focal region of their beam was prolate spheroid with a Gaussian distribution of intensity, these authors assumed the source to be either spherical or cylindrical in shape in their computations. In addition, the role of blood flow was analyzed by using an effective thermal conductivity term, similar to equation (**2**). Despite many assumptions made in their analysis, these authors obtained adequate agreement between their analysis and data on a cat's brain.

Recently, Parker and Lele (1980) have calculated the temperature distribution in a cat's brain insonated with one or more moving, focused beams of ultrasound. These authors have assumed that the insonation creates a temperature "forcing function" along its axis, with highest temperatures occurring in the focal plane. Using the BHTE, these authors obtained the steady state temperature profile that reproduced their data adequately.

While some investigators (Marmor et al. 1978; Lele 1980) have reported tumor temperature distributions caused by unfocused and focused beams, theoretical analyses of these data have not been carried out (see chapter 16).

Disadvantages	Reference
Use of high-resistance leads requires highly sophisticated electronics and a short lead length to keep the total resistance within bounds. Probes are large in diameter.	Larsen, Moore, and Acevedo 1974; Bowman 1976
Instability and hysteresis in the liquid crystal material require frequent calibration. Probes are large in diameter.	Rozzell et al. 1974; Deficis and Prion 1977
Because of source instabilities, it is necessary to monitor and correct for the source fluctuations. Probes are large in diameter.	Cetas 1976; Cetas and Connor 1978
Source instabilities need to be monitored and corrected for.	Christensen 1977, 1979
Requires a fairly sophisticated optical and electronic system.	Christensen 1975; James, Quick, and Strahan 1979; Wickersheim and Alves 1979; Samulski and Shrivastava 1980

Microwaves

Although considerable effort has been put into the analysis of energy absorbed by model tissues during microwave diathermy (see section on absorption of thermal energy), only limited studies have been done on the measurement or prediction of resulting temperature distributions in vivo.

The efforts to monitor temperature distributions in tissues exposed to electromagnetic fields by thermocouples and thermistors have been limited for the following reasons: (1) they scatter incident electromagnetic radiation into the surrounding tissue, causing a perturbation of the electromagnetic field and modifications in the thermal pattern; (2) the incident electromagnetic field induces an electric field in the probe material and, as a result, leads to ohmic heating, and (3) the current induced in the connecting wires of a probe interferes with the signal from the probe.

Investigators using MW and RF heating have avoided some of these problems by (1) turning off the field for a short time period for periodic temperature measurements, (2) orienting the probe and connecting leads properly so that the interference is minimum, (3) using a probe as small as possible, and (4) using nonconducting probe and leads (table 2.12).

Recently, as a result of several developments in the optical fiber technology, several new probes have been developed and tested that are electrically isolated and nonperturbing to the electromagnetic field. Table 2.12 compares several probes, many of which are currently available commercially. While these probes differ from each other in several ways, they all modify the intensity, wavelength, polarization, emission, or reflection of incident light as a function of their temperature (Christensen 1979; Wickersheim and Alves 1979). This change in the optical data is conducted through optical fibers to an optoelectronic processor to obtain analog or digital signals for recording the temperature.

Despite these new developments, only limited data have been reported in the literature on the detailed temperature distributions during MW diathermy (Sandhu, Kowal, and Johnson 1976). Figure 2.29 shows the steady-state temperature distribution in a pig exposed to microwaves; 915 MHz is more effective in heating at less than a 2-cm depth, while 430 MHz is effective at greater depths. The depth of heating can be increased by skin cooling (Subjeck et al. 1980).

Assuming the exponential decay in the electromagnetic field (cf. equation 11), Chan and associates (1973) recently have calculated the temperature distribution in a three-layer (fat-muscle-bone) model of human thigh. To account for the thermal regulation of the tissue, these authors introduced a nonlinear convective heat transfer term with a delay function in

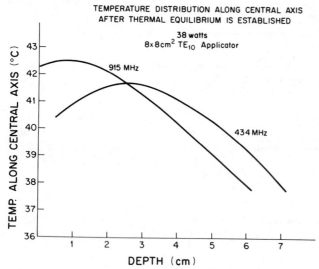

Figure 2.29. Steady-state temperatures measured as a function of tissue depth in pigs exposed to 434- and 915-MHz microwaves. Note that 434 MHz is more effective for heating at depths greater than 2 cm. (Subjeck et al. 1980. Reprinted by permission).

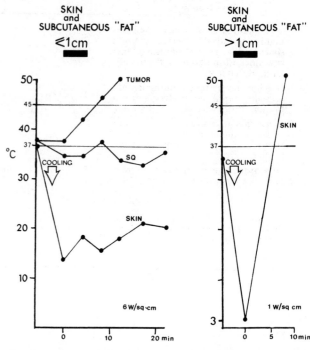

Figure 2.30. Temperature changes in normal (skin and subcutaneous fat) and neoplastic tissues of a human subject undergoing localized hyperthermia with RF capacitance heating. Note that surface cooling appears to be effective only if skin and subcutaneous tissue are less than 1 cm in thickness. (Storm et al., in press. Reprinted by permission.)

the BHTE. Their results are in qualitative agreement with their data obtained for 915-MHz microwaves. Kritikos and Schwan (1979) also have calculated the temperature rise in a spherical region simulating a hot spot in the central region of a human head; however, the assumption that the energy absorbed is uniform in a spherical active region imbedded in an infinite region is unrealistic, and makes this analysis of limited use in designing heating protocols. Trembly and coworkers (1980) recently have calculated the temperature distributions around implanted antennas in homogeneous tissues. Volpe and Jain (1982) have simulated temperature distributions in patients during microwave heating.

More effort is needed in analyzing the temperature distributions in tissues exposed to electromagnetic radiation.

Radiofrequency Heating

As in the investigations of MW diathermy, analyses of heat transfer and temperature distributions during RF heating have been limited. Temperature measurements in vivo have been hindered by the problems listed in the previous section. Figures 2.30 and 2.31 show the results of two groups investigating the temperature distributions in tumors and normal tissues exposed to capacitive (13.56 MHz) and inductive (27.12 MHz) heating, respectively (Storm et al., in press; Hahn and Kim 1980). Note that despite surface cooling, the high specific absorption by the subcutaneous fat limits the use of capacitive heating to nonobese patients.

The recently developed magnetic-loop induction applicator, in which the patient can be placed, leads to elevated temperatures in deep-seated tumors (fig. 2.32). While theoretical results are just recently becoming available for these devices, the experimental results on the temperature distribution look promising, especially for large tumors ($\geq$ 5 cm in diameter) (see chapter 14).

SUMMARY AND RECOMMENDATIONS

The objective of this chapter was to present various theoretical frameworks that may be used to predict temperature distributions in normal and neoplastic

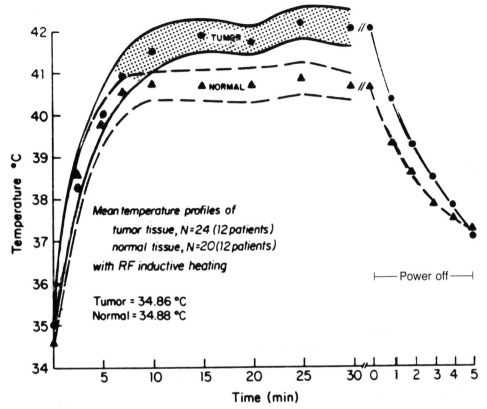

Figure 2.31. The average temperature changes in tumor and adjacent normal tissue of patients undergoing localized hyperthermia with RF inductive heating. (The *shaded areas* indicate the standard error.) (Hahn and Kim 1980. Reprinted by permission.)

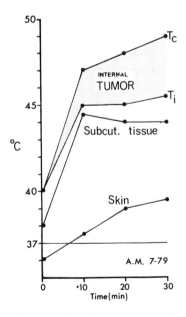

Figure 2.32. Temperature distributions in an internal solid tumor (gastric carcinoma; 10 × 6 × 4 cm) and normal tissues (skin and subcutaneous tissue) of a patient undergoing regional hyperthermia with a magnetic-loop induction applicator. No cooling of any kind was used. (T_c = tumor center; T_i = tumor/normal tissue interface.) (Storm et al., in press. Reprinted by permission.)

tissues during hyperthermia. Because the various biological, physiologic, and immunologic factors (fig. 2.33) are discussed extensively in the accompanying chapters, they have been mentioned only briefly (Jain 1982b). For this reason, the role of hyperthermia as an adjuvant to radiotherapy or chemotherapy was also not discussed (Overgaard 1980; Hahn 1980; Jain, Weissbrod, and Wei 1980; Jain 1982a). Instead, the emphasis in this review has been on the physical factors involved in hyperthermia. To this end, two approaches—distributed and lumped—were presented for modeling the thermal distribution between the normal and neoplastic tissues of various mammals. Evaluation of thermal energy produced during US, MW, and RF heating was discussed in depth. The experimental and theoretical results on the temperature distributions were summarized according to the method used for heating. Several important and poorly understood problems were pointed out at various places in the text in the hope of stimulating interest in the investigators involved in this multidisciplinary research.

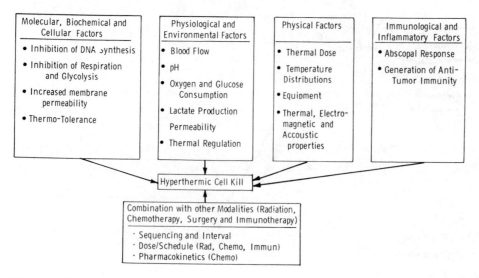

Figure 2.33. Various physical, physiologic, biochemical, and immunologic factors that determine the effectiveness of hyperthermia, alone or in combination with other modalities, in cancer treatment.

Some heat transfer related areas that deserve further attention are listed here:

1. While the BHTE, as it stands, appears to give adequate results in several applications, a precise description of heat transfer in tissues remains a tedious but challenging problem.
2. As pointed out early in the chapter, data on the thermal, electrical and acoustic properties, intratumor blood flow rates, and metabolism of tumors are limited. These parameters should be measured over a wide range of temperatures and stages of growth and regression.
3. It is essential to have and provide accurate measurements of intratissue temperature distribution during normothermia and hyperthermia. New developments in the noninvasive thermometric techniques are now needed for optimal use of hyperthermia.
4. Physiologic studies in animals and humans are needed to understand thermoregulation during hyperthermia. This information should then be incorporated into the mathematical models of the type described in this chapter.
5. Analysis of heat transfer and temperature distributions during US, MW, and RF heating is imperative for the proper use of these techniques. Such efforts in the past have been limited at best.
6. Quantitative comparison of various methods of inducing hyperthermia should be made on the basis of temperatures reached in various parts of normal and neoplastic tissues. This type of analysis will help in deciding the most efficacious modality of heating a given tumor and will help in designing feedback control mechanisms and heating protocols for maintaining desired temperatures in normal and neoplastic tissues.

Current research is directed toward answering many of these questions.

References

Aisenberg, A. C. *The glycolysis and respiration of tumors.* New York: Academic Press, 1961, pp. 159–179.

Algire, G. H., and Legallais, F. Y. Vascular reactions of normal and malignant tissues *in vivo*; effect of peripheral hypotension on transplant tumors. *J. Natl. Cancer Inst.* 12:399–408, 1951.

Allen, N. et al. Topographic blood flow in experimental nervous system tumors and surrounding tissues. *Trans. Am. Neurol. Assoc.* 100:157, 1975.

Altman, P., and Ditmer, D. Biology data book. *Fed. Am. Soc. Exptl. Biol.* Bethesda, Md.: Federation of American Society for Experimental Biology, 1964.

Amalric, R. et al. Thermography in diagnosis of breast diseases. *Bibl. Radiol.* 6:65–76, 1975.

Anne, A. "Scattering and absorption of microwaves by dissipative dielectric objects: the biological significance and hazard to mankind." Ph.D. dissertation, University of Pennsylvania, 1963.

Arcangeli, G. et al. Biological rationale for an optional scheduling of heat and ionizing radiation: clinical results on neck node metastases. In *Proceedings of the Third International Symposium on Cancer Therapy Hyperthermia, Drugs, and Radiation*, eds. L. A. Dethlefsen and W. C. Dewey. Bethesda, Md.: *J. Natl. Cancer Inst.* (special journal supplement), in press.

Barber, P. W. Electromagnetic power deposition in prolate spheroidal models of man and animals at resonance. *IEEE Trans. BME* 24:513–521, 1977.

Bicher, H. I. et al. Physiological mechanism of action of localized microwave hyperthermia. In *Proceedings* of the Third International Symposium on Cancer Therapy by Hyperthermia, Drugs, and Radiation, eds. L. A. Dethlefsen and W. C. Dewey. Bethesda, Md.: *J. Natl. Cancer Inst.* (special journal supplement), in press.

Bischoff, K. Some fundamental considerations of the applications of pharmacokinetics to cancer chemotherapy. *Cancer Chemotherapy Reports* 59:777–793, 1975.

Bladel, J. van. *Electromagnetic fields.* New York: McGraw-Hill, 1964.

Bowman, F. Heat transfer mechanisms and thermal dosimetry. In *Proceedings of the Third International Symposium on Cancer Therapy by Hyperthermia, Drugs, and Radiation*, eds. L. A. Dethlefsen and W. C. Dewey. Bethesda, Md.: *J. Natl. Cancer Inst.* (special journal supplement), in press.

Bowman, H. F.; Cravalho, E. G.; and Woods, M. Theory, measurements and applications of thermal properties of biomaterials. *Annu. Rev. Biophys. Bioeng.* 4:43–80, 1975.

Bowman, R. R. A probe for measuring temperature in radiofrequency-heated material. *IEEE Biomed. Eng. MTT* 24:43–45, 1976.

Burdette, E. C.; Cain, F. L.; and Seals, J. *In vivo* probe technique for determining dielectric properties at VHF through microwave frequencies, *IEEE Trans. MTT* 28(Suppl.):414–429, 1980.

Caldwell, W. E., and Durand, R. E., eds. *Proceedings of the Conference on Clinical Prospects of Hypoxic Cell Sensitizers and Hyperthermia*, Madison, Wis.: Wisconsin Press, 1978.

Cavaliere, R. et al. Selective heat sensitivity of cancer cells, biochemical and clinical studies. *Cancer* 20:1351–1381, 1967.

Cavaliere, R. et al. Heat transfer problems during local perfusion in cancer treatment. *Ann. N.Y. Acad. Sci.* 335:311–326, 1980.

Cetas, T. C. A birefringent crystal optical thermometer for measurements in electromagnetically induced heating. In *Proc. 1975 USNC/URSI Symp.*, eds. C. C. Johnson and J. L. Shore. Rockville, Md.: Bureau of Radiological Health, DHEW Publ. (FDA)77-8011, 1976, p. 338.

Cetas, T. C., and Connor, W. G. Thermometry considerations in localized hyperthermia. *Med. Phys.* 5:79–91, 1978.

Chan, A. K. et al. Calculation by the method of finite differences of the temperature distribution in layered tissues. *IEEE Trans. BME* 20:86–90, 1973.

Chan, A. K.; Sigelmann, R. A.; and Guy, A. W. Calculations of therapeutic heat generated by ultrasound in fat-muscle-bone layers. *IEEE Trans. BME* 21:280–284, 1974.

Chato, J. C. Heat transfer in bioengineering. In *Advanced heat transfer*, ed. B. T. Chao. Urbana, Ill.: University of Illinois Press, 1969, p. 395.

Chen, K. M., and Guru, B. S. Internal EM field and absorbed power density in human torsos induced by 1–500 MHz EM waves. *IEEE Trans. MTT* 25:746–756, 1977a.

Chen, K. M., and Guru, B. S. Induced EM fields inside human bodies irradiated by EM waves of up to 500 MHz. *J. Microwave Power* 12:173–183, 1977b.

Chen, M. M., and Holmes, K. R. Microvascular contributions in tissue heat transfer. *Ann. N.Y. Acad. Sci.* 335:137–150, 1980.

Christensen, D. A. Optical etalon temperature sensor of microwave tissue heating applications. Paper read at *USNC/URSI* Symposium on Biological Effects of E. M. Waves, Boulder, Colo., 1975.

Christensen, D. A. A new nonperturbing temperature probe using semiconductor band edge shift. *J. Biomedical Engineering* 1:541–545, 1977.

Christensen, D. A. Thermal dosimetry and temperature measurements. *Cancer Res.* 39:2325–2327, 1979.

Chrysanthopoulos, G. "Thermal interactions between normal and neoplastic tissues in mammalian systems during hyperthermia." Master's thesis, Columbia University, 1979.

Chrysanthopoulos, G., and Jain, R. K. Thermal interactions between normal and neoplastic tissues in the rat, rabbit, swine, and dog during hyperthermia. *Med. Phys.* 7:529–539, 1980.

Coley, W. B. The treatment of malignant tumor by repeated inoculations of erysipelas with a report of original cases. *Am. J. Med. Sci.* 105:487–511, 1893.

Cooney, D. O. *Biomedical Engineering Principles.* New York: Marcel Dekker, 1976, pp. 93–156.

Cooper, T. E., and Trezak, G. J. Correlation of thermal properties of some human tissues with water content. *Aerospace Medicine* 42:24–28, 1971.

Cunningham, D. J. An evaluation of heat transfer through skin in the human extremity. In *Physiological and behavioral temperature*, eds. J. D. Hardy, P. A. Gagge, and J. A. Stolwijk. Springfield, Ill.: Charles C. Thomas, 1970, pp. 302–315.

Deficis, A., and Prion, A. Non-perturbing microprobes for measurement in electromagnetic fields. *Microwave J.* 20:55, 1977.

Dewey, W. C., and Dethlefsen, L. A., eds. *Proceedings of the Third International Symposium on Cancer Therapy by Hyperthermia, Drugs, and Radiation.* Bethesda, Md.: *J. Natl. Cancer Inst.* (special journal supplement), in press.

Dickson, J. A. The effects of hyperthermia in normal animal tumour systems. *Recent Results Cancer Res.* 59:43–111, 1977.

Dickson, J. A., and Calderwood, S. K. Temperature range and selective sensitivity of tumors to hyperthermia: a critical review. *Ann. N.Y. Acad. Sci.* 335: 180–205, 1980.

Dickson, J. A., and Suzanger, M. *In vitro-vivo* studies on the susceptibility of the solid Yoshida sarcoma to drugs and hyperthermia (42°). *Cancer Res.* 34:1263–1274, 1974.

Dietzel, F. *Tumor und temperatur.* Berlin: Urban & Schwarzenberg, 1975.

Dobson, J. Equipment for local and regional hyperthermia. Report prepared by WSA, Inc., San Diego, Calif., for the Radiotherapy Development Branch, DCT, NCI, NIH, February, 1980.

Dudar, T. E., and Jain, R. K. Microcirculatory changes during hyperthermia in normal and neoplastic tissues. Paper presented at the 75th Annual Meeting of A.I.Ch.E., Los Angeles, 1982.

Durney, C. H. et al. *Radiofrequency radiation dosimetry handbook, 2nd edition.* Report SAM-TR-78-22, USAF School of Aerospace Medicine, Brooks AFB, Texas, 1978.

Durney, C. H.; Massoudi, H.; and Johnson, C. C. Long wavelength analysis of plane wave irradiation of a prolate spheroid model of man. *IEEE Trans. MTT* 23: 246–253, 1975.

Eberhart, R. E., and Shitzer, A., eds. *Heat transfer in biological systems: analysis and applications.* New York: Plenum Press, 1982, in press.

Eddy, H. A. Alterations in tumor microvascular during hyperthermia. *Radiology* 137:515–521, 1980.

Eddy, H. A.; Sutherland, R. M.; and Chmielewski, G. Tumor microvascular response: hyperthermia, drug, radiation combinations (abstr.). *Proceedings of the Third International Symposium on Cancer Therapy by Hyperthermia, Drugs, and Radiation*, eds. L. Dethlefsen and W. C. Dewey. Bethesda, Md.: National Cancer Institute, 1980, p. 88.

Emami, B. et al. Histopathological study on the effects of hyperthermia on microvasculature. *J. Radiat. Oncol. Biol. Phys.* 7:343–348, 1981.

Endrich, B. et al. Tissue perfusion in homogeneity during early tumor growth in rats. *J. Natl. Cancer Inst.* 62:387–395, 1979a.

Endrich, B. et al. Quantitative studies of microcirculatory function in malignant tissue: influence of temperature on microvascular hemodynamics during the early growth of the BA 1112 rat sarcoma. *Int. J. Radiat. Oncol. Biol. Phys.* 5:2021–2030, 1979b.

Esmay, M. L. *Principles of animal environment.* Westport, Conn.: AVI Publishing, 1969, pp. 77–80.

Fan, L. T.; Hsu, F. T.; and Hwang, C. L. A review of mathematical models of the human thermal system. *IEEE Trans. BME* 18:218, 1971.

Gandhi, O. P.; Hagerman, M. J.; and D'Andrea, J. A. Partbody and multibody effects on absorption of radiofrequency electromagnetic energy by animals and by models of man. *Radio Sci.* 14:15–21, 1979.

Gautherie, M. Thermopathology of breast cancer: measurement and analysis of *in vivo* temperature and blood flow. *Ann. N.Y. Acad. Sci.* 335:383–415, 1980.

Gautherie, M. et al. Puissance thermogène des épithéliomas mammaires I. Détermination par thérmométrie Intratumorale et thérmographie infrarouge cutanée. *Rev. Europ. Études Clin. et Biol.* 17:776–781, 1972.

Gemmill, C. L., and Brobeck, J. R. Energy exchange. In *Medical physiology*, ed. V. B. Mountcastle. St. Louis: C. V. Mosby, 1968, pp. 485–488.

Gershon-Cohen, J. et al. Advances in thermography and mammography. *Ann. N.Y. Acad. Sci.* 121:283–300, 1964.

Goldacre, R. J., and Sylvén, B. A rapid method of studying tumour blood supply using systemic dyes. *Br. J. Cancer* 16:306, 1962.

Goss, S. A.; Johnston, R. L.; and Dunn, F. Comprehensive compilation of empirical ultrasonic properties of mammalian tissues. *J. Acoust. Soc. Am.* 64:423–457, 1978.

Gullino, P. M. *In vitro* perfusion of tumors. In *Organ perfusion and preservation*, eds. J. C. Norman et al. New York: Appleton-Century-Crofts, 1968, pp. 887–898.

Gullino, P. M. *In vivo* utilization of oxygen and glucose by neoplastic tissue. In *Oxygen transport to tissue*, vol 2, eds. J. Groste, D. Reneau, and G. Thews. New York: Plenum, 1976, pp. 521–536.

Gullino, P. M. Influence of blood supply on thermal properties and metabolism of mammary carcinomas. *Ann. N.Y. Acad. Sci.* 335:1–21, 1980.

Gullino, P. M. et al. Relationship between oxygen and glucose consumption by transplanted tumors *in vivo*. *Cancer Res.* 27:1041–1052, 1967.

Gullino, P. M., and Grantham, F. H. Studies on the exchange of fluids between host and tumor. I. A method for growing "tissue-isolated" tumors in laboratory animals. *J. Natl. Cancer Inst.* 27:679–693, 1961a.

Gullino, P. M., and Grantham, F. H. Studies on the exchange of fluids between host and tumor. II. The blood flow of hepatomas and other tumors in rats and mice. *J. Natl. Cancer Inst.* 27:1465–1491, 1961b.

Gullino, P. M.; Grantham, F. H.; and Courtney, A. H. Utilization of oxygen by transplanted tumors *in vivo*. *Cancer Res.* 27:1020–1030, 1967a.

Gullino, P. M.; Grantham, F. H.; and Courtney, A. H. Glucose consumption by transplanted tumors *in vivo*. *Cancer Res.* 27:1031–1040, 1967b.

Gullino, P. M.; Jain, R. K.; and Grantham, F. H. Temperature gradients and local perfusion in a mammary carcinoma. *J. Natl. Cancer Inst.* 68:519–533, 1982.

Gullino, P. M.; Yi, P. N.; and Grantham, F. H. Relationship between temperature and blood supply or consumption of oxygen and glucose by rat mammary carcinomas. *J. Natl. Cancer Inst.* 60:835–847, 1978.

Gump, F. E., and White, R. L. Determination of regional tumor blood flow by krypton-85. *Cancer* (Philadelphia) 21:871–875, 1969.

Guru, B. S., and Chen, K. M. Experimental and theoretical studies on electromagnetic fields induced inside finite biological bodies. *IEEE Trans. MTT* 24:433–440, 1976.

Guttner, W. Ultraschall in menschlichen korper. *Acustica* 4:547–554, 1954.

Guy, A. W. Electromagnetic fields and relative heating patterns due to a rectangular aperture source in direct contact with bilayered biological tissue. *IEEE Trans. MTT* 19:214–223, 1971.

Guy, A. W., and Lehman, J. F. The determination of an optimum microwave diathermy frequency for a direct contact applicator. *IEEE Trans. BME* 13:76–87, 1966.

Guy, A. W.; Lehman, J. F.; and Stonebridge, J. B. Therapeutic applications of electromagnetic power. *Proc. IEEE* 62:55–75, 1974.

Hagerman, M. J., and Gandhi, O. P. Numerical calculation of electromagnetic energy deposition in models of man with grounding and reflector effects. *Radio Sci.* 14:23–29, 1979.

Hahn, E. W., and Kim, J. H. Clinical observations on the selective heating of cutaneous tumors with the radiofrequency inductive method. *Ann. N.Y. Acad. Sci.* 335:347–355, 1980.

Hahn, G. M. Studies on drug-hyperthermia interaction (abstr.). In *Proceedings of the Third International Symposium on Cancer Therapy by Hyperthermia, Drugs, and Radiation*, eds. L. A. Dethlefsen and W. C. Dewey. Bethesda, Md.: National Cancer Institute, 1980, p. 104.

Hahn, G. M. et al. Some heat transfer problems associated with heating by ultrasound, microwaves, or radiofrequency. *Ann. N.Y. Acad. Sci.* 335:327–346, 1980.

Hales, J. R. S. Physiological responses to heat. In *MTP Int. Rev. Ser. Physiology Series*, vol. 7, ed. D. Robertshaw. London: Butterworth, 1974, pp. 107–162.

Hand, J. W. Microwave heating patterns in simple tissue models. *Phys. Med. Biol.* 22:981–987, 1977.

Hardy, J. D.; Gagge, A. P.; and Stolwijk, J. A. J. *Physiological and behavioral thermal regulation.* Springfield, Ill.: Charles C. Thomas, 1970.

Har-Kedar, I., and Bleehen, N. J. Experimental and clinical aspects of hyperthermia applied to the treatment of cancer with special reference to the role of ultrasonic and microwave heating. *Adv. Radiat. Biol.* 6:229–266, 1976.

Henderson, M. A., and Pettigrew, R. Induction of controlled hyperthermia in treatment of cancer. *Lancet* 1:1275–1277, 1971.

Herman, T. S.; Zukoski, C. F.; and Anderson, R. M. Whole-body hyperthermia via blanket technique (abstr.). In *Proceedings of the Third International Symposium on Cancer Therapy by Hyperthermia, Drugs, and Radiation*, eds. L. A. Dethlefsen and W. C. Dewey. Bethesda, Md.: National Cancer Institute, 1980, p. 112.

Hizal, A., and Baykal, Y. K. Heat potential distribution in an inhomogeneous spherical model of a cranial structure exposed to microwaves due to loop or dipole antennas. *IEEE Trans. MTT* 26:607–612, 1978.

Ho, H. S. Contrast of dose distribution in phantom heads due to aperture and planewave sources. *Ann. N.Y. Acad. Sci.* 247:454–472, 1975a.

Ho, H. S. Dose rate distribution in triple layered dielectric cylinder with irregular cross-section irradiated by planewave sources. *J. Microwave Power* 10:421–431, 1975b.

Ho, H. S. et al. Electromagnetic heating patterns in circular cylindrical models of human tissue. *Proc. Eighth Int. Conf. Med. & Biol. Eng.*, sessions 2–4, July, 1969.

Ho, H. S. et al. Microwave heating of simulated human limbs by aperture sources. *IEEE Trans. MTT* 19:224–231, 1971.

Holmes, K. R., and Chen, M. M. Local thermal conductivity of para-7 fibrosarcoma in hamster. *Adv. in Bioeng.* 147:149, 1979.

Holt, J. A. G. The use of V.H.F. radiowaves in cancer therapy. *Australas. Radiol.* 19:223–241, 1975.

Huckaba, C. E., and Tam, H. S. Modeling of human thermal system. In *Advances in biomedical engineering*, part 1, ed. D. O. Cooney. New York: Marcel Dekker, 1980, pp. 1–58.

Hunt, J. W. Application of microwave, ultrasound and radiofrequency heating *in vivo*. In *Proceedings of the Third International Symposium on Cancer Therapy by Hyperthermia, Drugs, and Radiation*, eds. L. A. Dethlefsen and W. C. Dewey. Bethesda, Md.: National Cancer Institute, 1980, p. 53.

Hwang, C. L., and Konz, S. A. Engineering models of the human thermoregulatory system—a review. *IEEE Trans. BME* 24:309–315, 1977.

Jain, R. K. Effect of inhomogeneities and finite boundaries on temperature distributions in a perfused medium with application to tumors. *J. Biomech. Eng. Trans., ASME* 100:235–241, 1978.

Jain, R. K. Transient temperature distributions in an infinite perfused medium due to a time dependent, spherical heat source. *J. Biomech. Eng. Trans., ASME* 101:82–86, 1979.

Jain, R. K. Temperature distributions in normal and neoplastic tissues during normothermia and hyperthermia. *Ann. N.Y. Acad. Sci.* 335:48–66, 1980a.

Jain, R. K. Heat transfer in tumors—characterization and applications to thermography and hyperthermia. In *Advances in biomedical engineering*, part 1, ed. D. O. Cooney. New York: Marcel Dekker, 1980b, chap. 2.

Jain, R. K. Mass and heat transfer in tumors. In *Advances in transport processes*. 1982a, in press.

Jain, R. K. Analysis of heat transfer and temperature distributions in tissues during local and whole body hyperthermia. In *Heat transfer in biological systems: analysis and applications*, eds. R. Eberhart and A. Shitzer. New York: Plenum, 1982b, in press.

Jain, R. K.; Grantham, F. H.; and Gullino, P. M. Blood flow and heat transfer in Walker 256 mammary carcinoma. *J. Natl. Cancer Inst.* 62:927–933, 1979.

Jain, R. K., and Gullino, P. M., eds. Thermal characteristics of tumors: applications in detection and treatment. *Ann. N.Y. Acad. Sci.* 335:48–64, 1980a.

Jain, R. K., and Gullino, P. M. Analysis of transient temperature distribution in a perfused medium due to a spherical heat source with application to heat transfer in tumors—homogeneous and perfused medium. *Chemical Eng. Comm.* 4:95–118, 1980b.

Jain, R. K., and Wei, J. Dynamics of drug transport in solid tumors distributed parameter model. *J. Bioeng.* 1:313–330, 1977.

Jain, R. K.; Weissbrod, J.; and Wei, J. Mass transfer in tumors: characterization and applications in chemotherapy. *Adv. Cancer Res.* 33:251–311, 1980.

James, K. A.; Quick, W. H.; and Strahan, W. H. Fiber optics: the way to time digital sensors. *Control Eng.* 26:30–33, 1979.

Johnson, C. C. et al. Electromagnetic power absorption in anisotropic tissue media. *IEEE Trans. MTT* 23:529–532, 1975.

Johnson, C. C.; Durney, C. H.; and Massoudi, H. Long-wavelength electromagnetic power absorption in prolate spheroidal models of man and animals. *IEEE Trans. MTT* 23:739–747, 1975.

Johnson, C. C., and Guy, A. W. Nonionizing electromagnetic wave effects in biological materials and systems. *Proc. IEEE* 60:692–717, 1972.

Joines, W. T., and Spiegel, R. J. Resonance absorption of microwaves by the human skull. *IEEE Trans. BME* 21:46–48, 1974.

Kallman, R. F.; de Nardo, G. L.; and Stasch, M. J. Blood flow in irradiated mouse sarcoma as determined by the clearance of xenon-133. *Cancer Res.* 32:483, 1972.

Kantor, G. New types of microwave diathermy applicators: comparison of performance with conventional types. In *Proceedings of a Symposium on Biological Effects and Measurement of Radiofrequency/Microwaves*, ed. D. G. Hazzard. DHEW Publication #FDA-77-8026, July, 1977.

Keller, K. H., and Seiler, L. An analysis of peripheral heat transfer in man. *J. Appl. Physiol.* 30:779, 1971.

Kjartansson, I. E. Tumor circulation, an experimental study in the rat with a comparison of different methods for estimation of tumour blood flow. *Acta Chir. Scand. (suppl.)* 471:1–74, 1976.

Klinger, H. G. Heat transfer in perfused biological tissue. I: General theory. *Bull. Math. Biol.* 36:403, 1974.

Klinger, H. G. The description of heat transfer in biological tissue. *Ann. N.Y. Acad. Sci.* 335:133–136, 1980.

Kritikos, H. N., and Schwan, H. P. Hot spots generated in conducting spheres by EM waves and biological implications. *IEEE Trans. BME* 19:53–58, 1972.

Kritikos, H. N., and Schwan, H. P. The distribution of heating potential inside lossy spheres. *IEEE Trans. BME* 22:457–463, 1975.

Kritikos, H. N., and Schwan, H. P. Formation of hot spots in multilayer spheres. *IEEE Trans. BME* 23:168–172, 1976.

Kritikos, H. N., and Schwan, H. P. Potential temperature rise induced by electromagnetic field in brain tissues. *IEEE Trans. BME* 26:29–34, 1979.

Larsen, L. E.; Moore, R. A.; and Acevedo, J. A. A microwave decoupled brain temperature transducer. *IEEE Trans. BME* 22:438–444, 1974.

Law, H. T., and Pettigrew, R. T. Heat transfer in whole-body hyperthermia. *Ann. N.Y. Acad. Sci.* 335:298–310, 1980.

Lele, P. P. A transient thermal pulse technique for measurement of tissues thermal diffusivity *in vivo*. *Ann. N.Y. Acad. Sci.* 335:83–85, 1980.

Lett, J. T., and Alder, H., eds. *Adv. Radiat. Biol.*, vol. 6, 1976.

Levin, W.; Wasserman, H.; and Blair, R. M. Tumor temperature augmentation utilizing 433 MHz microwaves in patients undergoing whole body hyperthermia (abstr.). In *Proceedings of the Third International Symposium on Cancer Therapy by Hyperthermia, Drugs, and Radiation*, eds. L. A. Dethlefsen and W. C. Dewey. Bethesda, Md.: National Cancer Institute, 1980, p. 20.

Lin, J. C. et al. Microwave selective brain heating. *J. Microwave Power* 8:275–286, 1973.

Lin, J. C.; Guy, A. W.; and Johnson, C. C. Power deposition in a spherical model of man exposed to 1–20 MHz electromagnetic fields. *IEEE Trans. MTT* 21: 791–797, 1973.

Livesay, D. E., and Chen, K. Electromagnetic fields induced inside arbitrary shaped biological bodies. *IEEE Trans. MTT* 22:1273–1280, 1974.

Mäntylä, M. J. Regional blood flow in human tumors. *Cancer Res.* 39:2304–2306, 1979.

Mäntylä, M. J.; Kuikka, J.; and Rekonen, A. Regional blood flow in human tumours with special reference to the effect of radiotherapy. *Br. J. Radiol.* 49:335–338, 1976.

Marmor, J. B. et al. Treating spontaneous tumors in dogs and cats by ultrasound-induced hyperthermia. *Int. J. Radiat. Oncol. Biol. Phys.* 4:967–973, 1978.

Massoudi, H.; Durney, C. H.; and Johnson, C. C. Long wavelength electromagnetic power absorption in ellipsoidal models of man and animals. *IEEE Trans. MTT* 25:47–52, 1977.

Massoudi, H.; Durney, C. H.; and Johnson, C. C. A geometrical-optics and an exact solution for internal fields in and energy absorption by a cylindrical model of man irradiated by an electromagnetic plane wave. *Radio Science* 14:35–42, 1979.

Milder, J. W., ed. Proceedings of a Conference on Hyperthermia in Cancer Treatment. *Cancer Res.* 39: 2231–2340, 1979.

Mitchell, J. W., and Myers, G. E. An analytical model of the countercurrent heat exchange phenomena. *Biophys. J.* 8:897, 1968.

Moller, U., and Bojsen, J. Temperature and blood flow measurements in and around 7,12-dimethylbenz(A)anthracene-induced tumors and Walker 256 carcinomas in rats. *Cancer Res.* 35:3116–3121, 1975.

Neuder, S. M. et al. Microwave power density absorption in a spherical multilayered model of the head. In *Biological effects of electromagnetic waves*, vol. II, eds. C. C. Johnson and M. L. Shore. Washington: HEW Publication #FDA-77-8011, December, 1976, pp. 199–210.

Neuder, S. M., and Meijer, P. H. E. Finite element variational calculus approach to the determination of electromagnetic fields in irregular geometry. In *Biological effects of electromagnetic waves*, vol. II, eds. C. C. Johnson and M. L. Shore. Washington: HEW Publication #FDA-77-8011, December, 1976, pp. 193–198.

Nugent, L., and Jain, R. K. Diffusional transport and permeability of macromolecules in normal and neoplastic tissues. Paper read at the *74th Annual Meeting of A.I.Ch.E.*, New Orleans, 1981.

Overgaard, J. Influence of sequence and interval on the biological response to combined hyperthermia and radiation (abstr.). In *Proceedings of the Third International Symposium on Cancer Therapy by Hyperthermia, Drugs, and Radiation*, eds. L. A. Dethlefsen and W. C. Dewey. Bethesda, Md.: National Cancer Institute, 1980, pp. 105–108.

Overgaard, J., and Nielson, D. S. The role of tissue environmental factors on the kinetics and morphology of tumor cells exposed to hyperthermia. *Ann. N.Y. Acad. Sci.* 335:245–280, 1980.

Parker, K., and Lele, P. P. Discussion. *Ann. N.Y. Acad. Sci.* 335:64, 1980.

Peloso, R. Master's thesis. Carnegie-Mellon University, 1981.

Pennes, H. H. Analysis of tissue and arterial blood temperatures in the resting human forearm. *J. Appl. Physiol.* 1:93, 1948.

Peterson, H. I. et al. Studies on the circulation of experimental tumours. I. Effect of induced fibrinolysis and antifibrinolysis on capillary blood flow and the capillary transport function of two experimental tumors in the mouse. *Eur. J. Cancer* 5:91, 1969.

Plengvanit, U. et al. Regional hepatic blood flow studied by intrahepatic injection of xenon in normals and in patients with primary carcinoma of the liver, with particular reference to the effect of hepatic artery ligation. *Aust. N.Z. J. Med.* 1:44, 1972.

Poppendiek, H. F. et al. Thermal conductivity measurements and predictions for biological fluids and tissues. *Cryobiology* 3:318, 1967.

Reinhold, H. S. et al. Utilization of the Siemens unit techniques (abstr.). In *Proceedings of the Third International Symposium on Cancer Therapy by Hyperthermia, Drugs, and Radiation*, eds. L. A. Dethlefsen and W. C. Dewey. Bethesda, Md.: National Cancer Institute, 1980, p. 114.

Reinhold, H. S.; Blachiewicz, B.; and van den Berg-Blok, A. Decrease in tumor microcirculation during hyperthermia. In *Proceedings of the Second International Symposium on Cancer Therapy by Hyperthermia and Radiation*, eds. C. Streffer et al. Baltimore and Munich: Urban & Schwarzenberg, 1978, pp. 231–232.

Reinhold, H.S., and van den Berg-Blok, A. Enhancement of thermal damage to sandwich tumors by additional treatment (abstr.). In *Proceedings of the Third International Symposium on Cancer Therapy by Hyperthermia, Drugs, and Radiation*, eds. L. A. Dethlefsen and W. C.

Dewey. Bethesda, Md.: National Cancer Institute, 1980, p. 96.

Robert, J.; Martin, J.; and Burg, C. Évolution de la vascularisation d'une tumeur isologue solide de la souris au cours de sa croissance. *Strahlentherapie* 133:621, 1967.

Robinson, T. C., and Lele, P. P. An analysis of lesion development in the brain and in plastics by high-intensity focused ultrasound at low megahertz frequencies. *J. Acoustical Soc. Ann.* 51:1333–1351, 1969.

Robinson, J. E.; McCulloch, D.; and McCready, W. A. Blood perfusion of Murie tumor at normal and hyperthermal temperatures (abstr.). In *Proceedings of the Third International Symposium on Cancer Therapy by Hyperthermia, Drugs, and Radiation*, eds. L. A. Dethlefsen and W. C. Dewey. Bethesda, Md.: National Cancer Institute, 1980, p. 77.

Robinson, J. E., and Wizenberg, M. J., eds. *Proceedings of the First International Symposium on Cancer Therapy by Hyperthermia and Radiation*. Chicago: American College of Radiology, 1976.

Rogers, W. Tissue blood flow in transplantable tumors of the mouse and hamster. *Diss. Abstr. Int. B* 28:5185, 1968.

Rossi-Fanelli, A. et al., eds. *Selective heat sensitivity of cancer cells*. (*Recent results in cancer research*, vol. 59.) New York: Springer-Verlag, 1977.

Rowlinson, G. J., and Barber, P. W. Absorption at higher-frequency RF energy to biological models: calculations based on geometrical optics. *Radio Science* 14:43–50, 1979.

Rozzell, T. C. et al. A nonperturbing temperature sensor for measurements in electromagnetic fields. *J. Microwave Power* 9:241–249, 1974.

Rukspollmuang, S., and Chen, K. M. Heating of spherical versus realistic models of human and infrahuman heads by electromagnetic waves. *Radio Science* 14:51–62, 1979.

Samulski, T., and Shrivastava, P. A. Photoluminescent thermometer probes: temperature measurements in microwave fields. *Science* 208:193–194, 1980.

Sandhu, T. S.; Kowal, H.; and Johnson, R. J. The development of hyperthermia microwave generators and thermometry. *Int. J. Radiat. Oncol. Biol. Phys.* 1(suppl.):100, 1976.

Schepps, J. L., and Foster, K. R. The UHF and microwave dielectric properties of normal and tumor tissue. *Phys. Med. Biol.* 25:1149–59, 1980.

Schwan, H. P. Heating of fat-muscle layers by electromagnetic and ultrasonic diathermy. *Proc. AIEE* 72:483–487, 1953.

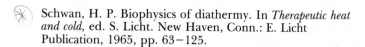

Schwan, H. P. Biophysics of diathermy. In *Therapeutic heat and cold*, ed. S. Licht. New Haven, Conn.: E. Licht Publication, 1965, pp. 63–125.

Schwan, H. P.; Carstensen, E. L.; and Li, K. Electric and ultrasonic deep heating diathermy. *Electronics* March, 1954, pp. 172–175.

Schwan, H. P., and Li, K. Hazards due to total body irradiation by radar. *Proc. IRE* 44:1572–1581, 1956.

Schwan, H. P., and Piersol, G. M. The absorption of electromagnetic energy in body tissues. *Am. J. Phys. Med.* 33:371, 1954.

Schwan, H. P., and Piersol, G. M. The absorption of electromagnetic energy in body tissues. I. Biophysical aspects. II. Physiological and clinical aspects. *Am. J. Phys. Med.* 33:371–404, 1954; 34:425–448, 1955.

Shah, S. A. et al. Modification of blood flow in W256 carcinoma by hyperglycaemia and hypervolemia: a thermal probe method. *Proc. Am. Assoc. Cancer Res.* 22:60, 1981.

Shah, S., and Jain, R. K. Effects of hyperthermia and hyperglycaemia on the metastasis formation and on survival of rats bearing W256 carcinosarcoma. *Proc. NAHG/ISOTT Conf.*, 1981.

Shapiro, A. R.; Lutomirski, R. F.; and Yura, H. T. Induced heating within a cranial structure irradiated by an electromagnetic plane wave. *IEEE Trans. MTT* 19:187–196, 1971.

Shibata, H. R., and MacLean, L. D. Blood flow to tumors. *Prog. Clin. Cancer* 11:33–47, 1966.

Sien, H. P. Dynamics of temperature distributions in normal and neoplastic tissues during hyperthermia. Master's thesis, Columbia University, 1978.

Sien, H. P., and Jain, R. K. Temperature distributions in normal and neoplastics tissues during hyperthermia: lumped parameter analysis. *J. Therm. Biol.* 4:157–164, 1979.

Sien, H. P., and Jain, R. K. Intratumor temperature distributions during hyperthermia. *J. Therm. Biol.* 5:127–130, 1980.

Slotman, G. J. et al. Quantitative changes in tumor blood flow with expanding tumor mass in the VX2 carcinoma (abstr. 203). *Proc. Am. Assoc. Cancer Res.* 21:51, 1980.

Song, C. W. Effect of hyperthermia on vascular functions of normal tissues and experimental tumors. *J. Natl. Cancer Inst.* 60:711–713, 1978.

Song, C. W. Role of blood flow and pH change in hyperthermia. In *Proceedings of the Third International Symposium on Cancer Therapy by Hyperthermia, Drugs, and Radiation*, eds. L. Dethlefsen and W. C. Dewey. Bethesda,

Md.: *J. Natl. Cancer Inst.* (special journal supplement), in press.

Song, C. W. et al. Effect of hyperthermia on vascular function in normal and neoplastic tissues *in vivo*. *Ann. N.Y. Acad. Sci.* 335:35–47, 1980.

Song, C. W.; Payne, J. T.; and Levitt, S. H. Vascularity and blood flow in x-irradiated Walker carcinoma 256 of rats. *Radiology* 104:693–697, 1972.

Stehlin, J. S. Hyperthermic perfusion with chemotherapy for cancer of the extremities. *Surg. Gynecol. Obstet.* 129:305–308, 1969.

Stehlin, J. S. Hyperthermic perfusion for melanoma of the extremities: experience with 65 patients, 1967 to 1979. *Ann. N.Y. Acad. Sci.* 335:352–355, 1980.

Stolwijk, J. A. J. Mathematical models of thermal regulation. *Ann. N.Y. Acad. Sci.* 335:98–106, 1980.

Stoner, E. K. Luminous and infrared heating. In *Therapeutic heat and cold*, 2nd edition, ed. S. Licht. New Haven, Conn: E. Licht Publisher, 1965, pp. 252–265.

Storm, F. K. et al. Normal tissue and solid tumor effects of hyperthermia in animal models and clinical trials. *Cancer Res.* 39:2245–2251, 1979.

Storm, F. K. et al. Clinical local hyperthermia by radiofrequency: a review. *Proceedings of the Third International Symposium on Cancer Therapy by Hyperthermia, Drugs, and Radiation*, eds. L. A. Dethlefsen and W. C. Dewey. Bethesda, Md.: *J. Natl. Cancer Inst.* (special journal supplement), in press.

Strauss, A. A. *Immunologic resistance to carcinoma produced by electrocoagulation. Based on fifty-seven years of experimental and clinical results.* Springfield, Ill.: Charles C Thomas, 1969.

Straw, J. A. et al. Distribution of anticancer agents in spontaneous animal tumors. I. Regional blood flow and methotrexate distribution in canine lymphosarcoma. *J. Natl. Cancer Inst.* 52:1327–1331, 1974.

Streffer, C. et al., eds. *Proceedings of the Second International Symposium on Cancer Therapy by Hyperthermia and Radiation.* eds. C. Streffer et al. Baltimore and Munich: Urban & Schwarzenberg, 1978, pp. 151–153.

Subjeck, J. et al. Cell survival dependence on heating method (abstr.). In *Proceedings of the Third International Symposium on Cancer Therapy by Hyperthermia, Drugs, and Radiation*, eds. L. A. Dethlefsen and W. C. Dewey. Bethesda, Md.: National Cancer Institute, 1980, p. 46.

Sugaar, S., and LeVeen, S. H. A histopathologic study on the effects of radiofrequency thermotherapy on malignant tumors of the lung. *Cancer* 43:767–783, 1979.

Sutton, C. H. Tumor hyperthermia in the treatment of malignant gliomas of the brain. *Trans. Am. Neurol. Assoc.* 96:195–199, 1971.

Sutton, C. H. Discussion. *Ann. N.Y. Acad. Sci.* 335:45–47, 1980.

Takacs, L.; Debreczeni, L. A.; and Farsang, C. S. Circulation in rats with Guérin carcinoma. *Appl. Physiol.* 38:696, 1975.

ter Haar, G. R. Computed temperature profiles in tissues resulting from ultrasonic irradiation (abstr.). In *Proceedings of the Third International Symposium on Cancer Therapy by Hyperthermia, Drugs, and Radiation*, eds. L. A. Dethlefsen and W. C. Dewey. Bethesda, Md.: National Cancer Institute, 1980, p. 39.

Thauer, R. Circulatory adjustments to climatic requirements. In *Handbook of physiology*, sec. 2, vol. III. eds. W. F. Hamilton and P. Dow. Washington: *Am. Physiol. Soc.*, 1965, pp. 1921–1966.

Townsend, J. G.; Jain, R. K.; and Eddy, H. Adriamycin pharmacokinetics in normal and neoplastic tissues of hamsters. Paper read at *Second World Congress of Chemical Engineers*, Montreal, 1981.

Trembly, B. S. et al. Hyperthermia induced by an array of invasive microwave antennas (abstr.). In *Proceedings of the Third International Symposium on Cancer Therapy by Hyperthermia, Drugs, and Radiation*, eds. L. A. Dethlefsen and W. C. Dewey. Bethesda, Md.: National Cancer Institute, 1980, p.74.

Trezak, G. J., and Jewett, D. L. Nodal network of transient temperature fields from cooling sources in anesthetized brain. *IEEE Trans. Biomed. Eng.* 4:281, 1970.

Turner, P., and Kumar, L. Computer solution for applicator heating patterns (poster). In *Proceedings of the Third International Symposium on Cancer Therapy by Hyperthermia, Drugs, and Radiation*, eds. L. A. Dethlefsen and W. C. Dewey. Bethesda, Md.: National Cancer Institute, 1980.

Turner, P. F. Deep heating of cylindrical or elliptical tissue masses (abstr.). In *Proceedings of the Third International Symposium on Cancer Therapy by Hyperthermia, Drugs, and Radiation*, eds. L. A. Dethlefsen and W. C. Dewey. Bethesda, Md.: National Cancer Institute, 1980, p. 74.

Tuttle, E. P., and Saddler, J. S. Measurement of renal tissue fluid turnover rates by thermal washout technique. *Hypertension* 13:3, 1964.

Vaupel, P. Atemgaswechsel und glucosestoffwechsel von implantationstumoren (DS-carcinosarkom) *in vivo*. In *Akademie der wissenschaften und der literatur*, Mainz, 1974.

Vaupel, P. Interrelationship between mean arterial blood pressure, blood flow and vascular resistance in solid tumor tissue of DS-carcino-sarcoma. *Experience* 31:587, 1975.

Vaupel, P. et al. Impact of localized hyperthermia on the cellular microenvironment in solid tumors (abstr.). In *Proceedings of the Third International Symposium on Cancer Therapy by Hyperthermia, Drugs, and Radiation*, eds. L. A.

Dethlefsen and W. C. Dewey. Bethesda, Md.: National Cancer Institute, 1980, p. 100.

Volpe, B. T. Master's thesis. Carnegié-Mellon University, 1981.

Volpe, B. T., and Jain, R. K. Temperature distributions and thermal response in humans. I. Simulations of various modes of whole-body hyperthermia in normal subjects. *Med. Phys.* 9:506–513, 1982.

Weil, C. M. Absorption characteristics of multilayered sphere models exposed to UHF-microwave radiation. *IEEE Trans. BME* 22:468–476, 1975.

Weinbaum, S. General discussions. *Ann. N.Y. Acad. Sci.* 335:173–175, 1980.

Westermark, N. The effect of heat upon rat-tumors. *Skand. Arch. Physiol.* 52:257–322, 1927.

Wickersheim, K. A., and Alves, R. B. Recent advances in optical temperatures measurements. *Luxtron Corp.* 408:727–2221, 1979.

Wu, C., and Lin, J. C. Absorption and scattering of electromagnetic waves by prolate spheroidal model of biological structures. *Digest of International Symposium*, Stanford University, Calif., June 20–22, 1977, pp. 142–145.

Wu, T. K., and Tsai, L. L. Electromagnetic fields induced inside arbitrary cylinders of biological tissues. *IEEE Trans. MTT* 25:61–65, 1977.

Wulff, W. The energy conservation equation for the living tissue. *IEEE Trans. BME* 21:494, 1974.

CHAPTER 3
Arrhenius Analysis of Thermal Responses

Kurt J. Henle

Introduction and Rationale
Historical Background of Thermodynamic Formulas
Concepts and Principles of the Arrhenius Equation
Thermodynamic Analysis of Data
Conclusions

INTRODUCTION AND RATIONALE

The Arrhenius equation is based on an empirical relationship between the rates of chemical reactions and the absolute temperature (Snell et al. 1965). The equation has been used extensively in the study of the temperature dependence of biological processes, particularly for bacterial systems (Johnson, Eyring, and Polisar 1954; Farrell and Rose 1967) and isolated enzyme-substrate reactions (Johnson, Eyring, and Polisar 1954; Brandts 1967). More recently, Arrhenius analysis has been applied also to heat-induced killing of mammalian cells in vitro (Westra and Dewey 1971; Dewey et al. 1977), as well as to heat damage of mammalian tissues in vivo (Field 1978). The following sections examine the potential utility of an Arrhenius analysis in the study of heat-induced cell killing, potential alternatives to an Arrhenius analysis, and the meaning of the Arrhenius parameters in the thermodynamic interpretation of cell killing at various temperatures.

HISTORICAL BACKGROUND OF THERMODYNAMIC FORMULAS

In 1889, Arrhenius empirically formulated the increase of chemical reactions with increasing temperature in an equation that was capable of representing reaction data in terms of a straight line whenever the temperature range was not too large. Arrhenius also applied this equation to biological phenomena, but the temperature range over which straight lines could be obtained was generally smaller than that for chemical reactions; the actual range was dependent not only on the specific process, but also on the specific chemical environment. Arrhenius proposed the concept of an activated state for reacting molecules, and this hypothesis was further developed by Eyring and colleagues into a quasi-equilibrium transition state concept, the so-called *rate theory* (Johnson, Eyring, and Polisar 1954). Rate theory attempted to provide a thermodynamic basis to the Arrhenius equation and has been applied to the analysis of numerous biological phenomena. In particular, bacterial luminescence received much attention since it is based on a simple enzyme system (luciferase) that can be assayed well in isolation and in the intact bacterium. The enzyme-substrate reaction may be considered reversible up to a temperature of about 35°C, and the temperature de-

This study was supported by grant PDT-98 from the American Cancer Society and by U.S. Public Health Service grant CA−20333 from the National Institutes of Health.

pendence of the irreversible destruction of luciferase above 35°C can be extrapolated from that of luciferase activity near the optimal temperature (Johnson, Eyring, and Polisar 1954). Using rate theory, the activation energy (related to the change in enthalpy) and the activation entropy for the enzymatic reaction can be calculated; however, for the irreversible destruction of luciferase, a similar calculation provides only estimates of the thermodynamic state variables, since the underlying theory is based on equilibrium (reversible) thermodynamics. Johnson Eyring, and Polisar (1954) have attempted to justify the use of rate theory for irreversible processes by arguing that many of the processes that appear irreversible are in fact reversible, and this can be demonstrated by the partial recovery of enzymatic activity when a heat-denatured enzyme is returned to colder temperatures. Separately, other investigators have tried to analyze irreversible biological phenomena thermodynamically by substituting either steady-state (Bertalanffy 1950) or near-equilibrium conditions (Iberall 1977; Yates 1978) for the reversibility requirement in classical thermodynamics. At present, however, adequate theoretical approaches to the analysis of highly irreversible phenomena, such as those that may be involved in heat-induced cell death, are not available.

CONCEPTS AND PRINCIPLES OF THE ARRHENIUS EQUATION

The Arrhenius equation is commonly written as:

$$v = A \exp(-\mu/RT), \qquad (1)$$

where v represents the velocity of the reaction; A, the Arrhenius constant; μ, the proportionality constant corresponding to v (activation energy); R, the gas constant (1.99 cal/mole °K); and T, the absolute temperature of the reaction (Johnson, Eyring, and Polisar 1954). The Maxwell-Boltzmann equation, which pertains to the equilibrium distribution of molecular energies (kinetics, vibrational, rotational), has a form similar to that of the Arrhenius equation:

$$\ln(n/n_o) = -E/RT. \qquad (2)$$

This equation defines the fraction of molecules that possess an energy value of E or greater, and thus, by analogy, the constant μ in the Arrhenius equation can be interpreted as the energy that individual molecules must attain in order to react (activation energy) (Snell et al. 1965). The transition state concept is based on the ideas of molecular collisions and probabilities of molecular events, and requires that molecules pass through a transition state prior to the formation of products in a specific reaction. The transition state is a molecular configuration with an average energy greater than that of the reactants by an amount equal to the activation energy. Molecules in the activated state may either return to the reactant or the product state (reversible reaction), and the rate of the reaction is proportional to the number of molecules in this transition state at any time. The rate constant v can then be related to the steady-state constant K, which reflects the transition state. If K is considered to be an equilibrium constant, classical thermodynamics provides an expression for the Gibbs' free energy in terms of the equilibrium constant K,

$$\Delta G° = RT \ln(K), \qquad (3)$$

where $\Delta G°$ is the free energy change associated with the reversible process.
Since

$$\Delta G° = T\Delta S° - \Delta H°, \qquad (4)$$

it can be shown that the rate constant is

$$v = \gamma kT/H \times \exp(\Delta S°/R) \times \exp(-\Delta H°/RT), \qquad (5)$$

where $\Delta S°$ and $\Delta H°$ are the entropy and enthalpy changes associated with the reversible formation of the transition state molecules; k, the Boltzmann constant; h, Planck's constant; and γ, the relative probability of a molecule in the transition state proceeding to form products or reactants.

Equations 3 to 5 represent an outline of the derivation of the Eyring extension of the Arrhenius equation; more detail is available elsewhere (Johnson, Eyring, and Polisar 1954; Snell et al. 1965). The application of this theory to heat-induced cell killing must be based on certain underlying assumptions. Specifically, heat-induced cell death may well constitute an irreversible process that cannot be approximated by near-equilibrium conditions; thus, the specific values of activation entropies and enthalpies would not be meaningful.

In a thermodynamically well-defined system, the rate-limiting step in a series of sequential reactions can, in theory, be identified from a knowledge of ΔH and ΔS for both the complete process and the individual steps. It is thus tempting to identify the "critical" heat target for cell death by comparing values of ΔH and ΔS for heat-induced cell killing with a number of temperature-sensitive molecular events (for example, glycolysis, DNA synthesis, etc.) or general processes (such as protein denaturation, DNA melting, etc.). For such an approach, however, one must consider the uncertainty in the calculated values of ΔH and ΔS for cell death and the possibility that molecular events leading to cell death are not sequential, but may involve coupled reactions with feedback. Thus, thermodynamic parameters for cell death may never reflect those of a single rate-limiting step, but that of the complete circuit of linked reactions. An additional problem with such an approach arises from the strong dependence of the values of the thermodynamic parameters for enzyme systems in isolation on experimental conditions, such as pH, ionic environment, concentrations of reactants, enzymes, products, and so forth. It is unlikely that reaction conditions in isolation reflect actual conditions in the intact cell (Johnson, Eyring, and Polisar 1954; Brandts 1967). Therefore, the identification of critical heat targets, based on thermodynamic analyses, requires advances both in the theory of nonequilibrium thermodynamics and in the biochemistry of the heat-injured cell.

Without the immediate prospects of a meaningful thermodynamic analysis of heat-induced cell death, the Arrhenius plot becomes essentially a representational device. It provides a convenient "map" for the temperature dependence of heat-induced killing after either hyperthermia alone, or in conjunction with other cytotoxic agents. The Arrhenius plot is not the only means of mapping the temperature dependence of cell killing, however; as shown in the next section, a time-temperature plot is equally practical as a representational device.

THERMODYNAMIC ANALYSIS OF DATA

In Vitro

Cell survival after hyperthermia can be parameterized in terms of the slope ($1/D_o$) and the shoulder (extrapolation number, n) of the heat survival curve (Westra and Dewey 1971). The temperature dependence of heat-induced cell killing usually is represented in terms of the D_o on the Arrhenius plot ($\ln(1/D_o)$ vs $1/T$) only. A summary of the in vitro data from the literature for a number of mammalian cell lines was published in a recent review by Henle and Dethlefsen (1980). For this summary, the published survival curves were all analyzed by the same fitting program after making an appropriate temperature transient correction for the heating times (Roti-Roti and Henle 1979). A similar summary is shown in figure 3.1, but

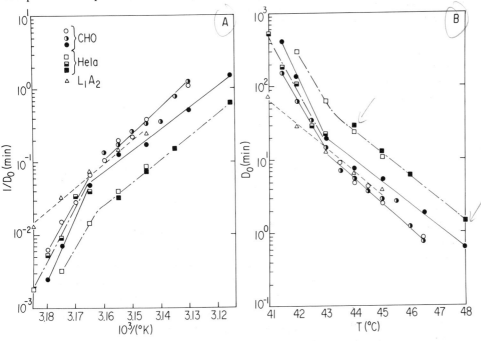

Figure 3.1. The temperature dependence of heat-induced cell killing for three independent data sets of CHO and HeLa cells and one set for L_1A_1 cells. The data are graphed on the Arrhenius plot ($\ln(1/D_o)$ versus $1/T(°K)$) in panel A and on the time-temperature plot ($\ln D_o$ versus $T(°C)$) in panel B. Survival data from the literature were analyzed identically; the data were taken from Henle and Dethlefsen (1980) with the exception of the data for L_1A_2 cells, which are from Nielsen and Overgaard (1979). CHO data points are connected by *solid lines*; HeLa, by *alternating long and short lines*; and L_1A_2, by *short dashed lines*.

for only three cell lines, CHO, HeLa, and L_1A_2. These were chosen because three independent data sets are available for both CHO and HeLa cells to show the variability that may appear for the same cell line assayed in different laboratories and the variability in temperature response of different cell lines, particularly the lack of an inflection point at 43°C for L_1A_2 cells. The summary of the temperature dependence of these cells is made on the Arrhenius plot in panel A and on a time-temperature plot ($\ln(D_o)$ vs T) in panel B. The $\ln(D_o)$ versus T representation in panel B can be superimposed onto the data in panel A if panel B is rotated by π radians about the x-axis even though, in principle, a straight line on the plot of $\ln(D_o)$ versus T should not correspond to a straight line fitted to the same data on the Arrhenius plot. In the narrow temperature range of 39°C to 48°C, however, both functions provide a good fit to the biological data, and the implied nonlinearity in transferring data on a straight line from panel A to panel B is small. For example, an activation energy of 140 kcal/mole across the entire temperature range in figure 3.1 would correspond to an exponential slope in panel B of 0.702, 0.694, and 0.681 at 43°C, 45°C, and 48°C, respectively.* Conversely, a constant slope s = 0.694 in panel B would correspond to activation energies of 138, 140, and 143 kcal/mole at the same respective temperatures. The biological data cannot differentiate between the linear and the reciprocal temperature fit in either representation; thus, without overriding a priori reasons for choosing one specific representation, both representations are equally valid. Data in panel B remain generally more accessible and the exponential slope constant is more readily converted to simple time-temperature conversion rules. For example, the exponential slope constant of −0.693 is equivalent to the frequently cited rule (Suit and Shwayder 1974; Dewey et al. 1977) of reducing the heating time by a factor of two per degree increase in temperature for equal biological effects ($e^{-0.693} = 0.5$).

The temperature dependence of the other survival curve parameter, n, is more difficult to define. Figure 3.2 shows a plot of $\ln(n)$ versus T for the survival data from the same cell lines shown in figure 3.1. Particularly for HeLa and L_1A_2 cells, the value of n remains close to unity, and even for CHO cells, n remains generally below 20. Since n is an exponential number, the effect on survival is probably less than $\ln(20) \simeq 3$ over the temperature range of interest, whereas the D_o can vary by as much as a factor of 7 per degree (fig. 3.1). In principle, however, even though

*The slope, s, is calculated by the equation $s = \ln\left(\dfrac{D_o(T+1)}{D_o(T)}\right) = \mu/RT(T+1)$.

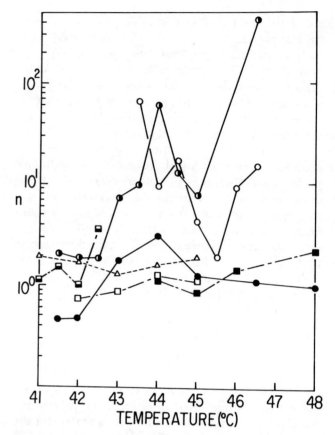

Figure 3.2. The temperature dependence of the extrapolation number, n, from the same survival curves used in figure 3.1. Symbols are identical to those in figure 3.1.

the description of cell killing in terms of the (n, D_o) survival model may be adequate using either the Arrhenius or the time-temperature plot, it is incomplete. A complete parameterization of cell survival after hyperthermia is possible with the linear-quadratic survival model, where both the linear (α) and the quadratic (β) rate constants are linear on the Arrhenius plot without any evidence of inflection points (Roti-Roti and Henle 1979). For additional detail, the reader is referred to the cited reference.

The large role of thermal history in modifying the cellular heat sensitivity for CHO cells is shown concisely in figure 3.3. The circles, which represent the control heat response of cells grown at 37°C, except for a single heating period at constant temperature, show the same inflection point near 43°C as the cells in figure 3.1. The inflection point can be removed by step-down heating (SDH), which consists of an acute heat shock immediately prior to heating at a lower assay temperature (Henle and Leeper 1976; Henle, 1980). For the data in figure 3.3, the acute heat shock consisted of 10 minutes at 45°C. The sensitization to heat killing below 43°C suggests that some of the heat lesions that lead to cell death at 39°C to 43°C are

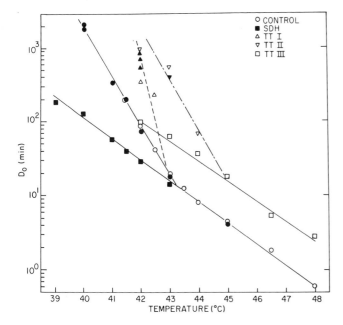

Figure 3.3. The effect of the cellular thermal history on the temperature response of CHO cells. Control cells were heated once at a constant temperature (*circles*) and step-down heating (SDH, *squares*) changed the slope of the control line at temperatures below 43°C. Thermotolerance (*TT*) appears in three distinct classes, each marked by a characteristic temperature dependence (see text). *Open symbols* are data taken from Bauer and Henle (1979) and *solid symbols* from Henle (1980).

reversible at 37°C. Cooling of cells to 0°C or 20°C after 45°C hyperthermia can also partially reverse lethal heat damage, but the effect is small (Henle and Leeper 1979). Such heat damage, which potentially leads to cell death but can be reversed, may be discernible by thermodynamic analysis, but a clear separation of the reversible and irreversible components of heat damage becomes necessary.

The phenomenon of thermotolerance has three distinct classes, each characterized by a different exponential slope constant. Thermotolerance I (TT I) represents the D_o's of the resistant phase of the biphasic survival curves that appear with prolonged (more than three to four hours) heating at temperatures below 43°C (Bauer and Henle 1979). Similarly, thermotolerance II (TT II) represents the D_o of the resistant phase of the biphasic survival curves that appear after acute heat conditioning (10 minutes at 45°C) and an eight-hour period (37°C) for the development of thermotolerance. The initial phase of these same thermotolerance survival curves is denoted by TT III (Bauer and Henle 1979).

In spite of the large variations in heat sensitivity induced by different thermal histories, the highly geometric picture of the temperature response of CHO cells is dominated by only two exponential slope constants, and these same slope constants are similar for a wide variety of different cell lines (fig. 3.1). Whereas this observation cannot be appreciated yet in terms of underlying mechanisms, such information can form the basis for correlation attempts with biochemical heat-induced lesions, which may result eventually in the identification of "critical" targets for heat-induced cell death.

In Vivo

Heat effects in vivo can be analyzed, in principle, in the same manner as data in vitro. The measurement of heat effects in vivo is based, however, almost exclusively on fixed endpoints (50% necrosis, TCD_{50}, etc.) rather than on a dose response (survival curve). The Arrhenius analysis requires the measurement of a rate, and thus in vivo data cannot be represented on the Arrhenius plot without assuming that the accumulation of heat damage in vivo is a purely exponential function of heating time. This assumption is unnecessary for showing in vivo data on the time-temperature plot.

Available heat response data for a large number of different tissues, measured by different endpoints, were summarized using the time-temperature plot (Henle and Dethlefsen 1980). Figure 3.4 shows this summary in diagrammatic form; the cross-hatched area represents the cluster of in vivo data, and the brace to the top line shows the complete range of time-temperature combinations for in vivo effects. Much of the data in the upper half of the unshaded in vivo range comes from the older literature, when hyperthermia production probably was inadequate, and temperature measurements reflected more the temperature of the heating medium than of the tissue of interest.

The vertically hatched area represents the data from figure 3.1, panel B for CHO and HeLa cells, and it overlaps considerably with the in vivo data. Superficially, this suggests that either relatively few cells are killed for a tissue reaction (corresponding to D_o; i.e., 63% or less), or that cells in vivo are relatively more heat sensitive, assuming that most cells must be killed for tissue necrosis. In some cases, however, a tissue reaction can be achieved without measurable cell killing. For example, the murine small intestine has an $LD_{50/7}$ response to hyperthermia without significant killing of the crypt stem cells, while the quantitative cellular heat response of this tissue (crypt survival) can be measured only with massive antibiotic support (Henle and Dethlefsen 1980).

Comparisons of the heat response for different

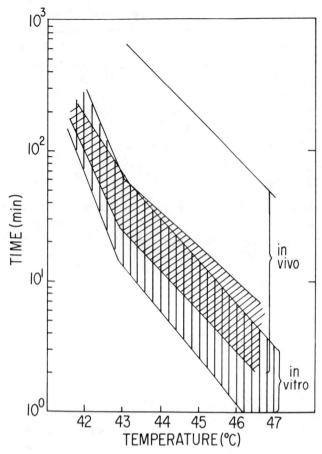

Figure 3.4. Diagrammatic representation of the time-temperature relationship for heat-induced tissue damage *(diagonal hatching)* and cell killing in vitro *(vertical hatching)*. The in vivo data for this diagram were summarized by Henle and Dethlefsen (1980). The majority of the available in vivo data fall into the diagonally hatched area; however, some other tissue data appeared more heat resistant and the complete range of in vivo data is indicated by the *brace*. The area marked *in vitro* is the envelope of the data in figure 3.1 for CHO and HeLa cells.

tissues, particularly malignant versus normal, may not reflect differential cellular sensitivities. For example, nonimmunogenic tumors may be able to regrow from single surviving cells; thus a TCD_{50} assay may appear extraordinarily heat resistant compared to normal tissue heat reactions, which may be achieved by only 50% cell killing, in spite of equal cellular heat sensitivity.

CONCLUSIONS

At present, the value of an Arrhenius analysis, as applied to heat-induced cell killing, lies in its representation of the temperature dependence of thermal death. A simple time-temperature representation is equally satisfactory for this purpose, however, for in vivo data it constitutes the representation of choice. Although the meaning of activation enthalpies and entropies for heat-induced cell death, calculated from the Eyring extension of the Arrhenius equation, is obscure at present, advances in irreversible thermodynamics and a better understanding of the biochemical perturbations from hyperthermia may justify a thermodynamic analysis of cell death.

References

Bauer, K. D., and Henle, K. J. Arrhenius analysis of heat survival curves from normal and thermotolerant CHO cells. *Radiat. Res.* 78:251–263, 1979.

Bertalanffy, L. V. The theory of open systems in physics and biology. *Science* 111:23–29, 1950.

Brandts, J. F. Heat effects on proteins and enzymes. In *Thermobiology*, ed. A. H. Rose. London: Academic Press, 1967, pp. 39–70.

Dewey, W. C. et al. Cellular responses to combinations of hyperthermia and radiation. *Radiology* 123:463–474, 1977.

Farrell, J., and Rose, A. H. Temperature effects on micro-organisms. In *Thermobiology*, ed. A. H. Rose. London: Academic Press, 1967, pp. 150–154.

Field, S. B. Some effects of hyperthermia on normal tissues. In *Clinical prospects of hypoxic cell sensitizers and hyperthermia*, eds. W. L. Caldwell and R. E. Durand. Madison, Wis.: Radiation Oncology Press, 1978, pp. 159–173.

Henle, K. J. Sensitization to hyperthermia below 43°C induced by step-down heating. *J. Natl. Cancer Inst.* 64:1479–1483, 1980.

Henle, K. J., and Dethlefsen, L. A. Time-temperature relationships for heat-induced killing of mammalian cells. *Ann. NY Acad. Sci.* 335:234–253, 1980.

Henle, K. J., and Leeper, D. B. Combination of hyperthermia (40°, 45C°) with radiation. *Radiology* 121:451–454, 1976.

Henle, K. J., and Leeper, D. B. Interaction of sublethal and potentially lethal 45°-hyperthermia and radiation damage. *Eur. J. Cancer* 15:1387–1394, 1979.

Iberall, A. S. A field and circuit thermodynamics for integrative physiology. I. Introduction to the general notions. *Am. J. Physiol.* 233:R171–R180, 1977.

Johnson, F. H.; Eyring, H.; and Polisar, M. J. *The kinetic basis of molecular biology.* New York: John Wiley & Sons, 1954, pp. 4, 19, 123, 198, 236, 438.

Nielsen, O. S., and Overgaard, J. Effect of extracellular pH on thermotolerance and recovery of hyperthermic damage *in vitro*. *Cancer Res.* 39:2772–2778, 1979.

Roti-Roti, J. L., and Henle, K. J. Comparison of two mathematical models for describing heat-induced cell killing. *Radiat. Res.* 78:522–531, 1979.

Snell, F. M. et al. *Biophysical principles of structure and function*. Reading, Mass.: Addison-Wesley, 1965, pp. 342–346.

Suit, H. D., and Shwayder, M. Hyperthermia: potential as an antitumor agent. *Cancer* 34:122–129, 1974.

Westra, A., and Dewey, W. C. Variation in sensitivity to heat shock during the cell-cycle of chinese hamster cells *in vitro*. *Int. J. Radiat. Biol.* 19:467–477, 1971.

Yates, F. E. Thermodynamics and life. *Am. J. Physiol.* 234:R81–R83, 1978.

CHAPTER 4
Thermosensitivity of Neoplastic Cells In Vitro

Beppino C. Giovanella

Introduction
Historical Background
Thermosensitivity of Neoplastic Cells In Vitro
Discussion and Conclusions

INTRODUCTION

The study of the in vitro effect of supranormal temperatures on neoplastic and normal cells has produced many important contributions to our knowledge of the mechanism of action of hyperthermia. It is the use of in vitro methodologies that enables investigators to probe for definitive answers to questions such as: why is the cancer cell more thermosensitive than the normal cell? We are faced with a vast and fascinating field of activity. The identity of cellular processes that are affected more and those that are affected less by heat is still quite uncertain; there are tantalizing hints that some phases of the cell cycle are more thermosensitive than others—but solutions to such problems are elusive. Many areas of investigation in hyperthermia await exploration.

HISTORICAL BACKGROUND

The first recorded observation that cultured neoplastic cells have higher thermosensitivity than normal cells was made in 1912. Lambert, using one of the first tissue culture systems, found that mouse and rat sarcoma cells are more thermosensitive than normal mesenchymal cells. For each temperature tested in the range 42°C to 47°C, it was possible to find a time of exposure that would suffice to kill sarcoma cells and yet allow mesenchymal cells to survive. Survival was measured by the capacity of cells to migrate from fragments of tissue maintained in a plasma clot. Curiously enough, this crucial observation elicited no interest and no follow-up.

In 1928, Friedgood exposed hanging drop cultures of Walker rat tumor No. 1 to various temperatures for 10 to 30 minutes, incubated them at 38°C for 48 hours, and then assessed their viability. Viability was once again determined by the presence or absence of cells migrating from the explant. The author assumed that only spindle-shaped cells were tumor cells, whereas the round cells observed were considered normal (mononuclear cells, probably macrophages). He did not find any difference in thermosensitivity between the two cell types. The obvious criticism of this experiment is that to differentiate between neoplastic and nonneoplastic cells in the same culture on purely morphologic grounds is impossible. In 1941, Vollmar compared the survival at supranormal temperatures of normal (spleen cells, presumably fibroblasts) and neoplastic (Ehrlich ascites carcinoma and Jensen's sarcoma) rodent cells. Tumor cells were killed at temperatures between 40°C and 42°C, whereas normal cells resisted exposure up to 43°C. Her experiments were carefully executed on a large group of cultures; unfortunately, quantitation of the results was very primitive, being based on subjective assessments of the rate of growth.

In 1966, Bender and Schramm published the results obtained after heating for 30 minutes at various temperatures cultures of 46 different animal and human tumors. As controls, cultures of whole embryos and of various organs from six species were also tested. Survival was measured by further proliferation at 37°C. It was found that tumor cells were 1°C to 2°C more thermosensitive than normal cells. Unfortunately, the authors did not count the cells before and after treatment, making it difficult to estimate the quantitative effect of heating.

Auersperg (1966) heated cultures of human fibroblasts and epithelial cells from carcinoma lines C4 and C27 at temperatures varying from 44°C to 48°C for 30 minutes. She found that the neoplastic epithelial cells survived longer than did fibroblasts. It must be noted, however, that all the determinations of viability were by the indirect methods of vital staining and protein content, which later were demonstrated to be unreliable for the determination of cell killing caused by exposure to supranormal temperatures (Giovanella, Lohman, and Heidelberger 1970; Harris 1966; Muckle and Dickson 1971).

Ossovski and Sachs in 1967 found that hamster cells transformed by polycyclic hydrocarbons had the same thermosensitivity as the cell of origin, and that cells transformed by SV40 virus were more thermoresistant than both. None of these cells was tested for their tumor-producing ability.

Kachani and Sabin in 1969 repeated the experiments of Ossovski and Sachs and found that hamster cells transformed in vitro by SV40 or polyoma virus were as thermosensitive as normal cells, but if they were injected into hamsters and produced tumors, cultures derived from these tumors were more thermosensitive than normal hamster embryo cells. This observation can be easily explained if we postulate that a large number of the cells morphologically transformed in vitro were non-tumor-producing, and that the in vivo passage was selective for the tumor-producing cells, which were also highly thermosensitive.

In 1969, Chen and Heidelberger found that mouse prostate cells cultured in vitro became more thermosensitive after undergoing malignant transformation after treatment with polycyclic aromatic hydrocarbons. In this communication, the definition of neoplastic cell is supported by biological experiments. Ten transformed cells produced malignant tumors when injected into syngeneic mice, whereas 10^3 "normal" cells did not.

Levine and Robbins in 1969 also compared the thermosensitivity of normal and neoplastic cells in cultures and found that neoplastic cells were much more adversely affected by supranormal temperatures. Unfortunately, their findings were based on an experimental set-up that could give only ambiguous results. Before heating, they allowed cells to reach confluence. At this point, normal (contact-inhibited) cells stop multiplying, whereas neoplastic cells continue to divide. Accordingly, the comparison was between cycling neoplastic cells and normal cells stopped in the G_0 phase. It has been abundantly demonstrated that the same cell changes its thermosensitivity according to the phase of the cell cycle during which the heating is performed. Both normal and neoplastic cells possess this property, and the difference in thermosensitivity exhibited by the same cell heated during G_0 or cycling is very large. In order to have meaningful comparisons between a normal cell and a tumor cell, both must be heated when in logarithmic growth, when their thermosensitivity is maximal. Theoretically, comparisons made with both cells in G_0 are also valid, but in practice it is difficult to determine whether all the tumor cells are in this stage.

In 1970 Giovanella, Lohman, and Heidelberger extensively studied the quantitative effects of supranormal temperatures on L1210 leukemia cells. The neoplastic cells were heated in vitro and their viability was assessed by inoculation into mice and determination of the survival of the inoculated animals. Using this direct method to assess the viability of the treated cells, it was possible to establish firmly the unreliability of dye exclusion tests, at least for hyperthermia experiments. The survival curves at various temperatures were determined, and it was found that the critical temperature range for these cells was between 40°C and 43°C. Below 40°C, the cells could survive indefinitely; above 43°C, their survival was so short—1×10^4 killing in less than an hour—that meaningful quantitative studies were difficult. Two interesting facts were also established. It was found that if cells were first heated at a low temperature (40°C) for four hours and then treated at higher temperature (42°C for two hours), they became relatively resistant to the second treatment. On the contrary, if cells were first heated at high temperature (42°C for one hour), they became sensitized to a further exposure to low temperature (40°C for four hours) (see chapter 6).

Giovanella, Lohman, and Heidelberger (1970) also studied the association of hyperthermia with various chemotherapeutic agents. It was found that L-erythro-α,β-dihydroxybutyraldehyde (DHBA) is clearly synergistic with hyperthermia, and phenylalanine mustard is at least additive. Actinomycin D was not effective in hyperthermic conditions. Such results are particularly interesting in view of the main pharmacologic activity of DHBA, protein synthesis inhibition.

In 1973, Giovanella and colleagues compared the thermosensitivity of normal and neoplastic mouse cells. The definition of normal and neoplastic cells was supported by inoculation into syngeneic hosts. The neoplastic cells (cultured from chemically induced fibrosarcomas) produced tumors when injected at doses of 1×10^6 cells in adults and 1×10^5 in newborns. The normal cells (fibroblasts cultured from mouse embryos of the same inbred strain C57 BL/6) did not. All the cultures were heated during their logarithmic growth phase and the doubling times of some of the normal cells were much shorter than those of the neoplastic cells. Without exception, all the normal cultures proved to be more thermoresistant than the neoplastic ones. One of the normal cultures developed into an established cell line, MEF 1, lacking neoplastic properties, fast growing, but strictly contact inhibited. From this line, a neoplastic subline, MEF 1T, was derived. These two lines have approximately the same doubling time (20 to 22 hours), but whereas MEF 1T can produce tumors when 1×10^3 cells are injected in adult C57 BL/6 mice, 1×10^7 cells of MEF 1 did not produce any. Two hours of exposure at 42.5°C will kill 98% of MEF 1T cells and only 37% of MEF 1. These results, together with those of Chen and Heidelberger (1969) indicate that mouse cells, as soon as they acquire malignant properties, become more thermosensitive than the cells of origin.

In 1974, Kim, Kim, and Hahn published the results of their work on the synergism of action of ionizing radiation with heat. HeLa cells exhibited significant enhancement of radiosensitivity when the irradiated cells were exposed to elevated temperatures, although 3T3 cells did not show such an enhancement. It is obviously not enough to have such a difference in behavior between two cell lines to conclude that hyperthermia enhances the radiosensitivity of tumor cells, but not of normal cells. Such an observation, however, should stimulate further work in the same direction. Strangely, this has not been the case.

In 1975, Kase and Hahn found that human nonneoplastic fibroblasts can be transformed by SV40 virus into cells with neoplastic potential, as demonstrated by their ability to produce tumors when injected into "nude" (thymus-deficient) mice. It has been effectively demonstrated (Giovanella, Stehlin, and Williams 1972, 1974; Giovanella and Stehlin 1974) that only human cultured cells from malignant neoplasms will grow as tumors when inoculated in nude mice; human or murine cells derived from normal tissues of adults or embryos will not, even if growing vigorously in vitro. Inoculation of tissue-cultured cells in nude mice is, consequently, a sensitive test for their malignant potential. The human cells transformed by SV40 virus increased their thermosensitivity considerably at the same time that they developed neoplastic capability. Such difference in thermosensitivity, however, was evident only when cells were studied at low population densities, disappearing at high densities when both cell systems became highly thermoresistant (Kase and Hahn 1976). This variability stresses the great care necessary for the correct evaluation of results obtained under different experimental conditions. Another note of caution was sounded in 1977 by Li, Hahn, and Shiu, who noticed that at 43°C organic solvents used to dissolve water-insoluble chemotherapeutic agents become themselves lethal to cells at the concentrations needed to make drug solutions. The logical conclusion is that in every experiment with such drugs, the results are invalid if there has not been a control with the solvent alone.

In 1975, Stehlin and associates determined that human normal melanocytes derived from the uvea of the eye were more thermoresistant than human melanoma cells. The same investigators (Giovanella, Stehlin, and Morgan 1976) then extended these findings to a wide variety of human nonneoplastic and neoplastic cells—the definition *neoplastic* being given only to cells cultured from human neoplasms that are aneuploid and capable of producing malignant tumors when injected into nude mice. The tumors were identified as human by karyotypic analysis. In this way, it was demonstrated that human colon carcinoma cells are more thermosensitive than normal epithelial cells derived from fetal intestine, that fibrosarcoma cells are more thermosensitive than normal fibroblasts, and so on. All the normal and neoplastic cells studied had similar doubling times; sometimes the normal cells divided faster than their neoplastic counterparts.

In 1975, Gerner and Schneider found that HeLa cells exposed continuously to 44°C are killed more rapidly than cells exposed to 44°C for one hour, cooled to 37°C for two hours, and then returned to 44°C. This suggests that the early part of thermal damage is reversible and can be repaired to some extent. These results were confirmed in 1976 by Henle and Leeper and afterward by many other investigators.

Also in 1975, Hahn, Braun, and Har-Kedar studied the effect of hyperthermia associated with adriamycin and bleomycin on Chinese hamster cells (CHO) and EMT6 mouse cells. They found synergism at 43°C, but not at 41°C. In 1978, however, Donaldson, Gordon, and Hahn established that exposure to hyperthermic temperatures reduces the toxic effect of actinomycin D on CHO cells.

Sapareto and colleagues in 1978, also working with Chinese hamster ovary cells, established that in the temperature range 42°C to 46°C the survival of the treated cells is very sensitive to small variations in temperature: an increment of 0.1°C can change survival by as much as a factor of 10.

By 1968, Giovanella and Heidelberger, working with L1210 leukemia cells, had established that cycling cells are much more sensitive to the lethal effect of hyperthermic temperatures than nondividing cells. In 1969, Giovanella, Mosti, and Heidelberger, working with HeLa cells, found that some phases of the cell cycle are more thermosensitive than others, with maximum thermosensitivity in the late S-phase. These studies were extended and confirmed by Palzer and Heidelberger (1973a, 1973b), and Kim, Kim, and Hahn (1976).

Many investigators have studied the effect of treating cells with both heat and ionizing radiation (Westra and Dewey 1971; Ben-Hur, Bronk, and Elkind 1972; Ben-Hur, Elkind, and Bronk 1974; Li, Evans, and Hahn 1976). Although not yet completely defined, there seems to be a consensus that the maximum lethal effect is obtained when cells are first irradiated and then subjected to hyperthermia. The best explanation for this seems to be that repair mechanisms to radiation damage are quite thermosensitive.

Very little is known about what causes the difference in thermosensitivity of normal and tumor cells. This is not surprising since we do not know why mammalian cells (both normal and neoplastic) die after exposure for a certain time to supranormal temperatures of up to 44°C to 45°C. Whether the limiting factor is the same for both normal and neoplastic cells, with only a quantitative difference, or if in the tumor cell a new, more thermosensitive vital function is present, is unknown at present.

Studies on the possible mechanism of action of heat have so far been conducted mostly by observing which cell function is affected first or more extensively by exposure to supranormal temperatures. Unfortunately, such an approach suffers from the drawback that it is impossible to decide if the loss of the function (for example, DNA synthesis) causes the death of the cells or the death of the cells is the cause of the loss of function.

THERMOSENSITIVITY OF NEOPLASTIC CELLS IN VITRO

Concepts, Principles

It is conceptually essential to clarify two major points when considering the results obtained in the field of hyperthermia. First, any study of the effects of supranormal temperatures on tumor cells is meaningless if there are no adequate controls, such as normal cells that are comparable (because of embryologic origin and/or rate of division) to the neoplastic cells used in the experiment. This is not to say that neoplastic (or normal) cells cannot be studied alone when mechanisms of action are being investigated. When the usefulness of a certain hyperthermic treatment (or, for that matter, of any treatment) for cancer is in question, however, it is totally useless to experiment with tumor cells alone. They can be killed by many agents, both chemical and physical. It is useless to list such agents if they do not have a selective action against tumor cells. Heavy metal salts, cyanide, and other toxic substances can kill cancer cells effectively and efficiently. Unfortunately, such substances kill normal cells no less efficiently. In vivo experiments have a built-in safety valve provided by the evident toxicity to the host animal. In vitro experiments do not have such a safety valve, and it is the responsibility of the researcher to plan the experimental design in a way that will produce meaningful results. When working with physical agents (such as heat), or chemical agents (such as chemotherapy) to ascertain anticancer capabilities, it is esssential to have an experimental design that allows determination of the presence or absence of a selective lethal action of the agent(s) chosen on the tumor cells. Such selectivity can be *qualitative* (e.g., asparaginase will kill tumor cells that are asparagine-dependent and will not kill normal cells or other tumor cells that are glutamine-dependent) or *quantitative* (e.g., 41°C or 42°C applied for a certain length of time will kill most tumor cells and few normal cells; however, if the time of exposure to heat is lengthened, the selectivity of the killing diminishes and then disappears altogether). The second point is that even if a selectivity of action has been established using a rational control, this is not enough without further studies to justify the use of a treatment. The safety of a treatment is directly proportional to the weakest component in the system. As a practical example, consider the administration of 43°C for four hours; there is a good differential effect on malignant melanoma cells versus human normal melanocytes, but this does not authorize systemic heating of melanoma patients for four hours at 43°C. Patients will die long before the end of the four hours because there are normal cells essential to survival that are more thermosensitive than melanocytes. Moreover, what is not lethal for isolated cells can be quite deadly for organs and complete organisms, owing to such factors as increased permeability of membranes (which causes pulmonary and cerebral edemas) and hyperproduction of lactic acid in the muscle masses (which exhausts the buffering capacity of the blood, causing a fatal acidosis). Also, hyperthermia too high or too prolonged causes a progressive increase in the heart rate, leading to fibrillation and death.

Cell kill by hyperthermic treatment, alone or associated with other agents, can be detected and measured by cell count or colony count. Such direct methods are reliable and precise when properly used. Indirect methods, such as dye exclusion tests, are imprecise and unreliable for measuring hyperthermic cell kill (Harris 1966, 1967, 1969; Giovanella, Lohman, and Heidelberger, 1970). This seems to contrast with the reliable performance of indirect methods used for other cytotoxic agents. It must be remembered, however, that different agents kill cells in different ways. The dye exclusion test does not detect directly the viability of a cell; it detects alterations in the cell membrane, which in many cases are temporary and probably caused by the death of the cell, if the cell has been killed by certain mechanism(s). Depending upon the cytotoxic agent, however, cells may acquire such permeability without dying, or not acquire it for some time after their apparent death. It is important to remember that the concept of death is different when applied to a whole animal or to a cell. A mammal is considered dead when its nervous system is no longer functional. Obviously, the viability of cells cannot be established by such a criterion. The most commonly used criterion of viability for cells in culture is their ability to reproduce. This can bring about some paradoxes, in that certain differentiated normal cells, such as neurons, do not reproduce at all. In their case, other criteria, such as the growth of neurites, are substituted. For work on experimental therapy of tumors, however, irreversible block of cell division is an excellent indication of the therapeutic effectiveness of a cytotoxic agent. A tumor cell that has permanently lost the ability to divide does not constitute a danger to the host. Whether such a cell is dead is purely a question of semantics.

Cell counts and colony counts after heat treatment are used to measure different parameters, and it is important to keep this in mind when interpreting results. *Cell counts* give a picture of the effect of heat treatment on the size of the total cell population regardless of the contribution to the total number of individual cells. In other words, suppose 100 cells are plated in two flasks, with one kept as control and the other heated at 43°C for two hours 24 hours after plating, and the cells are counted a week later, and 1000 cells are found in the control flask and 100 in the treated flask. The conclusion will be that the heat exposure has inhibited growth of cells by 90%, but there will be no information on the distribution of the damage within the population affected. The same result can be achieved if there is 100% plating efficiency (which in practice seldom, if ever, occurs), having none of the 100 cells dividing and none disappearing because of lysis, or having 60 of them lysed, 10 dividing twice, and 30 once. The total final number will be in both cases the same—100—as it will be with an almost infinite number of combinations. On the contrary, colony counting provides much valuable information on the subpopulations affected.

Colony counts give us the number of clonogenic cells (stem cells) present in the population under consideration. In other words, we come to know the number of cells actually capable of proliferating and the distribution of subpopulations according to growth rate. For example, if we plate 100 cells, wait two weeks and count the colonies formed (after defining the smallest number of cells that can be called a colony, perhaps 32, then 26 colonies are found, of which 15 contain 32 to 64 cells each, 7 colonies contain 64 to 128 cells, and 4 colonies contain more than 128 cells. Such observation yields the following information: (1) 26 of 100 cells in our population are capable of dividing more than five times every two weeks; (2) of these cells, 15 divided an average of five or six times, seven divided six or seven times, and four divided more than seven times. Obviously, in any definition of a colony there is an element of arbitrariness introduced by the cut-off point above which a group of cells is a colony, below which it is not. The problem is further complicated by our ignorance of the factors that cause a group of cells not to exceed such a threshold. A group of 16 cells, in our example, could be formed by a cell and its progeny dividing four times, once a day for four days, then lying dormant for ten days. A group of 32 cells could be formed by a cell and its progeny dividing five times, once a day again for five days, then lying dormant for nine days. Or, both may have a long cell cycle, the first dividing once every 3.5 days, the second dividing once every 2.8 days. The 16-cell group will not be counted as a colony; the 32-cell group will be. The implication is that the cell originating the first colony is nonclonogenic while the second cell is clonogenic. In both cases, a relatively minor quantitative difference translates itself into a major qualitative one because of our methodologic approach. However imprecise such an approach, it is the best available at present. Its use requires awareness of its limitations.

Research Methodology

For studies of sensitivity to hyperthermia in vitro, two basic approaches can be followed. The first and simplest is to have the cells under study growing in a flask and, when they are in the logarithmic phase of growth, expose them to the desired temperature for as long as required. The cells are then trypsinized and a given number plated in fresh medium and incubated at 37°C. After an appropriate time interval, depending on the rate of growth of the cells, their number or the number of colonies formed is counted. If the cells grow in suspension, trypsinization can be omitted.

In the second approach, the cells are plated in flasks, dishes, or microplates in the appropriate numbers, left to attach, and counted. The cells are then heated, incubated at 37°C for the length of time necessary for the development of the proper size colonies, and fixed. Colonies are then stained and counted.

The first method is obviously much simpler, but it introduces an element of variability—namely, the trypsinization after treatment. It is well known that proteolytic enzymes can lyse cells that have been partially damaged by a variety of agents, leaving cells intact that are not so damaged. This selective digestion of damaged cells is actually used to detect the percentage of such cells present in a population (McCormick et al. 1980). Some of the damaged cells, if not trypsinized, can recover. In other words, trypsinization tends to magnify the killing action of any pretreatment with a noxious agent of the cells under experiment.

Plating before heating has the great advantage of leaving the cells free of manipulations during and after the treatment. It is technically more difficult, however. The main problem encountered is that of cell multiplication after plating. If cells are plated as single cell suspensions and then the investigator waits too long, some of them can divide, detach during heating, and replate individually. This will give rise to an excess of colonies, which will lead to an underestimation of the killing effect of heat. This pitfall can be avoided easily if it is firmly established that cells should be heated within six hours of plating, and comparative controls also are fixed and counted at the end of the six hours. For cells that do not attach to solid supports, such as lymphoblastoid cell lines and suspension cultures, there is generally no problem of trypsinization. The killing effect of heat can be assessed by cell count after some time or by count of colonies growing in semisolid media (agar, agarose, methyl-cellulose, etc.).

Heat treatment is performed usually by sealing hermetically the tissue culture vessels in plastic bags and submerging the whole in water baths with precise regulation of the temperature (within 0.1°C to 0.2°C) and water circulation (to avoid temperature gradients). At least two precision thermometers periodically calibrated against reliable standards should be used to monitor the water-bath temperature.

DISCUSSION AND CONCLUSIONS

It is clear that the use of cultured cells has been useful for both qualitative and quantitative studies of the effects of hyperthermia on normal and neoplastic tissues. Without such studies, it would have been impossible to quantitate the effect of supranormal temperatures on different cell types or to establish the exact temperatures and duration of exposure necessary to achieve a kill of four logs or more. Even more important, it was only by means of tissue culture techniques that it was possible to demonstrate the selective lethal effect of supranormal temperatures on neoplastic cells, since only by in vitro techniques is it possible to compare side by side normal and neoplastic cells of the same embryologic origin growing at the same rate and to quantitate such results. The cultured cells constitute an ideal material for the study of mechanisms of action of heat on the living cells, both normal and neoplastic. Such investigations are still in their infancy and need prolonged study, but if pursued to their ultimate conclusions—the elucidation of the mechanisms of cell killing by heat and of the differences of such mechanisms between normal and neoplastic cells—they may provide the clue to the selective killing effect of heat on neoplastic tissues.

Another area in which in vitro techniques are irreplaceable is the investigation of the effect of heat combined with other agents upon living matter. Already it has been possible to find chemicals that act additively with heat, some possibly even synergistically. The main question has yet to be answered, however: what are the drugs that in conjunction with heat give the best selective tumor killing effect? Again, it is not the toxic effect of a drug or of a combination therapy that counts, but the selective, specific antitumor effect. This requirement leads to a major investigatory need: good controls in order to establish differential toxic effects. A beginning has been made using as controls normal melanocytes for melanoma cells, fetal epithelial intestinal cells for colon carcinoma cells, and normal fibroblasts for fibrosarcoma cells. These controls provide a reliable way to evaluate the increased thermosensitivity of neoplastic cells over their normal counterparts under well controlled and equal conditions. More is needed, however, especially good in vitro models of the human tissues more vulnerable to the action of supranormal temperatures applied alone or associated with other agents. This is rendered very difficult by our almost total ignorance about the relative thermosensitivities of the different normal human tissues, both in vivo and in vitro, in the dividing and in the resting state. Another difficulty is that, although it is known what tissues are most sensitive to the majority of chemotherapeutic agents (bone marrow and intestinal mucosa), a good in vitro model system for bone marrow is still lacking. Once all these controls are available, it may be possible to use an in vitro system to ascertain which human malignancies offer the best targets for hyperthermic treatments. Such a system could supply quantitative data on the length of time necessary to apply a certain temperature in order to kill a given number of tumor cells (i.e., a tumor of a corresponding size). Use of the model also may reveal how to best fractionate treatment and which drugs will have the maximum effect when administered in conjunction with heat. Similar information is needed about ionizing radiation associated with hyperthermia.

References

Auersperg, N. Differential heat sensitivity of cells in tissue culture. *Nature* 209:415–416, 1966.

Bender, E., and Schramm, T. Untersuchungen zur thermosensibilitat von tumor—und normalzellen in vitro. *Acta Biol. Med. Ger.* 17:527–543, 1966.

Ben-Hur, E.; Bronk, B. V.; and Elkind, M. M. Thermally enhanced radiosensitivity of cultured Chinese hamster cells. *Nature; New Biology* 238:209–211, 1972.

Ben-Hur, E.; Elkind, M. M.; and Bronk, B. V. Thermally enhanced radio-response of cultured Chinese hamster cells: inhibition of repair of sublethal damage and enhancement of lethal damage. *Radiat. Res.* 58:38–51, 1974.

Chen, T. T., and Heidelberger, C. Quantitative studies on the malignant transformation of mouse prostate cells by carcinogenic hydrocarbons in vitro. *Int. J. Cancer* 4:166–178, 1969.

Donaldson, S. S.; Gordon, L. F.; and Hahn, G. M. Protective effect of hyperthermia against the cytotoxicity of actinomycin-D on Chinese hamster cells. *Cancer Treat. Rep.* 62:1489–1495, 1978.

Friedgood, H. B. Thermal death point of sarcoma and normal mononuclear cells (Walker rat tumor no. 1). *Arch. Exp. Zellforsch.* 7:243–248, 1928.

Gerner, E. W., and Schneider, M. J. Induced thermal resistance in HeLa cells. *Nature* 256:500–502, 1975.

Giovanella, B. C. et al. Selective lethal effect of supranormal temperatures on mouse sarcoma cells. *Cancer Res.* 33:2568–2578, 1973.

Giovanella, B. C. and Heidelberger, C. Biochemical and biological effects of heat on normal neoplastic cells. *Proc. Am. Assoc. Cancer Res.* 9:24, 1968.

Giovanella, B. C.; Lohman, W. A.; and Heidelberger, C. Effects of elevated temperatures and drugs on the viability of L1210 leukemia cells. *Cancer Res.* 30:1623–1631, 1970.

Giovanella, B. C.; Mosti, R.; and Heidelberger, C. Further studies of the lethal effects of heat on tumor cells. *Proc. Am. Assoc. Cancer Res.* 10:29, 1969.

Giovanella, B. C., and Stehlin, J. S. Assessment of the malignant potential of cultured cells by injection in "nude" mice. In *Proceedings, first international workshop on nude mice*, eds. I. Rygaard, and C. O. Povlsen. Stuttgart: Fischer-Verlag, 1974, pp. 279–284.

Giovanella, B. C.; Stehlin, J. S.; and Morgan, A. C. Selective lethal effect of supranormal temperatures on human neoplastic cells. *Cancer Res.* 36:3944–3950, 1976.

Giovanella, B. C.; Stehlin, J. S.; and Williams, L. J., Jr. Development of invasive tumors in the "nude" mouse after injection of cultured human melanoma cells. *J. Natl. Cancer Inst.* 48:1531–1533, 1972.

Giovanella, B. C.; Stehlin, J. S.; and Williams, L. J., Jr. Heterotransplantation of human malignant tumors in "nude" thymusless mice. II. Malignant tumors induced by injection of cell cultures derived from human solid tumors. *J. Natl. Cancer Inst.* 52:921–930, 1974.

Hahn, G. M.; Braun, J.; and Har-Kedar, I. Thermochemotherapy: synergism between hyperthermia (42-43) and adriamycin (or bleomycin) in mammalian cell inactivation. *Proc. Natl. Acad. Sci. USA* 72:937–940, 1975.

Harris, M. Criteria of viability in heat-treated cells. *Exp. Cell Res.* 44:658–661, 1966.

Harris, M. Temperature-resistant variants in clonal populations of pig kidney cells. *Exp. Cell Res.* 46:301–314, 1967.

Harris, M. Growth and survival of mammalian cells under continuous thermal stress. *Exp. Cell Res.* 56:382–386, 1969.

Henle, K. J., and Leeper, D. B. Interaction of hyperthermia and radiation in CHO cells. Recovery kinetics. *Radiat. Res.* 66:505–518, 1976.

Kachani, Z. F. C., and Sabin, A. B. Reproductive capacity and viability at higher temperatures of various transformed hamster cell lines. *J. Natl. Cancer Inst.* 43:469–480, 1969.

Kase, K., and Hahn, G. M. Differential heat response of normal and transformed human cells in tissue culture. *Nature* 255:228–230, 1975.

Kase, K. R., and Hahn, G. M. Comparison of some response to hyperthermia by normal human diploid cells and neoplastic cells from the same origin. *Eur. J. Cancer* 12:481–491, 1976.

Kim, J. H.; Kim, S. H., and Hahn, E. Thermal enhancement of the radiosensitivity using cultured normal and neoplastic cells. *Am. J. Roentgenol.* 121:860–864, 1974.

Kim, S. H.; Kim, J. H.; and Hahn, E. W. The enhanced killing of irradiated HeLa cells in synchronous culture by hyperthermia. *Radiat. Res.* 66:337–345, 1976.

Lambert, R. A. Demonstration of the greater susceptibility to heat of sarcoma cells. *JAMA* 59:2147–2148, 1912.

Levine, E., and Robbins E. R. Differential temperature sensitivity of normal and cancer cells in culture. *J. Cell. Physiol.* 76:373–380, 1969.

Li, G. C.; Evans, R. G.; and Hahn, G. M. Modification and inhibition of repair of potential lethal x-ray damage by hyperthermia. *Radiat. Res.* 67:491–550, 1976.

Li, G. C.; Hahn, G. M.; and Shiu, E. C. Cytotoxicity of commonly used solvents at elevated temperatures. *J. Cell. Physiol.* 93:331–334, 1977.

McCormick, K. J. et al. An enzymatic assay for the detection of natural cytotoxicity. *J. Immunol. Methods* 35:83–90, 1980.

Muckle, D. S., and Dickson, J. A. The selective inhibitory effect of hyperthermia on the metabolism and growth of malignant cells. *Br. J. Cancer* 25:771–778, 1971.

Ossovski, L., and Sachs, L. Temperature sensitivity of polyoma virus: induction of cellular DNA synthesis and multiplication of transformed cells at high temperature. *Proc. Natl. Acad. Sci. USA* 58:1938–1943, 1967.

Palzer, R. J. and Heidelberger, C. Influence of drugs and synchrony on the hyperthermic killing of HeLa cells. *Cancer Res.* 33:422–427, 1973a.

Palzer, R. J., and Heidelberger, C. Studies on the quantitative biology of hyperthermic killing of HeLa cells. *Cancer Res.* 33:415–421, 1973b.

Sapareto, S. A. et al. Effects of hyperthermia on survival and progression of Chinese hamster ovary cells. *Cancer Res.* 38:393–400, 1978.

Stehlin, J. S. et al. Results of hyperthermia perfusion of melanoma of the extremities. *Surg. Gynecol. Obstet.* 140:338–349, 1975.

Vollmar, H. Über den einfluss der temperatur auf normales gewebe und auf tumorgewebe. *Z. Krebsforsch.* 51:71–99, 1941.

Westra, A., and Dewey, W. C. Variation in sensitivity to heat shock during the cell cycle of Chinese hamster cells in vitro. *Int. J. Radiat. Biol.* 19:467–477, 1971.

CHAPTER 5

Thermosensitivity of Neoplastic Tissues In Vivo

John A. Dickson
Stuart K. Calderwood

Introduction
Historical Background
Concepts and Principles
Methods and Materials
Analysis of Data
Tumor Volume and Thermal Response
Tumor Cell Kinetics and Thermosensitivity
Role of the Tumor Microenvironment in Heat Sensitivity
Potentiators of Hyperthermia
Perspective

INTRODUCTION

All living tissues are temperature sensitive. There is mounting evidence that tumors are selectively vulnerable to heat treatment (Wizenberg and Robinson 1976; Rossi-Fanelli et al. 1977; Streffer et al. 1978; Dethlefsen and Dewey, in press). Our current knowledge of the thermal treatment of cancer is, however, still rudimentary; for instance, adequate answers are not available to the following simple questions: to what temperature are tumors sensitive? What heating dose (in terms of temperature and time) will be needed to eradicate a tumor? To what extent can normal tissues be spared while tumors are destroyed? Even more basic, what are the best methods of tumor heating? (The plural noun is used, since different techniques will almost certainly be required for heating different sites, and maybe even for heating different tumor types and volumes.) These questions may be answered by careful empirical study. To have a greater measure of control over treatment design, however, it is essential to understand the mechanism(s) of thermal cell killing. Knowledge of the factors involved in heat sensitivity may permit a rational approach to therapy. Moreover, these factors may cast light on weak points in the tumor defenses and allow the selection of suitable agents to use in combination with heat. In the present chapter, therefore, factors influencing the heat sensitivity of tumors in vivo are discussed with the aim of providing practical information to aid in the development of hyperthermia as a treatment for cancer.

HISTORICAL BACKGROUND

The use of heat, in the form of a red-hot iron to cauterize malignant tumors, dates back at least to 2000 B.C. (Wolff 1907). There are many reports from the late 1800s and early 1900s describing the regression of both primary and secondary tumors after infection with pyrogenic bacteria (Bruns 1887; Coley 1893; Rohdenburg 1918).

These infections caused an increase in body temperature to 40.0°C to 40.6°C for several days. Clinical work using the physical application of heat also gave indications that hyperthermia might have a place in cancer therapy (Westermark 1898; Percy 1916). In this early era, however, the most widely applied technique clinically was the induction of pyrexia using bacterial toxins (Nauts, Fowler, and Bogatko 1953). In spite of the considerable success achieved by Coley in curing various types of malignancy over many years, the toxins proved unpredictable in effect and difficult to standardize biologically. As a result, the technique fell into general abeyance following Coley's death in 1936 (Dickson 1977). This early work, although suggesting the potential of hyperthermia as a treatment modality, offered little quantitative data on dose response to heat or the mode of thermal cell killing. One pioneer of the systematic approach to the evaluation of hyperthermia was Westermark (1927). In the investigations of Westermark, rat tumors were treated at measured intratumor temperatures of 44°C and 45°C; dose response in vivo and in vitro, differential effects of heat on tumors and normal tissues, and the effects of hyperthermia on tumor metabolism were evaluated.

The modern era of hyperthermia began with the work of Crile (1961, 1962, 1963), some of whose investigations are only now being evaluated. Crile also discovered the phenomena of delayed cell killing after hyperthermia, thermal tolerance, and increased heat sensitivity (1963) after blood vessel clamping.

The stimulus that led to the current renaissance of interest in hyperthermia was provided by the work of Cavaliere and colleagues (1967), who treated 22 patients with limb tumors by regional perfusion with prewarmed blood (tumor temperature 41.5°C to 43.5°C). The treatment was extremely effective, with tumor regression occurring in 16 of 22 patients. These workers also investigated the mechanisms of hyperthermic cell killing, concentrating on tumor metabolism in vitro. Findings included a specific inhibition of tumor cell respiration (Mondovi et al. 1969a) and decreased incorporation of labeled precursors into DNA, RNA, and proteins of malignant cells at elevated temperatures (Mondovi et al. 1969b); results have been confirmed by other workers (Muckle and Dickson 1971; Dickson and Shah 1972; Dickson and Suzangar 1976a). The changes occurring in cellular biochemistry in different phases of the cell cycle (Mitchison 1971) were also found to have a bearing on thermal cell killing; cells were found to be specifically heat sensitive in the phase of DNA replication (S phase) (Westra and Dewey 1971; Palzer and Heidelberger 1973b). Palzer and Heidelberger (1973a) also reported that hyperthermic cell killing was enhanced by drugs that stimulated DNA or protein synthesis or inhibited RNA synthesis.

More recently, the effects of hyperthermia on cell structure have been shown to be of importance. Determinants of the response of cells to hyperthermia appear to be the membrane constituents that contribute to membrane fluidity and permeability—the phospholipids, proteins, and the fatty acids (proportions of saturated to unsaturated acids); in particular, reduction in membrane cholesterol concentration has been found to be closely correlated with thermal cell killing (Gerner et al. 1980). More empirical work carried out in vitro has confirmed the dose-response relationship between heat and cell destruction found by Westermark (1927) and Crile (Bhuyan et al. 1977; Gerweck 1978). A general finding, however, has been that tumors are more heat sensitive in vivo than in vitro (Dickson 1978), which indicates that there may be additional factors in the in vivo environment that increase tumor thermal sensitivity (Overgaard 1976a; Marmor, Hahn, and Hahn 1977; Dickson and Calderwood 1980; Dickson, Calderwood, and Jasiewicz 1977).

Two of these in vivo factors may be the interrelated host-defense mechanisms, the inflammatory and immune responses, the latter including the mononuclear phagocyte system (macrophages). Crile (1963) investigated the use of serotonin, a chemical mediator of inflammation, and the effects of heat on immunogenic tumors have been explored by others (Dickson 1978; Dickson and Shah 1980; Szmigielski and Janiak 1978).

More recently, the interrelationships of tumor volume, cell kinetics, and the physiologic properties of the tumor extracellular microenvironment in thermal sensitivity have come to the fore (Dickson and Calderwood 1976, 1979, 1980; Calderwood and Dickson 1980b; Urano et al. 1980; Gerweck 1978). The tumor microenvironment has been implicated also in the intriguing phenomena of thermal resistance and step-down heating, which are now attracting increasing interest. The induction of thermal tolerance in tumors and in normal tissues by a wide range of hyperthermic temperatures has been well documented (Gerner et al. 1980). With cells in vitro, low pH has been shown to inhibit the development of this resistance (Gerweck 1977; Overgaard and Nielson 1980). Step-down heating is the phenomenon whereby marginally lethal or nonlethal temperatures become toxic when preceded by acute heat treatments at higher temperatures. This was first described for L1210 leukemia cells in 1970 by Giovanella, Lohman, and Heidelberger and has been confirmed by other workers (Henle and Leeper 1976; Henle, Karamuz, and Leeper 1978). From studies of this data on Arrhenius plots, Henle and Dethlefsen (1980) have pointed out that the acute heat conditioning of cells at 45°C has similarities to the sensitization of cells by exposure to an external pH of 6.7.

It is apparent that the response of cancer cells to heat encompasses a number of components; one postulated interaction of such governing factors in tumor response in vivo is illustrated in figure 5.1. It is to be noted that not only local considerations pertaining to both tumor and normal tissues are operative,

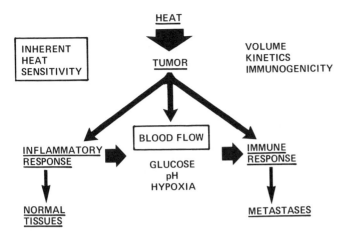

Figure 5.1. Interplay of factors involved in the response of a solid tumor in vivo to heat. Blood flow in the tumor during and after hyperthermia plays a central role in determining the metabolic status (e.g., pH, oxygenation) and uniformity of heat distribution in the tumor cells. An immune response to heating would require blood flow into the tumor for access of lymphocytes and/or macrophages and egress of such cells and possibly egress of breakdown products from the heated tumor. Changes in blood flow characterize the inflammatory reaction that occurs in normal tissue subjected to the physical injury of heat; a biochemical feature of inflammation is the acid pH that develops at the site (Menkin 1956). Local tissue reaction, which may vary in different anatomic sites, could therefore exert a major sensitizing influence on tumor response to hyperthermia. Inherent (metabolic) heat sensitivity of the cancer cells could constitute another major determinant in the response to heat, and other properties of the tumor (e.g., volume, cell population kinetics, tumor immunogenicity) may be of importance in the local and systemic consequences of hyperthermia. (Dickson and Calderwood 1980. Reprinted by permission.)

but the extremely important participation of the host at the systemic level must be kept in mind, even when the heating is applied locally to the tumor. Tumor blood flow is emerging as the major mediator in tumor response to heat, since it governs not only the local tumor environment (nutrient supply, oxygen level, pH) but is also the key link in the host-tumor relationship. In this latter role, blood flow affords access of components of the host defense system (leukocytes, lymphocytes, macrophages) to the tumor, and provides egress of tumor breakdown products following hyperthermia; it is also probably the major connection in the important but little understood relationship operating between a primary tumor and its metastases. It has been surmised for many years that a tumor as it grows conditions the host for its further spread, and there is evidence that the primary tumor may exert an influence (stimulatory or inhibitory) on the number and rate of growth of its metastases. The association between a primary tumor and its satellites and the host-tumor relationship have been reviewed in detail elsewhere (Dickson 1977; Dickson and Shah, in press).

These considerations mean that tumor heating and the various factors influencing it (fig. 5.1) cannot be viewed simply in terms of the physical application of heat to a mass of malignant cells. Normal local tissue reaction and systemic host response may be major determinants in the destruction of a tumor and its metastases following hyperthermia (Dickson and Calderwood 1980). More attention is now being focused on the effects of heat on normal tissues, but most of the experimental work to date on hyperthermia has ignored the host. This disregard makes questionable the value of many of the results in relation to human neoplasia, since it constitutes the difference between tumor cure and host cure in a disease characterized par excellence by metastasis. In humans this difference is vital, as succinctly expressed in Black's maxim, "Around every tumor there is a patient" (1972).

Again in the context of the host-tumor relationship, it must be emphasized that most of the information on factors involved in tumor response and reviewed in this chapter refers to local hyperthermia, since this technique has been most studied to date. There are, however, at least two other major approaches to cancer thermotherapy—regional hyperthermia and total body (systemic) hyperthermia, and each embodies a major effect upon the host.

Regional hyperthermia is most conveniently applied by arterial perfusion with heated blood. Tumor temperatures of 40°C to 43°C are employed (Stehlin et al. 1977; Moricca et al. 1977), although 45°C has been achieved in limbs (Cavaliere et al. 1967). This technique is noteworthy in that it induces a marked and probably highly significant (Dickson and Shah 1980; Dickson 1978) reaction (inflammatory response) in the normal tissues relating to the tumor. Little data is available on factors involved in tumor response to heating by this method. Tumor hypoxia should not occur, since the heated blood passes through a pump oxygenator. It is of interest, however, that a metabolic feature of inflammation is the acid pH that develops at the site (Menkin 1956). There is circumstantial evidence that this decreased pH may be important, since Moricca and co-workers (1977) found that if it was inhibited by perfusion of the limb with alkaline solution after hyperthermia, there was an increased incidence of tumor recurrence. Whether the acid pH is important in the tumor response to heat or for the inflammatory reaction is unknown. A host immune response has also been implicated in the tumor regression following hyperthermic perfusion (Stehlin et al. 1977; Moricca et al. 1977), but the evidence in favor of this is not convincing (Dickson and Shah 1980; this volume, chapter 23).

Total-body hyperthermia can be applied by a variety of techniques (see chapter 19), the important common denominator being that 42°C is the highest core temperature that can be safely induced at present. Because of this temperature restriction, and because in this form of therapy all the host normal tissues (including the cells of the immune and mononuclear phagocyte systems) are heated, it has been suggested that local and total body heating should be regarded as distinct and separate approaches to the treatment of malignant disease (Dickson 1977). Again, as with regional hyperthermia, it is not known to what extent the various factors detailed in figure 5.1 may participate in tumor response to total-body heating. The subject is much more complex than with local hyperthermia and encompasses the host-tumor and primary tumor-metastases relationships, possible differential heat sensitivity of primary tumor and its satellites, and the considerable impact of systemic hyperthermia on homeostasis, especially fluid and electrolyte balance (this volume, chapter 19; Larkin 1979, 1980; MacKenzie et al. 1976) and enzyme levels (Larkin 1979; Pettigrew et al. 1974). In addition, the maximum temperature of 42°C imposes restrictions in relation to tumor heating. It is becoming apparent that 42°C may not be a high enough temperature to shut down the microcirculation in some tumors (Jain 1980); this inhibition of blood flow is believed currently to play a major role in tumor heat sensitivity. An overview of clinical results also indicates that the majority of human solid tumors are not significantly damaged by 42°C (Dickson 1978), but require temperatures in the region of 45°C to 50°C (LeVeen et al. 1976; Storm et al. 1979; in press). Whether this reflects immunity of the tumor microcirculation or an inherent resistance of the cancer cells at 42°C is not known. It does, however, emphasize the importance of acquiring information on factors governing tumor cell thermosensitivity, and of seeking potentiators of the heat response. The possibility of increasing the thermal tolerance of normal tissues—the opposite side of the coin, as it were—should be borne in mind (see chapter 6). It has been mentioned that recent data reveal a correlation between thermotolerance and membrane composition, resistance to elevated temperature being associated with an increase in the proportion of saturated fatty acids or an increase in cholesterol. There is some evidence that the composition of cell membrane lipids can be altered by controlling the proportion of saturated fatty acids in the diet (Henle and Dethlefsen 1978), and that cell membrane cholesterol content can be increased by agents such as alcohol (Chin, Parsons, and Goldstein 1978). The possibility of increasing thermotolerance of host normal tissues without a similar effect on the tumor cells is an exciting prospect that warrants intense investigation.

CONCEPTS AND PRINCIPLES

Heat Sensitivity And Thermal Death Time

Quantitation of appropriate treatment doses for cancer therapy is based on both dose-response data and knowledge of the overall biological effects of the treatment. Cytotoxic agents used in tumor therapy (drugs [Hill and Baserga 1975] and radiation [Hall 1973]) usually are directed against single or a limited number of targets in tumors. Thus, at least a theoretic dose can be set for inactivation of the molecular or biological process at which therapy is aimed. In the case of hyperthermia, myriad effects have been observed on metabolic (Strom et al. 1977), macromolecular and ultrastructural (Mondovi 1976), cytokinetic (Bhuyan 1979), extracellular and vascular (Dickson and Calderwood 1980; Song, in press), and immunologic phenomena (Dickson and Shah 1980). It is not known whether any single effect is of primary importance, and it may be that hyperthermia, particularly in vivo, is a multitarget therapy. It is also not clear whether the nature of thermal cell killing remains the same across the whole spectrum of temperatures currently used for cancer treatment (40°C to 50°C).

The present study focuses on factors involved in heat sensitivity of tumors in vivo. One of the major aims of the work was to examine the effects induced in tumors by heat and to try to discover if these effects are important in tumor destruction or merely irrelevant side-effects (Appelgren 1979)—therapeutic "red herrings." It may be of value to examine the strategy evolved for this purpose. The study was designed to look for answers to the following types of questions:

1. Is the magnitude or incidence of the property under examination related to heat sensitivity?
2. Do changes that occur in the property after heating coincide in timing and magnitude with the degree of tumor destruction?
3. Can heat sensitivity be increased by altering the property in question? If the property is important, changing it should alter the response to heat.

An example of this approach was the evaluation of the role of blood flow in hyperthermia (described later in this chapter), in which the following questions were posed: (1) Is the heat sensitivity of tumors related to the amount of blood flowing through them?

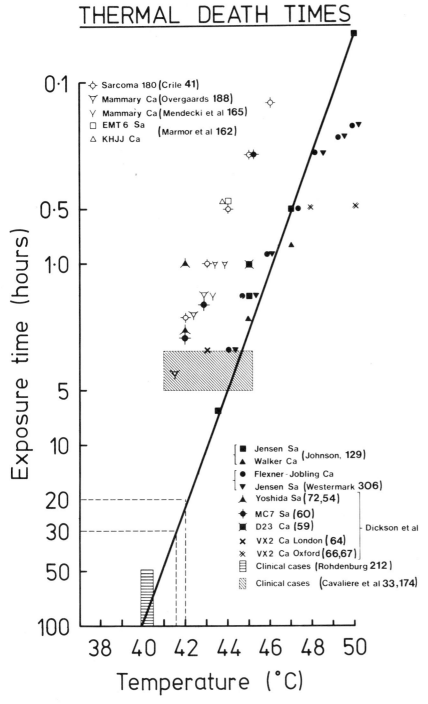

Figure 5.2. Thermal death times for animal and human cancers determined in vivo. *Solid symbols* refer to tumors of the rat, *open symbols* to tumors of the mouse. *Crosses* refer to the VX-2 carcinoma in rabbits, a tumor strain obtained from London, England, and a strain obtained from Oxford, England. The graph points relate to a cure rate of 50% (London VX-2) and 75% (Oxford VX-2) of all tumors treated. Data for the Jensen's, Walker, Flexner-Jobling, Yoshida, and MC-7 rat tumors, and the EMT-6 and mammary tumors in mice, refer to regression obtained in 100% of the tumors heated; in the S180 mouse data, "most" of the tumors were cured; the D-23 rat and KHJJ mouse points refer to a 50% cure rate, and a 25% (approximate) cure rate was reported by Overgaard and Overgaard (1972) for the mouse mammary carcinoma. The clinical results of Rohdenburg (1918) refer to systemic hyperthermia (pyrexia); those of Cavaliere and colleagues (1967) refer to tumors of the limb treated by regional hyperthermic perfusion. The *line* drawn through the rat and rabbit values extrapolates to the data for human tumors that underwent spontaneous regression with sustained pyrexia (Rohdenburg 1918; Dickson 1978). The overall figure *line* represents the LD_{100} of the majority of tumors studied to 1977. (Dickson and Calderwood 1980. Reprinted by permission.)

(2) To what extent is heat destruction of tumors caused by changes in tumor blood flow after hyperthermia? (3) Can tumors be sensitized to heat by increasing or decreasing blood flow? After this preliminary work, the findings were consolidated by determining what influence tumor volume, tumor site, tumor type, or species of the host may have on the factor under investigation and its role in thermal cell killing. With this data it may be possible to understand the mechanisms of thermal destruction of tumors and to accurately define the role of hyperthermia in the treatment of clinical cancer.

The temperature-time relationships of cell killing can be readily analyzed with cells in vitro using the well-established techniques of radiobiology, the cell survival curve, and the Arrhenius plot. Quantitation of cell killing in vivo is significantly more difficult than in cell culture. In many cases, the types and even the location of the target cells that determine the measured response is unknown. Since the shape of the heat survival curve for the target cells cannot be defined, an Arrhenius analysis, comparable to the data obtained in vitro, is impossible without major assumptions (Henle and Dethlefsen 1980). Another approach to the in vivo situation depends upon an understanding of the ways in which tissue, normal or neoplastic, is capable of accommodating energy input in the form of heat. This process is governed by the laws of thermodynamics, which in this case are embodied in the form of a partial differential equation, known as the *bioheat equation* (Cravalho, Fox, and Kan 1980). Although this is a comprehensive equation that attempts to take into account energy transfer from the environment, energy liberation from metabolic processes, and energy transport by the circulatory system, it does have sources of error (chiefly related to blood flow) in relation to its application under in vivo conditions (Cravalho, Fox, and Kan 1980; Eberhart, Shitzer, and Hernandez 1980), besides being a somewhat specialized approach to the problems of heat transfer.

In the present study, the end point chosen for quantitation of thermal response was thermal death time. *Thermal death time* is the minimum time of heating that produces irreversible damage to the tissue at a given temperature. Thermal death time has been determined in a number of tumor types in both humans and rodents by several groups of workers, and the results of these studies are shown in figure 5.2. When logarithm of exposure time is plotted against temperature (fig. 5.2), a straight line relationship is observed, indicating that for each degree of temperature increase, thermal death time is halved. This appears to be true of both the aggregate data shown in the figure and when single tumors are analyzed individually

(Johnson 1940; Westermark 1927; Crile 1962; 1963; Dickson and Calderwood 1980; Henle and Dethlefsen 1980).

Similar studies have also been carried out on a limited number of normal tissues using such criteria as loss of peripheral organs or skin necrosis as end points (fig. 5.3). Above 44°C, the time-temperature relationship for normal tissues (human and pig skin, rat testis,

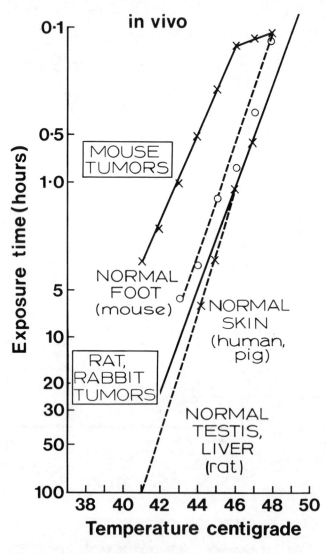

Figure 5.3. Thermal death times for animal and human normal tissues in vivo compared to the thermal death times for tumors given in figure 5.2. Data for normal human and pig skin begin at 44°C and extend to higher temperatures, as do Westermark's values for rat testis germinal epithelium and liver. The data of Fukui for germinal epithelium of testis extend from 41°C to 48°C. For detailed references for the data collated in this figure, see Dickson 1978. (Dickson and Calderwood 1980. Reprinted by permission.)

and liver) is similar to that for tumors, the respective lines fusing at about 46°C. Below 44°C, the time base for normal tissues is significantly increased compared to that for tumors. Thus, a change of 1°C is equivalent to altering the heating time by a factor of more than 2 (Dickson 1977), indicating thermal resistance in normal tissues up to 44°C. Henle and Dethlefsen (1980) have recently reported an in vivo thermal death time plot that includes some animal tumors and normal tissues not plotted in figure 5.2. The graph is essentially similar to figure 5.2, with the tumors occupying a linear and well-defined time-temperature response band, and the normal tissues being placed above the resistant edge of this band, since at a given temperature above 42°C the normal tissues withstand damage for a longer time (Henle and Dethlefsen 1980). An exception to this finding of relative heat resistance in normal tissues is the small intestine, which was found to be more heat-sensitive (in studies on mice) than most tumors (Henle and Dethlefsen 1980). This may not represent a true response to heat at the cellular level, however, but instead probably reflects breakdown of intestinal barriers to bacterial infection (Henle and Dethlefsen 1980).

The data presented in figures 5.2 and 5.3 suggest a range of temperatures over which selective treatment of tumors may be possible in most anatomic sites. At temperatures above 44°C, the thermal death times of normal and malignant tissues converge to such a degree as to make exploitation of any remaining time difference impracticable. Below 42°C, the prolonged heating times required for tumor destruction likewise make such temperatures unsuitable for therapy purposes. A selective temperature range of 42°C to 44°C is thus indicated. A number of studies have indicated that at 42°C in particular, there is a selective and irreversible inhibition of metabolism in several types of rodent, rabbit, and human tumors; the inhibition is correlated with loss of biological malignancy in the tumors assessed by bioassay (Dickson 1977, 1978). Furthermore, at 42°C the metabolism of normal organs and tissues from these hosts under comparable experimental conditions is not inhibited (Dickson 1977, 1978). Above 42°C, inhibition of metabolism is more rapid in the tumor cells, but normal tissues also become progressively affected. At such temperatures, the incidence of complications rises sharply with local hyperthermia in animals (Dickson 1977) and humans (Moricca et al. 1977), and host tolerance considerations preclude the use of temperatures greater than 42°C for total-body hyperthermia (Dickson 1977; Pettigrew 1976). At temperatures below 42°C, but above body temperature, there is evidence for a stimulatory range, for at these temperatures the metabolism and dissemination of tumor cells may be accelerated (De Duve and Wattiaux 1966; Dickson 1977; this volume, chapter 23). Thus, the temperature range that offers greatest selectivity in tumor destruction and has the widest applicability in cancer treatment would seem to be 41.5°C to 42°C (Dickson 1976, 1977, 1978; Dickson and Calderwood 1980). This temperature range represents a reasonable compromise between losing the differential susceptibility between malignant and normal cells to heat damage on the one hand and prolonged heating times on the other.

It must be emphasized that the thermal death times shown in figure 5.2 refer to conditions that are by no means standardized or even comparable for the tumor types. Tumors of different volume in various anatomic sites were heated by a range of techniques; the tumor cure rates achieved by the various workers ranged from 25% to 100%. Only the physical factor of heat is considered, and no account is taken of other factors such as the role of the immune response or other components of the host reaction to tissue damage.[1] A clear relationship between time and temperature has nevertheless been demonstrated in many tumors; this dose-response relationship and the overall heat sensitivity appear to be similar in many tumors. When the thermal death time graph is used for predictive rather than descriptive purposes, the formulation at first sight appears less satisfactory. For example, one might predict from the curve a thermal death time of approximately 20 hours at 42°C for human tumors, but use of total-body heating for periods in excess of the predicted 20 hours has not led to permanent tumor regression: an initial objective response rate of 47% in a series of patients with different advanced malignancies was followed by recurrence in all cases by about three months after treatment (Pettigrew et al. 1974). Using hyperthermic regional perfusion for tumors of the limb, Moricca and associates (1977) have obtained remarkable and massive necrosis in melanomas and sarcomas heated at 42°C to 43°C for two to five hours. Figure 5.2 predicts a heating time of 10 hours for tumor destruction at 43°C. Again, however, recurrence was a problem following tumor heating (Moricca et al. 1977). Viewed in terms of the thermal death times line, tumor recurrence in these two series may indicate that the degree of heating employed was inadequate for tumor destruction. On the other hand, at the upper end of the temperature range, the 75% cure rate of the VX-2 carcinoma in the rabbit by radio-

---

[1] For a more detailed analysis of the thermal death times graph (construction, tumor volumes, methods of heating) the reader is referred to previous publications (Dickson and Calderwood 1980; Dickson 1977, 1978).

frequency (RF) heating at 47°C for 30 minutes was not improved by increasing the degree of heating (Dickson and Shah 1977). These findings may reflect that the exponential nature of the temperature time graph does not hold at higher temperatures; indeed, although the points in figure 5.1 are limited in number, the data do suggest than an increase in temperature above 47°C gives little further reduction in the exposure time required for destruction of several tumor types. With human cancer, a notable feature in practically all series of cases treated at temperatures up to 45°C, even with prolonged heating, has been tumor recurrence. This has led to the proposal that tumors fall into two zones of thermal sensitivity, 42°C to 43°C and 45°C to 50°C, and that while immunogenic tumors like melanoma and sarcoma may be placed in the lower 42°C to 43°C zone, the majority of human solid tumors may require temperatures in excess of 45°C for adequate destruction (Dickson and Calderwood 1980; Dickson 1978). The necessity for such high temperatures would appear to be borne out by the recent work of LeVeen and colleagues (1976) and of Storm and associates (1979, 1980) with RF heating; there are also reports that osteogenic sarcoma in humans has proved resistant to 50°C for 30 to 60 minutes achieved by microwave or RF heating (Dickson 1979). With animal tumors and heating temperatures of 42°C to 50°C, most workers have found difficulty in achieving 100% tumor cure rates, especially using nonimmunogenic tumors. Although in this context, and also in the case of human cancer, inhomogeneous tumor heating must be considered, it has been believed since the era of Coley toxins that factors other than the heat itself are involved in the regression of tumors after hyperthermia (Dickson 1977). It is becoming apparent also that there are a number of these factors, that they pertain to both the tumor and the host, and that they operate at local and systemic levels (fig. 5.1).

In the treatment of human tumors, a wide range of temperatures, from approximately 42°C to over 50°C, is currently employed. Tumor destruction at these temperatures probably varies from subtle alterations in tumor biology at 42°C to gross coagulation necrosis at 50°C (Storm et al. 1979, 1980). The present study is concerned more with factors involved in tumor response at the lower end of the temperature range (42°C to 43°C), although where data is available on heating at high temperatures it is reviewed. As mentioned, the higher temperature range (45°C to 50°C) has recently been shown to be effective in treating human cancer with a marked degree of specificity (LeVeen et al. 1976; Storm et al. 1979, 1980). This is new terrain in the field of hyperthermia, and probably it will expand rapidly. Work in this area is, however, more likely to be operational and empirical rather than basic and analytical, owing to the overwhelming nature of the energy levels involved. Because of limited host tolerance, the lower temperature range probably will continue to be employed, especially for regional and systemic hyperthermia. These techniques involve the heating of large volumes of normal tissue and are thus critically dependent on any factor causing preferential tumor destruction.

Tumor Growth, Cell Kinetics, and Cancer Treatment

The nature of malignant neoplasms is to grow progressively at the expense of the host; the aim of cancer treatment is to halt tumor growth and ultimately to eradicate the tumor mass with preservation of the host. Tumor therapy may be aided by knowledge of the underlying mechanisms that determine the dynamics of tumor growth.

Tumor growth rate, commonly expressed as *doubling time* (T_D) or *tumor doubling time* (TDT), usually is determined in solid tumors by serial measurement of tumor volume. Increase in tumor volume is not caused entirely by tumor cell proliferation and, particularly in larger tumors, may be due in part to increased extracellular fluid volume (Gullino 1975) and an increased volume of necrotic tissue (Steel 1975), and decreasing proportion of space occupied by blood vessels may occur (Vaupel 1979). Enlargement of tumor volume has been shown, however, to be proportional to increase in cell number (Steel 1975). Tumor doubling time is thus, by definition, proportional to the rate of production of new cells, or birth rate. The latter parameter, birth rate, is itself dependent on two more basic properties of the cell population: *cell cycle time* (T_C), the amount of time taken for a cell to prepare for and undergo cell division, and *growth fraction* (GF), the number of cells in the population engaging in cell division. The cell cycle of eukaryotic cells has been shown (Howard and Pelc 1953) to be divided into four distinct phases (fig. 5.4). S phase, when DNA replication takes place; M (or mitosis), the division phase; and the gap phases G_1 and G_2. The biochemical events occurring in G_1 and G_2 differ from each other and from the events of the M and S phases and are not biochemically homogeneous phases (Mitchison 1971; Prescott 1976). Tumors thus contain a heterogeneous fraction of proliferating (P) cells in addition to a population of quiescent, nonproliferating (Q) cells (Lala 1971). The Q cell compartment also appears to be a heterogeneous group of cells. This compartment may include genuine resting (G_O) cells, which leave the cell

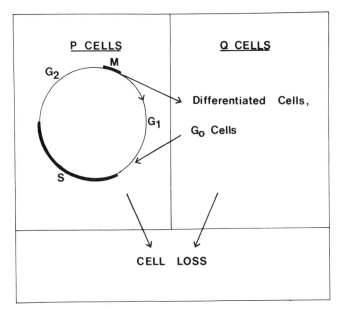

Figure 5.4. Schematic representation of the major parameters controlling the population kinetics of mammalian cells in vivo.

cycle after mitosis and may, by the appropriate stimulus, be recalled into the proliferative compartment (Van Putten 1974). In addition, there may be cells blocked in any of the cell cycle phases from further progression round the cycle (Gelfant 1977; Sarna 1974). The third main factor that determines tumor growth rate is cell loss from the P or Q cell compartment (Steel 1967, 1968, 1975; Cooper 1973); *cell loss* is envisaged as a consequence of exfoliation in peripheral tumors (Lala 1971), cell migration and metastasis (Weiss 1967), and cell death, particularly at the center of large, poorly vascularized tumors (Cooper 1973). These factors, which are the principle determining factors of tumor growth, are summarized in figure 5.4. Changes in tumor growth rate occur by alteration of one or all of the three controlling parameters in cell population kinetics—cell cycle time, growth fraction, and cell loss factor. Thus, for instance, deceleration in the growth of experimental tumors with increasing volume has been found in most cases to be due to a decrease in growth fraction and increase in cell loss (Algire, Legallais, and Park 1947; Calderwood and Dickson 1980a; Watson 1976). Cell cycle time in most studies has remained relatively constant, although there have been some reports of tumors in which T_C increases with tumor volume (Simpson-Herren 1977). Knowledge of the cell population kinetics of normal and malignant tissues has led to a more rational approach to the use of ionizing radiation and drugs in cancer therapy. Cytotoxic drugs have been classified according to their effects on lymphoma cells and normal hemopoietic stem cells (Bruce and Valeriote 1966) into

1. Nonspecific agents, which indiscriminately kill P or Q cells,
2. Cycle-specific agents, which selectively destroy P cells and spare Q cells, and
3. Phase-specific agents, which are selectively toxic to cells at one stage of the cell cycle.

Studies in vitro indicate that drugs and x-rays exert two main effects on target cells. They exert a cell killing effect, which is maximal in a certain phase or phases of the cell cycle (Hill and Baserga 1975; Tannock 1978). The original concept of drugs being truly phase-specific now appears to need modification, since for most agents, several phases may be susceptible (Hill and Baserga 1975). Secondly, cytotoxic agents may exert a blocking effect on the progression of cells through the cell cycle (Frindel and Tubiana 1971; Madoc-Jones and Mauro 1968). Unfortunately, these agents tend to be as toxic to rapidly proliferating normal tissues, such as bone marrow and crypt cells in the intestinal epithelium, as they are to tumor cells. A differential effect may be achieved by careful timing of dose applications to exploit differences in the kinetic properties of normal and malignant tissues (Hill and Baserga 1975; Tannock 1978); however, especially with systemic agents such as chemotherapeutic drugs, the effectiveness of treatment is limited by toxicity to dividing normal cells (Hill and Baserga 1975).

The differential sensitivity of resting and cycling cells to cytotoxic agents has been observed in vitro when comparing cells in the stationary growth phase with those growing exponentially in culture (Thatcher and Walker 1969; Twentyman 1976). The relevance of such data to the prediction of response of Q cells to treatment in vivo is not clear, as variations in duration and conditions of culturing, and especially the presence or absence of serum growth factors, can radically alter the response observed (Hill and Baserga 1975; Hahn 1974).

This knowledge has enabled therapists to combine agents with contrasting specificity in order to kill a wide spectrum of the tumor cell population and to time dose applications to spare normal tissues (Hill and Baserga 1975).

Studies of the effects of drugs or radiation in vivo have indicated that response to a given agent in terms of tumor volume may be misleading. Treatments that merely arrest the overall tumor growth rate or cause slight regression may, in fact, lead to a reduction in the clonogenic cell population by more than 99% (Hermends and Barendsen 1969; Wilcox et al. 1965). These findings indicate that many decades of cell killing may be required for the eradication of residual tumor (Skipper 1971; Steel 1975). A common finding in solid tumors in vivo is that cell killing by a single agent reaches a limit because of the existence of a resistant subpopulation of cells (Fowler, Thomlinson, and Howes 1970; Valeriote, Bruce, and Meeker 1968). This has led to the use of adjuvant and/or multidose therapy in the attempt to eradicate resistant cells. Use of multiple-dose treatment is complicated by the finding that a single application of therapy not only kills sensitive cells but perturbs the proliferation of the surviving population and influences the response to a second dose (Denekamp and Fowler 1977). For instance, following radiotherapy, response to further fractions of treatment is conditioned by what has been designated "the four Rs" of radiobiology (Denekamp and Fowler 1977): repair, redistribution, reoxygenation, and repopulation. Thus, in planning repeated dose schedules, the following factors need to be considered:

1. Intervals between fractions must not be too long or the repair of the various types of nonlethal damage in surviving cells may occur (Denekamp and Fowler 1977).
2. The agent may kill cells in one specific cell cycle phase and/or inhibit cell cycle progression in other phases. This may leave cells from other phases partially synchronized and thus affect response to a second treatment. As many studies have shown, however, cell populations have great heterogeneity in the duration of T_C (Steel 1977), and this factor tends to cause rapid desynchronization after release of cell cycle blockade (Lala 1971; Denekamp and Fowler 1977).
3. The treatment may alter the interrelationship of tumor cells and extracellular microenvironment. After radiotherapy, cell killing causes a decreased O_2 demand, followed by oxygenation of previously hypoxic cells, resulting in increased sensitivity of the previously radioresistant hypoxic cells to irradiation (Hall 1973). With other types of treatment, such as hyperthermia (Bicher et al. 1980), impairment of blood supply may result, with consequent tumor hypoxia.
4. Repopulation by resistant cells may occur, with cells reentering the cell cycle. This phenomenon of repopulation may be used to the benefit of the host, as the recruited cells, now actively proliferating, may become sensitive to cell cycle specific agents (Hall 1973; Mauer, Murphy, and Hayes 1976).

Thus, knowledge of cell population kinetics in tumors may be useful in identifying cancers sensitive to a particular type of therapy. Furthermore, investigations of the effect of cytotoxic agents on cytokinetic processes may increase the effectiveness of adjuvant and multiple-dose treatment.

Analysis of Tumor Growth and Cell Population Kinetics

Tumor Growth and Regression

The growth of solid tumors is generally studied by serial estimation of volume (Mendelsohn 1963; Steel 1977). For internal tumors, volume is estimated by x-ray photography (Schwartz 1961) and when lesions are superficial, by caliper measurement (Steel 1977). The linear dimensions may be related to tumor weight by means of a calibration curve, thus measurements in one dimension (length, breadth, or width of tumor), two dimensions (area), or three dimensions (volume) may be used to plot growth curves, provided the measurement can be related to tumor weight. In the present study, tumor volume was studied primarily in subcutaneous foot tumors. Tumor dimensions were fitted to a number of equations of likely three-dimensional geometric figures and the computed volumes plotted against tumor weight over the full range of tumor growth. Tumor volume calculated using the equation for an oblate sphere gave the best correlation with tumor weight in the tumor lines used (Yoshida, MC-7, D-23). Volume calculated in this way was used routinely for monitoring tumor growth. The equation is

$$V = \frac{1}{6} \pi a b^2, \qquad (1)$$

where a = length of longitudinal axis (cm), and b = length of vertical axis.

A number of mathematical functions have been proposed to describe tumor growth (Mendelsohn 1963; Laird 1964), with the aim of deriving fundamental information on the growth dynamics of neoplastic tissues. The present study was concerned with growth rate, and data were processed in such a way as to obtain accurate values for tumor doubling time (T_D). Plotting tumor growth on semilogarithmic graph paper gave in most cases a straight line over most of the tumor growth period. When linearity did not result, values for T_D were obtained from a tangent to the curve.

In studying tumor response to treatment, *tumor cure* is the most clinically relevant end point (Kallman and Rockwell 1977). Cure implies disappearance of both primary and secondary tumors that do not recur during the lifetime of the host. Another criterion, which may be more useful in some systems, is *tumor control*, which implies complete tumor disappearance from the primary site and absence of local recurrence over an adequate period of observation. A convenient parameter in measuring therapeutic effectiveness is the tumor control dose 50 (TCD$_{50}$), which is the dose of treatment necessary to cure 50% of animals (Kallman and Rockwell 1977; Suit and Shalek 1963; Kallman and Tapley 1964). Each of these methods is time-consuming, requiring many months of observation before a tumor in a host can be designated cured or controlled.

A more simple and rapid approach to assessing response to treatment is by analyzing changes in tumor growth. Some studies have involved analysis of tumor regression rate and compared the relative effectiveness of therapy doses in terms of the rapidity of tumor disappearance. In these circumstances, it is not possible to differentiate between effects of cell killing, changes caused by differences in resorption of dead tissues, or alterations in the tumor stroma (Kallman and Rockwell 1977). A more informative approach is to measure regrowth time either to the initial volume or some larger volume; changes in doubling or tripling time are sometimes studied (Hill and Denekamp 1979). In the present study, doubling time (T$_D$) was used to compare the effects of hyperthermia treatments on tumor growth (fig. 5.5). This parameter was chosen rather than regrowth to initial treatment volume, as tumors often increased rather than decreased in volume after heating at 42°C or 43°C because of edema (fig. 5.5), and in many cases no actual decrease in tumor volume occurred. In the treatment doses used in this study, edema subsided before tumors doubled in volume (fig. 5.5).

The information gained by regrowth experiments is highly empirical and gives no precise indication of events occurring in the cell population (Hermens and Barendsen 1969; Wilcox et al. 1965). Such information may be provided by the various approaches to the study of cell population kinetics.

Cell Population Kinetics

Cell population kinetics is largely concerned with the behavior of proliferating (P) cells; nonproliferating (Q) cells are of interest only because they constitute part of the bulk of the tumor and are able to reenter the cell cycle. Proliferating tissues fall into two types, characterized by steady state or expanding cell populations. In actively growing tumors, there is by definition an expanding population of cells. This affects the age distribution of cells in the tumor (age, in this context, is in relation to the cell cycle; newly born cells at the beginning of G$_1$ are deemed young and cells in G$_2$ relatively old). In a population with 100% proliferating cells and no cell loss, there are twice as many cells entering G$_1$ as old cells leaving telophase; there is a truly exponential age distribution as in figure 5.6. In a steady state population, where cell loss equals cell birth, if cell loss occurs at an even rate through the cell cycle, then a rectangular age distribution as in figure

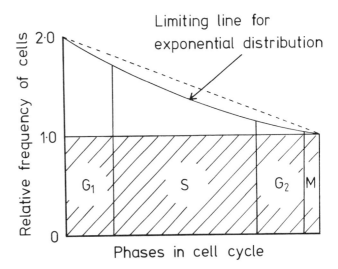

Figure 5.6. Age distribution of cells along the cell cycle for two model cell populations. The rectangle *(shaded)* represents the expected distribution for a steady-state population. The total area, bound above by the continuous line, and below by the abscissa passing through zero, represents the distribution for an exponentially growing population, where all cells are in cycle, and there is no cell loss. For computational purposes, the upper limiting continuous line may be replaced by a straight line *(broken)* without introducing any significant error. Relative number of cells in any phase is given by the area limited on two sides by the vertical lines raised at the boundaries (adapted from Lala 1971).

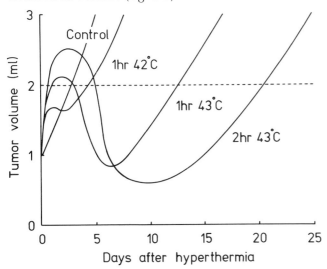

Figure 5.5. Regrowth of the D-23 carcinoma in the feet of rats after hyperthermia; tumors were treated at a mean volume of 1.0 ml. Each regrowth line was plotted using mean values for tumor volume measured at two-day intervals in 10 to 20 tumors.

5.6 is obtained (Lala 1971). In practice, neither of these model age distributions describes the situation in a solid tumor in which only a fraction of the population proliferates and varying degrees of cell loss occur, possibly at different rates in discrete phases of the cell cycle. It has been shown, however, that in fast, exponentially growing tumors, the assumption of an exponential age distribution is acceptable without much error (Lala 1971). Age distribution is important as it affects the calculations of parameters derived from primary data, such as cell loss factor and growth fraction. Equations shown in this text all assume exponential age distribution.

Cell Cycle Time

Cell cycle time (T_C), the time elapsing between consecutive mitoses, has been measured by direct observation in vitro using time-lapse cinematography (Showacre 1968). In general, however, the mean or median T_C of a cell population is measured. A number of techniques have been used to measure T_C in vivo (Lala 1971). The most widely used and informative method for the study of the cell cycle is the *labeled mitoses technique* (Quastler and Sherman 1959). In this technique, the passage of a cohort of cells labeled in S phase is followed through successive mitotic divisions. The method depends on the use of a labeled DNA precursor with an availability time that is short compared with the duration of the cell cycle phases, as is normally the case with in vivo administration of [^{3}H]-TdR (Cleaver 1967). For an asynchronous cell population with no spread in the duration of the different phases, a determination of the fraction of mitotic figures labeled at various times after [^{3}H]-TdR injection will show a periodic variation, as in figure 5.7A. The cell cycle phase times can be measured as follows (Lala 1971):

T_{G_2} = time between [^{3}H]-TdR injection and appearance of first labeled mitosis

T_M = time occupied by ascending or descending limb of curve

T_S = interval occupied by the plateau plus interval occupied by ascending/descending limb

$T_{G_1+G_2}$ = interval between zero points of the descending limb of the first wave and ascending limb of the second wave.

In practice, percentage labeled mitoses (PLM) curves (often referred to as fraction labeled mitoses curves [FLM]) generally depart from this ideal form owing to spread in phase durations, resulting in a damped periodic variation, as in figure 5.7B. Basic information on the cell cycle and phase times can be obtained by

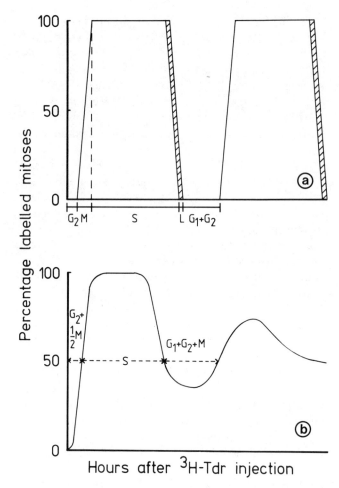

Figure 5.7. Percentage of labeled mitoses as a function of time after a pulse label of [^{3}H]-TdR. *A*, ideal case for a population with no variation between cells in the rate of progression through the cell cycle (L = length of pulse labeling) (Lala 1971). This is short enough to be ignored in practice. *B*, the type of PLM curve often encountered in solid tumors in vivo (Steel 1972), showing a damped periodic variation. Median durations of the cell cycle and its phases are measured from the 50% intercepts in the waves.

visual inspection of the PLM curve at the 50% points, as in figure 5.7.

In the present study, PLM curves were constructed from data obtained from the Yoshida sarcoma at two different stages of growth (fig. 5.8). Values for T_C and other parameters derived from the curves are presented later (table 5.5).

For study of variation in T_C and age distribution of the population, various mathematical analyses based on computer modeling have been devised (Barrett 1966; Gilbert 1972; Mendelsohn and Takahashi 1971; Steel and Hanes 1971).

Growth Fraction

Growth fraction (GF) is the proportion of proliferating cells in the population. Some concern has been

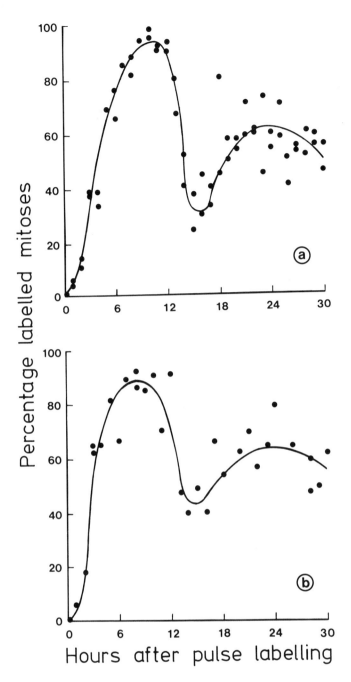

Figure 5.8. Percentage labeled mitoses (PLM) curves of the Yoshida sarcoma in the feet of rats at tumor volumes of A, 1.0 to 1.5 ml and B, 3.0 to 3.5 ml. Each point on the graphs represents a separate tumor. The PLM curves were fitted to the points after visual inspection. (Calderwood and Dickson 1980a. Reprinted by permission.)

expressed at the possible underestimation of GF owing to the presence of cells in the population with a cell cycle too long to be measured by conventional means (Lala 1971). The extent of errors caused by such cells is unknown, but the effect is unlikely to cause large discrepancies in measurements on fast-growing animal tumors. GF may be represented as follows:

$$GF = \frac{\text{number of cycling cells } (N_C)}{\text{total number of cells } (N)}. \quad (2)$$

This equation may be rewritten as

$$GF = \frac{\text{no. cells in S phase } (N_S)/\text{total no. cells } (N)}{\text{no. cells in S phase } (N_S)/\text{no. cycling cells } (N_C)}. \quad (3)$$

The upper and lower portions of the equation may be determined individually. The upper portion (N_S/N) is given by the flash [^{3}H]-TdR labeling index. This parameter is the proportion of cells labeled with [^{3}H]-TdR a short time (one hour) after a single injection of the label. The lower portion of the equation (N_S/N_C) represents the fraction of cycling cells in S phase. This parameter may be determined from values of T_C and other cell cycle phase times derived from the PLM curve (fig. 5.7) and the age distribution of cells within the cycle (Lala and Patt 1966) as follows:

$$N_{S/N_C} = (\exp(T_S{}^{\ln 2}/T_C) - 1) \\ \exp(T_{G_2} + M{}^{\ln 2}/T_C). \quad (4)$$

Another method is available for measurement of N_S/N_C based on the PLM technique and is described by Mendelsohn (1962).

It was formerly thought that repeated or continuous labeling of a cell population with [^{3}H]-TdR should lead to a plateau of labeled cells, the height of which is equal to GF. It is now known, however, that this method seriously overestimates the magnitude of the GF, owing to transit of labeled cells into the Q cell compartment (Dyson and Heppelston 1976; Lala 1971). The use of this technique for the study of proliferation in tumors after thermotherapy is discussed in the section on cytokinetic studies later in this chapter.

Cell Loss Factor

Loss of cells owing to exfoliation, migration, metastasis, and death is now acknowledged as a major factor in determining tumor growth rate (Steel 1967, 1968). Two main types of determination have been used to measure cell loss:

A biochemical method is based on the loss of iodine-125 (^{125}I) activity from a population in which the constituent cells were previously labeled with ^{125}I-iododeoxyuridine (^{125}I-IUdR) (Hofer, Prensky, and Hughes 1969; Dethlefsen 1971; Kumar et al. 1974). An ingenious variation on this method was employed by Weber and colleagues (1978) to investigate the differential effects on hypoxic and oxygenated cells of combined heat and γ irradiation. Tumors were labeled with ^{125}IUdR three days prior to treatment, so that ^{125}I-labeled cells permeated the whole tumor; this included areas close to capillaries and areas distant from the blood supply, which might be expected to be hypoxic. Immediately prior to treatment, animals were injected with ^{131}I-IUdR, which labeled only cells close to the capillaries. Radiation from the two isotopes used (^{125}I and ^{131}I) could be distinguished by differences in energy spectrum; thus the differential loss of ^{125}I (whole tumor) and ^{131}I (oxygenated area) could be calculated, and some estimate of the effect of oxygenation on sensitivity to treatment could be made. The most widely used method is based on the discrepancy between measured rates of population growth and potential growth computed from the birth rate of new cells (Steel 1967, 1968).

Steel (1968) derived the following equation from calculation of the cell loss factor (ø):

$$\text{ø} = \frac{\text{cell loss rate } (K_L)}{\text{cell birth rate } (K_B)}. \quad (5)$$

The cell loss rate (K_L) can be calculated from knowledge of the tumor doubling time, cell cycle time, and growth fraction:

$$K_L = \frac{\ln(1 + GF)}{T_C} - \frac{\ln 2}{T_D}. \quad (6)$$

Birth rate (K_B) can be calculated from estimates of T_C and GF when the age distribution of the cell population is known (Lala 1971). Birth rate also can be measured directly as the rate of entry into mitosis by using vincristine, which blocks cells in metaphase. Birth rate is calculated from the slope of the mitotic accumulation line (fig. 5.9). A more detailed account of the use of vincristine is given later in this chapter.

For a comparison of the ^{125}I-IUdR technique and Steel's method, see Dethlefsen (1971).

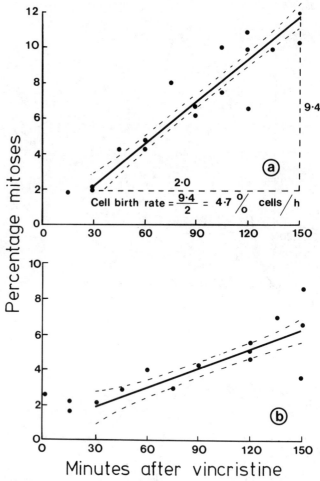

Figure 5.9. Mitotic accumulation after vincristine arrest in untreated Yoshida sarcomas at volumes of A, 1.0 to 1.5 ml; B, 3.0 to 3.5 ml. Cell birth rate in A, is calculated from the slope of the percentage mitoses line (9.4/2.0 = 4.7% cells per hour). Unlabeled broken lines indicate 95% confidence limits. (Calderwood and Dickson 1980a. Reprinted by permission.)

Cell Kinetics and Hyperthermia

Measurement of cell kinetic parameters in a highly perturbed system, such as a tumor after treatment, is a difficult and complex problem. A number of different approaches have evolved from the study of tumors after therapy.

After relatively mild treatments, tumor cell kinetics may be studied using the techniques described earlier to measure the major parameters of cell proliferation: cell cycle time, cell loss factor, and growth fraction (Denekamp and Fowler 1977).

After more severe treatments, such as curative heating, which produce a highly perturbed cell population, the assumptions on which computation of GF and ø are made will no longer be justified. In these circumstances, a number of in vivo and in vitro techniques have been investigated in an attempt to simplify measurement of tumor response. All these techniques suffer from an inherent drawback. Solid tumors need to be cut up and digested to establish a single cell suspension before they can be used in tissue culture in vitro. The effect of the disaggregation process on cells already damaged by cytotoxic therapy is unknown and adds extra variables to compound the difficulties in interpretation of tumor response. Second, the effects of treatment, especially hyperthermia, are not all immediate; assessment of cytokinetic response in cells transferred to the in vitro environment immediately after treatment may obscure delayed alterations occurring in the cell population in vivo (Crile 1963; Kang, Song, and Levitt 1980). The techniques do, however, possess some major advantages.

Flow Microfluorometry

In this technique, cellular DNA is combined quantitatively with a fluorescent dye and the DNA content of the cells is determined by fluorometry. Flow of cells into the fluorometer compartment and recording of data are automated in modern instruments (Crissman, Mullaney, and Steinkamp 1975). Thus, using this technique the distribution of DNA content in the cell population may be determined (Crissman, Mullaney, and Steinkamp 1975). Cells undergo a gradual doubling in DNA content as they progress around the cell cycle. Thus, DNA content gives an indication of position in the cell cycle. The technique is useful in examining the effects of treatment on cell cycle progression; synchrony is observed as an altered DNA distribution. Kal and Hahn (1976) used the method to study the effect of heat (43°C) and x-rays on cell kinetics in mouse tumors. The method has two major drawbacks:

1. Q cells cannot be distinguished from cells in G_1.
2. The existence of a large number of dead or dying cells in which the DNA content is not known will complicate the interpretation of the DNA distribution pattern.

Thymidine Labeling In Vitro

This technique may be of use in analyzing clinical samples, but in animal systems offers no apparent advantage over in vivo labeling and has many disadvantages (Steel 1977).

Survival and Growth of Clonogenic Cells after In Vivo Treatment

The effect of tumor therapy in vivo may be assayed by disaggregating the tumor and assessing growth and survival of the cells in vitro (Steel 1977), or measuring the ability to form colonies in the spleen (Till and McCulloch 1961) or lungs (Steel 1977) of recipient animals. This is a purely empirical measurement and gives no direct indication of cell cycle effects or of the relative rates of cell killing and repopulation in vivo. The advantage of this technique is that only the most important component of the cell population, clonogenic cells, is studied. In the classical analyses of cell population kinetics using [^{3}H]-TdR labeling technique, the behavior of such cells may be obscured by the presence of a large number of nonclonogenic cells (Steel 1977). These cell survival techniques are the most widely used methods of studying the effects of treatment on cell proliferation.

The Microenvironment of Tumor Cells

Tumor cells are more sensitive to heat when growing in the host than in tissue culture (Kim and Hahn 1979; LeVeen et al. 1976; Overgaard 1977a; Storm et al. 1979). Attention has recently been focused on the properties of the tissues and fluids in the immediate environment of tumor cells and the impact of these properties on heat sensitivity (Gerweck 1978; Overgaard and Nielsen 1980; Dickson and Calderwood 1980). The key factor in the maintenance and growth of a tissue is an adequate blood supply. In normal tissues, blood flows into the organ via one or more main arteries. Once inside the organ, vessels branch successively until even the most remote cells are supplied with blood by the capillaries forming the tissue microcirculation (Richardson 1976). Used blood is then collected into a main vein and carried away. In healthy tissues the extracellular fluid bathing the cells is kept relatively rich in oxygen and nutrients, and waste products are removed (Richardson 1976).

In malignant tissues in the early stages of growth, nutrition, and excretion are catered for by the microcirculation of the host tissue (Folkman 1975). Tumor growth is made possible by the infiltration of newly formed capillaries, stimulated by the secretion of tumor angiogenesis factor, into the neoplastic mass (Folkman 1975; Warren 1979). Angiogenesis appears to occur by random attraction of capillaries, which grow into the tumor (Gullino 1975), and the resulting microcirculation is formed both from the new capillaries and by vessels surviving from the vasculature of the invaded tissue (Gullino 1975). The vessels themselves become dilated and tortuous, displaying redundant curves (Warren 1979). As the tumor grows, the system as a whole becomes increasingly overextended, tumor growth exceeding angiogenesis (Tannock 1970; Karlsson et al. 1980).

Gullino, Grantham, and Smith (1965) found that the growth of solid tumors was associated with the incorporation of a large volume of fluid into the tumor mass. This fluid, which accumulates in the space between the endothelial wall of capillaries and the plasma membrane, is the *tumor interstitial fluid* (TIF). The volume of interstitial fluid in tumors was found to be 30% to 60% of tumor volume, compared to less than 10% in most normal tissues (Peterson 1979a). The raised concentration of TIF probably is caused by the "leaky" endothelium of blood vessels in neoplastic tissues and the absence of lymphatic drainage in tumors (Gullino 1975). The raised hydrostatic pressure of the interstitial fluid combined with the pressure of expanding tissue causes additional microcirculatory collapse (Gullino 1975). The large pool of TIF appears to be perfused at a relatively slow rate by the bloodstream, often at $\frac{1}{10}$ to $\frac{1}{100}$ the rate of host tissues (Cataland, Cohen, and Sapirstein 1962; Gullino and Grantham 1961; Gullino 1966; Peterson 1979b). Tumor cells thus exist in an extracellular microenvironment, which is to some extent independent of the host tissues, and this independence is reflected in the different biochemical consistency of the TIF compared to plasma and normal interstitial fluid (Gullino 1966, 1975). In contrast to normal body fluids, the TIF is deficient in O_2 and glucose (Gullino, Grantham, and Courtney 1967a, 1967b; Shapot 1972) and contains increased concentrations of CO_2 and the products of glycolysis, including lactic acid (Gullino 1975; Voegtlin, Fitch, and Kahler 1935; Burgess and Sylvén 1962; Gullino et al. 1965). It is not clear to what extent different areas of the TIF are in equilibrium. Mixing by convective currents may occur owing to the relatively low hyaluronic acid content of the TIF (Gullino 1975). Large gradients in both pH (Eden, Haines, and Kahler 1955) and pO_2 (Vaupel et al. 1973a) exist between different zones in tumors, however; thus the biochemistry of the TIF not only differs from normal tissue fluids but varies considerably in different parts of the tumor. The uneven distribution of nutrients and catabolites probably contributes in large measure to the increasing incidence of central necrosis in tumors as volume increases (Gullino 1975).

One of the most distinctive metabolic properties of tumors is rapid aerobic glycolysis and lactic acid production (Bodansky 1975; Warburg 1930, 1956). This property, combined with a poor and uneven microcirculation (Peterson 1979b), might be expected to lead to an acidic tumor microenvironment (Eden, Haines, and Kahler 1955; Overgaard 1976a). Low tumor pH has recently come to be associated with thermal sensitivity because of three main areas of speculation.

First, low pH in the tumor microenvironment is thought by some workers to be at least partially responsible for the heat sensitivity of tumors in vivo on the basis of a marked thermal potentiation by low extracellular pH in vitro (Overgaard and Nielsen 1980; Freeman, Dewey, and Hopwood 1977; Suit and Gerweck 1979; Urano et al. 1980).

Second, thermal killing per se has been attributed to a lowering of pH in tumor cells or extracellular fluid after heat. Mechanisms to account for this pH change are (1) thermal inhibition of respiration and compensatory increase in glycolysis and lactic acid production (Overgaard 1976a) and (2) inhibition of blood flow by heat, with a resultant accumulation of acidic catabolites (Song, in press).

Third, Von Ardenne (1971, 1972, 1978) has put forward a hypothesis suggesting that hyperglycemia in vivo produces tumor lactic acidosis because of stimulation of glycolysis (fig. 5.10). According to this scheme, lactic acidosis leads to a decrease in tumor pH and tumor sensitization to the effects of hyperthermia owing to activation of lysosomal enzymes. This approach to tumor thermal sensitization by selective alteration of the tumor microenvironment, and other relevant aspects of tumor pH, are discussed in detail later in this chapter.

Measurement of pH In Vivo

Knowledge of the pH of tissue fluids is of value since the rates of biochemical reactions and the stability of structural cell components are all more or less pH

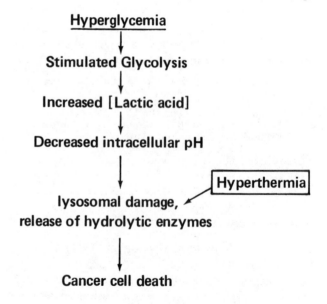

Figure 5.10. Chain of biochemical and cellular reactions in tumors after hyperglycemia and hyperthermia proposed by Von Ardenne (1971, 1972).

dependent. By definition, pH is a scale of values describing the relative acidity and alkalinity of solutions and is related to the chemical potential of the hydrogen ion. It is an operational scale of values, the pH of test solutions being derived by reference to primary standards (Waddell and Bates 1969). Hydrogen ion concentration is related to pH, but a value for H^+ concentration cannot be uniquely determined without knowledge of the activity coefficient of the H^+ ion under the test conditions.

Techniques for measuring pH are based on either (a) chemical indicators, which undergo a change in absorbance of visible or ultraviolet light in solutions of different pH, or (b) using pH sensitive materials that change in potential in solutions of differing pH. Such a material is pH-sensitive glass, which can be calibrated so as to directly measure pH with high specificity over a wide range of acidity and alkalinity (Bates 1964). All methods of pH measurement in biological materials involve the use of pH sensitive electrodes at some point.

There are many problems involved in the measurement of pH in vivo. One of the difficulties is the heterogeneity of tissues, which are composed of two main fluid compartments, the intracellular and extracellular water. Measurements of pH in vitro have shown that the relationship between pH in the two compartments is complex (Poole, Butler, and Waddell 1964; Dickson and Oswald 1976; Thomas 1978). The pH of one compartment cannot be predicted solely from a measurement of pH in the other compartment. A number of techniques have thus been devised for pH measurement in vivo.

Methods of Measuring Extracellular pH (pH_e)

Miniature Glass Electrodes

Electrodes with a diameter too large to enter cells but small enough to be inserted into the tumor have been used to measure tissue pH. These electrodes inevitably cause tissue damage (Voegtlin, Kahler, and Fitch 1935), and the electrode tip probably is immersed in a mixture of interstitial fluid, intracellular fluid from damaged cells, and blood from ruptured capillaries. Most workers have found that, after an initial period of instability, probably caused by injury and the mixing of tissue fluids described, a stable reading of pH is obtained within 5 to 40 minutes (Voegtlin, Kahler, and Fitch 1935; Ashby 1966; Dickson and Calderwood 1979). This method has the advantage of providing continuous pH monitoring. The disadvantage is that the probes cause gross perturbation of the tumor microenvironment. One solution to the problem of tissue perturbation by electrodes is to implant the probes in situ and grow the tumor around the implanted probe. Perturbation may be minimized by using electrodes of as small a tip diameter as possible, although such fine-tipped electrodes introduce practical problems. Tip size is proportional to electrical resistance (Waddell and Bates 1969), and measurement of the tiny output obtained from fine-tipped electrodes is prone to interference from stray electrical signals (Voegtlin, Kahler, and Fitch 1935). A pH measured by miniature electrodes appears to be a function of extracellular pH, however, and changes thus registered are assumed to reflect pH changes in the interstitial fluid (Voegtlin, Fitch, and Kahler 1935; Eden, Haines, and Kahler 1955; Dickson and Calderwood 1979).

Measurement of pH on Interstitial Fluid from Micropore Chambers Incorporated into Tissues

Gullino, with Clark and Grantham in 1964 and later with Grantham and Smith (1965), incorporated micropore chambers into tumors (tumor cells grew around the chambers) and into normal subcutaneous (SC) tissues. Interstitial fluid was free to pass through the walls of the micropore chamber, and could be sampled directly from the chamber. Acid-base parameters were then determined on the tumor or SC fluid samples using a blood-gas analyzer. Compared with invasive electrodes, the method has the advantage that gross tissue damage is not involved, and pH is measured on a true fluid, without interference from the presence of solid tissues and blood vessels. The effect of the presence of the chamber on tumor pH is, however, not known (Gullino, Grantham, and Smith 1965). Also, measurements using electrodes indicate that pH in the TIF may be extremely heterogeneous (Eden, Haines, and Kahler 1955); the TIF probably does not constitute a "well-stirred" homogeneous solution. Thus, TIF samples using the micropore chamber method may be unrepresentative of the whole tumor. The method is not of value when rapid, transient pH changes are to be investigated by sequential sampling over a time period of minutes or even hours. Time taken for the chamber to fill with TIF (a period of several hours) is the limiting factor (Gullino, Clark, and Grantham 1964).

Methods of Measuring Intracellular pH (pH_i)

Glass Microelectrodes

Reliable electrodes now are available for measurement of pH_i. Owing to the size of the tip of these electrodes (approximately 1 μ [Thomas 1976]), they are currently appropriate only for pH_i measurement in large, robust cells (such as invertebrate neurones or mammalian muscle). Smaller, more delicate cells, such as most tumor cells, are irreversibly damaged by insertion of such electrodes (Calderwood 1978). Also, with the equipment currently available (Thomas 1976), pH_i measurement is practicable only in vitro because of the extreme delicacy of the electrodes. These problems, although serious, are technical in nature and pH-sensitive microelectrodes may at some future time be used for pH_i measurement in vivo. Microelectrodes have the advantage of fast response time and the direct measurement of pH, which is unambiguous in interpretation. Apart from the operational problems mentioned, intracellular microelectrodes have the disadvantage of being invasive and may damage cells. Measurement of pH_i in a single cell may be advantageous in that the response of a specific cell is determined, or the measurement may be undesirable in that overall effects may be missed.

Weak Acids and Bases
Biological membranes in general tend to be impermeable to ionized substances (Hall and Baker 1977). Thus with weak acids or bases, the undissociated form of the molecule may cross membranes, while the ionic form is excluded. Distribution of such compounds between intracellular and extracellular compartments under conditions of equilibrium can thus be used for indirect measurement of intracellular pH. The substance most used for this purpose is the weak, nonmetabolizable acid 5-5-dimethyloxazolidine-2,4-dione (DMO) (Waddell and Bates 1969; Waddell and Butler 1959; Calderwood and Dickson 1978; Hinke and Menard 1978).

As with any of these weak acids or bases, at equilibrium the undissociated form of DMO is partitioned in equal concentration on either side of the cell membrane. The concentration of the ionized species is then determined solely by the pH and apparent dissociation constant, pK', of the compound on each side of the membrane, the concentration of the ionic form being directly proportional to the hydrogen ion concentration. The total concentration determined analytically, which is the sum of undissociated and dissociated forms, will be higher on the side on which ionization is more extensive—the side of higher pH with an acid, the side of lower pH in the case of a base. Thus, if the total DMO concentration on both sides of the membrane at equilibrium and the pH on one side of the membrane can be measured, knowing the pK' of DMO, the pH on the other side of the membrane can be calculated. The DMO concentration is expressed in terms of tissue water, and a correction is made for the DMO present in the extracellular fluid. For tissue samples, this requires an estimation of total water content of the sample and of the volume of extracellular fluid present.

In 1959, Waddell and Butler first introduced DMO as an indicator for pH_i, demonstrating its special suitability for that purpose, and deriving a mathematical equation for calculating pH_i from values for DMO concentration in the intracellular and extracellular phases (extracellular space being determined by inulin), extracellular pH (pH_e), and pK' for DMO. Subsequent work in a variety of intracellular pH investigations has shown DMO to be a reliable indicator of pH_i, and the suitability and limitations of the acid in this respect have been discussed at length (Robson, Bone, and Lambie 1968; Waddell and Bates 1969). The introduction of ^{14}C-labeled DMO (Poole, Butler, and Waddell 1964) enabled the technique to be extended to measure pH_i in small amounts of tissue, and in 1965 Schloerb and Grantham used tritiated water for determination of total water, and sodium chloride-36 for measurement of extracellular water in the tissue. From the relationship between these three radioisotopes, as expressed in the equation of Waddell and Butler (1959), it became possible to calculate tissue pH_i in vivo without otherwise measuring water, chloride, or DMO in plasma or tissues. The method devised for pH_i measurement in the present study is a development of the earlier work of Schloerb and Grantham (1965).

The value for pH_i obtained using this technique is an overall pH for a tissue and may conceal large variations between the disparate zones and cell types that may be encountered in tissues. In addition, the method measures the overall pH of the cell water; pH differences, if any, between different parts of the cell may be obscured. Also, the technique is critically dependent on free equilibration between the vascular, extracellular, and intracellular compartments. Despite these drawbacks, the method has the advantages of wide applicability in most cell types in vitro and in tumors in vivo. Measurement of the overall response of tissue pH_i may be an advantage when investigating the effect of treatments used to cause tissue acidosis or alkalosis (Calderwood and Dickson 1978; Dickson and Calderwood 1979). The method also is noninvasive and does not damage cells or tissues. The advantages and disadvantages of these techniques for pH measurement in vivo are summarized in table 5.1.

In experiments described in the present work, pH_e was measured by tissue electrode. This technique gave reproducible values for pH_e in normal and malignant tissues under normal conditions. The electrode system also had the advantage of giving continuous monitoring of pH_e and rapid response to pH

Table 5.1
Approaches to Tumor pH Measurement In Vivo

Means of Measurement	Parameter Measured	Advantages	Disadvantages
1. Extracellular electrode (tip diameters ranging between 1 μ[a] and 1 mm[b] have been used)	pH_e	(a) Continuous monitoring (b) Rapid response	(a) Tissue perturbed (b) Measures undefined mixture of blood, ECF, cellular debris
2. Indwelling millipore chamber for sampling ECF[c]	pH_e	(a) Tissue unperturbed when sampling (b) Other parameters may be measured (c) Parameters assayed under controlled conditions in vitro	(a) Delay while samples accumulate (b) Nonphysiologic
3. Intracellular electrode (tip 1 μ)[d]	pH_i	(a) Direct measurement (b) Rapid response (c) Continuous monitoring	(a) Possible damage to cell membrane (b) Fine electrode easily snapped by movement (c) Measures pH of single peripheral cells only
4. Distribution[e] of weak acids and bases between intracellular and extracellular fluid (e.g., DMO method)	pH_i	(a) Easy to use in vivo (b) Measures overall pH_i of tissues	(a) Depends on free equilibration of indicator between blood, ECF, and cell water (b) Measures overall pH_i of cell, tissue

[a] Bicher et al. 1980.
[b] Dickson and Calderwood 1979.
[c] Gullino 1970.
[d] Thomas 1976.
[e] Calderwood and Dickson 1978; Hinke and Menard 1978; Waddell and Bates 1969.

changes in conditions of acid-base disturbance. Intracellular pH was measured by the DMO method. Reproducible values were obtained for pH_i in tumors and normal organs, and the method was sensitive to pH_i changes induced in tissues by systemic acidosis and alkalosis (Calderwood and Dickson 1978).

METHODS AND MATERIALS

Experimental Animal Tumors

The work described in this report was carried out using a spectrum of four different types of transplantable animal tumor. These included one rabbit tumor, the allogeneic VX-2 carcinoma, and three types of rat tumor, the allogeneic Yoshida sarcoma and the syngeneic MC-7 sarcoma and D-23 carcinoma. Details of the origin and biological properties of these tumors are shown in table 5.2. For investigation of the effects of hyperthermia in rat tumors, the neoplasms were grown subcutaneously on the dorsum of the foot. For biochemical and blood flow investigations, tumors growing in the thigh muscles (IM) and the flank (SC) were used in addition to foot tumors. For all work carried out on the VX-2 carcinoma, tumors were grown in the thigh (IM).

Transplantation

Rat Tumors (Dickson and Suzangar 1974)
After excision from donor animals, tumors were sliced into small fragments (less than 1 mm³) and mixed thoroughly with an antibiotic mixture at a concentration of 25 μl/gm slices (antibiotic solution contained 100 units penicillin/ml, 100 μg streptomycin/ml, and

Table 5.2
Origin, Biological Properties, and Malignancy of Transplantable Tumors Used in the Present Study

Cell Line	Yoshida Sarcoma	MC-7 Sarcoma	D-23 Carcinoma	VX-2 Carcinoma
Species	Rat	Rat	Rat	Rabbit
Tissue of origin	Unknown[a]	Subcutaneous tissue, cell of origin unknown	Liver (hepatocellular carcinoma)	Shope papilloma[a] cells
Carcinogenic agent that induced tumor line	O-aminoazotoluol[a] (chemical)	3-methylcholanthrene (chemical)	4-dimethylaminoazo-benzene (chemical)	Shope virus[a]
Histologic type of tumor and genetic heterogeneity of host animals	Undifferentiated, allogeneic	Syngeneic, anaplastic	Syngeneic, anaplastic	Anaplastic, allogeneic[a]
Sex and weight of host animal	Male noninbred 200 g	Female inbred 200 g	Male inbred 250 g	Male, 2.5 kg
Anatomic site of tumor used (a) for passage	Thigh (IM)	Flank (SC)	Flank (SC)	Thigh (IM)
Site (b) used for hyperthermia experiments	Foot (SC)	Foot (SC)	Foot (SC)	Thigh (IM)
Target tissues for metastasis from primary site (b)	Regional and distant lymph nodes and viscera[b]	Regional and distant lymph nodes; lungs	Regional lymph nodes and viscera	Regional and distant nodes and lungs
Mean death time (days after tumor inoculum into site (b)	45 ± 7 days[c]	43 ± 8 days	49 ± 10 days	70 ± 6 days[d]

[a] Stewart et al. 1959.
[b] Dickson and Ellis 1976.
[c] Calderwood and Dickson 1980a.
[d] Muckle and Dickson 1973.

mycostatin at 100 units/ml). Slices were then injected into recipient animals, in 0.1 ml aliquots using a 14-g trocar.

Rabbit Tumors (Muckle and Dickson 1971)

The VX-2 carcinomas were excised and minced with scalpel blades in Ca^{++}- and Mg^{++}-free Rinaldini saline (1959) at 4°C. After a further wash in Rinaldini solution, tumor slices were placed in a 50-ml flask with a magnetic stirrer. The tumor pieces were then disaggregated into single cells by a 20-minute digestion in 30-ml Rinaldini solution containing 1% trypsin, 0.1% deoxyribonuclease, 0.01% $MgSO_4$ (Mg^{++} is the activator of DNAse), and penicillin (100 units/ml). The cell suspension was filtered successively through 80- and 200-mesh stainless steel meshes, and then centrifuged at 200 g for five minutes. The cell pellet was washed in Waymouth culture medium containing 10% human pooled AB serum. The cells were resuspended in fresh medium at 2×10^6 cells/ml, counted by hemocytometer and viability determined using trypan blue dye; the viability of the tumor cell population was always in excess of 80%. Male New Zealand white rabbits weighing approximately 2.5 kg were inoculated with 1×10^6 viable tumor cells in 0.5-ml medium, 1 cm deep into the muscles of the left hind limb.

Anesthesia

Rats

Rats were anesthetized with intraperitoneal (IP) pentobarbitone sodium (Sagatal, May & Baker Ltd., Dagenham, England). Sagatal was given at a dose of 24 mg/kg body weight (0.1 ml of a 12-mg/ml solution of Sagatal in 0.9% NaCl per 50 g rat weight). Narcosis was maintained by additional small doses of the barbiturate as required.

Rabbits

Rabbits were anesthetized with either Sagatal, 0.6 mg/kg IV or with Hypnorm, 0.5 ml/kg IM (Hypnorm, Janssen Pharmaceutica; Fentanyl base, 0.2 ml, from Crown Chemical Co. Ltd., Lamberhurst, Kent, England).

Hyperthermia

In the majority of the studies on hyperthermia, the normal or tumor-bearing feet of animals were heated by water-bath immersion.

Temperature Monitoring

Animals were anesthetized with pentobarbitone as described in the discussion of experimental animal tumors earlier in this chapter. Temperature-monitoring probes were then placed in the tumor, and hyperthermia was applied by water-bath immersion. Temperature was measured to ±0.1°C at 10-minute intervals by means of a multiprobe 12-channel direct reading electric thermometer with a scale range of 36°C to 46°C (model 3GID, Light Laboratories, Brighton, England). The instrument had a fast response time of four seconds, recorded temperature with an accuracy of ±0.05°C, and was unaffected by changes in ambient temperature.

For intratumor and intraabdominal temperature measurement, the thermistor probes were 5 cm long, needle type 1H, 0.88 mm in diameter, recording temperature only at the needle tip. Polythene-covered probes were used for rectal and water-bath temperature measurement. The intratumor sensor was inserted 1.0 cm into the foot tumor, along the line of the limb, which acted as a splint, and the probe was immobilized by a nonrestricting tape bandage around the leg. Each tumor had an indwelling thermistor during heating. It was established previously that the presence of a temperature probe in tumors (volume 1–10 ml) did not significantly alter the biological behavior or response to heat of the Yoshida tumor (Dickson and Ellis 1974, 1976). Central body, or core, temperature was monitored by the intraabdominal needle introduced to a depth of 2.5 cm below the liver in a right paramedian position. Core temperature was also measured by a rectal probe inserted 3 cm into the anus. Taping the rectal probe to the base of the tail prevented dislodgement of the sensor during heating.

Thermocouples (Bailey Instruments Inc., Saddle Brook, New Jersey) were also used for temperature monitoring. The sensors were embodied in 23-gauge needles, or the metal junction and leads were enclosed in a fine, clear polyethylene sleeve (tissue-implantable probes). The latter were introduced into the tissue via the lumen of a stainless steel needle, which acted as a trocar. The needle was then withdrawn, leaving the sensor in position. The thermocouples were used in association with five-channel digital readout meters (Doric Instruments, San Diego, California).

Before use, temperature sensors were calibrated against a mercury in glass thermometer of the National Standards Laboratory, Hemel Hempstead, Hertfordshire, England. During heating, each tumor contained at least one temperature sensor; often two or more probes were inserted, depending on tumor volume.

Heating Technique

The heating bath consisted of a perspex tank (33 × 33 × 15 cm) containing 10 liters of water heated by a Circotherm II (Shandon Scientific Co., Ltd., London, England) constant-temperature unit with a 700-watt coil heater and circulating pump with an output of 12 liters per minute. At an ambient temperature of 25°C, this unit maintained the bath temperature constant to ± 0.05°C. The rat was placed on a perspex platform resting over the bath, and the tumor-bearing foot was immersed in the water through a 10-cm diameter padded opening. The foot was supported in the bath at a depth that permitted complete submersion of the tumor. Immediately after heat therapy, each rat was given 1.0 ml of 4% dextrose in 0.18% NaCl to replace fluid loss. The animal was then wrapped in a blanket and placed under an infrared heater for 10 to 15 minutes; this helped to control the return of body temperature to normal without an overswing to subnormal temperature, which can occur rapidly in rats following hyperthermia (Dickson 1977).

Work also has been carried out on hyperthermia in the VX-2 carcinoma at 47°C to 50°C (Dickson and Shah 1977) and on heating the Yoshida, MC-7, and D-23 tumors at 42°C to 45°C (Dickson, Calderwood, and Jasiewicz 1977), using radiofrequency heating at 13.56 MHz. As this work does not form a major part of the study described in this report, radiofrequency heating is not described here, and readers are referred to earlier publications for details of heating technique and temperature monitoring (Dickson, Calderwood, and Jasiewicz 1977; Dickson and Shah 1977; Dickson et al. 1977; Dickson 1979).

Cytokinetic Studies

In the present study, the effects of hyperthermia were investigated in the Yoshida sarcoma using only in vivo techniques. In this tumor, the treatment used (one hour at 42°C) is curative, and would thus be expected to perturb grossly the age distribution of cells in the tumor population. Therefore, no attempt was made to compute derived parameters such as GF or ø from the measurements. The following techniques were found useful in studying cytokinetic response.

Flash [³H]-TdR Labeling Index

The flash labeling index is the proportion of cells labeled with [³H]-TdR a short time (one hour) after a single injection of the label. This parameter gives an

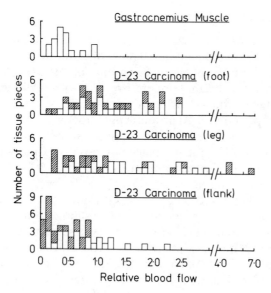

Figure 5.11. Distribution of blood flow between discrete zones of gastrocnemius muscle and in the D-23 carcinoma in the foot, leg, or flank of rats.

Relative blood flow was determined in 50- to 100-mg slices of tumor or muscle excised after IV injection of 100 μCi of ^{86}Rb. Values for relative blood flow were calculated as cpm ^{86}Rb per g wet weight of tissue. In gastrocnemius muscle, blood-flow distribution was determined in 24 slices obtained from four rats (six slices per rat). For each tumor site, relative blood flow was measured in 48 slices obtained from four rats (12 slices per rat). Of the 12 slices, 6 were from areas that appeared necrotic on macroscopic examination (represented as *hatched rectangles* in the histogram). The six remaining slices were from viable areas *(clear rectangles)*.

Mean tumor volumes in the foot, leg, and flank of rats were, respectively, 3.0, 8.2, and 12.4 ml.

estimate of the number of cells in the population in S phase. The technique is rapid and simple and was used to investigate changes in the fraction of cells in S phase after heating.

Repeated [^{3}H]-TdR Labeling

The repeated [^{3}H]-TdR labeling technique, in which thymidine was injected into animals every six hours, proved effective in examining situations of rapid change in the cell population. The almost continuous availability of [^{3}H]-TdR means that any cell entering DNA synthesis will be labeled. Also, any decline in the proportion of labeled cells (over the short time-periods used in this study) is unambiguous proof of cell killing. The method is thus an indicator of cell killing and repopulation, and was used to investigate these phenomena (figs. 5.11 and 5.12).

Thymidine Studies: Technical Details

[^{3}H]-TdR (specific activity 21 Ci/mM) was diluted with 0.9% NaCl before injection into rats in the following

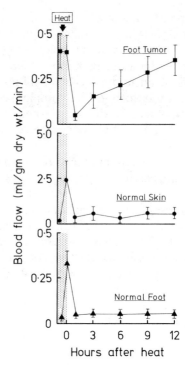

Figure 5.12. Blood flow in the Yoshida foot tumor (1.0 to 1.5 ml) and in the skin and tissues of the normal rat foot during the 12 hours after hyperthermia) 42°C for one hour). Blood flow was measured by the ^{86}Rb distribution technique. (Dickson and Calderwood 1980. Reprinted by permission.)

doses: animals to be killed for the PLM and flash-labeling studies were given a single injection of 2 μCi/g body weight in 1 ml 0.9% NaCl; animals for repeated tumor labeling experiments received 0.5 μCi/g [^{3}H]-TdR every six hours and a further 0.5 μCi/g one hour before killing.

Autoradiographs of 4 μm paraffin-embedded hemisections of the tumors were prepared using the dipping technique (Baserga and Malamud 1969) with Ilford K-2 emulsion (Ilford Ltd., Ilford, Essex, England). After 14 days of exposure, the autoradiographs were processed with D-19 developer and F-5 fixer (Rogers 1967) and stained with hematoxylin and eosin. For the PLM curves, 500 anaphases and metaphases were counted "blind," and for the flash and repeated labeling indices (% labeled cells) a total of 2000 cells was scored.

When determining the fraction of labeled cells in the tumor population, the following three questions posed by Baserga and Malamud (1969) were considered:

1. Is the group of cells to be counted a homogeneous population that can be identified on the basis of histologic criteria?
2. What are the criteria for classifying a cell as labeled?

3. What are the limits of accuracy in determining the fraction of labeled cells?

The Yoshida sarcoma consisted of sheets of histologically similar round cells embedded in a small amount of connective tissue and was considered to be a single population. To establish whether a cell was labeled, a tumor-bearing animal was injected with [^{3}H]-TdR and the tumor fixed 30 minutes later. This period is too brief for mitoses to become labeled (Lala 1971). Therefore, any silver grains over such a mitosis in an autoradiograph would be due to background radiation. Cells with a number of grains greater than the mean grain count over these mitoses were considered labeled. Sections of the tumor from the animal given the 30-minute [^{3}H]-TdR label were prepared in large numbers and used as standards for processing with each batch of autoradiographs.

To reduce inaccuracies in counting caused by sampling error, such as counting an unrepresentative area of tissue, counts were carried out on tumor hemisections. The sections were scanned from the surface to the interior of the tumor.

Metaphase Arrest with Vincristine

Agents that block cells in mitosis have been used for many years in the investigation of cell proliferation (Dustin 1936; Aherne, Camplejohn, and Wright 1977; Wright and Appleton 1980). In a study comparing these stathmokinetic drugs, Tannock (1965) found that vincristine was a suitable agent for the quantitative study of mitosis in vivo. Vincristine blocked cells in metaphase almost immediately after injection and gave an accurate value for birth rate in a rat mammary tumor (Tannock 1965). In the Yoshida sarcoma, as in other solid tumors, accumulation of cells in metaphase after administration of vincristine, although linear, showed a degree of variability (fig. 5.9). It was thus necessary to carry out multiple observations (at least seven in this study) before an accurate value for birth rate was obtained. Also, in each experimental condition in which the drug was used, a series of measurements of metaphase arrest was made between 30 and 150 minutes after vincristine injection to ensure linearity of metaphase accumulation under each condition.

The method proved a valuable complement to the thymidine labeling techniques. It is essential in assessing the effects of treatment to study more than one cell cycle phase to ensure that results obtained do not merely indicate changes in progression through one phase, rather than overall effects on cell proliferation.

Technical Details

Vincristine was injected into rats at an IP dose of 1 mg/kg in 10 ml 0.9% NaCl. Paraffin hemisections (4 μm) of tumors were stained with Harris's hematoxylin and the metaphase index (percentage of metaphases) determined by counting 4000 cells. Presence in the tumor section of cells in anaphase or telophase was an indication that the mitotic block was not complete. Under these circumstances, rate of entry into mitosis is underestimated (Tannock 1965). Therefore, data from animals in which incomplete metaphase blockade occurred were rejected.

Measurement of pH In Vivo

Technical Details and Calculations

Measurement of Extracellular pH with Miniature Glass Electrodes

Tissue pH was studied by means of miniature glass electrodes (type MI 400) with a 1-mm diameter tip and by a reference microelectrode (type MI 401) filled with 3M KCl saturated with AgCl (Microelectrodes Inc., Grenier Industrial Village, Londonderry, New Hampshire). The electrodes were coupled by high impedance amplifier to a digital pH meter (PHM 63: Radiometer, Copenhagen, Denmark). The electrodes were calibrated at 30°C, 38°C, and 42°C, with standard buffers of pH 7.40 to 6.40 and checked for drift and linearity before and after each experiment.

The anesthetized rat was immobilized on an electrically insulated board, the distal 1 cm of its tail was amputated, and the bleeding tail was immersed in a 250-ml Erlenmeyer flask containing 200 ml 0.9% NaCl and the reference microelectrode. This large volume of saline was found to be a stable reference as reported by earlier workers (Kahler and Robertson 1943). For measurement of tumor pH, the glass microelectrode was inserted 3 to 5 mm through a small incision made in the upper surface of the foot tumor with the point of a scalpel blade. The electrode was then secured vertically in position and connected to the digital pH meter, and the system was left to stabilize. When electrode stability was achieved, pH readings varied less than ±0.03 units/hour; this usually required 40 to 60 minutes. In experiments on hyperglycemia, an IP injection of 50% glucose (6 g/kg body weight) (Voegtlin, Fitch, and Kahler 1935) was given at this point, and pH was monitored for four to nine hours. At the end of the experiment, the animal was killed, and the capillary electrode track was examined to ensure that it had been situated in viable tumor.

For measurement of liver pH, a 2-cm incision was made in the abdominal wall immediately beneath the diaphragm, one lobe of the liver was exposed, and the incision was packed with gauze moistened with saline. The pH probe was inserted 3 to 5 mm into the exposed liver and fixed in position, and the liver was covered with saline-moistened gauze. The same procedure as with tumor was then followed.

Muscle pH was investigated in the gastrocnemius of the right leg. The skin was reflected, the pH probe was inserted approximately 5 mm via a small cut in the gastrocnemius muscle, and the probe was splinted to the rat's leg by elastic tape. As with the tumor, approximately one hour was required for the pH recording system to stabilize after insertion of the sensing electrode into the liver or muscle.

Measurement of Intracellular pH by the DMO Method

Following administration of the isotope mixture (legend, fig. 5.13), rats were sacrificed after a three-hour equilibration period (Calderwood and Dickson 1978). Samples (approximately 0.5 g) of tissue were then excised and stored at $-70°C$ prior to pH_i determination.

Determination of Intracellular pH

After equilibration of the isotopes with the tissues, a sample of blood was removed by cardiac puncture under anaerobic conditions and its pH determined, using a Radiometer Microelectrode Unit (PHA 931) and Acid-Base Analyzer (PHM 71) with a pCO_2 module (PHA 931) (Radiometer, Copenhagen, Denmark). Temperature was maintained at 38°C by a Water Thermostat (Radiometer model VTS 13), and the pH measuring equipment was standardized at 38°C using commercial (BDH) buffers for this temperature. Animals then were sacrificed by cervical dislocation, another blood sample was taken by cardiac puncture, and the tissue samples were removed as above for pH_i measurements, plus samples from lung, kidney, and both heart ventricles. Blood was spun at 3000 rpm and the plasma removed to a universal container. Solid tissue samples were blotted, coarsely minced, and placed in universal containers.

After adding 1.0 ml of distilled water to each sample, solid tissues were homogenized (Polytron Microhomogenizer, Northern Media Supply Ltd., Hull, England) at full speed for two minutes. Four ml of 5M NaH_2PO_3 were added, the samples mixed, and 4 ml of the mixture pipetted into containers. Sixteen ml of 1:1 ethyl acetate/toluene was added, the samples gently rotated for 20 minutes and then centrifuged at 3000 rpm for 15 minutes. The overlying ethyl acetate/toluene layer was removed, and duplicate 1-ml samples of the underlying phosphate buffer layer transferred to 10 ml of toluene based scintillation fluid (Patterson and Greene 1965). To these vials were added 6 ml distilled water, the contents thoroughly mixed and left overnight at 0°C in the dark to form thixotropic gels. Duplicate 4-ml samples of the ethyl acetate/toluene layer were added to 10 ml of scintillation fluid and left at 0°C overnight with the phosphate buffer samples. Activity in the vials was counted in three channels using an Intertechnique SL30 Liquid Scintillation Spectrometer (Intertechnique Ltd., Portslade, Sussex, England). The settings on the log-linear scale of this instrument were:

	Instrument Setting	Kev	Isotope Counted	Sample Layer
Channel 1	2.5–5.5	2–50	^3H	Lower
Channel 2	6.5–8.0	90–800	^{36}Cl	Lower
Channel 3	4.0–7.0	5–200	^{14}C	Upper

Calculation of Intracellular pH

Calculation of pH_i was done by inserting the counts in the following formula as used by Schloerb and Grantham (1965). This formula represents the original equation of Waddell and Butler (1959) expressed in the form proposed by Poole, Butler, and Waddell (1964) for use with isotopic indicators:

$pH_i = 6.13 +$ log of the expression

$$\left\{ \left[\frac{^{14}C_t \, ^3H_p}{^{14}C_p \, ^3H_t} \left(1 + \frac{^{36}Cl_t \, ^3H_p}{1.05 \, ^{36}Cl_p \, ^3H_t - ^{36}Cl_t \, ^3H_p} \right) - \left(\frac{^{36}Cl_t \, ^3H_p}{1.05 \, ^{36}Cl_p \, ^3H_t - ^{36}Cl_t \, ^3H_p} \right) \right] \times [1 + 10^{(pH_e - 6.13)}] - 1 \right\}, \quad (7)$$

where

6.13 = pK' for DMO;
pH_i = intracellular pH;
pH_e = extracellular pH = blood pH + 0.02;
^3H_p = HTO counts in plasma;
$^{14}C_p$ = DMO counts in plasma;
$^{36}Cl_p$ = ^{36}Cl counts in plasma;
^3H_t = HTO counts in tissue;
$^{14}C_t$ = DMO counts in tissue;
$^{36}Cl_t$ = ^{36}Cl counts in tissue.

The great convenience of this approach is that the relationship among the three isotopes, as incorporated in this equation, enables pH_i to be calculated from the ratio of counts in the tissue to the corresponding counts in the plasma, and the exact amount of each isotope introduced is therefore not critical.

Operational Guidelines

Determination of Extracellular pH

Before and after each experiment, electrodes were calibrated using standard buffers as described previously. Electrodes were selected for use by observing their performance in terms of accuracy and speed of response in blood and plasma of known pH (pH range 6.0–7.5) in vitro. Electrodes that gave reasonably rapid (1 to 5 minutes) and accurate (±0.02 pH units) measurements were chosen. These test conditions were used because they resembled the conditions inside tissues more closely than did standard buffers. After repeated use, electrodes began to give erratic readings, probably because of clogging with tissue debris and protein. This malfunction usually could be reversed by soaking the electrodes in 1N HCl for one to five minutes. The effects, if any, on the accuracy of pH_e measurement of proteins adhering to electrodes in vivo are not known. Any effect probably would be, at least initially, to increase response time.

Changes in temperature alter the performance of glass electrodes (Bates 1964). The pH electrodes therefore were calibrated at the desired intratumor temperature before pH_e measurement in tumors under hyperthermic conditions.

The effect of treatments such as hyperthermia and hyperglycemia or systemic acidosis/alkalosis on tumor pH_e were examined by two types of experiment: (1) The glass electrode was inserted into the tumor and the system left to equilibrate prior to perturbation. Experiments were then carried out with the probe in situ as in figure 5.14. In this type of arrangement, the animal acted as its own control and the effects of treatment could be directly observed. (2) Probes were inserted into the tumors after treatment and pH values compared to pH_e in untreated animals. When the two approaches were compared in the same animal, they were found to give similar results.

Determination of Intracellular pH

For pH_i measurement in vivo, it is necessary for the radioisotopes used in the determination to equilibrate between the tissues and between the vascular, interstitial, and intracellular fluids of each tissue. In the normal and malignant tissues used in this study, HTO, ^{14}C-DMO, and ^{36}Cl had reached apparently stable levels by 70 minutes after injection (fig. 5.13). These levels were maintained until at least three hours after isotope injection (fig. 5.13). When pH_i was calculated using these isotope concentrations, however, reproducible pH values were obtained only after two and one-half hours of equilibration. It is thus apparent that although isotopes are evenly distributed between tissues by 70 minutes, a longer period (150 minutes) is required for complete equilibration in tissue compartments and accurate determination of pH_i. A similar observation was made by Schloerb and Grantham (1965) using the same isotope combination for pH_i determination; a three-hour equilibration period for the isotopes was indicated by their study. Schloerb and Grantham (1965) also calculated the redistribution time for the isotopes after animals were subjected to $NaHCO_3$ infusion or CO_2 inhalation. Two hours was required for this redistribution process. These various data indicate that a lag period of at least two hours is required for adequate equilibration of the isotopes in vivo; we routinely employ an equilibration period of three hours (Calderwood and Dickson 1978; Dickson and Calderwood 1979). A limitation to the DMO technique is that pH_i can only be measured in tissues with an intact blood supply. In situations in which blood supply is interrupted, as in the Yoshida sarcoma during glucose infusion (Dickson and Calderwood 1979) or the MC-7 sarcoma heated for three hours at 42°C, the isotopes do not equilibrate with the tissue, and pH_i cannot be measured.

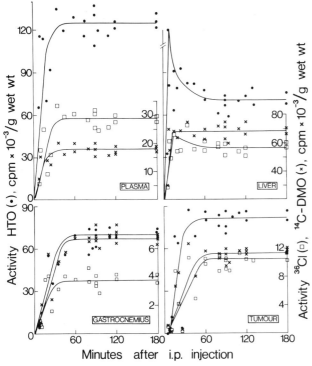

Figure 5.13. Levels of HTO, DMO-^{14}C and Na^{36}Cl in plasma, liver, gastrocnemius muscle, and Yoshida sarcoma of rats after IP injection of isotopes. Each rat received an IP injection of 1.0 ml of 0.9% NaCl containing 50 μCi of HTO, 1 μCi of Na^{36}Cl and 1 μCi of DMO-2-^{14}C. (Calderwood and Dickson 1978. Reprinted by permission.)

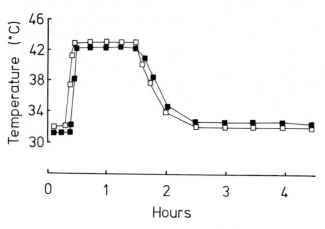

Figure 5.14. Simultaneously recorded bath (□) and intratumor (■) temperatures and pH_e (●) for 1.5-ml Yoshida sarcoma treated by hyperthermia. The tumor-bearing foot with a capillary glass pH electrode and thermistor probe inserted in the tumor was placed in a bath of 0.9% NaCl at 30°C. After immersion of the foot, the reference electrode was placed in the saline. When tumor pH, bath temperature, and intratumor temperature had stabilized, the tumor was heated for one hour at 42°C by elevating the bath temperature. The bath thermostat was then switched to 30°C and cooling accelerated by addition of solid CO_2 to the bath fluid.

Blood-Flow Measurement

Techniques for blood-flow measurement are reviewed elsewhere in this volume and this account will be restricted to methods used in the present study. Two techniques were employed.

Fractional Distribution of ^{86}Rb

This method was devised by Sapirstein (1958, 1962) and is based on the following premise. A foreign substance, after a single, rapid intravenous injection, will be distributed initially to the tissues in proportion to their blood flow and then be carried away by venous drainage. For a certain period of time, however, the venous drainage will be negligibly small compared to arterial delivery. During this time, the fractional distribution of the substance among the organs will correspond to the fractional distribution of the cardiac output among them (Sapirstein 1958, 1962). The time of minimum venous drainage will be greatest for substances that are completely transferred from the vascular system to the tissues with least hindrance and with a large volume of distribution within a tissue. The radioactive isotopes ^{42}K and ^{86}Rb are such substances (Sapirstein 1958). In the present study, ^{86}Rb was chosen for blood-flow measurement. This choice was made mainly because the longer half-life of ^{86}Rb (18.7 days compared to 12.4 hours for ^{42}K) allowed it to be stored for a longer period and experiments carried out over a number of days.

Sapirstein (1958) found that after IV injection, the time : concentration curve of the labeled indicator in all organs except the brain showed a steep rise during the first 10 to 15 seconds, a flattening from 20 to 60 seconds, and variable behavior beyond this point. The plateau of the time : concentration curve was interpreted as an indication that the amount of indicator found in each organ for about 40 seconds was approximately the amount brought by blood flow. During this time period, the percentage of distribution of indicator in the tissues was proportional to the fraction of the cardiac output.

In the present study, anesthetized animals were given injections of 100 μCi of $^{86}RbCl$ in 0.1 ml of 0.9% NaCl solution into the right femoral vein. Animals were killed by cutting through the thorax just above the diaphragm with a guillotine. Tissues were removed and radioactivity determined as cpm/g dry weight.

Time-Course of ^{86}Rb Uptake

To determine whether ^{86}Rb uptake followed similar kinetics to those described by Sapirstein (1958) in normal tissues and by Gullino and Grantham (1961) in tumors, animals with Yoshida sarcomas in the foot, leg, and flank were sacrificed at intervals after ^{86}Rb injection. Normal tissues showed a plateau of ^{86}Rb uptake values in the 30- to 60-second period (table 5.3) as previously shown by Sapirstein (1958). Uptake of ^{86}Rb by tumors was more erratic (table 5.3), probably owing to blood-flow differences between tumors. ^{86}Rb uptake in tumors was therefore investigated by external counting of individual tumors (fig. 5.15) as suggested by Gullino and Grantham (1961). Tumors or normal tissues investigated in this way were shielded from the remainder of the body by at least a 5-cm thickness of lead. As shown in figure 5.15, ^{86}Rb uptake in the Yoshida foot tumor reached a plateau level in 20 to 30 seconds and was maintained at this level for at least 60 seconds. A similar finding, of stable ^{86}Rb level for 30 to 60 seconds after injection, was

Table 5.3
Time-Course of ^{86}Rb Uptake by Normal and Malignant Rat Tissues

Time after Isotope Injection (sec)	^{86}Rb Uptake (cpm/g × 10^{-3})					Yoshida Sarcoma		
	Liver	Kidney	Spleen	Small Intestine	Gastrocnemius Muscle	Foot	Thigh	Flank
5	14.7	156.1	12.8	31.3	2.2	1.9	2.8	0.9
15	22.4	211.0	31.2	48.1	5.1	8.8	7.6	4.6
30	25.1	277.7	25.1	56.7	4.5	6.7	11.9	22.2
45	32.6	343.3	31.2	63.2	4.1	8.4	12.0	6.5
60	26.7	282.7	28.7	70.1	3.3	8.2	4.1	13.0
90	29.8	339.3	24.6	47.7	4.3	10.3	11.9	2.8

NOTE: Each figure represents the mean value obtained from the tissues of three rats.

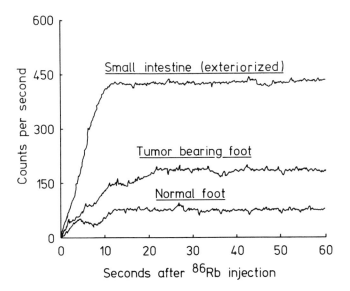

Figure 5.15. Radioactivity in normal and malignant rat tissues after injection of 100 μCi of ^{86}Rb in 0.1 ml of 0.9% NaCl into the right femoral vein. Radioactivity was measured in tissues shielded by at least 5-cm thickness of lead. Activity was detected with a NaI scintillation crystal coupled to a ratemeter and chart recorder.

observed in the Yoshida, MC-7, and D-23 tumors growing in the foot, leg, or flank of rats and for the VX-2 carcinoma in the thigh muscles of rabbits.

Blood-Flow Determination

Tissues were removed 40 seconds after IV ^{86}Rb injection. Blood-flow values were obtained by multiplying the cardiac output fraction (fraction of injected dose) by total cardiac output in a pentobarbitone-anesthetized 200-g rat (Vidt et al. 1959) as shown below:

$$\text{Blood flow (ml/g dry wt./min)} = \frac{\text{cpm (tissue)} \times 100 \times \text{total cardiac output (ml/min)}}{\text{cpm (total)} \times \text{fraction dry wt.}}. \quad (8)$$

Blood-Flow Measurement with ^{86}Rb in Experimental Conditions

Rubidium, like potassium, is maintained in intracellular water at a much higher concentration than in interstitial fluid or plasma (intracellular [Rb$^+$] = 20 − 30 × extracellular [Rb$^+$]) (Tietz and Siggaard-Anderson 1976). Maintenance of the Rb$^+$ gradient is an active process (Davies 1973), and thus treatments aimed at damaging tumors (e.g., hyperthermia or hyperglycemia) might alter this gradient. Under such circumstances, erroneous values for blood flow would be obtained. For instance, abolition of the Rb$^+$ gradient in muscle would lead to a 90% to 98% underestimation of blood flow. Therefore, the effects on Rb$^+$ uptake of in vivo hyperthermia and hyperglycemia carried out both singly and in combination were investigated. Uptake of Rb$^+$ was measured in tumor slices immediately after in vivo treatment (table 5.4). These treatments had no significant effect on ^{86}Rb uptake by tumor slices. This indicated that ^{86}Rb could be used to investigate blood flow under the hyperthermic and hyperglycemic conditions described, and that the measurements were not invalidated by alterations in ^{86}Rb$^+$ distribution at the cellular level (Calderwood and Dickson 1980b).

^{133}Xe Clearance Technique

This technique is based on the assumption that an inert gas such as ^{133}Xe (or ^{85}Kr) injected locally into a tissue will mix homogeneously with the part of the tissue close to the injection site. The ^{133}Xe then equilibrates with venous blood, and the rate of disappearance of the isotope from the tissue is proportional to blood flow (Appelgren 1979; Kety 1951).

In the present study, xenon clearance was measured after intratissue injection of 50 μCi of ^{133}Xe in 0.05 ml of 0.9% NaCl solution. Radioactivity was then determined by external counting using a scintillation crystal positioned 1 to 2 cm above the tissue, which was shielded to avoid detection of ^{133}Xe in the lungs of the animals. Radioactive decay followed a multiexponential function (fig. 5.16); the half-time for clearance (t½) was calculated from the initial part of the curve and a value for blood flow obtained using the equation

$$\text{Blood flow (ml/min)} = \frac{\log_e 2 \times \lambda}{t\frac{1}{2} \text{ (min)}}, \quad (9)$$

where λ is the partition coefficient for ^{133}Xe between tumor cells and blood (Lam et al. 1979; Appelgren 1979).

This technique has certain drawbacks, which have been discussed by Appelgren (1979). In the present study, values for blood flow obtained using the ^{133}Xe clearance technique were of a similar order to blood-flow levels observed with the ^{86}Rb distribution method. Furthermore, serial estimations carried out on the same tumor gave very reproducible estimates for blood flow.

These two methods were found to complement each other. The ^{86}Rb technique was of value when simultaneous comparison of blood flow in a number of tissues was required. The ^{133}Xe technique was used for sequential measurements when investigating the time course of blood-flow changes after treatment.

Table 5.4
Effect of In Vivo Hyperglycemia and Hyperthermia on ^{86}Rb Uptake by Yoshida Sarcoma Slices

Condition	^{86}Rb Incorporation (cpm/g dry wt)	% of Control	t	P
(a) Normoglycemic	12,375,050	—	—	—
(b) Hyperglycemic	10,633,685	85.93	0.677	> 0.10
(c) Normoglycemic + 1 hr/42°C	9,987,605	80.71	0.723	> 0.10
(d) Hyperglycemic + 1 hr/42°C	9,956,586	80.46	0.714	> 0.10

NOTE: Tumors were from four groups of Sagatal-anesthetised animals with six animals in each group. Each group received a different treatment, and animals were killed after (a) four-hour Sagatal anesthesia (control group), (b) four-hour hyperglycemia (glucose injection, 6g/kg), (c) four-hour anesthesia prior to heating for one hour at 42°C, and (d) four-hour hyperglycemia prior to one hour at 42°C. Immediately upon removal, thin tumor sections (less than 1 cu mm) were prepared (Dickson and Suzangar 1976b). The slices were washed in 5-ml Waymouth's medium containing 0.1% albumin, and 100-mg aliquots were weighed into 5-cm Petri dishes.

To the tumor slices were added 3 ml of medium containing ^{86}Rb (2.5 μCi/ml), and the slices were incubated at 37°C for two hours (Kimelberg and Mayhew 1975).

The time elapsing between removal of tumor from the animal and addition of ^{86}Rb to culture medium was less than 45 minutes. After equilibration (two hours), slices were washed three times in 5 ml of ice-cold phosphate-buffered saline (pH 7.4). Radioactivity was expressed as cpm/g dry weight. ^{86}Rb incorporation in slices from treated animals was compared to incorporation in control slices using a paired t-test (Walpole 1974).

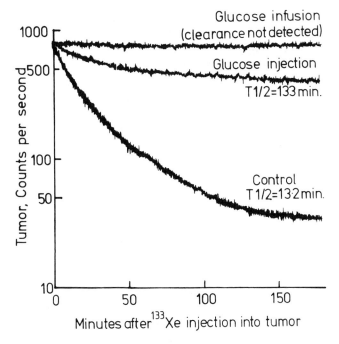

Figure 5.16. Clearance of ^{133}Xe after intratumor injection into the Yoshida foot sarcoma. Sample chart recordings are shown of ^{133}Xe clearance of individual tumors four hours after glucose injection, after a four-hour infusion, and in a control (normoglycemic) animal. T½ is the time taken for ^{133}Xe activity to decline to 50% of the initial value.

ANALYSIS OF DATA

Tumor Volume and Thermal Response

Tumors may be more (Crile 1963; Storm et al. 1979; in press, Urano et al. 1980) or less (Dickson and Ellis 1976) heat sensitive as they increase in volume. Few detailed studies on this topic have been reported. With the Yoshida sarcoma growing subcutaneously on the feet of rats, Dickson and Ellis (1976) found a dose-response relationship in tumors up to 3 ml in volume. After heating at 42°C for one hour, small tumors (1.0 to 1.5 ml) regressed completely in 11 to 13 days, with a tumor volume halving time of 2.6 days. Following two-hour hyperthermia (42°C), the tumors disappeared in 6 to 7 days, and tumor volume was halved every 1.2 days. Tumors of 2- to 3-ml volume were cured by heating for two hours at 42°C, even though they had left the exponential phase of the growth curve. Tumors in excess of approximately 3 ml were not affected by heating for two hours at 42°C, as judged by tumor volume measurements. As discussed earlier, volume measurements can underestimate tumor cell kill by several orders of magnitude (the Wilcox phenomenon) (Wilcox et al. 1965). In addition, following effective treatment of a tumor, variables, such as the rate of dead cell lysis and absorption and the kinetics of multiplication of resistant cells, govern measurable regression rate (Skipper 1971). In a series of patients with solid tumors treated at 50°C or more by RF energy, Storm and associates (in press) found that selective tumor heating was possible more often with the larger lesions (tumors greater than 5 cm in size). The authors pointed out that while standard methods of cancer therapy (surgery, irradiation, chemotherapy) were most effective for smaller tumors, their results suggested that hyperthermia may be uniquely effective against larger tumors.

Dickson and Ellis (1976) also reported that inadequate heating of the Yoshida sarcoma at 40°C, instead of the curative temperature of 42°C, led to enhanced tumor cell dissemination from the heated primary. The dissemination became more marked with increase in volume of the tumor beyond 3.0 ml. The authors postulated that at these larger volumes, a contributory factor to the augmented spread may have been an altered host "soil," owing to conditioning of the host by the well-established neoplasm (Dickson and Ellis 1976). Increased tumor dissemination from inadequate heating is discussed in chapter 23 of this volume. Its importance relates to the great difficulty in avoiding inadequate heating in larger tumors, caused by nonuniform heating, even with temperatures believed to be lethal. Another form of inadequate heating of a large tumor volume may be total-body hyperthermia in the patient with advanced cancer (large tumor burden). The potential of this approach is limited by the relatively low temperatures that must be used (41°C to 42°C) to ensure host survival (this volume, chapter 23; Henle and Dethlefsen 1980). The importance of the host component of the host-tumor relationship as the tumor volume increases has been alluded to earlier.

Tumor growth is accompanied by an alteration in the kinetic parameters of the cell population. The rate of cell production tends to decrease with increase in volume owing largely to decreased growth fraction and increased cell loss (Aherne, Camplejohn, and Wright 1977; Calderwood and Dickson 1980a; Watson 1976); with a few exceptions, cell cycle time alters little with increase in tumor volume. The importance of such changes is denoted by work on animal tumors and in the clinic, demonstrating that as a tumor enlarges its response to drugs alters, and it becomes more difficult to influence the course of the disease (Dickson and Ellis 1976). Hyperthermia has similarities to cytotoxic drugs and radiotherapy in that it is most effective against cells in cycle (in the 42°C to 45°C range). At temperatures greater than 45°C, such as employed by Storm and colleagues (1979, in press) and LeVeen and associates (1976), coagulation necrosis occurs; cell kinetics and tumor volume therefore become less relevant to thermal response.

Cater, Silver, and Watkinson (1964) emphasized that there is a limit to the volume of tumor tissue that it is safe to destroy within a given time. These authors state that after rapid destruction of large tumors there is the danger that absorption of toxic products will kill the patient; no illustrative cases are cited, however. In a series in 1967 by Cavaliere and co-workers, 15 of 22 patients treated by hyperthermic perfusion for limb tumors responded with rapid massive tumor necrosis. Total tumor necrosis occurred in 8 of the 15 patients, and one of these developed a "crushed limb syndrome." The patient had a huge fibrosarcoma of the femur and it was felt that the rapid lysis may have overloaded the host system with necrotic tumor debris. Crile (1961) reported that dogs with spontaneous lymphomas died within 18 hours of hyperthermia; the animals had extensive involvement of mesenteric and mediastinal nodes by tumor, and death may have resulted from "sudden destruction of massive amounts of tumor tissue." The problem of pathologic effects resulting from tumor breakdown after heat, therefore, appears to be associated with rapid lysis of very large amounts of tumor, especially sarcomas and lymphomas. With improved treatment protocols, including adequate IV fluids and stimulated diuresis, the problem would appear to be avoidable (Moricca et al. 1977; Pettigrew et al. 1974). Heating temperature also may be important, since the complication of tumor breakdown products has not been reported following tumor heating at more than 45°C. Storm and colleagues (1979) attribute this to the vascular necrosis and thrombosis that occur at these high temperatures, thus preventing absorption of the tumor and resulting in fibrous replacement over a prolonged period. It is of interest in this context that in several thousand rats and several hundred tumor-bearing rabbits treated at temperatures ranging from 42°C to 52°C by various techniques in our laboratory, no evidence of damaging effects of tumor breakdown products has been encountered. The treated tumor volumes were equivalent in terms of host size (1.5 ml in a 250-g rat, 15.0 ml in a 2.5-kg rabbit; these volumes would be equivalent to a 400-g tumor in a 70-kg human) to render comparable the host load of tumor breakdown products after heating.

Overall tumor blood flow decreases with the volume of tumors (Peterson 1979a) while heterogeneity in perfusion rates between discrete zones of the tumor increases (Kjartansson 1976) (fig. 5.11). It may be argued that larger tumors are heated at a more homogeneous temperature because of lower overall blood flow. The converse also may be argued—that heating in larger tumors may be more heterogeneous because of the increased intratumor differences in perfusion found in larger tumors. This is a question more likely to be solved by experiment than by theoretical argument, and tumor type, site, and heating temperature are probably of paramount importance. Larger tumors also may pose greater problems in treatment because of their irregular shape, especially when treated by radiant heat. In outgrowing the original site, tumors may follow devious routes in the invasion of normal tissues; these areas of invading tumor, probably with the highest proportion of P cells, may be outside the main tumor mass and may obtain a lower treatment dose. Temperature gradients in tumors also may be influenced by the effect of heat on blood flow in adjacent normal tissues (Dickson and Calderwood 1980). Some of these factors are discussed in the section in this chapter on the role of tumor microenvironment in heat sensitivity.

Response to treatment may alter with tumor volume because of changes in the extracellular microenvironment. There has been much speculation on the possibility of a decrease in tumor pH with increased volume (Gerweck, in press; Urano et al. 1980), although there is no direct data to indicate such a relationship. If such volume-related acidosis does occur, then it might contribute to modifying the thermal sensitivity of larger tumors. An extracellular factor related to tumor volume is hypoxia; pO_2 decreases with increase in tumor volume (Vaupel 1979). Hypoxia sensitizes tumor cells to heat in vitro (Suit and Gerweck 1979) and in vivo (this chapter, section on nutrient levels and hyperthermia), and this volume-related increase in hypoxia may sensitize larger tumors to heat. The role of pH, pO_2, and other microenvironmental factors in thermal sensitivity is discussed more fully in the section in this chapter on tumor microenvironment.

The relationship of tumor volume to thermal sensitivity is thus a highly complex one, and probably involves most or all of the factors discussed in this chapter as influencing tumor thermal sensitivity. Heat sensitivity at any volume appears to be a balance between factors increasing response to heat with volume (microenvironmental factors, blood flow decrease) and those which may decrease sensitivity (cytokinetic factors, increased tumor burden). Not only factors operating in the tumor itself, but also the local response of normal tissues (blood flow, inflammatory response) and systemic host reaction (which may include the immune system and mononuclear phagocyte system) may exert a modifying influence (fig. 5.1). The result may even be influenced by the manner in which treatment is assessed. Quantitation of response based on tumor regrowth analysis or cell kinetics may indicate an enhancement of the effects of hyperthermia at larger tumor volume, while the opposite conclusion may be drawn from the same study if response is measured by cure rate of animals. For example, the cell population kinetics of large Yoshida sarcomas are more severely disrupted by heat than are the kinetics of small tumors (see section on tumor cell kinetics and thermosensitivity); after heating, animals with small tumors were cured, while those with the large tumors had a significantly reduced lifespan because of increased metastasis (Dickson and Ellis 1976).

Tumor Cell Kinetics and Thermosensitivity

Studies in vitro and in vivo indicate that the kinetics of the tumor cell population may be involved in thermal cell killing. Hyperthermia shows cell cycle phase specificity, with selective killing of cells in S phase (Bhuyan 1979; Palzer and Heidelberger 1973b; Westra and Dewey 1971). This phase specificity is complementary to that of radiation, which kills cells in phase G_1 (as well as in M) and may be a factor in the reported additive interaction between the modalities (Westra and Dewey 1971). Hyperthermia also has been shown to inhibit progression through each phase of the cell cycle in vitro (Westra and Dewey 1971; Sapareto et al. 1978) and, in particular, to cause accumulation of cells in phase $G_2 + M$ (Kal, Hatfield, and Hahn 1975; Kase and Hahn 1975; Schlag and Lücke-Hühle 1976). Some in vivo data also indicate a possible role for cell kinetics in tumor destruction. The heat sensitivity of tumors alters with tumor volume (Dickson and Ellis 1976; Urano et al. 1980), and this alteration may be caused by volume-related changes in population kinetics of tumors (see previous section, this chapter, on tumor volume and thermal response). Also, Muckle and Dickson (1971) showed that multiple dose heating of the VX-2 carcinoma within the T_C of the tumor (one hour at 42°C on three successive days) was more effective than one single dose of heating for three hours at 42°C. This indicated some modulation of tumor parameters by the initial heating dose, inducing sensitivity to the subsequent doses. Altered distribution of cells in the mitotic cycle or recruitment of Q cells might be indicated. The role of cytokinetic factors and volume in tumor sensitivity to one hour at 42°C therefore was investigated in the Yoshida sarcoma at two different growth stages.

Cytokinetic Control Mechanisms in Yoshida Foot Sarcoma Growth

The growth curve of the Yoshida sarcoma is shown in figure 5.17. Following an initial lag period until five

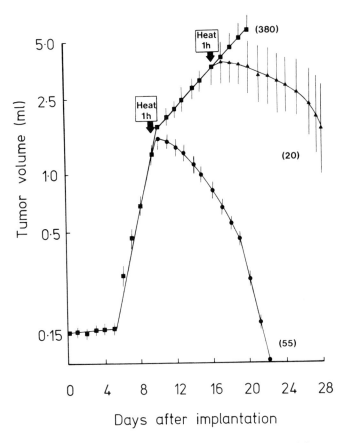

Figure 5.17. Growth curve of the Yoshida sarcoma (■) and tumor-volume changes after curative hyperthermia (intratumor temperature 42°C for one hour) on day 9 after implantation (1.0 to 1.5 ml, ●) and on day 16 (3.0 to 3.5 ml, ▲). Vertical bars indicate SD. During heating of the 1.0- to 1.5-ml tumors, intraabdominal temperature of the rats remained within the normal range (36.8°C to 39.5°C); with the 3.0- to 3.5-ml tumors, body temperature usually increased to 39.5°C to 40.5°C during hyperthermia. (Calderwood and Dickson 1980a. Reprinted by permission.)

days after implantation, a period of rapid exponential growth ensued between days 5 and 10, when the tumor volume increased from 1.5 to 2.0 ml, with a doubling time (T_D) of 36 hours. Between days 10 and 20, slower exponential growth occurred as tumor volume increased from 2.0 to 5.0 ml, with a T_D of 144 hours. Cell kinetic studies were carried out at two different volumes: 1.0 to 1.5 ml (rapid exponential growth), and at 3.0 to 3.5 ml (slower exponential growth).

The cause of the deceleration of tumor growth was investigated in relation to the main factors controlling the dynamics of cell population increase: cell cycle time (T_C), growth fraction (GF), and cell loss factor (ø). These factors were determined using the PLM and stathmokinetic techniques described in the section on cell population kinetics.

Cell cycle and phase times in the large and small tumor were essentially similiar, with a cell cycle time of approximately 14 hours in each case (table 5.5). The decreased growth rate of the 3.0- to 3.5-ml tumor compared to the rate of the smaller tumor (doubling time 144 hours compared to 36 hours) appeared to be mainly a consequence of a reduced proportion of proliferating cells, with the growth fraction reduced by almost half (from 67.8% to 39.6%). An increase in cell loss also occurred in the larger tumor (75.9% compared to 59.0% in the smaller tumor). The maintenance of T_C as a constant factor has been observed previously in several studies (Watson 1976; Aherne, Camplejohn, and Wright 1977; Steel 1977), although in some tumors an increase in T_C with volume occurs, caused mainly by an increase in duration of the G_1 period (Simpson-Herren 1977).

Increase in tumor volume in Yoshida sarcoma from between 1.0 and 1.5 ml to between 3.0 and 3.5 ml was accompanied also by a greater degree of heterogeneity in [^{3}H]-TdR labeling between different parts of the tumor (table 5.6). Labeling index in the 1.0 to 1.5-ml tumor underwent an approximate 50% decrease from 73.7% labeled cells at the periphery to 37.6% in the center of the tumor (table 5.6). In the larger tumor, a progressive decline in labeling was observed, from 49% at the periphery to less than 10% of this value (3.9% labeled cells) in areas adjacent to the central necrotic zones. The 3.0- to 3.5-ml tumor was characterized by broad areas of central necrosis bordered by those zones largely devoid of [^{3}H]-TdR-labeled cells or mitoses. In the smaller tumor no necrosis could be seen on macroscopic examination. Small areas of up to six pyknotic cells per field were, however, observed in some sections examined at higher magnification.

Table 5.5
Cell Kinetic Parameters of the Yoshida Sarcoma

Cell Cycle Times

Tumor Volume (ml)	Cell Cycle Time (hr)	Cell Cycle Phase Times			
		T_{G2}	T_S	T_{G1}	T_M
1.0–1.5	14.1	4.0	9.7	—	0.4
3.0–3.5	13.8	3.2	10.2	—	0.6

Cell Kinetic Parameters

Tumor Volume (ml)	Tumor Doubling Time (hr)	Growth Fraction (GF)	Cell Loss Factor (ø)
1.0–1.5	36	67.8	59.0
3.0–3.5	144	39.6	75.9

NOTE: Cell cycle times were determined by visual inspection of percentage labeled mitoses curves shown in figure 5.8. Computed kinetic parameters were derived as follows:
Growth fraction (Lala 1971)

$$= \frac{LI}{\left(\left\{\exp\left(T_S \frac{\ln 2}{T_C}\right) - 1\right\} \exp\left(T_{G_2+M} \frac{\ln 2}{T_C}\right)\right)},$$

where growth fraction = percentage proliferating cells in the population; LI = thymidine labeling index or percentage of cell population incorporating thymidine one hour after injection. Cell loss factor (ø), or percentage of cell loss from the proliferating population (Steel 1968), was determined as follows:

$$ø = \frac{K_L}{K_B},$$

where K_L = rate of cell loss from population

$$= \frac{\ln (1 + GF)}{T_C} - \frac{\ln 2}{T_D};$$

K_B = cell birth rate, measured as rate of entry into mitosis using vincristine for mitotic arrest (fig. 5.9).

Hyperthermia and Cell Kinetics in the Yoshida Sarcoma

The effect of hyperthermia (one hour at 42°C) on tumor volume is shown in figure 5.17. Treatment of tumors of 1.0 to 1.5 ml caused a decrease in tumor volume and complete regression within 14 days. There was a 96% cure rate of the animals. Treatment of the tumor at 3.0 to 3.5 ml produced a restraint in growth followed by partial regression, although variation in response between tumors was considerable (fig. 5.17). No tumor increased in volume after heating, however. Hyperthermia significantly reduced the lifespan of rats bearing these larger tumors (survival time

Table 5.6
Distribution of [³H]-TdR Labeled Cells Between Areas of the Yoshida Sarcoma at 1.0–1.5 and 3.0–3.5 ml

Tumor Volume (ml)	% [³H]-TdR Labeled Cells		
	Zone 1	Zone 2	Zone 3
1.0–1.5	73.7 (±7.6)	43.5 (± 6.5)	37.6 (±8.7)
3.0–3.5	49.0 (±5.8)	24.5 (±10.5)	3.9 (±3.1)

NOTE: Flash-labeling index was determined in square areas of tumor sections (sides of square 200 μm each). Tumor hemisections were scanned from surface to interior. Zones selected for determination of percentage of [³H]-TdR labeled cells were located in
zone 1, immediately beneath normal skin;
zone 2, halfway between zones 1 and 3;
zone 3, in 1.0- to 1.5-ml tumors that included no large areas of central necrosis, zone 3 constituted the final 200 μm of tissue at the tumor interior. In the 3.0- to 3.5-ml Yoshida sarcoma, zone 3 was situated in the section of tumor immediately preceding the central necrotic mass.

Means (± standard deviations [SD]) are from four tumors at each of the volumes studied (1.0–1.5 and 3.0–3.5 ml). In each tumor, at least six regions were counted at each of the three designated zones in the tumor.

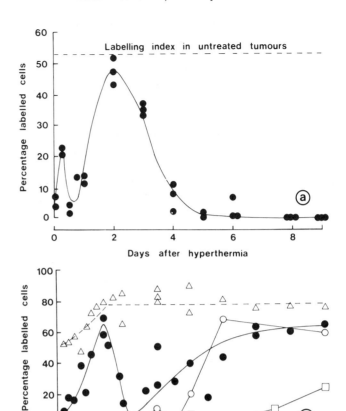

Figure 5.18. [³H]-TdR labeling in 1.0- to 1.5-ml Yoshida sarcoma after hyperthermia. A, flash (one hour) labeling index in tumors after curative heating; B, repeated labeling in unheated Yoshida tumors (△), tumors labeled 0 to 48 hours after hyperthermia (●), tumors in which labeling was terminated 14 hours after heating (□), or tumors in which labeling began at 14 hours and continued to 48 hours after heating (○). (Calderwood and Dickson 1980a. Reprinted by permission.)

27.4 ± 5.5 days compared to 44.6 ± 7.23 days in untreated tumor-bearing controls [$P < 0.001$]). At autopsy, animals that died at 27 days showed enhanced spread of tumor locally and to distant sites; the role of heat in this enhanced dissemination has been reported previously (Dickson and Ellis 1976).

The 1.0- to 1.5-ml Tumor
The effect of hyperthermia on the cell population kinetics of the 1.0- to 1.5-ml tumor was investigated by [³H]-TdR and mitotic arrest techniques (figs 5.18 and 5.19). The fraction of cells in S phase, as indicated by the flash-labeling index (fig. 5.18A), underwent a complex sequence of alterations after heating. Following initial decrease in the S phase fraction, an increase to control levels occurred by two days after heat, and this preceded a decrease to zero by seven to nine days after heat, at which time no viable cells were seen in the tumor. Cell proliferation from 0 to 48 hours after heating was examined in more detail using repeated [³H]-TdR labeling (fig. 5.18B). In controls repeatedly labeled every six hours, the labeling index increased to a plateau level of almost 80% labeled cells after eight hours. With curative hyperthermia, a complex sequence of events ensued. Labeling was depressed immediately after heating but increased rapidly to a maximum of 60% after eight hours; an equally rapid decline in labeling index to a minimum of 5% at 14 hours then preceded a final perturbed recovery to a plateau of approximately 60% labeled cells 36 hours after heating. The decrease in [³H]-TdR labeling immediately after heat appeared to be a consequence of a two- to three-hour block in the progression of cells into S phase. After release of the block, the partially synchronized cells then entered S phase, causing the maximum in labeling to occur at eight hours.

The decline in labeling to 5% at 14 hours after heat can only be due to cell death. Thus, more than 90% of cells proliferating (P cells) after hyperthermia had been lost by 14 hours, the median cell cycle time of the tumor. This would imply that cells damaged at the time of heating had progressed through one cell cycle and died, possibly in mitosis. It would seem unlikely that the rapid recovery in labeling from 14 to 36 hours is caused by cells in cycle at the time of heating. A more likely explanation is the entry into S phase of a population of cells not proliferating (Q cells) at the time of heat treatment. Further evidence in favor of this is shown in figure 5.18B. When repeated labeling was terminated at 14 hours after heat, the time of maximum cell death, only a small increase in labeling was subsequently observed. When repeated labeling was commenced at 14 hours, labeling index recovered to the 60% region by 30 hours after heat. These two experiments support the hypothesis that the recovery in proliferation from 14 hours after heat in this tumor is due mainly to cells in the Q cell compartment at the time of heat treatment (Dickson and Calderwood 1976; Calderwood and Dickson 1980a). A similar effect was noted by Lücke-Hühle and Dertinger (1977) using V-79 spheroids in vitro. Cells in the center of the spheroids had a low proportion of P cells under normal conditions. Heating at 42°C for four hours caused extensive loss from the population of cells, with a high growth fraction at the periphery of the spheroids, and appeared to stimulate the entry of Q cells into cycle. This caused an increase in the proportion of cells in S phase at the interior of the spheroids for 12 to 24 hours after heating. Kal and Hahn (1976), using a flow fluorometry technique to examine cytokinetic factors in EMT-6 tumors, also postulated recruitment of Q cells into cycle after heating in vivo.

In order to check that the effects observed on [^{3}H]-TdR labeling indicated overall changes in cell proliferation in the tumor and not merely some alteration in the properties of S phase (such as duration or rate of entry into and transit through the phase), entry into mitosis was studied also (fig. 5.19). The stathmokinetic study gave supportive evidence to the inferences drawn from the thymidine experiments. Immediately after heat, the rate of entry into mitosis (fig. 5.19) was lower and more variable than in controls (fig. 5.9). This initial blockade of entry into mitosis paralleled the inhibition of entry into S phase at this time (fig. 5.18B). This blocking effect of heat on progression around the cell cycle is a frequent finding in vitro (Bhuyan 1979). The mitotic block appeared to be released by six hours, and mitotic rate recovered partially (fig. 5.19). Recovery in M phase at this time paralleled the entry into S phase (fig. 5.18B). The high base-line mitotic index of 4% to 6% at six hours after heating (compared to 2.25% in controls) may indicate cells blocked in mitosis, possibly in the early stages of mitotic death. The decline in mitotic rate to approximately 0% at 12.0 to 14.5 hours and at 18.0 to 20.5 hours (fig. 5.19) is consistent with destruction of the P cell population from 8 to 14 hours after heating, as inferred from the repeated labeling study. By 24 hours (fig. 5.19), mitotic rate began to increase, al-

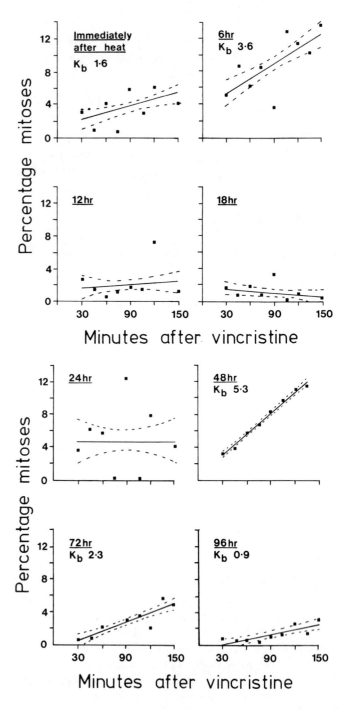

Figure 5.19. Mitotic accumulation in the 1.0- to 1.5-ml Yoshida sarcoma after vincristine arrest at different times after hyperthermia. Vincristine was injected at each time indicated (0 to 96 hours) and mitotic accumulation studied over the following 2.5 hours. The least-squares best-fit line ± SD is plotted. K_b is the cell birth rate as percentage of cells per hour, calculated as in figure 5.9. After the 12-, 18-, and 24-hour intervals, the increase in mitotic index was not significant at the 5% level, so K_b was indistinguishable from zero. (Calderwood and Dickson 1980a. Reprinted by permission.)

though great variation between tumors was observed. Recovery of mitotic rate to control levels by 48 hours (fig. 5.19) supports the inference drawn from the thymidine studies (fig. 5.18)—that proliferation in the Yoshida sarcoma had recovered essentially to control levels by 48 hours after curative heating. The progressive decline in mitotic rate after 48 hours (fig. 5.19) to 0% at six days, as the tumor regressed, followed a similar time course to the decline in S phase indicated by the flash [^{3}H]-TdR studies (fig. 5.18A). By day nine, the tumor was totally necrotic, and mitoses and [^{3}H]-TdR-labeled cells were no longer observed in sections of such tumors. It is not clear whether the reduction in the fraction of S phase cells or in the mitotic rate from two to nine days was due to inhibition of cell division and wastage by cell loss or to delayed thermal cell killing. The mitotic rate and [^{3}H]-TdR labeling index of cells in viable areas of the tumor 72 and 96 hours after heating (figs. 5.18 and 5.19) was considerably reduced compared to controls, and thus a reduced rate of cell production appeared to be at least partially implicated in destruction of the tumor population two to nine days after heating. From nine days, tumor regression was largely a reflection of resorption of dead material.

The PLM curve of the tumor 30 hours after heat treatment provided durations for the cell cycle components that were not significantly different from those of the untreated tumor (table 5.7). This indicates that changes in the tumor after curative hyperthermia are in population size and the relationship of the P and Q cell compartments, rather than in cell generation times (Calderwood and Dickson 1980a).

The 3.0- to 3.5-ml Tumor
The cytokinetic changes in the 3.0- to 3.5-ml Yoshida sarcoma after one hour at 42°C (fig. 5.20) followed a pattern similar to the alterations in the smaller tumor. Following [^{3}H]-TdR injection immediately after heating, no labeled cells were seen in the tumor (fig. 5.20A). The labeling index slowly increased from 1.0% at 12 hours to 2.5% at 24 hours, reaching a maximum of 16% labeled cells in S phase at 48 hours. There was then a progressive decrease, to a mean of 8% labeled cells at 72 hours and to 0% seven days after heating. Tumors at seven to eight days after hyperthermia were completely necrotic and contained no [^{3}H]-TdR-labeled cells or mitoses. In untreated animals repeated [^{3}H]-TdR labeling led to an increase from 28% in one hour to a plateau of about 55%-labeled cells by 18 to 24 hours (fig. 5.20B). Following hyperthermia, labeling index varied between 6% and 8% from 0 to 24 hours. After 24 hours, a slow increase occurred to 32% at 48 hours.

Table 5.7
Cell Cycle Times in the 1.0- to 1.5-ml Yoshida Sarcoma under Normal Conditions and in Cells Repopulating the Tumor after One Hour at 42°C

	T_C	T_S	T_{G_1}	T_{G_2}	T_M
Untreated tumor	14.1	9.7	—	4.0	0.4
30–54 hr after 1 hr/42°C	13.0	8.0	—	4.4	0.6

Despite the similarity in time sequence of the repeated labeling patterns after heat at each tumor volume, the cell population in the 3.0- to 3.5-ml tumor appeared to have suffered more damage. In the larger tumor, labeling did not increase significantly until 48 hours after heating, and there was no peak in labeling index at six to eight hours as in the 1.0- to 1.5-ml tumor (figs. 5.18B and 5.20B). The larger tumor at all times after hyperthermia contained large areas of heat-killed cells characterized by absence of intercellular connections and pyknotic nuclei. In the 1.0- to 1.5-ml tumor, such areas of heat-induced necrosis were absent from approximately 30 to 60 hours after heating. It is apparent, therefore, that the larger tumor with a smaller growth fraction (39.6% vs 67.8%; table 5.5) was at least as heat-sensitive as the 1.0- to 1.5-ml tumor. This would indicate that only a fraction of the Q cell population in the larger tumor was able to enter the cell cycle and repopulate the tumor. The capacity of the Q cell fraction to be recruited into the cell cycle was investigated by hemisection of the larger tumor and study of the ensuing [^{3}H]-TdR-labeling pattern (fig. 5.21).

Tumor Hemisection
The effect of hemisection (surgical resection of half the tumor) was used in the present work to study tumor repopulation after a perturbation. The use of cytotoxic drugs and radiation to produce partial destruction of the cell population was avoided in order to minimize complicating effects on the cell cycle associated with these agents (Hill and Baserga 1975; Denekamp and Fowler 1977). Animals were kept under light pentobarbitone-induced anesthesia for the duration of the experiment in order to minimize stress associated with surgery and pain. Systemic stress has been shown to enhance the growth of the Walker-256 carcinoma in rats (Van Den Brenk et al. 1976). The prolonged anesthesia had no effect on proliferation in the Yoshida tumor, as indicated by absence of any alteration in [^{3}H]-TdR labeling (Calderwood 1978). In the flash-labeling study, a gradual increase was

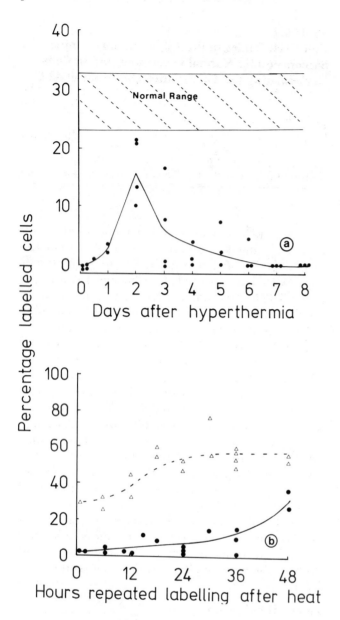

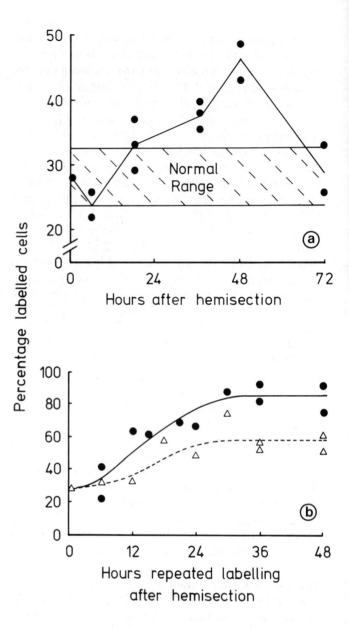

Figure 5.20. [^{3}H]-TdR labeling in the 3.0- to 3.5-ml Yoshida sarcoma after hyperthermia. *A*, flash-labeling index in tumors after curative heating. The normal range represents the mean LI (28.2% ± 4.6% labeled cells) calculated from 10 control tumors. *B*, repeated labeling in control tumors (△----△) and in tumors after heating for one hour at 42°C (●——●). (Calderwood and Dickson 1980a. Reprinted by permission.)

Figure 5.21. [^{3}H]-TdR labeling in the 3.0- to 3.5-ml Yoshida sarcoma after hemisection (surgical resection of the distal half of the tumor mass). Animals were anesthetized throughout the surgical and subsequent procedures and were fed by injections every six hours of 4.0 ml of 4% glucose in 0.9% NaCl. *A*, flash-labeling index in tumors after hemisection. The normal range represents the mean LI (28.2% ± 4.6% labeled cells) calculated from 10 control tumors. *B*, animals with untreated (△----△) and hemisected (●——●) tumors were given IP injections of 100 µCi of [^{3}H]-TdR at six-hour intervals. Control animals were anesthetized throughout the [^{3}H]-TdR labeling procedure and fed with 4% dextrose as in the treated animals. The dextrose doses were given between the [^{3}H]-TdR injections, so that labeled thymidine was injected at 0, 6, 12, 18 hours, and so on, while glucose was given at 3, 9, 15, 21 hours, and so on.

observed in the proportion of labeled cells by 24 hours after hemisection (fig. 5.21A); labeling index increased to significantly elevated levels by 36 hours (38%) and 48 hours (46%), compared to 28% in controls (fig. 5.21A). Labeling index declined to control values by 72 hours after hemisection. The repeated labeling study (fig. 5.21B) showed essentially the same effect, a gradual increase in labeling index after hemisection, reaching a plateau value of approximately 85% labeled cells by 30 hours (compared to 60% in controls).

The increase in the proportion of S phase cells in the tumor after hemisection could be due to three possible mechanisms:

1. Recruitment of Q cells into cycle
2. Decrease in the proportion of P cells decycling after mitosis
3. Decrease in cell loss from the P cell compartment

Evidence for recruitment of cells into cycle after hemisection is that cells in the interior of tumors, near the necrotic areas, which were mainly unlabeled with [^{3}H]-TdR under normal conditions (table 5.6), began to show a large increase in labeling by 30 to 48 hours after surgery.

These studies, therefore, indicate that a large number of Q cells in the 3.0 to 3.5-ml tumor possess the potential to enter the cell cycle and progress into S phase following a perturbation. Such effects have been reported following surgical excision of tumor tissue (Simpson-Herren, Sanford, and Holmquist 1976; Simpson-Herren et al. 1977). The labeling index of 46% in these tumors 48 hours after hemisection was not significantly different from the 52.5% of the smaller 1.0- to 1.5-ml tumor. The decrease in Q cell recruitment in the larger tumor after hyperthermia would seem to indicate, therefore, an increased heat sensitivity of the Q cell fraction with increasing volume. The large necrotic areas in the center of the 3.0- to 3.5-ml tumor would be likely to be bordered by zones of cells remote from the tumor microcirculation and deficient in oxygen and nutrients. Such conditions have been shown in vitro to sensitize some types of tumor cells to heat (Gerweck, Gillette, and Dewey 1974; Bass and Moore 1978; Overgaard and Nielson 1980; Gerweck, in press), and may contribute to the increased heat sensitivity sometimes found in larger tumors (Urano et al. 1980).

Thermal Cell Killing In Vivo

The killing of cells in the Yoshida sarcoma after one hour at 42°C is a complex process and may be due to a number of factors. Some of the processes believed to be responsible for thermal cell killing are

1. Direct damage to cells by physical heat. The expression of such thermal damage in terms of cell killing may be immediate (Bhuyan 1979) or delayed for several cell generations (Palzer and Heidelberger 1973a).
2. Delayed cell killing owing to effects of heat on the tumor blood supply and extracellular microenvironment (Crile 1963); Kang, Song, and Levitt 1980).
3. Long-term cell killing owing to immunologic (Crile 1963; Szmigielski and Janiak 1978) and inflammatory responses (Crile 1963; Dickson and Calderwood 1980) in tumors after heat.

It would be of benefit, especially when planning multiple-dose hyperthermic treatment, to know the effects of individual doses on the kinetics of the cell population, as this may influence subsequent response to therapy. The perturbing effects of radiotherapy on cell populations were listed by Denekamp and Fowler (1977) as redistribution, repopulation, repair, and reoxygenation. Hyperthermia in the Yoshida sarcoma led to redistribution and repopulation. Initial inhibition of progression of cells through S and M phase was followed by entry of cells into cycle (figs. 5.18 and 5.19). Although this redistribution was observed in a doomed population of cells, similar findings have been observed in vitro in cell populations that survive heat treatment (Bhuyan 1979).

Repopulation occurred in the Yoshida sarcoma, at both tumor volumes studied, from 24 to 48 hours after heating. Similar findings have been reported on V-79 spheroids in vitro (Lücke-Hühle and Dertinger 1977) and EMT-6 cells in vivo (Kal and Hahn 1976). In the Yoshida sarcoma at both 1.0 to 1.5 ml and 3.0 to 3.5 ml, tumor repopulation was initiated by the cells surrounding blood vessels. At about 15 to 20 hours after heating, when tumor repopulation was commencing (figs. 5.18–5.20), [^{3}H]-TdR-labeled cells and mitoses were observed in a circular arrangement around blood vessels. This indicates that a limiting factor in heat treatment by a single dose of hyperthermia may be tumor repopulation from Q cells from the better vascularized zones of tumors. The importance of proliferation of surviving cells around blood vessels in the recurrence of tumors after hyperthermia has been emphasized also by Overgaard and Nielsen (1980).

Tumor repopulation after other treatment modalities is associated with the other processes of repair and reoxygenation. Repair of various types of cell damage has been reported to occur after heating in vitro, although little is known about such processes in vivo. Heating does not appear to lead to reoxygenation of tumor cells in vivo, but, on the contrary, causes hypoxia, the extent and duration of which depends on the degree of heating (Bicher et al., in press). The role of extracellular effects of heat in delayed cell killing is discussed later in this chapter. Multiple-dose heating may have the dual advantage of sensitizing cells distant from blood vessels to further heating doses from microenvironmental effects of heat (ischemia, low pO_2 and pH) and destroying repopulating Q cells that border blood vessels. Another factor that may be involved in tumor response to repeated heating doses is thermal tolerance, which may be induced in tumors by an initial mild heating dose (Crile 1963; Jung, in press; Henle and Dethlefsen 1980). Induction of thermal tolerance may be prevented by a more severe initial dose of hyperthermia (Henle and Dethlefsen 1980; Jung, in press). Such a severe heat dose may in fact sensitize the tumor cells to subsequent damage at a lower temperature—the so-called *step-down phenomenon* (Henle and Dethlefsen 1980). The role of cell kinetics in these phenomena is unknown.

Value of Cell Kinetic Data in Cancer Treatment by Hyperthermia

While basic knowledge of the kinetics of cell killing by hyperthermia may be of use in deciding overall approaches to treatment, cytokinetic studies to date have given no real practical guidelines for thermotherapy. The data are mainly descriptive, and any predictions made from them are highly tentative. At most, the evidence of tumor repopulation after heat may form a basis for empirical studies on multiple dose heating.

From the point of view of predictions of thermal sensitivity of tumors, the cytokinetic approach is similarly limited. The data on the 1.0- to 1.5-ml Yoshida sarcoma indicated that P cells might be more heat sensitive than Q cells. It became apparent in the study on the larger tumor that some Q cells, perhaps those in zones remote from the microcirculation, were at least as heat sensitive as the P-cell fraction. The situation is thus highly complex, and the simple cytokinetic measurements feasible on human tumors, such as mitotic or flash-labeling index, would not be of value in prediction of thermal response.

Preferential destruction of P cells followed by recruitment of Q cells into cycle has also been reported after radiation or chemotherapy to solid tumors, and the timing of recovery has been used to plan fractionated therapy regimens (Steel 1977). Again, however, because of the problems of repeated tumor sampling and the complex nature of the situation, the approach has had little impact on the therapy of human tumors (Tannock 1978).

Role of the Tumor Microenvironment in Heat Sensitivity

The tumor microenvironment is the large, mainly stagnant, volume of interstitial fluid between the malignant cells and tumor microvasculature. The properties of the interstitial fluid ultimately are dependent on tumor blood flow. Tumor blood flow is thus the major key to changes in the tumor microenvironment.

Blood Flow and Hyperthermia

The major processes preventing temperature increase in tissues are direct heat loss into air or surrounding tissue and heat dissipation by circulating blood (Jain 1980). While the first of these factors is predictable, the second, blood flow, is likely to vary in magnitude and response to heat from tissue to tissue (Peterson 1979b). Thus, from a practical point of view, it is essential to know the relationship between tumor blood flow and thermal sensitivity.

Blood flow in rat tumors (0.41 to 1.08 ml/g dry wt/min) was of a lower order of magnitude than flow in normal tissues of high metabolic activity such as kidney, liver, and spleen (blood flow measured respectively at 12.51, 1.36, and 1.35 ml/g dry wt/min; table 5.8). Tumor blood flow, however, was higher than the perfusion rates in normal peripheral tissues such as resting gastrocnemius muscle (0.19), skin from the normal foot (0.13), and the residual tissues of the foot after removal of skin (0.04 ml/g dry wt/min; table 5.8). The overall values shown in table 5.8 probably conceal considerable intratumor variation in blood flow rates. In the D-23 carcinoma in the leg, for example, differences in flow between discrete parts of the tumor were as much as 35-fold compared with a maximal 10-fold difference in muscle (fig. 5.11). Blood flow in the larger subcutaneous D-23 flank tumor tended to be of lower values than those in the smaller subcutaneous foot tumor. In the intramuscular thigh tumor very large differences in blood flow were observed, possibly because of the persistence of normal blood vessels in

Table 5.8
Blood Flow in Normal and Malignant Animal Tissues

Tissue	Tumor Site and Volume (ml)	Rat		Rabbit	
Normal					
Liver		(15)	1.36 (±0.44)	(5)	0.84 (±0.30)
Kidney		(16)	12.51 (±4.70)	(5)	11.68 (±4.56)
Spleen		(17)	1.35 (±0.60)	(5)	4.06 (±1.07)
Gastrocnemius muscle		(16)	0.19 (±0.04)	(5)	0.38 (±0.08)
Skin		(10)	0.13 (±0.05)		—
Normal foot		(11)	0.04 (±0.02)		—
Malignant					
Yoshida sarcoma	foot, SC (1.0–1.5)	(13)	0.41 (±0.15)		
	flank, SC (7–10)	(10)	0.61 (±0.17)		
	leg, IM (5–8)	(10)	0.62 (±0.41)		
MC-7 sarcoma	foot, SC (1.0–1.5)	(6)	0.43 (±0.13)		
D-23 carcinoma	foot, SC (1.0–1.5)	(6)	0.78 (±0.19)		
	leg, IM (6–8)	(6)	1.08 (±0.32)		
	flank, SC (8–11)	(6)	0.85 (±0.38)		
VX-2 carcinoma	leg, IM (15–20)			(5)	1.52 (±0.76)

NOTE: Blood flow (ml/g dry weight/min) was measured in pentobarbitone-anesthetized animals at room temperature (20°C) using the ^{86}RB distribution technique. Mean values are given ±1 SD. Numbers of estimations are shown (in parentheses) preceding each blood flow value.

this site. Variation in muscle blood flow was considerably lower than variation observed in tumors (fig. 5.11). This nonhomogeneous flow distribution has been observed in a wide variety of tumors (Kjartansson 1976), with overall blood flow rate decreasing with increasing tumor volume (Peterson 1979b). The overall values for blood flow in table 5.8 are similar to those described in previous reports, which indicate that the range of blood flow values in experimental tumors generally occupies an intermediate position between the higher values found in actively metabolizing organs and the lower values in resting peripheral tissues (Peterson 1979b; Robinson, McCulloch, and McCready, in press). Blood flow in some tumors at the upper end of the range encountered in experimental tumors may be as high as, or higher than, flow in some actively metabolizing tissues (Straw et al. 1974; Zanelli and Fowler 1974). In the present study, the VX-2 rabbit carcinoma had a blood perfusion rate twice that of normal liver (table 5.8). Blood flow values reported for human tumors (15 to 35 ml/100 g/min) (Mäntylä 1979) are similar to the values obtained in rodent tumors and lagomorphs.

When the tumor blood flow values determined in the present study (table 5.8) were examined in relation to hyperthermia, an inverse relationship between blood flow and thermal sensitivity was observed (table 5.9). The tumors with the lowest blood flow, the Yoshida and MC-7 sarcomas, were most heat sensitive (100% cure rate with 42°C/60 min, 75% cure rate with 43°C/120 min), and the VX-2 carcinoma with the highest blood flow was most heat resistant, requiring 47°C for cure. More data is required to determine whether this relationship between blood flow and heat sensitivity is a general phenomenon. Although blood flow in the normal foot and skin was less than the perfusion rate in the Yoshida sarcoma (table 5.8), these tissues were not heat sensitive at 42°C for one hour.

The reason for this thermal resistance of normal tissues is shown in figure 5.22. Heating the normal foot at 42°C by water-bath immersion gave a 20-fold increase in skin blood flow and a 10-fold increase in flow in the residual tissues of the foot. With continued hyperthermia, blood flow was diminished, but remained four to seven times greater than the resting values. Blood flow through the 1.0- to 1.5-ml Yoshida sarcoma remained at normal levels for the first hour of heating. When the heating was continued beyond this point, a progressive decrease in blood flow to zero by three hours occurred. This finding of heat-induced vasodilation in normal tissues and its absence in tumors has been noted in several recent reports. Song

Table 5.9
Tumor Blood Flow and Thermosensitivity

Tumor Type	Host Animal	Tumor Volume	Blood Flow (ml/g dry wt/min)	Curative Heating Dose
Yoshida sarcoma (foot/SC)	rat	1.0–1.5 ml	0.41 (±0.15)	42°C/60 min (100%)
MC-7 sarcoma (foot/SC)	rat	1.0–1.5 ml	0.43 (±0.19)	43°C/120 min (75%)
D-23 carcinoma (foot/SC)	rat	1.0–1.5 ml	0.78 (±0.18)	45°C/60 min (50%)
VX-2 carcinoma (leg/IM)	rabbit	15–20 ml	1.52 (±0.76)	47°C/30 min (75%)

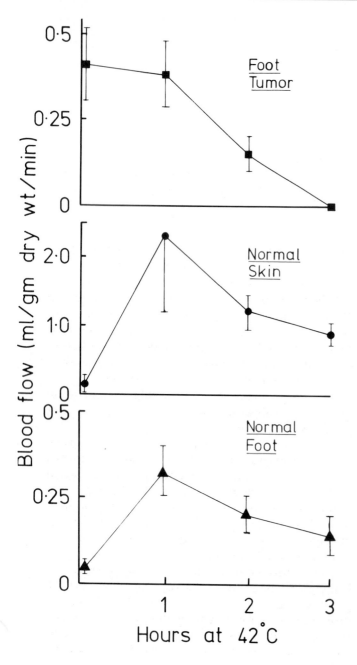

(1978, in press) and Song, Rhee, and Levitt (1980) have noted an increase in blood flow, intravascular volume, and vascular permeability in skin and muscle of Sprague-Dawley rats heated for one hour at 43°C. No blood flow increase was seen in tumors. It appears from recent data that this differential heat response of tumors and normal tissues may hold only for the range of temperatures 41°C to 45°C. At the lower end of the range, Bicher and co-workers (1980) noted a stimulation in blood flow of murine mammary adenocarcinomata at 41°C. At higher temperatures, Song (in press) found decreased blood flow and hemorrhage in skin heated to 46°C for one hour.

The recent results of Gullino (1980) are of note, since they are apparently at variance with the concept of a tumor behaving as a heat reservoir at elevated temperature. In a sophisticated tissue-isolated preparation, the tumor was implanted within the ovary of rats in vivo. The ovary was destroyed by the developing tumor, which then took over the ovarian artery and vein as its exclusive blood supply. By cannulating the artery and vein, Gullino was able to manipulate

Figure 5.22. Blood flow through the Yoshida tumor (1.0 to 1.5 ml) growing on the dorsum of the foot in rats versus blood flow through normal skin and normal foot. Blood flow for the normal foot represents values obtained from the remaining tissues of the foot (muscle, bone, footpad) after the skin was removed for measurement. Blood-flow values for the skin and other normal tissues of the tumor-bearing foot were similar to those in the normal foot, except when these tissues became involved by tumor. In this case, the normally augmented blood flow was reduced by an extent that depended upon the degree of tumor involvement. Regional blood flow was measured by the fractional distribution of ^{86}Rb in the tissues. Except where indicated, the symbols in this and the following figures represent the mean ± SD for at least five or six values at each point. (Dickson and Calderwood 1980. Reprinted by permission.)

and measure directly the tumor blood flow; oxygen and glucose use were monitored also. With Walker-256 tumors of 1.7 to 13.2 g in this system, it was found that during heating at 40°C to 43°C a consistent blood-flow response did not occur; in 10 of 20 tumors flow decreased, in 3 it increased, and in 7 it was unchanged. Nor did these flow changes correlate with tumor size. During hyperthermia, the vascular network of the tumors remained remarkably efficient, with oxygen and glucose consumption being as high as in unheated tumors. Similar findings were obtained with a mammary adenocarcinoma. The temperature of the unheated tumors was usually higher than that of the surrounding normal tissues and could not be reduced by increasing tumor blood flow. Gullino (1980) concluded that these results were contrary to the hypothesis that reduced heat dispersion owing to poor blood supply was the cause of higher tumor temperature. In these studies tumor damage was not reported, so the temperatures used may not have been tumor lethal. It would be useful to determine whether tumor regression rate and histologic damage were related to differences in blood flow in equivalent sized tumors at the same heating temperature. There are tumors that have a higher blood flow than normal tissues (VX-2 carcinoma) (Straw et al. 1974; Zanelli and Fowler 1974), and tumor blood flow may not alter greatly during heating. The critical factor for therapy is whether the normal tissues can maintain a nondamaging temperature at the elevated temperature required for tumor damage, as illustrated by destruction of the rabbit VX-2 carcinoma at 47°C (Dickson et al. 1977). Another decisive factor is whether the tumor blood flow is inhibited soon after heating (figs. 5.12 and 5.23). Gullino (1980) found tumor blood flow to be unchanged at 18 hours postheating, and the tumors were necrotic in a cortical zone of about one-third of the tumor diameter. No blood-flow data were provided on the crucial time period immediately after hyperthermia.

The pattern of blood flow following heating also differed in the Yoshida sarcoma and normal tissues. After one hour at 42°C, blood flow through the tumor fell almost to zero within an additional 60 minutes (fig. 5.12). Flow then returned slowly toward normal values over the next 12 hours. In the normal tissues, blood flow returned to resting levels by 60 minutes after heating at 42°C, and no inhibition was found at the time points studied. This long-term decrease in tumor blood flow after heating has been observed in a number of studies. Kang, Song, and Levitt (1980) found a 90% decrease in vascular volume of SCK mammary adenocarcinomata in mice five hours after heating at 43.5°C for 30 minutes, recovering to approximately

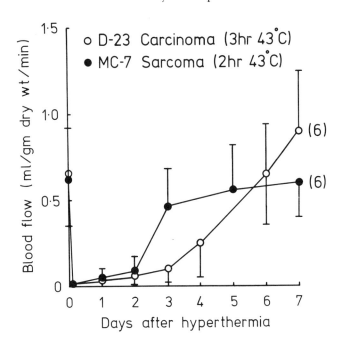

Figure 5.23. Blood flow in the D-23 carcinoma and MC-7 sarcoma determined serially after hyperthermia in two series of six rats. Blood flow was measured in each tumor at one- or two-day intervals by ^{133}Xe clearance. Mean values are plotted ± SD.

50% normal levels by 48 hours. Eddy, Sutherland, and Chmielewski (in press) found that vascular stasis and damage in a hamster tumor increased as the temperature for a 30-minute heating period was raised from 41°C to 45°C; at 45°C there was total stasis and marked hemorrhage. Blood flow inhibition in tumors after heat seems to be a general finding (Robinson, McCulloch, and McCready, in press; LeVeen et al. 1976; Storm et al. 1979; Reinhold and Van den Berg-Blok, in press; Vaupel et al., in press; Dickson and Calderwood 1980; Song, in press; Vaupel, Ostheimer, and Muller-Klieser 1980), and has been associated by Kang, Song, and Levitt (1980) with delayed thermal cell killing. This delay in cell death in tumors heated in vivo was first observed by Crile (1963), who found that cells of a mouse sarcoma could be transplanted with 100% take immediately after heating for 30 minutes at 44°C. Leaving tumors in the animal for four hours or more after heat caused a 97% decrease in take rates. Crile (1963) attributed this effect to the development of a postheating inflammatory reaction.

In the Yoshida sarcoma, one hour at 42°C cured 100% of animals but caused only a temporary (12-hour) decrease in blood flow (fig. 5.12). A similar finding was observed in the D-23 carcinoma and MC-7 sarcoma (fig. 5.23). Heating for three hours at 43°C caused an 82% regression rate of D-23 foot tumors and produced immediate 99% to 100% inhibition of blood flow. Flow resumed gradually, increasing to 50% of control levels by four days and to normal levels in all tumors by seven days, regardless of whether regression ensued. Similarly, in the MC-7 sarcoma, two hours at 43°C, which resulted in 69% cure of animals, caused initial 98% to 100% inhibition of blood flow. Blood flow in all animals returned to control levels in three to five days. Thus the effect of heat on tumor blood flow, although of importance in cell killing in the first few days after heating, does not appear to be decisive in cure of the rat tumors used in the present study.

The postheating period also has been associated with decreased extracellular pH and decreased pO_2 in some tumors (Song, in press; Vaupel et al., in press). Another effect that may be related to delayed cell killing is increased activity of lysosomal enzymes in tumors after heat (Overgaard 1976a; Overgaard and Nielsen 1980). Effects of microenvironmental changes on cell membranes after heating also could be involved (Gerner et al. 1980). The participation of normal tissue response (inflammation) in tumor regression after heating is considered later in this chapter. These various factors are linked to blood flow, but the precise role of blood flow inhibition and delayed killing in tumor regression after heating is not clear.

Tumor pH and Hyperthermia

There is now a large amount of evidence indicating that cells in vitro exposed to buffers of low pH are sensitized to hyperthermia (Overgaard and Nielsen 1980; Overgaard 1976b; Freeman, Dewey, and Hopwood 1977; Gerweck 1977, 1978, in press). Gerweck (1977) showed that the sensitizing effect took place over a range of temperatures (41°C to 44°C) and increased with decreasing pH; the effect was particularly pronounced at 42°C, the sensitization being less evident at 43°C and 44°C. Maintenance of tissue culture cells at low pH after heat also was found to increase the cytocidal effects of hyperthermia (Gerweck 1978) and to inhibit the onset of thermal tolerance (Overgaard and Nielson 1980). Recently it has been shown that the key parameter determining thermal sensitivity in vitro may be intracellular pH rather than extracellular pH (Hofer and Mivechi 1980). In cultured cells, pH_i has been shown to depend on both the pH and the composition of the incubation buffer (Hofer and Mivechi 1980; Poole, Butler, and Waddell 1964; Dickson and Oswald 1976). Thus, no assumption can be made about pH_i based merely on measurement of pH_e. Studies in which pH_e rather than pH_i is compared with thermal sensitivity may therefore result in difficulties in interpretation when different buffer systems are used. Dickson and Oswald (1976) found no sensitization of SDB rat mammary cells to 42°C at low pH_i (6.2 to 6.6). It seems, therefore, that low pH may not have a uniform effect on the heat sensitivity of malignant cells in vitro. The effect of low pH may vary between cell types at different temperatures, different pH values, and between buffer systems (Gerweck, in press; Hofer and Mivechi 1980; Dickson and Oswald 1976).

In vivo, pH values of tumors and normal tissues measured in this laboratory are shown in table 5.10. Extracellular pH (range 7.19 to 6.99) in tumors was slightly lower than pH_e in normal liver (7.32) and muscle (7.21) (table 5.9). Eden, Haines, and Kahler (1955), also using glass electrodes, found pH_e values in eight types of rat tumor to range between 6.9 and 7.3 compared to pH 7.4 in liver and muscle. Mouse tumors had pH_e values in the 6.5 to 6.9 range. Subsequent workers confirmed that in a wide spectrum of rodent tumors (Naesblund and Swenson 1953; Tagashira et al. 1953, 1954) and human cancers (Ashby 1966; Pampus 1963), tumor pH_e measured by tissue electrode was slightly, but persistently, lower than pH_e in normal tissues. Evidence of low pH_e in tumors was also provided by the studies of Gullino and colleagues (1965), on TIF samples obtained from millipore chambers implanted in rat tumors in vivo. In five tumor types pH_e ranged from 6.95 to 7.19, compared to pH 7.34 in subcutaneous interstitial fluid. As in the case of

Table 5.10
Extracellular and Intracellular pH in Tumors and Normal Tissues

Tissue Type/ Species	Extracellular pH (pH_e)	Intracellular pH (pH_i)
Tumors		
Yoshida sarcoma (rat)	(20) 7.19 (±0.13)	(48) 7.21 (±0.16)
MC-7 sarcoma (rat)	(22) 7.17 (±0.08)	(34) 7.19 (±0.15)
D-23 carcinoma (rat)	(10) 7.13 (±0.04)	(12) 6.94 (±0.29)
VX-2 carcinoma (rabbit)	(6) 6.99 (±0.15)	(12) 6.98 (±0.15)
Normal Tissues		
Diaphragm (rat)	—	(38) 6.95 (±0.16)
Gastrocnemius muscle (rat)	(20) 7.21 (±0.15)	(34) 6.85 (±0.18)
Liver (rat)	(38) 7.32 (±0.11)	(36) 7.11 (±0.04)

NOTE: Extracellular pH was measured by miniature electrode, and pH_i was determined using the DMO method. Mean values of determinations, each carried out on a different animal, are presented ±1 SD. Numbers in parentheses indicate the number of estimations (animals).

blood flow, pH_e values differed considerably between different parts of tumors (Bicher et al. 1980; Vaupel et al., in press; Eden, Haines, and Kahler 1955). The data of Eden, Haines, and Kahler (1955) indicate that hydrogen ion concentrations in some areas of tumors may be over 10 times those in areas of neutral pH.

Tumor pH_i, measured in the current study by the partitioning agent DMO, was of a similar order to or higher than pH_i in normal tissues (table 5.10). A similar observation was made by Schloerb and associates (1965), who showed pH_i in the Walker-256 carcinoma in rats to be 7.21 compared to pH_i of 6.8 in skeletal muscle. The intracellular pH of tumors thus seems at least as well regulated as pH_i in normal tissues (Schloerb et al. 1965; Calderwood and Dickson 1978; Dickson and Calderwood 1979). The extent of pH_i regulation is indicated in figure 5.24, which shows the correlation between pH_i and pH_e in the Yoshida sarcoma and normal tissues. The gradient of pH_i on pH_e (given by the first term of the equation), which gives an approximate indication of the degree of pH homeostasis in the cell water, was lower in the tumor than in the normal tissues, indicating a more effective pH_i regulation in the tumor. A similar finding of more effective buffering in tumor cells was obtained by Schloerb and colleagues (1965) comparing the Walker-256 carcinoma and muscle. Recent in vitro studies of pH_i regulation in mammalian muscle indicate an extremely tenacious system for pH_i maintenance based on a proton carrier system. The increased buffering efficiency of tumor cells in vivo may be an adaptation to existence in the adverse conditions of the tumor microenvironment (Gullino 1975). Intracellular pH in large VX-2 carcinomas (15 to 20 ml) was similar in central zones adjacent to necrotic regions and peripheral areas (table 5.11). This implies efficient pH_i maintenance even in areas remote from the microcirculation.

Table 5.12 records data on tumor pH_i and pH_e and response to hyperthermia; the two tumors with the lowest pH_i and pH_e, the D-23 rat carcinoma and VX-2 rabbit carcinoma, were the most thermally resistant, but the differences in pH values of the tumors are too small to attach significance to this. Little published information on this subject is available, and it is probably premature to judge whether tumor resting pH and thermal sensitivity are related. The subject of induced decrease in tumor pH as a thermosensitizer is treated later in this chapter.

The effect of curative heating on pH in the Yoshida and MC-7 sarcomas is shown in table 5.13. In the Yoshida sarcoma, one hour at 42°C, although curative, led to no decrease in pH_e or pH_i during heating or in the subsequent three hours. In the MC-7 sarcoma, pH_e declined from 7.17 to 6.72 after three hours. Song (in press) observed a decrease in pH_e of SCK

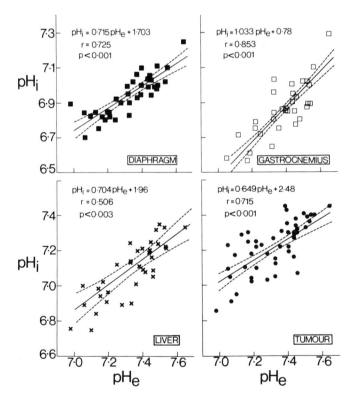

Figure 5.24. Relationship between intracellular pH (pH_i) and extracellular pH (pH_e) in diaphragm, gastrocnemius muscle, liver, and Yoshida tumor of rats. The regression line ± 1 SE, the equation to the line, and the correlation coefficient (r) are given in each case. (Calderwood and Dickson 1978. Reprinted by permission.)

Table 5.11
Intracellular pH in Necrotic and Viable Areas of the VX-2 Carcinoma in the Rabbit

	Tumor Zone	pH_i Value			
		A	B	C	Mean
Periphery	1	7.21	6.76	6.84	6.94
	2	6.89	6.73	7.12	6.91
	3	7.21	6.97	6.62	6.93
Center	4	7.16	7.22	7.00	7.13
Blood pH		7.33	7.25	7.38	7.32

NOTE: Intracellular pH was determined in 0.7- to 1.0-cm³ pieces of tumor excised from areas between the periphery (zone 1) and center of the tumor (zone 4) adjacent to necrotic tissue. Estimations were carried out in duplicate on tumors from three conscious, unanesthetized animals, A, B, and C maintained at room temperature.

Table 5.12
Tumor pH and Thermal Sensitivity

Tumor	Volume (ml)	pH_e	pH_i	Curative Heating Dose
Yoshida sarcoma (foot, SC)	1.0–1.5	7.19	7.21	42°C/60 min (100%)
MC-7 sarcoma (foot, SC)	1.0–1.5	7.17	7.19	43°C/120 min (75%)
D-23 carcinoma (foot, SC)	1.0–1.5	7.13	6.94	45°C/60 min (50%)
VX-2 carcinoma (leg, IM)	15–20	6.99	6.98	47°C/30 min (75%)

tumors in mice from 7.05 to 6.65 after one hour at 43°C; a further one hour at 43°C, 30 minutes after the first heating, caused an additional pH_e decrease to pH 6.58. A similar split-dose treatment in the Walker tumor in rats led to decline in pH_e from 7.0 to 6.65. In human tumors, Bicher and colleagues (1980) reported pH_e decreases of 0.5 to 1 unit at temperatures above 42°C. It thus seems that the magnitude of pH_e change after heating varies according to tumor type and heating conditions. The cause of the pH_e decrease is not clear but may be related to postheating reductions in the blood flow and pO_2 (Kang, Song, and Levitt 1980); pH_e decrease after hyperthermia has been implicated in delayed thermal cell killing, as discussed earlier in this chapter. Tumor ischemia after even curative heating, although prolonged, has not been permanent in the tumors studied in this laboratory (figs. 5.24 and 5.25). The exact role of decreased blood flow or any of the microenvironmental changes that stem from it (acidosis, hypoxia, nutrient depletion) has yet to be unraveled.

Tumor Nutrient Levels and Hyperthermia

There is considerable evidence to indicate that tumors are poorly oxygenated compared with normal tissues (Vaupel et al. 1973a, 1973b, in press; Thomlinson and Gray 1955). Tumor hypoxia probably is caused by sluggish blood flow (Peterson 1979b) and the high oxygen affinity of cancer cells (Gullino, Grantham, and Courtney 1967a; Vaupel et al., in press). The concentration of the other major nutrient, glucose, in the tumor environment is, under normal conditions, approximately zero (Gullino, Grantham, and Courtney 1967b). This zero glucose concentration in the tumor extracellular fluid (ECF) is maintained by almost immediate removal of free glucose by tumor cells, which have an extremely high glucose consumption (Gullino, Grantham, and Courtney 1967b; Racker 1976). No evidence is available on intratumor glucose distribution. It is, however, reasonable to assume that the sugar has a similar distribution to oxygen, tumor cells having a high affinity for both molecules. Cells

Table 5.13
Effect of Curative Hyperthermia on Tumor pH

Tumor Type	Heating Time/Temp.	Hours after Hyperthermia	No.	pH_e	No.	pH_i
Yoshida sarcoma (rat)	control	—	20	7.19 (0.13)	48	7.21 (0.16)
	1 hr/42°C	0	5	7.15 (0.10)	3	7.47 (0.04)
	1 hr/42°C	3	5	7.17 (0.13)	3	7.33 (0.14)
MC-7 sarcoma (rat)	control	—	22	7.17 (0.08)	34	7.19 (0.15)
	3 hr/42°C	0	4	6.97 (0.07)		—
	3 hr/42°C	3	3	6.72 (0.18)		—

NOTE: Mean values are given ± 1 SD. Numbers of estimations (animals) are shown in parentheses. Intracellular pH could not be measured in the MC-7 sarcoma after three hours at 42°C owing to exclusion from the tumor of the isotope mixture used in pH_i measurement. Isotope exclusion probably was caused by 99% to 100% inhibition of blood flow in such tumors.

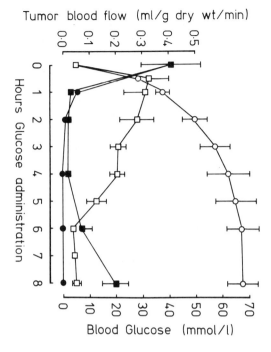

Figure 5.25. Blood glucose and blood flow after an IP glucose injection (6 g/kg; □, ■) and after glucose injection combined with IV glucose at 2 g/kg/hr (glucose infusion; ○, ●). Blood flow was measured by ^{86}Rb distribution and glucose assayed by a glucose oxidase technique (Dickson and Calderwood 1979).

remote from the tumor capillaries, especially in poorly perfused tumors, are thus likely to be deficient in both oxygen and glucose.

Early studies showed hyperthermia to have an advantage over other therapeutic modalities in that oxygenated and acutely hypoxic cells were equally heat sensitive in vitro (Gerweck, Gillette, and Dewey 1974; Schulman and Hall 1974; Kim, Kim, and Hahn 1975). It has since been shown in vitro that hypoxia maintained over a prolonged period has a significant heat sensitizing effect (Born and Trott 1978). Thermal sensitization was shown to be proportional to the degree of hypoxia (Gerweck 1978). The role of glucose deprivation in thermal sensitivity has been less extensively studied. Starved fibroblasts were shown to be more heat sensitive than fed cells in vitro (Rottinger 1976). There is also evidence that glucose deprivation acts synergistically with hypoxia in sensitizing cells to heat (Gerweck, in press). Additionally, 5-thio-D-glucose, a D-glucose analog, potentiates hyperthermia in hypoxic tumor cells, presumably by competitive binding to carrier proteins or enzymes involved in energy production from glucose (Kim et al. 1978).

No direct evidence links hypoxia with heat sensitivity in vivo. Indirect evidence is provided by arterial vascular occlusion, which has been shown to sensitize some tumors (Hill and Denekamp 1979; Suit 1976) and normal tissues (Morris, Myers, and Field 1977) to heat. Also, the increase in heat sensitivity with tumor volume observed by several workers (Crile 1963; Storm et al. 1979, in press; Urano et al. 1980) may be related to the reported increase in the hypoxic cell fraction with tumor size (Vaupel et al. 1973b; Vaupel 1979). Neither of these strands of indirect evidence is conclusive and the thermosensitizing effects observed could be due to decreased blood flow, acidosis, or metabolite deprivation. The synergistic action of heat and radiation on tumors in vivo may be due partially to heat destruction of radioresistant hypoxic cells (Overgaard 1977b; Dewey, Thrall, and Gillette 1977).

Evidence that delayed cell killing after heat may be related to oxygen deprivation has been provided by Bicher and colleagues (1980). Tumor pO_2 was found to be reduced to near zero levels in murine and human tumors after heating at temperatures in excess of 42°C. The change in pO_2 was accompanied by inhibition of blood flow and a 0.5 to 1.0 unit decrease in pH_e. Parallel results were obtained by these workers in normal tissues, although at higher temperatures (≥ 44°C). It is of note that these results are in keeping with the thermal death times graph (fig. 5.3), in which the lines for tumor and normal tissue heat damage converge in the region of 44°C.

Tumor Microcirculation[2] and Microenvironment in Thermal Response

There is thus ample evidence that the distinctive properties of the tumor microenvironment have an influence on heat sensitivity of cancer cells. Low pH, pO_2, and glucose act alone or combine together to sensitize cells to heat in vitro, particularly at 42°C (Gerweck 1978), a temperature innocuous to most normal tis-

---

[2] The generic term *microcirculation* is defined as the microscopic subdivisions of the vascular system that come to lie within the tissue proper and are exposed to the tissue milieu (extracellular fluid). The range and size of blood vessels so classified varies somewhat from tissue to tissue but in general encompasses all vessels less than 100 μm in diameter. This includes arterioles, capillaries, and venules (Richardson 1976).

sues (Dickson and Calderwood 1979). The relative importance of individual parameters, such as blood flow, pH, pO_2, and glucose supply, cannot be distinguished in most studies performed so far in vivo. The significance of the extracellular microenvironment in cancer treatment by heat may be analyzed by considering three questions: first, can the thermal sensitivity of tumors be predicted from measurements of microenvironmental factors? The initial data seems to indicate an inverse relationship between blood flow and heat sensitivity. Much more information is needed on this subject, which may be of crucial importance in selection of patients for thermotherapy. No correlation has yet been shown between any of the other parameters (pO_2, pH, glucose level) and heat sensitivity in vivo. Related to this question is the influence of tumor volume on thermal sensitivity, which was discussed earlier in this chapter.

Second, do heat-induced changes in the tumor microenvironment offer a basis for selective treatment of cancer? The finding of increased blood flow in heated normal tissues and no such response in tumors offers a basis for *selective heating* of cancer cells. This selectivity occurs over a limited range of heating doses since normal tissue circulation has an upper limit in its capacity for thermal dissipation (Jain 1980). Storm and colleagues (1979), however, were able to heat dog tumors at temperatures above 45°C, while normal tissue temperatures remained at physiologic levels. In patients, intratumor temperatures in the region of 50°C have been achieved with minimal damage to normal tissues (Dickson 1979; Storm et al., in press; Rossi-Fanelli et al. 1977).

The differential response to heat of the microcirculation of normal and malignant tissues permits a judgement of the relative merits of current heating techniques. The three main methods of heating are (1) local perfusion with heated blood (Cavaliere et al. 1967), (2) immersion in warm fluid (Henderson and Pettigrew 1971), and (3) heat induction by radiant energy, as by RF (Dickson 1979). The two main routes of heat dissipation permitting tissue cooling are thermal conduction into surrounding matter and heat removal by circulating blood. Thus, local perfusion is inferior to the other heating modes in that no cooling can take place via circulating blood. This in effect means that normal tissues are selectively perfused with heated blood because of their increased blood flow (see fig. 5.22); tumors may be spared if there is a stable or decreased blood flow during heating. The tumor still will receive heat from the surrounding normal tissues, but even if blood flow is impaired at the operating temperature, the tumor may not heat differentially. The end result with this type of hyperthermia is that selective tumor heating is unpredictable, and normal tissue temperature is the limiting factor. These considerations are borne out in practice. An analysis of the temperatures achieved by Cavaliere and associates (1967) in hyperthermic limb perfusion indicate that tumor temperature may run at a level slightly higher or lower than, or be similar to, that achieved in surrounding normal muscle (for a detailed analysis of these temperature differentials see Dickson 1979). In immersion techniques, heat cannot be lost to the air by skin and other tissues surrounding treated tumors, and again the simultaneous elevation of temperature in these normal tissues is a limiting factor, both for host damage and in preventing selective tumor heating. The use of electromagnetic or mechanical radiations, although still at an experimental stage, has advantages over the other modes, offering the prospect of true selectivity in tumor heating because of the possibility of normal tissue cooling via both conduction and blood flow.

Third, does thermal destruction of the tumor population occur by destruction of tumor microcirculation and microenvironmental changes? Although hyperthermia has been shown to cause inhibition of blood flow and vascular damage, these changes may only be temporary, as illustrated above for rat tumors. Cell killing after the resumption of blood flow is likely to be due to host defense factors. In tumors of doubtful immunogenicity, such as the majority of human cancers (Dickson and Shah 1980), it would thus seem that a single heating dose of 42°C to 43°C would not eradicate all or even most tumor cells. This may be the reason for the reports of recurrence of tumors apparently destroyed by local or total-body hyperthermia (Dickson 1978). It seems probable, therefore, that in nonimmunogenic tumors, the major damage is caused by initial heat-shock killing, and delayed death is caused by microenvironmental alterations in the heated, ischemic tumor. On the other hand, Storm and colleagues (1979, in press) found that most of the human tumors that could not be heated to 45°C displayed physiologic adaptation to heat similar to that of adjacent normal tissues. At temperatures of 45°C and above, in contrast to 42°C to 43°C (Dickson 1977; Overgaard and Overgaard 1976), striking changes occur in the stroma of tumors, with destruction of the microvasculature and thrombosis (LeVeen et al. 1976, 1980; Storm et al. 1979, in press); repeated heatings have produced total necrosis of the vasculature in a range of human cancers (LeVeen et al. 1980; Storm et al. 1979). Multiple heatings at temperatures of 45°C and above may therefore offer a means of more adequate or even total tumor control. Alternatively, combination of moderate hyperthermia (42°C to 43°C) with other modalities, such as radiotherapy (Overgaard and Nielsen 1980) or chemotherapy (Hahn 1978; Stehlin 1980) might be effective for eradication of the common solid human tumors.

It is informative to note that we were preempted by numerous earlier workers in our current ideas concerning the effects of heat on the dynamics and structure of tumor vasculature. It was well recognized in the era of tumor biotherapy that Coley's toxins contained a polysaccharide component (tumor-necrotizing fraction) that caused tumor hemorrhage and necrosis in animals and in humans. The vascular damage was limited to tumors, normal tissues not being affected (see review by Dickson 1977). From studies on mouse tumors injected with the polysaccharide, Klyuyeva and Roskin (1963) concluded that the necrosis produced in cancer cells was the result of breakdown of the tumor vascular network. Algire, Legallais, and Park (1947) believed that the hemorrhage and destruction that occurred in sarcomas following bacterial polysaccharide injection resulted from slowing of blood flow and development of stasis in vessels of both the tumor and the surrounding tissues of the host.

It is apparent from the conclusions of these various workers that the tumor destruction following injection of bacterial polysaccharide was secondary to an impaired microcirculation; that is, the effector of cancer cell necrosis was deleterious change(s) in the tumor microenvironment.

Potentiators of Hyperthermia

The major role of hyperthermia in the treatment of clinical cancer is currently envisaged as an adjuvant to radiotherapy and chemotherapy. These two combinations form subjects in themselves and are discussed elsewhere. This section will concentrate on potentiators of heat that exploit some of the factors governing tumor response to hyperthermia in vivo. These factors may operate chiefly in the tumor itself (*local* factors), or they may be generated systemically and thus represent more directly the participation of the host (*systemic* factors).

Local Factors

Cell Kinetics and Multiple-Dose Heating

Hyperthermia kills cells preferentially in S phase (Sisken, Morasca, and Kibby 1965; Westra and Dewey 1971; Palzer and Heidelberger 1973b). One theoretical approach to treatment thus might be to heat tumors after the use of a synchronizing agent that interferes with cell cycle progression by metabolic blockade. The most effective agents cause an accumulation of cells at the G_1/S barrier, with subsequent synchronization of the population as it enters S phase. This apparently would favor the use of hyperthermia. There is, however, little convincing evidence that response to cancer treatment in animals can be manipulated by altering the age distribution of cells in a tumor (Tannock 1978); nor are there reliable techniques available for inducing synchrony in vivo (Steel 1977). Even with effective methodology, the broad range of intermitotic times in solid tumors and the rapid occurrence of posttreatment desynchrony militate against this approach (Steel 1977). The application of cell synchrony in vivo has been reviewed by Tubiana, Frindel, and Vassort (1975). It may be concluded that while there is some evidence that synchrony procedures can usefully increase the proportion of cells vulnerable to therapy in leukemias, this has been harder to demonstrate in solid tumors, even in rapidly proliferating animal tumors. One feasible approach that does exploit cell kinetic factors is to combine the cycle specificity of hyperthermia (S, early G_2, and M) (Palzer and Heidelberger 1973b; Bhuyan 1979) with that of radiation (G_1 and M) (Dewey, Furman, and Miller 1970; Dawson et al. 1973). The synergism between these two modalities thus may be a consequence of complementary phase specificities; however, thermal killing of hypoxic radioresistant Q cells probably is involved also.

A common finding in studies of radiotherapy and chemotherapy has been the existence of resistant subpopulations of cells (Valeriote, Bruce, and Meeker 1978; Fowler, Thomlinson, and Howes 1970). From these resistant cells tumors may be repopulated (Hill and Baserga 1975). Recruitment of cells into cycle occurs after hyperthermia in vivo (figs. 5.18 and 5.20). Therefore, repeated heating doses could have the advantage of killing formerly resistant cells recruited into the cell cycle. Repeated heating was more effective in treating the VX-2 carcinoma in rabbits than the same treatment dose given singly (Muckle and Dickson 1973, 1971). A similar finding in mouse mammary tumors has recently been reported by Henle and Dethlefsen (in press). Some of the data, however, which showed that one treatment of 20 minutes at 45°C followed by 40 minutes at 42°C (step-down heating) was more effective than three fractionated doses of either 180 minutes at 42°C or 20 minutes at 45°C, indicated a complex situation; differential effects of the treatments on the development of thermal tolerance, repair of hyperthermic damage, and the state of the extracellular microenvironment may be involved. It must be remembered that hyperthermia differs from other modalities in killing Q cells from poorly perfused zones of tumors (Calderwood and Dickson 1980a; Overgaard and Nielsen 1980). Thus, multiple-dose heating may be advantageous in both killing cells recruited into S phase and in destruction of the tumor circulation with resulting ischemia. Tumor ischemia may be cytotoxic in itself (Kang, Song, and Levitt 1980) and may sensitize tumors to further heating doses. These considerations also bear upon the approach of multiple fractions of heat combined with radiotherapy, since although heat may remain effective in repeated doses, the damage to the microcirculation and consequent tumor cell hypoxia would be expected to render successive fractions of irradiation less efficient for tumor eradication.

The Tumor Microenvironment

Hypoxia, acidosis, and low glucose level have been shown by studies in vitro to sensitize tumor cells to heat (Gerweck, in press). The nature of the extracellular fluid of tumors in vivo at any given time depends in large measure on blood flow (Gullino 1975; Vaupel 1979); reduction in blood flow thus is likely to lead to the conditions above, thereby potentiating hyperthermia (Gerweck, in press). Since blood flow is also the major vehicle of heat dissipation from tumors, the most direct methods of potentiating the effects of heat upon tumors concern inhibition of tumor blood flow, the rationale being to selectively convert the tumor into a heat reservoir.

Alteration of Tumor Blood Flow

Elevated Blood Sugar Level It has recently been shown that high blood sugar levels lead to a selective inhibition of tumor blood flow (Calderwood and Dickson 1980b, in press; Dickson and Calderwood 1980; Dickson et al. 1980; Von Ardenne and Kruger 1980). The effect of glucose administration on blood glucose and blood flow in the Yoshida sarcoma is shown in figure 5.25. Injection of glucose at a dose of 6 g/kg body weight caused an immediate increase in blood glucose concentration, which remained elevated (20 to 30 mM/L) for four hours, returning to control levels in six hours. The curve of tumor blood flow inhibition was roughly the mirror image of glucose increase, flow decreasing to near zero levels within one hour and remaining inhibited until four to six hours after glucose injection. Using glucose infusion, blood glucose could be maintained at levels above 50 mM/L for as long as required (fig. 5.25). Blood flow in tumors declined rapidly to zero by two hours, and remained inhibited for the duration of the infusion. Hyperglycemia had no effect on blood flow in normal tissues (Dickson and Calderwood 1980). Blood flow was also investigated by arteriography, and the selective decrease in tumor blood flow induced by hyperglycemia was confirmed by the results illustrated in figure 5.26. Inhibition of blood flow at high blood glucose levels also occurred in other rat tumors (fig. 5.27). Glucose infusion under the conditions described for figure 5.25 caused total inhibition of blood flow in the MC-7 sarcoma and D-23 carcinoma after two-hour hyperglycemia; flow remained at zero level for the duration of the infusion. Hyperglycemia caused blood flow inhibition in each tumor type at a range of sites (subcutaneous tumors in foot or flank; intramuscular tumors in thigh). Blood flow also was inhibited to a similar extent at a range of volumes (1.5 to 10.0 ml) in each of the tumors. This effect was not confined to tumors in rats, but also occurred in the rabbit VX-2 carcinoma. Glucose infusion for four to six hours (blood glucose values ranging between 70 and 100 mM/L) caused an 85% decrease in blood flow in the VX-2 carcinoma. Blood flow in the rabbit tumor was thus less susceptible to blockade than were the rat tumors in which complete inhibition of blood flow occurred; this might indicate some species difference in the effect.

To determine whether the hemostatic effects of hyperglycemia were due to tumor glucose metabolism, the experiments were repeated using galactose, a sugar not metabolized by the tumors. Galactose injection or infusion had a marked inhibitory effect on blood flow in the Yoshida tumor, with the increase in blood galactose mirroring the decline in blood flow (fig. 5.28). The effects of galactose on blood flow were essentially similar both qualitatively and quantitatively to those of glucose. High blood galactose had a similar inhibitory effect on blood flow in the MC-7 sarcoma and D-23 carcinoma. These effects of high levels of blood sugar on tumor blood flow, therefore, were specific to tumors, and were independent of tumor type, site, volume, and the species of the host. The cause of the blood-flow inhibition, although not precisely known, is probably related to alterations in blood viscosity in the tumor microcirculation during hyperglycemia. This aspect has been extensively discussed elsewhere (Calderwood and Dickson 1980b; Dickson and Calderwood 1980). Sugars, therefore, are potent inhibitors of tumor blood flow and represent a useful adjuvant of hyperthermia; glucose, as a physiologic substance, has further appeal. The effect of hyperglycemia and hyperthermia (one hour at 42°C), applied singly and in combination, on blood flow in the Yo-

Thermosensitivity of Neoplastic Tissues In Vivo

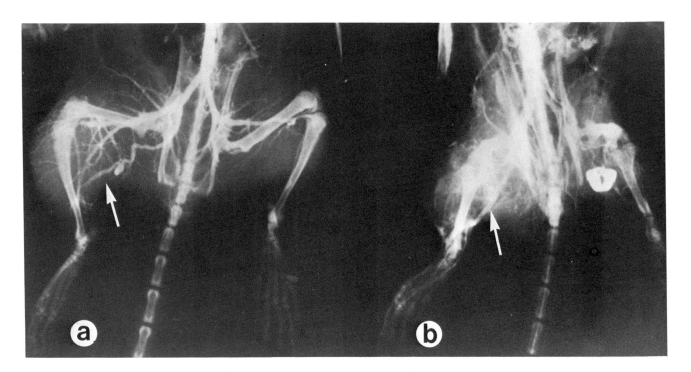

Figure 5.26. Arterial blood supply in normal and Yoshida sarcoma-bearing *(arrow)* hind legs of untreated rat, *A*, and after four-hour hyperglycemia (blood glucose 52 mM/L), *B*. The network of vessels supplying the tumor is visualized in *A*, while only the main arterial trunk is clearly identified in *B*. Blood vessels were given injections of contrast medium (Hypaque) prior to x-ray photography. The metal clip used to seal the left femoral vein, used for the infusion, is seen in *B*. The different anatomic position of the pelvis in *A* and *B* resulted from uncontrollable reflex spasm of the animals' hindquarters owing to the irritant nature of the contrast medium.

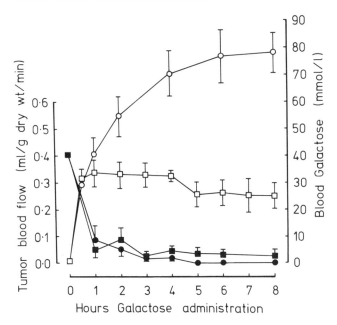

Figure 5.27. Inhibition of blood flow in the MC-7 sarcoma and D-23 carcinoma during glucose infusion (infusion conditions and assays as described in legend, fig. 5.25).

Figure 5.28. Blood galactose and blood flow after an IP galactose injection (6 g/kg; □, ■) and after galactose injection combined with IV galactose infusion at 2 g/kg/hr (galactose infusion; ○, ●). Blood flow was measured by the ^{86}Rb technique and blood galactose by oxidation with galactose dehydrogenase.

111

Table 5.14
Effect of Hyperglycemia and Hyperthermia on Blood Flow in the (1.0- to 1.5-ml) Yoshida Sarcoma

Treatment	Blood Flow (ml/g/min)
Control	0.405 (±0.151)
1 hr/42°C	0.311 (±0.110)
Hyperglycemia	0.030 (±0.011)
Hyperglycemia + 1 hr/42°C	0.036 (±0.010)

NOTE: Blood flow in hyperglycemic animals was measured four hours after IP injection of glucose (6 g/kg). In the combined treatment, tumors were heated for one hour at 42°C after four hours of hyperglycemia.

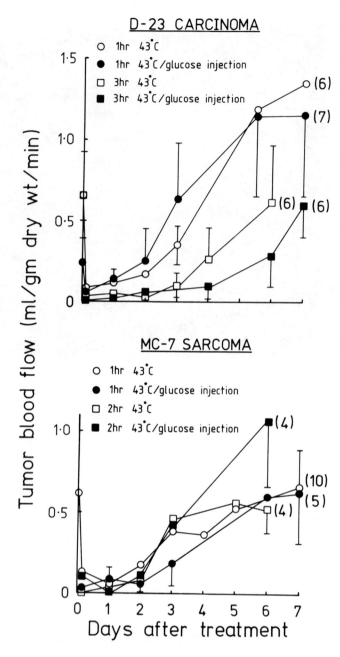

Figure 5.29. Blood flow in the D-23 carcinoma and MC-7 sarcoma after hyperthermia applied singly or after a four-hour period of hyperglycemia (6 g glucose/kg rat). Blood flow was measured sequentially in groups of tumor-bearing animals *(numbers indicated in parentheses)* at one- to two-day intervals after treatment using the ^{133}Xe clearance technique.

shida sarcoma immediately after treatment is shown in table 5.14. Hyperthermia had no significant effect on tumor blood flow. Hyperglycemia inhibited blood flow by 90%, and this decrease was not altered by addition of hyperthermia. It has been claimed that hyperglycemia combined with moderate hyperthermia (25 mM/L glucose and two to three hours at 43°C) leads to irreversible inhibition of blood flow (measured by Evans blue infusion technique) in the DS-carcinosarcoma (Von Ardenne and Reitnauer 1980). Such an irreversible effect was not observed with the Yoshida sarcoma (in which one hour at 42°C cured 100% of treated animals). The curative dose of heat with or without hyperglycemia (fig. 5.12) caused a prolonged (12- to 24-hour) but reversible inhibition of blood flow. Nor did irreversible inhibition of blood flow occur in the MC-7 sarcoma or D-23 carcinoma heated for longer periods (two to three hours at 43°C after hyperglycemia (fig. 5.29). In the MC-7 sarcoma, one hour at 43°C and two hours at 43°C with or without hyperglycemia caused short-term (two- to three-day) but reversible decrease in blood flow (fig. 5.29). In the D-23 carcinoma, one hour and three hours at 43°C caused a more persistent blood-flow inhibition, which was not markedly affected by hyperglycemia, but again was followed by resumed normal blood flow in three to six days (fig. 5.29).

Tumors in hyperglycemic animals required a lower heating dose for hyperthermia and were maintained at a more homogeneous temperature during therapy (figs. 5.30 and 5.31). In the Yoshida sarcoma in normoglycemic rats, the temperature gradient between the water bath at 43°C and tumor was approximately 1°C (i.e., tumor temperature at 42°C) for the first 40 minutes of heating (fig. 5.30). This gradient decreased to 0.7°C as the heat treatment was contin-

ued to 50 and 60 minutes. In tumors heated at four hours after glucose injection at a dose of 6 g/kg, the temperature gradient was reduced to a mean of approximately 0.2°C throughout the 60-minute heating period. Following glucose infusion for four hours, the temperature gradient between tumor and bath was

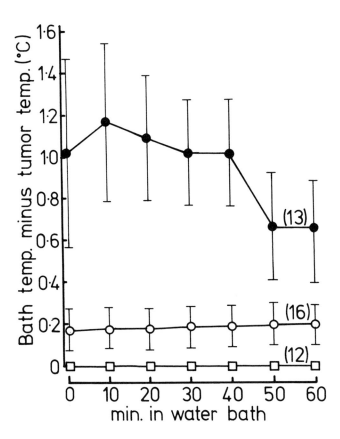

Figure 5.30. Temperature gradient between Yoshida sarcoma and water bath (at 43°C) in control, normoglycemic animals (●), after a single IP glucose injection at 6 g/kg (○), or after IV glucose infusion (□). (Calderwood and Dickson 1980b. Reprinted by permission.)

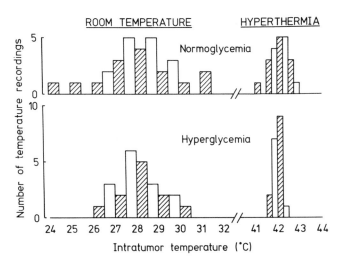

Figure 5.31. Distribution of temperature in the 1.0- to 1.5-ml Yoshida foot sarcoma in normoglycemic and hyperglycemic (6 g glucose/kg rat for four hours) animals. Temperature was measured in unheated animals and after local tumor heating at 42°C (intratumor temperature) for 30 minutes. Thermocouples were used to measure temperature in three regions of each tumor. The sensors were implanted vertically to a depth of approximately 3 to 4 mm, in distal, central, and proximal zones of the tumor and separated by approximately 3 mm along the longitudinal axis of the neoplasm. Two groups of rats (normoglycemic and hyperglycemic group), each consisting of 10 animals, were used. The three temperature values from each animal were pooled; this resulted in a total of 30 intratumor temperature readings per treatment group, and from this aggregate data the temperature distribution histograms were plotted.

abolished (fig. 5.30). The range of temperatures observed in Yoshida sarcomas under normal and hyperglycemic conditions is shown in figure 5.31. In unheated tumors, a wide range of values (approximately 26°C to 30°C) was obtained under normal and hyperglycemic conditions. In hyperglycemic animals, this variation was due largely to differences between animals (intertumor difference). The mean temperature difference between discrete parts of the same tumor[3] (intratumor difference) was 0.36°C compared with 1.36°C in normoglycemic hosts. For hyperthermia at a mean intratumor temperature of 42°C, hyperglycemic tumors required a lower bath temperature (fig. 5.30). Heating under hyperglycemic conditions was more uniform, tumors being heated at 42°C ± 0.09°C, compared with 42.2°C ± 0.39°C in normoglycemic animals

(fig. 5.30). Variation in intratumor temperature could be completely abolished if glucose was infused under the conditions described in figure 5.30. Uniformity of tumor heating, particularly in the 41°C to 43°C temperature range, may be important in determining the response to hyperthermia. It has been shown in vitro that in the 41°C to 43°C range, an increase in temperature of the order of 0.1°C causes a large increase in cell killing (Gerweck 1978). It is recognized that inhomogeneous heating is a serious drawback with virtually all current heating techniques. Thus, heating under more uniform conditions might allow the elimination of cells in the cooler zones around blood vessels (Overgaard and Nielsen 1980). The use of hyperglycemia to inhibit blood flow also might be indicated in heating tumors with a rapid blood flow or in the use of higher temperatures. Under the conditions of specifically reduced tumor blood flow during hyperglycemia, the heat input could be reduced, giving greater selectivity in tumor treatment.

---

[3] This intratumor temperature difference was calculated by subtracting the lowest recorded intratumor temperature from the other recordings and taking the mean.

Decreased Blood Pressure Tumor blood supply is composed of newly formed vessels and the host vessels surviving in the tumor (Gullino 1975). The remaining host vessels may retain contractile and nervous apparatus capable of responding to physiologic stimuli (Gullino and Grantham 1961). Tumors in general, however, are characterized by an absence of adrenergic innervation (Mattson et al. 1979). Tumor blood flow can be decreased using vasoconstrictor drugs such as adrenalin and noradrenalin (Cater, Grigson, and Watkinson 1962; Kjartansson 1976; Mattson et al. 1979). The general effect, however, seems to be a passive one, depending on blood-flow decreases through surrounding normal tissues; tumor blood vessels per se appear to be relatively inert channels (Mattson et al. 1979). Similarly, vasodilator drugs by reducing blood pressure may decrease circulation through the tumor (Cater, Grigson, and Watkinson 1962). Such effects, however, may be counteracted by physiologic responses—for example, vasodilation leads to a compensatory increase in pulse rate to maintain cardiac output. The tumor vasculature is usually regarded as a high resistance bed that is sensitive to some extent to changes in perfusion pressure, and the precise response to a given drug will be governed by several unknowns, including the contribution of host vessels. The most apparent drawback to employing such drugs is that the effect on a tumor is unpredictable. In older patients, strong vasoactive agents also may be hazardous, owing to systemic hypo- or hypertension. Little work has been described on the use of vasoactive agents in clinical hyperthermia, and although LeVeen and co-workers (1980) claim that in human tumors the heating effect can be intensified by hypotension produced by administration of sodium nitroprusside, no supporting data is provided.

Tumor Embolization A promising approach to the manipulation of blood flow in some types of tumor is embolization of arteries afferent to the tumor site. A technique for this purpose was recently described by Wallace and colleagues (in press), who used gelfoam particles to embolize the peripheral arterial bed and stainless steel coils to occlude the central vessel of renal carcinomata in human patients. This procedure resulted in tumor ischemia. The procedure has proved valuable in providing a relatively bloodless operating field for the surgeon removing highly vascular tumors of the kidney. These workers pointed out, however, that it was impossible to completely infarct the kidney or renal carcinoma, which continued to be supplied by some peripheral vessels. The usefulness of this technique to manipulate blood flow in organs other than the kidney is, however, largely untested. Neoplasms growing in uncapsulated tissues are unlikely to be supplied by a single major blood vessel, and blood perfusion from a multiplicity of sources appears to be the rule in most tumors (Warren 1979). Embolization of such tumors might entail nonspecific occlusion of blood supply to the whole region of tissue in the vicinity of the tumor, but may prove suitable in selected patients for reducing tumor blood flow prior to hyperthermia.

Tumor pH

It has been known for many years that tumor pH_e can be reduced from normal levels (7.1 to 6.9) to the 6.7- to 6.0-region by injection of large doses of glucose (Voegtlin, Fitch, and Kahler 1935; Kahler and Robertson 1943; Eden, Haines, and Kahler 1955). Thermosensitivity measurements in vitro indicate that low pH is a powerful sensitizer of cells to damage by hyperthermia (Gerweck, in press). Von Ardenne (1971, 1972) has proposed the combination of hyperglycemia, to reduce pH, with hyperthermia for the treatment of cancer in vivo (fig. 5.10).

The effect of glucose (6 g/kg) injection on pH_e and pH_i in normal and malignant animal tissues studied in this laboratory is summarized in figure 5.32. No decrease in pH_i or pH_e occurred in the normal tissues following glucose injection. Three of the four tumor types investigated, the Yoshida sarcoma, MC-7 sarcoma, and D-23 carcinoma, showed a decrease in pH_e after glucose injection (fig. 5.32). Extracellular pH declined to a minimum of 6.63 in the Yoshida sarcoma in four hours, and to 6.1 and 6.05 in the D-23 and MC-7 tumors in five and six hours, respectively. In each tumor, pH_e returned to normal levels within 24 hours of glucose administration.

In only one of the tumors, the MC-7 sarcoma, was a decrease in pH_i observed; pH_i fell from 7.18 to 6.52 in seven hours, returning to normal levels in 24 hours. In the other tumors, pH_i remained within normal limits, with some fluctuation, or increased (D-23) during the period of hyperglycemia. Maintenance of pH_i at normal levels during hyperglycemia was observed previously in the Walker-256 carcinoma in vivo (Schloerb et al. 1965). The same tumor in vitro showed a marked increase in pH_i in the presence of high glucose in the culture medium (Hult and Larsen 1976). In Ehrlich ascites cells in vitro, addition of glucose (11 mM/L) caused rapid accumulation of lactic acid in the medium; pH_e decrease (7.35 to 6.70) was more than twice the decrease in pH_i (7.17 to 6.95). Thus, although hyperglycemia has led to a decrease in pH_e in the majority of tumors (Eden, Haines, and Kahler 1955; Von Ardenne 1972; Dickson and Calderwood 1979), intracellular acidosis is not a common

Thermosensitivity of Neoplastic Tissues In Vivo

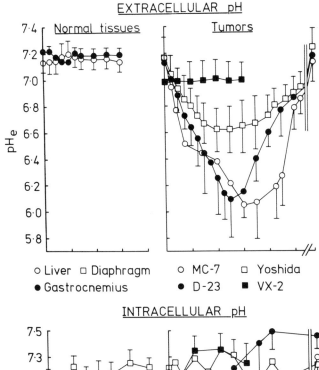

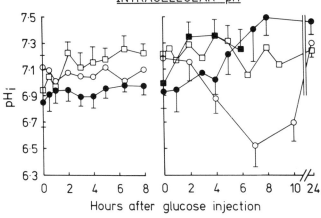

Figure 5.32. Extracellular (pH_e) and intracellular pH (pH_i) in normal rat tissues and in the Yoshida sarcoma, MC-7 sarcoma, D-23 carcinoma, and VX-2 carcinoma after hyperglycemia. Rats received 6 g glucose/kg IP, and rabbits were given 2 g glucose/kg/hr by IV infusion. Rat tumors (Yoshida, D-23, MC-7 grown subcutaneously in the foot) were used at a volume of 1.0 to 1.5 ml, and the rabbit VX-2 tumor was studied at a volume of 15 to 20 ml.

finding in these conditions (Schloerb et al. 1965; Poole 1967; Dickson and Calderwood 1979) (fig. 5.32). Intracellular pH homeostasis may be due to an efficient H^+ pumping system (Thomas 1978) in the membranes of most malignant cells. Galactose, which is not metabolized by most tumors, did not decrease pH_e in the Flexner-Jobling carcinoma (Voegtlin, Fitch, and Kahler 1935) or pH_i in the Walker-256 carcinoma (Schloerb et al. 1965). Preliminary results using the Yoshida sarcoma also indicate that galactose injection (6 g/kg) did not decrease pH_e.

The mechanism whereby glucose administration causes tumor acidosis is not known in detail; effects of the glucose on glycolysis and tumor blood flow appear to be involved, however. That the effect is not wholly caused by blood-flow inhibition is illustrated by the data on galactose administration; hypergalactosemia inhibits Yoshida tumor blood flow but does not decrease pH_e. The relative concentrations of glucose and lactate in the Yoshida sarcoma after glucose and galactose administration (fig. 5.33) indicate a mechanism for the pH_e decrease. After injection, glucose accumulated in the tumor for one to two hours at five to ten times the normal concentration; tumor glucose then fell to zero by three hours as blood perfusion was strongly inhibited. Tumor lactate concentration increased twofold by two hours and was maintained at this ceiling level until four hours after glucose injection (fig. 5.33). Lactate in the tumor returned to normal values by six hours, when blood glucose levels

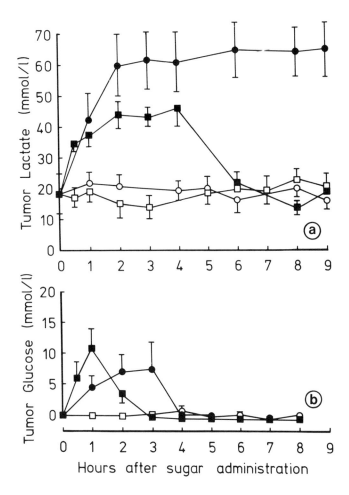

Figure 5.33. Levels of A, lactate, and B, glucose in the 1.0- to 1.5-ml Yoshida sarcoma. Tumor glucose and lactate were measured after injection (■) or infusion (●) of glucose and after injection (□) and infusion (○) of galactose.

115

were again normal (fig. 5.25). After galactose injection, no such rise in tumor glucose, or lactate, was observed (fig. 5.33). When glucose was given by infusion (fig. 5.33), the pattern of changes in tumor glucose level was similar to those after glucose injection; a transient (three-hour) rise in tumor glucose preceded a decline to zero levels, which were maintained for the remainder of the infusion. Tumor lactate tripled in concentration, reaching a ceiling of 60 mM/L by two hours of infusion; this level was maintained for the duration of the infusion. Galactose infusion had no effect on tumor glucose or lactate. The sequence of alterations in tumor physiology after glucose injection is thus (figs. 5.25 and 5.33):

1. Increased blood and tumor glucose levels within 30 minutes
2. Rapid inhibition of tumor blood flow, reaching 50% to 60% by 30 minutes and more than 90% by one hour
3. Metabolism of the excess tumor glucose with resulting increase in lactate concentration
4. Lactate production to a ceiling level at two hours owing to diminishing substrate glucose
5. A stable level of high lactate (~40 mM/L) and zero glucose by three hours owing to interruption of molecular exchange between tumor ECF and blood
6. Return to steady-state levels of metabolites on resumption of blood flow

Further evidence for this scheme is shown in figure 5.34, which shows the effect of hyperglycemia on exchange of labeled substances between the Yoshida sarcoma and plasma. At four hours after glucose injection, glucose uptake and lactate egress were inhibited by 80% to 90% in the tumor. When the blood glucose concentration was increased by infusion, almost total inhibition of glucose and lactate exchange occurred (99% to 100%). Under these conditions, the exchange of labeled water, chloride, and the weak nonmetabolizable acid DMO was completely inhibited. This indicates complete isolation of the tumor from the host bloodstream. Recovery of tumor pH_e (fig. 5.32) lagged behind clearance of lactate (fig. 5.33) and resumption of blood flow (fig. 5.25). Normal pH_e was not attained until 18 to 24 hours after the injection of glucose. The reason for this delay is not clear, but may be due to persistence of some other acidic catabolite(s).

Hyperglycemia, Tumor Blood Flow, and Tumor pH The concomitant changes in blood supply and pH_e found in the Yoshida sarcoma after hyperglycemia suggest that the three effects are interrelated. The evidence for inhibition of blood flow in the tumor is strong; both tumor uptake and clearance of a wide range of chemical species was inhibited during hyperglycemia (figs. 5.34 and 5.35, table 5.15), and quantitation of the

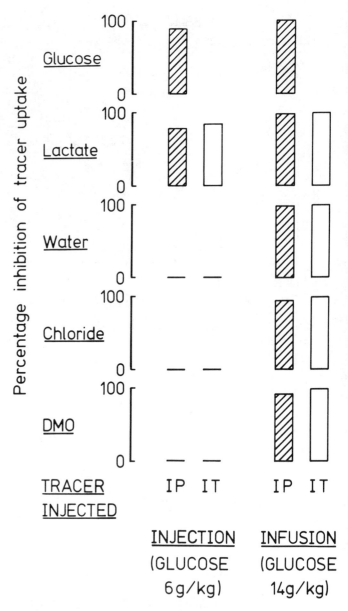

Figure 5.34. Effect of hyperglycemia on exchange of ^3H-2-deoxyglucose, ^{14}C-lactate, HTO, ^{36}Cl, and ^{14}C-DMO between the Yoshida sarcoma in the foot (1.0 to 1.5 ml) and host. Glucose was administered by either single IP injection or IV infusion as described in legend, figure 5.25.

% inhibition of tracer uptake =

$$\frac{cpm/g \text{ in control tissue} - cpm/g \text{ in experimental tissue}}{cpm/g \text{ in control tissue}} \times 100$$

% inhibition of tracer egress =

$$\frac{cpm/g \text{ (experimental)} - cpm/g \text{ (control)}}{cpm/g \text{ (experimental)}} \times 100$$

IP = intraperitoneal; IT = intratumor.

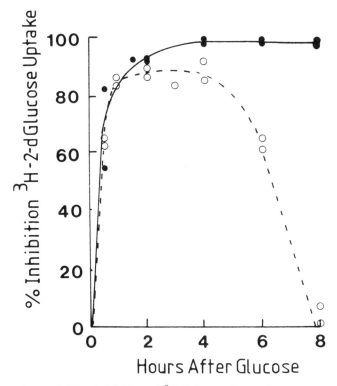

Figure 5.35. Inhibition of ^{3}H-2-deoxyglucose incorporation into the Yoshida foot tumor after glucose injection (6 g glucose/kg) or glucose infusion (as described in legend, fig. 5.25). The analog 2-deoxyglucose has similar transport properties to glucose, but is only slowly metabolized and was used in the present study to examine glucose transport (see Dickson and Calderwood 1980). (Calderwood and Dickson 1980b. Reprinted by permission.)

inhibition by different methods (^{86}Rb distribution and ^{133}Xe clearance) yielded comparable results (figs. 5.16 and 5.25). The results confirm an earlier finding of Algire and Legallais (1951), who found that hyperglycemia (blood sugar level unspecified) inhibited blood circulation in tumors growing in transparent chambers implanted in mice. In rat tumors, inhibition of blood flow during hyperglycemia has been reported more recently by Von Ardenne (1978), who has proposed a hypothesis to account for the concomitant fall in tumor blood flow and pH$_e$. Hyperglycemia, it is postulated, leads to a stimulation of glycolysis with lactic acidosis in the tumor; erythrocytes entering the acidified tumor would be expected to undergo a pH-mediated change in membrane conformation, causing a decreased flexibility. Low pH has been shown to alter erythrocyte membrane structure (Verma and Wallach 1976) and to increase blood viscosity (attributed to increased erythrocyte rigidity) (Murphy 1967). Such erythrocytes, it is argued, would lack the flexibility needed to pass through narrow capillaries and would physically block the tumor vessels (Von Ardenne 1978). Von Ardenne's theory therefore implicates a decrease in pH as the initiating event, with a resulting inhibition of tumor blood flow.

The present data do not support this hypothesis. Tumor blood flow decreased rapidly after glucose injection, and the curve of blood flow inhibition (fig. 5.25) was almost a mirror image of the blood glucose curve (fig. 5.25); blood flow decreased as blood glucose increased, and flow increased as glucose decreased. Tumor pH$_e$ declined progressively but much more slowly than blood flow, reaching a minimum in three and one-half to four hours (fig. 5.32). Previous work also showed that glycolysis (both aerobic and anaerobic) in the Yoshida tumor was inhibited by 35% to 60% during hyperglycemia (Dickson and Calderwood 1979), rather than stimulated as predicted by Von Ardenne (1972). Within 30 minutes after glucose injection, however, there was a rapid accumulation of lactate in the tumor to two to three times the normal level (fig. 5.33). Again, the rapidity of this accumulation paralleled the elevation of blood sugar and the decrease in blood flow and was well in advance of the fall in tumor pH$_e$. The time sequence of events there-

Table 5.15
Effect of Glucose (6 g/kg) on [^{14}C] Lactate Efflux from Tumors

Tissue	Control		Hyperglycemia	
	cpm/g	%	cpm/g	%
Tumor	101,219 ± 25,196*	100.0	365,980 ± 52,165	100.0
Plasma	24,332 ± 1,655	24.0	18,650 ± 4,858	5.1
Liver	6,319 ± 1,965	6.2	2,198 ± 796	0.6
Diaphragm	25,126 ± 5,316	24.8	10,950 ± 4,965	3.0
Gastrocnemius	18,926 ± 3,168	18.7	7,916 ± 1,615	2.16

NOTE: [^{14}C] lactate (1 μCi in 0.1 ml of 0.9% NaCl solution) was injected into tumors two hours after IP administration of glucose at a dose of 6 g/kg. Animals were sacrificed 30 minutes after the lactate injection, and ^{14}C activity was determined in the tumor and normal organs. In controls, tissue radioactivity was determined 30 minutes after an IT injection of [^{14}C] lactate. The ^{14}C activity in the tissues is presented both as cpm/g wet weight and as percentage of tumor radioactivity, the tumor value being taken at 100%.
*Mean ± SD of four determinations (animals).

fore favors inhibition of blood flow as the initial event, with a subsequent decrease in tumor pH_e secondary to arrest of lactate egress from the tumor. The maintenance of tumor pH_i at control levels despite a 0.6-pH unit fall in pH_e (fig. 5.32) probably is due to intracellular buffering and active transport of protons out of the cell (Roos 1975; Thomas 1978), as well as to the self-limiting effect of hyperglycemia preventing access of glucose to the cells.

The report by Gullino, Grantham, and Courtney (1967b) of a considerable increase in glucose use by the Walker-256 carcinoma, hepatoma 5123, and fibrosarcoma 4956 rat tumors in the first few hours after hyperglycemia (blood glucose $\geq$ 20 mM/L) would seem to be at variance with the present data. The use did reach a saturation level after six to seven hours of hyperglycemia, although no values for tumor blood flow were quoted by the authors. Increased glucose use in tumors after hyperglycemia would imply that no rapid inhibition of tumor blood flow occurred. The discrepancy between the findings of Gullino, Grantham, and Courtney (1967b) and the present study may be due to differences between experimental tumors used in the studies. Moreover, the tumors used in the investigations of Gullino, Grantham, and Courtney (1967b) were grown in ovarian tissue isolated from the other normal tissues and connected to the host blood supply by a single artery and vein. This is a situation considerably different from that of the tumors grown by simple subcutaneous or intramuscular implantation as used in the present study. In these circumstances, the tumor blood supply is connected by an agglomeration of tumor-induced new vessels to the vascular beds of surrounding normal tissues (Warren 1979). It is conceivable that blood flow in such tumors might be more susceptible to disruption than in tumors supplied by a single large artery and vein that constituted the original vessels to a normal organ.

The effect of glucose on tumor pH would thus seem to be complex, depending on rates of glucose influx into the tumor, lactate production and efflux, and buffering power. The balance between these processes in a tumor and the effect of hyperglycemia upon $pH_i : pH_e$ ratio may hinge on whether the glucose inhibits tumor blood flow and at what glucose concentration this inhibition occurs.

Our concept, therefore, envisages an initial effect of hyperglycemia on tumor blood flow, and this may encompass afferent and efferent vessels and/or microvasculature, with subsequent effects on tumor pH. At tumor temperatures of 42°C to 43°C, the effects on blood flow are reversible. According to Von Ardenne's hypothesis, the primary effect is a postulated decrease in pH within tumor capillaries, with subsequent interruption of tumor blood flow, and the consequences for blood flow are irreversible (Von Ardenne and Reitnauer 1980).

Hyperglycemia, Tumor pH, and Tumor Cure Rate after Hyperthermia The effect of combination hyperglycemia and hyperthermia on volume growth and cure of rat tumors is shown in table 5.16. Glucose alone had no effect on growth in any of the tumors. In each tumor, the combination of hyperglycemia and hyperthermia caused a significant increase in doubling time (T_D). In the Yoshida sarcoma, combination of hyperglycemia with heat for 45 minutes at 42°C increased T_D from 2.5 days after heat alone to 4.1 days after the combined treatments. In the D-23 carcinoma, hyperglycemia led to small but significant increases in T_D when combined with heat for one hour at 42°C (4.5 days to 6.0 days) or one hour at 43°C (12.9 days to 20.5 days). In neither of these tumors did hyperglycemia cause any increase in cure rate of animals. In the MC-7 sarcoma, however, hyperglycemia increased cure rate of animals given heat for one hour at 43°C from 20% to 69%.

Possible causes of tumor thermal sensitization are enhanced uniformity of heating (fig. 5.30), low pH_e (fig. 5.32), low level of glucose in the tumor (fig. 5.32), or hypoxia (on which no data is available). By far the greatest sensitization was found in the MC-7 sarcoma, which was also the tumor in which hyperglycemia led to a significant fall in intracellular pH (fig. 5.32). Therefore, low pH_i is strongly implicated as a sensitizer of hyperthermia in vivo. A similar conclusion was made recently by Hofer and Mivechi (1980) from studies on cells in vitro. These workers maintained BP-8 murine sarcoma cells in either bicarbonate buffer or the organic buffer system HEPES-MES. Cells acidified in the HEPES-MES buffer retained a pH_i 0.4 to 0.5 pH units above the buffer pH. In contrast, addition of the weak acid DMO to cells in HEPES-MES, or of carbon dioxide to cells in bicarbonate, caused rapid equilibration between pH_i and pH_e. This was due to rapid penetration of the cell membranes by both DMO and carbon dioxide, while HEPES and MES, which are zwitterionic buffers, were largely excluded from biological membranes (Hofer and Mivechi 1980). At a pH_e of 6.5, tumor cells were considerably more heat sensitive in bicarbonate and HEPES-MES plus DMO than in HEPES-MES, indicating that pH_i probably is a more important parameter in heat sensitivity than pH_e. The reason for the less pronounced response of the D-23 and Yoshida tumors to hyperglycemia may be efficient pH_i regulation. This regulation of intracellular pH may be a limiting factor in the usefulness of low pH as a potentiator of heat.

Although heating at 42°C for 45 minutes or hy-

Table 5.16
Effect of Hyperthermia and Hyperglycemia on Tumor Regression and Regrowth

Treatment	Yoshida Sarcoma			MC-7 Sarcoma			D-23 Carcinoma		
	N	T_D	% Cure	N	T_D	% Cure	N	T_D	% Cure
Control		1.50 (0.50)	—		3.4	—	20	2.64 (0.95)	—
45 min/42°C	6	2.51 (1.34)	—						
45 min/42°C + hyperglycemia	6	4.09 (2.91)	—						
1 hr/42°C	12	—	100				6	4.45 (2.04)	—
1 hr/42°C + hyperglycemia	12	—	100				6	5.93 (3.09)	
1 hr/43°C				15	13.9 (7.1)	20.0	20	12.89 (6.42)	5
1 hr/43°C + hyperglycemia				16	18.8 (4.61)	68.8	10	20.50 (10.25)	—
2 hr/43°C				11	18.9 (7.7)	63.6			
2 hr/43°C + hyperglycemia				6	—	100.0			
3 hr/43°C							10	18.7 (11.6)	80
3 hr/43°C + hyperglycemia							6	21.6 (12.4)	67
Hyperglycemia only	15	1.27	—	8	3.5 (1.1)	—	3	2.9	—

perglycemia are individually noncurative for the Yoshida sarcoma, we previously reported that combination of these two regimens was synergistic in effect. When the heat was given four hours after glucose injection, 80% of 1.0- to 1.5-ml Yoshida tumors growing on the feet of rats regressed with host cure, while the remainder resumed exponential growth with a prolonged T_D of 14 days (Dickson and Calderwood 1980). It was subsequently found that this effect was confined to female tumor-bearing hosts. The synergistic response has proved highly reproducible, but at present we have no explanation for its occurrence only under these restricted conditions. It apparently was not related to pH_i, since the response of pH_i and pH_e to hyperglycemia and heat was similar in tumors borne by both sexes.

The Concept of Lysosome Labilization In a series of investigations that began in 1935 and spanned 20 years, Kahler and Robertson (1943) studied pH and its modification by glucose loading of the host. The work encompassed 12 types of transplantable rodent carcinomas and sarcomas as well as spontaneous tumors in rats and mice. Following intraperitoneal injection of 6 g glucose/kg body weight, decreases in tumor pH (measured by capillary glass electrode) of 1 unit or more to the region of 6.2 to 6.6 were obtained, while the pH of normal tissues remained unchanged.

In discussing their earlier results, Voegtlin, Fitch, and Kahler (1935) postulated that because necrosis in tumors is the result of deficient nutrition and accumulation of waste products induced by inadequate blood supply, glucose loading of the host might establish chemical conditions (high lactic acid and low pH) that would favor such necrosis. No evidence in support of this hypothesis ever was presented by these or other workers in spite of the large number and variety of tumors subjected to prolonged low pH by glucose administration (Kahler and Robertson 1943; Voegtlin, Fitch, and Kahler 1935; Eden, Haines, and Kahler 1955).

In a succession of reports beginning in 1965, Von Ardenne (1971, 1972) supported the concept of inducing tumor cell damage by glucose saturation of the tumor. He postulated that hyperglycemia of 300 to 500 mg/100 ml (17 to 28 mM/L) in the tumor-bearing host leads to a selective increase in glycolysis in the tumor, with a resultant rise in concentration of tumor lactic acid and decrease in pH_i. Labilization of the lysosomal membrane ensues, and activation of the released lysosomal enzymes leads to destruction of the cancer cells. This process begins at pH 6.7 to 6.8, and for optimal tumor acidification, pH values of 6.0 to 6.5 were recommended. The destructive effect becomes progressively greater as the pH decreases, and values as low as 5.8 (measured by glass microelectrode) occurred in rodent tumors following three to five hours of hyperglycemia in the host. The lysosomal damage is potentiated by hyperthermia, the optimal tumor acidification acting as a preconditioner that enables an elevated temperature as low as 40°C to be effective in tumor destruction. This two-pronged attack forms the basis of Von Ardenne's multistep approach to cancer therapy (so-called Krebs-Mehrschritt Therapie), in which heat plus hyperacidity are supplemented with one or more of seven other modalities ranging from radiation to Calmette-Guérin bacillus (BCG). The results indicate that in the few experiments in which heat plus optimal tumor acidity were employed without other adjuvants for the therapy of rat solid tumors, no tumor cures were obtained (Von Ardenne 1971, 1972).

More recently, the concept (Overgaard and Overgaard 1972; Overgaard 1976a) has been propagated that increased lysosomal enzyme activity is responsible for the destruction of cancer cells by heat. Electron microscopy on the HB mouse mammary carcinoma heated in vivo revealed that during the early hyperthermia reaction (up to 24 hours after treatment) the number of small and large lysosomes increased in the cells; these acid phosphatase-positive lysosomes often contained a residue of digested heterogeneous material. Overgaard believes an outflow of the activated lysosomal hydrolases into the cytoplasm may then be triggered by a decrease in pH within the cells as postulated by Von Ardenne (1971, 1972). The decrease in pH is envisaged by Overgaard to result from "a relative increase in anaerobic glycolysis and a higher amount of lactic acid" (1977a), occasioned by a selective inhibition of respiration in the heated cancer cells (1976a, 1977a). The findings that most lysosomal enzymes have an acid pH optimum (DeDuve and Wattiaux 1966), and that mouse ascites cells in vitro showed an increased number of lysosomes and augmented heat damage when incubated in medium at pH 6.4 (Overgaard 1976b), are adduced as evidence in favor of a low pH-mediated lysosomal enzyme destruction of the cell as the primary agent of hyperthermic cell damage (Overgaard 1976a). The increased in vitro heat sensitivity of some cell lines under conditions of hypoxia (Gerweck, Gillette, and Dewey 1974; Schulman and Hall 1974) or inadequate nutrition (Hahn 1974) are also attributed to an increased acidity in the tumor cell milieu (Overgaard 1976b), because such conditions may lead to increased glycolysis and production of lactic acid by tumor cells in vitro (Warburg, Posener, and Negelein 1924; Burk, Woods, and Hunter 1967). Inasmuch as hypoxia and poor nutrition are common features of tumors in vivo, the concept of pH-induced lysosomal enzyme activation as the crucial means of hyperthermic cell damage has been extrapolated to apply to tumors heated in the host (Overgaard 1977a).

The hypothesis of lysosomal enzyme activation has been difficult to examine because it embodies a number of assumptions derived from widely different studies in vitro and in vivo, and some of these suppositions have become insidiously entrenched in the literature of cancer biochemistry. Results that enable the various steps comprising the postulate to be evaluated in a specific tumor system have not been available to date. Central to the hypothesis is the belief that in cancer cells undergoing glycolysis, lactic acid production leads to an intracellular acidosis and that this decrease in pH can be augmented by further stimulation of glycolysis. Data supporting this theory are sparse, and some evidence to the contrary has been found. Using DMO to measure pH_i of the Walker-256 carcinoma in vivo, Schloerb and colleagues (1965) found that intracellular acidosis could not be produced by glucose administration. Indeed, three hours after glucose loading of the host rats (6 g glucose/kg body weight), a slight rise of tumor pH_i from 7.19 to 7.36 was observed. With slices of the same tumor in vitro, Hult and Larson (1976) recorded a significant increase in pH_i from 7.12 to 7.42 within five minutes of the addition of glucose (5 mM) to the medium, and to 7.44 after 40 minutes; 20 minutes after addition of 25 mM glucose, the pH_i of the Walker carcinoma was 7.51. Poole (1967), using DMO with Ehrlich ascites cells maintained in vitro in phosphate or bicarbonate media supplemented with different sugars, has reported that the relationships between pH_e and pH_i are similar regardless of whether cells are undergoing glycolysis. With glucose and other fermentable sugars, a marked decrease in pH of the medium was noted within 15 to 30 minutes of the addition of the sugar, and the fall in pH agreed closely with that calculated from the accumulation of lactate. Extracellular pH values of about 6.0 were obtained when phosphate was the incubation buffer, and pH_i decreased concomitantly to approximately 6.5 under these conditions. With a more physiologic bicarbonate buffer, pH_e decreased to approximately 6.6, and pH_i remained relatively unaffected, decreasing from 7.1 to 6.9. These in vitro data of Poole (1967) therefore agree closely with our in vivo results with the Yoshida sarcoma. In their earlier work on glucose loading of tumor-bearing rats, Eden, Haines, and Kahler (1955) and others (Kahler and Robertson 1943; Voegtlin, Fitch, and Kahler

1935) were careful to emphasize that the observed tumor pH values of 6.5 did not represent pH_i, inasmuch as the capillary electrode was in contact with both cells and intercellular fluid. Reference has already been made to the remarkable stability of pH_i in the tumors examined in this laboratory; only in the MC-7 sarcoma has it been possible to decrease pH_i. Although the findings agree with the view that the microenvironment (interstitial compartment) of tumors is amenable to extensive modification of the acid-base status as detailed by Gullino and colleagues (1965) and by Gullino in 1966, our work indicates that the intracellular compartment of tumors may be highly resistant to such changes.

Far from increasing tumor glycolysis, as postulated by Von Ardenne, hyperglycemia significantly decreased both aerobic and anaerobic glycolysis in the Yoshida tumor (Dickson and Calderwood 1979). Tumor lactate level did increase, however, and pH_e decreased because of metabolite accumulation during blood-flow inhibition. Heating the tumors at the sublethal temperature of 40°C in association with hyperglycemia had no significant effect on Yoshida foot tumor volume or host survival, whereas tumor regression and host cure resulted from hyperthermia at 42°C. The results do not, therefore, support the postulate of Von Ardenne that low tumor pH acts as a sensitizer, enabling the tumor-lethal temperature to be reduced to 40°C. The results of Orth, Swidler, and Zarakov (1977) are of interest in this context. These investigators reported a 40% increase in survival time for mice treated by combination hyperglycemia (400 to 500 mg glucose/100 ml maintained for three hours) and one hour total-body hyperthermia at 40°C. Tumor response to hyperglycemia may vary with tumor type, and the role of glucose as a sensitizer for heat may be governed by host as well as tumor factors, depending on the method of heating (local or systemic).

In the present work, hyperthermia did not alter pH in the Yoshida sarcoma, but did cause a decrease in pH_e (7.17 to 6.72 units) in the MC-7 sarcoma after heating for three hours at 43°C. Song (in press) also has reported a decrease in tumor pH_e after heating. Results with a considerable number of different types of cancer cells, both animal and human, have shown that heating in vitro (Dickson 1977; Dickson and Suzangar 1976b; Strom et al. 1977) or in vivo (Dickson and Suzangar 1974; Dickson and Muckle 1972) induced an irreversible inhibition of respiration; glycolysis, however, remained unaffected or was decreased by the heating. In three types of transplantable rodent tumors, Gullino (1975) found no evidence to support the view that in vivo tumor metabolism shifted from respiration to glycolysis when the oxygen supply was deficient. Therefore, little evidence exists for the concept of Overgaard (1976b, 1977a) that lactic acid produced by increased glycolysis as a result of hyperthermia or hypoxic conditions decreases pH_i and is a crucial agent in hyperthermic cell damage. An increase in the number of lysosomes following in vivo heating of the HB mouse mammary carcinoma has been reported (Overgaard 1976a), but there is little evidence that within intact cells supranormal temperatures lead to significant release of hydrolytic enzymes (De Duve and Wattiaux 1966). Although low pH_e potentiates the destructive effects of heat on several types of cancer cell in vitro, recent study indicates that this effect may involve a low pH_i. Even if a low pH_i is operative in such cells, the stability of pH_i in tumor cells in vivo serves to emphasize the difference between the in vitro and in vivo situations.

In summary, low-pH-mediated labilization of the lysosomes with release of hydrolytic enzymes, which autodigest the cells of tumors following hyperthermia, or the induction by hyperglycemia of such a chain reaction as a preconditioner for tumor heating, has undoubted appeal. There is currently little evidence that the lysosome can be exploited in this way to potentiate hyperthermia. Because of the central role of lysosomal enzymes in cell autolysis and necrosis, however, the concept is attractive and may represent a line of attack for the future.

Caveat Because environmental acidity is a powerful modifier of hyperthermic response in vitro, there has been a ready assumption that pH must be of similar importance in vivo (Overgaard and Nielsen 1980; Dewey and Freeman 1980), on the basis that tumors are characterized by a high glycolytic rate and other features favoring an acidic milieu (hypoxia, inadequate nutrition). Table 5.17 represents a collation of published data on pH in tumors and normal tissues in various types of host. It is seen that the results, in general, support the belief that tumors have a lower pH than normal tissues. In most cases, however, the pH of tumors is in the region 7.0 to 7.2; the tumors are not markedly acidic. This finding is apparent especially in the results of those workers who have examined pH in very large numbers of tumors (Eden, Haines, and Kahler 1955; Gullino et al. 1965; Calderwood and Dickson 1980b, in press). In one series (Tagashira et al. 1953, 1954), very acidic values in the region of 6.5 were reported for tumors, with values of 6.65 to 6.95 for the Yoshida sarcoma, for example. In an extensive series of investigations, we have never obtained pH values in this very low range for the Yoshida tumor. Little data is available on the pH of human tumors. The results of Ashby (1966) were from

Table 5.17
Normal Tissue and Tumor pH In Vivo

Tissue	Species	Methods	Reference	pH
Normal	Humans	Microelectrode	Ashby 1966	7.30 – 7.54
Malignant	"	"		6.63 – 7.00
C_3H mammary Ca	Mouse	"	Bicher et al. 1980	6.75
SC tissue	Rat	Micropore chamber, tissue interstitial fluid (TIF)	Gullino et al. 1965	7.325
Plasma				7.369
Novikoff hep.				7.001
Walker CA				7.044
Hep. 3683				7.107
Fibro. 4956				7.091
Hep. 5123				6.913
Liver	Mouse	Microelectrode	Naesblund and Swenson 1953	7.65
SC tissue				7.65
Tumor (type unstated)	"	"	"	6.9
Gastrocnemius muscle	Rat	Miniature electrode	Calderwood and Dickson, in press	7.21 (0.15)
Liver				7.32 (0.11)
Yoshida sarcoma				7.19 (0.13)
MC-7 sarcoma				7.17 (0.08)
D-23 carcinoma				7.13 (0.04)
VX-2 carcinoma	Rabbit	"	"	6.99 (0.15)
SC tissue	Rat	Microelectrode	Tagashira et al. 1953, 1954	7.0 – 7.2
Liver				6.95 – 7.05
Blood				7.3
Fibrosarcoma				6.65 – 7.0
Reticulosarcoma				6.55 – 6.7
Yoshida sarcoma (SC)				6.65 – 6.95
OAT hepatoma				6.95
Muscle	Rat	Miniature electrode	Eden, Haines, and Kahler 1955	7.39
Liver				7.39
Hepatoma				6.96
"				6.95
"				7.13
"				7.09
Lymphosarcomas				6.98
E2730 sarcoma				7.04
Carcinoma 2226				7.00
Sarcoma 1643				7.01
Fibrosarcoma ACMCA2				6.83
Hepatoma 3924A				7.06

NOTE: Most of these are overall values. Where pH distribution has been studied (Eden, Haines, and Kahler 1955; Bicher et al. 1980; results, this chapter) wide variation has been observed.

a very small number of tumors. For technical reasons it is difficult to obtain measurements on human tumors, and it is not known how representative the values obtained by Ashby may be.

The crucial point, of course, is not necessarily the resting pH of the tumors, but whether this pH is decreased by heat and/or hyperglycemia, and if so, whether such a decrease is of relevance to tumor thermal sensitivity. In animals, the situation is extremely complex with tumor pH, hypoxia, and blood flow being inextricably intertwined. There is little evidence, however, in the limited number of animal tumors studied to date, that low pH per se is a conditioner for hyperthermia. In human tumors, the field of pH, oxygenation, and blood flow is an uncharted sea. Although Von Ardenne for some years (1971, 1972) has been a protagonist of low tumor pH (achieved by hyperglycemia) as a sensitizer for hyperthermia, data—animal or human—supporting this claim are difficult to find. It is also of note that several workers have included hyperglycemia at levels suggested by Von Ardenne in their protocols of total-body hyperthermia in humans (R. T. Pettigrew, personal communication; this volume, chapter 19; Krementz 1980; LeVeen et al. 1980). Although it is difficult to assess the effect of this procedure, owing to the problems of establishing well-controlled series in patients with advanced malignancy who had already received several types of orthodox treatment, no striking results have been reported in these patients. In addition, total-body hyperthermia results in host metabolic acidosis (Larkin 1979, 1980; MacKenzie et al. 1976; this volume, chapter 19), a condition that might be expected to further the cause of decreased pH as a thermosensitizer in both primary and metastatic tumors. Therefore, the beneficial effects of low tumor pH in vivo are to date unconvincing.

Systemic Acidosis and Alkalosis Another approach to the induction of low tumor pH is the use of systemic agents such as $NaHCO_3$, NH_4Cl, CO_2 (Gullino et al. 1965; Anghileri 1975; Calderwood and Dickson 1975), or HCl (Harguindey, Henderson, and Naeher 1979). Systemic acidosis caused by chronic ingestion of HCl led to decreased growth and to regression of sarcoma 180 in mice (Harguindey, Henderson, and Naeher 1979). Gullino and colleagues (1965) investigated the effects of some of these systemic agents on tumor pH_e using fluid aspirated from millipore chambers implanted in rat tumors. In animals breathing 10% CO_2 for five hours, pH in tumor interstitial fluid of Novikoff hepatomas declined from pH 7.00 to 6.67. In subcutaneous interstitial fluid, pH fell from 7.33 to 7.13, and plasma pH declined from 7.38 to between 7.08 and 7.24. After chronic NH_4Cl ingestion, pH in the Walker carcinoma reached 6.5, compared with approximately pH 7 in subcutaneous tissue and plasma. A selective reduction in the pH of TIF to 6.5 (approximately one pH unit lower than plasma pH) was also produced by bicarbonate infusion (Gullino et al. 1965). The effects of these changes on tumor growth were not reported.

The effect of NH_4Cl ingestion and $NaHCO_3$ infusion in the concentrations used by Gullino and associates were investigated in the Yoshida sarcoma in relation to pH_i (table 5.18); pH_i increased in the Yoshida sarcoma and normal tissues after $NaHCO_3$ infusion. Ingestion of NH_4Cl caused pH_i decrease to approximately 6.95 in the tumor and normal tissues. Thus, neither treatment gave a specific decrease in tumor pH_i. Schloerb and colleagues (1965), investigating the effect of CO_2 inhalation, found a pH_i decrease from 7.2 to 6.9 in the Walker-256 carcinoma, accompanied by a pH_i fall in rat skeletal muscle from 6.8 to

Table 5.18
Intracellular pH Following Administration of NH_4Cl or $NaHCO_3$

Tissue	Control	NH_4Cl	$NaHCO_3$
Blood pH	(48) 7.35 ± 0.15	(6) 6.90 ± 0.05	(6) 7.57 ± 0.03
Diaphragm	(38) 6.95 ± 0.16	(6) 6.95 ± 0.09	(6) 7.19 ± 0.07
Gastrocnemius	(34) 6.85 ± 0.18	(6) 6.96 ± 0.10	(6) 7.15 ± 0.05
Liver	(36) 7.11 ± 0.24	(6) 6.96 ± 0.02	(6) 7.17 ± 0.06
Yoshida tumor	(48) 7.21 ± 0.18	(6) 6.94 ± 0.06	(6) 7.42 ± 0.11

NOTE: Figures in parentheses indicate the number of animals examined in each case. Mean pH_i values are given ±1 SD.
  NH_4Cl was given orally to six rats as four doses each of 7.5 mM/kg body weight at hourly intervals, and this regimen was repeated 18 hours after the last dose; $NaHCO_3$ was infused IV into the femoral vein of six rats at 4 mM/kg/hr for three hours (Gullino et al. 1965). At the end of NH_4Cl or $NaHCO_3$ therapy, pH_i was measured, the isotope mixture being injected three hours before to allow equilibration.

6.5. The effects of systemic acidosis on pH_i were thus similar in normal and malignant tissues. In one recent report, however, inhalation of 15% CO_2 by mice led to increased cell killing in sarcomas heated at 43.5°C (Urano et al. 1980).

The use of systemic acidosis combined with heat has not been extensively studied and additional data is required to assess its potential. The approach is severely limited, however, by host toxicity. A departure of the blood pH from its normal range of 7.35 to 7.45 represents a considerable upset in acid-base balance, and inhalation of carbon dioxide at concentrations in excess of that normally present in the alveolar air (5.6%) overcomes the host compensatory mechanisms and can be rapidly fatal (Wright 1952).

Hypoxia
There is no data directly linking hypoxia with thermal sensitivity in vivo, although there is a considerable body of indirect evidence (Overgaard and Nielsen 1980). It has been shown that clamping the tumor blood supply increases thermal sensitivity (Crile 1963; Suit 1976), and that this sensitizing effect increases with duration of clamping before heating (Hill and Denekamp 1978). This indicates that chronic hypoxia is more important than acute hypoxia, and is thus in keeping with in vitro findings on hypoxia and the response to heat. The physical clamping is an analogous approach to using high blood sugar levels to inhibit tumor blood flow, except that it is not generally applicable, since it is not possible in most cases to clamp or ligate the tumor blood supply without impairing the circulation to the normal tissues in the region. The increased heat sensitivity with duration of clamping indicates the effect involves more than just an increased uniformity of heating; potentiation could, however, be due to changes in PO_2, glucose, catabolite level, pH, or other biochemical parameters altered under ischemic conditions.

The reported synergism between hyperthermia and hypoxic cell radiosensitizers appears to be due to thermal enhancement of the cytotoxic properties of the drugs (Stratford, Watts, and Adams 1978). These drugs may be of particular significance because of their sensitization to heat of chronically hypoxic cells, which may possess thermal tolerance (Adams et al., in press). More data is required on the differential effects of these drugs on response to heat of normal and malignant tissues before an assessment can be made of their usefulness as potentiators of hyperthermia.

Pretreatment of tumors by embolization or sugar loading to inhibit blood flow and facilitate heating results in hypoxia. The use of induced hypoxia as a preconditioner of heat is linked with other effects, such as pH change, on the tumor microenvironment.

Membrane Permeability Modulators
It is known that cholesterol is an effective regulator of cell membrane permeability, and a recent development has been the finding that membrane lipid composition is related to thermal sensitivity. It has been reported that agents that increased the permeability or fluidity of the membrane act as preconditioners for hyperthermia. Two groups of agents are currently under investigation: polyamines (Gerner et al. 1980) and local anesthetic drugs, such as procaine (Yatvin 1977). In vitro, polyamines have been found to inhibit the development of thermal resistance and to overcome thermal tolerance in CHO cells. Preliminary results indicate that the effect of heat on rat tumors is potentiated by intratumor injection of procaine. The goal of increasing the sensitivity of tumor cells to heat, or decreasing the sensitivity of normal tissues, by manipulation of cell membrane permeability would appear to be feasible. The subject is discussed further in this volume (see chapter 6).

The Inflammatory Response
Inflammation has been defined (Movat 1979) as the reaction of living tissue to injury and comprises the series of changes of the terminal vascular bed, of the blood, and of the connective tissue that tends to eliminate the injurious agent and to repair the damaged tissue. The rapid onset of acute inflammation in tumors after hyperthermia is the most obvious visible change when external tumors are heated. In the initial period after heat, tumors are red in color, hot, painful to the host, and host tissues show loss of function, the classic characteristics of the inflammatory response (Dickson and Calderwood 1980). Swelling is a more persistent symptom and may last for several days after treatment (fig. 5.5). The changes in tumor appearance are mainly a reflection of alterations in the tumor capillary bed; the initial redness probably is due to hyperemia, which occurs in normal tissues in the first few hours after injury (Movat 1979; Majno 1964) and is observed in the skin over the tumor. The hyperemia seems to be more common in normal tissues and was not observed in the Yoshida sarcoma (fig. 5.17) or in several other tumors (Gullino, Yi, and Grantham 1978; Song 1978; Song, Rhee, and Levitt 1980) after hyperthermia; tumor vessels appear to be largely incapable of increasing blood flow in response to stimuli (Mattson et al. 1979; Kruuv, Inch, and McCredie 1967).

Because of increased viscosity, the initial hyperemic period is followed in normal tissues by slowing of the blood (Movat 1979). Raised viscosity appears to be due to an increase in vascular permeability (Movat 1979; Song 1978; Sevitt 1958, 1964) in the injured tissue and loss of plasma into the interstitial space (Movat 1979), which results in the observed swelling. Eventually, if injury is sufficiently severe, blood flow stops altogether. Stasis may be reversible, in which case blood cells remain intact, or irreversible, when vessels are filled with a red, homogeneous paste composed of lysed blood cells (Movat 1979). Hemostasis caused by increased vascular permeability was observed in research on burns induced in animal normal tissues heated for short periods between 47°C and 60°C (Movat 1979; Sevitt 1958, 1964). The degree of vascular permeability induced was proportional to the degree of injury (Sevitt 1958). More recently, Song (in press) has reported increased vascular permeability in animal tumors after heating, in conjunction with decreased blood flow and a fall in extracellular pH. In the extensive studies of Menkin (1956), a decrease in pH to values as low as 6.0 to 6.5, associated with lactic acidosis in the inflamed tissue, was a common finding in normal tissues undergoing an acute inflammatory response. Thus, many of the microenvironmental changes in heated tumors are encountered in normal tissues undergoing acute inflammation. Malignant tissues may be more susceptible to the induction of inflammation owing to the tortuous, inefficient (Gullino 1975; Warren 1979), microcirculation and slow perfusion (Gullino and Grantham 1961; Peterson 1979a) rate found in many tumors. Blood flowing sluggishly through such a system is likely to be more viscous than fast-flowing blood, owing to the increased aggregation of blood cells known to occur at low perfusion rates (Schmid-Schonbein et al. 1975; Richardson 1976). In addition, tumor capillaries and sinusoids often have a discontinuous endothelium and sometimes lack even a basement membrane (Gullino 1975; Warren 1979). Such a system would seem particularly sensitive to disruption by inflammatory processes. The selective inhibition of blood flow in many tumors by hyperthermia (Dickson and Calderwood 1980; Song 1978; Song, Rhee, and Levitt 1980; Bicher et al. 1980) appears to confirm this hypothesis.

The importance of the inflammatory process in heat treatment of cancer was emphasized by Crile (1963, 1966). Crile found that lethal damage in mouse tumors did not occur until the development of a postheating inflammatory response. In addition, tumors could be sensitized to heat by chemical mediators of the inflammatory response, such as serotonin and histamine. These compounds administered alone produced some growth delay in tumors, but unlike hyperthermia, resulted in few complete regressions (Crile 1963, 1966; Sokoloff 1968). This illustrates that in tumors, as well as in normal tissues (Movat 1979), the acute inflammatory reaction probably involves direct damage to the tissue microvascular system, as well as chemically mediated effects (Movat 1979; Crile 1963). The importance of inflammation in tumor response to heating was emphasized further by Moricca and colleagues (1977), who stated that the inflammatory reaction following perfusion of human limb tumors with heated blood should not be avoided but constituted part of the hyperthermic treatment. The cellular reactions occurring in tissues after the acute inflammatory response also may play an important part in tumor treatment by hyperthermia (Strauss 1969), since the hyperemia can be expected to bring more effector cells of the immune system (lymphocytes) and also cells of the mononuclear phagocyte system (macrophages) into the region. These cellular responses are discussed in relation to immunologic phenomena in chapter 23.

Apart from the results of Crile on mouse tumors, little information is available on the use of chemical mediators of inflammation as potentiators in tumor heating. With the dosage of serotonin used by Crile (1963) on a body weight basis in rats with foot tumors, we found that the rats died from systemic toxicity shortly after intratumor injection of the drug. The approach certainly warrants further study, however, in view of the striking tumor regressions obtained by Cavaliere and his group (1967) with regional hyperthermic perfusion (Moricca et al. 1977), in which the inflammatory response probably was an integral component in tumor destruction.

Systemic Factors

Systemic factors influencing tumor response to heat involve participation of the host defense mechanisms: the immune system and the mononuclear phagocyte system. Although it occurs locally, the inflammatory response also has a systemic component, since the cells of the circulating blood participate, and a severe response may be accompanied by fever.

Defense Mechanisms

In animal tumor systems there is a considerable amount of data indicating that the immune system is a participant in the regression of both a primary tumor and its metastases following curative hyperthermia to the primary. Total body heating, on the other hand, abrogates these beneficial immune reactions by a direct depressive action on the lymphoid tissues. Even in immunogenic animal tumors, however, it has been difficult to demonstrate that the immune response to heating is in any way tumor-specific. Rather, the reaction is nonspecific, involving a major macrophage component. Current data indicate that the magnitude of the immune response following tumor heating is an artifact, caused by the artificial nature (chemical or virus induced) of transplantable animal cancers.

In humans, there is no convincing evidence for the generation of an antitumor immune response following tumor heating—even a nonspecific immune response. A nonspecific reaction on the part of the mononuclear phagocyte system (macrophages) appears much more likely and would be in keeping with the characteristics of tumor regression induced in the era of cancer biotherapy with extracts of bacteria (Coley's toxin) or protozoa (trypanosomes).

The occurrence of a nonspecific host reaction after heating may have advantages from the point of view of exploiting the reaction to potentiate the effects of hyperthermia. This is because there are a number of macrophage stimulants, such as *Corynebacterium parvum*, that can be given to boost host defenses, while our knowledge of how to stimulate an effective antitumor immune response is elementary.

For a detailed consideration of the immune system and hyperthermia, see chapter 23.

PERSPECTIVE

It is apparent that a number of factors are emerging that may influence the response of cancer cells to heat. Because of the intense efforts now being channeled into hyperthermia research, the field is advancing rapidly, and some of these modifying factors were unknown or were deemed unimportant in reviews published as recently as three years ago (Rossi-Fanelli et al. 1977; Streffer et al. 1978). In vitro experiments, because of the convenience and relative simplicity of the approach, have featured prominently in these advances. It is becoming increasingly evident that there are significant differences between the thermal response of cancer cells in vitro and tumors in vivo. These differences relate to the three-dimensional and histologic structures of tumors and the tumor milieu (Overgaard and Nielsen 1980; Gullino 1980), to blood flow (Jain 1980; Song, in press) and to the very important influence of host normal tissues, including the defense systems (immune system and mononuclear phagocyte system) (Dickson and Shah 1980; this volume, chapter 23).

The relevance of some of the seemingly important in vitro findings to tumors in vivo is unknown at present, and great caution should be exercised in extrapolation from the in vitro to the in vivo situation. The example of environmental pH has been cited. Although pH is a powerful modifier of hyperthermic response in vitro, and the characteristics of tumors (high potential for glycolysis, hypoxia, inadequate nutrition) supposedly favor an acidic milieu, there is little evidence that pH is a thermal sensitizer in vivo for the majority of tumors investigated to date.

Almost all available data on factors that may modify tumor response to heat have been obtained from animal systems. Such systems differ from the human tumor-bearing host in a number of significant ways that pertain to both the host (laboratory animals have a lower heat tolerance than humans) and the tumor (animal tumors have a different, usually more distinct, antigenicity and a shorter cell cycle time than human neoplasms; transplantable tumors are more homogeneous and more easily cured, by cytotoxic drugs, for example, than human cancers) (this volume, chapter 23; Dickson 1977). Even accepting these differences, no data is available on the heat response of animal tumors that could be considered equivalent (at least anatomically) to the common solid internal tumors of man: carcinoma of the bronchus, stomach, and colon. It must be remembered, therefore, that the majority of findings discussed in this chapter refer to artificially induced, transplantable tumors that are convenient for experimental work, but may be poor models of human cancer in etiology, anatomic site, pathology, pathogenesis, and in the response to heat. The increasing volume of results with animal tumor models can provide guidance for clinical research, but the problems of using hyperthermia for the treatment of human cancer remain formidable. It is reasonable to ask how the role of factors (some of them, such as inflammation, complex and multifaceted) governing the response of human tumors to heat can be evaluated when little is known about the basic degree of heating (temperature × time) required to produce regression of such cancers. The difficulties are compounded by the currently elementary state of temperature monitoring systems, the paucity of knowledge of heat dosimetry, and the primitive nature of heating technology. Furthermore, even if this information was available, the final temperature distribution for a given tumor: normal tissue geometry may vary appreciably, depending on the mode of induction of the hyperthermia (Hahn et al. 1980). Such temperature profiles could be vitally important from the point of view of normal tissue damage (the distribution of hyperthermic damage as distinct from thermal distribution) (Suit and Blitzer 1980) and repair and the generation of an inflammatory response, for example.

Another question is whether the design of our experiments in animal systems provides the type of information likely to be of benefit in heating human tumors. The tumors examined in this laboratory and most of the animal tumors used by other workers have regressed after very modest degrees of hyperthermia—heating durations in the region of one hour on one or two occasions at 42°C to 43°C. Human tumors have proved much less obliging in their response to heat. It is to be expected that with longer heating times and/or higher temperatures, such as will be necessary for human cancers, factors like hypoxia and acidity would assume greater importance. More data is needed, therefore, on tumors like the MC-7 sarcoma, which required three hours of hyperthermia to produce regression, and in which there occurred a decrease in pH_i after hyperglycemia.

Tumor blood flow currently is attracting much attention as a major determinant in tumor response to heating. The data indicate a possible inverse relationship between blood flow and heat sensitivity. The response of tumor blood flow to heat has two components: (1) an inability to increase during heating to cope with the imposed heat load, thus creating a heat reservoir effect; and (2) shut-down of the microcirculation with duration of heating, probably caused by intravascular stasis and clumping of blood elements. These changes in blood flow have been quantitated at 42°C to 43°C. It is certain there are tumors in which the blood flow will be adequate to cope with a heat load in this temperature range. This may be unimportant, provided the tumor microcirculation shuts down or can be closed after heating, assuming 42°C to 43°C is the tumor lethal temperature and will damage a considerable fraction of the tumor cell population. Rather than hope that the tumor circulation will close down with heating, it seems more logical to short-circuit the chain of events by selectively inhibiting the circulation via elevated blood glucose, and then heating the tumor. The length of time during which the tumor circulation should be inhibited may prove critical, even with total inhibition of blood flow. Results with hyperthermia indicate that irreversible damage to all cells of a tumor is necessary for cure (see chapter 23), and it is known that malignant cells have an exceptional resistance to deficiency of oxygen. Tumor takes occur regularly when necrotic areas of carcinomas or sarcomas are transplanted (Goldacre and Sylvén 1962); after perfusion for over 50 hours with a solution containing only 10% of the oxygen normally consumed in vivo, mammary carcinomas grew as well as control transplants (Gullino 1968). Many types of tumor, including the VX-2 carcinoma (Shah and Dickson 1978) and numerous tissue culture cell lines, form tumors readily in the host after continuous storage for several years at liquid nitrogen temperature ($-196°C$).

These considerations may, therefore, favor multiple dose hyperthermia, while tumor hypoxia or anoxia is maintained by hyperglycemia. Radiotherapy would be best applied initially to kill oxygenated cells. Heat and radiotherapy could be combined or used sequentially in several sessions to the maximum tolerated dosage of radiation, with elevated blood sugar level to potentiate the heat fractions. This could then be followed by the more prolonged hypoxic multiple-dose hyperthermia-hyperglycemia protocol. Manipulation of tumor blood flow is potentially one of the best adjuvants to hyperthermia now recognized and has implications for investigation and therapy of tumors beyond the field of hyperthermia. Blood flow inhibition by hyperglycemia would enable tumors to be treated or studied in isolation from the normal tissues, promoting increased specificity in cancer therapy. Examples of this application could be the deposition of high concentrations of drugs in glucose-isolated tumors by the techniques of interventional radiology and the ischemic infarction of tumors in a manner akin to embolization of the arterial supply.

It will be necessary to ascertain the lowest level of elevated blood glucose required to cause inhibition of tumor blood flow. It also may be feasible to use nonmetabolizable sugars or other substances for this purpose and so decrease the metabolic impact on the host of prolonged elevation of blood glucose. If decrease in tumor pH proves to be of value in the response of tumors to heat, then an initial period of high-glucose infusion could be used to provide the tumor with enough substrate to produce the required lactic acid; blockade of the tumor microcirculation could then be maintained by a more inert agent.

In view of the great importance of blood flow in tumor thermal response and the variability of human tumors, it would seem appropriate to monitor tumor blood flow routinely, in addition to temperature, during and after hyperthermia. A convenient method for doing so is already available in the ^{133}Xe clearance technique. Its use would enable tumor heating not only to be put on a firm physiologic basis, but would enable the heating to be assessed, monitored, and adjusted for individual cancers, thus providing data for a custom-designed tumor heating protocol.

It is intriguing to wonder whether, and to what extent, a curative heat dose at 42°C to 43°C to a tumor would be effective in the presence of an uninterrupted blood flow. This would be another approach to separating the effects of hypoxia from direct damage to a tumor by heat. Two other components of the heating response that cloud the issue in animal tumors are the immune response and the inflammatory reaction. The parts played by these factors make it even more difficult to evaluate the role of hypoxia and pH in tumor sensitivity. We require more prolonged heating experiments on tumors that are only slightly or nonimmunogenic; the use of reagents, such as steroids, to block the local inflammatory reaction may be instructive.

Participation of the immune system (Hersh, Mavligit, and Gutterman 1976) in tumor regression and the intriguing phenomenon of disappearance of distant tumor following curative heating of the primary cancer (abscopal response) appear to be confined to animal tumor systems. Indeed, the practice in human hyperthermia trials of heating a superficial tumor and using another similar tumor as an untreated control implies acceptance that an abscopal response does not occur.

Bearing in mind the limitations of our systems, and the consequent limited applicability of the conclusions, what practical guidelines are emerging in relation to the treatment of human cancer by heat?

First, data on tumor cell kinetics have been of value for planning drug therapy in animal tumors (Skipper 1971) (e.g., L1210 leukemia in mice), but such data have proved less valuable in human cancer (Hall 1971). With hyperthermia, the situation is very complex, and even if we could obtain repeated samples of tumor from patients, there is little indication from the results with animal tumors that kinetic data would be of value for therapy planning. The results do suggest that large tumors are equally (or more) vulnerable to heat damage as small tumors, possibly owing to an increased heat sensitivity of Q cells in larger tumors, rendering them less able to reenter cycle. This is in agreement with the clinical findings of Storm and associates (1979, in press) that human tumors greater than 5 cm in diameter respond better to heating than do cancers of lesser diameter. The results also may be associated with the poorer blood flow associated with larger tumors. The finding that tumor hemisection brings Q cells into cycle favors surgical debulking of accessible large tumors prior to hyperthermia. This would present the therapist with a larger growth fraction to destroy, and would minimize the hazards from rapid tumor lysis and liberation of breakdown products following heating.

Second, as pointed out elsewhere (Dickson and Calderwood 1980), results to date indicate that tumors fall into two zones of thermal sensitivity, 42°C to 43°C and 45°C to 50°C. From the therapeutic point of view, these zones are of great importance, since only in the lower range can the selective heat sensitivity of cancer cell biochemistry be exploited. Only the lower range is suitable for regional hyperthermia (e.g., by perfusion) and total-body hyperthermia. The effect of inhibiting tumor blood flow by elevated blood sugar level should now be tested by clinical trial in these two types of hyperthermia.

For the common types of human solid tumor, such as cancer of the lung or bowel, a rational approach would be the use of the higher temperature range, 45°C to 50°C, for local heating as approved methods of achieving such temperature fields become available. The limited clinical results already obtained support the view that tumor temperatures in this range are an attainable goal. Again, the inhibition of tumor blood flow by hyperglycemia should make for selective and more uniform tumor heating and may enable the heating temperature to be lowered; maintenance of the hyperglycemia will prolong the tumor hypoxia if the 45°C to 50°C heating does not totally destroy the tumor microcirculation.

Third, the value of membrane solubilizers, such as procaine and the polyamines, should be energetically investigated. The great potential of these agents is that they could be employed locally in tumors as well as systemically to sensitize metastases, but we must ascertain the extent to which such substances act selectively on cancer cells. The merits of alcohol as a membrane thermostabilizer for normal cells warrant analysis; again the agent would have systemic application, but we have little knowledge of its selectivity for normal versus malignant cells in this respect. We must keep in mind also the possibility of using local heating for defined tumor masses in association with total-body hyperthermia. This approach has been described in the literature, but there is insufficient data to comment upon its value.

Fourth, as stressed in this chapter, participation of the host is an important component in tumor response to heating, whether for local, regional, or total-body hyperthermia. At the local or regional level, inducing an aggressive normal tissue reaction by injection of a chemical mediator (such as serotonin or perhaps another vasoactive substance) may be a reasonable course. On the systemic level, mustering the host defenses in the form of macrophages and possibly a nonspecific immune response to the heated tumor has precedent in the results obtained years ago with Coley's toxins and recently, to a more limited extent, with immunostimulators like BCG and *C. parvum.* While little is known about how best to employ such agents, the undoubted lethal potential to cancer cells of cells like macrophages should not be overlooked.

References

Adams, G. E. et al. Hyperthermic enhancement of the hypoxic cell toxicity of electron-affinic agents and other drugs—a review. In *Proceedings of the Third International Symposium on Cancer Therapy by Hyperthermia, Drugs and Radiation,* eds. L. A. Dethlefsen and W. C. Dewey. Bethesda, Md.: National Cancer Institute (special journal supplement), in press.

Aherne, W. A.; Camplejohn, R. S.; and Wright, N. A. *An introduction to cell population kinetics.* London: Arnold, 1977.

Algire, G. H., and Legallais, F. Y. Vascular reactions of normal and malignant tissues *in vivo.* IV. The effect of peripheral hypotension on transplanted tumors. *J. Natl. Cancer Inst.* 12:399–421, 1951.

Algire, G. H.; Legallais, F. Y.; and Park, H. D. Vascular reactions of normal and malignant tissues *in vivo.* II. The vascular reaction of normal and neoplastic tissues of mice to a bacterial polysaccharide from Serratia Marcescens culture filtrates. *J. Natl. Cancer Inst.* 8:53–62, 1947.

Anghileri, L. J. Tumor growth inhibition by ammonium chloride induced acidosis. *Int. J. Clin. Pharmacol. Biopharm.* 12:320–326, 1975.

Appelgren, L. K. Methods of recording tumor blood flow. In *Tumor blood circulation, angiogenesis, vascular morphology and blood flow of experimental and human tumors,* ed. H.-I. Peterson. Boca Raton, Fla.: CRC Press, 1979, pp. 97–98.

Ashby, B. S. pH studies in human malignant tumours. *Lancet* 2:312–315, 1966.

Barrett, J. C. A mathematical model of the mitotic cycle and its application to the interpretation of percentage labelled mitoses data. *J. Natl. Cancer Inst.* 37:443–450, 1966.

Baserga, R., and Malamud, D. *Autoradiography.* New York: Harper & Row, 1969.

Bass, H., and Moore, J. L. Lethality in mammalian cells due to hyperthermia under oxic and hypoxic conditions. *Int. J. Radiat. Biol.* 33:57–67, 1978.

Bates, R. G. *Determination of pH: theory and practice.* New York: John Wiley & Sons, 1964, pp. 364–367.

Bhuyan B. K. Kinetics of cell kill by hyperthermia. *Cancer Res.* 39:2277–2284, 1979.

Bhuyan, B. K. et al. Sensitivity of different cell lines and of different phases in the cell cycle to hyperthermia. *Cancer Res.* 37:3780–3784, 1977.

Bicher, H. I. et al. Effects of hyperthermia on normal and tumor microenvironment. *Radiology* 137:523–530, 1980.

Bicher, H. I. et al. Physiological mechanism of action of localized microwave hyperthermia. In *Proceedings of the Third International Symposium on Cancer Therapy by Hyperthermia, Drugs and Radiation,* eds. L. A. Dethlefsen and W. C. Dewey. Bethesda, Md.: National Cancer Institute (special journal supplement), in press.

Black, M. M. General discussion of interaction of humoral and cellular mechanisms in tumor immunity. *Natl. Cancer Inst. Monogr.* 35:276, 1972.

Bodansky, O. *Biochemistry of human cancer.* New York: Academic Press, 1975, pp. 33–60.

Born, R., and Trott, K. R. The influence of hyperthermia on chronically hypoxic cells. In *Proceedings of the Second International Symposium on Cancer Therapy by Hyperthermia and Radiation,* eds. C. Streffer et al. Baltimore and Munich: Urban & Schwarzenberg, 1978, p. 177.

Bruce, W. R., and Valeriote, F. A. Comparison of the sensitivity of normal hematopoietic and transplanted lymphoma colony-forming cells to chemotherapeutic agents administered *in vivo. J. Natl. Cancer Inst.* 42:1015–1023, 1966.

Burgess, E. A., and Sylvén, B. Glucose, lactate and lactic dehydrogenase activity in normal interstitial fluid and that of solid mouse tumors. *Cancer Res.* 22:581–588, 1962.

Burk, D.; Woods, M.; and Hunter, J. On the significance of glycolysis for cancer growth, with special reference to Morris rat hepatomas. *J. Natl. Cancer Inst.* 38:839–863, 1967.

Calderwood, S. K. pH studies on the Yoshida sarcoma *in vivo.* In *Studies on population kinetics and pH in the response of the Yoshida sarcoma in the rat to hyperthermia.* Ph.D. thesis, University of Newcastle-upon-Tyne, England, 1978, p. 83.

Calderwood, S. K., and Dickson, J. A. Rapid method for measuring intracellular pH *in vivo. Cell Biol. Int. Rep.* 2:327–337, 1978.

Calderwood, S. K., and Dickson, J. A. Influence of tumour volume and cell kinetics on the response of the solid Yoshida sarcoma to hyperthermia (42°C). *Br. J. Cancer* 41:22–32, 1980a.

Calderwood, S. K., and Dickson, J. A. Effect of hyperglycemia on blood flow, pH and response to hyperthermia (42°C) of the Yoshida sarcoma in the rat. *Cancer Res.* 40:4728–4733, 1980b.

Calderwood, S. K., and Dickson, J. A. Inhibition of tumor blood flow at high blood sugar levels; effects on tumor pH and hyperthermia. In *Proceedings of the Third International Symposium on Cancer Therapy by Hyperthermia, Drugs and Radiation*, eds. L. A. Dethlefsen and W. C. Dewey. Bethesda, Md.: National Cancer Institute (special journal supplement), in press.

Cataland, S.; Cohen, C.; and Sapirstein, L. A. Relationship between size and perfusion rate of transplanted tumors. *J. Natl. Cancer Inst.* 29:389–394, 1962.

Cater, D. B.; Grigson, C. M. B.; and Watkinson, D. A. Changes in oxygen tension in tumors induced by vasoconstrictor and vasodilator drugs. *Acta Radiol. Ther. Phys. Biol.* 58:401–434, 1962.

Cater, D. B.; Silver, I. A.; and Watkinson, D. A. Combined therapy with 220 KV roentgen and 10 cm microwave heating in rat hepatoma. *Acta Radiol. Ther. Phys. Biol.* 2:321–336, 1964.

Cavaliere, R. et al. Selective heat sensitivity of cancer cells: biochemical and clinical studies. *Cancer* 20:1351–1381, 1967.

Chin, J. H.; Parsons, L. M.; and Goldstein, D. B. Increased cholesterol content of erythrocyte and brain membranes in ethanol-tolerant mice. *Biochim. Biophys. Acta* 513:358–363, 1978.

Cleaver, J. E. *Thymidine metabolism and cell kinetics.* Amsterdam: North Holland, 1967.

Cooper, E. H. The biology of cell death in tumours. *Cell Tissue Kinet.* 6:87–95, 1973.

Cravalho, E. G.; Fox, L. R.; and Kan, J. C. The application of the bioheat equation to the design of thermal protocols for local hyperthermia. *Ann. NY Acad. Sci.* 335:86–96, 1980.

Crile, G., Jr. Heat as an adjunct to the treatment of cancer; experimental studies. *Cleve. Clin. Q.* 28:75–89, 1961.

Crile, G., Jr. Selective destruction of cancers after exposure to heat. *Ann. Surg.* 156:404–407, 1962.

Crile, G., Jr. The effects of heat and radiation on cancers implanted into the feet of mice. *Cancer Res.* 23:372–380, 1963.

Crile, G., Jr. Inhibition of growth of mouse tumors by injections of serotonin or serotonin and histamine combined. *Cleve. Clin. Q.* 33:25–29, 1966.

Crissman, H. A.; Mullaney, P.F.; and Steinkamp, J. A. Methods and applications of flow systems for analysis and sorting of mammalian cells. In *Methods in cell biology*, vol. 9, ed. D. M. Prescott. New York: Academic Press, 1975, pp. 179–246.

Davies, M. *Functions of biological membranes. Outline studies in biology.* London: Chapman & Hall, 1973.

Dawson, K. B. et al. Studies on the radiobiology of a rat sarcoma treated *in situ* and assayed *in vitro*. *Eur. J. Cancer* 9:59–68, 1973.

De Duve, C., and Wattiaux, R. Functions of lysosomes. *Annu. Rev. Physiol.* 28:436–492, 1966.

Denekamp, J., and Fowler, J. F. Cell proliferation kinetics and radiation therapy. In *Cancer, a comprehensive treatise*, vol. 6, ed. F. F. Becker. New York and London: Plenum, 1977, pp. 101–138.

Dethlefsen, L. A. An evaluation of radioiodine labelled 5-iodo-2'-deoxyuridine as a tracer for measuring cell loss from solid tumors. *Cell Tissue Kinet.* 4:123–138, 1971.

Dethlefsen, L. A., and Dewey, W. C., eds. *Proceedings of the Third International Symposium on Cancer Therapy by Hyperthermia, Drugs and Radiation.* Bethesda, Md.: National Cancer Institute (special journal supplement), in press.

Dewey, W. C., and Freeman, M. L. Rationale for use of hyperthermia in cancer therapy. In *Thermal characteristics of tumors: applications in detection and treatment*, eds. R. K. Jain and P. M. Gullino. *Ann. NY Acad. Sci.* 335:373–378, 1980.

Dewey, W. C.; Furman, S. C.; and Miller, H. H. Comparison of lethality and chromosomal damage induced by x-rays in synchronized Chinese hamster cells *in vitro*. *Radiat. Res.* 43:561–581, 1970.

Dewey, W. C.; Thrall, D. E.; and Gillette, E. L. Hyperthermia and radiation—a selective thermal effect on chronically hypoxic tumor cells *in vivo*. *Int. J. Radiat. Oncol. Biol. Phys.* 2:99–103, 1977.

Dickson, J. A. Hazards and potentiators of hyperthermia. In *Proceedings of the First International Symposium on Cancer Therapy by Hyperthermia and Radiation*, eds. M. J. Wizenberg and J. E. Robinson. Chicago: American College of Radiology, 1976, pp. 134–150.

Dickson, J. A. The effects of hyperthermia in animal tumor systems. In *Selective heat sensitivity of cancer cells. Recent Results in Cancer Research*, vol. 59, eds. A. Rossi-Fanelli et al. Berlin and New York: Springer-Verlag, 1977.

Dickson, J. A. The sensitivity of human cancer to hyperthermia. In *Proceedings of Conference on Clinical Prospects for Hypoxic Cell Sensitizers and Hyperthermia*, eds. W. L. Caldwell and R. E. Durand. Madison Wis.: University of Wisconsin Press, 1978, pp. 174–193.

Dickson J. A. Destruction of solid tumors by heating with radiofrequency energy. *Radio Science* 14:6S, 285–295, 1979.

Dickson, J. A. et al. Immune regression of metastases after hyperthermic treatment of primary tumors. In *Proceedings EORTC Metastasis Conference*, eds. K. Hellman, P. Hilgard, and S. Eccles. The Hague: Martinus Nijhoff, 1980, pp. 260–265.

Dickson J. A. et al. Tumor eradication in the rabbit by radiofrequency heating. *Cancer Res.* 37:2162–2169, 1977.

Dickson, J. A., and Calderwood, S. K. In vivo hyperthermia induces entry of non-proliferating cells into cycle. *Nature* 263:772–774, 1976.

Dickson, J. A., and Calderwood, S. K. Effects of hyperglycemia and hyperthermia on the pH, glycolysis and respiration of the Yoshida sarcoma *in vivo*. *J. Natl. Cancer Inst.* 63:1371–1381, 1979.

Dickson, J. A., and Calderwood, S. K. Temperature range and selective sensitivity of tumors to hyperthermia: a critical review. In *Thermal characteristics of tumors: applications in detection and treatment*, eds. R. K. Jain and P. M. Gullino. *Ann. NY Acad. Sci.* 335:180-207, 1980.

Dickson, J. A.; Calderwood, S. K.; and Jasiewicz, M. L. Radio-frequency heating of tumours in rodents. *Eur. J. Cancer* 13:753–763, 1977.

Dickson, J. A., and Ellis, H. A. Stimulation of tumour cell dissemination by raised temperature (42°C) in rats with transplanted Yoshida tumours. *Nature* 248:354–358, 1974.

Dickson, J. A., and Ellis, H. A. The influence of tumor volume and the degree of heating on the response of solid Yoshida sarcoma to hyperthermia (40–42°C). *Cancer Res.* 36:1188–1195, 1976.

Dickson, J. A., and Muckle, D. S. Total body hyperthermia versus primary tumor hyperthermia in the treatment of the rabbit VX2 carcinoma. *Cancer Res.* 32: 1916–1923, 1972.

Dickson, J. A., and Oswald, B. E. The sensitivity of a malignant cell line to hyperthermia (42°C) at low intracellular pH. *Br. J. Cancer* 34:262–271, 1976.

Dickson, J. A., and Shah, D. M. The effect of hyperthermia (42°C) on the biochemistry and growth of a malignant cell line. *Eur. J. Cancer* 8:561–571, 1972.

Dickson, J. A., and Shah, S. A. Technology for the hyperthermic treatment of large solid tumours at 50°C. *Clin. Oncol.* 3:301–318, 1977.

Dickson, J. A., and Shah, S. A. Hyperthermia and the immune response in cancer therapy: a review. *Cancer Immunology and Immunotherapy* 9:1–10, 1980.

Dickson, J. A., and Shah, S. A. Hyperthermia, the immune response and tumor metastasis. In *Proceedings of the Third International Symposium on Cancer Therapy by Hyperthermia, Drugs and Radiation*, eds. L. A. Dethlefsen and W. C. Dewey. Bethesda, Md.: National Cancer Institute (special journal supplement), in press.

Dickson, J. A., and Suzangar, M. *In Vitro-in vivo* studies on the susceptibility of the solid Yoshida sarcoma to drugs and hyperthermia. *Cancer Res.* 34:1263–1274, 1974.

Dickson, J. A., and Suzangar, M. The *in vitro* response of human tumors to drugs and hyperthermia (42°C) and its relevance to clinical oncology. In *Organ culture in biomedical research*, eds. M. Balls and M. Monnickendam. Cambridge and New York: Cambridge University Press, 1976a.

Dickson, J. A., and Suzanger, M. A predictive *in vitro* assay for the sensitivity of human solid tumors to hyperthermia (42°C) and its value in patient management. *Clin. Oncol.* 2:141–155, 1976b.

Dustin, A. P., Sr. La colchicine réactif de l'imminence caryocinétique. *Archives of Portuguese Science and Biology* 5:38–43, 1936.

Dyson, P., and Heppelston, A. G. Cell kinetics of urethane-induced murine pulmonary adenomata. II. The growth fraction and cell loss factor. *Br. J. Cancer* 33:105–111, 1976.

Eberhart, R. C.; Shitzer, A.; and Hernandez, E. J. Thermal dilution methods: estimation of tissue blood flow and metabolism. *Ann. NY Acad. Sci.* 335:107–131, 1980.

Eddy, H. A.; Sutherland, R. M.; and Chmielewski, G. Tumor microvascular response: hyperthermia, drug, radiation combinations. In *Proceedings of the Third International Symposium on Cancer Therapy by Hyperthermia, Drugs and Radiation*, eds. L. A. Dethlefsen and W. C. Dewey. Bethesda, Md.: National Cancer Institute (special journal supplement), in press.

Eden, M.; Haines, B.; and Kahler, H. The pH of rat tumors measured *in vivo*. *J. Natl. Cancer Inst.* 16:541–556, 1955.

Folkman, J. Tumor angiogenesis. In *Cancer, a comprehensive treatise*, vol. 3, ed. F. F. Becker. New York and London: Plenum, 1975, pp. 355–385.

Fowler, J. F.; Thomlinson, R. F.; and Howes, E. E. Time-dose relationships in radiotherapy. *Eur. J. Cancer* 6:207–221, 1970.

Freeman, M. L.; Dewey, W. C.; and Hopwood, L. E. Effect of pH on hyperthermic cell survival. *J. Natl. Cancer Inst.* 58:1837–1839, 1977.

Frindel, E., and Tubiana, M. Radiobiology and the cell cycle. In *The cell cycle and cancer*, ed. R. Baserga. New York: Marcel Dekker, 1971, pp. 389–447.

Gelfant, S. A new concept of tissue and tumor cell proliferation. *Cancer Res.* 37:3845–3862, 1977.

Gerner, E. W. et al. Factors regulating membrane permeability after thermal resistance. In *Thermal characteristics of tumors: applications in detection and treatment*, eds. R. K. Jain and P. M. Gullino. *Ann. NY Acad. Sci.* 335:254–280, 1980.

Gerweck, L. E. Modification of cell lethality at elevated temperatures: the pH effect. *Radiat. Res.* 70:224–235, 1977.

Gerweck, L. E. Influence of microenvironmental conditions on sensitivity to hyperthermia or radiation for cancer therapy. In *Proceedings of Conference on Clinical Prospects for Hypoxic Cell Sensitizers and Hyperthermia*, eds. W. L. Caldwell and R. E. Durand. Madison, Wis.: University of Wisconsin, 1978, pp. 113–126.

Gerweck, L. E. Effect of microenvironmental factors on the response of cells to single and fractionated heat treatments. In *Proceedings of the Third International Symposium on Cancer Therapy by Hyperthermia, Drugs and Radiation*, eds. L. A. Dethlefsen and W. C. Dewey. Bethesda, Md.: National Cancer Institute (special journal supplement), in press.

Gerweck, L. E.; Gillette, E. L.; and Dewey, W. C. Killing of Chinese hamster cells *in vitro* by heating under hypoxic or aerobic conditions. *Eur. J. Cancer* 10:691–693, 1974.

Gilbert, C. W. The labeled mitoses curve and the estimation of the parameters of the cell cycle. *Cell Tissue Kinet.* 5:53–63, 1972.

Giovanella, B. C.; Lohman, W. A.; and Heidelberger, C. Effects of elevated temperatures and drugs on the viability of L1210 leukemia cells. *Cancer Res.* 30:1623–1631, 1970.

Goldacre, R. J., and Sylven, B. On the access of blood-borne dyes to various tumor regions. *Br. J. Cancer* 16:306–311, 1962.

Gullino, P. M. The internal milieu of tumors. *Prog. Exp. Tumor Res.* 8:1–25, 1966.

Gullino, P. M. *In vitro* perfusion of tumors. In *Organ perfusion and preservation*, eds. J. C. Norman et al. New York: Appleton-Century-Crofts, 1968, pp. 877–898.

Gullino, P. M. Techniques for the study of tumor physiopathology. *Methods in Cancer Research* 5:45–91, 1970.

Gullino, P. M. Extracellular compartments of solid tumors. In *Cancer, a comprehensive treatise*, vol. 3, ed. F. Becker. New York and London: Plenum, 1975, pp. 327–350.

Gullino, P. M. Influence of blood supply on thermal properties and metabolism of mammary carcinomas. In *Thermal characteristics of tumors: applications in detection and treatment*, eds. R. K. Jain and P. M. Gullino. *Ann. NY Acad. Sci.* 335:1–18, 1980.

Gullino, P. M. et al. Modifications of the acid-base state of the internal milieu of tumors. *J. Natl. Cancer Inst.* 34:857–869, 1965.

Gullino, P. M.; Clark, S. H.; and Grantham, F. H. The interstitial fluid of solid tumors. *Cancer Res.* 24:780–798, 1964.

Gullino, P. M., and Grantham, F. H. Studies on the exchange of fluids between host and tumor. II. The blood flow of hepatomas and other tumors in rats and mice. *J. Natl. Cancer Inst.* 27:1465–1491, 1961.

Gullino, P. M.; Grantham, F. H.; and Courtney, A. H. Utilization of oxygen by transplanted tumors *in vivo*. *Cancer Res.* 27:1020–1030, 1967a.

Gullino, P. M.; Grantham, F. H.; and Courtney, A. H. Glucose consumption by transplanted tumors *in vivo*. *Cancer Res.* 27:1031–1040, 1967b.

Gullino, P. M.; Grantham, F. H.; and Smith, S. H. The interstitial water space of tumors. *Cancer Res.* 25:727–731, 1965.

Gullino, P. M.; Yi, P. N.; and Grantham, F. H. Relationship between temperature and blood supply or consumption of oxygen and glucose by rat mammary carcinomas. *J. Natl. Cancer Inst.* 60:835–847, 1978.

Hahn, G. M. Metabolic aspects of the role of hyperthermia in mammalian cell inactivation and their possible relevance to cancer treatment. *Cancer Res.* 34:3117–3123, 1974.

Hahn, G. M. Interactions of drugs and hyperthermia *in vitro* and *in vivo*. In *Proceedings of the Second International Symposium on Cancer Therapy by Hyperthermia and Radiation*, eds. C. Streffer et al. Baltimore and Munich: Urban & Schwarzenberg, 1978, pp. 72–79.

Hahn, G. M. et al. Some heat transfer problems associated with heating by ultrasound microwaves or radiofrequency. In *Thermal characteristics of tumors: applications in detection and treatment*, eds. R. K. Jain and P. M. Gullino. *Ann. NY Acad. Sci.* 335:327–396, 1980.

Hall, E. J. *Radiobiology for the radiobiologist*. New York: Harper & Row, 1973, pp. 49–62.

Hall, J. L., and Baker, D. A. *Cell membranes and ion transport*. London and New York: Longman, 1977, pp. 1–24.

Hall, T. C. Limited role of cell kinetics in clinical cancer chemotherapy. *J. Natl. Cancer Inst.* 34:15–17, 1971.

Harguindey, S.; Henderson, E. S.; and Naeher, C. Effects of systemic acidification of mice with sarcoma 180. *Cancer Res.* 39:4364–4371, 1979.

Henderson, M. A., and Pettigrew, R. T. Induction of controlled hyperthermia in the treatment of cancer. *Lancet* 1:1275–1277, 1971.

Henle, K. J., and Dethlefsen, L. A. Heat fractionation and thermotolerance: a review. *Cancer Res.* 38:1843–1851, 1978.

Henle, K. J., and Dethlefsen, L. A. Time-temperature relationships for heat induced killing of mammalian cells. In *Thermal characteristics of tumors: applications in detection and treatment*, eds. R. K. Jain and P. M. Gullino. *Ann. NY Acad. Sci.* 335:234–252, 1980.

Henle, K. J., and Dethlefsen, L. A. Heat fractionation and stepdown heating of murine mammary tumors in the foot. In *Proceedings of the Third International Symposium on Cancer Therapy by Hyperthermia, Drugs and Radiation*, eds. L. A. Dethlefsen and W. C. Dewey. Bethesda, Md.: National Cancer Institute (special journal supplement), in press.

Henle, K. J., and Leeper, D. B. Combinations of hyperthermia (40, 45°C) with radiation. *Radiology* 121:451–454, 1976.

Hermens, A. F., and Barendsen, G. W. Changes of cell proliferation characteristics in a rat rhabdomyosarcoma before and after X-irradiation. *Eur. J. Cancer* 5:173–189, 1969.

Hersh, E. V.; Mavligit, G. M.; and Gutterman, J. U. Immunodeficiency in cancer and the importance of immune evaluation of the cancer patient. In *Symposium on immunotherapy in malignant disease*, ed. W. D. Terry. *Med. Clin. North Am.* 60(3):623–640, 1976.

Hill, B. T., and Baserga, R. The cell cycle and its significance for cancer treatment. *Cancer Treat. Rep.* 2:159–175, 1975.

Hill, S. A., and Denekamp, J. The effect of vascular occlusion on the thermal sensitization of a mouse tumor. *Br. J. Radiol.* 51:997–1002, 1978.

Hill, S. A., and Denekamp, J. The response of six mouse tumors to combined heat and X-rays; implications for therapy. *Br. J. Radiol.* 52:209–218, 1979.

Hinke, J. A. M., and Menard, M. R. Evaluation of the DMO method for measuring intracellular pH. *Respir. Physiol.* 33:31–40, 1978.

Hofer, K. G., and Mivechi, N. F. Tumor cell sensitivity to hyperthermia as a function of extracellular and intracellular pH. *J. Natl. Cancer Inst.* 65:621–625, 1980.

Hofer, K. G.; Prensky, W.; and Hughes, W. L. Death and metastatic distribution of tumor cells in mice monitored with ^{125}I-Iododeoxyuridine. *J. Natl. Cancer Inst.* 43:763–773, 1969.

Howard, A., and Pelc, S. R. Synthesis of deoxyribonucleic acid in normal and irradiated cells and its relation to chromosome breakage. *Heredity* (suppl). 6:261–273, 1953.

Hult, R. L., and Larsen, R. E. Dissociation of 5-fluorouracil uptake from intracellular pH in Walker 256 carcinosarcoma. *Cancer Treat. Rep.* 60:867–873, 1976.

Jain, R. K. Temperature distributions in normal and neoplastic tissues during normothermia and hyperthermia. In *Thermal characteristics of tumors: applications in detection and treatment* eds. R. K. Jain and P. M. Gullino. *Ann. NY Acad. Sci.* 335:48–66, 1980.

Johnson, H. J. The action of short radio waves on tissues. III. A comparison of the thermal sensitivities of transplantable tumors *in vivo* and *in vitro*. *Am. J. Cancer* 38:533–550, 1940.

Jung, H. Induction of thermotolerance and sensitization in CHO cells by combined hyperthermic treatments at 40°C and 43°C. In *Proceedings of the Third International Symposium on Cancer Therapy by Hyperthermia, Drugs and Radiation*, eds. L. A. Dethlefsen and W. C. Dewey. Bethesda, Md.: National Cancer Institute (special journal supplement), in press.

Kahler, H., and Robertson, W. V. Hydrogen ion concentration of normal liver and hepatic tumors. *J. Natl. Cancer Inst.* 3:495–501, 1943.

Kal, H. B.; Hatfield, M.; and Hahn, G. M. Cell cycle progression of murine sarcoma cells after X-irradiation or heat shock. *Radiology* 117:215–217, 1975.

Kal, H. B., and Hahn, G. M. Kinetic responses of murine sarcoma cells to radiation and hyperthermia *in vivo* and *in vitro*. *Cancer Res.* 36:1923–1929, 1976.

Kallman, R. F., and Rockwell, S. Effects of radiation on animal tumor models. In *Cancer, a comprehensive treatise. Radiotherapy, surgery and immunotherapy*, vol. 6, ed. F. F. Becker. New York and London: Plenum, 1977, pp. 225–280.

Kallman, R. F., and Tapley, N. D. V. Radiation sensitivity and recovery patterns of spontaneous and isologously transplanted mouse tumors. *Acta Unio Internationale Contra Le Cancrum* 20:1216–1221, 1964.

Kang. M.-S.; Song, C. W.; and Levitt, S. H. Role of vascular function in response of tumors *in vivo* to hyperthermia. *Cancer Res.* 40:1130–1135, 1980.

Karlsson, L. et al. Intratumor distribution of vascular and extravascular spaces. *Microvasc. Res.* 19:71–79, 1980.

Kase, K., and Hahn, G. M. Differential heat response of normal and transformed human cells in tissue culture. *Nature* 255:228–230, 1975.

Kety, S. S. Theory and applications of the exchange of inert gas at the lungs and tissues. *Pharmacol. Rev.* 3:1–15, 1951.

Kim, J. H. et al. 5-thio-D-glucose selectively potentiates hyperthermic killing of hypoxic tumor cells. *Science* 200:206–207, 1978.

Kim, J. H., and Hahn, E. W. Clinical and biological studies of localized hyperthermia. *Cancer Res.* 39:2258–2261, 1979.

Kim, S. H.; Kim, J. H.; and Hahn, E. W. Enhanced killing of hypoxic tumor cells by hyperthermia. *Br. J. Radiol.* 48:872–874, 1975.

Kimelberg, H. K., and Mayhew, E. Increased ouabain-sensitive $^{86}Rb^+$ uptake and sodium and potassium ion-activated adenosine triphosphate activity in transformed cell lines. *J. Biol. Chem.* 250:100–104, 1975.

Kjartansson, I. Tumour circulation; an experimental study in the rat with a comparison of different methods for estimation of tumor blood flow. *Acta Chir. Scand.* (suppl.) 471:503–509, 1976.

Klyuyeva, N. G., and Roskin, G. I. *Biotherapy of malignant tumors.* Translated by J. J. Oliver. London and New York: Pergamon Press, 1963, pp. 9–14.

Krementz, K. T. In discussion of paper by Cavaliere et al. *Ann. NY Acad. Sci.* 335:325–326, 1980.

Kruuv, J. A.; Inch, W. R.; and McCredie, J. A. Blood flow and oxygenation of tumors in mice. *Cancer* 20:51–70, 1967.

Kumar, A. R. V. et al. Comparative rates of dead tumor cell removal from brain, muscle, subcutaneous tissue and peritoneal cavity. *J. Natl. Cancer Inst.* 52:1751–1755, 1974.

Laird, A. K. The dynamics of tumour growth. *Br. J. Cancer* 28:490–502, 1964.

Lala, P. K. Studies on tumour cell population kinetics. *Methods in Cancer Research* 6:3–95, 1971.

Lala, P. K., and Patt, H. M. Cytokinetic analysis of tumor growth. *Proc. Natl. Acad. Sci. USA* 56:1735–1742, 1966.

Lam, P. H. M. et al. A simple technique of measuring liver blood flow—intrasplenic injection of ^{133}Xenon. *Acta Chir. Scand.* 145:95–100, 1979.

Larkin, J. M. A clinical investigation of total body hyperthermia as cancer therapy. *Cancer Res.* 39: 2252–2254, 1979.

Larkin, J. M. In discussion of paper by Law and Pettigrew. *Ann. NY Acad. Sci.* 335:309, 1980.

LeVeen, H. H. et al. Tumor eradication by radiofrequency therapy. Response in 21 patients. *JAMA* 235:2178–2200, 1976.

LeVeen, H. H. et al. Radiofrequency therapy: clinical experience. In *Thermal characteristics of tumors: applications in detection and treatment,* eds. R. K. Jain and P. M. Gullino. *Ann. NY Acad. Sci.* 335:362–371, 1980.

Lücke-Hühle, C., and Dertinger, H. Kinetic response of an *in vitro* "tumour model" (V-79 spheroids) to 42°C hyperthermia. *Eur. J. Cancer* 13:23–28, 1977.

MacKenzie, A. et al. Total body hyperthermia: techniques and patient management. In *Proceedings of the First International Symposium on Cancer Therapy by Hyperthermia and Radiation,* eds. M. J. Wizenberg and J. E. Robinson. Chicago: American College of Radiology, 1976, pp. 272–281.

Madoc-Jones, H., and Mauro, F. Interphase action of vinblastine and vincristine: differences in their lethal action through the mitotic cycle of cultured mammalian cells. *J. Cell. Physiol.* 72:185–196, 1968.

Majno, G. Mechanisms of abnormal vascular permeability in acute inflammation. In *Injury, inflammation and immunity,* eds. L. Thomas, J. W. Uhr, and L. Grant. Baltimore: Williams & Wilkins, 1964, pp. 58–93.

Mäntylä, M. J. Regional blood flow in human tumors. *Cancer Res.* 39:2304–2306, 1979.

Marmor, J. B.; Hahn, N.; and Hahn, G. M. Tumor cure and cell survival after localized radiofrequency heating. *Cancer Res.* 37:879–883, 1977.

Mattson, J. et al. Tumor vessel innervation and influence of vasoactive drugs on tumor blood flow. In *Tumor blood circulation,* ed. H.-I. Peterson. Boca Raton, Fla.: CRC Press, 1979, pp. 129–136.

Mauer, A. M.; Murphy, S. B.; and Hayes, F. A. Evidence for recruitment and synchronization in leukemia and solid tumors. *Cancer Treat. Rep.* 60:1841–1844, 1976.

Mendecki, J.; Friedenthal, E.; and Botstein, C. Effects of microwave-induced local hyperthermia on mammary carcinoma in C_3H mice. *Cancer Res.* 36:2113–2114, 1976.

Mendelsohn, M. L. Autoradiographic analysis of cell proliferation in spontaneous breast cancer of C_3H mouse. III. The growth fraction. *J. Natl. Cancer Inst.* 28:1015–1029, 1962.

Mendelsohn, M. L. Cell proliferation and tumor growth. In *Cell proliferation,* eds. L. F. Lamerton and R. J. Fry. Oxford, Eng.: Blackwell, 1963, pp. 190–210.

Mendelsohn, M. L., and Takahashi, M. A critical analysis of the fraction of labelled mitoses method as applied to the analysis of tumor and other cell cycles. In *The cell cycle and cancer,* ed. R. Baserga. New York: Marcel Dekker, 1971, pp. 58–95.

Menkin, V., ed. The role of the hydrogen ion concentration and the cytology of an exudate. Glycolysis in inflammation. Some aspects of the chemistry of exudates. In *Biochemical mechanisms in inflammation.* Springfield, Ill.: Charles C. Thomas, 1956, pp. 120–144.

Mitchison, J. M. *The biology of the cell cycle.* Cambridge, Eng.: Cambridge University Press, 1971.

Mondovi, B. Biochemical and ultrastructural lesions. In *Proceedings of the First International Symposium on Cancer Therapy by Hyperthermia and Radiation*, eds. M. J. Wizenberg and J. E. Robinson. Chicago: American College of Radiology, 1976, pp. 3–15.

Mondovi, B. et al. The biochemical mechanism of selective heat sensitivity of cancer cells. I. Studies on cellular respiration. *Eur. J. Cancer* 5:129–136, 1969a.

Mondovi, B. et al. The biochemical mechanism of selective heat sensitivity of cancer cells. II. Studies on nucleic acids and protein synthesis. *Eur. J. Cancer* 5:137–146, 1969b.

Moricca, G. et al. Hyperthermic treatment of tumours: experimental and clinical applications. In *Selective heat sensitivity of cancer cells. Recent results in cancer research*, vol. 59, eds. A. Rossi-Fanelli et al. Berlin and New York: Springer-Verlag, 1977, pp. 112–152.

Morris, C. C.; Myers, R.; and Field, S. B. The response of the rat tail to hyperthermia. *Br. J. Radiol.* 50:576–580, 1977.

Movat, H. Z., ed. The acute inflammatory reaction. In *Inflammation, immunity and hypersensitivity.* New York: Harper & Row, 1979, pp. 1–162.

Muckle, D. S., and Dickson, J. A. The selective inhibitory effect of hyperthermia on the metabolism and growth of malignant cells. *Br. J. Cancer* 25:771–778, 1971.

Muckle, D. S., and Dickson, J. A. Hyperthermia (42°C) as an adjuvant to radiotherapy and chemotherapy in the treatment of the allogeneic VX2 carcinoma in the rabbit. *Br. J. Cancer* 27:307–315, 1973.

Murphy, J. R. The influence of pH and temperature on some physical properties of normal erythrocytes from patients with hereditary spherocytosis. *J. Lab. Clin. Med.* 69:758–775, 1967.

Naesblund, J., and Swenson, K. E. Investigations on the pH of malignant tumors in mice and humans after the administration of glucose. *Acta Obstet. Gynecol. Scand.* 32:359–367, 1953.

Nauts, H. C.; Fowler, G. A.; and Bogatko, F. H. A review of the influence of bacterial infections and bacterial products (Coley's toxin) on malignant tumors in man. *Acta Med. Scand.* (suppl.) 276:1–103, 1953.

Orth, R. E.; Swidler, H. J.; and Zarakov, M. S. Survival of pleomorphic sarcoma 37 transplanted into virgin female DBA/2J mice: hyperthermia and hyperglycemia, alone and in combination with drugs. *J. Pharm. Sci.* 66:437–438, 1977.

Overgaard, J. Ultrastructure of a murine mammary carcinoma exposed to hyperthermia *in vivo. Cancer Res.* 36:983–995, 1976a.

Overgaard, J. Influence of extracellular pH on the viability and morphology of tumor cells exposed to hyperthermia. *J. Natl. Cancer Inst.* 56:1243–1250, 1976b.

Overgaard, J. Effect of hyperthermia on malignant cells *in vivo. Cancer* 39:2637–2646, 1977a.

Overgaard, J. Effect of sequence and time intervals of combined hyperthermia and radiation treatment of a solid mouse mammary adenocarcinoma *in vivo. Br. J. Radiol.* 50:763–765, 1977b.

Overgaard, J., and Nielsen, O. S. The role of tissue environmental factors on the kinetics and morphology of tumor cells exposed to hyperthermia. In *Thermal characteristics of tumors: applications in detection and treatment*, eds. R. K. Jain and P. M. Gullino. *Ann. NY Acad. Sci.* 335:254–280, 1980.

Overgaard, K., and Overgaard, J. Investigations on the possibility of a thermic tumor therapy. I. Short-wave treatment of a transplanted isologous mouse mammary carcinoma. *Eur. J. Cancer* 8:65–78, 1972.

Overgaard, K., and Overgaard, J. Pathology of heat damage: studies on the histopathology in tumor tissue exposed "in vivo" to hyperthermia and combined hyperthermia and Roentgen irradiation. In *Proceedings of the First International Symposium on Cancer Therapy by Hyperthermia and Radiation*, eds. M. J. Wizenberg and J. E. Robinson. Chicago: American College of Radiology, 1976, pp. 115–127.

Palzer, J. A., and Heidelberger, C. Studies on the quantitative biology of hyperthermic killing of HeLa cells. *Cancer Res.* 33:415–421, 1973a.

Palzer, J. A., and Heidelberger, C. Influence of drugs and synchrony on the hyperthermic killing of HeLa cells. *Cancer Res.* 33:422–427, 1973b.

Pampus, F. Die wasserstoffionenkonzentration des hirngewebes bei raumfordernden intracraniellen. *Prozessen Acta Neurochirurgie* 11:305–318, 1963.

Patterson, M. S., and Greene, R. C. Measurement of low energy beta-emitters in aqueous solution by liquid scintillation counting of emulsions. *Anal. Chem.* 37:854–857, 1965.

Percy, J. F. Heat in the treatment of carcinomas of the uterus. *Surg. Gynecol. Obstet.* 22:77–79, 1916.

Peterson, H.-I., ed. Vascular and extravascular spaces in tumors: tumor vascular permeability. In *Tumor blood circulation.* Boca Raton, Fla.: CRC Press, 1979a, pp. 77–86.

Peterson, H.-I., ed. Tumor blood flow compared with normal tissue blood flow. In *Tumor blood circulation.* Boca Raton, Fla.: CRC Press, 1979b, pp. 103–114.

Pettigrew, R. T. Cancer therapy by whole body heating. In *Proceedings of the First International Symposium on Cancer Therapy by Hyperthermia and Radiation*, Washington, D.C., 1975, eds. M. J. Wizenberg and J. E. Robinson. Chicago: American College of Radiology, 1976, pp. 282–288.

Pettigrew, R. T. et al. Circulatory and biochemical effects of whole body hyperthermia. *Br. J. Surg.* 61:727–730, 1974.

Poole, D. T. Intracellular pH of the Ehrlich ascites tumor cell as it is affected by sugars and sugar derivatives. *J. Biol. Chem.* 242:3731–3736, 1967.

Poole, D. T.; Butler, T.C.; and Waddell, W. J. Intracellular pH of the Ehrlich ascites tumor cell. *J. Natl. Cancer Inst.* 32:939–946, 1964.

Prescott, D. M. *Reproduction of eukaryotic cells.* New York: Academic Press, 1976.

Quastler, H., and Sherman, F. G. Cell population kinetics in the intestine of the mouse. *Exp. Cell Res.* 17:420–438, 1959.

Racker, E. Why do tumor cells have a high aerobic glycolysis? *J. Cell. Physiol.* 89:697–700, 1976.

Reinhold, H. S., and Van den Berg-Blok, A. Enhancement of thermal damage to "sandwich" tumors by additional treatment. In *Proceedings of the Third International Symposium on Cancer Therapy by Hyperthermia, Drugs and Radiation*, eds. L. A. Dethlefsen and W. C. Dewey. Bethesda, Md.: National Cancer Institute (special journal supplement), in press.

Richardson, D. R. *Basic circulatory physiology.* Boston: Little, Brown & Co., 1976.

Rinaldini, L. M. J. An improvement for the isolation and quantitative cultivation of embryonic cells. *Exp. Cell Res.* 16:477–505, 1959.

Robinson, J. E.; McCulloch, D.; and McCready, W. A. Blood perfusion of murine tumors at normal and hyperthermal temperatures. In *Proceedings of the Third International Symposium on Cancer Therapy by Hyperthermia, Drugs and Radiation*, eds. L. A. Dethlefsen and W. C. Dewey. Bethesda, Md.: National Cancer Institute (special journal supplement), in press.

Robson, J. S.; Bone, J. M.; and Lambie, A. T. Intracellular pH. *Adv. Clin. Chem.* 11:213–274, 1968.

Rogers, A. W. *Techniques of autoradiography.* Amsterdam: Elsevier, 1967, p. 253.

Rohdenburg, G. L. Fluctuations in the growth of malignant tumors in man, with special reference to spontaneous recession. *J. Cancer Res.* 3:193–225, 1918.

Roos, A. Intracellular pH and distribution of weak acids across cell membranes. A study of D and L lactate in rat diaphragm. *J. Physiol.* 249:21–25, 1975.

Rossi-Fanelli, A. et al., eds. *Selective heat sensitivity of cancer cells. Recent results in cancer research*, vol. 59. Berlin and New York: Springer-Verlag, 1977.

Rottinger, E. M. Response of starved fibroblasts to hyperthermia. In *Proceedings of the First International Symposium on Cancer Therapy by Hyperthermia and Radiation*, eds. M. J. Wizenberg and J. E. Robinson. Chicago: American College of Radiology, 1976, p. 233.

Sapareto, S. A. et al. Effects of hyperthermia on survival and progression of Chinese hamster ovary cells. *Cancer Res.* 38:393–400, 1978.

Sapirstein, L. A. Regional blood flow by fractional distribution of indicators. *Am. J. Physiol.* 193:161–168, 1958.

Sapirstein, L. A. Fractionation of the cardiac output of rats with isotopic potassium. *Circ. Res.* 4:689–692, 1962.

Sarna, G. The resting cell: a chemotherapeutic problem, part 1. *Biomedicine* 20:322–326, 1974.

Schlag, H., and Lücke-Hühle, C. Cytokinetic studies on the effect of hyperthermia on Chinese hamster lung cells. *Eur. J. Cancer* 12:827–831, 1976.

Schloerb, P. R. et al. Intracellular pH and buffering capacity of the Walker 256 carcinoma. *Surgery* 58:5–11, 1965.

Schloerb, P. R., and Grantham, J. J. Intracellular pH measurement with tritiated water, carbon-14 labelled 5,5-dimethyl-oxazolidine-2,4-dione and chloride-36. *J. Lab. Clin. Med.* 65:669–676, 1965.

Schmid-Schonbein, H. et al. Effect of O-(β-hydroxyethylrutosides on the microrheology of human blood under defined flow conditions. *Vasa* 4:263–270, 1975.

Schulman, N., and Hall, E. J. Hyperthermia: its effect on proliferative and plateau phase cell cultures. *Radiology* 113:207–209, 1974.

Schwartz, M. A biomathematical approach to clinical tumor growth. *Cancer* 14:1272–1294, 1961.

Sevitt, S. Inflammatory changes in burned skin: reversible and irreversible effects and their pathogenesis. In *Injury, inflammation and immunity*, eds. L. Thomas, W. H. Uhr, and L. Grant. Baltimore: Williams & Wilkins, 1964, pp. 183–210.

Sevitt, H. S. Early and delayed oedema and increase in capillary permeability after burns of the skin. *J. Pathol.* 75:27–40, 1958.

Shah, S. A., and Dickson, J. A. Preservation of enzymatically prepared rabbit VX2 tumor cells *in vitro. Eur. J. Cancer* 14:447–448, 1978.

Shapot, V. S. Some biochemical aspects of the relationship between the tumor and the host. *Adv. Cancer Res.* 15:253–289, 1972.

Showacre, J. L. Staging of the cell cycle with time-lapse photography. In *Methods in cell physiology*, vol. 3, ed. D. M. Prescott. New York: Academic Press, 1968, pp. 147–159.

Simpson-Herren, L. M. Growth kinetics as a function of tumor size. In *Growth kinetics and biochemical regulation of normal and malignant cells*, eds. B. Drewinko and R. M. Humphrey. Baltimore: Williams & Wilkins, 1977, pp. 547–559.

Simpson-Herren, L. M. et al. Kinetics of metastasis in experimental tumors. In *Cancer invasion and metastasis: biologic mechanisms and therapy*, eds. S. B. Day et al. New York: Raven Press, 1977, pp. 117–133.

Simpson-Herren, L. M.; Sanford, A. H.; and Holmquist, J. P. Effects of surgery on the cell kinetics of residual tumor. *Cancer Treat. Rep.* 60:1749–1760, 1976.

Sisken, J. E.; Morasca, L.; and Kibby, S. Effects of temperature on the kinetics of the mitotic cycle of mammalian cells in culture. *Exp. Cell Res.* 39:103–116, 1965.

Skipper, H. E. Kinetic behaviour versus response to chemotherapy. *Natl. Cancer Inst. Monogr.* 34:2–14, 1971.

Skipper, H. E. Cell cycle and chemotherapy of cancer. In *The cell cycle and cancer*, ed. R. Baserga. New York: Marcel Dekker, 1971, pp. 358–387.

Sokoloff, B., ed. Oncostatic activity of serotonin. In *Carcinoid and serotonin, Recent Results in Cancer Research*, vol. 15. Berlin and New York: Springer-Verlag, 1968, pp. 20–30.

Song, C. W. Effect of hyperthermia on vascular functions of normal tissues and experimental tumours; brief communication. *J. Natl. Cancer Inst.* 60:711–713, 1978.

Song, C. W. Physiological factors in hyperthermia. In *Proceedings of the Third International Symposium on Cancer Therapy by Hyperthermia, Drugs and Radiation*, eds. L. A. Dethlefsen and W. C. Dewey. Bethesda, Md.: National Cancer Institute (special journal supplement), in press.

Song, C. W. et al. Effect of hyperthermia on vascular function, pH and cell survival. *Radiology* 137:795–803, 1980.

Song, C. W.; Rhee, J. G.; and Levitt, S. H. Blood flow in normal tissues and tumors during hyperthermia. *J. Natl. Cancer Inst.* 64:119–124, 1980.

Steel, G. G. Cell loss as a factor in the growth rate of human tumours. *Eur. J. Cancer* 3:381–387, 1967.

Steel, G. G. Cell loss from experimental tumors. *Cell Tissue Kinet.* 1:193–207, 1968.

Steel, G. G. The cell cycle in tumors: an examination of data gained by the technique of labelled mitoses. *Cell Tissue Kinet.* 5:87–100, 1972.

Steel, G. G. The growth kinetics of tumors in relation to their therapeutic response. *Laryngoscope* 85:359–370, 1975.

Steel, G. G. *Growth kinetics of tumours*. Oxford, Eng.: Clarendon Press, 1977.

Steel, G. G., and Haines, S. The technique of labelled mitoses: analysis by automatic curve fitting. *Cell Tissue Kinet.* 4:93–105, 1971.

Stehlin, J. S. Hyperthermic perfusion for melanoma of the extremities: experience with 165 patients, 1967–1979. In *Thermal characteristics of tumors: applications in detection and treatment*, eds. R. K. Jain and P. M. Gullino. *Ann. N.Y. Acad. Sci.* 335:352–355, 1980.

Stehlin, J. S. et al. Hyperthermic perfusion of extremities for melanoma and soft tissue sarcomas. *Recent Results Cancer Res.* 59:171–185, 1977.

Stewart H. L. et al. *Transplantable and transmissible tumors of animals*. Washington, D.C.: Armed Forces Institute of Pathology, 1959.

Storm, F. K. et al. Normal tissue and solid tumor effects of hyperthermia in animal models and clinical trials. *Cancer Res.* 39:2245–2251, 1979.

Storm, F. K. et al. Hyperthermia therapy for human neoplasms: thermal death time. *Cancer* 46:1849–1854, 1980.

Storm, F. K. et al. Clinical radiofrequency hyperthermia: a review. In *Proceedings of the Third International Symposium on Cancer Therapy by Hyperthermia, Radiation and Drugs*, eds. L. A. Dethlefsen and W. C. Dewey. Bethesda, Md.: National Cancer Institute (special journal supplement), in press.

Stratford, I. J.; Watts, M. E.; and Adams, G. E. The effect of hyperthermia on the differential cytotoxicity of some electron-affinic hypoxic cell radiosensitizers on mammalian cells *in vitro*. In *Proceedings of the Second International Symposium on Cancer Therapy by Hyperthermia and Radiation*, eds. C. Streffer et al. Baltimore and Munich: Urban & Schwarzenberg, 1978, pp. 267–270.

Strauss, A. A. *Immunologic resistance to carcinoma produced by electrocoagulation. Based on fifty-seven years of experimental and clinical results*. Springfield, Ill.: Charles C Thomas, 1969.

Straw, J. A. et al. Distribution of anticancer agents in spontaneous animal tumors. I. Regional blood flow and methotrexate distribution in canine lymphosarcoma. *J. Natl. Cancer Inst.* 52:1327–1331, 1974.

Streffer, C. et al., eds. *Proceedings of the Second International Symposium on Cancer Therapy by Hyperthermia and Radiation*. Baltimore and Munich: Urban & Schwarzenberg, 1978.

Strom, R. et al. Biochemical aspects of heat sensitivity of tumour cells. In *Selective heat sensitivity of cancer cells*, eds. A. Rossi-Fanelli et al. Berlin and New York: Springer-Verlag, 1977, pp. 7–35.

Suit, H. D. Hyperthermia in the treatment of tumors. In *Proceedings of the First International Symposium on Cancer Therapy by Hyperthermia and Radiation*, eds. M. J. Wizenberg and J. E. Robinson. Chicago: American College of Radiology, 1976, pp. 107–114.

Suit, H. D., and Blitzer, P. H. Thermally induced resistance to hyperthermic damage. In *Thermal characteristics of tumors: applications in detection and treatment*, eds. R. K. Jain and P. M. Gullino. Ann. NY Acad. Sci. 335: 379–382, 1980.

Suit, H. D., and Gerweck, L. E. Potential for hyperthermia and radiation therapy. Cancer Res. 39:2290–2298, 1979.

Suit, H. D., and Shalek, R. J. The response of anoxic C$_3$H mouse mammary carcinoma isotransplants (1–25 mm^3) to X-irradiation. J. Natl. Cancer Inst. 31:479–484, 1963.

Szmigielski, S., and Janiak, M. Reaction of cell-mediated immunity to local hyperthermia of tumors and its potentiation by immunostimulation. In *Proceedings of the Second International Symposium on Cancer Therapy by Hyperthermia and Radiation*, eds. C. Streffer et al. Baltimore and Munich: Urban & Schwarzenberg, 1978, pp. 80–88.

Tagashira, Y. et al. Continual pH measuring by means of inserted microglass electrode in living normal and tumor tissues (1st report). Gan 44:63–64, 1953.

Tagashira, Y. et al. Continuous pH measuring by means of microglass electrode inserted in living normal and tumor tissue (2nd report), with an additional report on interaction of SH-group of animal protein with carcinogenic agent in the carcinogenetic mechanism. Gan 45:99–101, 1954.

Tannock, I. F. A comparison of the relative efficiencies of various metaphase arrest agents. Exp. Cell Res. 47: 345–356, 1965.

Tannock, I. F. Population kinetics of carcinoma cells, capillary endothelial cells, and fibroblasts in a transplanted mouse mammary tumor. Cancer Res. 30:2470–2476, 1970.

Tannock, I. F. Cell kinetics and chemotherapy: a critical review. Cancer Treat. Rep. 62:1117–1133, 1978.

Thatcher, C. J., and Walker, I. G. Sensitivity of confluent and cycling embryonic hamster cells to sulphur mustard 1,3-bis (2-chloroethyl)-1-nitrosourea and actinomycin D. J. Natl. Cancer Inst. 42:363–368, 1969.

Thomas, R. C. Construction and properties of recessed-tip microelectrodes for sodium and chloride ions and pH. In *Ion and enzyme electrodes in biology and medicine*, eds. M. Kessler et al. Baltimore and Munich: Urban & Schwarzenberg, 1976, pp. 141–148.

Thomas, R. C. Comparison of the mechanisms controlling intracellular pH and sodium in snail neurones. Respir. Physiol. 33:63–73, 1978.

Thomlinson, R. H., and Gray, L. H. The histological structure of some human lung cancers and the possible implications for radiotherapy. Br. J. Cancer 9:539–549, 1955.

Tietz, N. W., and Siggaard-Anderson, O. Acid-base and electrolyte balance. In *Fundamentals of clinical chemistry*, ed. N. W. Tietz. Philadelphia: W. B. Saunders, 1976, pp. 947–948.

Till, J. E., and McCulloch, E. A. A direct measurement of the radiation sensitivity of normal mouse bone marrow. Radiat. Res. 14:213–222, 1961.

Tubiana, M.; Frindel, E.; and Vassort, F. Critical survey of experimental data on the *in vivo* synchronization by hydroxyurea. Recent Results Cancer Res. 52:187–205, 1975.

Twentyman, P. R. Comparative chemosensitivity of exponential versus plateau phase cells in both *in vitro* and *in vivo* model systems. Cancer Treat. Rep. 60: 1719–1722, 1976.

Urano, M. et al. Response of a spontaneous murine tumor to hyperthermia: factors which modify the thermal response *in vivo*. Radiat. Res. 83:312–322, 1980.

Valeriote, F. A.; Bruce, W. R.; and Meeker, B. E. Synergistic action of cyclophosphamide and 1,3-bis(2-chloroethyl)-1-nitrosourea on a transplanted murine lymphoma. J. Natl. Cancer Inst. 40:935–944, 1968.

Van Den Brenk, H. A. S. et al. Lowering of innate resistance of the lungs to the growth of blood-borne cancer cells in states of topical and systemic stress. Br. J. Cancer 33:60–77, 1976.

Van Putten, L. M. G$_o$ a useful term? *Biomedicine* 20:5–8, 1974.

Vaupel, P. Oxygen supply to malignant tumors. In *Tumor blood circulation*, ed. H.-I. Peterson. Boca Raton, Fla.: CRC Press, 1979, pp. 143–168.

Vaupel, P. et al. PO$_2$ histograms and PO$_2$ profiles in tumor tissue (DS-carcinosarcoma) during different stages of growth. In *Oxygen supply—theoretical and practical aspects of oxygen supply and microcirculation of tissue*, eds. A. Kessler et al. Baltimore and Munich: Urban & Schwarzenberg, 1973a. pp. 189–192.

Vaupel, P. et al. Oxygen supply of tumor tissue (DS-carcinosarcoma) *in vivo*. In *Oxygen supply—theoretical and practical aspects of oxygen supply and microcirculation of tissue*, eds. A. Kessler et al. Baltimore and Munich: Urban & Schwarzenberg, 1973b, pp. 285–287.

Vaupel, P. et al. Impact of localized hyperthermia on the cellular microenvironment in solid tumors. In *Proceedings of the Third International Symposium on Cancer Therapy by Hyperthermia, Drugs and Radiation*, eds. L. A. Dethlefsen and W. C. Dewey. Bethesda, Md.: National Cancer Institute (special journal supplement), in press.

Vaupel, P.; Ostheimer, K.; and Muller-Klieser, W. Circulatory and metabolic responses of malignant tumors during localized hyperthermia. *J. Cancer Res. Clin. Oncol.* 98:15–29, 1980.

Verma, S. P., and Wallach, D. F. Erythrocyte membranes undergo co-operative pH-sensitive state transitions in the physiological temperature range: evidence from Raman spectroscopy. *Proc. Natl. Acad. Sci. USA* 73:3558–3561, 1976.

Vidt, D. G. et al. Effect of ether anesthesia on the cardiac output, blood pressure and distribution of blood flow in the albino rat. *Circ. Res.* 7:759–764, 1959.

Voegtlin, C.; Fitch, R. H.; and Kahler, H. The influence of the parenteral administration of certain sugars on the pH of malignant tumors. *National Institute of Health Bulletin* 164:1–14, 1935.

Voegtlin, C.; Kahler, H.; and Fitch, R. H. II. The estimation of the hydrogen ion concentration of tissues in living animals by means of the capillary glass electrode. *National Institute of Health Bulletin* 164:15–27, 1935.

Von Ardenne, M. *Theoretische und experimentelle grundlagen der Krebs-Mehrschritt-Therapie*, 2nd edition. Berlin: VEB Verlag Volk und Gesundheit, 1971.

Von Ardenne, M. Selective multiphase cancer therapy: conceptual aspects and experimental basis. *Adv. Pharmacol.* 10:339–380, 1972.

Von Ardenne, M. On a new physical principle for selective local hyperthermia of tumor tissues. In *Proceedings of the Second International Symposium on Cancer Therapy by hyperthermia and radiation*, eds. C. Streffer et al. Baltimore and Munich: Urban & Schwarzenberg, 1978, pp. 96–104.

Von Ardenne, M., and Kruger, W. The use of hyperthermia within the frame of cancer multistep therapy. In *Thermal characteristics of tumors: applications in detection and treatment*, eds. R. K. Jain and P. M. Gullino. *Ann. NY Acad. Sci.* 335:356–361, 1980.

Von Ardenne, M., and Reitnauer, P. G. Selective occlusion of cancer tissue capillaries as the central mechanism of the cancer multistep therapy. *Japanese Journal of Clinical Oncology* 10:31–48, 1980.

Waddell, W. J., and Bates, R. G. Intracellular pH. *Physiol. Rev.* 49:285–329, 1969.

Waddell, W. J., and Butler, T. C. Calculation of intracellular pH from the distribution of 5,5-dimethyloxazolidine-2,4-dione (DMO). Application to skeletal muscle of the dog. *J. Clin. Invest.* 38:720–729, 1959.

Wallace, S. et al. Embolization of renal carcinoma: experience with 100 cases. *Radiology*, in press.

Walpole, R. E. *Introduction to statistics*. New York: Macmillan, 1974.

Warburg, O. *Metabolism of tumors*. London: Constable, 1930.

Warburg, O. On the origin of cancer cells. *Science* 123:309–313, 1956.

Warburg, O.; Posener, K.; and Negelein, E. Über den stoffwechsel der carcinomzelle. *Biochem. Z.* 152:309–344, 1924.

Warren, B. A. Tumor angiogenesis. In *Tumor blood circulation*, ed. H.-I. Peterson. Boca Raton, Fla.: CRC Press, 1979, pp. 49–76.

Watson, J. V. The cell proliferation kinetics of the EMT6/M/AC mouse tumour at four volumes during unperturbed growth *in vivo*. *Cell Tissue Kinet.* 9:147–156, 1976.

Weber, H.-J. et al. *In vivo* analysis of the influence of combined hyperthermia and gamma irradiation on oxic and hypoxic tumor cells. In *Proceedings of the Second International Symposium on Cancer Therapy by Hyperthermia and Radiation*, eds. C. Streffer et al. Baltimore and Munich: Urban & Schwarzenberg, 1978, pp. 276–277.

Weiss, L. *The cell periphery, metastasis and other contact phenomena*. Amsterdam: North Holland, 1967.

Westermark, F. Über die behandlung des vecerirended cercixcarcinoms mittel konstanter wärme. *Zbl. Gynak.* 1335–1339, 1898.

Westermark, H. The effect of heat upon rat tumors. *Skand. Arch. Physiol.* 52:257–322, 1927.

Westra, A., and Dewey, W. C. Variation in sensitivity to heat shock during the cell cycle of Chinese hamster cells *in vitro*. *Int. J. Radiat. Biol.* 19:467–477, 1971.

Wilcox, W. S. et al. Experimental evaluation of potential anticancer agents. XVII. Kinetics of growth and regression after treatment of certain solid tumors. *Cancer Chemother. Rep.* 47:27–39, 1965.

Wizenberg, M. J., and Robinson, J. E., eds. *Proceedings of the First International Symposium on Cancer Therapy by Hyperthermia and Radiation*. Chicago: American College of Radiology, 1976.

Wolff, J. *Die lehre von der krebskrankheit.* Jena, 1907.

Wright, N. A., and Appleton, D. R. The metaphase arrest technique: a critical review. *Cell Tissue Kinet.* 13:643–663, 1980.

Wright, S. *Applied physiology*, 9th edition. Oxford: Oxford University Press, 1952, pp. 87–103, pp. 394–395.

Yatvin, M. B. The influence of membrane lipid composition and procaine on hyperthermic death of cells. *Int. J. Radiat. Biol.* 32:513–521, 1977.

Zanelli, G. D., and Fowler, J. F. The measurement of blood perfusion in experimental tumors by uptake of ^{86}Rb. *Cancer Res.* 34:1451–1456, 1974.

CHAPTER 6
Thermotolerance

Eugene W. Gerner

Introduction
Operational Definitions of Thermotolerance
Factors Affecting the Expression of Thermotolerance
Thermotolerance: Single Mechanism or Multiple Phenomena?
Effect of Thermotolerance on Radiosensitization
Effect of Thermotolerance on the Interaction of Hyperthermia and Chemotherapeutic Drugs
Clinical Considerations
Conclusions

INTRODUCTION

The response of biological systems, including cell lines as well as organized tissues, to temperatures greater than 37°C has been widely studied for a variety of reasons. These include studies on the mechanisms of growth of thermophilic organisms (see Ingraham 1962 for review); attempts to develop temperature-resistant mutants with the goal of investigating gene regulation (see Simchen 1978 for review); the production of temperature-sensitive mutants, which are cells that express one phenotype at a permissive, usually hypothermic, temperature while displaying a second phenotype at a nonpermissive, usually hyperthermic, temperature (see Simchen 1978 for review); and research on the potential of elevated temperatures as a tumoricidal agent (Gerner et al. 1975; Leith et al. 1977; Suit and Shwayder 1974). While these areas tend to have widely diverse goals, from understanding evolutionary processes to unraveling the mysteries of genetic regulation to developing new modes of cancer treatment, a common finding has been the ability of biological systems to express remarkable degrees of resistance to temperatures in excess of 37°C.

In some cases, resistance to heat is a heritable trait. *Thermus thermophilus* HB-8 and *Caldariella acidophila* grow at 75°C and 87°C, respectively (De Rosa et al. 1976; Oshima 1975), and heat resistant mutants of mammalian cells are killed at reduced rates compared to the parental cell populations from which they were isolated (Harris 1967a, 1967b, 1969; Reeves 1972; Selawry, Goldstein, and McCormick 1957). In other examples, heat resistance has been shown to develop in cell cultures in a nonheritable manner (Gerner et al. 1976; Gerner and Schneider 1975). This nongenetic type of resistance, which is regulated as a complex function of temperature, time of exposure to hyperthermia, growth state, environmental conditions, and other factors, may be responsible for several clinical observations regarding resistance during human cancer therapy employing hyperthermia (see Henle and Dethlefsen 1978 for review). The possibility of regulating the development of hyperthermia therapy so as to maximize tumor sensitivity and minimize normal tissue damage has created considerable interest. The molecular and cellular mechanisms underlying the ef-

The author gratefully acknowledges the significant contributions of Anne E. Cress, Paul W. Holmes, James A. McCullough, Max Costa, David J. M. Fuller, Terence S. Herman, Max L. M. Boone, William G. Conner, John A. Hicks, John T. Leith, Robert C. Miller, Diane H. Russell, M. Robert Boone, Patrick S. Culver, David K. Holmes, Julie A. Noterman, and Donna G. Stickney. This work was supported in part by U.S. Public Health Service grants CA-17343 and CA-18273 from the National Institutes of Health.

fect(s) are still poorly understood. This chapter summarizes our existing knowledge of this phenomenon and proposes a working hypothesis to explain its regulation.

OPERATIONAL DEFINITIONS OF THERMOTOLERANCE

Survival Responses

Thermotolerance develops and is expressed in several ways, depending on a variety of parameters detailed in the following section. Classically, this type of heat resistance is defined as a decrease in the rate of cell killing in response to an initial exposure to hyperthermia (Gerner et al. 1976; Gerner and Schneider 1975; Henle and Dethlefsen 1978). Figure 6.1 shows that when HeLa cells in culture are exposed to 44°C for one hour, then returned to 37°C for two hours, the response of these cells to subsequent graded doses of heat is different, displaying more resistance, when compared with cultures treated continuously at 44°C. Figure 6.2 demonstrates that the change in sensitivity, or increased resistance, of cells subjected to an initial heat dose is related to the magnitude of that heat dose. Cultures treated for 30 minutes at 44°C, then returned to 37°C for two hours show a D_o^1 of 66 minutes, while cells treated with an initial heat dose of 44°C for 60 minutes, then incubated at 37°C for two hours, show a D_o of 90 minutes for subsequent 44°C exposures.

Figure 6.3 combines the results of studies using several cell lines and shows that the degree to which thermotolerance develops is proportional to the logarithm of the damage, or cell killing, produced by the initial thermal dose (Leith et al. 1977). There does appear to be an optimum change in the slope of the survival curve as a function of time that is specific for the amount of damage produced by a given thermal dose (Henle, Karamuz, and Leeper 1978) and possibly cell type, since there is a wide range of D_os for various tissue culture cells at any given temperature (Bhuyan et al. 1977; Cress and Gerner 1980). Thermotolerance also can develop when no cell killing is produced by the initial hyperthermic damage (Henle, Karamuz, and Leeper 1978; Freeman, Raaphorst, and Dewey 1979). In this case, there is little or no change in the D_o when these thermotolerant cells are exposed to lethal hyperthermic temperatures; however, there is a dramatic increase in the shoulder region of the survival response. In any event, survival is greatly increased at

---

[1] D_o is the standard radiobiological parameter describing the survival curve slope and is defined as the dose required to reduce survival to 1/e of an initial value in the exponential region of the curve; units are of a reciprocal slope.

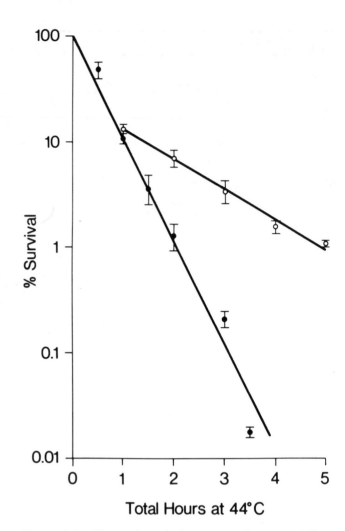

Figure 6.1. Thermal survival response of exponentially growing HeLa cells ●. Survival response of cells treated with single exposures of 44°C for increasing time; ○, response of cells treated at 44°C for one hour returned to the 37°C incubator for two hours and given second graded doses of 44°C for increasing times. Standard errors (SE) of survival determinations are as shown. (Gerner and Schneider 1975. Reprinted by permission.)

all exposure times in comparison to cells not previously exposed to the nonlethal, initial hyperthermic challenge. This type of response is seen in figure 6.4, taken from the work of Henle, Karamuz, and Leeper (1978).

The preceding paragraphs describe thermotolerance developing after an initial hyperthermic dose, followed by an incubation period at normal mammalian body temperature (37°C). When cells thus treated were reexposed to hyperthermia, thermotolerance was expressed. It was shown that if cells were incubated between hyperthermic doses at 0°C to inhibit cellular metabolism, thermotolerance was not expressed (Gerner et al. 1976; Gerner and Schneider 1975; Henle, Karamuz, and Leeper 1978). This led to

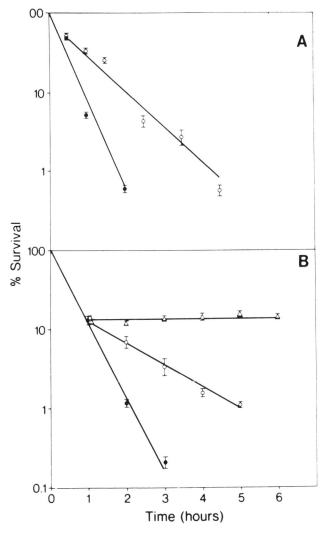

Figure 6.2. Closed symbols in both *A* and *B* represent the response of HeLa cells to continuous 44°C exposures. At the same time as these exposures, cells were treated with 44°C for either 0.5 hour (○, *A*) or one hour ○, *B*), incubated at 37°C for two hours, and then returned to 44°C for additional times as shown. The figure shows the total exposure time to 44°C but not the incubation period at 37°C. △, *B*, results of experiments in which cells were treated with 44°C for one hour. Cells were then plated for colony formation at hourly intervals after the end of this thermal dose. (Gerner et al. 1976. Reprinted by permission.)

the conclusion that cellular metabolism was required for the development of this resistance.

A number of reports have described several aspects of cellular metabolism that are inhibited to various degrees at hyperthermic temperatures (Mondovi et al. 1969b; Reeves 1972; Strom et al. 1973; Turano et al. 1970). These findings were puzzling, then, when several investigators (Gerweck, Nygaard, and Burlett 1979; Gerweck and Rottinger 1976; Harisiadis, Sung, and Hall 1977; Dewey et al. 1977; Sapareto et al. 1978)

showed that prolonged continuous exposures at hyperthermic temperatures also resulted in the expression of thermotolerance. This type of tolerance, as shown in figure 6.5, develops after continuous exposures in excess of 200 to 300 minutes at elevated temperatures. The rate of cell killing is dramatically reduced as thermotolerance is expressed, indicated as a plateau in the survival response. Note that this is similar to the results seen with the split-graded dose experiments already described. It is indeed remarkable that this tolerance can develop at temperatures in excess of 41°C without returning cultures to normal temperatures. Equally remarkable are the findings of Sapareto and colleagues (1978) that this tolerance can result in differences in survival of up to 4 logs for temperature differences of only 0.5°C (42.0°C vs 42.5°C).

Three experimental methods have been described that produce thermotolerance in cell cultures. These include initial exposures to lethal and nonlethal hyperthermic temperatures, followed by a return to nonlethal temperature in the former case, prior to the heat exposure showing thermotolerance. These two methods expressed tolerance primarily as slope and shoulder changes, respectively, in the survival responses. The third method did not require return to a nonlethal temperature, but showed thermotolerance after 200 to 300 minutes of continuous exposure to lethal hyperthermic temperatures. A fourth and more complex method, reported by Bauer and Henle (1979) and Henle, Bitner, and Dethlefsen (1979), produces tolerance in a manner that combines the first and third methods. These investigators combined split-graded dose exposures with extended, continuous second exposures to show that both slope changes and the plateau response can be produced in tolerant cells.

The question of whether these four different ways in which thermotolerance is expressed, based on survival curve analysis, actually represent different types of tolerance or similar phenomena, remains unanswered. Bauer and Henle (1979) have proposed multiple types of tolerance, based in part on thermodynamic arguments, while we have suggested that all these thermotolerant responses may be regulated by common mechanisms (Gerner et al. 1980a). Clarification of this point will await a detailed understanding of the mechanisms underlying the phenomenon(a).

Distinction between Thermotolerance and Genetic Heat Resistance

Since 1975, there has been considerable investigation of thermotolerance produced by all of the methods covered in the preceding section. Often, these types of experiments are discussed in terms of previous work

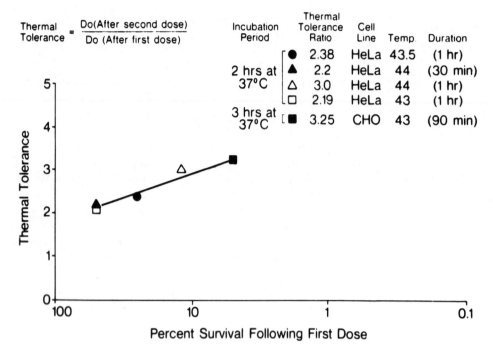

Figure 6.3. Schematic plot of the degree of induced thermal tolerance (defined as the ratio of the survival curve slope after a second thermal treatment to the survival curve slope after the initial thermal treatment). Cells were incubated for two to three hours at 37°C before administration of the second treatment. Data are shown for CHO and HeLa cells, with cells treated at different levels of survival (following the initial thermal treatment). (Leith et al. 1977. Reprinted by permission.)

dealing with heat resistance of a genetic nature, but without adequate distinctions drawn. Several investigators have isolated and studied heat resistant mutants of various cell lines (Harris 1967a, 1967b, 1969; Reeves 1972; Selawry, Goldstein, and McCormick 1957). These mutants usually are genetically altered from the parental populations from which they were isolated and pass their heat resistance on to their progeny. Often, procedures used to isolate these mutants include multiple exposures to temperatures in the 45°C to 50°C range and require the exposure of several cell generations before a stable mutant is isolated.

This type of heat resistance is very much in contrast to the resistance now commonly referred to as thermotolerance, produced as described in the preceding section. Thermotolerance can be produced in a large fraction of a cell population within 200 to 300 minutes by heat exposures that may kill only a few, or even none, of the initially treated population. As shown in figure 6.6, thermotolerance is a transient, nonheritable type of heat resistance. This phenomenon is fully expressed within several hours after an initial heat dose, remains evident for from 48 to 72 hours after the initial exposure, then disappears, and cells reacquire their normal heat sensitivity (Gerner et al. 1976; Henle and Leeper 1976). Thus, heat resistance in mutant cells is a different phenomenon from that referred to here as thermotolerance. The differences include the time course for the development of resistance, the number of cells in a population affected, and the transient effect of thermotolerance, while heat resistance in mutants is a genetic effect and is passed on to progeny.

Several characteristics of the transient, nonheritable thermotolerance phenomenon have been reported by various groups. First, while it is known that the sensitivity to heat-induced cell killing changes during the cell cycle (Bhuyan et al. 1977; Gerner, Holmes, and McCullough 1979; Palzer and Heidelberger 1973; Westra and Dewey 1971), it is now clear that thermotolerance does not develop because of accumulation of cells into a heat-resistant cell cycle phase. Hyperthermia does cause marked changes in normal cell cycle progression (Gerner, Holmes, and McCullough 1979; Gerner and Russell 1977; Kal, Hatfield, and Hahn 1975; Sapareto et al. 1978); however, thermotolerance has been shown to develop in treated synchronous G_1 cells (Dewey et al. 1977) and in plateau phase cultures (Gerner, Holmes, and McCullough 1979). In the latter example, all treatments occurred prior to stimulating the cultures to proliferate for colony formation, thus ruling out any argument invoking cell cycle redistribution as the basis for thermotolerance. Also, since thermotolerance can be produced by initial thermal

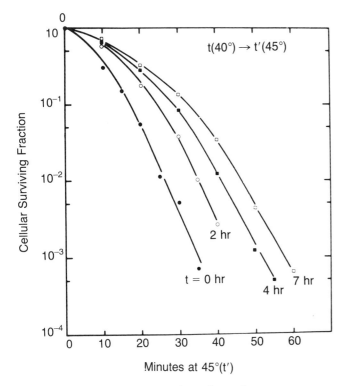

Figure 6.4. Surviving fraction of asynchronous CHO cells is plotted as a function of time of heating at 45°C. Cells were incubated at 40°C for various intervals immediately prior to 45°C hyperthermia. The primary effect of prior incubation at 40°C is one of widening the shoulder of the 45°C hyperthermia survival curve. The calculated values of the survival curve parameters are taken from Henle, Karamuz, and Leeper 1978. (Reprinted by permission.)

doses that in themselves are not cytotoxic, the phenomenon cannot be due to the existence of heat-resistant subpopulations within the original culture. Finally, it is becoming widely recognized that there is a wide range of thermal sensitivities among various cell lines in culture (Bhuyan 1977; Cress and Gerner 1980). These differences in sensitivities are reflected by D_os, which may vary by factors of 10 or more (Connor et al. 1977; Westra and Dewey 1971). Thermotolerance, however, develops in a similar manner in cells of widely different heat sensitivities (e.g., D_o at 43°C for Chinese hamster ovary cell (CHO) is 9 minutes and for HeLa is 66 minutes), as seen in figure 6.3. Thus, the cell line's characteristic thermal sensitivities that have been reported cannot be due to differential expression of thermotolerance in the untreated populations.

A Phenomenon with Many Names

Early investigators studying the genetic inheritance of heat resistance in mutants simply referred to their cells as heat-resistant cells (Harris 1967a, 1967b, 1969;

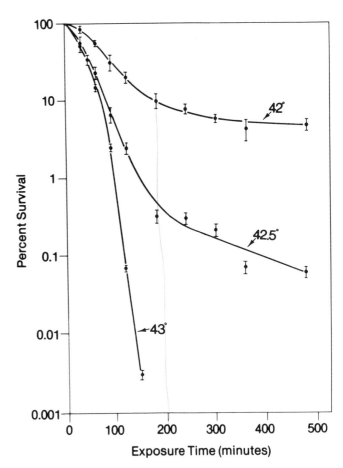

Figure 6.5. Survival response of log phase CHO cells following continuous hyperthermic exposures. Cultures were treated at 42°C, 42.5°C, and 43°C for times as indicated in the figure. Following thermal exposures, cells were harvested and plated for colony formation (Bhuyan et al. 1977). *Error bars* in this and subsequent figures represent the SE of the survival values. (Gerner et al. 1980a. Reprinted by permission.)

Reeves 1972; Selawry, Goldstein, and McCormick 1957). The phenomenon of thermotolerance, the nonheritable, transient phenomenon discussed here, has been referred to by many names, and this has provoked considerable discussion. Various reports have used terms such as induced thermal resistance (Gerner and Schneider 1975), thermotolerance (Gerner et al. 1976; Henle and Dethlefsen 1978; Henle, Karamuz, and Leeper 1978; Joshi and Jung 1979), thermal tolerance (Harisiadis, Sung, and Hall 1977), and various combinations of thermal/heat/hyperthermic resistance. While a specific name may not be of great importance, it is important to differentiate between heat resistance in mutants and this transient resistance, which can be expressed by most cells in a population in response to a single hyperthermic dose. In order to facilitate this distinction, and because of current widespread usage, the term *thermotolerance* is preferred.

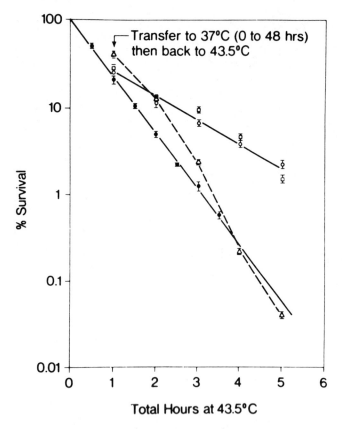

Figure 6.6. Survival response of HeLa cells either exposed continuously to 43.5°C (●) for times shown in the figure or given an initial thermal dose of 43.5°C (one hour) and then transferred to 37°C for two (○), 24 (□), or 48 (△) hours. Immediately after these incubation periods at 37°C, these cultures were reexposed to 43.5°C for increasing times. The values shown on the *abscissa* include the initial one hour at 43.5°C exposure prior to 37°C incubation plus the times of the second 43.5°C treatments. After the total thermal treatment, cells were removed from the monolayers by trypsin treatment, and single cells were plated for colony formation. (Gerner et al. 1976. Reprinted by permission.)

Induced thermal resistance is a misleading term, implying a mechanism for the effect. In the biological literature, to say that resistance is induced implies the induction of new repair enzymes or proteins via classic transcriptional or translational means. This possibility has been investigated, and results obtained suggest that this is not a primary mechanism. Studies on the recovery of macromolecular synthesis and experiments employing nucleic acid and protein synthesis inhibitors show that neither transcriptional nor translational events are absolutely required for the development of thermotolerance (Henle and Leeper 1979) (see Henle and Dethlefsen 1978 for review). In fact, protein synthesis inhibition may afford an additional degree of protection and lead to even greater hyperthermic resistance (Palzer and Heidelberger 1973).

Thus, while the phenomenon is certainly "induced" by an initial heat dose, the term *induced thermal resistance* conveys a misleading, and probably incorrect, view of the mechanism(s) underlying this phenomenon.

FACTORS AFFECTING THE EXPRESSION OF THERMOTOLERANCE

Many factors, physical, biological, and environmental, have been shown to affect the development and expression of thermotolerance in a wide variety of cell lines in culture. This section addresses these factors and, in addition, attempts to summarize data of relevance to the clinical use of hyperthermia in cancer therapy. Questions pertinent to this point include: does thermotolerance develop in vivo in experimental animal or human tissues? Does thermotolerance develop in a similar manner in normal and tumor tissues? Is there a correlation between thermotolerance as it is expressed in vitro and as it is expressed in vivo?

Temperature and Time of Exposure

Thermotolerance has been shown to develop in response to exposure to a wide range of temperatures. Measurement of cell survival indicates that thermotolerance is expressed somewhat differently in specific temperature regions. These differences are summarized in table 6.1. Three distinct temperature regions are evident for the development of thermotolerance. Interestingly, these regions correspond to distinct regions in the Arrhenius analysis of cell-killing kinetics of a large number of mammalian cell lines (Connor et al. 1977). Figure 6.7 shows this thermodynamic analysis of the rates of cell killing. For temperatures less than 41°C, little or no cell killing is produced, and activation energies are infinitely large. Between 41°C and 43°C, activation energies are cell-line dependent and exceed 200 kcal/mol. For temperatures greater than or equal to 43°C, activation energies are similar for all cell lines studied, with a value of 150 ± 25 kcal/mol. Thus, both 41°C and 43°C seem to be critical temperatures. Forty-one degrees Celsius is the threshold for cytotoxicity. Forty-three degrees Celsius is a critical temperature and designates a possible change in the mechanism of cell killing, as evidenced by the remarkable change in activation energy for cell killing. Using this thermodynamic argument, it may be reasonable to predict a fourth temperature range in which thermotolerance may be still differently expressed. Landry and Marceau (1978) recently have

Table 6.1
Summary of Cellular Phenomenology for Thermotolerance as a Function of the Temperature Range of the Initial Heat Dose

Temperature Range of Dose-Producing Thermotolerance	Characteristics of Thermotolerance Expression
T < 41°C	1. Initial dose produces little or no cytotoxicity
	2. Tolerance is evidenced primarily by a change in the shoulder region of the survival curve
	3. D_o is unaffected when cells are subsequently exposed to T < 41°C
	4. Return to 37°C is not required to elicit effect[a]
41°C ≤ T ≤ 43°C	1. Tolerance develops after continuous exposures lasting 200–300 minutes and is independent of heat-damage produced
	2. Return to 37°C is not required to elicit effect
	3. Tolerance is evidenced by a plateau in the survival response[b]
T ≥ 43°C	1. Development is dependent on return to T ≤ 41°C for some time after initial thermal dose
	2. Degree to which tolerance develops is dependent on the time during which cells are incubated at T ≤ 41°C after initial thermal dose, maximum tolerance develops, depending on cell line, between 120 and 720 minutes
	3. Cell metabolism is required for effect to develop
	4. Tolerance is evidenced both by a change in D_o and, with extended exposures, a plateau in the survival response
	5. The change in D_o is related to the magnitude of the initial hyperthermic damage
	6. Tolerance induced is not heritable[c]

[a]Henle, Karamuz, and Leeper 1978; Joshi and Jung 1979.
[b]Dewey et al. 1977; Harisiadis, Sung, and Hall 1977; Sapareto et al. 1978.
[c]Gerner, Holmes, and McCullough 1979; Gerner and Schneider 1975; Henle, Karamuz, and Leeper 1978; Joshi and Jung 1979; Leith et al. 1977.

confirmed that for temperatures in excess of 43°C, the activation energy is in the range of 150 kcal/mol; however, this value only holds for temperatures in the range of 43°C to 49°C. Above 49°C, the activation energy again decreases, to a value approaching 25 kcal/mol and may suggest a concomitant change in the expression of thermotolerance.

Influence of Initial Damage

The damage incurred as a result of the initial hyperthermic dose is critical in determining the degree to which thermotolerance develops. This has been shown in figures 6.2 and 6.3 and in table 6.1. Recent studies by Ben-Hur and Riklis (1979) also support this conclusion. These investigators found that the ubiquitous polycations, the polyamines, act significantly to sensitize cells to hyperthermic cell killing. Most interestingly, they report that when the cell-killing potential of the initial thermal dose is increased, not by changing exposure time or temperature but by utilizing the polyamines, the degree to which thermotolerance develops can be increased also. Thus, these results suggest that it is the amount of thermal, or thermal-like, damage produced that is critical in determining the degree of thermotolerance development, and that temperature and time are not the only parameters to consider.

Environmental Conditions

Cell environmental factors are known also to affect the development and expression of thermotolerance. Because it is known that tumors may contain areas that are at a lower pH than surrounding normal tissues (see Gullino et al. 1965 for example), pH has been one of the most extensively studied environmental parameters (Freeman, Dewey, and Hopwood 1977; Gerweck 1977; Gerweck, Nygaard, and Burlett 1979; Gerweck and Rottinger 1976; Overgaard 1976). In a series of elegant experiments, Gerweck, Jennings, and Wesolowski (1979) showed that, when pH is decreased as low as 6.7, the development of thermotolerance is inhibited in cultures exposed continuously to temperatures in the range of 42.0°C to 42.5°C. The degree of inhibition is directly related to the extracellular pH of the culture media. While low pH is very effective at inhibiting the development of thermotolerance, these investigators have shown recently that low pH cannot act to overcome thermotolerance once it has devel-

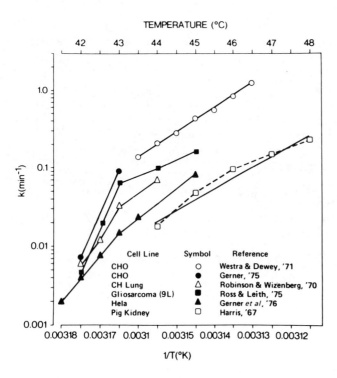

Figure 6.7. Arrhenius plot for cell killing at various temperatures above 41.5°C. Cell lines and references are listed in the key. The *dashed line* shows Harris's fit for his data. Our fit of the various data groups shows activation energies above 43°C ranging from 171 kcal for HeLa cells (Connor et al. 1977; Gerner et al. 1976) to 124 kcal for pig kidney cells (Harris 1967b). (Connor et al. 1977. Reprinted by permission.)

oped in Chinese hamster cells. This finding may be somewhat cell-line dependent, since Nielsen and Overgaard (1979) have reported that low pH can act in part to overcome thermotolerance in L_1A_2 cells, a hypotetraploid line.

George Hahn's group at Stanford University was among the first to recognize that environmental factors other than pH affect heat cell killing, repair of hyperthermic damage, and more recently, thermotolerance (Hahn 1974; Kase and Hahn 1974). They find that cells become heat sensitive in the absence of nutrients, either media or sera. They and others (Bass, Moore, and Coakley 1978; Gerweck, Gillette, and Dewey 1974; Harisiadis et al. 1974; Kim, Kim and Hahn 1975; Power and Harris 1977; Schulman and Hall 1974) find little or no dependence of oxygen concentration on heat cell killing, although in some of these studies techniques to produce hypoxia involved nutrient deprivation as well. Also, Hall (1976) has reported that hypoxia does not affect the heat sensitivity of thermotolerant cells. While several groups have shown that anisotonic media sensitize cells to hyperthermia (Hahn, Li, and Shui 1977; Raaphorst and Dewey 1978), Henle and Dethlefsen (1977) have suggested that thermotolerance does not result from an altered intracellular ionic environment, since tonicity changes sensitize both normal and thermotolerant cells to hyperthermia-induced cytotoxicity. Interestingly, Li and colleagues (1979) recently reported that the lack of nutrients primarily affects hyperthermic recovery mechanisms when nutrients are absent during, but not after, the hyperthermic exposures. This implies that repair mechanisms may play at best a minor role in modulating hyperthermic damage; rather, the limiting factor may be the environmental state of the cell during the interval in which cells are exposed to heat.

Growth State

Much of the early work done in the area of thermotolerance research was carried out using exponentially growing cells in culture. Since many tumors and most normal tissues contain nonproliferating cells, it was of some interest to know if thermotolerance developed only in dividing cells, or if it was an effect independent of growth. This issue was addressed using the in vitro model system of a nonproliferating population, that of plateau phase cells (Gerner, Holmes, and McCullough 1979). These cultures can be grown into and treated in a G_1-like state, similar to nonproliferating normal and tumor cells in vivo. Following treatments, the cultures are stimulated to proliferate by reducing the cell density and replating into fresh media to form colonies for quantitative assessment of thermal damage incurred when the cells were not proliferating. In this particular set of experiments, an attempt was made to minimize nutrient and pH effects by using fed plateau cultures, as described by Hahn and Little (1972). The results, shown in figure 6.8, demonstrate that thermotolerance is expressed in these plateau phase cultures, using split-graded dose experiments to detect changes both in shoulder and D_o. These results suggest that tolerance may also develop in nondividing cells in vivo.

Comparing the plateau phase results in figure 6.8 with similar results from log phase cultures, seen in figure 6.9, it is apparent that there are similarities and differences in the expression of tolerance as a function of growth state. Similarities include the development of tolerance in both cultures, following split-graded dose exposures. The degree of tolerance depends on the initial damage in both cases. Differences include the finding that tolerance develops in log phase but not in plateau phase cultures after continuous exposure to 43°C for three hours. An additional phenome-

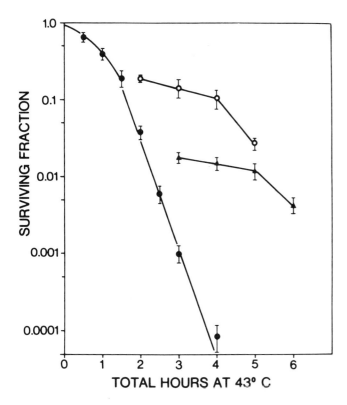
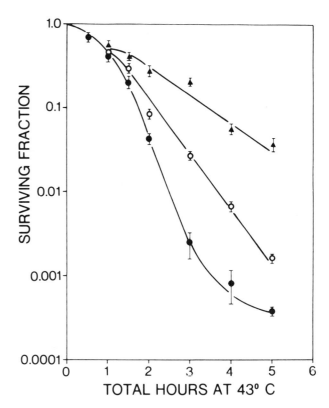

Figure 6.8. Fed plateau phase cultures were treated at 43°C for total times shown. Treatment protocols were continuous exposure to 43°C (●); 43°C for two hours followed by incubation at 37°C for two hours and subsequent treatment at 43°C (○); 43°C for three hours followed by incubation at 37°C for two hours and subsequent treatment at 43°C (▲). Bars, SE (Gerner, Holmes, and McCullough 1979. Reprinted by permission.)

Figure 6.9. Log phase EMT6/AZ cells were treated at 43°C for the total times shown. ●, surviving fractions during continuous 43°C exposures. ○, △, survival response for cultures treated at 43°C for one hour and then incubated at 37°C for either two or five hours, respectively, followed by retreatment at 43°C. Bars, SE (Gerner, Holmes, and McCullough 1979. Reprinted by permission.)

non seems to be occurring in plateau phase cultures and is not seen in log phase cultures. This effect is the recovery of heated plateau phase cells from potentially lethal heat damage, which occurs in addition to the development of thermotolerance in these cells (Gerner, Holmes, and McCullough 1979).

Rate of Heating

To date the majority of the information on biological responses to hyperthermia has been obtained using hot water baths as the method of heating cell cultures and experimental animals. Most investigators have taken great care to ensure that the time to equilibrate flasks or animal tissues, both tumor and normal, is brief (on the order of minutes), so that results can be expressed as a function of a specific temperature for varying times. Unfortunately, clinical applications of hyperthermia often deviate from procedures used to obtain experimental data, and the ability to draw analogies is restricted. A major case in point is the use of systemic hyperthermia in human cancer therapy. Using any of several current techniques to induce systemic hyperthermia, several hours usually are required to raise total body temperatures from 37°C to 41°C to 42.4°C (see Bull et al. 1979 for example). Since clinical results have not been overwhelming, it is possible that the long heat-up time, or slow heating rate, is producing thermotolerance. This idea is based on results showing that exposure to temperatures in the range of 38°C to 41°C could induce tolerance to subsequent heat doses using temperatures in excess of 41°C (Henle, Karamuz, and Leeper 1978; Joshi and Jung 1979).

Recent preliminary evidence has been reported that shows that a low heating rate actually produces a marked thermotolerant response (Herman, Stickney, and Gerner 1979). Using Chinese hamster cells and a transformed rat fibroblast cell line, we have shown that the degree of thermotolerance development is inversely proportional to the rate at which cultures are brought from 37.0°C to 42.4°C. Survival can be modified by a factor approaching 4 logs after three hours at 42.4°C, when rates of heating vary from three minutes to three hours to change temperatures from 37.0°C to 42.4°C. This phenomenon occurs in both normal and transformed cells.

Thermotolerance In Vivo

Little quantitative information is available regarding the development of thermotolerance in vivo. Evidence was available from earlier studies in both experimental animals and humans to suggest that thermotolerance can be expressed in organized tissues (see Henle and Dethlefsen 1978 for review). These data simply showed, however, that in certain instances fractionated thermal doses were less effective than a single large thermal dose. One of the most extensive recent in vivo studies has been done by Milligan and Leeper (1978). Using Chinese hamsters, these investigators studied time of exposure, temperature, and fractionation effects on the exteriorized small intestine, using $LD_{50/7}^2$ as their endpoint. The results produced in this study clearly show thermotolerance expression in vivo in this normal tissue. Of additional interest is the similarity of the parameters of thermotolerance development (time course, dependence on damage, degree of development) in this normal tissue to those parameters previously described by this group for thermotolerance development in Chinese hamster ovary cells in culture (Henle and Leeper 1976).

Other in vivo evidence for thermotolerance development comes from the work of Law, Ahier, and Field (1978) and Law, Coultas, and Field (1979). Several of their results found in vivo differ quantitatively from results found in vitro. A great deal more work must be done, however, before the relationship between thermotolerance in vitro and thermotolerance in vivo is well understood. For example, while in vitro studies have pointed to environmental factors (such as nutrients and pH) as being important in the development of tolerance, while hypoxia is of less significance, it is not clear how these factors will affect tumor responses when they are modulated in turn by vascular changes caused, perhaps directly, by heat. Also, while growth state has been studied, the effect of differentiation status on thermotolerance development has not. Thus, while gut may develop tolerance quite nicely, as shown by Milligan and Leeper (1978), other more differentiated normal tissues may not.

While some anecdotal clinical work has been interpreted to show thermotolerance development in

[2]$LD_{50/7}$ is the lethal dose to 50% of treated animals in seven days.

human cancer patients, the recent preliminary report by Hahn (1979) is the most convincing evidence to date. In their initial clinical studies, the Stanford University group has investigated the effects of several ultrasound-produced hyperthermic fractionation regimens on both cutaneous and subcutaneous tumor responses as well as normal skin response. Their results suggest that thermotolerance does develop in humans, and that the magnitude of the effect is such that it can increase the temperature required to produce a specific effect by as much as 2°C. Many questions still, however, remain unanswered. Is thermotolerance development organ specific? This may occur, at least in degree for a given temperature, since it is well documented that there is marked cell line specific sensitivity to hyperthermia-induced cytotoxicity (Bhuyan et al. 1977; Cress and Gerner 1980). Are the time courses and duration of thermotolerance similar or different in all normal tissues? The answer to this question will greatly influence our ultimate selection of proper fractionation schemes for clinical use. Finally, does thermotolerance develop in normal and malignant tissues similarly or differently? Important parameters include time to develop, extent of development, and duration of this effect in tumor and normal tissues. The determination of whether thermotolerance is a phenomenon that can be used to advantage by cancer therapists awaits the answer to these and related questions.

Transformation and the Malignant State

The important question of whether thermotolerance develops in tumors in a manner similar to its development in normal tissues is exceedingly complex. Because of the variety of cell-specific and environmental factors that influence thermotolerance, the answer to this question probably will not be definitive, but rather will require consideration of both cellular and physiologic aspects of the specific tumor and normal tissue in question. This may not be an impossible task, but assessment of individual tumors may require the use of either predictive assays of cell viability and heat sensitivity, or specific biochemical indicators of thermotolerance.

Limited data has begun to accrue, primarily from studies using cells in culture. A recent review by Bhuyan (1979) summarizes the current state of studies investigating the purported increased heat sensitivity of tumor cells compared to their normal tissue counterparts. While there is not complete uniformity in results, tumor cells seem to be, in general, more heat sensitive than their normal tissue counterparts growing under identical conditions and with similar proliferative status (e.g., growth fraction[3]). These results, however, must be considered parallel to more general studies demonstrating that there is a wide divergence of heat sensitivities among cell lines. No trend is evident suggesting that all tumor cell lines are more sensitive than all normal tissue cell lines (Bhuyan et al. 1977; Cress and Gerner 1980). Thus, while a tumor in a human patient may turn out to be more heat sensitive than the tissue from which it was derived, it may be much less sensitive than other limiting normal tissues in the treatment volume.

In addition to these results on differences in hyperthermic sensitivities between normal and tumor cells, some data is available regarding differences in thermotolerance development. Using the DNA tumor virus SV40 to transform human lung cells (WI38), Kase and Hahn (1975) showed that the transformed cells were more sensitive at 43°C than were the untransformed cells. The primary difference in the response of the transformed and untransformed cells was that the untransformed cell survival response showed a plateau after a time at 43°C, while the transformed cells did not. Thus, the difference in survival could have been caused by thermotolerance development in the untransformed line, but not in the transformed cells.

There is clear evidence that thermotolerance does develop in vitro in tumor cells capable of producing tumors in vivo. The mouse mammary sarcoma tumor cells, EMT6/AZ, show thermotolerance independent of their growth state and in response to fractionated as well as to continuous hyperthermic doses (Gerner, Holmes, and McCullough 1979). While the differences in sensitivity between the transformed and untransformed WI38 cells may have been due to differential expression of thermotolerance, it does not follow that transformed, or malignant, cells generally do not develop thermotolerance.

That the EMT6/AZ result is not an isolated instance is supported by preliminary studies on thermotolerance development in an RNA tumor-virus-transformed rat fibroblast cell system (Herman, Stickney, and Gerner 1979). Using differential heating rates to induce thermotolerance, it was shown, first, that the untransformed cells are more sensitive to hyperthermic doses between 42.0°C and 43.0°C than are the transformed cells, and second, that both transformed and untransformed cells develop thermotolerance. Together with the results from the DNA tumor virus transformed cells, these results support the contention that in some instances thermotolerance may be an asset that can be employed clinically, and in other cases a detriment to be overcome.

THERMOTOLERANCE: SINGLE MECHANISM OR MULTIPLE PHENOMENA?

As discussed earlier in this chapter, a thermotolerant response can be elicited in cells or tissues using fractionated thermal doses or continuous exposures, primarily at temperatures less than or equal to 43°C lasting in excess of 200 to 300 minutes. Using these methods and combinations of fractionated and extended exposures, Bauer and Henle (1979) have proposed the existence of three types of thermotolerance. Their proposal is based on differences in the observed survival responses and on thermodynamic arguments. The latter relates to differences in activation enthalpies, entropies, and Gibbs' free energies for cells treated under specific conditions. Similar arguments also have been used to suggest that the mechanism of heat cell killing may be different below 43°C than that above 43°C, since the activation energies seem different (> 200 kcal/mol and 150 ± 25 kcal/mol, respectively). This thermodynamic argument is somewhat incomplete in that a specific target(s) or cellular component(s) whose activation energy(ies) correspond to that observed for cell killing has not been identified.

Since thermotolerant responses are elicited only in response to initial heat or heatlike damage, it may be more appropriate to classify thermotolerant responses based on the thermodynamic characteristics of cells subjected to that initial damage. As discussed earlier in this chapter in the section on survival responses, the Arrhenius curve for cell killing shows three distinct regions for temperatures up to 49°C. These regions correspond to temperatures between 37° and 41°C, between 41°C and 43°C, and between 43°C and 49°C. The expression of thermotolerance, based on survival curve analysis, is then different when the initial heat dose triggering the response changes from one temperature region to another. For temperatures less than 41°C, thermotolerance is evident primarily as a shoulder change; for temperatures between 41°C and 43°C, tolerance is expressed primarily as a plateau in the survival response, and above 43°C, tolerance is expressed as a change in D_o.

Fractionated experiments using treatment of tissues in vivo contribute intriguing, yet inconclusive, data on this question. Overgaard (1978) has summarized several differences in thermotolerance expression for in vitro and in vivo systems. These differences

---

[3]Growth fraction is the fraction of cells in a population actively moving through the division cycle.

could have many interpretations. It is possible that many (at least three) types of thermotolerance can be expressed in isolated cells, but fewer, or different types, develop in organized tissues. There could be cell-type-specific differences in the types of thermotolerance expressed. While this list is not meant to be complete, a third possibility is that only one mechanism is responsible for all thermotolerant responses, and it is simply expressed differently under various conditions of time, temperature, cell type, and so forth.

This leads directly to the question, still unanswered, of the mechanism by which cells die from heat damage, and to the related issue of whether the mechanism of cell killing is the same in normal and thermotolerant cells. The latter question has been answered in part: Bauer and Henle's (1979) data show that the activation energy for cytotoxicity is similar in normal and thermotolerant cells. There are differences apparent in their thermodynamic analyses, such as the finding that the change in activation energy, which occurs at 43°C in normal cells, shifts to higher or lower temperatures in thermotolerant cells, depending on their treatment conditions. The suggestion remains that for specific conditions the mode of heat-induced cell death is the same in normal and thermotolerant cells. Table 6.2 summarizes some of the cellular components that have been implicated in cell death in response to hyperthermia. It is immediately apparent that heat produces marked and general changes in energy metabolism, nucleic acid metabolism, protein structure, protein compartmentalization, enzyme activity, organelle structure, and membrane structure and function. Note, however, that cell kill kinetics follow ordered functions of time and temperature and yield a specific activation energy for cell killing, which is cell-line independent in the range of 43°C to 49°C, supporting the interpretation that a specific event or sequence of events predominates in the determination of cell viability at hyperthermic temperatures. Furthermore, the work of Landry and Marceau (1978) suggests that the number of limiting events is not large and may be on the order of one to three related reactions spanning the temperature range from 41°C to 55°C.

Assuming that the mechanism of cell killing may be similar in normal and thermotolerant cells, it is appropriate to consider two recently proposed models for heat-induced cell killing. In 1979 Professor William C. Dewey, an important contributor to current

Table 6.2
Summary of Possible Mechanisms of Hyperthermic Cell Killing of Normal or Thermotolerant Cells

Proposed Mechanism	Primary Evidence
Protein denaturation	Correlation of thermodynamic parameters of cell killing and protein denaturation[a]
None proposed	Thermodynamic analysis of cell kill kinetics; activation energy is temperature dependent: $\Delta H+ > 150$ kcal/mol for $T < 43°C$ $\Delta H+ \simeq 150$ kcal/mol for $43°C \leq T \geq 49°C$ $\Delta H+ < 150$ kcal/mol for $T \geq 49°C$ Inflection points on Arrhenius curve shifted in thermotolerance cells[b]
RNA metabolism	Defect in ribosomal RNA processing[c]
Accumulation of nuclear proteins	Increased chromatin protein: DNA ratio[d]
Membrane structure and function; permeability and/or fluidity	Increased uptake of macromolecular precursors and cellular nutrients, heat sensitization by polytenic antibiotics, microscopic analysis of membrane preparations, use of drugs known to affect fluidity[e]
Incomplete replication of single-stranded regions in replicating DNA	Correlation of chromosomal aberrations with cell death in treated S phase but not mitotic and G_1 cells, increase in nuclear protein content, defects in DNA replication processes[f]
Membrane permeability modulated by the polyamines	Heat causes alterations in membrane permeability to the polyamines, polyamines sensitize cells to heat at the level of the membrane, polyamines inhibit increases in cholesterol that correlate with thermotolerance development[g]

[a]Bauer and Henle 1979; Dewey et al. 1977; Rosenberg et al. 1971.
[b]Bauer and Henle 1979; Connor et al. 1977; Landry and Marceau 1978.
[c]Strom et al. 1973.
[d]Roti-Roti and Winward 1978; Tomasovic, Turner, and Dewey 1978.
[e]Hahn, Li, and Shui 1977; Mondovi et al. 1969a, 1969b; Overgaard 1977; Yatvin 1977.
[f]Dewey 1979.
[g]Gerner et al. 1980a.

understanding of hyperthermic biology, suggested a mechanism for cell killing, specifically for cells in the process of replicating their DNA. Based on extensive morphologic, biochemical, chromosomal, and cellular studies, he suggested that cell killing involves accumulation of nuclear proteins, specifically around regions of single-stranded DNA. These proteins prevent normal synthesis of DNA and lead to the formation of chromosome aberrations. That these chromosome aberrations are responsible for cell death follows from studies correlating aberration frequency with cytotoxicity. Since G_1 and mitotic cells do not have these single-stranded DNA regions, which are formed as part of the process of DNA replication, the mechanism by which these cells are killed may be different than for S phase cells. According to Dewey's model, this is implied because they have shown that aberration frequency does not correlate with cell death in heated G_1 or mitotic cells (Dewey, Sapareto, and Betten 1978; Dewey et al. 1971).

A second comprehensive model was proposed by the authors and focuses on cell membrane changes (Gerner et al. 1980a). The model does not limit itself to cells in any particular phase of growth and is general for normal and thermotolerant cells. The specific aspect of the cell membrane that was proposed to be limiting was permeability, and included the action of the ubiquitously occurring polycations, the polyamines, as modulators of permeability. Table 6.3 shows a simplified pathway, with structure, of the polyamines in mammalian cells (Cohen 1971). Evidence for this model came from previous work showing immediate hyperthermic effects on intracellular polyamine levels (Gerner and Russell 1977), polyamine biosynthetic enzymes and their relation to cell growth (Fuller, Gerner, and Russell 1977), and more direct determinations of the polyamines' role in influencing cell survival at elevated temperatures (Gerner et al. 1980b; Gerner and Russell 1977). This latter evidence strongly suggested a membrane site of action and led to investigation of other factors affecting membrane permeability. These included the roles of cholesterol (Cress and Gerner 1980), more recently of phospholipids and membrane protein content, and a possible interaction between polyamines and membrane-bound cholesterol (Gerner et al. 1980a). The model is shown diagrammatically in figure 6.10. While this model shows a possible interaction of the polyamines with membrane-bound cholesterol, it is not meant to exclude other possible interactions. Since cholesterol is somewhat internal in the membrane, it is quite possible that polyamines interact more directly with either phospholipids or with membrane proteins to mediate similar changes in membrane permeability. Further work is needed to determine specific interactions with membrane components.

One of the most compelling pieces of evidence for a role of the polyamines in the mechanism of killing thermotolerant cells by heat is shown in figure 6.11. While several factors are known to inhibit the development of thermotolerance, such as low temperature (inhibiting metabolism) and low pH, as well as the polyamines, only the polyamines have been shown to be able to overcome thermotolerance once it is expressed (Gerner et al. 1980a). These results should be considered in view of the wide variety of biochemical investigations into the general mechanism of heat cell killing. While some authors have speculated that hyperthermia kills cells by mechanisms involving protein damage (Westra and Dewey 1971), the inhibition of protein synthesis does not potentiate heat damage (Palzer and Heidelberger 1973), nor does it markedly affect thermotolerance expression (Henle and Leeper 1979). RNA processing also has been implicated as a possible mode of heat cell killing (Strom et al. 1973),

Table 6.3
Polyamine Biosynthetic Pathway and Structure of the Polyamines Occurring Naturally in Mammalian Cells

Substrate	Enzyme	Structure
Ornithine		$\mathrm{NH_3^+ CH_2 CH_2 CH_2 CH_2 NH_3^+}$ with COOH
↓	Ornithine decarboxylase	
Putrescine		$\mathrm{^+NH_3 CH_2 CH_2 CH_2 CH_2 NH_3^+}$
↓ S-adenosylmethionine	S-adenosylmethionine decarboxylase	
Spermidine		$\mathrm{NH_3^+ CH_2 CH_2 CH_2^+ NH_2 CH_2 CH_2 CH_2 CH_2 NH_3^+}$
↓ S-adenosylmethionine	S-adenosylmethionine decarboxylase	
Spermine		$\mathrm{NH_3^+ CH_2 CH_2 CH_2 NH_2^+ CH_2 CH_2 CH_2 CH_2 NH_2^+ CH_2 CH_2 CH_2 NH_3^+}$

Source: Cohen 1971.

A. Heat Resistant State

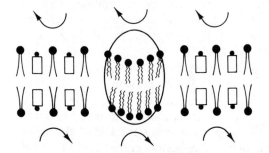

Membrane permeability reduced, in part due to restriction of motion of phospholipid molecule headgroups by
   a. cholesterol
   b. transmembrane proteins

B. Sensitization by Polyamines

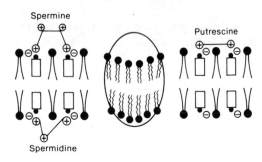

NH_3^+ groups on polyamines may interact with OH group on cholesterol.
   a. interaction may occur on inside or outside of membrane
   b. interaction may occur within the membrane

C. Heat Sensitive State

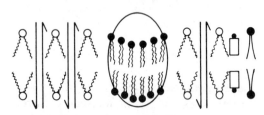

Membrane now permeable, unable to exert normal osmotic controls, partially due to reduced cholesterol levels in membrane.

Figure 6.10. Simplified model for how various membrane components could modulate resistance or sensitivity to hyperthermia induced cell killing and a possible role for the polyamines in overcoming thermotolerance. Panel A (heat resistant state), effect of cholesterol and other factors (e.g., transmembrane proteins) on membrane fluidity and membrane permeability. Here, fluidity is decreased by the action of either cholesterol or transmembrane proteins. The closed heads of the phospholipids (,) represent restricted motion. Panel B (sensitization by polyamines), proposed mechanism by which the polyamines might interact with cholesterol in the cell membrane through electrostatic interaction of the amino groups in the polyamines and the hydroxyl moiety of the cholesterol molecules (□). Panel C (heat sensitive state), conversion of the membrane from a heat resistant to sensitive state is accomplished by removal of cholesterol molecules via the action of the polyamines. Although not shown, transmembrane proteins also may be denatured and removed by hyperthermia, also resulting in increased fluidity and permeability. Not all cholesterol molecules need be removed, as depicted in the model, to achieve sensitization. The open heads on the phospholipids now indicate unrestricted head group motion, leading to increased fluidity and permeability of the membrane. (Gerner et al. 1980a. Reprinted by permission.)

yet inhibitors of nucleic acid metabolism, also, do not affect thermotolerant responses (Henle and Leeper 1979). There are unquestionable early modifications in membrane structure and function following exposure of cells to elevated temperatures (Gerner et al. 1980a, 1980b; Gerner and Russell 1977; Hahn, Li, and Shui 1977; Overgaard 1976; Reeves 1972; Turano et al. 1970; Yatvin 1977) coordinate with, or preceding, marked effects on general cellular metabolism. It is appropriate to consider a role of the polyamines in a general model for cell killing of normal and thermotolerant cells since they have biological characteristics consistent with a variety of these cellular responses at hyperthermic temperatures. First, the polyamines are ubiquitous, yet vary in metabolism depending on proliferative status and level of differentiation (see

Thermotolerance

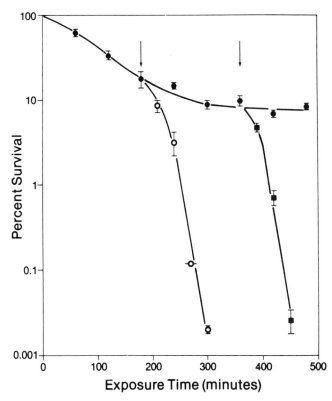

Figure 6.11. Effect of exogenous spermine addition on thermal tolerance. Log phase CHO cells were treated continuously with 42°C for up to 480 minutes. Replicate cultures were treated for up to 180 and 360 minutes. In the latter two cases, the culture media was adjusted to 10^{-3}M spermine at the end of the initial heating intervals. Cultures then were returned to 42°C for an additional 120 minutes of heating. The curves in this figure represent the survival of cells receiving 42°C alone (-●-), 42°C for 180 minutes followed by 42°C plus spermine for an additional 120 minutes (-○-), and cultures receiving 42°C alone for 360 minutes followed by 42°C plus spermine for an additional 120 minutes (-■-). All culture media were supplemented with 10^{-5}M aminoguanidine. (Gerner et al. 1980a. Reprinted by permission.)

Cohen 1971 for review). Second, polyamine receptors have been identified on cell membranes (Canellakis et al. 1978). Third, polyamine levels are regulated by enzymes, which by definition are proteins, and the rate limiting enzyme, ornithine decarboxylase (EC 4.1.1.17), has special characteristics. It has the shortest half-life of any known mammalian enzyme (approximately 15 to 40 minutes), and its activity is regulated by a heat-labile inhibitor protein, called an *antizyme* (Heller, Fong, and Canellakis 1976). Fourth, the polyamines are known to modulate DNA, RNA, and protein structure and synthesis (see Cohen 1971 for review). Fifth, thermophilic organisms have been shown to possess special polyamines not found in euthermic cells, and it has been speculated that these special polyamines are required for normal function of these thermophilic cells (De Rosa et al. 1976; Oshima 1975). Finally, it has been suggested that heat resistance in bacterial endospores is controlled by a mechanism involving osmotic regulation by divalent cations (Dring and Gould 1975; Gould and Dring 1975). These findings, in addition to the observations that the polyamines sensitize cells to heat damage and inhibit thermotolerance development as well as overcome it after development, strongly point to a central role for the polyamines in the mechanism of hyperthermic damage and thermotolerance. There appears to be, therefore, a single general mechanism underlying the phenomenon of thermotolerance. This mechanism, involving membrane permeability, would be similar for killing normal cells as well as heat-resistant mutants (Reeves 1972). This model would be consistent with different components of the membrane predominating in the mechanism for specific temperatures. In some instances, cholesterol levels may be prime regulators of heat sensitivity, while in others denaturation of membrane proteins or alterations in phospholipids may be the most important determinants in heat cell killing. Membrane permeability modifications, induced by any of several factors, would be an integral aspect of the mechanism of heat cell killing. For temperatures ranging up to 49°C, as discussed by Landry and Marceau (1978), other intracellular modifications could occur and participate in the overall mechanism responsible for cell death. These could include intracellular ionic changes, enzymatic alterations, or changes in protein compartmentalization.

EFFECT OF THERMOTOLERANCE ON RADIOSENSITIZATION

An important clinical application of hyperthermia will be its use in combination with radiation therapy in the treatment of human malignancies (see Streffer et al. 1978 for historical review and recent studies). Since conventional radiotherapy procedures involve fractionated dose techniques, protocols including radiation and hyperthemia therapy have been designed to use fractionation of both modalities. These designs introduce the consideration of thermotolerance, and especially the possibility that the development of thermotolerance may produce radiation resistance.

This question has been addressed most directly by Dewey and co-workers (1977). It is well known, based on this and the work of other groups, that hyperthermia interacts synergistically, or more than additively, with ionizing radiation. The synergistic interaction is most apparent for radiations of low linear energy transfer (LET)[4] and decreases with increasing LET (Gerner and Leith 1977). Heat interacts synergistically with radiation, in part, by inhibiting molecular and cellular repair mechanisms. In addition, heat is most effective at killing mitotic and S phase cells while radiation kills mitotic, G_1 and G_2 phase cells most efficiently (see Dewey et al. 1977 for review). In one series of experiments, Freeman, Raaphorst, and Dewey (1979) investigated possible effects of thermotolerance on heat and radiation interactions by combining 42°C for varying times with irradiation. These data are shown in figure 6.12. Knowing that thermotolerance would be expressed after 200 to 300 minutes, these investigators could ask whether resistance would be expressed also in cultures treated with both 42°C and x-rays. Their results demonstrated that no resistance developed in cultures treated with both heat and radiation, even after thermotolerance was expressed. Rather, results suggested that the degree of synergism most closely reflected the amount of thermal damage produced. Thus, when tolerance was expressed, longer exposures to 42°C and added radiation afforded only minor increases in cell killing. Resistance clearly was not expressed, however.

Studies on the production of chromosome aberrations by these two agents are also pertinent to a possible effect of thermotolerance on heat-radiation synergism. In a classic paper in 1971, Dewey, Miller, and Leeper demonstrated that chromosome aberrations produced by irradiation correlated with radiation cell killing. This group later showed that the production of chromosome aberrations correlated with heat cell killing only for S phase populations, but not in mitotic or G_1 cultures (Dewey, Sapareto, and Betten 1978; Dewey et al. 1971). When the two agents were combined in a synergistic fashion, the chromosome aberrations observed again correlated with cell killing, independent of the cell cycle phase in which treatment was administered (Dewey, Sapareto, and Betten 1978). Thus, this group emphasized that clear distinctions should be made between effects associated with hyperthermic cytotoxicity and hyperthermic radiosensitization.

Other in vitro and in vivo studies are also relevant to a discussion of the effects of thermotolerance on radiosensitivity. It was pointed out in the section on survival responses that a single thermal dose can produce tolerance in a population to subsequent thermal doses. Sequencing experiments by a large number of laboratories have shown that heat is most effective at

[4]Linear energy transfer describes the rate of energy deposition per unit distance by a specific radiation as it moves through a medium.

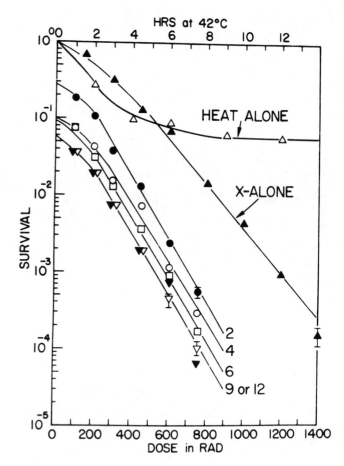

Figure 6.12. The effect of heat and/or radiation on cell survival. △:heat alone (42°C); ▲:radiation alone; ●, ○, □, ▽, ▼: heating for 2, 4, 6, 9, or 12 hours at 42°C followed by irradiation 0.5 minutes after the end of the heating period. The numbers of the curves represent the duration of heating. The plating efficiency for unheated and unirradiated cells was 0.46 ± 0.02 (mean ± SEM). (Freeman, Raaphorst, and Dewey 1979. Reprinted by permission.)

potentiating radiation damage when given concurrently or within a few hours of the radiation exposure (Dewey et al. 1977). Using an analogous split-graded heat dose experiment, which was used to assess thermotolerance, if radiation is substituted for the second heat treatment, no radiation resistance is apparent. The synergistic interaction is no longer evident, but the radiation response is similar to that seen if the culture had been exposed to radiation alone. Thus, thermotolerant cells express the same degree of heat-radiation synergism as do normal cells, and heat does not induce radiation resistance; nor does radiation induce thermotolerance.

While an effect of thermotolerance on radiosensitivity seems negligible, much additional work needs to be done in order to define the optimal application of these two agents in clinical human cancer therapy. In the treatment of local disease, temperatures used in therapy will usually exceed 43°C and be given in fractionated form. Hyperthermia will be employed as a cytotoxic agent and as a radiosensitizer. Thermotolerance probably will not be an issue in the latter case but will most certainly be in the former.

EFFECT OF THERMOTOLERANCE ON THE INTERACTION OF HYPERTHERMIA AND CHEMOTHERAPEUTIC DRUGS

The preceding section addressed the question of whether thermotolerance affected radiosensitivity in cells and tissues. Based on current results, the answer is negative. The mechanism of radiation-induced cell lethality is thought to be chromosomal and involve some aspect of DNA damage (Dewey, Miller, and Leeper 1971). On the other hand, the mechanism of heat cell killing in normal and thermotolerant cells is thought not to involve DNA damage directly, at least for temperatures up to 49°C (Connor et al. 1977; Dewey et al. 1977; Landry and Marceau 1978; Westra and Dewey 1971). Thermodynamic and other arguments point to protein damage and denaturation and to a possible involvement of the cell membrane. Thus, since the targets involved in cell killing for these two agents are different, it is not surprising that neither agent produces tolerance to the other.

Similar questions regarding the interaction of chemotherapeutic drugs and heat may not yield the same answers as did the study of heat-radiation. Classes of drugs exist that affect membrane (Hahn, Li, and Shui 1977) or protein structure and function. Thus, drugs that share mechanisms of action with heat may be expected to produce a thermotolerant state similar to that resulting from a heat dose. Since, however, the mechanism(s) of action of heat damage and thermotolerance is (are) unknown, studies of this type may prove helpful in identifying specific targets.

Hahn's group at Stanford University has been one of the most active in investigating the interaction of heat and chemotherapeutic agents. They have described heat potentiation of a number of drugs, including Adriamycin, bleomycin, the nitrosoureas, actinomycin D, amphotericin B, and cisplatin as a partial listing (Hahn 1979). In a 1979 study, Dr. Hahn addressed the question of heat-induced drug tolerance. Of all the drugs studied, only Adriamycin and actinomycin D seemed to show resistance in response to an initial heat dose. As Dr. Hahn pointed out, however, it is not clear whether the mechanism of heat-produced drug tolerance is the same as for hyperthermia-produced thermotolerance. For instance, drug tolerance could be expressed if transport mechanisms regulating cellular uptake of the drug were compromised. Another possibility might include true induction of metabolic enzymes that would degrade drugs into inactive species. The answer to the question of whether heat-produced thermotolerance and drug tolerance are similar remains to be determined.

Another related question is whether drugs can produce resistance to subsequent hyperthermia treatments. There are more questions than answers. Li and Hahn (1978) have demonstrated that alcohols can produce resistance to both heat and chemotherapeutic drugs. They argue that alcohols are known to affect cell membranes and thus are inducing thermotolerance in a manner similar to the mechanism by which heat damage is inflicted. Again, it is not clear whether the mechanism of chemically induced thermotolerance is the same as heat-produced thermotolerance.

Recent evidence suggests that the development of thermotolerance does affect the interaction of heat and the chemotherapeutic agent BCNU, 1,3-bis (2-chloroethyl)-1-nitrosourea (Herman, Stickney, and Gerner 1979). Table 6.4 abstracts some of these results. It is clear that when thermotolerance is expressed, by decreasing the rate at which temperature is raised from 37°C to 42.4°C, the ability of a specific heat dose to potentiate drug cytotoxicity decreases. It is important, however, to note that the degree of drug potentiation is related to the damage produced by the thermal dose. As in the case of radiation-heat interactions, resistance in the form of an increased D_o is not evident. The results simply show that when thermotolerance develops, continued heat exposures do not produce additional enhancement of drug damage.

CLINICAL CONSIDERATIONS

The rationale for hyperthermia as a cancer treatment modality is extensive and can be summarized as follows. Hyperthermia may have selective lethal effects on tumors compared to normal tissues, and heat enhances the cytotoxic effects of radiation and chemotherapeutic drugs. In the former case, heat may be acting selectively both at the cellular level and at the organized tissue level through presumed mechanisms involving the vascular nature of normal and tumor tissues. In the latter case, hyperthermia is acting at many levels, including affecting radiation and drug repair mechanisms, selectively killing cells resistant to drugs and radiation, and enhancing cellular drug up-

Table 6.4
Effect of Thermotolerance Development on the Sensitization of Chinese Hamster Cells to BCNU Cytotoxicity

BCNU Dose* (μg/ml for 1 hr)	Treatment Temperature (1 hr)	Time from Room Temperature to Treatment Temperature (min)	Surviving Fraction†
1	37°C	3	95
2.5	37°C	3	93
5	37°C	3	89
10	37°C	3	39
1	42.4°C	3	51
2.5	42.4°C	3	25
5	42.4°C	3	11
10	42.4°C	3	0.19
1	42.4°C	180	90
2.5	42.4°C	180	65
5	42.4°C	180	38
10	42.4°C	180	1.5

Source: Herman, Stickney, and Gerner 1979.
*Cells were treated with BCNU at the specified treatment temperature for one hour only.
†Surviving fraction is normalized to exclude cell killing caused by hyperthermia alone.

take, thereby overcoming one type of acquired drug resistance. Because of the many facets of hyperthermia, and since whole-body hyperthermia most probably will be limited to temperatures around 42°C (Bligh 1979), while local and regional hyperthermic temperatures from 43°C to 50°C are well tolerated, the clinician is faced with complicated decisions regarding thermal dose prescription. Depending on whether the hyperthermic treatment involves local or systemic therapy, or whether heat is used alone or in combination with drugs or radiation, and on knowledge of the thermosensitivity of the tumor and surrounding normal tissues, the physician must arrive at decisions including temperature, duration of heating, thermal dose fractionation, and sequence of application of heat with another modality.

A first consideration in prescribing a thermal dose schedule should include the probability that some, but not all, tumors have the potential to develop thermotolerance. Both in vitro and in vivo evidence exists to suggest thermotolerance can be expressed in tumor cells and tissues. As pointed out by several authors, however, certain characteristics of tumors, such as low pH, lack of essential nutrients, and hypoxia, may inhibit the expression of thermotolerance. Also, there may be specific tumor cell types that lack the ability to express thermotolerance or may express tolerance to a lesser degree than normal cells. Thus, in some cases it may prove efficacious to design a thermal dose schedule to produce thermotolerance in normal tissues when a certain tumor would not express tolerance, or would express it to a lesser degree. In other instances, it may be essential to derive a thermal dose prescription to avoid thermotolerance development, since other tumors may effectively express tolerance, thereby limiting the cytotoxic potential of hyperthermia by itself as well as its efficiency in potentiating the action of either radiation or drugs. While these considerations are complex, they are not new to either radiologists or chemotherapists, since they currently face the development of resistance during therapy. Basic research offers the promise of providing vital guidelines for making these decisions. Basic cellular studies may uncover generalizations regarding those tumors that might express thermotolerance and those that would not. In vitro cloning assays, such as those described by Salmon and colleagues (1978), may offer the possibility of determining before treatment whether a specific tumor has the ability to express tolerance and to what degree. Thus, for some types of cancer, daily or even more frequent fractionation schedules may be in order, while for others, fractions would be separated by two to three days in order to avoid thermotolerance expression during therapy.

A second major level of consideration in prescribing a thermal dose schedule should include factors of time and temperature. Based on currently available data, if temperatures less than 43°C are employed, as in whole-body hyperthermia, durations of heating in excess of three to four hours are not warranted since thermotolerance will be expressed by this time. This view must be considered in relation to the tumor-specific characteristics discussed in the preceding paragraph. Limitation of treatment times to three to four hours is not warranted when it is known, or suspected, that the tumor(s) being treated may be at low pH, so that they would not develop tolerance while normal tissues would. The general intent of whole-body hyperthermia, however, is to treat distant, often microscopic, metastasis. This type of disease usually is well vascularized and would not have the characteristics known to inhibit thermotolerance expression. Thus, whole-body hyperthermia should normally be restricted to treatment times of less than four hours. A notable exception would be tumor types unable to express thermotolerance because of some cellular mechanism.

A third aspect of thermotolerance involved in clinical decisions relates to damage produced by the thermal dose. It is known that the cytotoxic effects of a thermal dose are a function of time and temperature. Thermotolerance expression is a function of thermal dose. In cases where tolerance expression would be a desired component of treatment, a larger thermal dose would be in order at the beginning of treatment. An alternate approach would be to use a noncytotoxic thermal dose, including temperatures in the range of 38°C to 41°C, before administering primary therapy at toxic temperatures of 41°C and higher. Also, the rate of heating from normal body to hyperthermic temperatures could be reduced. Conversely, when thermotolerance is deemed detrimental to therapy, rapid heating rates from normal to therapeutic temperatures are essential, and fractionation schemes must be designed to allow this transient phenomenon to subside.

A final consideration would pertain to the effects of thermotolerance on the interaction of heat with drugs or radiation. Evidence currently available indicates that heat does not induce either radiation or drug resistance, with a few exceptions. These exceptions include drugs such as Adriamycin and actinomycin D. Also, a great deal of evidence suggests that neither radiation nor drugs produce thermotolerance. The sole exception here is the possibility that drugs, similar to ethanol, may produce thermotolerance. It is becoming apparent that the enhancement of radiation and drug-induced cytotoxicity by hyperthermia is directly related to the amount of cytotoxocity produced by the thermal dose. Thus, thermotolerance can be important in combined modality therapy in that radio- or chemosensitization may reach some maximum as thermotolerance is expressed.

CONCLUSIONS

Thermotolerance is an intriguing and complex biological phenomenon. It develops in in vitro as well as in vivo biological systems. It is expressed as a complicated function of physical, biological, and environmental factors. While the underlying mechanisms of thermotolerance remain to be elucidated, evidence has been presented to suggest that it shares aspects in common with mechanisms responsible for killing nontolerant cells and heat-resistant mutants. It may be as simple as a single mechanism, identical with that responsible for killing nontolerant cells, although some investigators have argued for up to three distinct mechanisms. Thermotolerance expression does not lead to radiation or chemotherapeutic drug resistance, but it does affect the efficacy of heat doses in enhancing the effects of these other modalities. Important generalizations regarding the relative degree to which tolerance develops in normal tissues and tumors remain to be formulated. Certain strategies may intentionally produce thermotolerance as a way of protecting normal tissues if certain tumors can be shown to express tolerance or express it to a lesser degree than normal tissues. In any event, it is clear that thermotolerance is a biological phenomenon that must be considered in the clinical application of hyperthermia in human cancer therapy.

References

Bass, H.; Moore, J. L.; and Coakley, W. J. Lethality of mammalian cells due to hyperthermia under oxic and hypoxic conditions. *Int. J. Radiat. Biol.* 33:57–67, 1978.

Bauer, K. D., and Henle, K. J. Arrhenius analysis of heat survival curves from normal and thermotolerant CHO cells. *Radiat. Res.* 78:251–263, 1979.

Ben-Hur, E., and Riklis, E. Enhancement of thermal killing by polyamines. III. Synergism between spermine and γ radiation in hyperthermic Chinese hamster cells. *Radiat. Res.* 78:321–328, 1979.

Bhuyan, B. K. Kinetics of cell kill by hyperthermia. *Cancer Res.* 39:2277–2284, 1979.

Bhuyan, B. K. et al. Sensitivity of different cell lines and of different phases in the cell cycle to hyperthermia. *Cancer Res.* 37:3780–3784, 1977.

Bligh, J. Aspects of thermoregulatory physiology pertinent to hyperthermic treatment of cancer. *Cancer Res.* 39:2307–2312, 1979.

Bull, J. M. et al. Whole body hyperthermia: a phase-I trial of a potential adjuvant to chemotherapy. *Ann. Intern. Med.* 90:317–323, 1979.

Canellakis, E. S. et al. Intracellular levels of ornithine decarboxylase, its half-life, and a hypothesis relating polyamine-sensitive membrane receptors to growth. In *Advances in polyamine research*, vol. 1, eds. R. A. Campbell, et al. New York: Raven Press, 1978, pp. 17–30.

Cohen, S. S. *Introduction to the polyamines.* Englewood Cliffs, N.J.: Prentice-Hall, 1971.

Connor, W. G. et al. Prospects for hyperthermia in human cancer therapy. II. Implications of biological and physical data for applications of hyperthermia to man. *Radiology* 123:497–503, 1977.

Cress, A. E. and Gerner, E. W. Cholesterol content inversely reflects the thermal sensitivity of mammalian cells in culture. *Nature* 382:677–680, 1980.

De Rosa, M. et al. Occurrence and characterization of new polyamines in the extreme thermophile *Caldariella acidophila. Biochem. Biophys. Res. Commun.* 69:253–261, 1976.

Dewey, W. C. Cell biology of hyperthermia and radiation. Paper read at the Sixth International Congress in Radiation Research, May 1979, Tokyo, Japan.

Dewey, W. C. et al. Heat induced lethality and chromosomal damage in synchronized Chinese hamster cells treated with 5-bromo-deoxyuridine. *Int. J. Radiat. Biol.* 20:505–520, 1971.

Dewey, W. C. et al. Cellular responses to combinations of hyperthermia and radiation. *Radiology* 123: 463–474, 1977.

Dewey, W. C.; Miller, H. H.; and Leeper, D. B. Chromosomal aberrations and mortality of x-irradiated mammalian cells; emphasis on repair. *Proc. Nat. Acad. Sci. USA* 68:667–671, 1971.

Dewey, W. C.; Sapareto, S. A.; and Betten, D. A. Hyperthermic radiosensitization of synchronous Chinese hamster cells: relation between lethality and chromosomal aberrations. *Radiat. Res.* 76:48–59, 1978.

Dring, G. J., and Gould, G. W. Reimposition of heat-resistance on germinated spores of bacillus cereus by osmotic manipulation. *Biochem. Biophys. Res. Commun.* 66:202–208, 1975.

Freeman, M. L.; Dewey, W. C.; and Hopwood, L. E. Effect of pH on hyperthermic cell survival. *J. Natl. Cancer Inst.* 58:1837–1839, 1977.

Freeman, M. L.; Raaphorst, G. P.; and Dewey, W. C. The relationship of heat killing and thermal radiosensitization to the duration of heating at 42°C. *Radiat. Res.* 78:172–175, 1979.

Fuller, D. J. M.; Gerner, E. W.; and Russell, D. H. Polyamine biosynthesis and accumulation during the G_1 to S phase transition. *J. Cell. Physiol.* 93:81–88, 1977.

Gerner, E. W. et al. The potential of localized heating as an adjunct to radiation therapy. *Radiology* 116:433–439, 1975.

Gerner, E. W. et al. A transient thermotolerant survival response produced by single thermal doses in HeLa cells. *Cancer Res.* 36:1035–1040, 1976.

Gerner, E. W. et al. Factors regulating membrane permeability alter thermal resistance. *Ann. N.Y. Acad. Sci.* 335:215–233, 1980a.

Gerner, E. W. et al. Enhancement of hyperthermia induced cytotoxicity by the polyamines. *Cancer Res.* 40:432–438, 1980b.

Gerner, E. W.; Holmes, P. W.; and McCullough, J. A. Influence of growth state on several thermal responses of EMT6/AZ tumor cells *in vitro. Cancer Res.* 39:981–986, 1979.

Gerner, E. W., and Leith, J. T. Interaction of hyperthermia with radiations of different linear energy transfer. *Int. J. Radiat. Biol.* 31:283–288, 1977.

Gerner, E. W., and Russell, D. H. The relationship between polyamine accumulation and DNA replication kinetics in synchronized CHO cells after heat shock. *Cancer Res.* 37:482–489, 1977.

Gerner, E. W., and Schneider, M. J. Induced thermal resistance in HeLa cells. *Nature* 256:500–502, 1975.

Gerweck, L. E. Modification of cell lethality at elevated temperatures: the pH effect. *Radiat. Res.* 70:224–235, 1977.

Gerweck, L. E.; Gillette, E. L.; and Dewey, W. C. Killing of Chinese hamster cells *in vitro* by heating under hypoxic or aerobic conditions. *Eur. J. Cancer* 10: 691–693, 1974.

Gerweck, L. E.; Jennings, M.; and Wesolowski, B. Influence of pH on hyperthermically induced thermal tolerant cells. Paper read at the Sixth International Congress in Radiation Research, May 1979, Tokyo, Japan.

Gerweck, L. E.; Nygaard, T. G.; and Burlett, M. Response of cells to hyperthermia under acute and chronic hypoxic conditions. *Cancer Res.* 39:966–972, 1979.

Gerweck, L., and Rottinger, E. Enhancement of mammalian cell sensitivity to hyperthermia by pH alteration. *Radiat. Res.* 67:508–511, 1976.

Gould, G. W., and Dring, G. J. Heat resistance of bacterial endospores and concept of an expanded osmoregulatory cortex. *Nature* 258:402–405, 1975.

Gullino, P. M. et al. Modifications of the acid-base status of the internal mileu of tumors. *J. Natl. Cancer Inst.* 34:857–869, 1965.

Hahn, G. M. Metabolic aspects of the role of hyperthermia in mammalian cell inactivation and their possible relevance to cancer treatment. *Cancer Res.* 34:3117–3123, 1974.

Hahn, G. M. Potential for therapy of drugs and hyperthermia. *Cancer Res.* 39:2264–2268, 1979.

Hahn, G. M. Radiofrequency, microwaves, and ultrasound in the treatment of cancer: some heat transfer problems. Paper read at the Conference on Thermal Characteristics of Tumors: Applications in Detection and Treatment, March 1979, New York, N.Y.

Hahn, G. M.; Li, G. C.; and Shui, E. Interaction of Amphotericin B and 43° hyperthermia. *Cancer Res.* 37:761–764, 1977.

Hahn, G. M., and Little, J. B. Plateau-phase cultures of mammalian cells: an *in vitro* model for human cancer. *Current Topics Radiation Research Quarterly* 8:39–83, 1972.

Hall, E. J. Response of cultured cells to hyperthermia with special reference to thermal tolerance and hypoxia. *Br. J. Radiol.* 49:806, 1976.

Harisiadis, L. et al. Hyperthermia: biological studies at the cellular level. *Radiology* 117:447–452, 1975.

Harisiadis, L.; Sung, D.; and Hall, E. J. Thermal tolerance and repair of thermal damage by cultured cells. *Radiology* 123:505–509, 1977.

Harris, M. Criteria of viability in heat-treated cells. *Exp. Cell Res.* 44:658–661, 1967a.

Harris, M. Temperature-resistant variants in clonal populations of pig kidney cells. *Exp. Cell Res.* 46:301–314, 1967b.

Harris, M. Growth and survival of mammalian cells under continuous thermal stress. *Exp. Cell Res.* 56:382–386, 1969.

Heller, J. S.; Fong, W. F.; and Canellakis, E. S. Induction of a protein inhibitor to ornithine decarboxylase by the end products of its reaction. *Proc. Natl. Acad. Sci. USA* 73:1858–1862, 1976.

Henle, K. J.; Bitner, A. F.; and Dethlefsen, L. A. Induction of thermotolerance by multiple heat fractions in Chinese hamster ovary cells. *Cancer Res.* 39:2772–2778, 1979.

Henle, K. J., and Dethlefsen, L. A. Sensitization to hyperthermia (45°C) of normal and thermotolerant CHO cells by anisotonic media. *Radiat. Res.* 70:632, 1977.

Henle, K. J., and Dethlefsen, L. A. Heat fractionation and thermotolerance: a review. *Cancer Res.* 38:1843–1851, 1978.

Henle, K. J.; Karamuz, J. E., and Leeper, D. B. Induction of thermotolerance in Chinese hamster ovary cells by high (45°) or low (40°) hyperthermia. *Cancer Res.* 38:570–574, 1978.

Henle, K. J., and Leeper, D. B. Interaction of hyperthermia and radiation in CHO cells: recovery kinetics. *Radiat. Res.* 66:505–518, 1976.

Henle, K. J., and Leeper, D. B. Effects of hyperthermia (45°C) on macromolecular synthesis in Chinese hamster ovary cells. *Cancer Res.* 39:2665–2674, 1979.

Herman, T. S.; Stickney, D. G.; and Gerner, E. W. Differential rates of heating influence hyperthermia induced cytotoxicity in normal and transformed cells *in vitro*. *Proc. Am. Assoc. Cancer Res.* 20:165, 1979.

Ingraham, J. L. The physiology of growth. In *The bacteria*, vol. 4, eds. I. C. Gunsalus and R. Y. Stanier. New York: Academic Press, p. 265.

Joshi, D. S., and Jung, H. Thermotolerance and sensitization induced in CHO cells by fractionated hyperthermic treatments at 38°–45°C. *Eur. J. Cancer* 15:345–350, 1979.

Kal, H. B.; Hatfield, M.; and Hahn, G. M. Cell cycle progression of murine sarcoma cells after x-irradiation or heat shock. *Radiology* 117:215–217, 1975.

Kase, K., and Hahn, G. M. Differential heat response of normal and transformed human cells in tissue culture. *Nature* 255:288–230, 1975.

Kase, K. R., and Hahn, G. M. Comparison of some responses to hyperthermia by normal human diploid cells and neoplastic cells from the same origin. *Eur. J. Cancer* 12:481–491, 1976.

Kim, S. H.; Kim, J. H.; and Hahn, E. W. Enhanced killing of hypoxic tumor cells by hyperthermia. *Br. J. Radiol.* 48:872–874, 1975.

Landry, J., and Marceau, N. Rate-limiting events in hyperthermic cell killing. *Radiat. Res.* 75:573–585, 1978.

Law, M. P.; Ahier, R. G.; and Field, S. B. The response of the mouse ear to heat applied alone or combined with x-rays. *Br. J. Radiol.* 51:132–138, 1978.

Law, M. P.; Coultas, P. G.; and Field, S. B. Induced thermal resistance in the mouse ear. *Br. J. Radiol.* 52:308–314, 1979.

Leith, J. T. et al. Hyperthermic potentiation: biological aspects and applications to radiation therapy. *Cancer* 39:766–779, 1977.

Li, G. C. et al. Recovery from potentially lethal hyperthermic damage: nutrient and pH dependence. Paper read at the Sixth International Congress in Radiation Research, May 1979, Tokyo, Japan.

Li, G. C., and Hahn, G. M. Ethanol-induced tolerance to heat and to adriamycin. *Nature* 274:699–700, 1978.

Milligan, A. J., and Leeper, D. B. The effect of hyperthermia on the Chinese hamster small intestine. *Radiat. Res.* 74:529, 1978.

Mondovi, B. et al. The biochemical mechanism of selective heat sensitivity of cancer cells. II. Studies on nucleic acids and protein synthesis. *Eur. J. Cancer* 5:137–146, 1969a.

Mondovi, B. et al. The biochemical mechanism of selective heat sensitivity of cancer cells. I. Studies on cellular respiration. *Eur. J. Cancer* 5:129–136, 1969b.

Nielsen, O. S., and Overgaard, J. Effect of extracellular pH on thermotolerance and recovery of hyperthermic damage *in vitro*. *Cancer Res.* 39:2772–2778, 1979.

Oshima, T. Thermine: a new polyamine from an extreme thermophile. *Biochem. Biophys. Res. Commun.* 63:1093–1098, 1975.

Overgaard, J. Influence of extracellular pH on the viability and morphology of tumor cells exposed to hyperthermia. *J. Natl. Cancer Inst.* 56:1243–1250, 1976.

Overgaard, J. Effect of hyperthermia on malignant cells *in vivo*. A review and a hypothesis. *Cancer* 39:2637–2646, 1977.

Overgaard, J. The effect of local hyperthermia alone, and in combination with radiation, on solid tumors. In Proceedings of the Second International Symposium on Cancer Therapy by Hyperthermia and Radiation, eds. C. S. Streffer et al. Baltimore and Munich: Urban & Schwarzenberg, 1978, pp. 49–61.

Palzer, R. J., and Heidelberger, C. Influence of drugs and synchrony on the hyperthermic killing of HeLa cells. *Cancer Res.* 33:422–427, 1973.

Power, J. A., and Harris, J. W. Response of extremely hypoxic cells to hyperthermia: survival and oxygen enhancement ratios. *Radiology* 123:767–770, 1977.

Raaphorst, G. P., and Dewey, W. C. Enhancement of hyperthermic killing of cultured mammalian cells by treatment with anisotonic NaCl or medium solutions. *J. Therm. Biol.* 3:177–182, 1978.

Reeves, O. R. Mechanisms of acquired resistance to acute heat shock in cultured mammalian cells. *J. Cell. Physiol.* 79:157–170, 1972.

Rosenberg, B. et al. Quantitative evidence for protein denaturation as the cause of thermal death. *Nature* 232:471–473, 1971.

Roti-Roti, J. L., and Winward, R. T. The effects of hyperthermia on the protein to DNA ratio of isolated HeLa cell chromatin. *Radiat. Res.* 74:159–169, 1978.

Salmon, S. E. et al. Quantitation of differential sensitivity of human-tumor stem cells to anticancer drugs. *N. Engl. J. Med.* 298:1321–1327, 1978.

Sapareto, S. A. et al. Effects of hyperthermia on survival and progression of Chinese hamster ovary cells. *Cancer Res.* 38:393–400, 1978.

Schulman, N., and Hall, E. J. Hyperthermia: its effect on proliferative and plateau phase cell cultures. *Radiology* 113:207–209, 1974.

Selawry, O. S.; Goldstein, M. N.; and McCormick, T. Hyperthermia in tissue-culture cells of malignant origin. *Cancer Res.* 17:785–791, 1957.

Simchen, G. Cell cycle mutants. *Annu. Rev. Genet.* 12:161–191, 1978.

Streffer, C. et al., eds. *Proceedings of the Second International Symposium on Cancer Therapy by Hyperthermia and Radiation.* Baltimore and Munich: Urban & Schwarzenberg, 1978.

Strom, R. et al. The biochemical mechanism of selective heat sensitivity of cancer cells. IV. Inhibition of RNA synthesis. *Eur. J. Cancer* 9:103–112, 1973.

Suit, H. D., and Shwayder, M. Hyperthermia: potential as an anti-tumor agent. *Cancer* 34:122–129, 1974.

Tomasovic, S. P.; Turner, G. N.; and Dewey, W. C. Effect of hyperthermia on non-histone proteins isolated with DNA. *Radiat. Res.* 73:535–552, 1978.

Turano, C. et al. The biochemical mechanism of selective heat sensitivity of cancer cells. III. Studies on lysosomes. *Eur. J. Cancer* 6:67–72, 1970.

Westra, A., and Dewey, W. C. Variation in sensitivity to heat shock during the cell-cycle of Chinese hamster cells *in vitro*. *Int. J. Radiat. Biol.* 19:467–477, 1971.

Yatvin, M. B. Influence of membrane lipid composition on hyperthermic and radiation killing of *E. coli* cells. *Radiat. Res.* 70:610–614, 1977.

Yatvin, M. J. The influence of membrane lipid composition and procaine on hyperthermic death of cells. *Int. J. Radiat. Biol.* 32:513–521, 1977.

CHAPTER 7
Histopathologic Effects of Hyperthermia

Jens Overgaard

Introduction
Morphologic Effects of Hyperthermia in Normal Tissues
Morphologic Effects of Hyperthermia in Solid Tumors
Influence of Environmental Factors on the Hyperthermic Response
Mechanism of Hyperthermic Cell Destruction
Conclusion
Clinical Implications

INTRODUCTION

The aim of hyperthermic cancer therapy is to obtain destruction of clonogenic tumor cells with minimal damage to the normal tissue. Therefore, most experimental and clinical hyperthermic studies have focused on the curative effect and regression pattern in tumors and physiologic responses in normal tissues; in cellular studies, survival curves have been the quantitative end point used.

Morphologic studies of heated cells and tissues may give important information on the heat-induced destructive mechanisms in malignant and normal cells and tissues. Briefly, such histopathologic examination may help to answer the following questions: *Why* are cells destroyed by moderate heat treatment? *What* are the mechanisms involved in the progressive cell destruction? *When* does the heat destruction occur? *Where* in the tissue is the heat damage present? A multi-focused investigation involving quantitative and qualitative end points is required to answer these questions. Histopathologic studies using the light and electron microscopes and histochemical procedures have given valuable knowledge, especially on the qualitative aspects of some of the parameters involved following hyperthermic treatment. Also, at a time when hyperthermic treatment technique is far from sufficiently developed, histologic examination can provide a tool for outlining areas of destruction and areas of treatment failures in tumors, as well as the location of damage to normal tissues. Thus, histopathologic investigations in experimental tumors have served as a sort of "biological dosimetry" and have demonstrated the location of treatment failures. (Overgaard 1978a, 1978b; Overgaard and Overgaard 1972a, 1976); it was by such methods that areas with intense vascular cooling were detected (Overgaard 1980; Overgaard and Nielsen 1980). Also, the well-established phenomenon of increasing heat sensitivity, which is present in nutritionally deprived environments dominated by increased acidity, is largely a consequence of histopathologic studies (Overgaard 1976a, 1976b, 1977b; Overgaard and Nielsen 1980; Overgaard and Overgaard 1972a, 1976). In spite of this, it is surprising that currently only a few investigators have included systematic histopathologic examinations in their experimental and clinical end points.

This chapter reviews the histopathologic effects of hyperthermia in normal tissues and provides a detailed analysis of the achievements of morphologic examinations of heat-treated tumors.

The author gratefully acknowledges the significant support of the Danish Cancer Society Grant No. 24/79 and of the Krista and Viggo Petersen Foundation.

MORPHOLOGIC EFFECTS OF HYPERTHERMIA IN NORMAL TISSUES

A surprisingly large amount of experimental and clinical information on the physiologic and morphologic effects of whole-body hyperthermia in normal tissue was published in the first part of this century (Bragdon 1947; Gore and Isaacson 1949; Heine et al. 1971). This attention resulted from the wide use of hyperthermia in the treatment of infectious diseases (especially venereal diseases) prior to the antibiotic era (Gore and Isaacson 1945; Heine et al. 1971; Wallace 1943). Unfortunately, the older data contained no information about specific heat-induced morphologic lesions, only general observations described as congestions, focal necrosis, and hemorrhage (Gore and Isaacson 1945; Hartman and Major 1935; Kew et al. 1967; Shibolet et al. 1967). Neither were any organs described to be especially sensitive to hyperthermia except for the testis, where an inhibition of spermatogenesis occurs following very moderate heat exposure (Hand et al. 1979). Autopsy records of fatal cases following early whole-body hyperthermia also described nonspecific changes in almost all organs, with lung or brain congestion as the most likely cause of death in the acute phase, and liver damage expressed as nonspecific hepatitis as one of the most significant findings in more prolonged lethal cases (Bianchi et al. 1972; Bragdon 1947; Gore and Isaacson 1949; Hartman and Major 1935; Shibolet et al. 1967). Nonspecific hepatitis also has been observed more recently in cancer patients treated with modern day whole-body hyperthermia. Furthermore, liver damage appears to be more frequent when the heat treatment is associated with anesthesia or treatment with drugs known to be hepatotoxic (e.g., barbiturates or phenytoin) (Pettigrew et al. 1974a; Willis, Findlay, and McManus 1976).

The histopathologic alteration following local hyperthermia to different normal tissues has been only sparsely described (von Heinz, Uerlings, and Grupe 1971; Gilchrist et al. 1965; Jacobsen and Hosol 1931; Okkels and Overgaard 1937), except in the skin, where Moritz (1947) and Moritz and Henriques (1947) extensively analyzed the physiologic and pathologic changes in human and pig skin following heating at different temperatures and times. They also described a time-temperature relation for thermal damage, which was confirmed experimentally in other normal tissues (Overgaard 1978a; Overgaard and Suit 1979; Overgaard and Overgaard 1977) and that provided the basis for the assumption that normal tissue injury in general follows the same time-temperature relation. Thus, local exposure to treatments up to about 44°C for 30 minutes, or an equivalent temperature and heating time, is tolerated without producing severe damage. This tolerance level is in the same range as the pain threshold. This normal tissue tolerance to heat treatment exists only if sufficient tissue vascularization is present; any inhibition of blood flow tends strongly to increase the heat damage and consequently to reduce the tolerance level (Baker and Wright 1976; Hill and Denekamp 1978; Overgaard and Nielsen 1980).

With such limited knowledge about the damage induced by local hyperthermia, it is important in the future to focus on transient and fixed damage in heated normal tissues in order to establish the heat response and tolerance in different organs.

MORPHOLOGIC EFFECTS OF HYPERTHERMIA IN SOLID TUMORS

Morphologic changes have been described in numerous experimental and human tumors exposed to various types of heat treatment (Cockett et al. 1967; Dickson and Suzangar 1974; Fajardo et al. 1980; Hall, Schade, and Swinney 1974; Hall et al. 1976; Holt 1975, Jares and Warren 1939; Ludgate et al. 1976; Lunglmayr et al. 1973; Marmor, Hahn, and Hahn 1977; Muckle and Dickson 1971; Okkels and Overgaard 1937; Overgaard 1978a, 1980; Overgaard and Nielsen 1980; Overgaard 1934; Overgaard and Okkels 1940a, 1940b; Overgaard and Overgaard 1972b, 1975; Pettigrew et al. 1974b; Stamm et al. 1975; Storm et al. 1979; Strauss, Appel, and Saphir 1962; Sugaar and LeVeen 1979; Warren 1935). Unfortunately, most observations are sporadic, and only a few investigators have focused on this specific subject—all in experimental systems (Fajardo et al. 1980; Okkels and Overgaard 1937; Overgaard 1976a, 1978a; Overgaard and Heyden 1974; Overgaard and Nielsen 1980; Overgaard and Overgaard 1972a, 1975, 1976). Nevertheless, the histopathologic changes in human and animal tumors agree and are so characteristic that they justify the following general description, based on an analysis of a variety of experimental tumors in mice (Overgaard 1976a; Overgaard and Nielsen 1980; Overgaard and Overgaard 1972a, 1975, 1976).

Early Response to Hyperthermia

Macroscopic Changes

It is evident that gross changes can be followed only in superficial tumors. Heat treatment with curative intent in the temperature range of 40.5°C to 45°C and given to tumors transplanted to the foot, leg, or flank of experimental animals resulted in the same macroscopic reaction in each (Crile 1961, 1963; Marmor, Hahn, and Hahn 1977; Okkels and Overgaard 1937; Overgaard 1978a; Overgaard and Suit 1979; Overgaard 1934; Overgaard and Overgaard 1972a, 1976). Immediately after treatment, the tumors and overlaying skin were edematous, with gradually increasing cyanosis that frequently turned into the formation of a dry black crust within one or two days. This crust detached normally within one to three weeks after treatment, leaving either a scar or a regrowing tumor.

Typically, the regrowth took place from the periphery, giving the recurring tumor a ring-shaped appearance. In tumors treated with techniques allowing an almost selective tumor heating, the crust formation was less common (Marmor, Hahn, and Hahn 1977; Overgaard and Overgaard 1972a, 1976), and the edema generally was followed by a regression of the tumor, resulting in permanent control, or a recurrent tumor. The recurrence was from the peripheral areas. When hyperthermia was compared with the effects of other treatment modalities (viz., radiation therapy), the time to a final response (a complete and permanent regression or recurrence) was rather rapid, and in most experimental tumors a definitive answer could be obtained within four weeks after treatment, whereas a similar assay following ionizing radiation would generally take three to four months (Crile 1961, 1963; Overgaard and Suit 1979; Overgaard and Overgaard 1972a, 1976). Such rapid and characteristic hyperthermia response typically was observed after local heating of human tumors, where prominent necroses have been described within hours or a few days after treatment (Cavaliere et al. 1967).

Light Microscopic Changes

In studies of 17 different experimental animal tumors, the histopathologic reaction following heat treatment in the temperature range 40.5°C to 45°C has been investigated at different times after treatment (Overgaard 1978b; Overgaard and Suit 1979; Overgaard and Overgaard 1972a, 1976). The tumors cover a wide range with regard to tissue of origin, tumor type, differentiation, and morphology. The therapeutic response to hyperthermia was different (Overgaard 1978b); however, the early histopathologic reaction after a hyperthermia treatment in the curative range showed characteristic morphologic changes in all experimental tumors despite variation in other characteristics, and apparently was not influenced by the heating technique or by the treatment temperature and time.

Hemorrhage and edema occurred in the tumor during and immediately after treatment, probably as a consequence of early vascular damage that resulted in stasis and increased vascular permeability combined

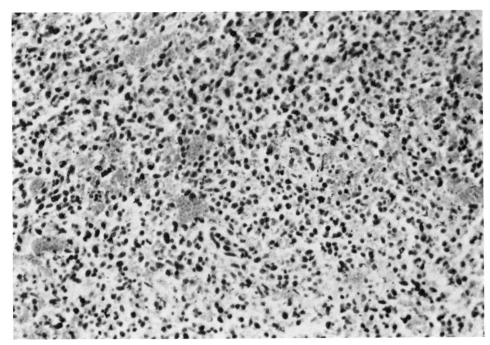

Figure 7.1. FSa I mouse sarcoma 24 hours after treatment for 80 minutes at 43.5°C. The tumor cells have shrunk remarkably and appear with dark pyknotic nuclei and a sparse amount of abrupt cytoplasm, which presents increasing eosinophilia. No mitotic activity is seen. The extracellular space is increased because of edema and hemorrhage (× 300).

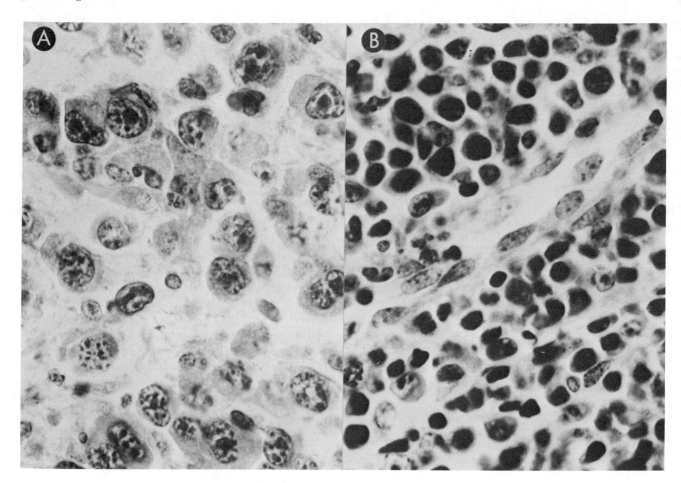

Figure 7.2. *A*, untreated mouse mammary carcinoma. Cells with large irregular hyperchromatic nuclei dominate (× 1200). *B*, same tumor one day after heat treatment at 42.5°C for one hour. Tumor cells show intense shrinkage of the cytoplasm, with small dark pyknotic nuclei. Note the apparent normal morphology of the fibroblasts (× 1200). (Overgaard and Overgaard 1977. Reprinted by permission.)

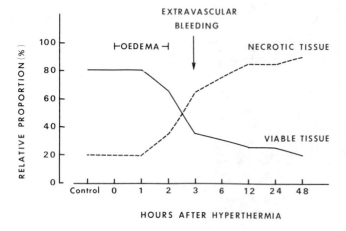

Figure 7.3. Relative proportion of viable and necrotic tissue based on morphologic examination of a mouse mammary carcinoma treated with 42.5°C for one hour.

with a reduction of the functional vascular volume. In the next few hours, the tumor cells became isolated. The cytoplasm shrank and became more acidophilic, and the cell borders were often indistinct and difficult to recognize with the light microscope. The nuclei initially showed condensation and granulation of the chromatin that subsequently became small, dark, and pyknotic (figs. 7.1 and 7.2). Typically, mitotic activity was blocked immediately after heat treatment. Cells in mitosis still were apparent, but underwent decay and no new mitoses were observed. Usually, a complete cell destruction with the presence of small, apparently destroyed tumor cells was observed within a few to 24 hours after heat treatment. This is illustrated in figure 7.3, which shows the time scale for a semiquantitative evaluation of necrotic versus viable tumor cells following moderate heat treatment in an experimental mammary carcinoma. In tumors treated with a heat dose able to induce permanent control, all cells showed the same specific destruction, independent of the localization in the tumor area. In contrast, the stroma and surrounding normal tissue were not generally primarily affected except for moderate edema and hemorrhage, and the stroma cells persisted with an apparently normal morphology (fig. 7.2). Some modifications of the reaction in the surrounding tissue did occur, however, depending on the location of the tumor.

Ultrastructural Changes

Most of the tumors studied were subjected also to examination by electron microscope. In general, treatment resulted in similar ultrastructural changes in the different tumors. A detailed study of the C3H mouse mammary carcinoma HB illustrates these findings (Overgaard 1976a). A mammary carcinoma before treatment is shown in figure 7.4. The first ultrastructural changes were observed immediately after treatment and were similar to the light microscopic observations, showing a separation of the tumor cells, with increasing shrinkage of cytoplasm and nuclei.

At the subcellular level, both cytoplasm and nuclei showed characteristic heat-induced damage. In the cytoplasm, the most prominent feature during the first few hours after treatment was the presence of a hypertrophic Golgi apparatus with an increasing amount of cisternae and small, coated vesicles associated with the occurrence of many small and larger acid phosphatase positive lysosomes, often with a heterogeneous content (fig. 7.5). Later, larger vacuoles and lipid droplets were observed, especially in the periphery of the cytoplasm (fig. 7.6). The mitochondria showed either a dense matrix with irregularly dilated intercristal space or, less frequently, a dilated matrix and ruptured cristae. No apparent abnormalities were observed initially in the rough endoplasmatic reticulum, but the polyribosomes showed a tendency to disaggregate into monoribosomes (fig. 7.7). At this stage, the plasma membrane in most tumors was still distinct and continuous; however, Fajardo and colleagues (1980) have recently described disruption of the plasma membrane as an early finding in the EMT6 tumor following hyperthermic treatment.

These changes occurred progressively during the

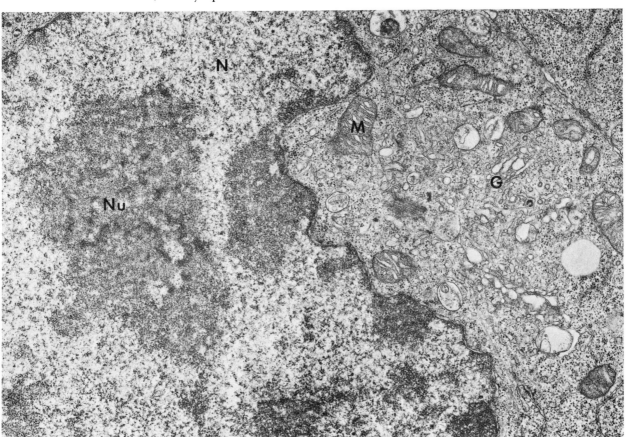

Figure 7.4. Ultrastructure of an untreated mouse mammary carcinoma cell. A part of the large nucleus *(N)* with the nucleolus *(Nu)* can be seen. The cytoplasm is sparse and contains only a few organelles, such as mitochondria *(M)* and a moderate Golgi apparatus *(G)*. The granular endoplasmic reticulum is very sparse, but free polyribosomes are abundant in the cytoplasm. ($\times$ 23,000).

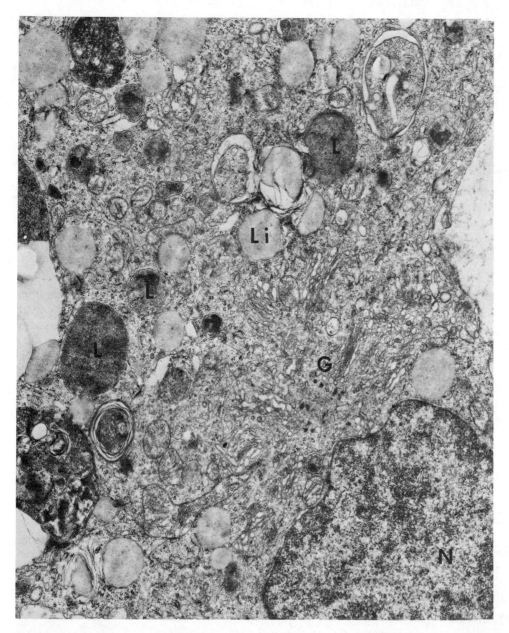

Figure 7.5. Electron micrograph showing part of a mammary carcinoma cell a few hours after treatment at 42.5°C for 60 minutes. The cytoplasm shows characteristic initial changes with a hypertrophic Golgi area *(G)*, with many cisternae and small coated vesicles (primary lysosomes). The surrounding cytoplasm is dominated by small and larger secondary lysosomes *(L)* and lipid droplets *(Li)*. The nucleus *(N)* is without major morphologic changes (× 25,000). (Overgaard and Overgaard 1977. Reprinted by permission.)

first six hours following treatment. Subsequently, the ultrastructural disintegration progressed: acid phosphatase positive lysosomes, with a residue of digested material, dominated the cytoplasm with an increasing number of lipid droplets, and the plasma membrane became more indistinct and seemed to be ruptured in several places (figs. 7.8 and 7.9).

Although the described changes were observed in almost all tumor cells, the time scale varied, and during the first 24 hours cells in different phases of destruction were present at the same time. After 24 hours, however, the destruction in all tumor cells was so severe that they appeared to be nonviable. They had shrunk further, especially those situated in the central parts of the tumor, and all demonstrated a peculiar burnt appearance. The cytoplasm was totally disar-

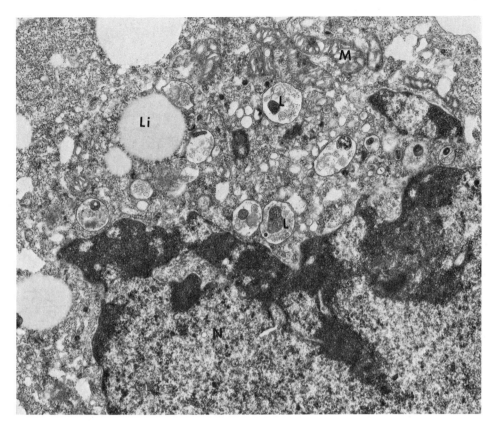

Figure 7.6. Same region as in figure 7.5 four hours after heating at 42.5°C for 30 minutes. A less distinct Golgi area with lysosomes (L) and a few lipid droplets (Li) appear in a slightly indistinct ground cytoplasm. The mitochondria (M) are irregular with a dense matrix and dilated intracristal spaces. The nucleus (N) is irregular in shape and has condensed heterochromatin ($\times$ 23,000). (Overgaard 1976a. Reprinted by permission.)

ranged and without any recognizable structures, and the granular ground cytoplasm contained more or less distinct vacuoles with broken membranes; the plasma membranes were disrupted, and no distinct cell borders could be found (fig. 7.10). Later the cytoplasmic remnant almost completely disappeared, leaving only a small pyknotic nucleus. At this time (two to three days after treatment), macrophages became numerous in the necrotic tumor tissue, and phagocytosis of the destroyed tumor material was prominent (fig. 7.11). It was characteristic, however, that no infiltration of macrophages or other reactive cells was observed in the early stages of the hyperthermic destruction.

In the same period, the nuclei also showed progressive destruction (figs. 7.7 and 7.12). Immediately after treatment the heterochromatin became more dense and had a coarse-appearing network, leaving the remaining karyoplasm with a pale, structureless appearance. In the nucleolus, a reduced amount of the granular component was found, whereas the fibrillary part seemed unchanged. Perichromatin granules were still visible, and the nuclei envelope did not show any morphologic alterations. The nucleolar changes became gradually more prominent with an almost total loss of the granular component and with only a fine fibrillary structure left after 24 hours. In some nucleoli, dense homogeneous spots were observed in the form of nucleolar microspherules. The nuclear envelope was also still continuous and almost unchanged 24 hours after treatment, except for a slight dilation of the perinuclear cisternae. This membranous structure appeared to be the most resistant in the heated tumor cells.

Cells in mitosis underwent the same cytoplasmic changes. The chromatin in the mitotic cells became dense and "clotted," similar to the interphase chromatin (figs. 7.10 and 7.13).

The described pattern of destruction apparently occurred in all tumor cells in solid tumors sufficiently heated in vivo. It was a typical finding, however, that the nonmalignant cells (such as fibroblasts and endothelial cells) did not manifest the same massive destructive changes (fig. 7.14). In fact, they showed only minor and reversible alterations in the cytoplasm, with a slightly increased amount of lysosomes and lipid

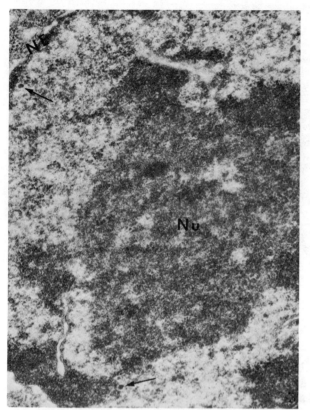 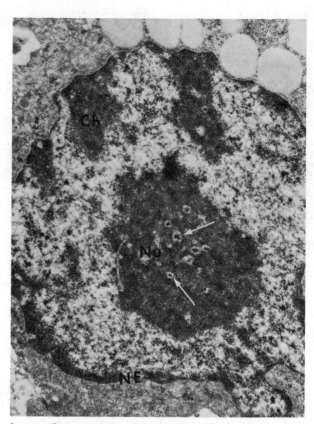

Figure 7.7. Changes in nucleoli and ribosomes. A, four hours after heating at 42.5°C for 30 minutes. Partly degranulated nucleolus (Nu); perichromatin granules (arrows) are visible in the nucleoplasm; nucleus envelope (NE) (× 38,000). B, tumor cell nucleus eight hours after treatment. Nearly all granular components have disappeared, and only fibrils and nucleolar microspherules (arrows) are left in the nucleolar (Nu) area; nuclear envelope (NE), heterochromatin (Ch) (× 15,500). C, two days after treatment. Part of nucleus

inclusions. After a few days, a dilated rough-endoplasmic reticulum could be observed, and collagen synthesis was sometimes present in the fibroblasts.

Within a few days after treatment, a large number of fibroblasts and macrophages gradually invaded the tumor periphery and replaced it by a fibrous scar within one or two weeks.

Although the changes described above might represent variations according to the treatment temperature, time, and tumor type, the pathologic reaction was in general of the same typical nature (Overgaard and Overgaard 1976) after treatment in a therapeutic range with temperatures up to 45°C. Treatment with lower heat doses in some cells showed a similar reaction, but a few of the tumor cells, especially in the periphery, revealed only minor, and probably reversible, damage that could be responsible for tumor recurrence within a few days or weeks after treatment (Overgaard and Overgaard 1972a).

The characteristic cytoplasmic reaction observed in tumor cells appears to be a specific reaction for hyperthermic damage. Not so much because of the actual changes, but the very rapid time scale in which they occur differs completely from what has been observed after other forms of cellular injury. Certainly, an increased lysosomal activity is a prominent finding in other forms of cellular injury and has been observed after exposure to cytostatic drugs, radiation, hypoxia, and during regression of hormone-dependent tumors (Kerr and Searle 1980; Kerr, Wyllie, and Currie 1972; Overgaard 1973, 1976a; Searle et al. 1974). In all these situations, however, the time scale for progressive lysosomal destruction ranges from several days to weeks, in contrast to the few hours observed after hyperthermia.

The early and prominent lysosomal reaction observed with the electron microscope is supported by biochemical and histochemical studies that show a high activity of lysosomal enzymes in heated tumors (Barratt and Wills 1979; Hofer, Brizzard, and Hofter 1979; Keech and Wills 1979; Overgaard and Heyden 1974; Overgaard and Nielsen 1980; Overgaard and Skovgaard Poulsen 1977; Overgaard and Overgaard 1972a, 1972b, 1975; Turano, Ferraro, and Strom 1970). The reason for this prominent lysosomal reaction probably is related to specific changes in the

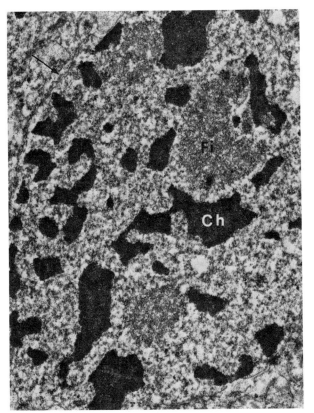
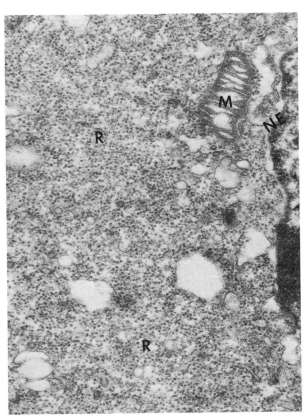

Figure 7.7. continued
with condensed heterochromatin *(Ch)* and remnants of nucleolar fibrils *(Fi)* and microspherules. The nuclear envelope is still fairly well preserved *(arrow)* (× 19,000). D, cytoplasm four hours after treatment. All the ribosomes have changed from polyribosomes into monoribosomes *(R)*. An irregular mitochondria with dilated intercristal spaces is present *(M)*, nuclear envelope *(NE)* (× 38,000). (Overgaard 1976a. Reprinted by permission.)

thermally treated tumor cells and may be attributed to a change in the tumor physiology, with increased acidity as the most prominent feature.

Whereas the cytoplasmic changes appear to be specific for tumors heated in vivo or in an acidic environment in vitro, the nuclear, and especially the nucleolar, changes are a phenomenon observed in all heated cells (Overgaard 1976a, 1977b). Thus, a degranulation of the nucleolus with formation of nucleolar microspherules or bundles of filaments associated with a reduced number of polyribosomes has been observed in several cell lines in vitro (Buckley 1972; Heine et al. 1971; Overgaard 1976a, 1976b, 1977b). Such morphologic changes in the nucleolus are consistent with the inhibition of the RNA synthesis that is known to occur after exposure to hyperthermia (Overgaard 1976a, 1976b, 1977b). This RNA synthesis inhibition is found to be related to both the synthesis of 45-S precursor RNA and to the conversion of 45-S RNA to the mature 28-S and 18-S RNA (Heine et al. 1971; Simard, Amalric, and Zalta 1969; Simard and Bernhard 1967; Warocquier and Scherrer 1969). Similar findings also have been observed in cells from different normal tissues, however, and are not likely to account for the marked destructive changes observed in vivo (Love, Soriano, and Walsh 1970; Simard and Bernhard 1967), especially since the effect of RNA synthesis appears to be transient, with complete recovery within one or two days if the cells are not otherwise damaged (Heine et al. 1971; Overgaard 1977b; Overgaard and Okkels 1940b).

Late Histopathologic Reaction

Whereas the initial histopathologic response to hyperthermia was similar in all tumors investigated, the subsequent elimination of the destroyed tumor cells and restoration of the tissue architecture were widely different. In general, tumors were replaced by ingrowth of fibroblasts and macrophages, ultimately resulting in the formation of a fibrotic scar (Overgaard and Overgaard 1972a, 1976). Some variations have been described in detail by Overgaard and Overgaard (1976), but they appear to be of minor importance in the evaluation and understanding of primary hyperthermic cellular damage.

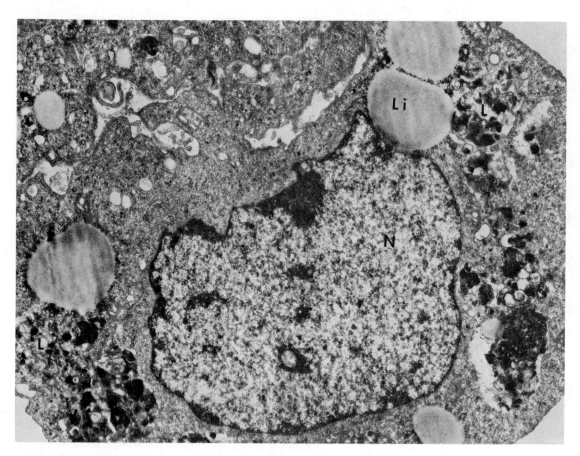

Figure 7.8. Shrunken tumor cell eight hours after treatment. Large lysosomes (L) with digested material are now observed in the cytoplasm together with lipid inclusions (Li); nucleus (N) (× 15,500). (Overgaard 1976a. Reprinted by permission.)

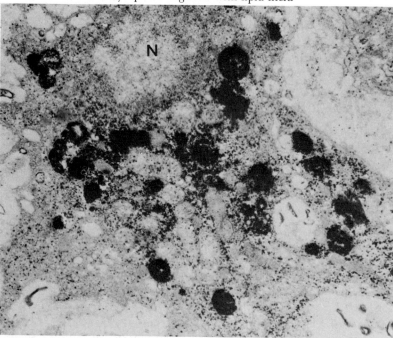

Figure 7.9. Intense positive acid phosphatase reaction in heated tumor-cell lysosomes; nucleus (N); Gomori lead citrate staining (× 8000). (Overgaard 1976a. Reprinted by permission.)

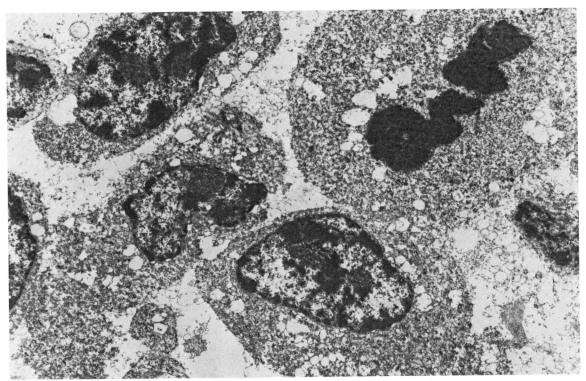

Figure 7.10. Tumor cells 24 hours after moderate heat treatment. All cells have complete loss of cytoplasmic structure without distinct membranes. The nuclei are small and dense with condensed chromatin. The nuclear envelope is still preserved in most cells, however. Note that the cell in mitosis is destroyed, similar to interphase cells ($\times$ 7200). (Overgaard 1976a. Reprinted by permission.)

In normal tissues, heat damaged tissue was either completely restored or a permanent fibroserous scar developed (Overgaard and Overgaard 1976). In contrast to ionizing radiation damage, the normal tissue damage observed in experimental animals following hyperthermia was "healed" or fixed within a short time after treatment, and no apparent changes or unexpected late effects have been observed (Overgaard and Suit 1979; Overgaard and Overgaard 1972a).

The Pattern and Mechanism of Recurrence

Treatment with low temperatures and/or short exposure times which did not yield a curative response resulted generally in the same overall histologic reaction, except that a few apparently morphologically undamaged cells occasionally could be found at the tumor periphery or around larger intact blood vessels (Overgaard 1977a, 1978a, 1978b, 1980; Overgaard and Nielsen 1980; Overgaard and Overgaard 1972a, 1976). Recurrences were apparent within a few days, and regrowth started typically around such blood vessels and/or at the tumor periphery (figs. 7.15 and 7.16).

There may be several explanations for the preservation of tumor cells located at the tumor periphery or close to intact blood vessels. Such cells may have been exposed to lower temperatures than the rest of the tumor, either because of cooling by the blood flow or because they may suffer from a heterogeneous heat distribution. In a previous study, it was demonstrated that the high incidence of local failures in tumors thermally treated by radiofrequency could be explained by insufficient peripheral heating (Overgaard 1978b). Under such circumstances, however, the pattern of recurrence was different from that seen with low heat doses, mainly because the regrowth was observed only in a single peripheral spot and normally was not found around the blood vessels. Furthermore, the pattern of recurrence observed after low heat doses can be seen in tumors heated with radiofrequency and by a much more uniform water bath immersion (Overgaard 1978b, 1980; Overgaard and Nielsen 1980). Heterogeneous tumor heating therefore is not likely to be the cause of the observed failure pattern in such tumors, although vascular cooling certainly may influence the heat distribution in deep-seeded tumors and in well-vascularized normal tissue (Hume, Robinson, and Hand 1979).

The explanation for this type of recurrence following low-dose heat treatment could be that malignant cells at the tumor periphery or close to blood

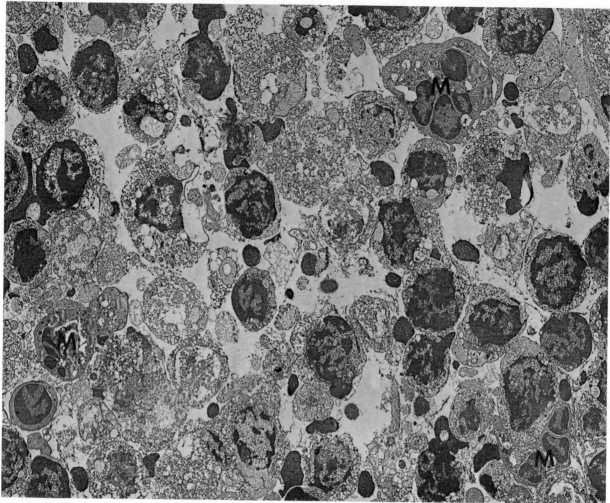

Figure 7.11. Central area of the tumor a few days after treatment. Small pyknotic tumor cells without any well-defined structure. Macrophages (M) are seen in the area (× 4500). (Overgaard 1976a. Reprinted by permission.)

vessels are situated in a more physiologic environment, with sufficient supply of oxygen and other nutrients. Therefore, such cells in a "normal" environmental milieu might be biologically more resistant to hyperthermia.

In favor of such a hypothesis is the observation that cancer cells heated during conditions in which their blood flow has been interrupted are considerably more heat sensitive than similar tumors with adequate vascularization (Baker and Wright 1976; Crile 1963; Hill and Denekamp 1978; Overgaard 1978a). Naturally, such mechanisms might be explained by an inhibition of vascular cooling, but Hill and Denekamp (1978) have in an elegant experiment shown that the hyperthermic sensitivity increases with increased clamping time. This indicates that cessation of the blood flow, per se, cannot fully account for enhanced heat sensitivity. This can more likely be explained as a consequence of accumulation of metabolites and changes in the environmental acidity during the clamping period.

Furthermore, heat also produces a characteristic necrosis in the chronically hypoxic and nutritionally deprived central part of spheroids heated in vitro, whereas the peripheral cells appear more resistant to hyperthermia (Durand 1976; Overgaard and Nielsen 1980). On this basis, it seems justified to conclude that cells in a physiologic environment appear to be more resistant to heat. Such cells are supplied with sufficient oxygen and nutrients and typically constitute the proliferating part of the tumor. On the other hand, as heat-sensitive cells in the tumors are subjected to deprived nutritional conditions and chronic hypoxia, their environment is likely to become increasingly acidic as a consequence of increased anaerobic metabolism and insufficient removal of metabolites. Cells situated in such an environment cannot normally contribute to proliferation. Figure 7.17 illustrates the pat-

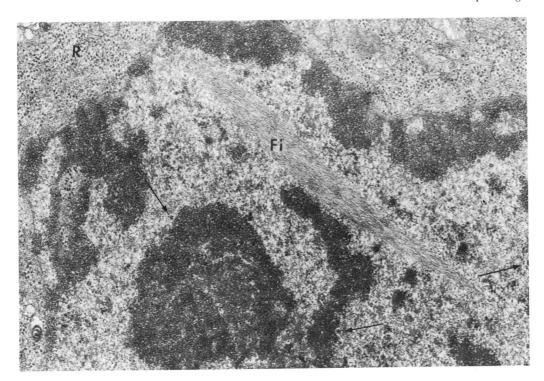

Figure 7.12. Part of L1A2 ascites cell six hours after heating. The nucleus contains a bundle of fine filaments *(Fi)* and an increased number of perichromatin granules *(arrows)*. Most ribosomes are in the form of monosomes *(R)* (× 39,000). (Overgaard 1976b.)

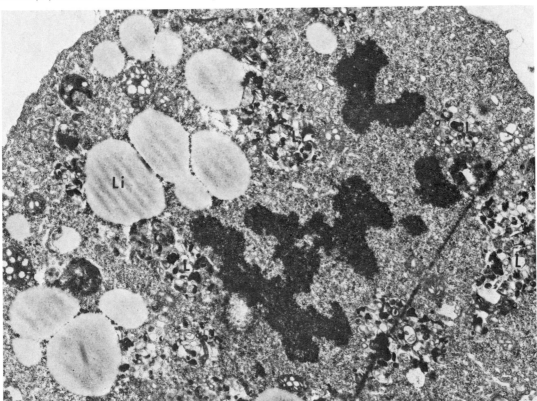

Figure 7.13. Same tumor as in figure 7.8. Mitotic cell eight hours after treatment. Intense cytoplasmic degeneration with lysosomes *(L)* and lipid droplets *(Li)* in the cytoplasm; similar to the findings in figure 7.8 (× 13,500). (Overgaard 1976a. Reprinted by permission.)

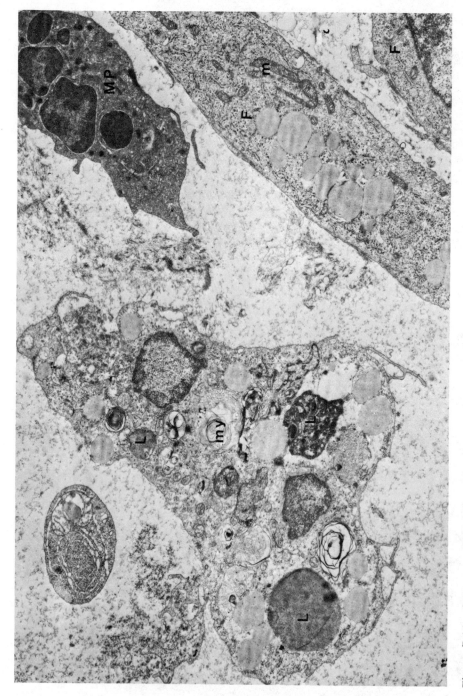

Figure 7.14. Sarcoma cell 24 hours after heating at 42.5°C for 60 minutes. Degenerative changes in cytoplasm, the lysosomes (*L*), and myelin figures (*my*) dominate. A well-preserved macrophage (*MP*) and parts of fibroblasts (*F*) are seen to the right. Note that the degenerative changes in these nonmalignant cells are only slight (lipid inclusions), but otherwise the cytoplasmic structures are well preserved with normal mitochondria (*m*) (× 9200). (Overgaard and Overgaard 1976. Reprinted by permission.)

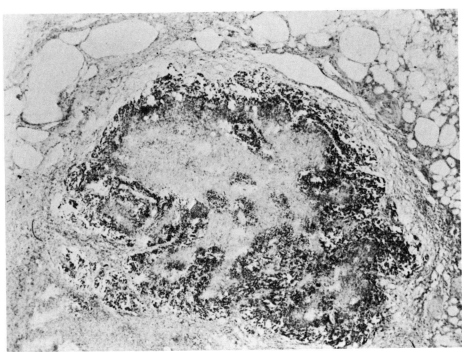

Figure 7.15. Mammary adenocarcinoma three days after a noncurative heat treatment. Recurrent tumor growth is visible at the tumor periphery and around a few of the larger intact blood vessels (× 50). (Overgaard 1980. Reprinted by permission.)

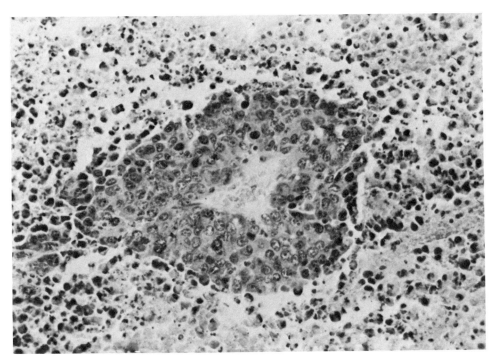

Figure 7.16. Higher magnification of figure 7.15 showing the typical recurrent tumor growth around a larger blood vessel. Note the destroyed small pyknotic cells in the less vascularized area (× 300). (Overgaard 1980. Reprinted by permission.)

EFFECT OF TUMOR ENVIRONMENT ON THE RESPONSE TO HYPERTHERMIA

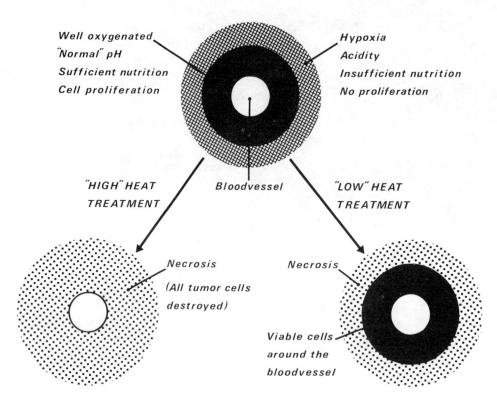

Figure 7.17. Schematic graph showing the effect of environment on the hyperthermic response in solid tumors. A "high" heat treatment results in necrosis of all tumor cells. A "low" heat treatment, however, will selectively destroy cells in the poorly vascularized hypoxic and acidic area. The cells surrounding the blood vessels are in a "normal" environment and therefore more resistant to hyperthermia.

tern of heat destruction in a tumor with respect to such differences in the cellular environment.

INFLUENCE OF ENVIRONMENTAL FACTORS ON THE HYPERTHERMIC RESPONSE

The concept that environmental factors may influence the hyperthermic response is far from new. It was suggested almost 50 years ago that characteristic changes in the tumor milieu may be responsible for the typical heat destruction observed in solid tumors (Overgaard and Nielsen 1980; Overgaard 1934). More recently, the historic morphologic observations have been extended, with quantitative studies of the heat response to treatment under well-defined environmental conditions, such as extracellular acidity, hypoxia, and nutritional deprivation.

Effects of Extracellular Acidity

In general, all quantitative in vitro studies have shown that increased extracellular acidity enhances the response to hyperthermia (Bichel and Overgaard 1977; Freeman, Dewey, and Hopwood 1977; Gerweck 1977; Gerweck and Rottinger 1976; Haveman 1979; Meyer, Hopwood, and Gillette 1979; Nielsen and Overgaard 1979; Overgaard and Bichel 1977; Overgaard and Nielsen 1980). A comparison between the different cell lines investigated, however, indicates that the effect of reduced pH may involve different mechanisms. An increase in direct heat sensitivity, a reduced capacity to accumulate sublethal hyperthermic damage, and a lowering of the threshold at which thermal resistance has been shown to occur during prolonged heating have been described (Fig. 7.18).

The mechanism by which the extracellular acidity affects heat-treated cells is not known. Based on mor-

Histopathologic Effects of Hyperthermia

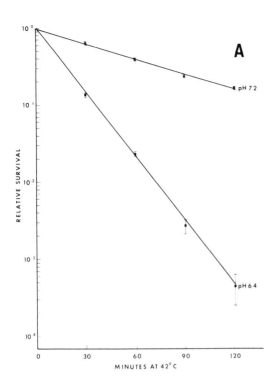

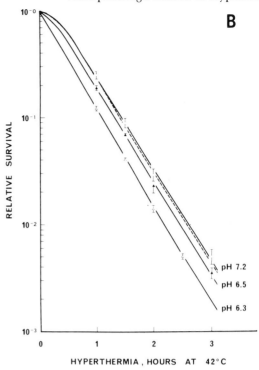

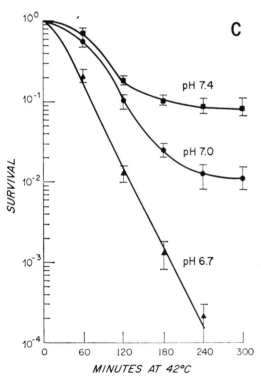

Figure 7.18. Hyperthermic cell survival curves showing three different mechanisms by which the effect of reduced extracellular pH may influence the survival. A, a direct increase in sensitivity, expressed by a decreased D_o (Overgaard 1977b); B, a reduced repair of sublethal damage by reducing the shoulder (Nielsen and Overgaard 1979); C, a reduced survival, caused mainly by lowering of the level at which thermal resistance occurs (Gerweck 1977). (Reprinted by permission.)

phologic and histochemical observations indicating an early and prominent lysosomal activity in most heated tumor cells (Barratt and Wills 1979; Keech and Wills 1979; Overgaard 1976a, 1976b, 1977b; Overgaard and Heyden 1974; Overgaard and Nielsen 1980; Overgaard and Overgaard 1972a, 1975, 1976), we have suggested that the reduced pH mainly accelerates a lysosomal destruction of malignant cells with a primary target in the cytoplasm. Such reaction could be intensified further by a relative increase in anaerobic glycolysis, which in turn will tend to accumulate lactic acid in the poorly vascularized tumor areas (Overgaard 1973, 1977b; Overgaard and Overgaard 1977). This hypothesis is supported by experiments indicating that incubation of cells under acidic conditions enhances both the amount of lysosomes and the ability of tumor cells to accelerate the lysosomal digestive processes (Overgaard 1976b; Overgaard and Skovgaard Poulsen 1977). It is likely, however, that the increased acidity also acts through other mechanisms. First, it has been questioned whether the extracellular pH is in itself the most important factor, or if the extracellular acidity is just modifying the intracellular pH (Dickson and Calderwood 1979; Dickson and Oswald 1976; Haveman 1979). This problem is still being debated. Second, the observation that a low pH affects the survival curve in different ways makes it apparent that the reduced pH also involves a number of unknown mechanisms, which may have targets in the nucleus or in the membranes (Gerweck 1977; Overgaard and Nielsen 1980; Suit and Gerweck 1979). Additional research on these mechanisms is needed to clarify the exact nature of this heat sensitizing effect; it is of utmost importance for the understanding of the nature of hyperthermic tumor cell destruction.

Effects of Hypoxia

Acute hypoxia (i.e., an incubation under low oxygen tension that does not apparently alter other parameters) does not appear to influence significantly the sensitivity to hyperthermia (Bass and Moore 1978; Durand 1978; Gerweck, Gillette, and Dewey 1974; Nielsen 1981; Overgaard and Bichel 1977; Overgaard and Nielsen 1980; Power and Harris 1977; Suit and Gerweck 1979). This is important because hypoxia is a characteristic environmental condition in large areas of solid tumors. Cells radiated under hypoxic conditions are known to be up to three times more resistant to therapy than well-oxygenated cells.

Cells heated after incubation for a prolonged period under hypoxic conditions (chronic), however, appear to increase their heat sensitivity when compared to well-oxygenated cells cultured in air (Born and Trott 1978; Gerweck, Nygaard, and Burlett 1979; Harisiadis et al. 1975; Kim, Kim, and Hahn 1975, 1978; Kim, Kim, and Hahn 1975, 1978; Schulman and Hall 1974). The techniques by which chronic hypoxia is obtained frequently involve metabolic depression of cells and consequently may often alter several other parameters, including the intra- and extracellular acidity. (Gerweck, Nygaard, and Burlett 1979; Overgaard and Bichel 1977). If such conditions are avoided, chronic hypoxia appears to enhance the effect of hyperthermia only to a relatively small extent (Gerweck, Nygaard, and Burlett 1979), and it is generally thought that hypoxia itself is of minor importance in the hyperthermic treatment (Myers and Field 1979; Overgaard 1977b, 1978a; Overgaard and Nielsen 1980).

Nevertheless, the observation that chronic hypoxic cells are most sensitive to heat if the hypoxia is achieved by metabolic depletion is clinically important, since this situation probably simulates the situation under which cells are located in poorly vascularized areas of solid tumors.

Effects on Plateau Phase Cells

Metabolically deprived tumor cells also show a reduced proliferation pattern and are to some extent comparable to unfed plateau phase cells in vitro. In contrast to the effects of increased acidity and hypoxia, the data differ enormously on the heat sensitivity of plateau phase cells, which are observed to be both more sensitive and/or more resistant than similar exponentially growing cells (Bhuyan 1979; Bhuyan et al. 1977; Bichel and Overgaard 1977; Durand 1978; Hahn 1974; Kase and Hahn 1975, 1976; Overgaard and Nielsen 1980; Schlag and Lücke-Hühle 1976; Schulman and Hall 1974; Suit and Gerweck 1979). The technical conditions for the experiments and the cell lines studied may be important, but no conclusion has yet been reached. It should, however, be mentioned that separating the influence of low pH from the nutritional deprivation indicated that increased acidity was much more important for the heat sensitivity than the state of growth per se (Bichel and Overgaard 1977).

MECHANISM OF HYPERTHERMIC CELL DESTRUCTION

Based on the morphologic evidence of intense cytoplastic damage and the importance of the environmental factors to hyperthermic cell destruction, a hypothesis of the mechanisms of hyperthermic cell destruction was proposed (Overgaard 1973, 1977b). Prominent and rapidly increased lysosomal activity observed in the cytoplasm of tumor cells in vivo seems in most tumors to be responsible for the cellular destruction. That such findings are present only in vivo may be explained by several factors. First, it is likely that the tumor cell environment in large areas of solid tumors, owing to metabolic depression and lack of oxygen, may be more acidic than normal tissues (Overgaard 1977b; Poole 1967; Suit and Gerweck 1979; Waddell and Bates 1969). Second, heat may inhibit the respiratory metabolism selectively in malignant cells, which may lead to a relative increase in anaerobic glycolysis and consequently to a higher amount of lactic acid, first in the cytoplasm and probably later in the extracellular space, owing in part to cellular membrane damage (Cavaliere et al. 1967; Johnson and Wiske 1976; Muckle and Dickson 1971; Overgaard 1977b). Third, heat causes a reduced blood flow in tumors compared to that in normal tissue, and it is likely that this in turn will increase the accumulation of acidic metabolic products, caused partly by a further reduction in oxygen supply (Kang, Song, and Levitt 1980). Therefore, large areas of the tumor environment are likely to become even more acidic. This increased acidity may in turn increase and intensify the activity of lysosomal enzymes, resulting in an accelerating cell destruction that starts in the cytoplasm but later involves the whole cell. Recent data have shown that leaving tumor cells in vivo following thermal therapy enhances the hyperthermic destruction by several orders of magnitude (Fajardo et al. 1980; Marmor,

Hahn, and Hahn 1977), supporting the hypothesis mentioned above. Nuclear changes, which occur in all cells, may not be fatal unless treatment has been given with extensive heat doses. This hypothesis has been questioned by several authors (Dickson and Calderwood 1979; Dickson and Oswald 1976; Fajardo et al. 1980; Hofer, Brizzard, and Hofter 1979), mainly because of its simplicity, and it is likely that the mechanism of hyperthermic cell destruction in vivo is more complex than was originally suggested. Thus, studies of the influence of extracellular acidity on hyperthermic destruction have shown that there may be targets outside the cytoplasm in which an increased acidity is likely to affect heated cells (Gerweck, Nygaard, and Burlett 1979). Damage to nuclei and cell membranes is likely to be involved, but the nature and significance of such additional factors need to be clarified and seem not to alter the importance of the prominent cytoplasmic effect (Bhuyan 1979; Suit and Gerweck 1979). At present, therefore, it can be concluded only that increased acidity is a factor of utmost importance in the hyperthermic destruction of cells, probably owing to an enhanced lysosomal destruction occurring primarily in the cytoplasm but also involving other targets, such as nuclei and membranes.

The protection of most normal cells during a moderate hyperthermic treatment in vivo may be explained by their location in the supporting tumor stroma, which is well vascularized, thereby creating a "normal environment." Furthermore, there are indications that heat may not inhibit the respiratory metabolism to the same extent in the normal cells as in tumor cells (Cavaliere et al. 1967; Overgaard 1977b). Therefore, stromal cells may suffer only transient nuclear damage, which will not be expressed to the extent as damage to tumor cells because of the low proliferation rate of these normal cells.

CONCLUSION

Histopathologic investigations indicate that hyperthermia causes a rapid and specific destruction of tumor cells in vivo. Ultrastructural, histochemical, and biochemical studies have shown that increased lysosomal activity is a prominent feature in early heat damage. Apparently, the damage to the cytoplasm is of primary importance for hyperthermic destruction in vivo, although nuclei and membranous structures also may be involved at an early stage.

The reaction is found almost selectively in malignant cells, and most normal tissues do not appear to be primarily damaged unless the hyperthermic treatment is given at a high temperature.

Characteristically, recurrences after low heat doses are located at the tumor periphery or around larger intact blood vessels. There is experimental evidence that cells situated in such a normal physiologic environment appear to be biologically more resistant to heat than cells situated under environmental conditions characterized by increased acidity, hypoxia, and nutritional depletion. Probably the most decisive factor for heat sensitivity is that the extracellular acidity and the influence of other environmental parameters act indirectly by changing the milieu into a more acidic state. Such environmental conditions are characteristic of large areas in solid tumors and may explain why tumor cells in general are much more sensitive in vivo than similar cells treated in vitro under normal conditions. Furthermore, tumors in vivo are subjected to secondary heat-induced changes in the vascularization, which again may alter the environmental conditions. Such variables emphasize the importance of using the proper experimental design in studies that attempt to evaluate the potential clinical use of heat treatment.

CLINICAL IMPLICATIONS

The implications for the role of hyperthermia in local cancer therapy may be that heat in "high" doses frequently is able to destroy solid tumors, but often with significant and unacceptable normal tissue damage if not given selectively to the tumor. Hyperthermia given in "low" doses, which apparently are tolerated by normal tissues, destroys preferentially and almost selectively cells in an environment of increased acidity, hypoxia, and insufficient nutrition. Such cells generally are the most resistant to radiation and chemotherapy, whereas the proliferating, well-oxygenated cells that are most resistant to heat are the most sensitive to these agents. An optimal local therapy would be to combine hyperthermia with other modalities. Among these, radiation therapy appears to be the most useful since this, like heat, is a local treatment that does not involve systemic toxicity and is known to interact with heat in a way that may be used to improve the therapeutic effect (Overgaard 1978a, 1980; Overgaard and Overgaard 1972b; Suit and Gerweck 1979).

Certainly such combined treatment involves numerous problems in terms of interaction between the two modalities, but early clinical experience has substantiated that local hyperthermia, particularly when combined with radiation, may have a great potential role in clinical cancer treatment. As with the application of histopathologic examinations to experimental tumors, decisions about future clinical treatments require biopsies of heated lesions. Biopsies are likely to yield information on the effectiveness of the heat treatment, on the heat distribution (e.g., occurrence of cold spots, etc.), and may be able to act as a form of biological dosimetry. Morphologic studies have contributed to the understanding of the nature of hyperthermic destruction by demonstrating characteristic qualitative changes. Therefore, in future experimental and clinical work, increased attention to comparative morphologic investigations should lead to a better understanding of the nature of hyperthermia.

References

Baker, G. M., and Wright, E. A. Effects of hypoxia and heat on mammary carcinoma in C_3H mice. *Br. J. Radiol.* 49:809, 1976.

Barratt, G. M., and Wills, E. D. The effect of hyperthermia and radiation on lysosomal enzyme activity of mouse mammary tumours. *Eur. J. Cancer* 15:243–250, 1979.

Bass, H. and Moore, J. L. Lethality in mammalian cells due to hyperthermia under oxic and hypoxic conditions. *Int. J. Radiat. Biol.* 33:57–67, 1978.

Bhuyan, B. K. Kinetics of cell kill by hyperthermia. *Cancer Res.* 39:2277–2284, 1979.

Bhuyan, B. K. et al. Sensitivity of different cell lines and of different phases in the cell cycle to hyperthermia. *Cancer Res.* 37:3780–3784, 1977.

Bianchi, L. et al. Liver damage in heatstroke and its regression. *Hum. Pathol.* 3:237–248, 1972.

Bichel, P., and Overgaard, J. Hyperthermic effect on exponential and plateau ascites tumor cells in vitro dependent on environmental pH. *Radiat. Res.* 70:449–454, 1977.

Born, R., and Trott, K.-R. The influence of hyperthermia on chronically hypoxic cells. In *Proceedings of the Second International Symposium on Cancer Therapy by Hyperthermia and Radiation*, eds. C. Streffer et al. Baltimore and Munich: Urban & Schwarzenberg, 1978, p. 177.

Bragdon, J. The hepatitis of hyperthermia. *N. Engl. J. Med.* 237:765–769, 1947.

Buckley, I. K. A light and electron microscopic study of thermally injured cultured cells. *Lab. Invest.* 26:201–209, 1972.

Cavaliere, R. et al. Selective heat sensitivity of cancer cells. *Cancer* 20:1351–1381, 1967.

Cockett, A. T. K. et al. Enhancement of regional bladder megavoltage irradiation in bladder cancer using local bladder hyperthermia. *J. Urol.* 97:1034–1039, 1967.

Crile, G., Jr. Heat as an adjunct to the treatment of cancer. *Cleve. Clin. Q.* 28:75–89, 1961.

Crile, G., Jr. The effects of heat and radiation on cancers implanted on the feet of mice. *Cancer Res.* 23:372–380, 1963.

Dickson, J. A., and Calderwood, S. K. Effects of hyperglycemia and hyperthermia on the pH, glycolysis and respiration of the Yoshida sarcoma in vivo. *J. Natl. Cancer Inst.* 63:1371–1381, 1979.

Dickson, J. A., and Oswald, B. E. The sensitivity of a malignant cell line to hyperthermia (42°C) at low intracellular pH. *Br. J. Cancer* 34:262–270, 1976.

Dickson, J. A., and Suzangar, M. In vitro-in vivo studies on the susceptibility of the solid Yoshida sarcoma to drugs and hyperthermia (42°C). *Cancer Res.* 34:1263–1274, 1974.

Durand, R. E. Effects of hyperthermia on the cycling, noncycling and hypoxic cells of irradiated and unirradiated multicell spheroids. *Radiat. Res.* 75:373–384, 1978.

Fajardo, L. F. et al. Effects of hyperthermia in a malignant tumor. *Cancer* 45:613–623, 1980.

Freeman, M. L.; Dewey, W. C.; and Hopwood, L. E. Effect of pH on hyperthermic cell survival. *J. Natl. Cancer Inst.* 58:1837–1839, 1977.

Gerweck, L. E. Modification of cell lethality at elevated temperatures. The pH effect. *Radiat. Res.* 70:224–235, 1977.

Gerweck, L. E.; Gillette, E. L.; and Dewey, W. C. Killing of Chinese hamster cells in vitro by heating under hypoxic or aerobic conditions. *Eur. J. Cancer* 10:691–693, 1974.

Gerweck, L. E.; Nygaard, T. G.; and Burlett, M. Response of cells to hyperthermia under acute and chronic hypoxic conditions. *Cancer Res.* 39:966–972, 1979.

Gerweck, L. E., and Rottinger, E. Enhancement of mammalian cell sensitivity to hyperthermia by pH alteration. *Radiat. Res.* 67:508–511, 1976.

Gilchrist, R. K. et al. Effects of electromagnetic heating on interval viscera: a preliminary to the treatment of human tumors. *Ann. Surg.* 161:890–896, 1965.

Gore, I., and Isaacson, N. H. The pathology of hyperpyrexia. Observations at autopsy in 17 cases of fever therapy. *Am. J. Pathol.* 25:1029–1059, 1949.

Hahn, G. M. Metabolic aspects of the role of hyperthermia in mammalian cell inactivation and their possible relevance to cancer treatment. *Cancer Res.* 34:3117–3123, 1974.

Hall, R. R. et al. Hyperthermia in the treatment of bladder tumours. *Br. J. Urol.* 48:603–608, 1976.

Hall, R. R.; Schade, R. O. K.; and Swinney, J. Effects of hyperthermia on bladder cancer. *Br. Med. J.* 2:593–594, 1974.

Hand, J. W. et al. Effects of hyperthermia on the mouse testis and its response to x-rays, as assayed by weight loss. *Int. J. Radiat. Biol.* 35:521–528, 1979.

Harisiadis, L. et al. Hyperthermia: biological studies at the cellular level. *Radiology* 117:447–452, 1975.

Hartman, F. W., and Major R. C. Pathological changes resulting from accurately controlled artificial fever. *Am. J. Clin. Pathol.* 5:392–410, 1935.

Haveman, J. The pH of the cytoplasm as an important factor in the survival of in vitro cultured malignant cells after hyperthermia. Effects of carbonylcyanide 3-chlorophenylhydrazone. *Eur. J. Cancer* 15:1281–1288, 1979.

Heine, U. et al. The behaviour of HeLa-S_3 cells under the influence of supranormal temperatures. *J. Ultrastruct. Res.* 34:375–396, 1971.

Hill, S. A., and Denekamp, J. The effect of vascular occlusion on the thermal sensitization of a mouse tumour. *Br. J. Radiol.* 51:997–1002, 1978.

Hofer, K. G.; Brizzard, B.; and Hofter, M. G. Effect of lysosome modification on the heat potentiation of radiation damage and direct heat death of BP-8 sarcoma cells. *Eur. J. Cancer* 15:1449–1457, 1979.

Holt, J. A. G. The use of V.H.F. radiowaves in cancer therapy. *Australian Journal of Radiology* 19:223–241, 1975.

Hume, S. P.; Robinson, J. E.; and Hand, J. W. The influence of blood flow on temperature distribution in the exteriorized mouse intestine during treatment by hyperthermia. *Br. J. Radiol.* 52:219–222, 1979.

Jacobsen, V. C., and Hosol, K. The morphologic changes in animal tissues due to heating by an ultrahigh frequency oscillator. *Archives of Pathology* 11:744–759, 1931.

Jares, J. J., and Warren, S. L. Physiological effects of radiation. A study of the in vitro effect of high fever temperatures upon certain experimental animal tumors. *Am. J. Roentgenol.* 41:685–708, 1939.

Johnson, H. A., and Wiske, P. S. Injury of the cell's respiratory system by heat and by formaldehyde. *Lab. Invest.* 35:179–184, 1976.

Kang, M. S.; Song, C. W.; and Levitt, S. H. Role of vascular function in response to tumors in vivo to hyperthermia. *Cancer Res.* 40:1130–1135, 1980.

Kase, K. R., and Hahn, G. M. Differential heat response of normal and transformed human cells in tissue culture. *Nature* 255:228–230, 1975.

Kase, K. R., and Hahn, G. M. Comparison of some response to hyperthermia by normal human diploid cells and neoplastic cells from the same origin. *Eur. J. Cancer* 12:481–491, 1976.

Keech, M. L., and Wills, E. D. The effect of hyperthermia on activation of lysosomal enzymes in HeLa cells. *Eur. J. Cancer* 5:1025–1031, 1979.

Kerr, J. F. R., and Searle, J. Apoptosis: its nature and kinetic role. In *Radiation biology in cancer research*, eds. R. E. Meyn and H. R. Withers. New York: Raven Press, 1980, pp. 367–384.

Kerr, J. F. R.; Wyllie, A. H.; and Currie, A. R. Apoptosis: a basic biological phenomenon with wide-ranging implications in tissue kinetics. *Br. J. Cancer* 26:239–257, 1972.

Kew, M. C. et al. The effects of heatstroke on the function and structure of the kidney. *Q. J. Med.* 36:277–300, 1967.

Kim, J. H.; Kim, S. H.; and Hahn, E. W. Enhanced killing of hypoxic tumour cells by hyperthermia. *Br. J. Radiol.* 48:872–874, 1975.

Kim, J. H.; Kim, S. H.; and Hahn, E. W. Killing of glucose-deprived hypoxic cells with moderate hyperthermia. *Radiat. Res.* 75:448–451, 1978.

Kim, S. H.; Kim, J. H.; and Hahn, E. W. The radiosensitization of hypoxic tumor cells by hyperthermia. *Radiology* 114:727–728, 1975.

Kim, S. H.; Kim, J. H.; and Hahn, E. W. Selective potentiation of hyperthermic killing of hypoxic cells by 5-Thio-D-glucose. *Cancer Res.* 38:2935–2938, 1978.

Love, R.; Soriano, R. Z.; and Walsh, R. J. Effect of hyperthermia on normal and neoplastic cells in vitro. *Cancer Res.* 30:1525–1533, 1970.

Ludgate, C. M. et al. Hyperthermic perfusion of the distended urinary bladder in the management of recurrent transitional cell carcinoma. *Br. J. Urol.* 47:841–848, 1976.

Lunglmayr, G. et al. Experimentelle untersuchungen über die wirkung temporärer hyperthermie auf blasentumore. *Urol. Int.* 28:314–321, 1973.

Marmor, J. B.; Hahn, N.; and Hahn, G. M. Tumor cure and cell survival after localized radiofrequency heating. *Cancer Res.* 37:879–883, 1977.

Meyer, K. R.; Hopwood, L. E.; and Gillette, E. L. The thermal response of mouse adenocarcinoma cells at low pH. *Eur. J. Cancer* 15:1219–1222, 1979.

Moritz, A. R. Studies of thermal injury III. The pathology and pathogenesis of cutaneous burns. *Am. J. Pathol.* 23:915–941, 1947.

Moritz, A. R., and Henriques, F. C., Jr. Studies of thermal injury II. The relative importance of time and surface temperature in the causation of cutaneous burns. *Am. J. Pathol.* 23:695–720, 1947.

Muckle, D. S., and Dickson, J. A. The selective inhibitory effect of hyperthermia on the metabolism and growth of malignant cells. *Br. J. Cancer* 25:771–778, 1971.

Myers, R., and Field, S. B. Hyperthermia and the oxygen enhancement ratio for damage to baby rat cartilage. *Br. J. Radiol.* 52:415–416, 1979.

Nielsen, O. S. Effect of fractionated hyperthermia on hypoxic cells in vitro. *Int. J. Radiat. Biol.* 39:73–82, 1981.

Nielsen, O. S., and Overgaard, J. Effect of extracellular pH on thermotolerance and recovery of hyperthermic damage in vitro. *Cancer Res.* 39:2772–2778, 1979.

Okkels, H., and Overgaard, K. Effect of high frequency currents upon normal tissues and malignant tumors in mice. *Arch. Exp. Zellforsch.* 19:466–470, 1937.

Overgaard, J. Ultrastructural changes in an irradiated carcinoma. *J. Ultrastruct. Res.* 44:446, 1973.

Overgaard, J. Ultrastructure of a murine mammary carcinoma exposed to hyperthermia in vivo. *Cancer Res.* 36:983–995, 1976a.

Overgaard, J. Influence of extracellular pH on the viability and morphology of tumor cells exposed to hyperthermia. *J. Natl. Cancer Inst.* 56:1243–1250, 1976b.

Overgaard, J. The effect of sequence and time intervals of combined hyperthermia and radiation treatment. *Br. J. Radiol.* 50:763–764, 1977a.

Overgaard, J. Effect of hyperthermia on malignant cells in vivo. *Cancer* 39:2637–2646, 1977b.

Overgaard, J. The effect of local hyperthermia alone, and in combination with radiation, on solid tumors. In *Cancer therapy by hyperthermia and radiation*, ed. C. Streffer. Baltimore and Munich: Urban & Schwarzenberg, 1978a, pp. 49–61.

Overgaard, J. Biological effect of 27.12-MHz short wave diathermic heating in experimental tumors. *IEEE Transactions on Microwave Theory and Techniques* 35:523–529, 1978b.

Overgaard, J. Simultaneous and sequential hyperthermia and radiation treatment of an experimental tumor and its surrounding normal tissue in vivo. *Int. J. Radiat. Oncol. Biol. Phys.* 6:1507–1517, 1980.

Overgaard, J., and Bichel, P. The influence of hypoxia and acidity on the hyperthermic response of malignant cells in vitro. *Radiology* 123:511–514, 1977.

Overgaard, J., and Heyden, G. Histological and histochemical studies on combined heat-roentgen treatment of a C_3H mouse mammary carcinoma. *Histochemistry* 42:47–59, 1974.

Overgaard, J., and Nielsen, O. S. The role of tissue environmental factors on the kinetics and morphology of tumor cells exposed to hyperthermia. *Ann. NY Acad. Sci.* 335:254–280, 1980.

Overgaard, J., and Skovgaard Poulsen, H. Effect of hyperthermia and environmental acidity on the proteolytic activity in murine ascites tumor cells. *J. Natl. Cancer Inst.* 58:1159–1161, 1977.

Overgaard, J., and Suit, H.D. Time-temperature relationship in hyperthermic treatment of malignant and normal tissue in vivo. *Cancer Res.* 39:3248–3253, 1979.

Overgaard, K. Über wärmetherapie bösartiger tumoren. *Acta Radiol.* 15:89–100, 1934.

Overgaard, K., and Okkels, H. Über den einflus der wärmbehandlung auf Woods sarkom. *Strahlentherapie* 68:587–619, 1940a.

Overgaard, K., and Okkels, H. The action of dry heat on Wood's sarcoma. *Acta Radiol.* 21:577–582, 1940b.

Overgaard, K., and Overgaard, J. Investigations on the possibility of a thermic tumour therapy I. Short-wave treatment of a transplanted isologous mouse mammary carcinoma. *Eur. J. Cancer* 8:65–78, 1972a.

Overgaard, K., and Overgaard, J. Investigations on the possibility of a thermic tumour therapy II. Action of combined heat-roentgen treatment on a transplanted mouse mammary carcinoma. *Eur. J. Cancer* 8:573–575, 1972b.

Overgaard, K., and Overgaard, J. Histologic and histochemical reactions in a mouse mammary carcinoma following exposure to combined heat-roentgen irradiation. *Acta Radiol.* 14:164–176, 1975.

Overgaard, K., and Overgaard, J. Pathology of heat damage. In *Proceedings of the First International Symposium on Cancer Therapy by Hyperthermia and Radiation*, eds. M. J. Wizenberg and J. E. Robinson. Chicago: American College of Radiology, 1976, pp. 115–127.

Overgaard, K., and Overgaard, J. Hyperthermic tumour-cell devitalization in vivo. *Acta Radiol.* 16:1–16, 1977.

Pettigrew, R. T. et al. Circulatory and biochemical effects of whole body hyperthermia. *Br. J. Surg.* 61:727–730, 1974a.

Pettigrew, R. T. et al. Clinical effects of whole body hyperthermia in advanced malignancy. *Br. Med. J.* 4:679–682, 1974b.

Poole, D. Intracellular pH of the Ehrlich ascites tumor cells as it is affected by sugars and sugar derivatives. *J. Biol. Chem.* 242:3731–3736, 1967.

Power, J. A., and Harris, J. W. Response of extremely hypoxic cells to hyperthermia: survival and oxygen enhancement ratios. *Radiology* 123:767–770, 1977.

Schlag, H., and Lücke-Hühle, C. Cytokinetic studies on the effect of hyperthermia on Chinese hamster lung cells. *Eur. J. Cancer* 12:827–831, 1976.

Schulman, N., and Hall, E. J. Hyperthermia: its effect on proliferative and plateau phase cell cultures. *Radiology* 113:207–209, 1974.

Searle, J. et al. An electron-microscope study of the mode of cell death induced by cancer chemotherapeutic agents in populations of proliferating normal and neoplastic cells. *J. Pathol.* 116:129–138, 1974.

Shibolet, S. et al. Heatstroke: its clinical picture and mechanism in 36 cases. *Q. J. Med.* 36:525–548, 1967.

Simard, R.; Amalric, F.; and Zalta, J.-P. Effet de la température supra-optimale sur les ribonucléprotéines et le RNA nucléolaire. *Exp. Cell Res.* 55:359–369, 1969.

Simard, R., and Bernhard, W. A heat-sensitive cellular function located in the nucleolus. *J. Cell Biol.* 34:61–76, 1967.

Song, C. W.; Rhee, J. G.; and Levitt, S. H. Blood flow in normal tissues and tumors during hyperthermia. *J. Natl. Cancer Inst.* 64:119–124, 1980.

Stamm, M. E. et al. Microwave therapy experiments with B-16 murine melanoma. *IRCS Medical Science* 3:392–393, 1975.

Storm, F. K. et al. Normal tissue and solid tumor effects of hyperthermia in animal models and clinical trials. *Cancer Res.* 39:2245–2251, 1979.

Strauss, A. A.; Appel, M.; and Saphir, O. Electro-coagulation of malignant tumors. *Am. J. Surg.* 104:37–45, 1962.

Sugaar, S., and LeVeen, H. A histopathologic study of the effects of radiofrequency thermotherapy on malignant tumors of the lung. *Cancer* 43:767–783, 1979.

Suit, H. D., and Gerweck, L. E. Potential for hyperthermia and radiation therapy. *Cancer Res.* 39:2290–2298, 1979.

Turano, C.; Ferraro, A.; and Strom, R. The biochemical mechanism of selective heat sensitivity of cancer cells. III. Studies on lysosomes. *Eur. J. Cancer* 6:67–72, 1970.

von Heinz, D.; Uerlings, I.; and Grupe, M. Strukturveränderungen von leberzellen bei supranormalen temperturen. *Exp. Pathol.* 5:2–10, 1971.

Waddell, W. J., and Bates, R. G. Intracellular pH. *Physiol. Rev.* 49:285–329, 1969.

Wallace, J. Physiological and biochemical changes following hyperthermia treatment. *Br. J. Vener. Dis.* 14:155–165, 1943.

Warocquier, R., and Scherrer, K. RNA metabolism in mammalian cells at elevated temperature. *Eur. J. Biochem.* 10:362–370, 1969.

Warren, S. L. Preliminary study of the effect of artificial fever upon hopeless tumor cases. *Am. J. Roentgenol.* 33:75–87, 1935.

Wills, E. J.; Findlay, J. M.; and McManus, J. P. A. Effects of hyperthermia therapy on the liver II. Morphological observations. *Clin. Pathol.* 29:1–10, 1976.

CHAPTER 8
Blood Flow in Tumors and Normal Tissues in Hyperthermia

Chang W. Song

Introduction
Development and Characteristics of Tumor Vasculatures
Vascular Physiology of Normal Tissue and Tumor
Vascular Changes by Hyperthermia
Mechanisms of Vascular Damage in Heated Tumors
Relationship between Blood Flow and Tissue Temperature During Heating
Implication of Vascular Damage to the Survival of Tumor Cells in Heated Tumors
Role of Blood Flow in Combination of Hyperthermia with Other Modalities

INTRODUCTION

The clinical benefit of hyperthermia can be anticipated only if heat preferentially affects neoplastic tissues, leaving surrounding normal tissues relatively unaffected. Circumstantial evidence appears to suggest that the damage by heat to malignant cells in vitro is not necessarily greater than that to normal cells (Bhuyan 1979; Raaphorst et al. 1979). A number of investigators have reported, however, that tumors in vivo are more susceptible to heat than are the normal tissues (Crile 1963; Dickson et al. 1977; Hornback et al. 1977; Kang, Song, and Levitt 1980; Kim and Hahn 1979; LeVeen et al. 1976; Muckle and Dickson 1971; Nelson and Holt 1977; Overgaard 1977; Song, Rhee, and Levitt 1980; Stehlin et al. 1979; Storm et al. 1979). The mechanism for this preferential effect of heat on the tumors in vivo is not yet clear, but the difference in blood circulation in tumors and normal tissues has been suggested to occupy the central role.

It is a general finding that tumors acquire higher temperatures than do the surrounding normal tissues during hyperthermia (Dickson et al. 1977; Kim and Hahn 1979; LeVeen et al. 1976; Song, Rhee, and Levitt 1980; Storm et al. 1979). The magnitude of the rise in temperature of a tissue during heating is a function of the rate of influx and efflux of heat (Guy 1976; Patterson and Strang 1979). Inasmuch as the efflux of heat is achieved mainly by circulating blood, it has been suggested that the higher temperature of tumors relative to that of surrounding normal tissues during heating may be due to languid dissipation of heat by sluggish blood circulation in the tumors (Song, Rhee, and Levitt 1980; Storm et al. 1979). In this context, a difference in temperature by as little as 0.5°C during hyperthermia has been known to result in a significant difference in the cell survival (Connor et al. 1977; Kang, Song, and Levitt 1980).

It has been increasingly clear in recent years that the differential rise in temperature may not be the sole causative factor for the differential heat sensitivity of tumors in vivo and of normal tissues to hyperthermia. Poor nutritional condition and acidic environment are known to expedite the killing of cells by heat (Dewey et al. 1979; Gerweck 1977; Hahn 1974; Overgaard 1976). Portions of tumor cells in solid tumors are believed to be nutritionally deprived, and the intratumor environment is acidic, owing possibly to an insufficient

The assistance of the American Cancer Society and the American Hospital Supply Corporation in the preparation of this review is gratefully acknowledged. The views expressed by the author in this work are entirely his own and are in no way endorsed by either the American Cancer Society or the American Hospital Supply Corporation.

oxygen supply, and thus to an enhanced formation of lactic acid (Gullino et al. 1965; Song et al. 1980). The poor tumor blood flow may be incriminated for this inadequate supply of nutrients and oxygen and the low pH in the tumors.

Indications are that tumor vascular beds are quite vulnerable to heat and more prone to develop occlusion and thrombosis after heating than are the vascular beds in normal tissues. A severe vascular occlusion would inevitably entail necrosis in the adjacent area. This suggests that the eradication of tumors by hyperthermia may result not only from the direct cytocidal effect of heat on tumor cells, but also from the tissue necrosis as a consequence of vascular damage (Kang, Song, and Levitt 1980; Song et al. 1980a, 1980b, 1980c; Song and Levitt 1971).

The present chapter attempts to describe the possible implication of blood circulation in the use of hyperthermia, alone or in combination with other modalities, for the control of malignant tumors.

DEVELOPMENT AND CHARACTERISTICS OF TUMOR VASCULATURES

The blood vasculatures are the principal integrating system between tumor and host. Without an adequate supply of nutrients from the host by the blood circulation, neoplastic tissues never would be able to establish themselves and maintain progressive growth. It is not surprising, therefore, that the earliest pathogenic phenomenon that can be observed in the normal tissues adjacent to tumors at the inception of tumor development is prominent changes in host vasculatures. These pathogenic vascular changes in the host tissues are believed to be initiated by diffusible substances released by the neoplastic cells. Folkman (1975) demonstrated that a wide variety of animal and human tumors release tumor angiogenesis factor (TAF), which is capable of eliciting angiogenesis from the host vascular system. Cavallo and colleagues (1973), however, showed that TAF was mitogenic not only on the capillary endothelium, but also on pericytes and surrounding connective tissues. Suddith and co-workers (1975) subsequently demonstrated that tumor cells produce a soluble mitogenic factor specific to endothelial cells and called this substance endothelial proliferation factor (EPF).

The process of neovascularization in tumors has been studied with a number of different methods (Gullino 1975; Peterson 1979). The direct observation of development of vasculatures in tumors growing in transparent chambers provided the most valuable information (Algire et al. 1946; Eddy and Casarett 1973; Goodall, Sanders, and Shubik 1965; Hubler and Wolf 1976; Warren and Shubik 1966). The first pathogenic changes in the normal tissues adjacent to the grafted tumor tissues in the transparent chamber are dilation and hyperemia of preexisting host vessels, and particularly venules, often with formation of telangiectasis. The endothelium of venules starts to proliferate actively, perhaps evoked by the TAF or EPF (Cavallo et al. 1973; Folkman 1975; Suddith et al. 1975). Out of the stimulated vessels, buds grow in three to four days after the tumor transplantation, and these buds soon begin to show lumen. Although red blood cells are abundant in the lumen of the buds, thorough circulation cannot be seen at this stage of growth. The buds grow in length and width with eventual formation of a network of sprouts. The growing sprouts or their branches randomly fuse, giving rise to loops and anastomose with the anterior end of host capillaries. When this is achieved, blood from the host arterial supply starts to flow through the tumor capillary network. In malignant neurilemoma growing in the cheek pouch of hamster, Eddy and Casarett (1973) observed that the development of buds and sprouts was seen only in the host venous vessels. This is in agreement with the earlier observation by Kligerman and Henel (1961) that the initial capillary beds of hepatoma were both supplied and drained by venules. There is, however, a possibility that some of the tumor capillaries grow out of the arteriolar site of host capillaries as well (Lindgren 1945).

Once the capillary network is established, the tumors begin to grow rapidly. As the tumor mass increases, more new capillaries are formed and incorporated into the tumor mass. The growth rate of the tumors at this stage may greatly depend on the rate of neovascularization. Willis (1960) observed that host arteries rarely are invaded by the growing tumor mass. In some tumors, however, the preexisting host vessels have been reported to incorporate into the tumor mass and constitute the tumor vascular network (Algire et al. 1946; Falk 1978; Gamill et al. 1976; Goodall, Sanders, and Shubik 1965; Gullino and Grantham 1962; Lindgren 1945; Warren and Shubik 1966). Gullino and Grantham (1962) found that the vascular tree of hepatoma and Walker-256 carcinoma of rat, growing as "tissue-isolated" tumor in kidney or ovary, consisted not only of the newly formed vessels, but also of the vessels of the kidney or ovary, and the vascular network of the host organs became the main branches of the vascular network of the growing tumors. The host vessels did not increase in number, however, and only the length and caliber increased. While part of the preexisting host vessels enclosed in the tumor mass are destroyed, the remaining part is functionally intact (Lindgren 1945). Using special staining methods, Gamill and associates (1976) could identify the presence of host vessels in several human tumors, including hemangioma of liver, cystadenoma of pancreas, and angiomyolipoma of kidney. The extent of incorporation of these vasculatures into the tumor varies, depending on the tumor type. For example, the host vessels may prevail in the vascular network of the tumors of the "central vascularization" type described by Falk (1978). The newly formed tumor vasculatures are known to respond to various stimuli differently from the normal tissue vessels (Abrams 1964; Bierman, Kelly, and Singer 1952; Edlich et al. 1966; Rockoff et al. 1966). Therefore, in studying the response of tumor vasculatures to vasoactive agents, including heat, the possibility that part of the tumor vasculatures are preexisting host vessels should be considered.

The patterns of tumor vascular networks are characteristic for different types of tumors (Falk 1978; Gullino 1975; Jirtle, Clifton, and Rankin 1978; Lindgren 1945; Peterson 1979; Rubin and Casarett 1966). Generally, it is believed that the vascular patterns reflect the histology of the tumor itself and not the stroma of its tissue origin (Milne et al. 1967). Even in the same tumors, a stable and stationary vascular network cannot be achieved both within the tumors and in the surrounding normal tissues when the tumors continuously grow.

The tumor capillaries are extremely coarse, irregularly constricted or dilated, and distorted with twisting and sharp bending (Algire et al. 1946; Eddy and Casarett 1973; Endrich et al. 1979b; Hubler and Wolf 1946; Kolstad 1965; LeServe and Hellman 1972). The irregular and coarse architecture of the capillary network may result from an uneven proliferation of the tumor cells and irregular enlargement of the mass from place to place in the tumors (Goodall, Sanders, and Shubik 1965). The blood coursing through the capillaries is drained into abundant collecting venules, which also are tortuous, irregularly dilated, and distended. Warren (1979) classified the tumor vessels into nine groups by morphologic features. The capillaries of most of the tumors are lined by a single layer of endothelial cells, although the vessels of some of the highly differentiated tumors have normal basement membrane (Peterson 1979). The thin-walled endothelium is often incomplete and lined in part by cords of neoplastic cells (Kolstad 1965; LeServe and Hellman 1972; Warren and Shubik 1966). Krylova (1973) reported that endothelial cells lining tumor vessels often protrude into the lumen and partly or completely occlude the capillary lumen. Tumor ischemia and necrosis may ensue from such obstruction of blood flow. In a mammary adenocarcinoma of mouse, Vogel (1965) observed that the relative proportion of sinusoidal vessels increased as the tumor size increased, and that many of the vascular areas occupied largely by sinusoidal vessels were on the verge of necrosis. Gamill and colleagues (1976) reported that new vessels in certain human tumors were large capillaries or sinusoids. The sinusoids were devoid of endothelial cells; they were simple holes in tumor parenchyma through which blood drained from the arteries to the capillaries or veins.

The histopathologic study of Thomlinson and Gray (1955) on the interrelationship among capillary density, oxygen tension, and necrosis in tumors is well known. They reported that all cords of carcinoma of human bronchus more than 200 μ in diameter contained central necrosis. It was concluded that the diffusion length of oxygen in tissue is about 100 μ, with the venous oxygen tension of 40 mm Hg, and necrosis develops when cell mass grows greater than this oxygen diffusion length. Kolstad (1965) observed that the intercapillary distance in human cervical cancer was 50 μ to 200 μ, averaging approximately 100 μ, and it increased with the advancing grade of the disease. Lindgren (1945), Rubin and Casarett (1966), and Tannock and Steel (1969) were able to observe the capillary in necrotic area in a variety of tumors, suggesting that the development of necrosis may not be due simply to an overgrowth of tumor cells. Arteries or arterioles seldom grow significantly (Algire et al. 1946). As the growth of the tumor mass proceeds, both the length and the number of ectatic capillaries fed by the same vessels increase. Such an increase in excess of the capacity of arterioles would lead to a decrease in arteriolar pressure, entailing a decline in the velocity of blood flowing through the capillaries (Eddy and Casarett 1973). The extravascular pressure around the lengthy capillary may sometimes exceed the arteriolar pressure and evoke vascular stasis in places. Such a stasis may or may not be a permanent one. Indeed, it is commonly observed in the tumors growing in transparent chambers that the blood flow in capillaries suddenly ceases for a moment, followed by a resumption of rapid blood flow, sometimes in the direction opposite to the previous one (Eddy and Casarett 1973; Endrich et al. 1979b; Reinhold 1979). Song and Levitt (1971) and Tannock and Steel (1969) found that complete equilibrium of intravenously injected ^{51}Cr-RBC between the tumor and the systemic circulation took more than 10 minutes, even though a complete circulation of blood through the body may take less than one minute. This rather sluggish mixing of newly injected radioactive red blood cells in tumors may be due, at least in part, to a temporal stasis of the tumor vasculatures. A prolonged obstruction of capillaries may result in local ischemia, necrosis of the vessels with fibrotic replacement and atrophy, and transformation of the vessels into hyaline core. Hemorrhage and necrosis in the tumor stroma would ensue from such vascular damages.

VASCULAR PHYSIOLOGY OF NORMAL TISSUE AND TUMOR

There has been considerable confusion as to the relative efficiency of the blood supply of neoplastic tissues and normal tissues. While some tumors have central vascular supply and drainage, others have a rich peripheral vascular network (Falk 1978). Tumors with such peripheral vascular supply could be mistakenly characterized as well vascularized, even though the inner part of the tumors may be avascular. Histologic studies often demonstrate rich vascularity in tumors (Rubin and Casarett 1966; Vogel 1965). It should be pointed out, however, that not all the vessels demonstrable in the histologic sections or those containing foreign materials, such as India ink or radiopaque materials, may be functional vessels under normal physiologic conditions (Tannock and Steel 1969). Tumors growing in transparent chambers also showed a rich vascularity. Goodall, Sanders, and Shubik (1965) reported that the vascular volume in hamster melanoma growing in transparent chambers amounted to about 40% to 60% of the total tumor volume. Such high vascular volume in tumors growing in transparent chambers was reported also by Eddy and Casarett (1973); the reasons for such rich vascularity are unclear. One conceivable explanation is that the concentration of the angiogenesis factors TAF or EPF in the chambers may be high due to accumulation within the chambers (Eddy and Casarett 1973).

Investigation with red blood cells labeled with radioisotopes showed that the functional vascular volume in the majority of rodent tumors is within 1% to 5% of the total tumor volume (Fujiwara 1974; Jirtle and Clifton 1971, 1973; Storey, Wish, and Furth 1951), although the vascular volume measured with ^{51}Cr-RBC method was 8.5% of the tumor volume in SCK tumor of A/J mice (Kang, Song, and Levitt 1980; Song et al. 1980b). The vascular volume in the tumors measured morphometrically ranged from 0.5% in large DS tumors of rat to 16.9% in mouse mammary carcinoma (Vaupel 1979).

Perhaps the role of blood flow may be more relevant to the hyperthermia of tumors than total vascular or blood volume. A number of different methods have been used to quantitate the blood flow in tissues, including tumors. Gullino (1975) and Gullino and Grantham (1961) transplanted a variety of tumors in kidney or ovary that were isolated from the hosts by a paraffin sack; the supply of blood to the tumor mass and draining of blood were carried out by a single artery and vein. By collecting the blood from the vein at a constant blood pressure, the blood flow in the tumors could be determined directly. One of the disadvantages of this method is the artificiality of the experimental setup. Furthermore, unless the total organ, such as kidney or ovary, is taken over completely by the tumor, the blood flow measured by this method would include the blood flow of normal tissue as well as that of tumor.

The blood flow can be measured indirectly using radioactive tracers. Sapirstein (1958) measured tissue blood flow of rats by ^{86}RbCl. When the tracer is injected intravenously into animals it is rapidly extracted by the tissues, and a negligible amount of the isotope leaves the tissues for certain periods of time, usually 20 to 120 seconds in animal tissues. It is thus possible to determine the fractional distribution of cardiac output in each tissue by counting the radioactivity in the tissues within a short period after the injection of the tracer. This method cannot be used, however, for tissues with a low extraction ratio, such as the brain. Since an increase in vascular permeability may influence the ^{86}RbCl uptake by the tissues (Lurie, Nintzel, and Rippey 1977; Whalen and Nair 1970), it may be that the ^{86}RbCl extraction method cannot be used for tissues under unusual physiologic conditions, such as after exposure to radiation or heat. Another disadvantage with this method is that the cardiac output should be known to calculate the tissue blood flow in terms of ml/min/unit weight. Nevertheless, Gullino and Grantham (1961) reported that blood flow in tumors measured by their "direct draining method" and by the indirect methods using ^{86}RbCl or ^{42}KCl were comparable under normal conditions.

The rate of clearance of inert gas, such as ^{85}Kr or ^{133}Xe, has been used to measure the capillary blood flow (Gump and White 1968; Kallman, DeNardo, and Stasch 1972; Song and Levitt 1971). Isotopes are injected into the tissues directly or delivered to the tissues by inhalation. The isotopes are cleared from the tissues at a rate proportional to the blood flow and almost completely excreted through the lungs in a single passage. The rate of decrease in radioactivity in tissues is monitored by an externally placed detector. When the isotopes are β emitters, the clearance rate determined by the β detector indicates mainly the blood flow in the surface of the tissues. The partition coefficient between blood and the tissues of interest must be determined for each isotope to convert the clearance rate into blood flow in ml/min/unit weight.

The newest development in the measurement of blood flow in animal tissues is the use of radioactive microspheres. When microspheres (3 to 50 μ in diameter) are injected into the left ventricle of the heart, they are transferred to and lodged in the vascular beds of each tissue in proportion to the cardiac output and the rate of blood flow in the tissue. Reference blood is withdrawn from a femoral artery at a constant rate beginning immediately before the injection of the microsphere and until its disappearance from the circulation at about one to two minutes after the injection. From the radioactivities in the tissues, and in the reference blood and the withdrawal rate of the reference blood, the blood flow in the tissues can be calculated in terms of ml/min/unit weight. The reliability of this method has been validated by a number of investigators and widely used to measure the blood flow in various tissues in experimental animals (Bickberg et al. 1971; Jirtle, Clifton, and Rankin 1978; Neutze, Wyler, and Rudolph 1968; Nishiyama, Nishiyama, and Frohlich 1977; Phibbs and Dong 1970; Tsuchiya, Walsh, and Frohlich 1977; Utley et al. 1974). With this method, Song and associates (1980a, 1980c) and Song, Rhee, and Levitt (1980) determined the effect of hyperthermia on the blood flow in tumor, skin, and muscle of rats.

Since the microspheres are firmly lodged in the vascular beds, the animals do not need to be killed immediately after the injection of microspheres. It is therefore possible to measure the blood flow in the same tissues repeatedly (i.e., before and after hyperthermia) by injecting microspheres labeled with different isotopes, such as ^{51}Cr, ^{125}I, ^{141}Ce, ^{95}Nb, or ^{45}Sc. Unfortunately, the measurement of blood flow with this method in animals smaller than the rat is arduous because of the delicate surgical procedures.

Another method to measure tumor blood flow has been devised by Hughes and co-workers (1979). Tumors are irradiated with 45 MeV x-rays to activate ^{16}O to ^{15}O, and the clearance rate of ^{15}O is monitored by an externally placed detector. It is assumed that the rate of clearance of ^{15}O from the tumors is proportional to the blood flow rate. The probable implication of physiologic changes by the radiation, such as changes in vascular permeability or oxygen consumption rate in the tumors, on the clearance rate of ^{15}O need to be clarified. Furthermore, the use of this method is limited by the availability of a high energy x-ray source.

It often has been suggested that the blood flow in tumors is small compared with the blood flow in normal tissues (Gullino 1975; Gullino and Grantham 1961; LeVeen et al. 1976). It appears that this notion is not always tenable. The blood flow varies enormously in different normal tissues and in different tumors. The blood flow in the resting human muscle has been reported to be 2.2 ml/min/100 g (Lassen, Lindbjerg, and Munch 1964). Song and co-workers (1980a, 1980c) and Song, Rhee, and Levitt (1980) found that the blood flow in normal skin and muscle of rat was 7.8 ml/min/100 g and 5.0 ml/min/100 g, respectively. On the other hand, the blood flow in the spleen of rabbit was as large as 897.2 ml/min/100 g (Neutze, Wyler, and Rudolph 1968). Gullino (1975) and Gullino and Grantham (1961) reported that the blood flow in rat mammary carcinoma growing in ovary or kidney was 3.3 to 13.3 ml/min/100 g, which was about 20-fold smaller than the blood flow in the corresponding normal tissues. There is abundant evidence, however, that the blood flow in some tumors is much higher (Bierman et al. 1952; Straw et al. 1974; Zanelli and Fowler 1974). The thyroid tumors of rat have been reported to have blood flow that approximates that in the normal thyroid (Zanelli and Fowler 1974). The blood flow in lymphoma of dog was as large as 1830 ml/min/100 g (Straw et al. 1974). The lymphomas of humans also appear to have relatively large blood flow (Mantyla, Kuikka, and Rekonen 1976). It is possible, however, that the large blood flow in some tumors may be due to the presence of arteriolar-venous shunts (Bierman et al. 1951; Bierman, Kelly, and Singer 1952).

The blood flow of tumors has been known to vary, depending upon the stage of growth and on the size of tumors (Algire et al. 1946; Cataland, Cohen, and Sapirstein 1962; Gullino 1975; Gullino and Grantham 1961; Kallman, DeNardo, and Stasch 1972; Song et al. 1980a, 1980c; Song, Rhee, and Levitt 1980; Vaupel 1979; Whalen and Nair 1970). Song and colleagues (1980a, 1980c) and Song, Rhee, and Levitt (1980) reported that the relationship between the weight of Walker-256 carcinoma growing in the leg of rat and the blood flow could be expressed by the following equation:

$$\log y = -0.1721 (\log X)^2 - 0.5382 (\log X) + 1.4745, \tag{1}$$

where X and Y are weight of tumor (g) and blood flow (ml/min/100 g), respectively (fig. 8.1). Kallman, DeNardo, and Stasch (1972) reported that the best-fitting polynomial curve for the blood flow in KHT sarcoma of C3H mice was

$$\log Y = -0.234 (\log X)^2 + 0.869 (\log X) + 0.525, \tag{2}$$

where X is tumor volume (cu mm), and Y is blood flow

(ml/min/100 g). The relationship between the blood flow and tumor weight of the DS carcinosarcoma of rat was reported by Vaupel (1979) to be

$$\log Y = -0.100X + 1.685, \qquad (3)$$

where X is tumor weight (g), and Y is blood flow (ml/min/100 g). The average blood flow in the 0.3- to 0.9-g Walker tumors was 48.1 ml/min/100 g, and it decreased to 15.7 ml/min/100 g in the tumors of 2 to 5 g (figs. 8.1 and 8.2). In the KHT tumors (Lassen, Lindbjerg, and Munch 1964), the blood flow was 22.1 and 6.7 ml/min/100 g in the 65- to 70-cu mm and 1121- to 2240-cu mm tumors, respectively. In the DS carcinoma of rats, the blood flow in 3-g tumors was 25 ml/min/100 g, and that in the 13-g tumors was 2.5 ml/min/100 g (Vaupel 1979). The cause of this diminished blood flow in the large tumors is obscure, but a number of factors may be involved. In light of the increase in necrosis as the growth of tumors proceeds, it is apparent that the main cause of the decrease in blood flow per unit weight or volume in the large tumors is the increase in the necrotic tissue volume in the large tumors. The increase in necrosis may not entirely account for the decrease in the blood flow, however, since significant necrosis would not develop in the tumors until the intercapillary distance becomes greater than the limit of oxygen diffusion length. One of the causes of the decrease in the blood flow during the early stage of tumor growth may be the progressive decrease in the capillary density. It is also probable that the coarse, twisted, branched, irregularly constricted, elongated, and dilated tumor capillaries in the large tumors offer a greater resistance to the blood flow. It may be concluded that a decrease in capillary density, an increase in necrotic tissue, and an increase in flow resistance are the major causes for the progressive decrease in the blood flow with the growth of tumors. In this context, it should be emphasized that when tumor blood flow is compared with normal tissue blood flow, the size of tumors should be taken into account. This is particularly true when the effect of hyperthermia on the blood flow is discussed, since the response of tumor vasculatures to heat may vary depending upon the size of tumors.

VASCULAR CHANGES BY HYPERTHERMIA

Normal Tissues

Most investigations so far reported on the thermal injury to blood vessels of normal tissues are limited to skin and muscle. Furthermore, the temperatures used in these studies were high compared with the temperatures relevant to hyperthermic treatment of tumors. Nevertheless, this information on the effect of relatively high temperatures still may be valuable for understanding of vascular changes in normal tissues during hyperthermia of tumors at moderate temperatures.

The most prominent changes in normal tissues following mild (first degree) thermal injuries are ery-

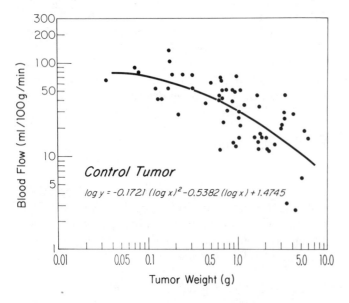

Figure 8.1. The blood flow in control Walker tumors of rats. The line is the best-fitting curve obtained by the least square method. (Song, Rhee, and Levitt 1980.)

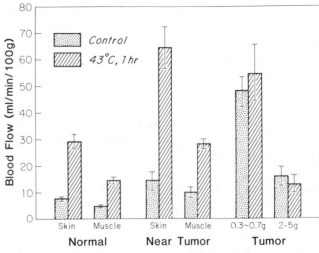

Figure 8.2. The blood flow in skin and muscle of normal leg of rat, and those adjacent to Walker tumor. The average blood flow in 0.3- to 0.7-g and 2- to 5-g tumors were obtained from the data shown in figure 8.1. Average ± 1 SE is shown. (Song et al. 1980c. Reprinted by permission.)

thema and edema. The erythema is due to vasodilation and lasts for a prolonged period after heating. The edema results from the accumulation, owing to an increase in vascular permeability, of water and plasma protein in the extravascular spaces. The mechanisms and kinetics of the increase in vascular permeability in the inflammatory skin, including thermal injury, are well known. A mild thermal injury, such as 54°C for 20 seconds, evokes a diphasic increase in vascular permeability consisting of an immediate and transient reaction, followed by delayed and prolonged reactions (Cotran and Majno 1964; Hurley, Ham, and Ryan 1967). The early increase subsides within 30 minutes, while the delayed reactions begin after 30 to 60 minutes and last for several hours after thermal injury. The early increase in vascular permeability is believed to be due to the formation of reversible gaps between endothelial cells of venules. These gaps are reported to result from active endothelial contraction mediated by histamine from the tissues and bradykinin from the blood. The delayed leakage of plasma protein appears to occur predominantly in capillaries (Cotran and Majno 1964; Hurley, Ham, and Ryan 1967), although venules may also be involved (Shea et al. 1973). Gabbiani and Badonnel (1975) reported that irregular dilation of interendothelial-clefts occurs in small vessels (arterioles, venules, capillaries) prior to the delayed increase in vascular permeability in the rat skin heated at 54°C for 20 seconds. Extensive thermal damage obviously would result in severe vascular damage leading to stasis and subsequent necrosis. The threshold temperature for the vascular changes marked by erythema and increased vascular permeability may depend upon the duration of heating and on the kind of tissues involved. Sevitt (1954) reported that the threshold temperature for an increase in vascular permeability in guinea pig dermis was in the range of 41°C to 45°C.

Song (1978) investigated the changes in vascular volume and permeability in the heated skin and muscle of leg of SD male rats. The vascular permeability and blood volume were measured with ^{125}I-plasma protein and ^{51}Cr-RBC, respectively. The vascular permeability in the control skin and muscle was 1.46 ml/hr/100 g and 0.49 ml/hr/100 g, respectively. After heating for one hour at 43°C, the vascular permeability increased 3.5-fold in the skin and 2.5-fold in the muscle. The heat-induced changes in the intravascular blood volume in the skin and muscle of rats are shown in figure 8.3. The blood volume in the control skin was 0.55 ml/100 g and that in the control muscle was 0.42 ml/100 g. The blood volume in the skin increased by about 3.0-fold when the legs were heated at 43°C for one hour. It remained at this level for one hour after

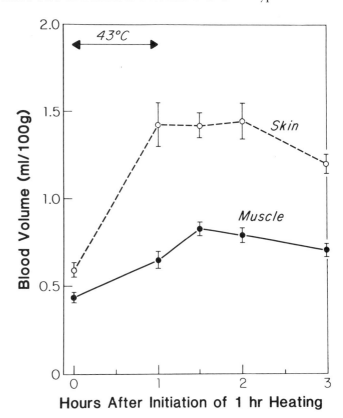

Figure 8.3. The blood volume changes in the skin and muscle of leg of rat heated at 43°C for one hour. Notice that the blood volume remained elevated after heating. Average ± 1 SE is shown.

the heating and then began to decline slowly. About 1.5-fold increase in the blood volume was observed in the muscle at the end of one hour of heating at 43°C. The blood volume in the heated muscle continued to increase to 1.8-fold of control during the 30-minutes postheating, and it remained at this level for more than two hours after the heating. This increase in the intravascular volume could be due to dilation of blood vessels or opening of more capillaries.

Investigators (Song et al. 1980a, 1980c; Song, Rhee, and Levitt 1980) reported that the blood flow in the skin and muscle of rat also increased remarkably when heated at 43°C. The blood flow in the control skin, measured with the radioactive microsphere method, was 7.82 ml/min/100 g, and that in the control muscle was 4.97 ml/min/100 g (figs. 8.2 and 8.4). When the rats' legs were heated with 43°C-water for one hour, the blood flow in the skin increased 3.7-fold and in the muscle by 2.9-fold. The blood flow started to decline rapidly to control level after the heating. Within one to two hours the blood flow in the skin and muscle returned to control level. This is in contrast to the vascular volume, which remained significantly elevated two hours after the heating (fig. 8.3). The blood flow rates in the skin overlying the tumor and that in

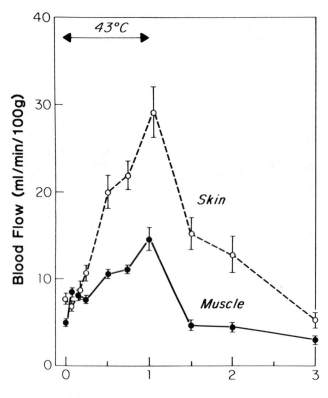

Figure 8.4. The blood flow rate in the skin (○) and muscle (●) of leg of rats heated at 43°C for one hour. The blood flow rate returned to control value within about two hours after heating. (Song et al. 1980c. Reprinted by permission.)

the muscle surrounding the tumor were 14.39 ml/min/100 g and 10.26 ml/min/100 g, respectively, which were about 2.0-fold greater than that in the normal skin and muscle (fig. 8.2). Heating for one hour with 43°C-water increased the blood flow in the skin and muscle near the tumors by 4.5- and 2.8-fold, respectively (fig. 8.2).

Tumors

Animals

Walker-256 Carcinoma of SD Rats

Song and associates (Song 1978; Song et al. 1980a, 1980c; Song, Rhee, and Levitt 1980) reported that the response to heat of vasculatures in the Walker-256 carcinoma growing subcutaneously in the leg of SD rat was quite different from that in the skin or muscle. Heating with water at 43°C for one hour had no noticeable effect on the blood volume and vascular permeability in the tumors (Song 1978).

The blood flow in the Walker tumors measured at the end of heating for one hour at 43°C and 45°C are shown as a function of tumor weight in figures 8.5 and

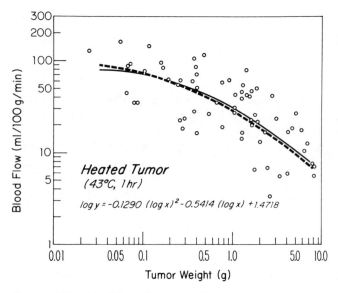

Figure 8.5. The blood flow in Walker tumors of rats. The *dotted line* is the best-fitting curve obtained by the method of least squares for 65 tumors heated at 43°C for one hour. The *solid line* is the blood flow in control tumors as shown in figure 8.1.

8.6 (Song et al. 1980a, 1980c; Song, Rhee, and Levitt 1980). The blood flow in the tumors measured three hours after heating at 45°C is shown also in figure 8.6 (closed circle). In figure 8.7, the blood flow curves for the control tumors (fig. 8.1), tumors heated at 43°C (fig. 8.5) and 45°C (fig. 8.6), are shown together for comparison. It is evident that no significant change in the tumor blood flow occurred upon heating at 43°C for one hour. It appeared, however, that the blood flow increased slightly in the small tumors at the end of heating at 45°C for one hour. The blood flow in the large tumors at the end of one hour of heating at 45°C was similar to that in the control tumors, but it decreased significantly when measured three hours after the heating (figs. 8.6 and 8.7). Gullino and Grantham (1961) reported that the blood flow in Walker tumors did not change when heated at temperatures up to 42°C, which is in agreement with the results herein described.

Mammary Adenocarcinoma of Mice

Song and colleagues (1980b, 1980c), Song and Levitt (1971), and Kang, Song, and Levitt (1980) observed that the blood volume of the SCK mammary adenocarcinoma growing subcutaneously in the leg of A/J mice was 8.8 ml/min/100 g when measured with the ^{51}Cr-RBC method. When the tumors were heated with water at 43.5°C, no noticeable change in the blood volume could be observed at the end of heating for 30 minutes. The blood volume, however, was found to be significantly decreased when measured three hours after heating at temperatures higher than 40.5°C. In

Blood Flow in Tumors and Normal Tissues in Hyperthermia

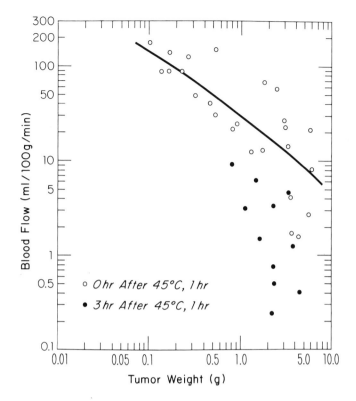

Figure 8.6. The blood flow in Walker tumors heated at 45°C. ○: measured at the end of heating; ●: measured at three hours after heating. The line is the best-fitting curve for the blood flow at the end of heating (○). (Song et al. 1980c. Reprinted by permission.)

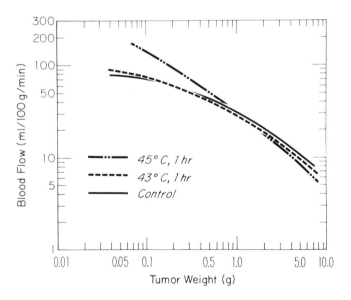

Figure 8.7. The curves in figures 8.1, 8.5, and 8.6 are shown together for comparison. (Song et al. 1980c. Reprinted by permission.)

figure 8.8, the changes in blood volume measured at various times after heating for 30 minutes or one hour at 43.5°C are shown. The minimal blood volume, about 5% of control, was observed seven hours after heating for one hour. The blood volume five hours after heating for 30 minutes was about 10% of that in the control tumors. The blood volume started to recover gradually thereafter, but it was still about 40% of control 48 hours after heating (Kang, Song, and Levitt 1980). The blood volumes measured three hours after heating at various temperatures are shown in figure 8.9. It is demonstrated that the decrease in blood volume was temperature dependent. Heating at 45.5°C decreased the blood volume to less than 10% of control blood volume three hours after heating. This decrease in blood volume indicated that severe vascular occlusion occurs in the SCK tumor following hyperthermia at temperatures above 40.5°C.

The effect of hyperthermia on blood flow in a C3H mouse mammary adenocarcinoma was measured by Bicher and co-workers (1980) using the hydrogen diffusion method. It was found that hyperthermia with microwave at a frequency of 2450 MHz has a dual effect on the tumor blood flow. Heating at temperatures up to 41°C increased the blood flow, while temperatures above 41°C caused a collapse in blood flow. As a consequence, the tumor pO_2 also increased at temperatures up to 41°C and then decreased at temperatures above 41°C.

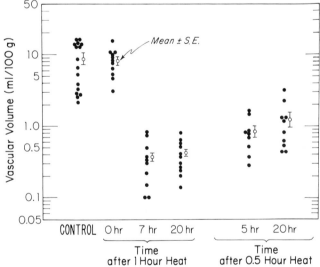

Figure 8.8. The blood volume (vascular volume) of SCK tumors of mouse at various times after heating at 43.5°C for 0.5 and 1.0 hours. The intratumor temperature during the heating was 42.9°C to 43.1°C. The vascular volume in control tumors also is shown. The *closed circles* are values for individual tumor, and the *open circles* are geometric mean for each group. The *bars* are 1 SE. (Song et al. 1980b. Reprinted by permission.)

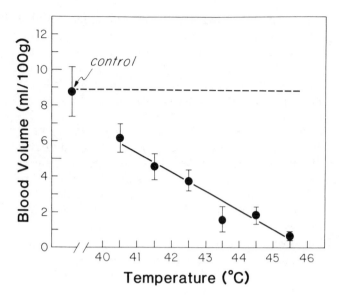

Figure 8.9. The blood volume of SCK tumor of mouse three hours after 30-minute heating at various temperatures. The averages of more than 10 tumors with 1 SE are shown. (Song et al. 1980c. Reprinted by permission.)

BA-1112 Rhabdomyosarcoma of WAG-RiJ Rats

Emami and associates (1981) studied the blood flow in the BA-1112 rhabdomyosarcoma of WAG-RiJ rat following hyperthermia with 1.25 MHz electric currents. The blood flow in the control tumors in scalp, measured by the clearance rate of ^{15}O, was 15 to 44 ml/min/100 g. Heating at 41°C for 40 minutes or at 42°C for 30 minutes reduced the blood flow by 50%, with some suggestion that the decreased blood flow following heating at 41°C was only temporary. Hyperthermia at 43°C for 40 minutes reduced the blood flow by more than 90%, and heating at 44°C for 60 minutes caused almost complete cessation of blood flow. The histopathologic investigation revealed a pattern of gradual change in tumor microvasculatures, with increasing temperature and increasing time intervals between heating and examination. Congestion of vessels could be seen soon after hyperthermia at 40.5°C for 40 minutes. This congestion was more pronounced at three hours but disappeared at 24 hours. This observation was in agreement with the measurement of blood flow in this tumor; the retarded blood flow in the tumor heated at 41°C was only a temporary reaction. At temperatures above 42.5°C, a marked dilation and congestion of the vessels, accompanied by a massive hemorrhage and necrosis, developed within a few hours.

The effect of hyperthermia on the velocity of red blood cells and vessel diameter in the vasculatures in BA-1112 rhabdomyosarcoma growing in transparent chambers has been studied by Endrich and colleagues (1979a). When the environmental temperature of rats was elevated from 25°C to 35°C, the velocity of red blood cells increased and, as a consequence, the arteriolar inflow increased more than twofold. The number of functional capillaries, however, remained unchanged at the elevated temperature, suggesting that the maximal number of capillaries were functional at 25°C. This is an apparent contrast to the propensity of a large proportion of capillaries to close at lower temperatures in normal tissues. When the tumors were heated at temperatures higher than 40°C, there was a slow, but steady, decline of the velocity of red blood cells. The number of perfused capillaries also decreased. It was of interest that leukocytes started to stick to the vascular wall in postcapillaries and collecting venules within five to ten minutes of heating. Petechial hemorrhage was seen in the venules and in the capillaries in which the blood flow could resume after a period of stasis during heating at mild temperatures. The circulatory disturbance by hyperthermia at 40°C to 42°C for 60 minutes was apparently irreversible.

Using a similar experimental device, Reinhold, Blachiewicz, and Berg-Blok (1978) observed that heating at 42.5°C for three hours impaired the microvasculatures in two-thirds of the rhabdomyosarcomas in the transparent chambers and caused necrosis in the central areas of the tumors within 24 hours. The tumor peripheries, as well as the surrounding normal tissues, were unaffected, however.

Squamous Cell Carcinoma of Syrian-Hamsters

Eddy (1980) used squamous cell carcinoma of Syrian hamster for the investigation of effect of heat on tumor vasculatures. The tumors were grown in transparent chambers in cheek pouches, and the blood flow and histologic changes were investigated during and after heating at 41°C, 43°C, and 45°C for 30 minutes. Within three to seven minutes after the beginning of heating at 41°C, the feeding arterioles constricted, and the blood flow decreased. The blood flow subsequently returned to preheating level and stayed unchanged until the end of heating. When examined histologically 15 to 24 hours after the heating, the blood vessels were found to be dilated and occasionally thrombosed (fig. 8.10). Hyperthermia at 43°C also caused a temporary vasoconstriction and reduction in blood flow within a few minutes after the beginning of treatment. The vessels then promptly returned to preheating conditions, but petechial hemorrhage and stasis of vessels started to occur during the heating at 43°C. The histologic sections taken during the 24 hours after heating at 43°C showed prominent hyperemia and hemorrhage. Endothelial linings were miss-

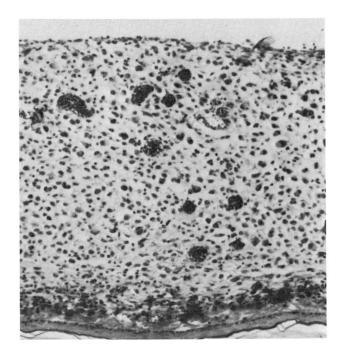

Figure 8.10. Squamous cell carcinoma growing in the transparent chamber in the cheek pouch of hamster. The tumor was heated at 41°C for 30 minutes and the tumors obtained 24 hours later. Notice the severe congestion of blood vessels.

ing in some vessels, and thrombosis was evident in others. Heating at 45°C also induced the early constriction of vessels, which was followed by dilation in some tumors. Within five to six minutes of heating, petechial hemorrhage became apparent. Vascular stasis also began at this time, and the circulation was completely ceased by 20 to 25 minutes. The hemorrhage was extensive, and the vessels were congested with red cells, thrombosed, and none of the vessels were functional 24 hours after the heating. The tumors demonstrated typical infarction and coagulation necrosis 24 to 48 hours after hyperthermia.

Ependymoblastoma of C57BL Mice

The effect of hyperthermia by 2450 MHz microwave on the blood flow in subcutaneous flank implants of ependymoblastoma of C57BL mice was investigated by Sutton (1979). The blood flow was measured by ^{133}Xe clearance method. When the tumors were heated at 40°C to 42°C, the blood flow began to increase after 15 minutes. The blood flow reached its maximum rate at 30 minutes and then decreased rapidly at 42°C and more slowly at 40°C. All the tumors showed a significant decrease in the blood flow, compared with the preheating level, from 75 minutes and 60 minutes at 40°C and 42°C, respectively. At 45°C, the blood flow decreased to half of the control value upon heating for only 15 minutes. Histologic studies of the tumors heated at 42°C for 60 minutes demonstrated a widespread coagulation and obstruction of blood vessels. It was concluded that hyperthermia at moderate temperatures increases blood flow for short periods, while excessive heating decreases the blood flow and ultimately produces vascular occlusion.

DS Carcinosarcoma of Rats

Von Ardenne, Böhme, and Kell (1979) reported that acidification of tumors by hyperglycemia and heating the tumors induced a complete cessation of blood flow in the DS carcinosarcoma of rats. Before heating the tumors, the animals were infused with glucose to elevate the blood glucose up to 500 ml/100 ml for several hours, which lowered the pH of the venules of tumors to 6.0 to 6.5. When the tumors were heated at 41°C with 27.12 MHz transmitter after achieving the acidification of the tumors, the blood flow decreased to 0.5% of the initial value. The histologic examination showed a complete occlusion of the capillaries (Von Ardenne et al. 1979). The heat dose for triggering vascular occlusion was 42°C for 60 to 120 minutes when the tumors were acidified as described.

Human Tumors

Sugaar and LeVeen (1979) investigated the histopathologic changes in malignant lung tumors of humans following hyperthermia with radiofrequency (RF) of 13.56 MHz. The early signs of vascular change within one day after a series of treatments were a reduced blood flow in the tumor's stromal capillaries and dilation of the vessels supplying the lung parenchyma. It appeared that the dilation of vessels in the surrounding normal tissues induced peripheral hypotension and reduced the blood flow in the tumors. Subsequently, the capillaries in some areas of tumor showed severe degenerative changes with loss of staining and marked fibrinoid necrosis of capillary wall. The damages in the capillaries progressed with time, and most of the capillaries of the heated tumors were functionally and anatomically destroyed. As a consequence, the tumor's stroma adjacent to the affected vessels became necrotic and infiltrated with small lymphocytes. It was postulated that the eradication of tumor by hyperthermia was, in part, enhanced by cell-mediated immune reaction by the small lymphocytes, which escaped through the ruptured stromal walls.

The histologic changes in the human tumors were described also by Storm and colleagues (1979). In human intraabdominal sarcoma, severe coagulation necrosis and vascular thrombosis could be observed within two weeks after five courses of hyperthermia at temperatures of 45°C to 50°C for 15 to 30 minutes using RF. The vessels were completely obliterated three months later.

MECHANISMS OF VASCULAR DAMAGE IN HEATED TUMORS

It has been observed that heating of normal tissues at moderate temperatures causes a profound increase in blood flow (Song et al. 1980a, 1980c; Song, Rhee, and Levitt 1980). In contrast, the blood flow in tumors does not increase but decreases upon heating (Emami et al. 1981; Endrich et al. 1979a; Gerweck 1977; Song et al. 1980b, 1980c; Song, Rhee, and Levitt 1980; Sugaar and LeVeen 1979; Sutton 1979; Von Ardenne, Böhme, and Kell 1979; Von Ardenne et al. 1979; Zanelli and Fowler 1974), although some of the tumors heated at 40°C to 42°C showed a slight and/or temporary increase in blood flow (Bicher et al. 1980; Sutton 1979). It is known that the neovasculatures in tumors are devoid of receptor or nervous apparatus to respond to various stimuli including heat stress (Hahn 1974, 1979; Hilmas and Gillette 1974; Hornback et al. 1977). It also is probable that tumor vasculatures are fully dilated and operated continuously at the maximum capacity, even at ambient temperature, to meet the nutritional demand by the progressively increasing tumor cell population (Endrich et al. 1979b). In the small Walker-256 carcinoma the slight increase in blood flow upon heating at 45°C (figs 8.6 and 8.7), as well as the temporary or slight increase in other tumors (Bicher et al. 1980; Sutton 1979), could be attributed to the presence of host vessels incorporated into tumors at the early stage of tumor growth, as discussed earlier in this chapter.

The tumor vascular beds apparently are more vulnerable to heat than are those in normal tissues, as indicated by the decrease in blood flow and histopathologic changes at temperatures significantly lower than the threshold temperature for the vascular damage in normal tissues. The heat sensitivity of blood vessels in different tumors appears to vary considerably among the tumors investigated. The blood vessels of Walker-256 carcinoma were the most heat resistant, since heating at 45°C for one hour was necessary to induce vascular damage and a decrease in blood flow (fig. 8.6). It should be noted that heating at 45°C for one hour is still significantly lower than the threshold heat dose for the vascular damages in skin and muscle of rat, which was about 46.5°C for one hour.

The reason for the greater sensitivity of tumor blood vessels than normal tissue blood vessels to heat is far from clear. As described in the section on vascular changes in tumors, Von Ardenne, Böhme, and Kell (1979) reported that acidification of tumors by hyperglycemia in animals enhanced the heat-induced occlusion of tumor vasculatures. It has been reported that red blood cells lose their membrane flexibility under acidic conditions (Schmid-Shönbein, Weiss, and Ludwig 1973), and that leukocytes adhere to the postcapillary wall during heating of tumors (Endrich et al. 1979a). Von Ardenne, Böhme, and Kell (1979) postulated, therefore, that the leukocytes attach on the walls of venules during heating and form a barrier where the rigid red blood cells lodge and occlude the blood flow.

It is well known that the intratumor environment is intrinsically acidic (Gullino et al. 1965; Song et al. 1980c). Song and colleagues (1980c) observed that tumor pH decreases rapidly upon heating of tumors. As shown in figure 8.11, the pH of SCK tumor of mice was 7.05 prior to heating. When heated with a 43.5°C water bath, the pH fluctuated at the beginning, but eventually decreased to 6.80 within 10 minutes and to 6.67 at the end of heating for 30 minutes. The pH decreased further to 6.58 when the tumor was heated again for 30 minutes at 43.5°C. Similar decreases in pH by heat were observed by Song and co-workers (1980c) in Walker-256 carcinoma of rat and also in mammary adenocarcinoma of C3H mouse by Bicher and colleagues (1980). This rapid decrease in pH in the heated tumors may result from an enhanced for-

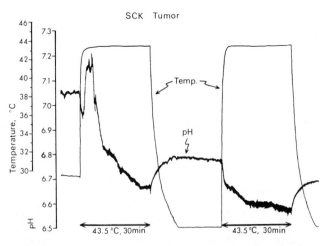

Figure 8.11. The change in pH in SCK tumor upon heating for 30 minutes at 43.5°C, 30 minutes apart. The temperature change of the tumor was recorded simultaneously with pH change. (Song et al. 1980c. Reprinted by permission.)

mation of lactic acid in the tumors, presumably by hypoxic cells, and the resultant acidity may enhance the vascular occlusion. Therefore, acidification of tumors by hyperglycemia prior to heating, as Von Ardenne, Böhme, and Kell (1979) proposed, may not be an essential condition of the induction of vascular occlusion in the heated tumors. In this connection, it should be noted that the deformation of red cells occurs at pH 6.7 to 6.9 (Schmid-Shönbein, Weiss, and Ludwig 1973), which can be attained easily by heating the tumors as described above. It was of interest that the intratumor pH remained low after heating (fig. 8.11). This may be ascribed to the further decrease in pO_2, accumulation of lactic acid, and lack of prompt drainage of the acidic fluid because of the vascular stasis. That is, the increase in acidity triggers the vascular occlusion, which in turn further increases the intratumor acidity. It is likely that when tumors are treated with fractionated hyperthermia the increased acidity in the tumors from the initial heating may render the tumor vasculatures more vulnerable to the subsequent heating.

Another conceivable mechanism of the cessation of blood flow in the heated tumors is that severe edema caused by an increased extravasation of plasma protein in the surrounding normal tissues, as well as in the tumor, may augment the extravascular pressure on the capillaries, which are devoid of supportive connective tissue, and invokes vascular occlusion (Eddy 1980). When the heat dose is moderate, the vascular damage is delayed. This delayed development of vascular damage could be attributed, in part, to the swelling and protruding of endothelial cells (Krylova 1973) and to the lysis of the endothelial cells, as well as the tumor cells, which line up with the endothelial cells along the walls of neovasculatures of tumors.

RELATIONSHIP BETWEEN BLOOD FLOW AND TISSUE TEMPERATURE DURING HEATING

In a well-perfused tissue most of the heat delivered by hyperthermia is dissipated by blood circulation, although part of the heat may be lost to surrounding tissues by a simple diffusion process. The role of blood flow in the rise of temperature in tissues during heating was defined by Peterson (1979) as follows:

$$\Delta T = \frac{I}{K S_t}(1 - e^{-kt}), \quad K = \frac{F\rho}{100\lambda}, \quad (4)$$

where I is heat input rate per g of tissue, S_t is specific heat of tissue ($J/g/°C$), F is tissue blood flow in ml/min/100 g, ρ is density of tissue (g/ml), and λ is ratio of solubility of heat in tissue and blood (i.e., ratio of specific heat). It can be seen that the increment of tissue temperature is inversely proportional to the rate of blood flow if it is assumed that (1) the equilibrium distribution of heat between blood and tissue is immediately attained, (2) heat is removed from the tissue only by the blood flow, (3) there is no recirculation of heat, and (4) the tissue is homogeneous.

Figure 8.12 shows the increment of temperature calculated using this equation in hypothetical tissues with the tissue specific heat of 4.2 J/g/°C. The curves A, B, and C are the calculated temperature profiles in tissues with no blood flow or with a constant blood flow of 10 ml and 50 ml/min/100 g when the heat input was 0.1 Wg^{-1}. The curve D is the temperature profile in a hypothetical tissue in which the blood flow was 10 ml/min/100 g prior to heating, and it increased to 50 ml/min/100 g within 10 minutes of heating. It is demonstrated that the temperature of the tissue declined as a consequence of the elevation of blood flow and an increased heat dissipation.

Actual measurement of tissue temperature demonstrated close relationships with the blood flow (Song et al. 1980c; Song, Rhee, and Levitt 1980). When the leg of rat was heated with 43°C water bath, the temperature of skin, measured by a thermocouple, rapidly reached a peak temperature of 42.8°C and then decreased slightly thereafter. Despite the large increase in blood flow (figs. 8.2. and 8.4), the temperature of skin was only 0.5°C below the water temperature. This high skin temperature could be anticipated since the skin was in direct contact with the heating water. Figure 8.13 shows the temperature profile of muscle in the leg and that of subcutaneous Walker-256 carci-

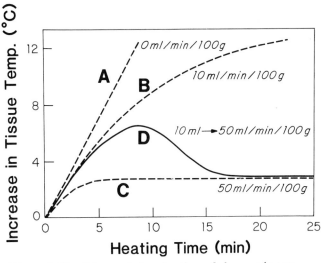

Figure 8.12. Theoretical estimation of changes in temperature in tissues with various blood flow rate.

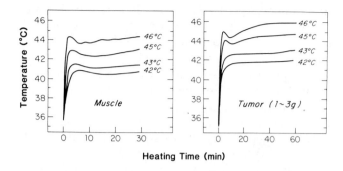

Figure 8.13. The temperature of muscles and Walker tumors of rats during 30 minutes of heating at various temperatures. The temperature reached was shown at the end of each curve.

noma growing in the leg of rat. When heated at 43°C the temperature of muscle rose to 41.5°C and then declined to 41.0°C. This considerable decrease in the muscle temperature could be attributed to an efficient heat dissipation by the increased blood flow (figs. 8.2 and 8.4). Upon heating at 43.0°C, the temperature in a Walker tumor rose to 42.5°C and remained at this temperature during the rest of the one-hour heating period. This temperature was significantly higher than that of muscle heated at the same temperature. The relatively small blood flow and the lack of increase in blood flow in the tumor upon heating at 43°C (figs. 8.2 and 8.5) may account for the high temperature in the tumor during the heating. It was of interest that the temperatures of Walker tumors (2.5 to 3.0 g) increased to 44°C and 45°C when heated at 45°C and 46°C, respectively. The temperatures then dropped rapidly by about 1°C and then rose again fairly rapidly. At the end of one hour of heating, the temperature in these tumors was almost the same as the temperature of the heating water. It appeared that when heated at the rather high temperatures, there was a temporal increase in blood flow in the Walker tumors that caused the decline in the intratumor temperatures. The blood flow soon decreased, however, and the tumor temperature rose to the temperature of heating water because of the decrease in the heat dissipation.

Storm and colleagues (1979) were able to heat liver metastases of patients up to 53°C while maintaining the temperature of liver and skin at 42°C and 39°C, respectively, during heating with RF waves of 13.56 MHz and a magnetic-loop applicator. The temperature of an intraabdominal tumor could be raised to 50°C, while the temperature of muscle and skin remained under tolerable levels. Such a selective heating of tumors up to 42°C to 50°C could be achieved in 75% of patients treated. It was postulated that the heat dissipation in the large tumors was languid owing to poor blood flow. Many tumors smaller than 5 cm in diameter could not selectively be heated, however, suggesting that the blood circulation in the small tumors was as effective as that in the normal tissues to dissipate the heat. LeVeen and associates (1976) also observed preferential rise of temperature and destruction of human neoplastic tissues upon heating with RF, and attributed this to the sluggish tumor blood flow.

It could be concluded that the blood flow in tumors, particularly in the large tumors, generally is intrinsically sluggish relative to that in the normal tissues. The differences in the blood flow in tumors and normal tissues become even larger during heating owing to the significant increase in the blood flow in the normal tissue and to the lack of such increase in tumor blood flow and/or to the damage in the tumor vessels. As a consequence, the temperature in tumors rises substantially higher than that in normal tissues during the heating and results in preferential destruction of tumor tissues.

IMPLICATION OF VASCULAR DAMAGE TO THE SURVIVAL OF TUMOR CELLS IN HEATED TUMORS

The vascular occlusion in the heated tumors would inevitably result in cessation of supply of nutrients, including oxygen. Further cell death as a result of vascular damage, in addition to the direct thermal killing of the cells, might then be expected to occur in the tumors after hyperthermia. This delayed or indirect killing of tumor cells after hyperthermia was first suspected by Crile (1962). It was observed that when the sarcoma-180 of Swiss mice was treated in vivo at 44°C for 30 minutes and transplanted to other mice immediately after the heating, tumors developed at most of the injected sites. When the tumors were left in situ for eight hours after hyperthermia and then transplanted to new hosts, tumor failed to develop in the new hosts. It was concluded that the inflammatory reactions in the heated tumors killed additional tumor cells when the tumors were left in situ after hyperthermia. Suit (1977) reported that half of the fibrosarcomas of C3H mice could be cured by heating at 43.5°C for 78 minutes. When the tumor cells were heated in vitro at the same temperature, however, heating for four hours at that temperature was necessary to inactivate the transplantability of the tumor cells. It therefore was suggested that the cure of tumor in vivo by hyperthermia is not achieved by direct thermal killing of cells. Marmor, Hilerio, and Hahn (1979) heated EMT6 tumor of mouse at 43.5°C to 44.0°C for 30 minutes with ultrasound waves and found that significant additional cell deaths occurred over a period of 2 to 48 hours after the heating. They concluded that immune reaction or vascular damage was responsible for the delayed cell deaths. A similar phenomenon was found by Kang, Song, and Levitt (1980) and by Song and co-workers (1980b) in SCK mammary carcinoma of A/J mice (fig. 8.14). When the tumors were left in situ after hyperthermia at 43.5°C for 30 minutes, there was a progressive decrease in the cell survival during 6 to 12 hours after the heating. No such delayed cell deaths occurred when the cells were heated and left in the culture flasks for several hours in vitro prior to subculture. Since SCK tumor is non-immunogenic to A/J mice, it was highly unlikely that immunogenic reactions were involved in the progressive cell death in the tumor left in situ. A similar time course in the reduction and recovery of the blood volume (fig. 8.8) and the cell survival (fig. 8.14) strongly suggested that the delayed cell death in the SCK tumors following hyperthermia was closely related to the vascular damage. As discussed, the intratumor pH significantly decreases upon heating, owing in part to the vascular damage (fig. 8.11). It is known that acidic environment renders mammalian cells sensitive to heat (Dewey et al. 1979; Gerweck 1977; Overgaard 1976). Furthermore, it has been observed by Dewey and colleagues (1979) that cell survival progressively decreases when the cells are maintained in acidic conditions after heating. It can be concluded that the progressive death of cells in tumor left in situ after heating results from lack of nutrients and an increase in the intratumor acidity as a consequence of vascular damage. Besides direct thermal killing, the delayed or indirect killing of tumor cells by vascular damage may play an important role in the tumoricidal effect of hyperthermia.

ROLE OF BLOOD FLOW IN COMBINATION OF HYPERTHERMIA WITH OTHER MODALITIES

It is known that hyperthermia synergistically enhances the effect of radiation (Ben-Hur, Bronk, and Elkind 1972; Dewey et al. 1977, 1979; Sapareto, Hopwood, and Dewey 1978) or chemotherapeutic drugs (Hahn 1979; Marmor 1979; Song, Guertin, and Levitt 1979) on mammalian cells. The sequence of radiation and hyperthermia significantly influences the magnitude of heat sensitization of radiation effect (Sapareto, Hopwood, and Dewey 1978). A series of experiments by Field, Hume, and Law (1979; Myers and Field 1977) on the combined effects of hyperthermia and radiation indicated that heating before radiation causes more damage to normal tissues than does the heating after radiation. It is known that the intracellular pO_2 in resting skeletal muscle is only 5 to 10 mm Hg (Coburn and Mayers 1971; Willis 1960), which can be regarded as radiobiologically hypoxic. When heat is applied first, the blood flow increases in normal tissues, and this would inevitably augment the pO_2 and render the normal tissues more sensitive to radiation.

Still undetermined is the sequence of heat and radiation to tumors that would exhibit the greatest therapeutic advantage. If hyperthermia at relatively low temperatures increases tumor blood flow, as Bicher and colleagues (1980) reported for temperatures under 41°C, the intratumor pO_2 may increase, which will render the tumors sensitive to subsequent irradiation. Inasmuch as the blood flow in normal tissues would also increase upon heating, the therapeutic gain of using heat prior to irradiation may not be significant. Furthermore, it should be stressed that an increase in tumor blood flow upon heating even at modest temperatures may not be a universal phenomenon. When the temperature is high enough to cause vascular damage and decrease in pO_2, the sensitivity of tumor cells to subsequent irradiation will inevitably

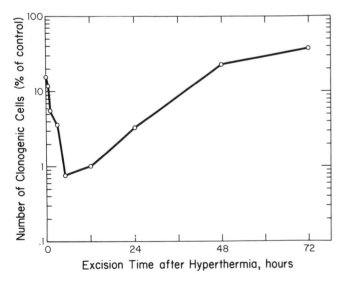

Figure 8.14. The change in number of clonogenic cells in SCK tumor as a function of time after heating at 43.5°C for 30 minutes. The clonogenic cell number was calculated by multiplying the number of cells recovered from 1 g of tumor with the plating efficiency of these cells. (Kang, Song, and Levitt 1980. Reprinted by permission.)

be reduced. Under such circumstances, applying hyperthermia prior to radiotherapy will unquestionably be less effective than the reversed sequence of these two modalities. Field, Hume, and Law (1979) suggested using heat at least several hours after irradiation in order to obtain the direct effects of both modalities and to avoid the thermal enhancement of radiation effects on tumors and normal tissues.

In combining hyperthermia and chemotherapeutic drugs, a crucial question is whether the thermal enhancement of damage by drug to tumors would be higher than that to normal tissues. The blood flow may play an important role in this regard as well. If the blood flow and vascular permeability in normal tissues become greater than those in tumors upon heating, the concentration of drug will become higher, and thus the damage will be greater in the normal tissues than in the tumors. It is apparent that the differential response of blood vessels of tumors and normal tissues to heat should be considered in the treatment of tumors with drugs in combination with hyperthermia.

References

Abrams, H. L. Altered drug response of tumor vessels in man. *Nature* 201:167–170, 1964.

Algire, G. H. et al. Vascular reactions of normal and malignant tissues in vivo. I. Vascular reactions of mice to wounds and to normal and neoplastic transplants. *J. Natl. Cancer Inst.* 6:73–85, 1946.

Ben-Hur, E.; Bronk, B. V.; and Elkind, M. M. Thermally enhanced radiosensitivity of cultured Chinese hamster cells. *Nature; New Biology* 238:209–210, 1972.

Bhuyan, B. K. Kinetics of cell killing by hyperthermia. *Cancer Res.* 39:2277–2284, 1979.

Bicher, H. I. et al. Effects of hyperthermia on normal and tumor microenvironment. *Radiology* 137:523–530, 1980.

Bickberg, G. D. et al. Some sources of error in measuring regional blood flow with radioactive microspheres. *J. Appl. Physiol.* 31:598–604, 1971.

Bierman, H. R. et al. Studies on the blood supply of tumors in man. III. Vascular patterns of the liver by hepatic arteriography in vivo. *J. Natl. Cancer Inst.* 12:107–131, 1951.

Bierman, H. R. et al. Studies on the blood supply of tumors in man. V. Skin temperature of superficial neoplastic lesions. *J. Natl. Cancer Inst.* 13:1–15, 1952.

Bierman, H. R.; Kelly, K. H.; and Singer, G. Studies on the blood supply of tumors in man. IV. The increased oxygen content of venous blood draining neoplasms. *J. Natl. Cancer Inst.* 12:701–707, 1952.

Cataland, S.; Cohen, C.; and Sapirstein, L. A. Relationship between size and perfusion rate of transplanted tumors. *J. Natl. Cancer Inst.* 29:389–394, 1962.

Cavallo, T. et al. Ultrastructural autoradiographic studies of the early vasoproliferative response in angiogenesis. *Am. J. Pathol.* 70:345–354, 1973.

Coburn, R. F., and Mayers, L. B. Myoglobin O_2 tension determined from measurements of carboxymyoglobin in skeletal muscle. *Am. J. Physiol.* 220:66–74, 1971.

Connor, W. G. et al. Prospects for hyperthermia in human cancer therapy, II. *Radiology* 123:497–503, 1977.

Cotran, R. S., and Majno, G. The delayed and prolonged vascular leakage in inflammation. I. Topography of the leaking vessels after thermal injury. *Am. J. Pathol.* 45:261–281, 1964.

Crile, G., Jr. Selective destruction of cancers after exposure to heat. *Ann. Surg.* 156:404–407, 1962.

Crile, G., Jr. The effect of heat and radiation on cancers implanted on the feet of mice. *Cancer Res.* 23:372–380, 1963.

Dewey, W. C. et al. Cellular responses to combinations of hyperthermia and radiation. *Radiology* 123:463–473, 1977.

Dewey, W. C. et al. Cell biology of hyperthermia and radiation. *Paper read at the Sixth International Congress of Radiation Research*, May 1979, Tokyo, Japan, eds. S. Okada et al.

Dickson, J. A. et al. Tumor eradication in the rabbit by radiofrequency heating. *Cancer Res.* 37:2162–2169, 1977.

Eddy H. A. Alteration in tumor microvasculature during hyperthermia. *Radiology* 137:515–521, 1980.

Eddy, H. A., and Casarett, G. W. Development of the vascular system in the hamster malignant neurilemmoma. *Microvasc. Res.* 6:63–82, 1973.

Edlich, R. F. et al. Effect of vasoactive drugs on tissue blood flow in the hamster melanoma. *Cancer Res.* 26:1420–1424, 1966.

Emami, B. et al. Histopathological study on the effects of hyperthermia on microvasculature. *Int. J. Radiat. Oncol. Biol. Phys.* 7:343–348, 1981.

Endrich, B. et al. Quantitative studies of microcirculatory function in malignant tissue: influence of temperature on microvascular hemodynamics during the early growth of the BA 1112 rat sarcoma. *Int. J. Radiat. Oncol. Biol. Phys.* 5:2021–2030, 1979a.

Endrich, B. et al. Hemodynamic characteristics in microcirculatory blood channels during early tumor growth. *Cancer Res.* 39:17–23, 1979b.

Falk, P. Patterns of vasculature in two pairs of related fibrosarcomas in the rat and their relation to tumour responses to single large doses of radiation. *Eur. J. Cancer* 14:237–250, 1978.

Field, S. B.; Hume, S. P.; and Law, M. P. The response of tissues to heat alone or in combination with radiation. *Paper read at the Sixth International Congress of Radiation Research*, May 1979, Tokyo, Japan. eds. S. Okada, et al.

Folkman, J. et al. Isolation of a tumor factor responsible for angiogenesis. *J. Exp. Med.* 133:275–288, 1971.

Fujiwara, K. Effects of irradiation on the fine vasculatures of normal and malignant tissues. *Jpn. J. Cancer Clin.* 20:52–54, 1974.

Gabbiani, G., and Badonnel, M. C. Early changes of endothelial clefts after thermal injury. *Microvasc. Res.* 10:65–75, 1975.

Gamill, S. L. et al. Roentgenology-pathology correlative study of neovascularity. *AJR* 126:376–384, 1976.

Gerweck, L. E. Modification of cell lethality at elevated temperatures: the pH effect. *Radiat. Res.* 70:224–235, 1977.

Goodall, C. M.; Sanders, A. G.; and Shubik, P. Studies of vascular patterns in living tumors with a transparent chamber inserted in hamster cheek pouch. *J. Natl. Cancer Inst.* 35:497–521, 1965.

Gullino, P. M. Extracellular compartments of solid tumors. In *Cancer 3—a comparative treatise*, ed. F. F. Becker. New York: Plenum Press, 1975.

Gullino, P. M. et al. Modification of the acid-base status of the internal milieu of tumors. *J. Natl. Cancer Inst.* 34:857–869, 1965.

Gullino, P. M., and Grantham, F. H. Studies on the exchange of fluids between host and tumor. II. The blood flow of hepatomas and other tumors in rats and mice. *J. Natl. Cancer Inst.* 27:1465–1491, 1961.

Gullino, P. M., and Grantham, F. H. Studies on the exchange of fluids between host and tumor III. Regulation of blood flow in hepatomas and other rat tumors. *J. Natl. Cancer Inst.* 28:211–229, 1962.

Gullino, P. M.; Yi, P.-N.; and Grantham, F. H. Relationship between temperature and blood supply or consumption of oxygen and glucose by rat mammary carcinomas. *J. Natl. Cancer Inst.* 60:835–847, 1978.

Gump, F. E., and White, R. L. Determination of regional tumor blood flow by Krypton-85. *Cancer* 21:871–875, 1968.

Guy, A. W. Physical aspects of the electromagnetic heating of tissue volume. In *Proceedings of the International Symposium on Cancer Therapy by Hyperthermia and Radiation*, eds. M. J. Wizenberg and J. E. Robinson. Chicago: American College of Radiology, 1976, p. 179.

Hahn, G. M. Metabolic aspects of the role of hyperthermia in mammalian cell inactivation and their possible relevance to cancer treatment. *Cancer Res.* 34:3117–3123, 1974.

Hahn, G. M. Potential for therapy of drugs and hyperthermia. *Cancer Res.* 39:2264–2268, 1979.

Hilmas, D. E., and Gillette, E. L. Morphometric analyses of the microvasculature or tumors during growth and after x-irradiation. *Cancer* 33:103–110, 1974.

Hornback, N. B. et al. Preliminary clinical results of combined 433 megahertz microwave therapy and radiation therapy on patients with advanced cancer. *Cancer* 40:2854–2863, 1977.

Hubler, W. R., Jr., and Wolf, J. E., Jr. Melanoma. Tumor angiogenesis and human neoplasia. *Cancer* 38:187–192, 1976.

Hughes, W. L. et al. Tissue perfusion rate determined from the decay of oxygen-15 activity after photon activation in situ. *Science* 204:1215–1217, 1979.

Hurley, J. V.; Ham, K. N.; and Ryan, G. B. The mechanism of the delayed prolonged phase of increased vascular permeability in mild thermal injury in the rat. *Journal of Pathology and Bacteriology* 94:1–12, 1967.

Jirtle, R., and Clifton, K. H. On carcinoma growth and vascular supply: a study of mouse mammary tumor strain MTG-B (35875). *Soc. Exp. Bio. Med.* 138:267–269, 1971.

Jirtle, R., and Clifton K. H. Effect of preirradiation of the tumor bed on the relative vascular space of mouse gastric adenocarcinoma 328 and mammary adenocarcinoma CA755. *Cancer Res.* 33:764–768, 1973.

Jirtle, R.; Clifton, K. H.; and Rankin, J. H. Measurement of mammary tumor blood flow in unanesthetized rats. *J. Natl. Cancer Inst.* 60:881–886, 1978.

Kallman, R. F.; DeNardo, G. L.; and Stasch, M. J. Blood flow in irradiated mouse sarcoma as determined by the clearance of Xenon-133. *Cancer Res.* 32:483–490, 1972.

Kang, M. S.; Song, C. W.; and Levitt, S. H. Role of vascular function in response of tumors in vivo to hyperthermia. *Cancer Res.* 40:1130–1135, 1980.

Kim, J. H., and Hahn, E. W. Clinical and biological studies of localized hyperthermia. *Cancer Res.* 39:2258–2261, 1979.

Kligerman, M. M., and Henel, D. K. Some aspects of the microcirculation of a transplantable experimental tumor. *Radiology* 76:810–817, 1961.

Kolstad, P. The development of the vascular bed in tumours as seen in squamous-cell carcinoma of the cervix uteri. *Br. J. Radiol.* 38:216–223, 1965.

Krylova, N. V. Endothelium characteristics of blood vessels in neoplasms. *Bibl. Anat.* 12:497–503, 1973.

Lassen, N. A.; Lindbjerg, J.; and Munch, O. Measurement of blood flow through skeletal muscle by intramuscular injection of Xenon-133. *Lancet* 1:686–689, 1964.

LeServe, A. W., and Hellmann, K. Metastases and the normalization of tumour blood vessels by ICRG 159: a new type of drug action. *Br. Med. J.* 1:597–601, 1972.

LeVeen, H. H. et al. Tumor eradication by radiofrequency therapy. Response in 21 patients. *JAMA* 235:2198–2200, 1976.

Lindgren, A. G. H. The vascular supply of tumours with special reference to the capillary angio architecture. *Acta Pathol. Microbiol. Scand.* 22:493–522, 1945.

Lurie, A. G.; Nintzel, B. B.; and Rippey, R. M. Vascular volume and perfusion in hamster cheek pouch carcinomas and tumor-bearing cheek pouches. *Cancer Res.* 37:3484–3489, 1977.

Majno, G.; Palade, G. E.; and Schoefl, G. I. Studies on Inflammation. II. The site of action of histamine and serotonin along the vascular tree: a topographic study. *J. Biophy. and Biochem. Cytology* 2:607–626, 1961.

Mäntyla, M.; Kuikka, J.; and Rekonen, A. Regional blood flow in human tumours with special reference to the effect of radiotherapy. *Br. J. Radiol.* 49:335–338, 1976.

Marmor, J. B. Interactions of hyperthermia and chemotherapy in animals. *Cancer Res.* 39:2269–2276, 1979.

Marmor, J. B.; Hilerio, F. J.; and Hahn, G. M. Tumor eradication and cell survival after localized hyperthermia induced by ultrasound. *Cancer Res.* 39:2166–2171, 1979.

Milne, E. N. C. et al. Histologic type-specific vascular patterns in rat tumors. *Cancer* 20:1635–1646, 1967.

Muckle, D. S., and Dickson, J. A. The selective inhibitory effect of hyperthermia on the metabolism and growth of malignant cells. *Br. J. Cancer* 25:771–778, 1971.

Myers, R., and Field, S. B. The response of the rat tail to combined heat and x-rays. *Br. J. Radiol.* 50:581–586, 1977.

Nelson, A. J. M., and Holt, J. A. G. The problems of clinical hyperthermia. *Australas. Radiol.* 21:21–30, 1977.

Neutze, J. M.; Wyler, R.; and Rudolph, A. M. Use of radioactive microspheres to assess distribution of cardiac output in rabbits. *Am. J. Physiol.* 215:486, 1968.

Nishiyama, K.; Nishiyama, A.; and Frohlich, E. D. Regional blood flow in normotensive and spontaneously hypertensive rats. *J. Physiol.* 3:691–698, 1977.

Overgaard, J. Influence of extracellular pH on the viability and morphology of tumor cells exposed to hyperthermia. *J. Natl. Cancer Inst.* 56:1243–1250, 1976.

Overgaard, J. Effect of hyperthermia on malignant cells in vivo. *Cancer* 39:2637–2646, 1977.

Patterson, J., and Strang, R. The role of blood flow in hyperthermia. *Int. J. Radiat. Oncol. Biol. Phys.* 5:235–241, 1979.

Peterson, H. I., ed. *Tumor blood circulation: angiogenesis, vascular morphology and blood flow of experimental and human tumors.* Boca Raton, Fla.: CRC Press, 1979.

Phibbs, R. H., and Dong, L. Nonuniform distribution of microspheres in blood flowing through a medium-size artery. *Can. J. Physiol. Pharmacol.* 48:415–421, 1970.

Raaphorst, G. P. et al. Intrinsic differences in heat and/or x-ray sensitivity of seven mammalian cell lines cultured and treated under identical conditions. *Cancer Res.* 39:396–401, 1979.

Reinhold, H. S. In vivo observations of tumor blood flow. In *Tumor blood circulation*, ed. H. I. Peterson. Boca Raton, Fla.: CRC Press, 1979.

Reinhold, H. S.; Blachiewicz, B.; and Berg-Blok, A. Decrease in tumor microcirculation during hyperthermia. In *Proceedings of the Second International Symposium on Cancer Therapy by Hyperthermia and Radiation*, eds. C. Streffer et al. Baltimore and Munich: Urban & Schwarzenberg, 1978, pp. 231–232.

Rockoff, S. D. et al. Variable response of tumor vessels to intra-arterial epinephrine. *Invest. Radiol.* 1:205–213, 1966.

Rubin, P., and Casarett, G. Microcirculation of tumors. Part I: anatomy, function and necrosis. *Clin. Radiol.* 17:220–229, 1966.

Sapareto, S. A.; Hopwood, L. E.; and Dewey, W. C. Combined effect of X-irradiation and hyperthermia on CHO cells for various temperatures and order of application. *Radiat. Res.* 73:221–233, 1978.

Sapirstein, L. A. Regional blood flow by fractional distribution of indicators. *Am. J. Physiol.* 193:161–168, 1958.

Schmid-Shöbein, H.; Weiss, J.; and Ludwig, H. A simple method for measuring red cell deformability in models of the microcirculation. *Blut. Band.* 26:369–379, 1973.

Sevitt, S. Local vascular changes in burned skin. Discussion: pathological sequelae of burns. *Proceedings of the Royal Society of Medicine* 47:225–228, 1954.

Shea, S. M. et al. Microvascular ultrastructure in thermal injury: a reconsideration of the role of mediators. *Microvasc. Res.* 5:87–96, 1973.

Song, C. W. Effect of hyperthermia on vascular functions of normal tissues and experimental tumors: brief communication. *J. Natl. Cancer Inst.* 60:711–713, 1978.

Song, C. W. et al. Effect of hyperthermia on vascular function in normal and neoplastic tissues. *Ann. NY Acad. Sci.* 355:35–47, 1980a.

Song, C. W. et al. Vascular damage and delayed cell death in tumors after hyperthermia. *Br. J. Cancer* 41:309–312, 1980b.

Song, C. W. et al. Effect of hyperthermia on vascular function, pH and cell survival. *Radiology* 137:795–803, 1980c.

Song, C. W., and Levitt, S. H. Quantitative study of vascularity of Walker Carcinoma 256. *Cancer Res.* 31:587–589, 1971.

Song, C. W., and Levitt, S. H. Vascular changes in Walker 256 carcinoma of rats following X-irradiation. *Radiology* 100:397–407, 1971.

Song, C. W.; Guertin, D. P.; and Levitt, S. H. Potentiation of cytotoxicity of 5-thio-D-glucose on hypoxic cells by hyperthermia. *Int. J. Radiat. Oncol. Biol. Phys.* 5:965–970, 1979.

Song, C. W.; Rhee, J. G.; and Levitt, S. H. Blood flow in normal tissues and tumors during hyperthermia. *J. Natl. Cancer Inst.* 64:119–124, 1980.

Stehlin, J. S., Jr. et al. Results of eleven years' experiences with heated perfusion for melanoma of the extremities. *Cancer Res.* 39:2255–2257, 1979.

Storey, R. H.; Wish, L.; and Furth, J. Organ erythrocyte and plasma volume of tumor-bearing mice. *Cancer Res.* 11:943–947, 1951.

Storm, A. K. et al. Normal tissue and solid tumor effects of hyperthermia in animal models and clinical trials. *Cancer Res.* 39:2245–2251, 1979.

Straw, J. A. et al. Distribution of anticancer agents in spontaneous animal tumors. I. Regional blood flow and methotrexate distribution in canine lymphosarcoma. *J. Natl. Cancer Inst.* 52:1327–1331, 1974.

Suddith, R. L. et al. In vitro demonstration of an endothelial proliferative factor produced by neural cell lines. *Science* 190:682, 1975.

Sugaar, S., and LeVeen, H. H. A histopathologic study on the effects of radiofrequency thermotherapy on malignant tumors of the lung. *Cancer* 43:767–783, 1979.

Suit, H. D. Hyperthermic effects on animal tissues. *Radiology* 123:483–487, 1977.

Sutton, C. H. Discussion. In *Tumor blood circulation*, ed. H. I. Peterson. Boca Raton, Fla.: CRC Press, 1979.

Tannock, I. F., and Steel, G. G. Quantitative techniques for study of the anatomy and function of small blood vessels in tumors. *J. Natl. Cancer Inst.* 42:771–782, 1969.

Thomlinson, R. H., and Gray, L. H. The histological structure of some human lung cancers and the possible implications for radiotherapy. *Br. J. Cancer* 9:539–549, 1955.

Tsuchiya, M.; Walsh, G. M.; and Frohlich, E. D. Systemic hemodynamic effects of microspheres in conscious rats. *Am. J. Physiol.* 233:H617–H621, 1977.

Utley, J. et al. Total and regional myocardial blood flow measurements with 25 micron, 15 micron, 9 micron, and filtered 1-10 micron diameter microspheres and antipyrine in dogs and sheep. *Circ. Res.* 34:391–405, 1974.

Vaupel, P. Oxygen supply to malignant tumors. In *Tumor blood circulation*, ed. H. I. Peterson. Boca Raton, Fla.: CRC Press, 1979.

Vogel, A. W. Intratumoral vascular changes with increased size of a mammary adenocarcinoma: new method and results. *J. Natl. Cancer Inst.* 34:571–578, 1965.

Von Ardenne, M. M.; Böhme, G.; and Kell, E. On the optimization of local hyperthermy in tumors based on a new radiofrequency procedure. *J. Cancer Res. Clin. Oncol.* 94:163–184, 1979.

Von Ardenne, M. et al. Histological proof for selective stop of microcirculation in tumor tissue at pH 6.1 and 41°C. *Naturwissenschaften* 66:59–60, 1979.

Warren, B. A. The vascular morphology of tumors. In *Tumor blood circulation*, ed. H. I. Peterson. Boca Raton, Fla.: CRC Press, 1979.

Warren, B. A., and Shubik, P. The growth of the blood supply to melanoma transplants in the hamster cheek pouch. *Lab. Invest.* 15:464, 1966.

Whalen, W. J., and Nair, P. Skeletal muscle PO_2: effect of inhaled and topically applied O_2 and CO_2. *Am. J. Physiol.* 218:937–980, 1970.

Willis, R. A. *Pathology of tumors*, 3rd edition. London: Butterworth, 1960.

Zanelli, G. D., and Fowler, J. F. The measurement of blood perfusion in experimental tumors by uptake of ^{86}Rb. *Cancer Res.* 34:1451–1456, 1974.

CHAPTER 9
Physiology and Morphology of Tumor Microcirculation in Hyperthermia

Haim I. Bicher
Fred W. Hetzel
Taljit S. Sandhu

Introduction
Background
Research Methodology
Analysis of Data
Discussion and Future Prospects

INTRODUCTION

The clinical use of hyperthermia as a cancer treatment modality is attracting considerable attention. Why does it work? Many in vitro studies have been conducted to determine if cancer cells are inherently more susceptible to hyperthermic damage than normal cells; the results are inconclusive at best. It is clear, however, that hyperthermia is effective in tumor eradication in vivo. It may be, therefore, the in vivo state, the morphology and physiology of the tumor *in its host*, that conveys to tumors a differential sensitivity to heat.

BACKGROUND

Tumors, as they grow, develop a vasculature that is quite different from that found in normal tissues. It was postulated by Folkman (1971) that the initial tumor vascular bed is derived from host venous vessels in response to tumor-cell-produced angiogenic factors. The resulting vascular network consists of thin-walled capillary-like vessels, usually lacking the degree of connective tissue support and regulatory mechanisms found in normal tissue (Eddy 1980). Their morphology and physiology also are abnormal, including dilated and tortuous veins, sinusoids, and arteriovenous shunting (Eddy 1980; Emami et al. 1980).

It is this vascular network that must support the maintenance and growth of the tumor by providing nutrients and oxygen while clearing metabolites. As demonstrated mathematically by Krogh (1918) and by Thomlinson and Gray (1955), this system becomes ineffective when intercapillary distances exceed 300 to 400 μ. At this separation, areas of anoxia and subsequent necrosis will develop because of the inability of oxygen (as well as of other substances) to diffuse.

The initial work of Thomlinson and Gray (1955) demonstrating necrotic regions in tumors at distances greater than 150 to 200 μ from a capillary and numerous reports of large hypoxic fractions in tumors (Bicher et al. 1980; Kallman 1972) confirm that the microvascular network in tumors is poorly developed and organized when compared to that of normal tissues. It is this physiologic difference between tumor and normal tissue that may provide a therapeutic advantage for a modality such as hyperthermia.

Several general physiologic responses to hyperthermia have been commonly observed. These include increased cellular metabolic activity in the heated region and flushing of the skin overlying the heated area (indicative of increased perfusion of the skin). The specific responses in the microenvironment of different organs and tissues (malignant and nonmalignant) to modifications in temperature have been studied recently in more detail (Bicher, Mitagvaria, and Hetzel 1980; Reinhold, Blachiewicz, and Berg-Blok 1979; Song 1978). Examination of several recent publications demonstrates that active investigations of the physiologic phenomena induced by hyperthermia are in progress (Bicher, Mitagvaria, and Hetzel 1980; Eddy and Casarett 1973; Reinhold, Blachiewicz, and Berg-Blok 1979; Storm et al. 1979; Streffer et al. 1978; Von Ardenne and Reitnauer 1978). The studies by Eddy (1980) and by Reinhold, Blachiewicz, and Berg-Blok (1978), employing "chamber systems" of different types, have shown changes in the microvascular network as a function of temperature and exposure time. The apparent sensitivity of the neovasculature is a critical observation by these investigators. Similar results in different test systems also have been observed by Emami and colleagues (1980) and by Dewhirst and Ozimek (1980).

Knowledge of the effect of hyperthermia on tumor and normal tissue blood flow and the subsequent effects on oxygen tension (pO_2) and pH is important not only for the effect of hyperthermia on hypoxic cells at the time of radiation, but also for differential tumor heating.

There is considerable evidence from plethysmography that elevation of normal tissue temperature to 41°C is accompanied by a considerable increase in blood flow (Lehman 1971). Cater and Silver (1960) reported on changes in tumor oxygen tension with hyperthermia but did not record changes in tumor temperature. They concluded that diathermy had not increased the oxygen tension in the tumor but, on the contrary, caused a decrease.

Bicher and co-workers (1980), studying a mouse leg tumor system, reported that tumor blood flow increased at temperatures up to 41°C and then progressively decreased. The oxygen tension in the tumor, as measured with a platinum electrode, generally followed the changes in tumor blood flow, but the exact tumor temperature at which the oxygen tension decreased was not determined. Similar changes in brain tissue oxygenation were reported earlier by the same author (1978).

Another possible explanation for the different response to heat of tumor tissue from that of normal tissue could be provided by a better understanding of the physiologic microenvironment of the tumor as it differs from the microenvironment of normal tissue. For example, several studies indicate that the pH of interstitial fluid in human and rodent solid tumors is 0.3 to 0.5 units lower than the normal tissue pH of about 7.4 (Eden and Kahler 1955; Gullino, Grantham, and Smith 1965; Meyer, Kammerling, and Amtman 1948; Naselund and Seenson 1953).

The reduced pH of tumor fluid may be due in part to elevated lactic acid production resulting from poor tumor vascularity and reduced oxygen tension. Reports also indicate that tumor cells produce lactic acid at an elevated rate under oxygenated conditions (Von Ardenne and Reitnauer 1978).

Gerweck (1977) has demonstrated that the lethal response of cells to hyperthermia was markedly enhanced by reduced pH at 42°C. In addition, chronically hypoxic cells are more sensitive to hyperthermic damage by at least a factor of 5, even if pH is controlled (Gerweck, Nygaard, and Burlett 1970). This is not the case for cells made acutely hypoxic, which are equally as sensitive as oxygenated cells at normal pH (Gerweck, Nygaard, and Burlett 1970). Reduced pH also has been shown to affect the transplantability of tumor cells heated in vitro (Overgaard 1976). Gerweck (1977) also has shown that there is a variable influence of pH according to temperature, and that there is a critical point to the increased lethality of heat below pH 6.7. The influence of some other parameters that could change the microenvironment of the tumor during heat production has just been mentioned. Such changes in glycolysis, respiration rate, and lactic acid production probably will influence also the response of the tissue by modifying the cellular environment.

Several other parameters may change and subsequently influence the response of cells or tissues to supranormal temperatures. Paramount among those are the vascular changes, blood flow responses, and the net result of these on tissue oxygenation that may change the effect of hyperthermia or radiation therapy when used in combination when treating tumors. Several authorities have reported an increase in blood flow during hyperthermia. Others have reported an increase in oxygen consumption during elevated local temperatures. The net result could, therefore, be either increase or decrease in local oxygen tension, depending on temperature and duration of heating.

England, Hallbrook, and Ling (1974) and Sutton (1976) have demonstrated that during hyperthermia there is an increase in the local blood flow in the tumor region and also in the organ hosting the tumor. In recent experiments, Bicher and co-workers (1980) have been able to demonstrate, measuring the local oxygen levels as well as the local blood flow to the tumor, that a net result of this process is an increase in the local oxygen tension. This increase is quite remarkable and leads to abolishing the local autoregulation processes that usually tend to keep oxygen levels constant in several organs, as well as in tumors (Bicher, Hetzel, and D'Agostino 1977; Cater and Silver 1960).

Recent results obtained by several research groups provide a considerable amount of information on the physiologic responses of tumors to hyperthermia; this chapter examines the methods employed and the results obtained.

RESEARCH METHODOLOGY

Tumor Systems

Measurements were performed on several different tumors and on two normal tissues as described here.

C$_3$H Mouse Mammary Adenocarcinoma

In situ studies were carried out in fourth generation transplants of C$_3$H mammary adenocarcinoma implanted in the hind legs of C$_3$H SED-BH mice (Bicher et al. 1980). The tumors were obtained from the Radiobiology Division, Massachusetts General Hospital (Suit, Sedlacek, and Faquez 1968). This is a syngeneic implantable tumor that is maintained using solid tissue transplants that are inoculated subcutaneously into recipient mice. Tumors used for experimentation were approximately 10 mm in diameter. The mice were anesthetized (Buelke-Sam et al. 1978) during microelectrode introduction with a combination of Ketamine™ 40µg/kg IM and Thorazine™ 50 mg/kg IM.

Human Tumors

Determinations were made in subcutaneous metastases in a group of 15 patients (Bicher et al. 1980). Tumors represented different histologies and locations but were grouped together as the responses were homogeneous. There were four melanomas, six chest wall recurrences of mammary adenocarcinomas, and five peripheral metastases of squamous cell carcinoma of the lung. The patients were not anesthetized.

SCK Tumor

This fast growing mammary adenocarcinoma arose spontaneously in a female A/J mouse in 1974 (Song 1978). This tumor has been maintained by alternate passage in vitro and in vivo. The investigators routinely subcultured the tumor cells in vitro several times before injecting back into animals. They used twenty-seventh to thirty-second generations for the study (i.e., cells that were cultured in vitro and grew in vivo 27 to 32 times). Cells obtained from in vivo tumors grew well in RPMI 1640 medium with 10% fetal calf serum (FCS).

For in vivo studies, the exponentially growing cells in culture were treated with 0.25% trypsin for 15 minutes, washed three times with RPMI 1640 medium, and the cells able to exclude trypan blue were counted. About 5×10^4 cells were injected subcutaneously into the right legs of A/J mice. When the tumors grew to 7 to 9 mm in diameter, which took about 8 to 10 days after the transplantation, experiments were conducted to study the effect of hyperthermia on the vascular volume, intratumor temperature, pH, and to analyze the effect of hyperthermia on the cell survival in vivo (Song 1978).

Squamous Cell Carcinoma

For studies of the tumor microvascular response to hyperthermia, the squamous cell carcinoma of the Syrian hamster grown in the transparent cheek pouch chamber has been employed (Song 1978; Eddy 1980). A specially designed, compartmented chamber allows control of the heating within ± 0.2°C, and the desired temperatures are achieved in the tissue within one minute. The use of this system results in the formation of a transparent "sandwich tumor," which enables repeated observations, and recording of events occurring during and after heating.

The initial studies have been qualitative in nature and are evaluated in terms of spatial and temporal changes in tumor vessel caliber and blood circulation, pathologic alterations in tumor vasculature, and recovery of circulation after return to normal temperatures. At present, observations have been made of changes in the tumor vasculature before, during, and after heating at either 41°C or 45°C. Tumor and pouch tissue were taken for histologic examination at 15 minutes or at 24 hours after the end of the heating period.

Normal Tissues

The normal tissues studied in these experiments were C$_3$H mouse muscle and cat brain.

C$_3$H Mouse Muscle

Employing the same animal system as in the study of C$_3$H mouse mammary adenocarcinoma, measurements were made in the muscle tissue of the hind leg (Bicher et al. 1980). Determinations were obtained in controls and in animals bearing an implanted tumor in the opposite leg. Since no difference was observed, no distinction is made in the results.

Cat Brain

All studies on brain were performed on cats. In each case the animal was anesthetized with Nembutal (30 mg/kg) prior to and during the procedure. After opening the scalp, a small opening (5 mm) was made in the skull with a dental hand drill, and the dura was carefully opened. Throughout the entire procedure, including microelectrode introduction and measurements, the opening was kept moist with isotonic saline.

Physiologic Determinations

Oxygen Ultra-microelectrodes

The oxygen *ultra*-microelectrodes used were of the "gold-in-glass" type described by Cater, Silver, and Wilson (1960). They were made by pulling a glass tube (KG-33, ID 1.0 mm, OD 2.0 mm, Garner Glass Co., Claremont, California), encasing a 20 μ gold wire (Sigmund Cohn Corp., Mt. Vernon, New York) in a David Kopf Model 700C vertical pipette puller. The exposed gold tip is about 10 μ in diameter, and is coated with a Rhoplex (Rhom Haas, Philadelphia, Pennsylvania) membrane (Bicher 1978). This probe is used as an external reference oxygen microelectrode (Bicher et al. 1980).

The performance of oxygen ultra-microelectrodes has proved to be in good agreement with what could be expected according to the theories of polarographic recording of partial pressures of oxygen in tissue. The adoption of industrial technology ensures good reproducibility between different electrodes in different experiments, and turns a hitherto complicated and sophisticated technique into a simple determination, potentially useful in a variety of physiologic and in vitro experiments.

The characteristic "quasi-plateau" shown by the microelectrodes in the current-voltage polarogram are similar to those described by Davies and Brink (1942) when using their open tip electrodes, or by Cater (1964) when testing flush-ended oxygen microelectrodes. This seems to be characteristic for non-recessed electrodes, and adequate coating of the open tip makes the voltage difference across the plateau very small, thus ensuring greater accuracy in pO_2 determination.

The electromotive force (EMF) to activate the oxygen cathode and the current output from the cathode are provided and measured with a Transidyne Model 1210 microsensor amplifier (Transidyne General Corp., Ann Arbor, Michigan) and recorded using an appropriate chart recorder.

Electrodes are calibrated as described by Silver (1963) in buffered saline solutions of known pO_2 values. The electrodes are conditioned by placing them in buffered saline and applying 0.8 V potential for two hours. After this treatment they are usually very stable. Stirring of the solution by a magnetic rotor does not change the oxygen current. The current reading at zero oxygen tension is very low (residual current), and the response of the microelectrode to changes of oxygen tension is very rapid.

The current voltage polarogram of these electrodes shows a quasi-plateau between 0.3 and 0.7 V. The plateau, however, is not absolute, and small differences can be detected at the extreme values. This type of plateau can be expected when using open tip electrodes (Davies and Brink 1942). In these experiments a polarizing voltage of 0.6 V has been used. The relation between current output and oxygen tension is linear, the current per mm Hg being of the order of magnitude of 0.6×10^{-11}A.

In human experiments, a platinum-iridium Teflon™-coated wire, 120 μ in diameter, was used as the oxygen electrode. Although the calibration was not as reliable to determine actual tissue oxygen tension (TpO_2) values, it was found in determining transients (response to oxygen breathing or hyperthermia) that the obtained values correlated well with those obtained using microelectrodes. The temperature artifact of both types of oxygen electrodes was determined and found to be 5% per degree Celsius. All results were corrected by taking this artifact into account.

Alternative Microelectrodes

Other microelectrodes are now available to measure K^+, Cl^-, pH, and so on. An antimony electrode has long been used in the measurement of pH, but investigators have found that the electrode potential is linear, with increasing pH only to pH 7.0. Consequently, these electrodes require frequent calibration. Bicher and Ohki (1972) successfully used an antimony pH microelectrode. Basically it consists of a very finely drawn (tip diameter 1 μ or less) glass micropipette that has a thin film overcoating of antimony. It is then coated with two layers of insulating epoxy resin, leaving an exposed tip of approximately 2 μ in length. A microcalomel electrode inserted into the same cell or tissue or solution serves as a reference. Results obtained in the squid giant axon were very satisfactory (Bicher and Ohki 1972).

Designs for glass pH microelectrodes have been developed, most notably by Hinke (1973) and Thomas (1970). The Thomas electrode consists of a Pyrex® glass micropipette drawn to a fine point, into which is inserted and fused a second pipette made of pH-sensitive glass. The tip of the pH-sensitive glass pipette is recessed in the tip of the Pyrex® glass pipette, and the electrode is filled with KCl electrolyte. The Hinke type electrode also consists of a pH-sensitive glass micropipette inside a Pyrex® glass pipette; the major difference is that the tip of the pH-sensitive micropipette is not recessed but extrudes from the Pyrex® glass pipette. A silver/silver-chloride electrode is inserted into the electrode stem, which is filled with 0.1 N HCl. The Hinke microelectrode then has an exposed tip, and its response time is instantaneous. This is an advantage over the Thomas microelectrode, in which the recessed tip may cause a response time of up to several minutes.

The pH of SCK tumor of mice was measured during and after hyperthermia (Song, Kang, and Rhee 1980). The pH electrode used was a commercially available needle electrode with a diameter of 0.8 mm (Micro Electrode, Inc., Londonderry, New Hampshire). Reference electrodes with a diameter of 0.1 mm were constructed in their laboratory. The pH electrode and reference electrode were carefully inserted into the center of the tumor. In some cases, a thermocouple also was inserted into the same tumor to monitor the relationship between the change in pH and temperature. The change in pH was monitored using a stripchart recorder. The electrode and thermocouple were immobilized by firmly attaching them to a stand placed above the tumor.

Microflow

Flow in microareas of tumor tissue was determined using the hydrogen diffusion method as described by Stosseck, Lubbers, and Cottin (1974). The method is based on the polarographic determination of the amount of hydrogen gas reaching a platinum electrode (hydrogen detector) from a hydrogen generating electrode located at a fixed distance. The amount of hydrogen reaching the reading electrode depends on the generation and diffusion rates, which are constant, and the blood flow clearance of hydrogen, which can be thus determined. In the present experiments, two platinum in Teflon® 100-μ wires placed 100 μ apart were used. The reading device was applied to the surface of the tumor. In the present experiments only relative changes in the rate of blood flow were determined. This method was used in the experiments on mouse tumors in situ (see earlier section this chapter on C_3H mouse mammary adenocarcinoma).

Histopathology Studies

To better understand the results of preclinical blood flow studies, the pathologic changes in the tumor microvasculature (i.e., arterioles, capillaries, and venules) caused by single applications of local hyperthermia have been investigated (Emami et al. 1980). The tumors studied were undifferentiated rhabdomyosarcoma (BA-1112) on the scalp of WAG/RiJ rats. In the present study, four- to five-week-old rats were used. Suspensions of 5000 to 20,000 tumor cells were prepared aseptically and inoculated in 0.05 ml into the subcutaneous tissue of the scalp between the ears. Tumors were used in experiments when they reached a mean base diameter of 10 to 12 mm.

The tumors were heated with 1.25-MHz electric currents. A temporary change, however, in the type of control potentiometer employed (on the signal generator) made it somewhat more difficult to hold the tumor temperature constant during the heating interval, resulting in variations about the respective, desired temperatures of up to 0.5°C.

Tumor-bearing animals were assigned to four different groups: a control group (no tumor heating), and groups receiving tumor hyperthermia at 40°C for 40 minutes, 42.5°C for 40 minutes, and at 44.5°C for 40 minutes. Within each group receiving tumor hyperthermia, animals were sacrificed either immediately after heating, three hours postheating, or 24 hours after local tumor heating. After sacrifice of animals in the respective groups, the tumors were removed and the tissue processed and subjected to detailed pathologic study.

Temperature Determinations

Tumor and mouse core temperatures were recorded using copper-constantan microthermocouples (tip diameter 30 to 100 μ) inserted into the tumoral tissue in close proximity to the oxygen microelectrode or in the animal's rectum for core measurements (Bicher et al. 1980). An Omega Engineering model 250 Digital Voltmeter amplifier was used as a link between the microthermocouple and the polygraph. Microwaves of a frequency 2450 MHz were produced by a Raytheon Magnetron™ and delivered through a specially designed 5-cm diameter circularly polarized applicator loaded with low loss dielectric material having a dielectric constant of 6 (Sandhu, Kowal, and Johnson 1978).

Heating

The mice were lightly anesthetized with pentobarbital sodium (0.04 mg/g), and the rats were anesthetized with Inactin® (sodium salt of ethyl-(1-methyl-propyl)-malonyl-thio-urea, BYK Gulden Konstanze, West Germany) (0.08 mg/g) (Bicher 1977). The hair of the legs with or without tumors was clipped prior to the heating. A Plexiglas® plate was placed over a preheated water bath (Thermomix™ 1480), and the animals were laid on the plate. The legs were immersed in water through the holes in the plate. The legs of mice were immobilized by anchoring a toe with thread to the Plexiglas® support adjacent to each hole. A 1-cm thick styrofoam sheet was placed between the animals and the Plexiglas® to minimize the rise in body temperature (Song, Kang, and Rhee 1980).

ANALYSIS OF DATA

The first parameter to be analyzed was tissue oxygenation. This parameter, as well as being of therapeutic and radiobiologic interest, is a strong indication of physiologic response in a region. For microelectrode studies in mouse tumors, figure 9.1 demonstrates a rise in TpO_2 that parallels the application of the microwaves and closely follows changes in tissue temperature. The response seen in the tissue was very fast, with TpO_2 increasing shortly after the rise in temperature and then decreasing as the tumor cools. This effect was present when heating was carried out up to 41°C. At higher temperatures (fig. 9.2) the same initial increase in TpO_2 was seen as the tissue temperature rose, but this was followed by a decrease to very low levels as the temperature was held constant at 46°C.

Similar effects were seen in both normal tissues studied, with one major difference: the temperature at which the TpO_2 begins to fall, or the "breaking point." In figure 9.3, the rise in oxygen tension in brain occurred up to 43°C, while it declined sharply at higher temperatures. A composite of the results obtained for brain, muscle, and tumor tissues is presented in figure 9.4. It is evident that the breaking point (in the pO_2 vs temperature curve) exists for each tissue and, at least for the tissues studied in these experiments, that the breaking point temperature is significantly lower for tumor.

Figure 9.5 shows an example of the effect of hyperthermia on local blood flow in mice. It is clear that blood flow increases significantly as the temperature increases. This figure illustrates several interesting details that must be examined carefully. First, since the tumor was located in the distal segment of the hind leg, the resting temperature of the tumor tissue near the surface (0.5 cm) was quite low. Second, the elevation of temperature to less than 40°C resulted in a dramatic increase in blood flow, which even necessitated instrument recalibration after five minutes. Finally, even at these relatively low temperatures of short duration, a pH drop of 0.3 units was evident.

The results presented in figure 9.6 were obtained in a series of experiments similar to that in figure 9.5. Values of pO_2 and relative blood flow were obtained at several temperatures in a series of animals. It is clear that a strong correlation exists between TpO_2 and blood flow as a function of temperature up to 45°C.

EFFECT OF HYPERTHERMIA ON T_pO_2

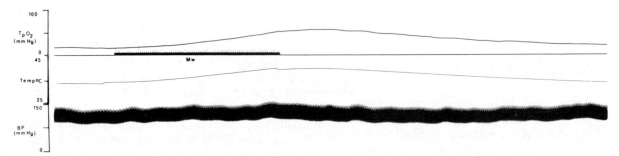

Figure 9.1. The effect of microwave hyperthermia on TpO_2 in mouse tumor. The *upper channel* records TpO_2 and the lower temperature in °C. Microwaves are on when indicated by the *timing pulse*. One minute is indicated by the space between two large peaks on the microwave tracing. Cooling and declining pO_2 can be seen when the microwaves are turned off.

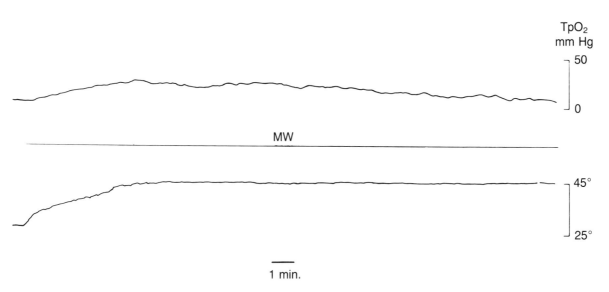

Figure 9.2. In this figure microwaves (*MW*) are indicated by the *solid center line*. Note that the temperature is constant at 46°C except for the first five minutes. The drop in TpO_2 is clear after only 10 minutes at 46°C.

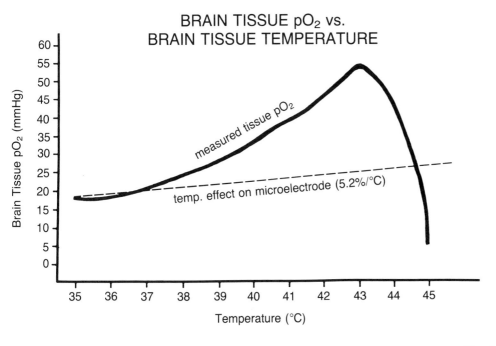

Figure 9.3. Brain tissue pO_2 is plotted as a function of brain tissue temperature in this composite figure. A breaking point is clearly seen at approximately 43°C. The temperature artifact induced in the microelectrode (5.2%/°C) is plotted on this graph to show clearly that the observed curve is not artifactual in nature.

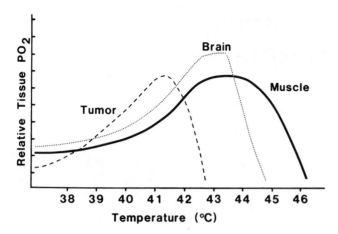

Figure 9.4. In this figure, relative tissue pO₂ is plotted as a function of temperature similar to the plot in figure 9.3. In this case the curves obtained for tumor, brain, and muscle are superimposed so that the differential effect and the different position of the breaking point can be seen more clearly.

In addition to blood flow measurements and tissue oxygen levels, tissue pH is also a good indicator of a tissue's physiologic status and response.

Studying eight different animals (C₃H mice) with implanted tumors, a large number of single point pH determinations were made both prior to and following hyperthermia. The results are plotted in histogram form showing frequency of values at different pH levels (figure 9.7). This method of representing microenvironmental tissue distribution was introduced by Stosseck, Lubbers, and Cottin (1974) for tissue oxygen levels.

The mean value of tissue pH was found to be 6.8 pH units in these mouse tumors. Upon heating for one hour at 43°C, there was a pH decrease of 0.5 to 1 pH unit to an average of 6.2. The individual variations seen from point to point within tumors are seen more clearly in figure 9.8. For each value of pH obtained at

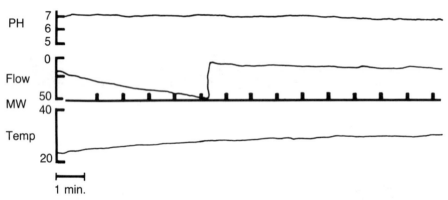

Figure 9.5. In this graph the effect of microwave hyperthermia on pH and blood flow are shown. The *upper tracing* showing pH indicates a decline in pH values as hyperthermia is continued, even though the temperature attained never reaches 40°C. The effect on blood flow is shown in the *second tracing* with increase in flow being in the down direction. The dramatic increase in blood flow required a recalibration after five minutes, which explains the vertical shift in the curve at that time.

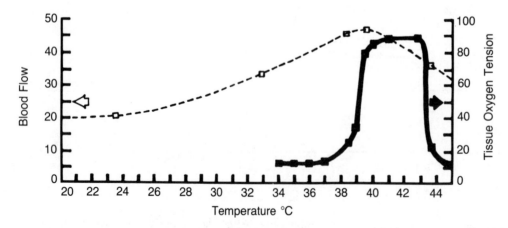

Figure 9.6. In this figure both blood flow (□) and tissue oxygen tension (■) are plotted as a function of temperature. All readings were taken in the same series of animals. It is clear that a strong correlation exists between change in blood flow with temperature and the change in tissue oxygen tension with temperature.

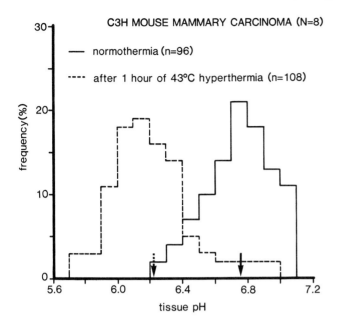

Figure 9.7. Histogram representation of tissue pH levels obtained in mouse tumor. A total of 96 separate determinations were made at normal temperature, while a total of 108 measurements were made following one hour of 43°C microwave hyperthermia. The mean value of pH obtained in the control group was approximately 6.75, while in the group after one hour of hyperthermia the mean value obtained was approximately 6.2 (Bicher et al. 1980. Reprinted by permission.)

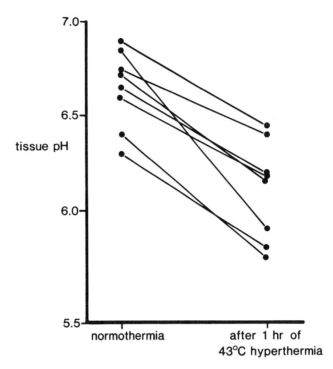

Figure 9.8. This figure shows changes in pH induced by one hour of 43°C hyperthermia examining single points within the tumor. In each case a reading was taken with the pH microelectrode, which was subsequently left in place throughout the period of hyperthermia, following which a second pH reading was obtained. It is clear that in each case the pH fell, the most dramatic being a drop of almost one complete unit of pH. (Bicher et al. 1980.)

normal temperatures, the pH microelectrode was left in place throughout the one hour of 43°C hyperthermia and a new value obtained. In each case, related data points are joined by a solid line.

The results obtained by Song, Kang, and Rhee (1980) in the SCK tumor system are qualitatively similar to those described above. The pH of an SCK tumor (1-cm diameter) was 7.05 before heating. Upon heating at 43.5°C, the pH increased to 7.18 during the first seven minutes, and then rapidly declined to 6.67 at the end of heating for 30 minutes. The pH of the tumor recovered to 6.9, but it decreased further to 6.58 after a second heating at 43.5°C for 30 minutes (fig. 9.9). The pH of muscle in the leg of the rat was 7.1 to 7.4. Contrary to the decrease in pH in the heated tumors, the pH in muscle significantly increased when heated to 43.5°C to 45.0°C. When the temperature was raised above 46.0°C, however, the pH rapidly decreased to as low as 6.60.

In a study by Emami and co-workers (1980), the effects of three levels of hyperthermia on the microvasculature were examined histologically following treatment. Table 9.1 shows the results obtained. The histology of the tumors in the control group is characterized by hypercellularity, pleomorphism, and anaplasticity with high mitotic index. The stroma is fine, reticular, and for the most part indistinct. Patent blood vessels are seen only occasionally, are quite small, and are concentrated more in the periphery than in the center of the tumor.

There were no apparent changes in tumor cells of tumors given modest hyperthermia (40°C ± 0.5°C for 40 minutes) and then removed immediately after heating. Moreover, the vasculature central to the tumor was still barely visible. At this time, the small blood vessels at the tumor periphery did show mild dilation and congestion, with the majority of dilated vessels having diameters of 10 to 12 μ. Occasional endothelial cells of increased prominence were seen projecting into the lumen. When these same tumors were examined three hours after heating, the bulk of the tumor, including both peripheral and central regions, displayed small vessels that were dilated and congested. By this time, the majority of the observed dilated blood vessels had diameters of 15 to 20 μ. The endothelium appeared to be intact with no areas of rupture visible. Vessel congestion was more pronounced than immediately after heating, and capillaries appeared to be packed with red blood cells (stasis). Projection of endothelial cells into the lumen of the capillaries was seen

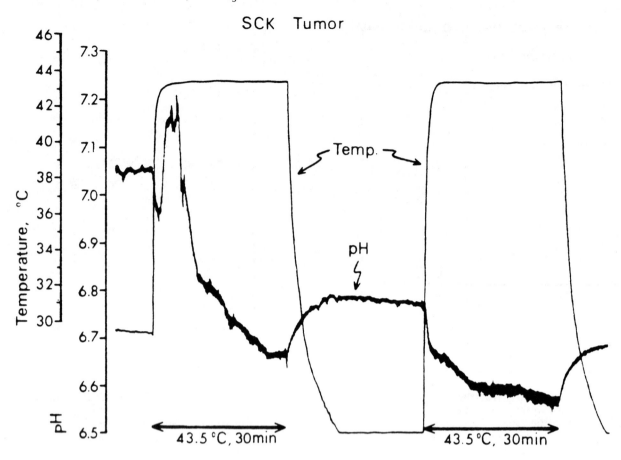

Figure 9.9. The change in pH in a SCK tumor heated twice at 43.5°C for 30 minutes, with 30 minutes between heatings. (Song, Kang, and Rhee 1980.)

Table 9.1
The Effects of Local Hyperthermia on Tumor Microvasculature

Treatment	Survival Hours	Changes in Blood Vessels		
		Congestion	Dilation	Rupture
40.5° ± 0.5°C/40 min	0	+	+	0
	3	+++	++	0
	24	±	+++	0
42.5° ± 0.5°C/40 min	0	++	++	±
	3	+++	+++	++
	24	±	++++	+++
44.5° ± 0.5°C/40 min	0	++	++	++
	3	++	+++	+++
	24	0	*	++++

SOURCE: Emami et al. 1980.
NOTES: 0 = none; ± = very few; + = mild; ++ = moderate; +++ = marked; ++++ = severe.
*Ghosts of previously dilated blood vessels.

occasionally. In tumors removed 24 hours after receiving modest hyperthermia, dilation of small blood vessels was much more pronounced, with a majority of dilated vessels having diameters of 20 to 50 μ. No significant congestion was visible. Endothelium was still intact with no signs of rupture. Occasionally (3 areas in 12 slides from 4 tumors), small foci of necrosis were seen, but the blood vessels in those areas appeared to be intact.

In tumors given intermediate hyperthermia (42.5°C ± 0.5°C for 40 minutes) and removed immediately after heating, marked dilation and congestion of the blood vessels, in all tumor regions, was seen. A few areas of hemorrhage and rupture of blood vessel walls were visible, but the blood vessel walls generally appeared to be intact. In tumors removed three hours after heating, a number of different effects were observed. In some regions of the tumor, the blood vessels showed numerous areas of rupture with associated hemorrhage. In other regions (mostly in the periphery of the tumor), there was marked congestion and dilation of blood vessels with intact endothelium. In some areas, the blood vessels were filled with a granular eosinophilic material compatible with hemolysis. Areas of marked necrosis were seen. In some areas (especially in the periphery of the tumor), however, the tumor cells appeared to remain undamaged. In tumors removed 24 hours after receiving intermediate hyperthermia there was seen to be marked hemorrhagic necrosis involving most of the central areas of the tumor mass. There were lakes of necrotic material within which no identifiable microvasculature existed. Vessels on the periphery of the tumor showed marked dilation and little congestion. Capillary endothelia in these areas remained essentially intact, although a few ruptured capillary walls were seen.

In tumors receiving extreme hyperthermia (44.5°C ± 0.5°C for 40 minutes), massive areas of hemorrhage and ruptured blood vessel endothelium were seen, even in tumors removed immediately after heating. In these tumors, lakes of hemorrhage and necrotic debris were present. The periphery of the tumors revealed areas of less severely damaged blood vessels. The same effects and processes, although much more pronounced and extensive, were observed in tumors subjected to extreme hyperthermia and removed either 3 hours or 24 hours after heating, similar to that subsequently observed in human tumors subjected to extreme heat (Storm et al. 1979).

Similar observations were made by Eddy (1980) in the hamster cheek pouch system. A summary of the time course of events he observed to occur at each temperature is given in table 9.2. At 41°C the constriction of feeding arterioles within three to seven minutes after the beginning of heating resulted in a reduction in vessel caliber and blood circulation through the tumor. Some tumors remained at this level while in others there was a progressive return of circulation to preheating levels by 15 minutes. At 10 to 15 minutes, occasional petechial hemorrhages occurred at the margins of the tumor, but leakage was well contained and bleeding did not progress. At this time also an occasional small tumor capillary became static and remained so throughout the heating period. During the remaining 15 minutes of heating no further alterations occurred. When the tumor temperature was reduced to preheating levels (34°C), the tumor bed dilated within two to three minutes and blood circulation increased. No notable changes occurred in the normal pouch tissue during the period of heating. Histologic sections taken at 15 minutes after the end of the experimental period showed dilated vessels with a few petechial hemorrhages within and at the margins of the tumor. At 24 hours hyperemia was prominent and an occasional vessel was thrombosed.

At 43°C Eddy (1980) observed constriction of feeding arterioles and moderate reduction in tumor vessel caliber and blood circulation within three minutes. The caliber of vessels subsequently increased so that by 15 minutes the tumor bed showed mild to moderate dilation. This degree of dilation remained during the rest of the heating period. Beginning petechial hemorrhages in the margins of the tumor and stasis of individual small vessels within the tumor occurred after 10 to 15 minutes of heating. During the remainder of the period of heating, petechiae continued to expand slowly, and a few additional vessels became static, while a few vessels developed bulbous segments. In general, the tumor vasculature remained mildly to moderately dilated when the temperature was returned to 34°C. Normal pouch tissue appeared unaltered during heating. Histologic sections taken at 15 minutes showed prominent hyperemia and an occasional thrombosed vessel. Petechial hemorrhages within the tumor were few and were well contained, while hemorrhaging was more extensive at the perimeter. At 24 hours, hyperemia continued to be prominent, and it appeared that hemorrhaging continued to occur, particularly at the tumor margins. Some vessels had lost their endothelial linings, while other vessels were thrombosed.

Eddy also examined the system at 45°C, where he noted that the tumor vascular network also reduced in caliber within three minutes, followed by dilation of some networks by six minutes. Petechial hemorrhaging in the margin of the tumor began at five to six minutes after the start of heating and progressed to obscure either the periphery alone or the entire tumor. Stasis in vessels also began at about this time and progressed, causing complete shutdown of the tumor circulation by 20 to 25 minutes. There was no change in this picture when the tumor was returned to 34°C. Vessels in the normal cheek pouch were also static and hemorrhaged by the end of the heating period. In these tumors, the vessels demonstrated three general types of responses. These usually occurred in combination with one another and included compression/occlusion, hemorrhage, and stasis/thrombosis. In some tumors or areas of tumors the tissue was compressed, perhaps as a result of an increased interstitial pressure or a reduction in hydrostatic pressure below interstitial pressure, and blood flow through the vessel network stopped. In such areas, hemorrhage did not occur since blood had been forced out of the network. In other areas of the tumor, where circulation persisted, hemorrhage into the tissue progressed to obscure the tumor. In still other areas, blood circulation through the vascular network became static for reasons that undoubtedly included thrombosis at key locations in the network. Histologic sections taken at 15 minutes showed erythrocytes packing the vessels and extensive hemorrhage within the tumor tissue, particularly at the margin. Nonoccluding thrombi and thrombosis were found throughout the tumor in large and small vessels and in host-feeding arterioles. At 24 hours there remained no functional vasculature within the tumor or pouch tissue. Vessels in the tumor perimeter were packed with erythrocytes, while within the tumor hemorrhaged erythrocytes and those in vessels showed progressive hypochromasia indicative of beginning lysis. The endothelial lining in some vessels appeared to have been lost, while others appeared to be degenerating. At 48 hours after heating at 45°C, tumors present a picture typical of coagulation necrosis.

DISCUSSION AND FUTURE PROSPECTS

Considerable controversy exists as to whether blood perfusion of tumors is greater than that of normal tissues. LeVeen, Wapnick, and Piccone (1976) stated that tumor blood flow, as measured by an isotope dilution technique from surgically excised material, was 2% to 15% of surgically excised normal material. They claimed that the success of radiofrequency (RF) energy in hyperthermia was due to differential cooling of the normal tissue by the improved normal tissue perfusion. Gullino and Grantham (1961), using transplanted tumors in rats' ovaries or kidneys, showed that the tumor blood flow per milligram of tissue was 10 to 20 times less than the blood flow per milligram of tissue of the host ovary or kidney. It must be pointed out, however, that in this experiment the tumor was 10 to 100 times greater in weight than the host ovary or kidney, and the actual total perfusion to the isolated ovary or kidney was increased following the growth of tumor.

Shibata and MacLean (1966), using radioactive microspheres in very large human tumors, showed microsphere normal tissue to tumor distribution ratios ranging from 3:1 to 20:1. Bierman, Kelly, and Singer (1952), using a differential between arterial and venous blood pO_2 as an index of blood flow, showed in 12 patients with metastatic lesions that there was a greater increase in blood flow through the tumors than in comparable normal tissue, possibly because of the presence of arterio-venous shunts. Bierman, Byron, and Kelly (1951) also showed an increase in the thermal index of superficial tumors as compared with the normal tissue.

The results discussed here indicate that localized microwave hyperthermia causes a rise in both blood flow and TpO_2 up to some temperatures, with a fall at higher temperatures. It also appears that the temperature at which the breaking point occurs is characteristic of the specific tissue. The mechanism of this effect seems to be mediated predominantly through the blood flow changes, any metabolic effects being secondary to a microcirculation that is activated at moderate hyperthermic temperatures and damaged at higher temperatures. These results also indicate that the normal tissue microcirculation is better able to cope with hyperthermic stress, as Storm and colleagues (1979) have observed during clinical trials.

The rise in the tumor temperature up to 41°C leads to a significant increase in tumor blood flow. This effect has also been demonstrated by England, Hallbrook, and Ling (1974) and by Sutton (1976) for both the tumor region and host organ. As to the cause of this increased flow, presumably different factors, including systemic and local autoregulation, must be considered. The oxygen partial pressures in several subcutaneous tumors in animals and in humans as measured with 100-μm-tip floating oxygen electrodes followed the change in blood flow (Bicher, Mitagvaria, and Hetzel 1980).

A further rise in tissue temperature up to 42°C results in a marked breakdown of tumor blood flow somewhat below the initial value. Similar results are obtained for in situ tumors, both in humans and in mice. It has been shown in metastatic lesions involving the skin that increases in flow occur owing to elevations of temperature up to 40°C. With tumor temperature elevated to 46°C, the tissue oxygen tension in microareas of the tumor decreases following a drop in tumor blood flow. This correlates with the findings of Reinhold and Berg-Blok (1980), who have shown that at 42°C the center of a sandwich tumor becomes necrotic because of a decrease of tumor microcirculation at this temperature. These results, however, do not correspond with those of Song (1978), who found that hyperthermia at 43°C did not change circulation in tumors, but that it did increase in normal tissues.

The restriction in blood flow at 42°C and the increase in total vascular resistance presumably result from a series of factors. As main determinants of the decline of blood flow, a reduction of red cell deformability, multiple microthrombi, as well as occlusions of microvessels, have to be taken into account. Many of these factors are seen in tables 9.1 and 9.2.

Reduced pH in tumors compared to that of normal tissues has been reported by several authors (Eden and Kahler 1955; Gullino, Grantham, and Smith 1965; Meyer, Kammerling, and Amtman 1948; Naselund and Seenson 1953). The reduced pH of tumor fluid may be due in part to elevated lactic acid production resulting from poor tumor vascularity and reduced oxygen tension. Reports also indicate that tumor cells produce lactic acid at an elevated rate even under oxygenated conditions (Von Ardenne and Reitnauer 1978). The significance of lowered pH to hyperthermia cell killing has been clearly demonstrated by several authors (Gerweck 1977; Gerweck, Nygaard, and Burlett 1970; Gerweck and Rottinger 1976; Overgaard 1976; Von Ardenne and Reitnauer 1978). Cell killing is increased by factors of 5 and above when chronically hypoxic cells are heated to 42°C at a pH only 0.5 units below normal.

The result of a pH drop in cancer tissue during or following hyperthermia is not surprising if one considers the familiar principle that temperature strongly influences the buffering processes and hence the pH. There usually is a shift to lower pH values if the temperature is elevated; in addition, the increase in the carbon dioxide partial pressure, induced by changes in cellular metabolic pathways during hyperthermia, enhances tumor tissue acidosis. These factors, together with the observed breaking point in blood flow and other vascular changes noted in this chapter, may provide a distinct therapeutic advantage for this modality.

The results discussed pose interesting questions and suggest possible areas of future investigation. Since the oxygen partial pressures in malignant tumors generally follow changes in blood flow, it can be expected that the radiosensitivity of cancer tissue may be improved during increased blood flow, thus pro-

Table 9.2
Time Course of Events Following Hyperthermia

Heating Period	41°C for 30 min	43°C for 30 min	45°C for 30 min
Vascular constriction (reduced flow rate)	3–7 min	< 3 min	< 3 min
Return of vessel caliber to preheating levels	15 min		
Return of vessel caliber to greater than preheating levels		+ to ++, 15 min	+, 6 min in some tumors
Petechial hemorrhage	±, 10–15 min	10–15 min ⟶ +	5–6 min ⟶ +++
Tumor vascular stasis	±, 10–15 min	10–15 min ⟶ +	5–6 min ⟶ ++++
Return to 34°, Observe for 10 Minutes			
Vascular dilation (increased flow rate)	++, 2–3 min	No change	No change, total stasis
Histology at 24 hrs postheating	±	Hyperthermia, few thrombosed vessels cells, continued petechial hemorrhage	No functional vasculature, some loss of endothelial cells, vessels packed with erythrocytes

SOURCE: Eddy 1980.
NOTES: ± = very few; + = mild; ++ = moderate; +++ = marked; ++++ = severe; ⟶ = progressive during the 30 minute heating period.

ducing a significant prolongation of survival time of tumor-bearing animals if they are treated with local hyperthermia in combination with radiation. This result would be based totally on the oxygen effect producing increased radiosensitivity of previously hypoxic areas and would not include any contribution to increased cell killing from the synergistic effects of combined hyperthermia and radiation (e.g., repair modification). Therefore, it is expected that hyperthermia can be a useful adjuvant during tumor therapy with ionizing radiation. The clinical utility of the simultaneous combination short term, low tumor hyperthermia (41°C) and radiation is currently being examined in vivo in several laboratories.

It also should be taken into consideration that in the range of maximum tumor blood flow (at moderate hyperthermia) the convective transport of substrates, of wastes, and of systemically introduced antiproliferative agents may be improved. This improvement is of special interest since it can achieve regionally higher concentrations of the antiproliferative agents in specific target tissues. In addition, by improving the substrate supply, a recruitment of the cancer cells belonging to the dormant G_o fraction may be obtained, thus enhancing the cancerostatic effect, especially of cycle-specific drugs.

Another important area for the future involves the apparent collapse of the microvasculature at a higher temperature. It appears that the breaking point in blood flow (the temperature at which blood flow begins to decrease) in tumors is lower than that in some normal tissues. Heating a region of the body that includes both tumor and normal tissues to a temperature above the breaking point for the tumor, but only approaching that of the normal tissue, could result in interesting consequences. With the increased metabolic activity and decreased flow, dramatic shifts in pH such as reported here would occur in the tumor (but not in the surrounding normal tissue). As previously described by Gerweck (1977) and others (Gerweck, Gillette, and Dewey 1974; Gerweck, Nygaard, and Burlett 1970), the combination of reduced pH and increased temperature is extremely cytotoxic. This effect alone should effectively eradicate large areas of the tumor, including the radioresistant hypoxic areas, prior to the start of any combination with radiation. This possibility must be explored in a variety of tumor and normal tissue systems.

We may conclude from the results presented here that the therapeutic effectiveness of hyperthermia may result, at least partially, from several induced physiologic modifications. First, moderate (41°C) hyperthermia in combination with ionizing radiation may result in improved tumor response by increasing oxygenation and hence, radiosensitivity coupled with a decrease in tumor pH. Second, higher levels of hyperthermia, 42°C and above, may be directly tumoricidal because of an elimination of tumor micro-blood flow and a concomitant sharp reduction in tumor pH, while normal tissue still is able to be properly perfused.

References

Bicher, H. I. Proceedings of the Second RTOG Hyperthermia Meeting, November 1977, Chicago, Illinois.

Bicher, H. I. *Increase in brain tissue oxygen availability induced by localized microwave hyperthermia.* New York: Plenum Press, 1978.

Bicher, H. I. et al. Effects of hyperthermia on normal and tumor microenvironment. *Radiology* 137:523–530, 1980.

Bicher, H. I.; Hetzel, F. W.; and D'Agostino, L. Changes in tumor tissue oxygenation induced by microwave hyperthermia. *Int. J. Radiat. Oncol. Biol. Phys.* 2:157, 1977.

Bicher, H. I.; Mitagvaria, N.; and Hetzel, F. W. Alterations in tumor tissue oxygenation by microwave hyperthermia. *Ann. NY Acad. Sci.* 335:20–21, 1980.

Bicher, H. I., and Ohki, S. Intracellular pH electrode experiments on the giant squid axon. *Biochim. Biophys. Acta* 255:900, 1972.

Bierman, H. R.; Byron, R. L.; and Kelly, K. R. Studies on the blood supply of tumors in man. III. Vascular patterns of the liver by hepatic arteriography in vivo. *J. Natl. Cancer Inst.* 12:107, 1951.

Bierman, H. R.; Kelly, K. H.; and Singer, G. Studies on the blood supply of tumors in man. IV. The increased oxygen content of venous blood draining neoplasm. *J. Natl. Cancer Inst.* 12:701–707, 1952.

Buelke-Sam, J. et al. Comparative stability of physiological parameters during sustained anesthesia in rats. *Lab. Anim. Sci.* 2:157–162, 1978.

Cater, D. B. *The measurement of pO_2 in tissue.* New York: Macmillan, 1964.

Cater, D. B., and Silver, I. A. Quantitative measurements of oxygen tension in normal tissues and in the tumors of patients before and after radiotherapy. *Acta Radiol.* 53:233–256, 1960.

Cater, D. B.; Silver, I. A.; and Wilson, G. M. Apparatus and technique for the quantitative measurement of oxygen tension in living tissues. *Proc. R. Soc. Lond.* 151:256–276, 1959–1960.

Davies, P., and Brink, F. Microelectrodes for measuring local oxygen tension in animal tissue. *Rev. Sci. Instrum.* 13:524–533, 1942.

Dewhirst, M. W., and Ozimek, E. J. Will hyperthermia conquer the elusive hypoxic cell? *Radiology* 137:811–817, 1980.

Eddy, H. A. Alterations in tumor microvasculature during hyperthermia. *Radiology* 137:515–521, 1980.

Eddy, H. A., and Casarett, G. W. Development of vascular system in the hamster malignant neurilemmoma. *Microvasc. Res.* 6:63–82, 1973.

Eden, M., and Kahler, H. The pH of rat tumors measured in vivo. *J. Natl. Cancer Inst.* 16:541–556, 1955.

Emami, B. et al. Physiological effects of hyperthermia: response of capillary blood flow and capillary under structure to local tumor heating. *Radiology* 137:805–809, 1980.

England, N. E.; Hallbrook, T.; and Ling, L. Skin and muscle blood flow during regional perfusion with hyperthermia perfusate. *Scand. J. Thorac. Cardiovasc. Surg.* 8:77–79, 1974.

Folkman, J. Isolation of a tumor factor responsible for angiogenesis. *J. Exp. Med.* 133:275–288, 1971.

Gerweck, L. E. Modification of cell lethality at elevated temperatures: the pH effect. *Radiat. Res.* 70:224–235, 1977.

Gerweck, L. E.; Gillette, E. L.; and Dewey, W. C. Killing of Chinese hamster cells in vitro by heating under hypoxic or aerobic conditions. *Eur. J. Cancer* 10:691–693, 1974.

Gerweck, L. E.; Nygaard, T. G.; and Burlett, M. Response of cells to hyperthermia under acute and chronic hypoxic conditions. *Cancer Res.* 39:966–972, 1970.

Gerweck, L. E., and Rottinger, E. Enhancement of mammalian cell sensitivity to hyperthermia by pH alteration. *Radiat. Res.* 67:508–511, 1976.

Gullino, P. M., and Grantham, F. H. Studies on the exchange of fluids between host and tumor. II. The blood flow of hepatomas and other tumors in rats and mice. *J. Natl. Cancer Inst.* 27:1465–1491, 1961.

Gullino, P. M.; Grantham, F. H.; and Smith, S. H. Modifications of the acid-base status of the internal milieu of tumors. *J. Natl. Cancer Inst.* 34:857–860, 1965.

Hinke, J. A. *Cation-selective microelectrodes for intracellular use. Glass electrodes for hydrogen and other cations.* New York: Marcel Dekker, 1973.

Kallman, R. F. The phenomenon of reoxygenation and its implications for fractionated radiotherapy. *Radiology* 105:135–142, 1972.

Krogh, A. The rate of diffusion of gases through animal tissues with some remarks on the coefficient of invasion. *J. Physiol.* 52:391–408, 1918.

Lehman, J. F. Diathermy. In *Handbook of physical medicine and rehabilitation*, eds. Krusen, Kottke, and Elwood. Philadelphia: W. B. Saunders, 1971, pp. 273–345.

LeVeen, H. H.; Wapnick, S.; and Piccone, V. Tumor eradication by radiofrequency therapy: response in 21 patients. *JAMA* 235:2198–2200, 1976.

Meyer, K. A.; Kammerling, E. M.; and Amtman, L. pH studies of malignant tissues in human beings. *Cancer Res.* 8:513–518, 1948.

Naselund, J., and Seenson, K. E. Investigations on the pH of malignant tumors in mice and humans after the administration of glucose. *Acta Obstet. Gynecol. Scand.* 32:359–367, 1953.

Overgaard, J. Influence of extracellular pH on the viability and morphology of tumor cells exposed to hyperthermia. *J. Natl. Cancer Inst.* 56:1243–1250, 1976.

Reinhold, H. S., and Berg-Blok, A. V. D. Features and limitations of the "in vivo" evaluation of tumor response by optical means. Paper read at the 9th L. H. Gray Memorial Conference, Cambridge, England: *Br. J. Cancer,* 1980.

Reinhold, H. S.; Blachiewicz, B.; and Berg-Blok, A. V. D. *Decrease in tumor microcirculation during hyperthermia.* Baltimore and Munich: Urban & Schwarzenburg, 1978.

Reinhold, H. S.; Blachiewicz, B.; and Berg-Blok, A. V. D. Reoxygenation of tumors in "sandwich" chambers. *Eur. J. Cancer* 15:481–489, 1979.

Sandhu, T. S.; Kowal, H. S.; and Johnson, R. Development of hyperthermia applicators. *Int. J. Radiat. Oncol. Biol. Phys.* 4:515–519, 1978.

Shibata, H. R., and MacLean, L. D. Blood flow in tumors. *Prog. Clin. Cancer* 2:33–47, 1966.

Silver, I. A. A simple microelectrode for measuring pO_2 in gas or fluid. *Med. Electron. Biol. Eng.* 1:547–551, 1963.

Song, C. W. Effect of hyperthermia on vascular functions of normal tissues and experimental tumors. *J. Natl. Cancer Inst.* 60:711–713, 1978.

Song, C. W.; Kang, M. S.; and Rhee, J. G. Effect of hyperthermia on vascular function, pH and cell survival. *Radiology* 137:795–803, 1980.

Storm, F. K. et al. Normal tissue and solid tumor effects of hyperthermia in animal models and clinical trials. *Cancer Res.* 39:2245–2251, 1979.

Stosseck, K.; Lubbers, D. W.; and Cottin, W. Determination of local blood flow (microflow by electrochemically generated hydrogen). *Pflugers Arch.* 348:225–238, 1974.

Streffer, C. et al. *Proceedings of the Second International Symposium on Cancer Therapy by Hyperthermia and Radiation.* Baltimore and Munich: Urban & Schwarzenberg, 1978, p. 344.

Suit, H. D.; Sedlacek, R.; and Faqudez, L. Tissue distribution recurrences of immunogenic and nonimmunogenic tumors following tumor irradiation. *Radiat. Res.* 73:251–266, 1968.

Sutton, C. H. Necrosis and altered blood flow produced by microwave induced tumor hyperthermia in a murine glioma (abstr.). *Proc. Am. Assoc. Cancer Res.* 17:63, 1976.

Thomas, R. C. A new design for a sodium sensitive plan microelectrode. *J. Physiol. (Lond.)* 210:82, 1970.

Thomlinson, R. H., and Gray, L. H. The histological structure of some human lung cancers and the possible implications for radiotherapy. *Br. J. Cancer* 9:539–549, 1955.

Von Ardenne, M., and Reitnauer, P. G. Amplification of the selective tumor acidification by local hyperthermia. *Naturwissenschaften* 65:159–160, 1978.

CHAPTER 10
Thermometry and Thermography

Douglas A. Christensen

Introduction
System Requirements
Difficulties with Temperature Measurements in an Electromagnetic Environment
Development of Nonperturbing Probes
Ultrasonic Computed Temperature Tomography
Conclusions

INTRODUCTION

As hyperthermia therapy becomes more intensely investigated, in clinical trials as well as in animal and phantom studies, the need for accurate and reliable thermometry has become particularly evident. In order to quantitate sensitive thermal effects where differences of only a few tenths of degrees Celsius may result in widely varying tissue response, the temperature measuring system must possess fine thermal precision and resolution. In addition, some techniques for delivering local hyperthermia are vulnerable to the production of hot spots—irregularities in the thermal pattern—which must be accurately detected and measured to assure safe, efficient heat dosage. This requires a spatial resolution of the system that adequately determines variations throughout the heated volume.

This chapter discusses the requirements placed upon the thermometry system by the special needs of hyperthermia, and outlines some presently available point-probe techniques for measurements in radiofrequency and microwave fields. Included also are details on current research in noninvasive temperature mapping schemes.

SYSTEM REQUIREMENTS

The desirable specifications of a thermometry system for hyperthermia monitoring may be classified in three main areas:

1. The precision with which the system measures the correct tissue temperature (thermometric accuracy and resolution)
2. The ability of the instrument to map accurately spatial variations in the thermal field (spatial resolution)
3. The fidelity with which the system follows rates of change in the detected temperature (temporal resolution)

Although a clearly understood picture of the mechanisms of hyperthermia therapy is yet to be completely defined, it is possible to set some general guidelines for the specifications in each of these three areas.

Thermometric Accuracy and Resolution

Clinical experiments to elucidate the sensitivity of in vivo tumor and normal tissues to different values of achieved temperatures have been sparse, possibly because of the many other variables involved (e.g., tumor type and location, time at temperature, other therapeutic modalities employed) in such studies in the population of patients available. In vitro cell studies, however, have shown high sensitivity of the heat inactivation rates to temperature, especially below 43°C, where a difference of only 0.1°C to 0.2°C can make a multifold difference in cell survival characteristics (Bhuyan 1979; Sapareto et al. 1978). Figure 10.1 shows the cell inactivation rate as a function of temperature for one study of Chinese hamster ovary cells (Sapareto et al. 1978). It can be seen that cell survival is sensitively dependent upon temperature, particularly below about 43°C. Based on these and similar data and on what is generally achievable in thermometry instruments, an *accuracy* and *resolution* of 0.1°C appears desirable. Resolution of 0.1°C is easier to realize than an absolute accuracy of 0.1°C, since long-term drifts and instabilities in thermometer systems, especially the state-of-the-art nonperturbing techniques described later in this chapter, may be on the order of a few tenths of a degree. Calibration therefore becomes a key component of a reliable system.

Calibration techniques are well documented in the literature (Cetas and Connor 1978; Quinn and Compton 1975). Regardless of the particular hyperthermia system used, some means for conveniently verifying its calibration must be provided. Careful hyperthermia researchers may even check their thermometric accuracy before and after each experiment. A gallium melting-point standard (Bowers 1977), used in conjunction with a water bath, is a convenient calibration technique at a temperature somewhat near those relevant to hyperthermia (29.77°C) and has the additional advantage of traceability to the National Bureau of Standards. Figure 10.2 depicts a typical gallium cell used for probe calibration.

Spatial Resolution

Guy (1971) and Cetas and Connor (1978) have performed microwave and radiofrequency (RF) heating pattern studies of phantom models by viewing the resulting patterns of exposed phantoms with infrared (IR) thermographic cameras. In the latter study, RF heating was produced by implanted needle electrodes; the unperfused models showed high spatial gradients, on the order of 5 to 10 degrees per centimeter, during the initial phases of heating. In order to map accurately the temperature spatial distribution to within a few tenths of a degree would require, in this case, spatial resolution to better than a millimeter. The high gradients, however, became much more gradual as heating progressed, owing to thermal diffusion, and the applied power subsequently was reduced to a lower steady-state value. There also is reason to expect that the gradients encountered in experiments with perfused tissue will be significantly less than those performed on static models, since blood flow will increase thermal diffusion to reduce inhomogeneities in temperature. Also, in the study described, the use of implanted needle electrodes, which tend to localize the applied power, undoubtedly contributed to the high thermal gradients. In experiments with larger applicators, such as parallel plate electrodes, induction coils, or microwave antennae, spatial variations in the heating patterns should be more gentle.

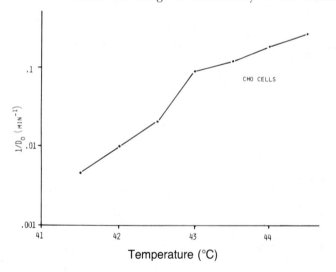

Figure 10.1. The inactivation rate for Chinese hamster ovary cells exposed to heat, showing the sensitivity of the cells to small temperature differences. (Data are from Sapareto et al. 1978.)

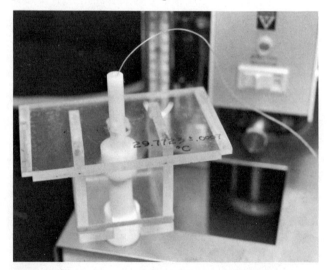

Figure 10.2. A gallium melting-point temperature reference with probe inserted for calibration. The melting point of gallium in this cell is 29.7723 ± 0.0007°C. During calibration, the cell is placed in the water bath shown in the background, where the temperature is held slightly above the gallium melting point.

Therefore, it appears that a spatial resolution of approximately 1 cm would be adequate for a realistic temperature monitoring system. If point probes are used, this would require excessive implantation unless each probe measured temperature at multiple sites along its length. For the noninvasive ultrasound techniques described later, resolution on the order of 1 to 2 cm may be possible.

Temporal Resolution

For normal hyperthermia monitoring, fast time response fortunately is not required. Owing to the relatively slow nature of the heat generation and the large thermal capacity of the regions to be heated, a time response of one second should be adequate. This temporal response is met by almost all of the point-probe systems presently being considered, but may prove a difficulty for the noninvasive ultrasound tomography schemes where computed reconstruction of collected data may require many seconds to several minutes of computation time, depending upon the level of sophistication of the computer. Should the scheme become a clinical reality, improvements would most likely be achieved through the use of dedicated, high-speed computational hardware.

These suggestions for resolution of a hyperthermia thermometry system can be summarized as

1. Thermometric accuracy and resolution: 0.1°C
2. Spatial resolution: approximately 1 cm
3. Temporal resolution: one second
4. Range: 20°C to 55°C

The temperature range 20°C to 55°C begins at a temperature low enough to allow the system to be useful for phantom studies carried out at room temperature and extends to values higher than those normally encountered in clinical hyperthermia trials.

DIFFICULTIES WITH TEMPERATURE MEASUREMENTS IN AN ELECTROMAGNETIC ENVIRONMENT

Conventional devices used for temperature measurements, such as thermistors, thermocouples, and resistance sensors, are very satisfactory regarding such factors as sensitivity, range, reliability, size, and cost. They all suffer, however, from one serious drawback—they contain metallic wires that can lead to large measurement errors when used in the presence of electromagnetic fields. Electromagnetic means is one of the most popular ways of generating hyperthermia and encompasses techniques such as RF electrode heating, microwave antenna and array applicators, and magnetic-induction heating. The ability to accurately monitor temperature in such an environment is, therefore, an important consideration.

The problems associated with the use of metal components in an electromagnetic field stem from the currents induced in the conductive parts by the incident fields. There are at least three related manifestations of this difficulty, as diagrammed in figure 10.3.

1. When a conducting material, such as a metal wire, is placed in a region of electromagnetic radiation, the incident electric field produces current flow on or near the surface of the metal. This is a consequence of the boundary conditions, which require that the tangential component of the incident electric field E_t be matched by an equal component of tangential electric field E_t internal to the boundary, and that the normal component of the incident electric field E_n be matched by an internal normal electric field with a value $E_n(\varepsilon_2/\varepsilon_1)$, where ε_1 and ε_2 are the complex permittivities of the outside and conductive materials, respectively (Ramo and Whinnery 1960). Figure 10.3 shows the case of an incident tangential field inducing an equal field at the surface of a metal, which in turn causes a current density $i_t = \sigma E_t$, where σ is the conductivity of the metal. One effect of this induced current is to cause reradiation of the incident fields, thereby perturbing the original radiation pattern. Such scattering modifies the heating distribution compared to what it would be in the absence of the probe, and tends to be most prevalent in the immediate neighborhood of the conductor, owing to the inverse square law dependence of scattered fields with respect to distance away from the scatterer.

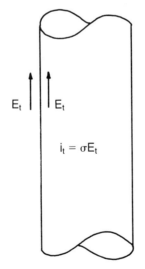

1. i_t reradiates fields

2. i_t causes heating:
   $P = i^2/\sigma$

3. i_t causes em interference

Figure 10.3. The three ways in which the current induced in a metallic conductor by an incident electric field may cause perturbation of the measured temperature. The electromagnetic boundary conditions require continuity of the tangential electric field components, setting up a surface or near-surface current density i_t.

The magnitude of this effect can be reduced somewhat by a careful alignment if the probe is long and thin. By orienting the long axis of the probe perpendicular to the known polarization direction of the incident electric field, the induced currents will be short in length, and the resulting short-dipole radiation is relatively inefficient. Unfortunately, it may not be possible to precisely determine the incident field orientation either because such orientation does not exist, as is the case in the near-field region of many applicators where all three vector components are present, or because intervening tissue refracts and reflects the radiation to disrupt the polarization to an unknown direction.

Scattering of the incident fields occurs whenever any foreign object, even a nonmetallic probe, is placed in the tissue if the intruding object has a complex permittivity that is different from that of the tissue. But the perturbation is significantly less for a purely dielectric object, such as glass or plastic optical fibers in which there are no conduction currents, than it is for metallic components with induced surface currents.

2. The induced currents described above lead to another difficulty. When current flows in a material with finite conductivity, ohmic losses produce heat in the material, whose power density is given by $P = i^2/\sigma$. Such losses also occur in the tissues surrounding the probe (these represent one of the contributions to heat generation in tissues by electromagnetic fields), but because the conductivity of metal is so much higher than that of tissue, the induced currents will be appreciably larger, leading to significantly more heating in the metal probe than in the tissue it displaces (Guy, Lehmann, and Stonebridge 1974). Furthermore, this heat production is localized in an area that has the greatest influence on the accuracy of the temperature measurement—in the temperature probe itself. Again, by judicious orientation of the long axis of the probe, the heating error can be minimized, but this is not always possible or convenient.

3. The presence of wires connecting the temperature sensor to its associated electronic signal processing and display circuitry may promote magnetic and electric coupling between the radiation fields and the temperature monitoring system. This source of interference can sometimes lead to inaccurate and fluctuating readings of temperature during the application of electromagnetic power.

DEVELOPMENT OF NONPERTURBING PROBES

The seriousness of the problem described in the previous section has spurred the development by several groups of so-called nonperturbing temperature probes. As discussed in the preceding section, these probes should more correctly be labeled "minimally perturbing" rather than "nonperturbing," but in comparison to conventional metallic devices, they are a significant improvement. Development of these probes has proceeded along two lines: the use of very high resistance lead wires to reduce the magnitude of the induced currents, and the use of optical fibers to eliminate conduction currents entirely. Examples of each category are given below.

High-Resistance Leads

From the discussion in the previous section, it should be apparent that the cause of most of the difficulty in using conventional probes in an electromagnetic environment is the abnormally high currents induced in the conductive pathways. Bowman (1976) and Larsen, Moore, and Acevedo (1974) have reduced the induced current densities in their devices by employing high resistance lead wires that attach to thermistor sensors. If the resistivity of the leads is made large enough (approaching even the tissue values of approximately 10^2 to 10^3 ohms−cm), the induced current densities are lowered to a level where reradiation of the incident fields and ohmic heating in the leads are within acceptable limits. To achieve the proper value of resistance, the plastic material of the leads is impregnated with carbon particles. Since a thermistor is used as the sensor, the device has relatively good accuracy, stability, and reproducibility. The lead diameter, however, is necessarily larger (overall diameters of about 1 mm have been achieved) than conventional fine-gauge wire, and probe length is limited by the requirement to keep total probe resistance manageable.

Fiberoptic Probes

Another approach has been to employ glass or plastic optical fibers as the connecting link between the sensor and the display electronics. Since the fibers are nonconducting, no currents are induced to cause errors in the temperature readings. As an added benefit, the fibers also possess low thermal conductivity, so they effectively eliminate any appreciable conduction of heat away from or toward the sensor that may cause reading errors with metal lead wires.

One of the first fiberoptic temperature probes was developed by Rozzell and colleagues in 1974 and by Johnson, Gandhi, and Rozzell in 1975. A capsule coated with a layer of cholesteric liquid-crystal material is affixed to the distal end of a bundle of plastic (polymethylmethacrylate) fibers. A portion of the fibers in the bundle carries red narrow-band light from a light-emitting diode (LED) source to the liquid crystal sensor, whose wavelength of maximum reflectivity changes as the effective spacing between its molecular layers changes with varying temperature. The reflected light captured by the remaining fibers is detected and, after appropriate scaling for offset and slope, is displayed as a direct temperature reading. Principal shortcomings of this device are its relatively large tip diameter (1.5 to 2.0 mm) and some chemical instability tendencies of the liquid crystal. After calibration, accuracy is approximately 0.1°C.

Cetas (1976) has pursued the development of an optical rotation sensor. A small birefringent sensor of $LiNbO_3$ rotates the polarization of light which passes through a polarizing filter immediately in front of the crystal. Upon reflection from the back face of the sensor and further rotation as it returns through the crystal, the amount of light passing again through the polarizing film is dependent upon the total degree of rotation, which in turn is a function of the temperature of the crystal. The nonlinear nature of the returned optical power is compensated by the electronics of the system. Minimum tip diameter to date is about 1.0 mm.

Wickersheim and Alves (1979) have developed a fluorescent-type sensor composed of a mixture of two phosphors. The ratio of the fluorescent intensities (each at unique wavelengths) may be detected and correlated with the temperature of the sensor. Advantages include the ability to separate input and output powers by narrow-band interference filters, insensitivity to exciting radiation intensity, and factory calibration of the sensors.

A sensor design using a small semiconductor crystal has been completed by Christensen (1977). The sensor is fabricated from GaAs, a semiconductor whose band-edge absorption lies in the near infrared region of the spectrum. When narrow-band radiation of this wavelength from an LED is passed through the sensor, a variable amount is absorbed by the process of exciting valence band electrons across the forbidden energy gap into the conduction band. Since the gap energy of GaAs varies sensitively with temperature, the amount of returned power is a function of tip temperature. The sensor is attached to the optical fibers (usually two) by transparent epoxy.

In this particular system, a microprocessor is employed to accomplish the linearization of the returned signals, thereby displaying temperature directly. The speed of the microprocessor actually allows the simultaneous use and display of four temperature probes in a single instrument. The prototype is shown in figure 10.4. Self-calibration of the probes is possible in this instrument through the use of a microprocessor-controlled calibration module integrated into the unit. When the probes to be calibrated are placed in a small

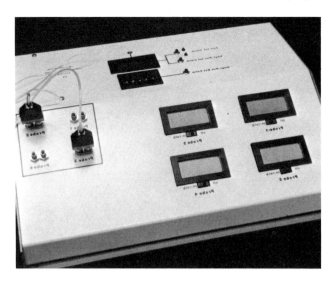

Figure 10.4. The prototype fiberoptic temperature probe system employing semiconductor (GaAs) sensors. Up to four probes may be used and displayed simultaneously. A microprocessor controls the display as well as the automatic calibration of the probes.

tube in the calibrator module, one of two possible automatic calibration modes can be selected by pushbutton: either a full-scale calibration, which covers the entire temperature range of the instrument in about 15 minutes, or a single-point correction (at the ambient module temperature) in about one to three seconds. Also provided is another single-point calibration mode for use with a precision external temperature standard such as the gallium cell mentioned earlier.

The tip diameter of the semiconductor probe, including the Teflon™ sheath employed for mechanical protection, is equivalent to 26 gauge (0.45 mm) over the distal 14 cm of the probe. Because the system was designed specifically for hyperthermia monitoring, an insertion set consisting of a blunt 25-gauge needle inside a Teflon™ catheter has been developed. A sample probe is shown with an insertion set in figure 10.5. After insertion in phantom or tissue, the needle is withdrawn and the probe is introduced into the catheter.

Accuracy for the semiconductor probe system is approximately ± 0.2°C over a range of 16.6°C to 50.0°C.

ULTRASONIC COMPUTED TEMPERATURE TOMOGRAPHY

A very different and noninvasive approach to monitoring internal tissue temperature is currently being

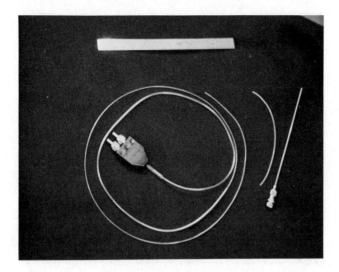

Figure 10.5. A sample fiberoptic probe with semiconductor sensor, shown alongside a 25-gauge insertion set. A Teflon™ catheter fits over the needle of the insertion set. After placement in phantom or tissue, the needle is withdrawn, and the probe is introduced into the catheter.

investigated by a small number of research teams. The basic technique uses computed tomography (CT) in a fashion similar to the radiographic CT scanner, but the energy source is ultrasound rather than radiation (Mueller, Kaveh, and Wade 1979). Figure 10.6 shows the fundamentals of the data collection instrument. A transmitting transducer array sends ultrasonic pulses through the region being scanned, whereupon they are detected by an array of receiving transducers. By rotating the arrays and taking advantage of the lateral sampling by the receiver elements, all possible paths are collected.

While the radiation CT scanner gives images of variations in radiopacity, the ultrasound scanner can produce images either of sonic opacity (measured by detecting the attenuation of ultrasound pulses transmitted through the tissue region over the multitude of different paths) or of the speed of ultrasound propagation (measured by the time-of-flight of the pulses over these paths). A reconstruction by computer techniques of sonic attenuation and velocity has shown promise as a means of detecting breast lesions (Greenleaf and Johnson 1978), but for temperature mapping the reconstruction of tissue speed characteristics is of primary interest. This is because all tissues investigated so far have shown a variation in sonic propagation speed as a function of temperature, and by using those known and cataloged dependencies, temperature data may be inferred from an image of velocity (Sachs and Tanney 1977; and Christensen 1979).

The obvious advantages of such a technique are that it does not employ indwelling probes and that it gives detail about the spatial distribution of heating patterns. Much more development, however, needs to be done before the method is proven as a practical clinical instrument. The computer algorithms are fairly well developed because of the great strides made in radiographic computed tomography, although ul-

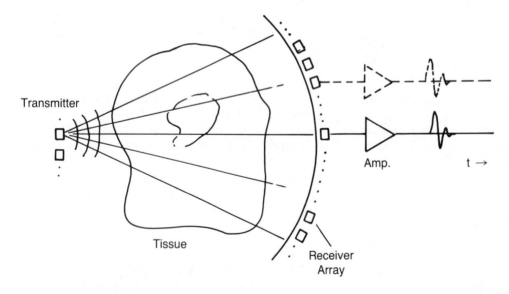

Figure 10.6. Configuration for the ultrasonic computed tomography instrument. Variations in the propagation speed of ultrasound pulses in the tissue may be related to temperature changes. All possible paths are measured by electronic selection of the receiver elements combined with possible mechanical rotation of the transducer arrays.

Thermometry and Thermography

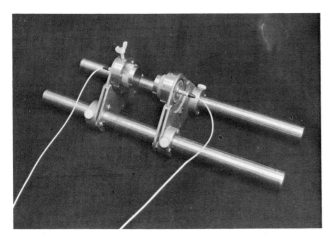

Figure 10.7. The transducer holder for a single-path measurement of tissue propagation speed. The tissue, in a saline-filled plastic bag, is placed between the transducers and the entire assembly immersed in a temperature-controlled water bath. Transducers shown here are 10 MHz.

trasonic wave propagation has a characteristic that is not important for x-ray propagation, the bending of wavefronts caused by variations of the speed of propagation, which makes the straight-ray assumption partially invalid for ultrasound. Modifications to the algorithms to correct for this effect in ultrasound are being investigated (Johnson et al. 1979). The hardware needed for the ultrasonic scanner, such as transducers, amplifiers, and time-of-flight detectors, are also reasonably well advanced. The one area receiving the most attention, and upon which the basis of temperature tomography rests, is the variation of ultrasonic velocity in various tissues as a function of temperature. Tissue characterization is therefore the focus of current interest.

Tissue Temperature Characterization

Jacobs and colleagues (1979) have developed a single-path laboratory technique for cataloging ultrasonic parameters for a variety of tissues. The method is a continuous-wave interferometric instrument using two opposed 10 MHz transducers. A tissue sample with known thickness is placed inside a thin polyvinylidene bag filled with a normal saline solution. This bag is then gently positioned between the transducers, and the entire assembly is placed in a temperature-controlled water bath. The transducers are accurately spaced by an invar rod arrangement to avoid variations in path length caused by thermal expansion of the holders (fig. 10.7).

Figure 10.8 is a schematic of the tissue-characterization scheme. The phase of the wave passing through the sample is compared with that of an electronic ref-

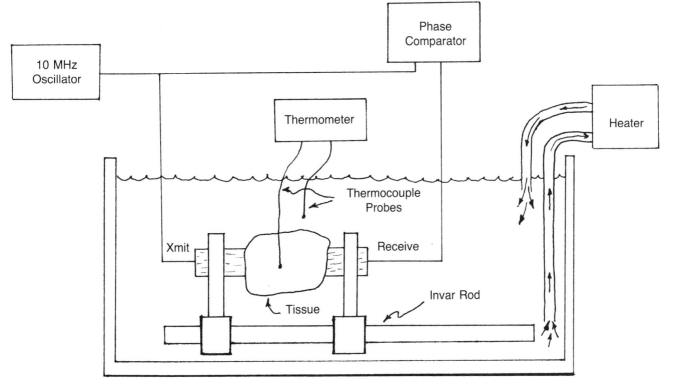

Figure 10.8. An interferometric technique for measuring the phase changes, and therefore the velocity changes, of ultrasound propagation through the tissue sample. The temperature of the water bath is varied in increments; the system is allowed to stabilize at each temperature before a measurement is made.

erence connected directly from the signal generator. The phase difference φ is given by

$$\phi = 2\pi \frac{\ell}{\lambda} = 2\pi \frac{\ell f}{c}, \qquad (1)$$

where ℓ = path length between transducers,
λ = wavelength in intervening tissue,
f = frequency of transmitted ultrasound,
c = speed of ultrasound in tissue sample.

If the frequency is held constant, as is the case when the signal generator is crystal-controlled, then changes in propagation speed c owing to temperature T may be accurately measured by detecting small changes in the phase difference as given by

$$\frac{d\phi}{dT} = -2\pi \frac{\ell f}{c^2} (dc/dT). \qquad (2)$$

Of course the length ℓ must be held steady, necessitating the invar rod mechanism.

The above technique yields an accurate measurement of the slope dc/dT. For absolute measurement of c at any temperature, the frequency is changed and the following equation holds:

$$\frac{d\phi}{df} = \frac{2\pi \ell}{c}. \qquad (3)$$

Thus, both c and dc/dT for various tissue samples may be conveniently measured with the interferometric technique. Clinical instruments now being planned, however, employ time-of-flight detectors to avoid multiple-path interference problems from the more complex volumes being scanned.

Early work at Mayo Clinic by Rajagopalan and co-workers (1979) showed that there was a nonuniformity in absolute speed between tissue types, but that, for nonfatty tissues, the magnitude of the variation of speed with temperature (dc/dT) was relatively independent of tissue classification. This discovery means that, although absolute temperature mapping requires a priori knowledge of the distribution of tissue type, temperature increases or decreases from a baseline measurement might be obtained from data regarding velocity changes. For example, figure 10.9 shows a summary of some typical data for various tissues (Nasoni et al. 1979; Jacobs et al. 1979; and

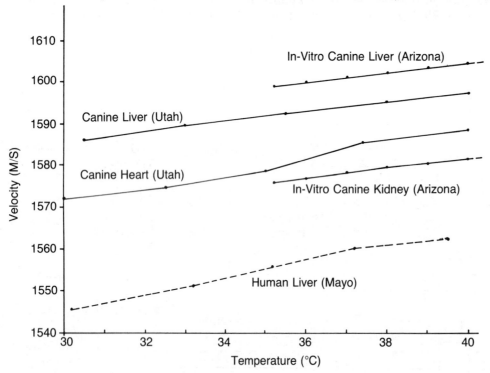

Figure 10.9. A summary of velocity data versus temperature for a few different tissue types. Data marked *Arizona* are from Nasoni and colleagues (1979), data marked *Mayo* are from Rajagopalan and co-workers (1979), and data marked *Utah* are from Jacobs and colleagues (1979). Slopes from these nonfatty tissues are reasonably uniform.

Rajagopalan et al. 1979). As can be seen, the temperature coefficients as given by the slopes for these examples are fairly uniform, although their absolute speeds vary considerably from tissue to tissue.

Further studies by Rajagopalan and associates (1979) and those by Bamber and Hill (1979) have found an exception to the rule of a uniform temperature coefficient characteristic of all tissues. The coefficient for fat-containing tissues (e.g., human breast fat and bovine peritoneal fat) appears to be either monotonically negative or at least piecewise negative. This makes the inference of temperature changes from velocity data more difficult, perhaps requiring an additional measurement of fat content of the tissues being probed. There is hope, however, that this additional information may be obtained ultrasonically by measuring other ultrasonic parameters of the tissues as well. Some combination of two or more of the parameters listed in table 10.1 might provide a tissue-independent characteristic that varies as a function of temperature only. More research currently is being pursued in this matter.

Instrument Design

The present apparatus for collecting tissue data employs a single transmitting transducer and a single receiving transducer, making scan times fairly long (5 to 10 minutes). Future plans call for multitransducer arrays similar to that in figure 10.6, in which the scanning will be achieved with all-electronic switching (or perhaps with a minimum of mechanical movement). This is expected to reduce scan times considerably. Dedicated electronic data processing hardware also will shorten the reconstruction times.

Theoretical improvements in the reconstruction algorithms, such as accounting for ray bending caused by refraction and for ray blocking in regions of bone or air as in the thorax or limbs, currently are being investigated. Although the remaining problems in the development of a clinically useful instrument are large, the promise of a noninvasive temperature mapping instrument has great appeal for hyperthermia monitoring.

Table 10.1
Measurable Ultrasonic Parameters of Tissues

c	α
dc/dT	$d\alpha/dT$
dc/df	$d\alpha/df$

NOTES: c = phase velocity; T = temperature; f = frequency; α = attenuation constant.

CONCLUSIONS

Accurate temperature monitoring is essential to the delivery of safe and effective thermal doses during hyperthermia therapy. The exact requirements placed upon the thermometry system depend upon the region being heated, the temperatures achieved, and the method of heat generation, but the suggestions for a general system as given in the discussion of resolution appear reasonable.

Each of the presently available temperature measurement techniques possesses certain advantages and shortcomings. If no electromagnetic radiation is involved in the delivery of the heat (such as with ultrasonic heating), thermistors or thermocouple devices work well. They cannot be used in microwave or RF fields, however, and they are invasive point measurements only. For use in electromagnetic fields, various nonconducting point probes have been developed, such as the high-resistance lead devices and the fiber-optic probes. These also are invasive point measurement techniques.

The noninvasive ultrasonic tomography scheme needs considerable development in order to become a practical clinical system. It also will be expensive and complex—but the attractiveness of noninvasive spatial temperature mapping is indeed high.

References

Bamber, J. C., and Hill, C. R. Ultrasonic attenuation and propagation speed in mammalian tissues as a function of temperature. *Ultrasound Med. Biol.* 5:149–157, 1979.

Bhuyan, B. K. Kinetics of cell kill by hyperthermia. *Cancer Res.* 39:2277–2284, 1979.

Bowers, G. N. The gallium melting-point standard. *Clin. Chem.* 23:709–710, 1977.

Bowman, R. R. A probe for measuring temperature in radio frequency-heated material. *IEEE Trans. MTT* 24: 43–45, 1976.

Cetas, T. C. A birefringent crystal optical thermometer for measurements in electromagnetically induced heating. In *Proceedings 1975 USNC/URSI Symposium (Bureau of Radiological Health)*, eds. C. C. Johnson and J. L. Shore. Rockville, Md.: Bureau of Radiologic Health, 1976.

Cetas, T. C., and Connor, W. G. Thermometry considerations in localized hyperthermia. *Med. Phys.* 5:79–91, 1978.

Christensen, D. A. A new nonperturbing temperature probe using semiconductor band edge shift. *J. Bioeng.* 1: 541–545, 1977.

Christensen, D. A. Thermal dosimetry and temperature measurements. *Cancer Res.* 39:2325–2327, 1979.

Greenleaf, J. F., and Johnson, S. A. Measurement of spatial distribution of refractive index in tissues by ultrasonic computer assisted tomography. *Ultrasound Med. Biol.* 3:327–339, 1978.

Guy, A. W. Analysis of electromagnetic fields induced in biological tissues by thermodynamic studies on equivalent phantom models. *IEEE Trans. MTT* 19:205–214, 1971.

Guy, A. W.; Lehmann, J. F.; and Stonebridge, J. B. Therapeutic applications of electromagnetic power. *Proc. IEEE* 62:55–75, 1974.

Jacobs, S. R. et al. Sound velocity versus temperature measurements in tissues. Paper read at the 98th meeting of the Acoustical Society of America, November 1979, Salt Lake City, Utah.

Johnson, C. C.; Gandhi, O. P.; and Rozzell, T. C. A prototype liquid crystal fiberoptic probe for temperature and power measurements in RF fields. *Microwave Journal* 18:55–59, 1975.

Johnson, S. A. et al. High spatial resolution ultrasonic measurement techniques for characterization of static and moving tissues. In *Ultrasonic tissue char. II*, eds. M. Linzer, NBS Special Publication 525:235–246, 1979.

Larsen, L. E.; Moore, R. A.; and Acevedo, J. A microwave decoupled brain temperature transducer. *IEEE Trans. MTT* 22:438–444, 1974.

Mueller, R. K.; Kaveh, M.; and Wade, G. Reconstructive tomography and applications to ultrasonics. *Proc. IEEE* 67:567–587, 1979.

Nasoni, R. L. et al. Temperature dependence of ultrasonic speed in tissue and its application to noninvasive temperature monitoring. *Ultrasonic Imag.* 1:34–43, 1979.

Quinn, T. J., and Compton, J. P. The foundations of thermometry. *Prog. Physics* 38:151–239, 1975.

Rajagopalan, B. et al. Variation of acoustic speed with temperature in various excised human tissues studied by ultrasound computerized tomography. In *Ultrasonic Tissue Char. II*, ed. M. Linzer, NBS Special Publication 525:227–233, 1979.

Ramo, S., and Whinnery, J. R. *Fields and waves in modern radio.* New York: John Wiley & Sons, 1960.

Rozzell, T. C. et al. A nonperturbing temperature sensor for measurements in electromagnetic fields. *J. Microwave Power* 9:241–249, 1974.

Sachs, T. D., and Tanney, C. S. A two-beam acoustic system for tissue analysis. *Phys. Med. Biol.* 22:327–340, 1977.

Sapareto, S. A. et al. Effects of hyperthermia on survival and progression of Chinese hamster ovary cells. *Cancer Res.* 38:393–400, 1978.

Wickersheim, K. A., and Alves, R. V. Recent advances in optical temperature measurement. *Indust. Res. and Dev.* 21:82, 1979.

CHAPTER 11
Hyperthermia Techniques and Instrumentation

E. Ronald Atkinson

Hyperthermia Technology Assessment
Thermometry
Hyperthermia Control Systems
Biophysical Data

HYPERTHERMIA TECHNOLOGY ASSESSMENT

Whole-Body Hyperthermia

Various techniques for the clinical practice of whole-body hyperthermia (WHB) have been reported (Byrne 1889; Cavaliere et al. 1969; Clowes 1906; Coley 1893; Grisham and Barnett 1973; Krementz and Ryan 1972). These techniques often have involved the use of overly complex technology. As a result of growing understanding, the means for producing and controlling WBH may be simplified to the point that clinical trials may be safely performed routinely at whole-body temperatures of 42°C for sessions in excess of four hours (Crile 1961; Hahn 1974). Patients have been treated in a wide variety of disease states. It is possible that WBH, so performed, may be simultaneously administered to several patients by paramedical personnel under the supervision of a trained physician. Approximately 90 minutes is required to attain 42°C nominal WBH (Grisham and Barnett 1973). Normothermia, without sequelae, is attained in approximately 30 minutes (Grisham and Barnett 1973; Lampert 1957b). Simple analgesia and electrolyte maintenance are the only concomitant requirements (Guy 1971; Lampert 1957b; Neymann and Osborne 1929).

WBH in the 39°C to 40°C range has the potential to facilitate local hyperthermia and possibly to improve tumor oxygenation through vasodilation and inhibition of oxygen vasoconstriction. These benefits require that the method by which WBH is produced does not decrease the subject's blood pressure and tumor blood flow (Crile 1961).

Local Hyperthermia

Local hyperthermia is produced by coupling energy to tissue through three commonly accepted modalities: radiofrequency coupling at frequencies ranging from 100 KHz to 100 MHz, microwave coupling at higher frequencies, and mechanical coupling by means of ultrasound.

Radiofrequency coupling

Radiofrequency coupling is under investigation since the technique appears to be applicable both to superficial sites and to certain deep tissue sites (Dickson and

The assistance of the American Cancer Society and the American Hospital Supply Corporation in the preparation of this review is gratefully acknowledged. The views expressed by the author in this work are entirely his own and are in no way endorsed by either the American Cancer Society or the American Hospital Supply Corporation.

Ellis 1976; Robinson and Wizenberg 1974). Workers at the University of California at Los Angeles have reported attaining local tumor temperatures over 50°C while maintaining acceptable normal surrounding tissue temperatures (Storm et al). Radiofrequency sources, amplifiers, and electrodes are simple and inexpensive (Lampert 1957a). Radiofrequency energy couples more readily to fatty tissue than to muscle (Goetze and Schmidt 1932); parenchyma (Goetze and Schmidt 1932; Michaelson et al. 1975), or connective tissues (Selawry, Carlson, and Moore 1958). This feature may be exploited to advantage or may be regarded as an inherent limitation to the method. Indwelling electrodes (Goodlad and Clark 1972; Michaelson 1969) may be used so that their geometric location can help to limit the tissue volume to be heated. The determination of temperature distributions resulting from the dissipation of radiofrequency energy in large volumes of perfused tissue at diverse anatomic sites is being investigated (Carlisle 1968; Comings 1973; Korb 1976; Loshek, MacKinnon, and Orr 1974; Rivière et al. 1965; Schmidt and Ott 1974; Warters, Winward, and Roti-Roti 1977). Electromagnetic interference with conventional monitoring equipment and thermometry is a factor that must be considered in the clinical use of radiofrequency-induced hyperthermia (Haam and Frost 1939; Korb 1976).

It appears at present that radiofrequency coupling techniques, using implanted electrodes for the production of local hyperthermia, are ideally suited to the treatment of small, well-defined tissue volumes that are readily accessible (Langendorff and Langendorff 1943; Sutton 1971). This technique may be expected to be of great immediate value as an adjunct to treatment in clinical situations where radioactive needle implants are used. It may be anticipated that radiofrequency coupling would be the method of choice in those special clinical situations where tumors are accessible through body cavities (Loshek, MacKinnon, and Orr 1974; Warters, Winward, and Roti-Roti 1977). Research is needed to identify the specific clinical situations that occur with sufficient frequency to warrant the development of appropriate electrode configurations, with consideration of contact impedances and the need for heat dissipation at the conductive interface to treat body cavity accessible lesions (DeLateur et al. 1970; Lehmann et al. 1965; Schwan and Ferris 1968).

Microwave Coupling

Microwave coupling complements radiofrequency coupling to a certain extent in that muscle, parenchyma, and connective tissue are more closely coupled than is fatty tissue (Cloudsley-Thompson 1963; Goetze and Schmidt 1932). The microwave coupling technique may be implemented by means of surface contacting applicators or by tuned indwelling coaxial lines. The electromagnetic interference problems associated with radiofrequency coupling are much less severe at these higher frequencies. Microwave coupling permits heating tissue volumes in the lung and gut that are inaccessible to mechanical coupling with ultrasound (Lehmann 1937; Lehmann et al. 1970). Tissues are moderately transparent to microwave radiation (Goetze and Schmidt 1932; McCaffrey et al. 1975). At a frequency of about 1 GHz, for example, tissue penetration to a depth of 3 to 5 cm is possible (McCaffrey et al. 1975; Schulman et al. 1961). Depths of this order, in many patients, should permit tissue heating in virtually any part of the body. Research is required to determine the degree of localization of heating that may be obtained by the use of multiple aperture, direct contact, and focused microwave applicators (Lehmann and Krusen 1955; Rahn, Reeves, and Howell 1975). Microwave coupling appears at present to be a technique ideally suited to the heating of larger regional tissue volumes and may be appropriate to WBH as a means of more rapidly elevating whole-body temperature than may be accomplished by direct skin or blood thermal conduction heating or endogenous metabolic heat confinement.

Potential nonthermal effects of microwaves on living tissues need to be delineated (Hahn et al. 1973; Overgaard 1977). Thermal hazards may be associated with resonance situations arising from special body and postural geometries that would be quite frequency specific (Dickson and Muckle 1972; Schamberg and Tseng 1927; Schulman et al. 1961). The confinement of microwave radiation is difficult because of leakage and reradiation from heated tissues. Such reradiation has been observed to occur at unexpected anatomic sites (Hahn, Alfieri, and Kim 1974).

Ultrasonic Coupling

Mechanical energy coupling by means of ultrasound to produce hyperthermia in tumors up to 5 cm beneath the skin surface has been clinically demonstrated at the Stanford University Medical Center (Gerner et al. 1975; Lampert 1957b, 1957c; Langendorff and Langendorff 1939, 1943; Lawson 1956; Lazarides 1975; Owen et al. 1967). As a result of scaled pilot studies, ultrasonic heating appears to be well suited for producing and sustaining controlled levels of uniform hyperthermia at deep tissue sites (Goldman, Green, and Iampietro 1965). Workers at the Massachusetts Institute of Technology have constructed prototype clinical systems on the basis of these studies (Goldman and Dreffer 1976; Goldman, Green, and Iampietro 1965; Goodlad and Clark 1972; Goss and Parsons 1972; Goulon et al. 1973; Grantham, Hill, and Gullino 1973). It is anticipated that a tissue volume may be heated to within a few millimeters of air-filled cavities or bone, and that such a system could be useful in producing hyperthermia in tissue volumes up to 8 cm in diameter (Goss and Parsons 1977; Krementz and Ryan 1972; Overgaard 1977). Multiple applicators, sequentially energized, may permit heating of deep tissue volumes that would be otherwise inaccessible because of acoustic shielding by internal interfaces (Gilchrist 1960; Krementz and Ryan 1972).

The short wavelength of ultrasound permits precise definition of the boundaries of the tissue volume in which energy is deposited (Goss and Parsons 1977; Shoulders et al. 1941). Contacting applicators are required by ultrasonic heating (Cabanac, Cunningham, and Stolwijk 1971). Although phased array beam steering may be used at ultrasonic frequencies (Lambert 1912), it is anticipated that the method would be difficult to apply to large tissue volumes (Lampert 1957a). Sites near or behind bone and other tissue interfaces present special problems associated with high ultrasonic absorption, cavitation, and mode conversion (Crile 1961; Overgaard and Poulson 1977).

THERMOMETRY

Since biological response is a critical function of temperature (Hall, Schade, and Swinney 1974; Overgaard and Bicher, in press) and temperature gradient (Overgaard and Bicher, in press; Selawry, Carlson, and Moore 1958), precision in vivo thermometry is essential for hyperthermia. Several types of nonperturbing implantable temperature transducers are either commercially available or in prototype development (Coley 1893; Overgaard and Bicher, in press; Selawry, Carlson, and Moore 1958). For the electromagnetic microwave and radiofrequency heating methods, temperature transducers must be nonconductive or of tissue-matching conductivity so that the transducers themselves will not perturb the electromagnetic field. Since it is difficult to predict the degree of depolarization of electromagnetic radiation that will result from tissue volume scattering (Krementz and Ryan 1972), direct transducer heating errors are also difficult to avoid, although multiplex techniques may resolve this problem without introducing unacceptable control system constraints.

The variation of sound velocity with temperature in a given tissue permits, in principle, the use of computed tomography techniques with ultrasonic imaging to map internal temperature changes noninvasively (Goldacre and Sylvén 1962; Saigusa et al. 1968). The sensitivity and accuracy of this method has not been determined (Sapozink, Deschner, and Hahn 1973). The ultimate utility of this method will depend upon the as yet unknown range of normal variation of tissue elastic moduli with temperature and degree of tissue perfusion. Another problem with this method may arise if the time required to execute the requisite computer algorithms should present unacceptable system constraints (Overgaard 1977). Any method of ultrasonic interrogation is subject to the same tissue accessibility limitations as were discussed above for ultrasonic heating.

HYPERTHERMIA CONTROL SYSTEMS

Hyperthermia treatment requires raising an assigned tissue volume temperature to a predetermined level within a given time and maintaining that temperature, with acceptable temperature gradients, for a predetermined time interval. From the present state of clinical knowledge (Cabanac and Massonnet 1974; Goodlad and Clark 1972; Gullino, Grantham, and Smith 1965), it is not possible to define with any degree of assurance exactly what final temperature will be desired (although it seems that 42°C to 45°C ± 0.2°C may be reasonable), what temperature gradients within surrounding normal tissues would be tolerable (although it seems that 40°C to 42°C ± 0.1°C, in normal tissue, may be reasonable), nor what heating rate is acceptable (although this tolerated rate presumably would depend upon the final temperature to be achieved). In spite of these gaps, it is essential to envision a control system that uses temperature transducers to control the heat-producing energy source to regulate temperature in a defined tissue volume and adjacent regions. Such a system would involve a distribution of temperature transducers whose data would be processed to control the energy deposition pattern. Special consideration must be given to the monitoring and control of any interfaces (fascia, etc.) that are known to be preferentially heated by the specific heating modality used. Evaluation criteria for such control systems would be based largely upon considerations of reliability, spatial homogeneity, and speed of response. Analytic models predicting tissue temperature distributions that would result from various energy deposition patterns under anticipated ranges of tissue perfusion rates are needed to gauge the performance of any such control system.

BIOPHYSICAL DATA

To permit the design and predict the range of performance and safety of a hyperthermia treatment system, certain biophysical data are needed. These include the properties of normal and tumor tissues such as electrical conductivity (Kim, Kim, and Hahn 1975a; Kluger, Ringler, and Anver 1975), dielectric constant (Knochel 1975; Overgaard and Suit 1977; Selawry, Goldstein, and McCormick 1977), thermal conductivity (Kim, Kim, and Hahn 1975a, 1975b; Overgaard and Overgaard 1972), heat capacity (Overgaard and Overgaard 1972), elastic moduli (Overgaard 1977), microvascular architecture (Lawson 1956; Pettigrew et al. 1974), perfusion rate, and temperature response.

Direct mode-mode coupling may take place in biological systems and support resonant excitation from low energy electromagnetic radiation (Gullino and Grantham 1964; Korb 1948). The mode-mode coupling hypothesis lends itself to verification by direct measurement of physical quantities. Observed ultrasonic attenuation constants in single axis ferromagnetics (Coley 1893; Korb 1948) and antiferromagnetics (Collins and Weiner 1968; Guy 1977) are in good agreement with the model. Particle diffusivity measurements in an isobutyric acid-water system (Horvath 1948), a nitrobenzene and n-hexane system (Marmor, Hahn, and Hahn 1977), and an analine-cyclohexane system (Crabb, Murdock, and Amelunxen 1975) also are in good agreement with the prediction of mode-mode coupling model. Frohlich (1977) has suggested direct verification of resonant mode-mode coupling between non-ionizing electromagnetic radiation and living tissue by means of spectral measurements. Biophysical data from such measurements could serve to define spectral distributions of excitation radiation having maximum biological interaction and could aid in the identification of susceptible tissues. Very little is known of these parameters, and now that the efficacy of hyperthermia treatment is becoming established, it is becoming extremely important that their values and range of normal and pathologic variation be established.

References

Byrne, J. A digest of 20 years' experience in the treatment of cancer of the uterus by galvanocautery. *Am. J. Obstet.* 22:1052–1053, 1889.

Cabanac, M.; Cunningham, D. J.; and Stolwijk, J. A. J. Thermoregulatory set point during exercise: a behavioral approach. *J. Comp. Physiol. Psychol.* 76:94–102, 1971.

Cabanac, M., and Massonnet, B. Temperature regulation during fever: Change of set point or change of gain? A tentative answer from a behavioural study in man. *J. Physiol.* 238:561–568, 1974.

Carlisle, H. J. Peripheral thermal stimulation and thermoregulatory behavior. *J. Comp. Physiol. Psychol.* 66:507–510, 1968.

Cavaliere, R. et al. Trattamento dei tumori in circolazione extracorporea regionale con ipertermia ed antiblastici. *Minerva Chir.* 24:1147–1148, 1969.

Cloudsley-Thompson, J. L. The mechanism of heat death. *New Scientist* 338:330–332, 1963.

Clowes, G. H. A. A study of the influence exerted by a variety of physical and chemical forces on the virulence of carcinoma in mice. *Br. Med. J.* 1548–1554, 1906.

Coley, W. B. The treatment of malignant tumors by repeated inoculations of erysipelas: with a report of ten original cases. *Am. J. Med. Sci.* 105:487–511, 1893.

Collins, K. J., and Weiner, J. S. Endocrinological aspects of exposure to high environmental temperatures. *Physiol. Rev.* 48:785–839, 1968.

Comings, D. E. A general theory of carcinogenesis. *Proc. Natl. Acad. Sci. USA* 70:3324–3328, 1973.

Crabb, J. W.; Murdock, A. L.; and Amelunxen, R. E. A proposed mechanism of thermophily in facultative thermophiles. *Biochem. Biophys. Res. Commun.* 62:627–633, 1975.

Crile, G., Jr. Heat as an adjunct to the treatment of cancer. *Cleve. Clin. Q.* 28:75–89, 1961.

DeLateur, B. J. et al. Muscle heating in human subjects with 915 MHz microwave contact applicator. *Arch. Phys. Med.* 51:147–151, 1970.

Dickson, J. A., and Muckle, D. S. Total-body hyperthermia versus primary tumor hyperthermia in the treatment of the rabbit VX-2 carcinoma. *Cancer Res.* 32:1916–1923, 1972.

Frohlich, H. Long-range coherence in biological systems. *Rivista del Nuovo Cimento* 7:399–418, 1977.

Gerner, E. W. et al. The potential of localized heating as an adjunct to radiation therapy. *Radiology* 116:433–439, 1975.

Gerner, E. W. et al. Biochemical aspects of hyperthermic damage. The effects of elevated temperatures on two enzymes involved in cell proliferation. *Radiat. Res.* 70:610, 1977.

Gilchrist, R. K. Potential treatment of cancer by electromagnetic heating. *Surg. Gynecol. Obstet.* 110:499–500, 1960.

Goetze, O., and Schmidt, K. H. Ortliche homogene uberwarmung gesunder und kranker gliedmassen. *Deutsche Zeitschrift fur Chirurgie* 234:623–670, 1932.

Goldacre, R. J., and Sylvén, B. On the access of blood-borne dyes to various tumor regions. *Br. J. Cancer* 16:306–322, 1962.

Goldman, L., and Dreffer, R. Microwaves, magnetic iron particles and lasers as a combined test model for investigation of hyperthermia treatment of cancer. *Arch. Dermatol. Res.* 257:227–232, 1976.

Goldman, R. F.; Green, E. B.; and Iampietro, P. F. Tolerance of hot, wet environments by resting men. *J. Appl. Physiol.* 20:271–277, 1965.

Goodlad, G. A. J., and Clark, C. M. Activity of gastrocnemius and soleus polyribosomes in rats bearing the Walker 256 carcinoma. *Eur. J. Cancer* 8:647–651, 1972.

Goss, P., and Parsons, P. G. The effect of hyperthermia and melphalan on survival of human fibroblast strains and melanoma cell lines. *Cancer Res.* 37:152–156, 1977.

Goulon, J. et al. Mésures interférométriques précises de la permittivité complex des liquides dans un large domaine de fréquences. *Révue de Physique Appliquée* 8:165–174, 1973.

Grantham, F. H.; Hill, D. M.; and Gullino, P. M. Primary mammary tumors connected to the host by a single artery and vein. *J. Natl. Cancer Inst.* 50:1381–1383, 1973.

Grisham, C. M., and Barnett, R. E. The role of lipid-phase transitions in the regulation of the (sodium + potassium) adenosine triphosphatase. *Biochemistry* 12:2635–2637, 1973.

Gullino, P. M., and Grantham, F. H. The vascular space of growing tumors. *Cancer Res.* 24:1727–1732, 1964.

Gullino, P. M.; Grantham, F. H.; and Smith, S. H. The interstitial water space of tumors. *Cancer Res.* 25:727–731, 1965.

Guy, A. W. Analyses of electromagnetic fields induced in biological tissues by thermographic studies on equivalent phantom models. *IEEE Transactions on Microwave Theory and Techniques* 19:205–214, 1971.

Guy, A. W. Biophysical characteristics of electromagnetic fields: problems of dosimetry and dosimetric techniques. *Neurosci. Res. Program Bull.* 15:81–88, 1977.

Haam, E. V., and Frost, T. T. Changes in the parenchymatous organs produced by artificially induced fever. *Proc. Soc. Exp. Biol. Med.* 42:99–103, 1939.

Hahn, E. W. et al. The radiation enhancement factor (REF) of hyperthermia in combination with fast neutrons on local tumor response. *Radiat. Res.* 59:141, 1973.

Hahn, E. W.; Alfieri, A. A.; and Kim, J. H. Increased cures using fractionated exposures of X irradiation and hyperthermia in the local treatment of the ridgway osteogenic sarcoma in mice. *Radiology* 113:199–202, 1974.

Hahn, G. M. Metabolic aspects of the role of hyperthermia in mammalian cell inactivation and their possible relevance to cancer treatment. *Cancer Res.* 34:3117–3123, 1974.

Hall, R. R.; Schade, R. O. K.; and Swinney, J. Effects of hyperthermia on bladder cancer. *Br. Med. J.* 2:593–594, 1974.

Horvath, J. Morphologisch untersuchungen uber die wirkung der ultraschallwellen auf das karzinomgewebe. *Strahlentherapie* 77:279–290, 1948.

Kim, S. H.; Kim, J. H.; and Hahn, E. W. Enhanced killing of hypoxic tumor cells by hyperthermia. *Br. J. Radiol.* 48:872–874, 1975a.

Kim. S. H.; Kim J. H.; and Hahn, E. W. The radiosensitization of hypoxic tumor cells by hyperthermia. *Radiology* 114:727–728, 1975b.

Kluger, M. J.; Ringler, D. H.; and Anver, M. R. Fever and survival. *Science* 188:166–168, 1975.

Knochel, J. P. Disseminated intravascular coagulation in heat stroke. *JAMA* 231:1053–1055, 1975.

Korb, H. Uber eine kombination der rontgenbestrahlung mit der kurzwellenbehandlung. *Strahlentherapie* 77:301–303, 1948.

Krementz, E. T., and Ryan, R. F. Chemotherapy of melanoma of the extremities by perfusion: fourteen years clinical experience. *Ann. Surg.* 175:900–917, 1972.

Lambert, R. A. Demonstration of the greater susceptibility to heat of sarcoma cells. *JAMA* 59:2147–2148, 1912.

Lampert, H. Hyperthermie und infektionskrankheiten als modell fur die abwehr beim carcinom. *Int. Rundschau Phys. Med.* 10:157–159, 1957a.

Lampert, H. Beeinflussung bosartiger geschwulste durch hyperthermie und hyperamie. *Int. Rundschau Phys. Med.* 10:150–151, 1957b.

Lampert, H. Schlusswort: zusammenfassung der ergebnisse der tagung. *Int. Rundschau Phys. Med.* 10:178, 1957c.

Langendorff, H., and Langendorff, M. Untersuchungen uber die ultrakurzwellenwirkung auf impftumoren. *Strahlentherapie* 64:512–519, 1939.

Langendorff, H., and Langendorff, M. Uber die wirkung einer mit ultrakurzwellen kombinierten rontgenstrahlenbehandlung auf das ehrlich-karzinom der maus. *Strahlentherapie* 72:211–219, 1943.

Lawson, R. Implications of surface temperatures in the diagnosis of breast cancer. *Can. Med. Assoc. J.* 75:309–310, 1956.

Lazarides, E. Immunofluorescence studies on the structure of actin filaments in tissue culture cells. *J. Histochem. Cytochem.* 23:507–528, 1975.

Lehmann, E. Nursing care in fever therapy. *Am. J. Nurs.* 37:1–7, 1937.

Lehmann, J. F., and Krusen, F. H. Biophysical effects of ultrasonic energy on carcinoma and their possible significance. *Arch. Phys. Med. Rehabil.* 36:452–459, 1955.

Lehmann, J. F. et al. Comparison of deep heating by microwaves at frequencies 2456 and 900 megacycles. *Arch. Phys. Med. Rehabil.* 46:307–314, 1965.

Lehmann, J. F. et al. Evaluation of a microwave contact applicator. *Arch. Phys. Med. Rehabil.* 51:143–146, 1970.

Loshek, D.; MacKinnon, D. J.; and Orr, J. S. Claims for a diathermy machine. *Lancet* 2:289, 1974.

McCaffrey, T. W. et al. Effect of isolated head heating and cooling on sweating in man. *Aviat. Space Environ. Med.* 46:1353–1357, 1975.

Michaelson, S. M. Biological effects of microwave exposure. *Biological effects and health implications of microwave radiation* 9:17–19, 1969.

Michaelson, S. M. et al. *Physiological responses to microwave exposure.* Rochester, N. Y., University of Rochester, Dept. of Electrical Engineering and Radiation Biology and Biophysics. *Proc. Natl. Electron. Conference* 30:233–234, 1975.

Neymann, C., and Osborne, S. Artificial fever produced by high frequency currents. *IMJ* 56:199–203, 1929.

Overgaard, J. Correspondence: ultrastructure of a murine mammary carcinoma exposed to hyperthermia in vivo. *Cancer Res.* 37:943–944, 1977.

Overgaard, J., and Bicher, P. Influence of hypoxia and acidity on the hyperthermic response of malignant cells in vitro. *Radiology*, in press.

Overgaard, J., and Overgaard, K. Some factors influencing the curability of transplanted tumors exposed to local hyperthermia. Radiumstationen, Denmark: The Institute of Cancer Research, 1975, pp. 1–17.

Overgaard, K., and Overgaard, J. Investigations on the possibility of a thermic tumor therapy—I. Short-wave treatment of a transplanted isologous mouse mammary carcinoma. *Eur. J. Cancer* 8:65–78, 1972.

Overgaard, J., and Poulsen, H. S. Effect of hyperthermia and environmental acidity on the proteolytic activity in murine ascites tumor cells. *J. Natl. Cancer Inst.* 58:1159–1161, 1977.

Overgaard, J., and Suit, H. D. Hyperthermia in vivo: time-temperature relation and split dose effect. *Radiat. Res.* 70:634, 1977.

Owen, O. E. et al. Brain metabolism during fasting. *J. Clin. Invest.* 46:1589–1595, 1967.

Pettigrew, R. T. et al. Circulatory and biochemical effects of whole body hyperthermia. *Br. J. Surg.* 61:727–730, 1974.

Rahn, H.; Reeves, R. B.; and Howell, B. J. Hydrogen ion regulation, temperature, and evolution; the 1975 J. Burns Amberson Lecture. *Am. Rev. Respir. Dis.* 112, 1975.

Redmann, K.; Burmeister, J.; and Jenssen, H. I. The influence of hyperthermia on the transmembrane potential, zeta-potential and metabolism of polymorphonuclear leukocytes. *Acta Biol. Med. Ger.* 33:187–196, 1974.

Rivière, M.-R. et al. Cancérologie—effets de champs électromagnétiques sur un lymphosarcome lymphoblastique transplantable du rat. *Comptes Rendus de l'Académie des Sciences* 260:2099–2102, 1965.

Robinson, J. E., and Wizenberg, M. J. Thermal sensitivity and the effect of elevated temperatures on the radiation sensitivity of Chinese hamster cells. *Acta Radiologica* 13:241–248, 1974.

Saigusa, K. et al. Influence of focused ultrasound on experimental tumor (Horie's sarcoma) (the 4th report). *Med. Ultrason.* 6:142–144, 1968.

Sapozink, M. D.; Deschner, E. E.; and Hahn, E. W. Induction of mitotic synchrony by intermittent hyperthermia in Ehrlich carcinoma in vivo. *Nature* 244:299–300, 1973.

Schmidt, K. L., and Ott, V. R. Immunreaktionen bei hoher korpertemperatur. *Med. Welt* 25:1963–1968, 1974.

Schwan, H. P., and Ferris, C. D. Four-electrode null techniques for impedance measurement with high resolution. *Rev. Sci. Instrum.* 39:481–485, 1968.

Selawry, O. S.; Carlson, J. C.; and Moore, G. E. Tumor response to ionizing rays at elevated temperatures. *Am. J. Roentgenol.* 80:833–839, 1958.

Selawry, O. S.; Goldstein, M. N.; and McCormick T. Hyperthermia in tissue-cultured cells of malignant origin. *Cancer Res.* 17:785–791, 1957.

Shoulders, H. S. et al. Preliminary report on the effect of combined fever and deep x-ray on the treatment of far-advanced malignant cases. *J. Tenn. Med. Assoc.* 34:9–15, 1941.

Shulman, R. G. et al. Ferromagnetic resonance in DNA samples. *Biochem. Biophys. Res. Commun.* 5:52–56, 1961.

Storm, F. K. et al. Normal tissue and solid tumor effects of hyperthermia in animal models and clinical trials. *Cancer Res.* 39:2245–2251, 1979.

Sutton, C. H. Tumor hyperthermia in the treatment of malignant gliomas of the brain. *Trans. Am. Neurol. Assoc.* 6:195–199, 1971.

Warters, R. L.; Winward, R. T.; and Roti-Roti, J. L. Effects of hyperthermia on chromatin composition and structure. *J. Cell Biol.* 75:126a, 1977.

Selected Readings

Adams, G. E. Chemical radiosensitization of hypoxic cells. *Br. Med. Bull.* 29:48–53, 1973.

Alfieri, A. A.; Hahn, E. W.; and Kim, J. H. The relationship between the time of fractionated and single doses of radiation and hyperthermia on the sensitization of an in vivo mouse tumor. *Cancer Chemother. Rep.* 58:296–298, 1974.

Alguire, G. H., and Legallais, F. Y. Vascular reactions of normal and malignant tissues in vivo. IV. The effect of peripheral hypotension on transplanted tumors. *J. Natl. Cancer Inst.* 12:399–421, 1951.

Allan, B. D., and Norman, R. L. Hyperthermia and cancerous tissue water structure. *Cancer Chemother. Rep.* 58:296–298, 1974.

Allan, B. D., and Norman, R. L. Letters: hyperthermia as an adjuvant in the treatment of osteogenic sarcoma. *Cancer Chemother. Rep.* 59:257–258, 1975.

Allen, F. M. Biological modification of effects of roentgen rays. II. High temperature and related factors. *Am. J. Roentgenol.* 73:836–848, 1955.

Allison, A. Therapeutic effects of lightning upon cancer. *Lancet* 1:77, 1880.

Almalric, F.; Simard, R.; and Zalta, J. P. Effet de la température supra-optimale sur les ribonucléoproteines et le RNA nucléolaire. II. Étude biochimique. *Exp. Cell Res.* 55:370–377, 1969.

Ardenne, M. V. Selective multiphase cancer therapy: conceptual aspects and experimental basis. *Adv. Pharmacol. Chemother.* 10:339–380, 1972.

Ardenne, M. V. et al. Zur hohe und bedeutung des effektiven glukosespiegels in tumoren. *Klin. Wochenschr.* 44:502–511, 1966.

Aschoff, V. J., and Wever, R. Kern und schale in warmehausfolt des menschen. *Naturwissenschaften* 20:477–485, 1958.

Aslan, E. E. Accuracy of a temperature-compensated precision RF power bridge. *IEEE Transactions on Instrumentation and Measurement* 18:232–236, 1969.

Aslan, E. E. Electromagnetic leakage survey meter. *J. Microwave Power* 6:169–177, 1971.

Aslan, E. E. Broad-band isotropic electromagnetic radiation monitor. *IEEE Transactions on Instrumentation and Measurement* 21:421–424, 1972.

Astrom, K. E. et al. An experimental neuropathological study of the effects of high-frequency focused ultrasound on the brain of the cat. *J. Neuropathol. Exp. Neurol.* 20:484–520, 1961.

Atkinson, E. R. Comments on Yerushalmi's article. *Eur. J. Cancer* 13:193–194, 1977.

Atkinson, E. R. Hyperthermia dose definition. *J. Bioeng.* 1:487–492, 1977.

Atkinson, E. R. Current status of whole body hyperthermia technology. In *Hyperthermia as an antineoplastic treatment modality*, NASA Publication 2051. Washington, D.C.: National Aeronautics and Space Administration, 1978.

Atkinson, E. R. Microwave-induced hyperthermia dose definition. *IEEE Trans. MTT* 26:595–598, 1978.

Atkinson, E. R. et al. *Protocol: phase I clinical trial of whole body hyperthermia.* Bethesda, Md.: National Institutes of Health, 1976.

Auersperg. N. Differential heat sensitivity of cells in tissue culture. *Nature* 209:415–416, 1966.

Bacchetti, S.; Cassandro, M.; and Mauro, F. Radiosensitivity in relation to the cell cycle and recovery from x-ray sublethal damage in diploid yeast. *Exp. Cell Res.* 46:292–300, 1967.

Baev, V. I., and Shcherbachev, I. P. Metabolism of certain phosphorus compounds of the brain the first minutes following hyperthermia and hypercapnia. *Patol. Fiziol. Eksp. Ter.* 14:20–24, 1970.

Baker, G. M., and Wright, E. A. Effects of hypoxia and heat on mammary carcinoma in C3H mice. *Br. J. Radiol.* 40:809, 1976.

Ballantine, H. T.; Bell, E.; and Manlapaz, J. Progress and problems in the neurological applications of focused ultrasound. *J. Neurosurg.* 17:858–876, 1960.

Ballard, B. E. Pharmacokinetics and temperature. *J. Pharm. Sci.* 63:9, 1974.

Barranco, S. C.; Novak, J. K.; and Humphrey, R. M. Studies on recovery from chemically induced damage in mammalian cells. *Cancer Res.* 35:1194–1204, 1975.

Barrett, A. H., and Myers, P. C. Microwave thermography. *Bibl. Radiol.* 6:45–56, 1975.

Barth, G., and Wachsmann, F. Uber den einfluss der temperatur auf die hautreaktion bei rontgenbestrahlungen. *Strahlentherapie.* 77:87–90, 1948.

Basauri, L., and Lele, P. P. A simple method for production of trackless focal lesions with focused ultrasound: statistical evaluation of the effects of irradiation on the central nervous system of the cat. *J. Physiol.* 160:513–534, 1962.

Bauer, K. D.; Henle, K. J.; and Dethlefsen, L. A. Induction of thermotolerance in CHO cells by multiple heat (45°C) fractions. *Radiation Res.* 70:631, 1977.

Bazett, H. C. Physiological responses to heat. *Physiol. Rev.* 7:531–599, 1927.

Beer, J. Z.; Lett, J. T.; and Alexander, P. Influence of temperature and medium on the x-ray sensitivities of leukemia cells in vitro. *Nature* 199:193–194, 1963.

Beisel, W. R.; Goldman, R. F.; and Joy, R. J. T. Metabolic balance studies during induced hyperthermia in man. *J. Appl. Physiol.* 24:1–10, 1968.

Beller, G. A., and Boyd, A. E. Heat stroke: a report of 13 consecutive cases without mortality despite severe hyperpyrexia. *Milit. Med.* 140:464–467, 1975.

Belli, J. A., and Bonte, F. J. Influence of temperature on the radiation response of mammalian cells in tissue culture. *Radiat. Res.* 18:272–276, 1963.

Bender, E., and Schramm, T. Untersuchungen zur thermosensibilitat von tumorund normalzellen in vitro. *Acta Biol. Med. Ger.* 17:527–543, 1966.

Bender, V. E., and Schramm, T. Weitere untersuchungen zur frage der ulterschiedlichen thermosensibilitat von tumor—und normalzellen in vitro and in vivo. *Archiv. Geschwulstforsch.* 32:215–230, 1968.

Ben-Hur, E.; Elkind, M. M.; and Bronk, B. V. Thermally enhanced radiosensitivity of cultured Chinese hamster cells. *Nature* 238:209–211, 1972.

Ben-Hur, E.; Elkind, M. M.; and Bronk, B. V. Thermally enhanced radioresponse of cultured Chinese hamster cells: inhibition of repair of sublethal damage and enhancement of lethal damage. *Radiat. Res.* 58:38–51, 1974.

Benzinger, T. H. Heat regulation: homeostasis of central temperature in man. *Physiol. Rev.* 49:671–759, 1969.

Berlin, R. D. et al. The cell surface. *N. Engl. J. Cancer* 292:515–520, 1975.

Berteaud, A. J. et al. Essai de corrélation entre l'évolution d'une affection par Trypanosoma équiperdum et l'action d'une onde électromagnétique pulsée et modulée. *Comptes Rendue de l'Académie des Sciences* 272:1003–1006, 1971.

Berteaud, A. J. et al. Action d'un rayonnement électromagnétique à longeur d'onde multimétrique sur la croissance bactérienne. *Comptes Rendue de l'Académie des Sciences* 271:843–846, 1975.

Beuchat, L. R., and Worthington, R. E. Relationships between heat resistance and phospholipid fatty acid composition of *vibrio Parahaemolyticus*. *Appl. Environ. Microbiol.* 31:389–394, 1976.

Biedert, P. (Vorlaufige) Heilung einer ausgebreiteten sarkomwucherung in einem kinderkopf durch erysipel. *Deutsche Medizinal-Zeitung* 7:45, 1886.

Bierman, H. R.; Kelly, K. H.; and Singer, G. Studies on the blood supply of tumors in man. IV. The increased oxygen content of venous blood draining neoplasms. *J. Natl. Cancer Inst.* 12:701–707, 1952.

Birkner, R., and Waschsmann, F. Uber die kombination von rontgenstrahlen und kurzwellen. *Strahlentherapie* 79:93–102, 1949.

Bleehen, N. M.; Honess, D. J.; and Morgan, J. E. The effect of hyperthermia and the hypoxic cell sensitizer Ro-07-0582 on the EMT6 mouse mammary tumor. *Br. J. Radiol.* 49:806, 1976.

Bleehen, N. M.; Honess, D. J.; and Morgan, J. E. Interaction of hyperthermia and the hypoxic cell sensitizer Ro-07-0582 on the EMT6 mouse tumor. *Br. J. Cancer* 35:299–306, 1977.

Bleiberg, I., and Sohar, E. The effect of heat treatment on the damage and recovery of the protein synthesis mechanism of human kidney cell line. *Virchows Arch. [Cell Pathol.]* 17:269–278, 1975.

Bligh, J. *Temperature regulation in mammals and other vertebrates.* Amsterdam: North Holland, 1973.

Block, J. B., and Zubrod, C. G. Adjuvant temperature effects in cancer therapy. *Cancer Chemother. Rep.* 57:373–382, 1973.

Boak, R. A.; Carpenter, C. M.; and Warren, S. L. The thermal inactivation time at 41.5 C of three strains of herpes simplex virus. *J. Exp. Med.* 71:169–173, 1940.

Bodel, P. Generalized perturbations in host physiology caused by localized tumors. *Ann. NY Acad. Sci.* 230:6–13, 1974.

Bodey, G. P. Antibiotic therapy of infections in patients undergoing cancer chemotherapy. *Antibiot. Chemother.* 18:49–88, 1974.

Bomford, R., and Christie, G. H. Mechanisms of macrophage activation by *Corynebacterium parvum*. II. In vivo experiments. *Cell. Immunol.* 17:150–155, 1975.

Bonard, E. C. La pyrétothérapie. Indications modernes d'une thérapeutique ancienne. Praxis. *Rev. Suisse Med.* 48:437–442, 1959.

Bonilla-Naar, A. 403 cancer patients treated with immunotherapy as a complement for orthodox methods of treatment (1963–1975). Paper presented at The First International Symposium on Cancer Therapy by Hyperthermia and Radiation, April 1975, Washington, D.C.

Boothby, W. M., and Sandiford, I. Summary of the basal metabolism data on 8614 subjects with especial reference to the normal standards for the estimation of the basal metabolic rate. *J. Biol. Chem.* 54:783–803, 1922.

Bowler, K. et al. Cellular heat injury. *Comp. Biochem. Physiol.* [A] 45:441–450, 1973.

Bradley, S. E., and Conan, N. J. Estimated hepatic blood flow and Bromsulfalein extraction in normal man during pyrogenic reaction. *J. Clin. Invest.* 26:1175, 1947.

Braun, J., and Hahn, G. M. Enhanced cell killing by bleomycin and 43 hyperthermia and the inhibition recovery from potentially lethal damage. *Cancer Res.* 35:2921–2927, 1975.

Bray, G. A., and Campfield, L. A. Metabolic factors in the control of energy stores. *Metabolism* 24:99–117, 1975.

Brett, D. E., and Schloerb, P. R. Intermittent hyperthermia on Walker 256 carcinoma. *Arch. Surg.* 85:1004–1007, 1962.

Bronk, B. V. Thermal potentiation of mammalian cell killing: clues for understanding and potential for tumor therapy. *Adv. Radiat. Biol.* 6:267–324, 1976.

Bronk, B. V.; Wilkins, R. J.; and Regan, J. D. Thermal enhancement of DNA damage by an alkylating agent in human cells. *Biochem. Biophys. Res. Commun.* 52:1064–1070, 1973.

Bruns, P. Die heilwirkung des erysipels auf geschwulste. *Beitraege zur Klinischen Chirurgie* 3:443–466, 1887.

Bucciante, L. Ulteriori ricerche sulla velocita della mitosi nelle cellule coltivate in vitro in funzione della temperatura. *Archiv fur Experimentelle Zellforschung* 5:1–24, 1928.

Bull, J. et al. *Protocol: a phase I-II trial of hyperthermia + methyl CCNU in malignant melanoma and other neoplasms.* Bethesda, Md.: National Institutes of Health, 1977.

Bull. J. et al. *Protocol: pilot study examining the effect of whole body hyperthermia combined with adriamycin in patients with metastatic sarcoma and other cancers.* Bethesda, Md.: National Institutes of Health, 1977.

Bullard, R. W.; Elizondo, R. S.; and Banerjee, M. Effect of local heating and arterial occlusion on sweat electrolyte. *J. Appl. Physiol.* 32:1–6, 1972.

Burger, F. J.; Malan, J.; and Bester, P. The effect of hyperthermia on serum enzymes' activity. *South Afr. Med. J.* 44:899–901, 1970.

Burk, D.; Woods, M.; and Hunter, J. On the significance of glucolysis for cancer growth, with special reference to Morris rat hepatomas. *J. Natl. Cancer Inst.* 38:839–863, 1967.

Busch, W. Uber den einfluss welchen heftigere erysipeln zuweilen auf organisierte neubildungen amiben. *Verhandl. Naturh. Preuss. Rhein Westphal.* 23:28–30, 1866.

Butler, T. P.; Grantham, F. H.; and Gullino, P. M. Bulk transfer of fluid in the interstitial compartment of mammary tumors. *Cancer Res.* 35:3084–3088, 1975.

Butler, T. P., and Gullino, P. M. Quantitation of cell shedding into efferent blood of mammary adenocarcinomas. *Cancer Res.* 35:512–516, 1975.

Byrne, J. Vaginal hysterectomy and high amputation or partial extirpation by galvano-cautery in cancer of cervix uteri. An inquiry into their relative merits. *Brooklyn Med. J.* 6:729–766, 1892.

Cann, J. R.; Stewart, J. M.; and Matsueda, G. R. A circular dichroism study of the secondary structure of bradykinin. *Biochemistry* 12:3780–3788, 1973.

Carlisle, H. J. Effect of preoptic and anterior hypothalmic lesions on behavioral thermoregulation in the cold. *J. Comp. Physiol. Psychol.* 69:391–402, 1969.

Castillo, J., and Goldsmith, H. S. Immunological competence of heat-treated extracts of tumor or lymph nodes. *Arch. Surg.* 106:322–324, 1973.

Cater, D. B.; Grigson, M. B.; and Watkinson, D. A. Changes of oxygen tension in tumors induced by vasoconstrictor and vasodilator drugs. *Acta Radiologica* 58:401–434, 1962.

Cater, D. B., and Silver, I. A. Quantitative measurements of oxygen tension in normal tissues and in the tumors of patients before and after radiotherapy. *Acta Radiologica* 33:233–256, 1960.

Cater, D. B.; Silver I. A.; and Watkinson, D. A. Combined therapy with 220 KV roentgen and 10 CM microwave heating in rat hepatoma. *Acta Radiologica* (new series) 2:321–336, 1964.

Cavaliere, R. et al. Selective heat sensitivity of cancer cells. *Cancer* 20:1351–1381, 1967.

Chaffee, R. R. J. et al. Temperature acclimation in birds and mammals. *Annu. Rev. Physiol.* 33:155–202, 1971.

Chan, A. K.; Sigelmann, R. A.; and Guy, A. W. Calculations of therapeutic heat generated by ultrasound in fat-muscle-bone layers. *IEEE Trans. Biomed. Eng.* 21:286–284, 1974.

Chang, C. H. Some new radiotherapeutic approaches and combined protocol trials in the management of malignant gliomas. *Recent Results Cancer Res.* 51:135–150, 1975.

Check, J. H. et al. Protection against spontaneous mouse mammary adenocarcinoma by inoculation of heat-treated syngeneic mammary tumor cells. *Int. J. Cancer* 7:403–408, 1971.

Chen, K., and Guru, B. Internal EM field and absorbed power density in human torsos induced by 1-500-MHz EM waves. *IEEE Trans. MTT* 25:746–756, 1977.

Chen, T. T., and Heidelberger, C. Quantitative studies on the malignant transformation of mouse prostate cells by carcinogenic hydrocarbons in vitro. *Int. J. Cancer* 4:166–178, 1969.

Chien Y., and Levine, L. Fluorescence quenching of adriamycin by specific antibodies. *Immunochemistry* 12:291–296, 1975.

Christiansen, E. N., and Kvamme, E. Effects of thermal treatment on mitochondria of brain, liver, and ascites cell. *Acta Physiol. Scand.* 76:472–484, 1969.

Clark, W. G.; Cumby, H. R.; and Davis, H. E. The hyperthermic effect of intracerebroventricular cholera enterotoxin in the unanaesthetized cat. *J. Physiol.* 240:493–504, 1974.

Clarke, P. R.; Hill, C. R.; and Adams, K. Synergism between ultrasound and x-rays in tumor therapy. *Br. J. Radiol.* 43:97–99, 1970.

Cleary, S. F. Uncertainties in the evaluation of the biological effects of microwave and radiofrequency radiation. *Health Phys.* 25:387–404, 1973.

Cockett, A. T. K. et al. Enhancement of regional bladder megavoltage irradiation in bladder cancer using local bladder hyperthermia. *J. Urol.* 97:1034–1039, 1967.

Connor, W. G. et al. Prospects for hyperthermia in human cancer therapy: part II. *Radiology* 123:497–503, 1977.

Copeland, E. S. Effect of selective tumor heating on the localization of 131I fibrinogen in the Walker carcinoma 256 (I. Heating by immersion in warm water.) *Acta Radiologica* 9:205–224, 1970.

Copeland, E. S., and Michaelson, S. M. Effect of selective tumor heating on the localization of 131I fibrinogen in the Walker carcinoma 256. (II. Heating with microwaves.) *Acta Radiologica* 9:323–336, 1970.

Corbit, J. D. Behavioral regulation of hypothalmic temperature. *Science* 166:256–258, 1969.

Corry, P. M.; Robinson, S.; and Getz, S. Hyperthermic effects on DNA repair mechanisms. *Radiology* 123:475–482, 1977.

Courtenay, V. D. et al. In vitro and in vivo radiosensitivity of human tumor cells obtained from a pancreatic carcinoma xenograft. *Nature* 263:771–772, 1976.

Cranston, W. I. et al. Noradrenergic mechanism in the central control of body temperature. *Int. J. Biometeorol.* 15:301–304, 1971.

Crile, G., Jr. Selective destruction of cancers after exposure to heat. *Ann. Surg.* 156:404–407, 1962.

Crile, G., Jr. The effects of heat and radiation on cancers implanted on the feet of mice. *Cancer Res.* 23:372–380, 1963.

Darzynkiewicz, Z. et al. Effect of divalent cations and alcohols. *J. Cell Biol.* 68:1–10, 1976.

Darzynkiewicz, Z. et al. Effect of 0.25 N sodium chloride treatment on DNA denaturation in situ in thymus lymphocytes. *Exp. Cell Res.* 100:393–396, 1976.

Debevec, M. Uber die verifizierung der lungentumoren vor der strahlentherapie. *Strahlentherapie* 147:149–158, 1974.

DeHoratius, R. J. et al. Immunologic function in humans before and after hyperthermia and chemotherapy for disseminated malignancy. *J. Natl. Cancer Inst.* 58:905–911, 1977.

Delario, A. J. Methods of enhancing roentgen-ray action. *Radiology* 25:617–627, 1935.

De La Torre, C. et al. The effect of thermal shock on the division cycle of meristematic cells. *Cell Tissue Kinet.* 4:569–575, 1971.

Devyatkov, N. D. Influence of millimeter-hand electromagnetic radiation on biological objects. *Soviet Physics, Uspekhi* 16:568–579, 1974.

Dewey, W. C. et al. Heat-induced lethality and chromosomal damage in synchronized Chinese hamster cells treated with 5-bromodeoxyurdine. *Int. J. Radiat. Biol.* 20:505–520, 1971.

Dewey, W. C. et al. *Radiology* 123:463–474, 1977.

Dickens, F.; Evans, S. F.; and Weil-Malherbe, H. The action of short radio waves on tissues. I. Effects produced in vitro. *Am. J. Cancer* 28:603–620, 1936.

Dickson, J. A. Hyperthermia in the treatment of cancer. *Cancer Chemother.* 58:294–296, 1974.

Dickson, J. A., and Calderwood, S. K. In vivo hyperthermia of Yoshida tumor induces entry of nonproliferating cells into cycle. *Nature* 263:772–774, 1976.

Dickson, J. A., and Ellis, H. A. Stimulation of tumor cell dissemination by raised temperature (42°C) in rats with transplanted Yoshida tumors. *Nature* 248:354–358, 1974.

Dickson, J. A., and Ellis, H. A. The influence of tumor volume and the degree of heating on the response of the solid Yoshida sarcoma to hyperthermia. *Cancer Res.* 36:1188–1195, 1976.

Dickson, J. A., and Oswald, B. E. The sensitivity of a malignant cell line to hyperthermia (42°C) at low intracellular pH. *Br. J. Cancer* 34:262–271, 1976.

Dickson, J. A., and Shah, D. M. The effects of hyperthermia (42°C) on the biochemistry and growth of a malignant cell line. *Eur. J. Cancer* 8:561–571, 1972.

Dickson, J. A., and Suzangar, M. In vitro sensitivity screening system for human cancers to drugs and hyperthermia (42°C). *Br. J. Cancer* 28:81, 1973.

Dickson, J. A., and Suzangar, M. In vitro-in vivo studies on the susceptibility of the solid Yoshida sarcoma to drugs and hyperthermia (42°C). *Cancer Res.* 34:1263–1274, 1974.

Dietzel, F. Tumor synchronization by microwaves: animal experiments concerning susceptibility of proliferation kinetics of the tumor in vivo to nonionizing radiation. *Strahlentherapie* 148:531–542, 1974.

Dietzel, F. Radiation sensitization of tumor cells by microwaves. End of the oxygen problem? *Naturwissenschaften* 62:44–45, 1975.

Dietzel, F. et al. Zur frage der tumorheilung durch alleinige hochfrequenz-hyperthermie (dezimeterwellen)—tierexperimentelle untersuchungen. *Biomed. Tech. (Berlin)* 16:213–220, 1971.

Dietzel, F. et al. Microwaves in radiotherapy of tumors—alternative to heavy particles? *Strahlentherapie* 149:438–441, 1975.

Dietzel, F.; Klobe, G.; and Seibert, G. Der einfluss einer hochfrequenz-hyperthermie auf den verlauf des Ehrlich-aszites-karzinoms der maus. *Strahlentherapie* 149:105–117, 1975.

Dilworth, N. M. The importance of changes in body temperature in paediatric surgery and anaesthesia. *Anaesth. Intensive Care* 1:480–485, 1973.

Doub, H. P. Artificial fever as a therapeutic agent. *Radiology* 25:360–361, 1935.

Doub, H. P. Osteogenic sarcoma of the clavicle treated with radiation and fever therapy. *Radiology* 25:355–356, 1935.

Draper, J. W., and Boag, J. W. Skin temperature distributions over veins and tumors. *Phys. Med. Biol.* 16:645–654, 1971.

Durand, R. E. Isolation of cell subpopulations from in vitro tumor models according to sedimentation velocity. *Cancer Res.* 35:1295–1300, 1975.

Eberhardt, A. et al. Immunological studies of individuals working in high environmental temperature. *Acta Physiol. Pol.* 23:291–297, 1972.

Ecker, H. A. Biomedical applications of EM radiation. *Microwave Journal* 18:47–50, 1975.

Edelman, G. M.; Yahara, I.; and Wang, J. L. Receptor mobility and receptor-cytoplasmic interactions in lymphocytes. *Proc. Natl. Acad. Sci. USA* 70:1442–1446, 1973.

Edidin, M. Rotational and translational diffusion in membranes. *Annu. Rev. Biophys. Bioeng.* 3:179–201, 1974.

Editorial. Fever in malignant disease. *Br. Med. J.* 1:591–592, 1974.

Elkind, M. M. et al. Sublethal and lethal radiation damage. *Nature* 214:1088–1092, 1967.

El Piner, I. E. Ultrasonics in experimental oncology (Russian). *Patol. Fiziol. Eksp. Ter.* 13:3–9, 1969 (Russ.).

El-Sayed, M. A. Double resonance and the properties of the lowest excited triplet state of organic molecules. *Ann. Rev. of Phys. Chem.* 26:235–258, 1975.

Esser, A. F., and Souza, K. A. Correlation between thermal death and membrane fluidity in bacillus stearothermophilus. *Proc. Natl. Acad. Sci. USA* 71:4111–4117, 1974.

Euler, J. et al. Wirkung von temperatur, pH, und thio-tepa auf angehraten und thymidineinbau von aszitestumorzellen. *Weiner Klinische Wochenschrift* 86:211–219, 1974.

Eyckmans, L.; Wouters, R.; and Vandenbroucke, J. Unexplained fever: seven-year experience. *Acta Clin. Belg.* 28:232–237, 1973.

Fahimi, H. D., and Cotran, R. S. Permeability studies in heat-induced injury of skeletal muscle using lanthanum as fine structural tracer. *Am. J. Pathol.* 62:143–152, 1971.

Fan, L. T.; Hsu, F. T.; and Hwang, C. L. A review on mathematical models of the human thermal system. *IEEE Trans. Biomed. Eng.* 18:218–234, 1971.

Ferguson, K. et al. Alteration of fatty acid composition of LM cells by lipid supplementation and temperature. *Biochemistry* 14:146–151, 1975.

Field, J.; Fuhrman, F. A.; and Martin, A. W. Effect of temperature on the oxygen consumption of brain tissue. *J. Neurophysiol.* 7:117–126, 1944.

Flanigan, W. F.; Bowman, R. R.; and Lowell, W. R. Nonmetallic electrode system for recording EEG and

ECG in electromagnetic fields. *Physiol. Behav.* 18: 531–533, 1977.

Fowler, J. F.; Adams, G. E.; and Denekamp, J. Radiosensitizers of hypoxic cells in solid tumors. *Cancer Treat. Rev.* 3:227–256, 1976.

Freeman, M. E. et al. Thermogenic action of progesterone in the rat. *Endocrinology* 86:717–720, 1970.

Frohlich, H. The extraordinary dielectric properties of biological materials and the action of enzymes. *Proc. Natl. Acad. Sci. USA* 72:4211–4215, 1975.

Frohlich, H. Dielectric theory and ion channels in nerve membranes. *Biosystems* 8:193–194, 1977.

Fry, W. J., and Fry, F. J. Fundamental neurological research and human neurosurgery using intense ultrasound. *IEEE Trans. Biomed. Eng.* ME-7:166–181, 1960.

Fung, B. M. Non-freezable water and spin-lattice relaxation time in muscle containing a growing tumor. *Biochem. Biophys. Acta* 362:209–214, 1974.

Furuyama, F. et al. Proceedings: body temperature equilibrium in hyperthermia and ambient temperature. *J. Physiol. Soc. Jpn.* 36:394–395, 1974.

Gale, C. C.; Mathews, M.; and Yound, J. Behavioral thermoregulatory responses to hypothalmic cooling and warming in baboons. *Physiol. Behav.* 5:1–6, 1976.

Gallo, R. C. et al. Relationships between components in primate RNA tumor viruses and in the cytoplasm of human leukemic cells: implications to leukemogenesis. *Cold Spring Harbor Symp. Quant. Biol.* 39:933–961, 1974–5.

Gautherie, M. Mechanism of local skin thermoregulation in man essentially controlled by a cooperative biosynthesis of bradykinin. *J. Physiol.* (Paris) 63:251–253, 1971.

George, K. C.; Hirst, D. G.; and McNally, N. J. Effect of hyperthermia on cytotoxicity of the radiosensitizer Ro-07-0582 in a solid mouse tumor. *Br. J. Cancer* 35:372–375, 1977.

Gericke, D. et al. In vitro thermosensibility of experimental tumors in small animals. *Naturwissenschaften* 58:155–156, 1971.

Gerner, E. W. et al. A transient thermotolerant survival response produced by single thermal doses in HeLa cells. *Cancer Res.* 36:1035–1040, 1976.

Gerner, E. W.; Leith, J. T.; and Boone, M. L. M. Mammalian cell survival response following irradiation with 4 MeV x-rays or accelerated helium ions combined with hyperthermia. *Radiology* 119:715–720, 1976.

Gerner, E. W., and Schneider, M. J. Induced thermal resistance in HeLa cells. *Nature* 256:500–502, 1975.

Gerweck, L. E. Modification of cell lethality at elevated temperatures. The pH effect. *Radiat. Res.* 70:224–235, 1977.

Gerweck, L. E.; Gillette, E. L.; and Dewey, W. C. Killing of Chinese hamster cells in vitro by heating under hypoxic or aerobic conditions. *Eur. J. Cancer* 10:691–693, 1974.

Gerweck, L. E.; Gillette, E. L.; and Dewey, W. C. Effect of heat and radiation on synchronous Chinese hamster cells; killing and repair. *Radiat. Res.* 64:611–623, 1975.

Gerweck, L., and Rottinger, E. Enhancement of mammalian cell sensitivity to hyperthermia by pH alteration. *Radiat. Res.* 67:508–511, 1976.

Gilbert, H. A., and Kagan, A. R. The immune response and cancer. A guide. *Radiologia Clinica et Biologica* 43:409–444, 1974.

Giles, U. The historic development and modern application of artificial heat. *New Orleans Med. Surg. J.* 91:655–670, 1938–9.

Gillette, E. L., and Thrall, D. E. Combination of heat and ionizing radiation on the C3H mouse mammary adenocarcinoma in vivo: effect of variable heat doses and repair of heat damage. *Radiat. Res.* 59:185–186, 1973.

Giovanella, B. C. et al. Selective lethal effect of supranormal temperatures on mouse sarcoma cells. *Cancer Res.* 33:2568–2578, 1973.

Giovanella, B. C., and Heidelberger, C. Biochemical and biological effects of heat in normal and neoplastic cells. *Proc. Am. Assoc. Cancer Res.* 9:24, 1968.

Giovanella, B. C., and Heidelberger, C. Mouse epidermal cells and carcinogenesis. I. Isolation of skin constituents. *Cancer Res.* 25:161–183, 1975.

Giovanella, B. C.; Lohman, W. A.; and Heidelberger, C. Effects of elevated temperatures and drugs on the viability of L1210 leukemia cells. *Cancer Res.* 30:1623–1631, 1970.

Giovanella, B. C.; Mosti, R.; and Heidelberger, C. Further studies of the lethal effects of heat on tumor cells. *Proc. Am. Assoc. Cancer Res.* 10:29, 1969.

Givoni, B., and Goldman, R. F. Predicting rectal temperature response to work, environment, and clothing. *J. Appl. Physiol.* 32:812–821, 1972.

Gold, J. Metabolic profiles in human solid tumors: I. A new technique for the utilization of human solid tumors in cancer research and its application to the anaerobic glycolysis of isologous benign and malignant colon tissues. *Cancer Res.* 26:695–705, 1966.

Gold, J. Cancer cachexia and gluconeogenesis. *Ann. NY Acad. Sci.* 230:103–110, 1974.

Goldenberg, D. M., and Langner, M. Direct and abscopal antitumor action of local hyperthermia. *Zeitschrift fur Naturforschung* 26B:359–361, 1971.

Grundler, W.; Keilmann, F.; and Frohlich, H. Resonant growth rate response of yeast cells irradiated by weak microwaves. *Physics Letters* 62:463–466, 1977.

Gullino, P. M.; Clark, S. H.; and Grantham, F. H. The interstitial fluid of solid tumors. *Cancer Res.* 24:780–797, 1964.

Gullino, P. M., and Grantham, F. H. Studies on the exchange of fluids between host and tumor. II. The blood flow of hepatomas and other tumors in rats and mice. *J. Natl. Cancer Inst.* 27:1465–1491, 1962.

Guru, B. S., and Chen, K. Hyperthermia by local EM heating and local conductivity change. *IEEE Trans. Biomed. Eng.* 24:473–477, 1977.

Guy, A. W. Electromagnetic fields and relative heating patterns due to a rectangular aperture source in direct contact with bilayered biological tissue. *IEEE Trans. MTT* 19:214–223, 1971.

Guy, A. W., and Lehmann, J. F. On the determination of an optimum microwave diathermy frequency for a direct contact applicator. *IEEE Trans. Biomed. Eng.* BME-13:76–87, 1966.

Guy, A. W.; Lehmann, J. F.; and Stonebridge, J. B. Therapeutic applications of electromagnetic power. *Proc. IEEE* 62:55–75, 1974.

Haddow, A. Addendum to molecular repair, wound healing, and carcinogenesis: tumor production, a possible overhealing? *Adv. Cancer Res.* 20:323–366, 1974.

Hahn, E. W.; Feingold, S. M.; and Kim, J. H. The effect of radiation and hyperthermia on growing bone. *Radiat. Res.* 59:186, 1973.

Hahn, E. W.; Feingold, S. M.; and Kim, J. H. Repair of normal tissue injury from local hyperthermia. *Radiat. Res.* 70:634, 1977.

Hahn, G. M. et al. Cell survival and repair of plateau-phase cultures after chemotherapy—relevance to tumor therapy and to the in vitro screening of new agents. *Cancer Chemother. Rep.* 57:473–475, 1973.

Hahn, G. M. et al. Repair of potentially lethal lesions in X-irradiated, density-inhibited Chinese hamster cells: metabolic effects and hypoxia. *Radiat. Res.* 55:280–290, 1973.

Hahn, G. M. et al. Response of solid tumor cells exposed to chemotherapeutic agents in vivo: cell survival after 2- and 24-hour exposure. *J. Natl. Cancer Inst.* 50:529–533, 1973.

Hahn, G. M.; Braun, J.; and Har-Kedar, I. Thermochemotherapy synergism between hyperthermia (42–43°C) and adriamycin (or bleomycin) in mammalian cell inactivation. *Proc. Natl. Acad. Sci. USA* 72:937–940, 1975.

Hahn, G. M., and Kal, H. B. Kinetic responses of murine sarcoma cells to radiation and hyperthermia in vivo and in vitro. *Cancer Res.* 36:1923–1929, 1976.

Hahn, G. M., and Kase, K. Differential heat response of normal and transformed human cells in tissue culture. *Nature* 255:228–230, 1975.

Hahn, G. M.; Li, G. C.; and Shiu, E. Interaction of amphotericin B and 43°C hyperthermia. *Cancer Res.* 37:761–764, 1977.

Hahn, G. M., and Little, J. B. Plateau-phase cultures of mammalian cells: an in vitro model for human cancer. *Curr. Top. Radiat. Res. Q.* 8:39–83, 1972.

Hales, J. R. S. et al. Thermoregulatory effects of prostaglandins E1, E2, F1, and F2 in the sheep. *Pfluegers Archiv* 339:125–133, 1973.

Hales, J. R. S., and Dampney, R. A. L. The redistribution of cardiac output in the dog during heat stress. *J. Therm. Biol.* 1:29–34, 1975.

Hall, E. J. Response of cultured cells to hyperthermia with special reference to thermal tolerance and hypoxia. *Br. J. Radiol.* 49:806, 1976.

Hammel, H. T. Regulation of internal body temperature. *Annu. Rev. Physiol.* 30:641–710, 1968.

Hardy, J. D. Control of heat loss and heat production in physiologic temperature regulation. *Harvey Lect.* 49:242, 1954.

Hardy, J. D. Thermoregulatory responses to temperature changes in the midbrain of the rabbit. *Fed. Proc.* 28:713, 1969.

Harisiadis, L. et al. Hyperthermia: biological studies at the cellular level. *Radiology* 117:447–452, 1975.

Harisiadis, L.; Sung, D.; and Hall, E. J. Thermal tolerance and repair of thermal damage by cultured cells. *Radiology* 123:505–509, 1977.

Har-Kedar, I. Effects of hyperthermia on bladder cancer. *Br. Med. J.* 3:345–346, 1974.

Har-Kedar, I., and Bleehen, N. M. Experimental and clinical aspects of hyperthermia applied to the treatment of cancer with special reference to the role of ultrasonic and microwave heating. *Adv. Radiat. Biol.* 6:229–266, 1976.

Harris, M. Temperature-resistant variants in clonal populations of pig kidney cells. *Exp. Cell Res.* 46:301–314, 1967.

Harris M. Growth and survival of mammalian cells under continuous thermal stress. *Exp. Cell Res.* 56:382–386, 1969.

Harris, M. Mutation rates in cells at different ploidy levels. *J. Cell. Physiol.* 78:177–184, 1971.

Hartman, J. T., and Crile, G., Jr. Heat treatment of osteogenic sarcoma. *Clin. Orthop.* 61:269–276, 1968.

Hayward, J. N., and Baker M. A comparative study of the role of the cerebral arterial blood in the regulation of brain temperature in five mammals. *Brain Res.* 16:417–440, 1969.

Hazel, J. R., and Prosser, C. L. Molecular mechanisms of temperature compensation in poikilotherms. *Physiol. Rev.* 54:620–677, 1974.

Healey, J. E., Jr. Vascular patterns in human metastatic liver tumors. *Surg. Gynecol. Obstet.* 120:1187–1193, 1965.

Heath, J. E.; Williams, B. A.; and Mills, S. H. Interactions of hypothalmic thermosensitivity and body size in vertebrates. *Int. J. Biometeorol.* 15:254–257, 1971.

Heckel, M. Ganzkorpererwarmung und steurerbare hyperthermie mittels tiefpenetrierender kurzwelliger infrarotstrahlung. *Med. Welt* 308–313, 1970.

Hegsted, D. M. Energy needs and energy utilization. *Nutrition Rev.* 32:33–39, 1974.

Heilbrunn, L. V. The colloid chemistry of protoplasm. IV. The heat coagulation of protoplasm. *Am. J. Physiol.* 69:190–199, 1924.

Heine, H. et al. The behavior of HeLa-S3 cells under the influence of supranormal temperatures. *J. Ultrastruct. Res.* 34:375–396, 1971.

Heinrich, V. M., and Osswald, H. Prufund der Krebs-Mehrschritt-Therapie (kombination der glukosevorbehandlung mit chemotherapie und ganzkorperhyperthermia), am Walker- und DS-karzinosarkom der ratte. *Arch. Geschwulstforsch.* 43:310–319, 1974.

Henderson, M. A., and Pettigrew, R. T. Induction of controlled hyperthermia in treatment of cancer. *Lancet* 19:1275–1277, 1971.

Henle, K. J., and Dethlefsen, L. A. Sensitization to hyperthermia (45°C) of normal and thermotolerant CHO cells by anisotonic media. *Radiat. Res.* 70:632, 1977.

Henle, K. J.; Karamuz, J. E.; and Leeper, D. B. Induction of thermotolerance in Chinese hamster ovary cells by high (45°C) or low (40°C) hyperthermia. *Cancer Res.* 38:570–574, 1978.

Henle, K. J., and Leeper, D. B. Interaction of hyperthermia and radiation in CHP cells: recovery kinetics. *Radiat. Res.* 66:505–518, 1976.

Henriques, F. C. Studies of thermal injury. *Archives of Pathology* 43:489–502, 1947.

Hensel, H. Neural processes in thermoregulation. *Physiol. Rev.* 53:948–1017, 1973.

Hill, R. P., and Stanley, J. A. The response of hypoxic B16 melanoma cells to in vivo treatment with chemotherapeutic agents. *Cancer Res.* 35:1147–1153, 1975.

Himms-Hagen, J. Cellular thermogenesis. *Annu. Rev. Physiol.* 38:315–351, 1976.

Hinshaw, J. R. Early changes in the depth of burns. *Ann. NY Acad. Sci.* 150:548–553, 1968.

Ho, H. S., and McManaway, M. Heat-dissipation rate of mice after microwave irradiation. *J. Microwave Power* 12:93–100, 1977.

Holm, D. A. The effects of non-thermal radio frequency radiation on human lymphocytes in vitro. *Experientia* 26:992–994, 1970.

Holman, R. A. Hyperthermia and cancer. *Lancet* 1:1027, 1975.

Holroyde, C. P. et al. Altered glucose metabolism in metastatic carcinoma. *Cancer Res.* 35:3710–3714, 1975.

Holt, J. A. G. The cure of cancer. *Australas. Radiol.* 18:15–16, 1974.

Holt, J. A. G. The use of VHF radiowaves in cancer therapy. *Australas. Radiol.* 19:223–241, 1975.

Hori, T., and Harada, Y. The effects of ambient and hypothalmic temperatures on the hyperthermic responses to prostaglandins E1 and E2. *Pfluegers Archiv* 350:123–134, 1974.

Isenberg, I. Some comments on broad electron spin resonance absorptions observed in nucleic acid preparations. *Biochem Biophys. Res. Commun.* 5:139–143, 1961.

Jares, J. J., and Warren, S. L. Some fundamental aspects of the cancer problem: the combined effects of roentgen-radiation and fever upon malignant tissue. In *Occasional publications of the American Association for the Advancement of Science. Science* 85(suppl. 4):225–226, 1937.

Jares, J. J., and Warren, S. L. Physiological effects of radiation. I. A study of the in vitro effect of high fever temperatures upon certain experimental animal tumors. *Am. J. Roentgenol.* 41:685–708, 1939.

Johnson, C. C., and Guy, A. W. Nonionizing electromagnetic wave effects in biological materials and systems. *Proc. IEEE* 60:692–718, 1972.

Johnson, H. A., and Pavelec, M. Thermal enhancement of thio-TEPA cytotoxicity. *J. Natl. Cancer Inst.* 50:903–910, 1973.

Johnson, H. J. The action of short radio waves on tissues. III. A comparison of the thermal sensitivities of transplantable tumors in vivo and in vitro. *Am. J. Cancer* 38:533–550, 1940.

Justensen, D. R. Diathermy versus the microwaves and other radio-frequency radiations: a rose by another name is a cabbage. *Radio Science* 12:355–364, 1977.

Juul, T., and Kemp, T. Uber den einfluss von radium- und rotgenstrahlen, ultraviolettem licht und hitze auf die zellteilung bei warmblutigen tieren. Studien an gewebekulturen. *Strahlentherapie* 48:457–499, 1933.

Kachani, Z. F., and Sabin, A. B. Reproductive capacity and viability at higher temperatures of various transformed hamster cell lines. *J. Natl. Cancer Inst.* 43:469–480, 1969.

Kal, H. B., and Hahn, G. M. Kinetic responses of murine sarcoma cells to radiation and hyperthermia in vivo and in vitro. *Cancer Res.* 36:1923–1929, 1976.

Kal, H. B.; Hatfield, M.; and Hahn, G. M. Cell cycle progression of murine sarcoma cells after X irradiation or heat shock. *Radiology* 117:215–218, 1975.

Kase, K. R., and Hahn, G. M. Differential heat response of normal and transformed human cells in tissue culture. *Nature* 255:228–230, 1975.

Kase, K. R., and Hahn, G. M. Comparison of some response to hyperthermia by normal human diploid cells and neoplastic cells from the same origin. *Eur. J. Cancer* 12:481–491, 1976.

Katsumi, S. The experimental and clinical study on the destructive action of intense focused ultrasound to the malignant tumor tissues. *Archiv für Japanische Chirurgie* 35:489–507, 1966.

Kidd, J. G., and Rous, P. A transplantable rabbit carcinoma originating in a virus-induced papilloma and containing the virus in masked or altered form. *J. Exp. Med.* 71:813–838, 1940.

Kim, J. H. et al. Local tumor hyperthermia in combination with radiation therapy. I. Malignant cutaneous lesions. *Cancer* 40:161–169, 1977.

Kim, J. H.; Kim, S. H.; and Hahn, E. W. The radiosensitization of hypoxic tumor cells by hyperthermia. *Radiat. Res.* 59:140, 1973.

Kim, J. H.; Kim, S. H.; and Hahn, E. Thermal enhancement of the radiosensitivity using cultured normal and neoplastic cells. *Am. J. Roentgenol.* 121:860–864, 1974.

Kiricuta, I. C., and Simplaceanu, V. Tissue water content and nuclear magnetic resonance in normal and tumor tissues. *Cancer Res.* 35:1164–1167, 1975.

Klastersky, J. Fever, its pathogenesis and its role in bacterial infections. *Acta Clin. Belg.* 28:266–277, 1973.

Klauder, J. V. Fever therapy of mycosis fungoides. *JAMA* 106:201–205, 1936.

Kligerman, M. M., and Henel, D. K. Some aspects of the microcirculation of a transplantable experimental tumor. *Radiology* 76:810–817, 1961.

Koch-Weser, J., and Blinks, J. R. The influence of the interval between beats on myocardial contractility. *Pharmacol. Rev.* 15:601–652, 1963.

Kopp, I., and Solomon, H. C. Shock syndrome in therapeutic hyperpyrexia. *Arch. Intern. Med.* 60:597–622, 1927.

Koppanyi, T., and Maling, H. M. The effects of pretreatment with reserpine, a-methyl-p-tyrosine, or prostaglandin E1 on adrenergic salivation (36553). *Proc. Soc. Exp. Biol. Med.* 140:787–793, 1972.

Kramer, S. Hyperthermia in the treatment of the cancer patient. *Cancer* 37:2075–2083, 1976.

Kruuv, J. A.; Inch, W. R.; and McCredie, J. A. Blood flow and oxygenation of tumors in mice. *Cancer* 20:60–65, 1967.

Kvetina, J., and Guaitani, A. A versatile method for the in vitro perfusion of isolated organs of rats and mice with particular reference to liver. *Pharmacology* 2:65–80, 1969.

Lampietro, P. F. et al. Exposure to heat: comparison of responses of dog and man. *Int. J. Biometeorol.* 10:175–185, 1966.

Lazarides, E. Tropomyosin antibody: the specific localization of tropomyosin in nonmuscle cells. *J. Cell Biol.* 65:549–561, 1975.

Leeper, D. B.; Karamuz, J. E.; and Henle, K. J. Hyperthermia-induced alterations of macromolecular synthesis. *Radiat. Res.* 70:610–611, 1977.

Lehmann, J. F. et al. Heating patterns produced by short-wave diathermy using helical induction coil applicators. *Arch. Phys. Med. Rehabil.* 49:193–198, 1968.

Lehmann, J. F.; DeLateur, B. J.; and Stonebridge, J. B. Selective muscle heating by short-wave diathermy with a helical coil. *Arch. Phys. Med. Rehabil.* 50:117–123, 1969.

Lehmann, J. F., and Johnson, E. W. Some factors influencing the temperature distribution in thighs exposed to ultrasound. *Arch. Phys. Med. Rehabil.* 39:347–356, 1958.

Leith, J. T. et al. Hyperthermic potentiation: biological aspects and applications to radiation therapy. *Cancer* 39:766–779, 1977.

Lele, P. P. A simple method for production of trackless focal lesions with focused ultrasound: physical factors. *J. Physiol.* 160:494–512, 1962.

Lele, P. P. Production of deep focal lesions by focused ultrasound—current status. *Ultrasonics* 5:105–112, 1967.

Lele, P. P. Application of ultrasound in medicine. *N. Engl. J. Med.* 286:1317–1318, 1972.

Lele, P. P., and Young, G. F. Focal lesions in the brain of growing rabbits produced by focused ultrasound. *Exp. Neurol.* 9:502–511, 1964.

Lerner, R. M., and Carstensen, E. L. Frequency dependence of thresholds for ultrasonic production of thermal lesions in tissue. *J. Acoust. Soc. Am.* 54:504–506, 1973.

Letter. General discussion: treatment of prostatic cancer. *Cancer Chemother. Rep.* 59:251–254, 1975.

LeVeen, H. H. et al. *JAMA* 235:2198–2224, 1976.

Levine, E. M., and Robbins, E. B. Differential temperature sensitivity of normal and cancer cells in culture. *J. Cell. Physiol.* 76:373–380, 1970.

Levinson, C., and Hempling, H. G. The role of ion transport in the regulation of respiration in the Ehrlich mouse ascites-tumor cell. *Biochim. Biophys. Acta* 135:306–318, 1967.

Lewin, S., and Pepper, D. S. Variation of the melting temperature of calf-thymus DNA with pH and type of buffer. *Arch. Biochem. Biophys.* 109:192–194, 1965.

Lindgren, A. G. H. The vascular supply of tumors with special reference to the capillary angioarchitecture. *Acta Pathol. Microbiol. Scand.* 22:493–522, 1945.

Linke, C. A. et al. Effects of marked hyperthermia upon the canine bladder. *J. Urol.* 107:599–602, 1972.

Linke, C. A. et al. Localized tissue destruction by high-intensity focused ultrasound. *Arch. Surg.* 107:887–891, 1973.

Linke, C. A. et al. Renal cortical necrosis: a model for the study of juxtamedullary nephron physiology. *J. Appl. Physiol.* 37:228–234, 1974.

Linke, C. A. et al. Response of rat testis to localized hyperthermia. *Urology* 5:76081, 1975.

Linke, C. A.; Lounsberry, W.; and Goldschmidt, V. Effects of microwaves on normal tissues. *J. Urol.* 88:303–311, 1962.

Linke, C. A.; Lounsberry, W.; and Goldschmidt, V. Comparison of tissue heating in ten second electrocoagulation and ten second soldering iron lesions. *Am. J. Med. Elec.* 2:110–114, 1964.

Linke, C. A.; Netto, I. C. V.; and Elbadawi, A. Marked hyperthermia: effect on male canine urinary bladder. *Urology* 1:347–350, 1973.

Lounsberry, W. et al. The early histologic changes following electrocoagulation. *J. Urol.* 86:321–329, 1961.

Love, R.; Soriano, R. Z.; and Walsh, R. J. Effect of hyperthermia on normal and neoplastic cells in vitro. *Cancer Res.* 30:1525–1533, 1970.

Lunglmayr, G. et al. Bladder hyperthermia in the treatment of vesical papillomatosis. *Int. Urol. Nephrol.* 5:75–84, 1973.

Lunglmayr, G. et al. Experimentelle untersuchungen uber die wirkung temporarer hyperthermie auf blasentumore. *Urologie Internationalis* 28:314–321, 1973.

Magbagbeola, J. A. D. The effect of atropine premedication of body temperature of children in the tropics. *Br. J. Anaesth.* 45:1139, 1973.

Maling, H. M. et al. Inflammation induced by histamine, serotonin, bradykinin, and compound 48/80 in the rat: antagonists and mechanisms of action. *J. Pharmacol. Exp. Ther.* 191:300–310, 1974.

Maling, H. M.; Williams, M. A.; and Koppanyi, T. Salivation in mice as an index of adrenergic activity. *Arch. Int. Pharmacodyn. Ther.* 199:318–332, 1972.

Manlapaz, J. S. et al. Effects of ultrasonic radiation in experimental focal epilepsy in the cat. *Exp. Neurol.* 10:345–356, 1964.

Marmor, J. B. Thermal denaturation of deoxyribonucleic acid isolated from a thermophile. *Biochim. Biophys. Acta* 38:342–343, 1960.

Marmor, J. B.; Hahn, N.; and Hahn, G. M. Tumor cure and cell survival after localized radiofrequency heating. *Cancer Res.* 37:879–883, 1977.

Marsili, M. et al. Bacterial invasion of the liver during the course of heatstroke in the dog heatstroke model. U.S. Army Research Institute on Environmental Medicine, Natick, Mass. 01760, Internal Report, 1974.

Marton, J. P. Conjectures in superconductivity and cancer. *Physiol. Chem. Phys.* 5:259–270, 1973.

Maskrey, M. Respiratory responses in heat-exposed rabbits: inhibition of tachypnoea offset by increase in tidal volume. *Experientia* 33:353–354, 1977.

Mayer, G. et al. Biologie—action de champs magnétiques associés à des ondes électromagnétiques sur l'orchite trypanosomienne du lapin. *C. R. Acad. Sci.* 274:3011–3014, 1972.

McCormick, W., and Penman, S. Regulation of protein synthesis in HeLa cells: translation at elevated temperatures. *J. Mol. Biol.* 39:315–333, 1969.

McSkimin, H. J. Velocity of sound in distilled water for the temperature range 20–75°C. *J. Acoust. Soc. Am.* 37:325–328, 1965.

Melzack, R., and Wall, P. D. Pain mechanisms: a new theory. *Science* 150:971–979, 1965.

Mendecki, J.; Friedenthal, E.; and Botstein, C. Effects of microwave-induced local hyperthermia on mammary adenocarcinoma in C3H mice. *Cancer Res.* 36:2113–2114, 1976.

Mendel, B. Uber die hitzeempfindlichkeit der krebszelle. *Klin. Wochenschr.* 7:457, 1928.

Merriman, J. R.; Holmquest, H. J.; and Osborne, S. L. A new method of producing heat in tissues: the inductotherm. *Am. J. Med. Sci.* 187:677–683, 1934.

Michaelson, S. M. Effects of exposure to microwaves: problems and perspectives. *Environ. Health Perspect.* 8:133–156, 1974.

Michaelson, S. M. Central nervous system responses to microwave-induced heating. *Neurosci. Res. Program Bull.* 15:98–100, 1977.

Michaelson, S. M. et al. Effects of electromagnetic radiations on physiologic responses. *Aerosp. Med.* 38:293–298, 1967.

Michaelson, S. M.; Thomson, R. A. E.; and Howland, J. W. Physiologic aspects of microwave irradiation of mammals. *Am. J. Physiol.* 211:351–356, 1961.

Miller, R. C. et al. Prospects for hyperthermia in human cancer therapy: part I. *Radiology* 123:489–495, 1977.

Miura, M. et al. A study of leukemic cell injury by physical agents. *Cancer Res.* 31:1451–1456, 1971.

Miura, O., and Usami, Y. Heat sensitivity of tumor cells. *Tohoku J. Exp. Med.* 113:291–297, 1974.

Mohindra, J. K., and Rauth, A. M. Increased cell killing by metronidazole and nitrofurazone of hypoxic compared to aerobic mammalian cells. *Cancer Res.* 36:930–936, 1976.

Moller, U., and Bojsen, J. Temperature and blood flow measurements in and around 7, 12-dimethylbenz (a) anthracene-induced tumors and Walker 256 carcinosarcomas in rats. *Cancer Res.* 35:3116–3121, 1975.

Mondovi, B. et al. Biochemical mechanism of selective heat sensitivity of tumor cells: preliminary results. *Ital. J. Biochem.* 17:101–106, 1968.

Mondovi, B. et al. The biochemical mechanism of selective heat sensitivity of cancer cells. I. Studies on cellular respiration. *Eur. J. Cancer* 5:129–136, 1969.

Mondovi, B. et al. The biochemical mechanism of selective heat sensitivity of cancer cells. II. Studies on nucleic acids and protein synthesis. *Eur. J. Cancer* 5:137–146, 1969.

Mondovi, B. et al. Increased immunogenicity of Ehrlich ascites cells after heat treatment. *Cancer* 30:885–888, 1972.

Moritz, A. R., and Henriques, F. C. Studies of thermal injury. II. The relative importance of time and surface temperature in the causation of cutaneous burns. *Am. J. Pathol.* 23:695–720, 1947.

Morrison, S. D. Partition of energy expenditure between host and tumor. *Cancer Res.* 31:98–107, 1971.

Muckle, D. S. The selective effect of heat in cancer. *Ann. R. Coll. Surg. Engl.* 54:72–77, 1974.

Muckle, D. S., and Dickson, J. A. The selective inhibitory effect of hyperthermia on the metabolism and growth of malignant cells. *Br. J. Cancer* 25:771–778, 1971.

Muckle, D. S., and Dickson, J. A. Hyperthermia (42°C) as an adjuvant to radiotherapy and chemotherapy in the treatment of the allogeneic VX2 carcinoma in the rabbit. *Br. J. Cancer* 27:307–315, 1973.

Muckle, D. S.; Dickson, J. A.; and Johnston, I. D. A. The effect of hyperthermia on the metabolic activity and growth rate of malignant cells. *Eur. Surg. Res.* 3:251–252, 1971.

Mulay, I. L., and Mulay, L. N. Effect on drosophila melanogaster and S37 tumor cells: postulates for magnetic field interactions. *Biol. Effects of Mag. Fields* 1:146–169, 1964.

Munson, A. E. et al. Antineoplastic activity of cannabinoids. *J. Natl. Cancer Inst.* 55:597–602, 1975.

Nagasawa, H., and Dewey, W. C. Effects of cold treatment on synchronous Chinese hamster cells treated in mitosis. *J. Cell. Physiol.* 80:89–106, 1972.

Nauts, H. C.; Fowler, G. A.; and Bogatko, F. H. A review of the influence of bacterial infection and of bacterial products (Coley's toxins) on malignant tumors in man. *Acta Med. Scand. [Suppl.]* 276:1–103, 1953.

Newburgh, L. H. et al. Further experiences with the measurement of heat production from insensible loss of weight. *J. Nutr.* 13:203–221, 1937.

Neymann, C. A. Historical development of artificial fever in the treatment of disease. *Medical Record* 150:89–92, 1939.

Nicolson, G. L. The interactions of lectins with animal cell surfaces. *Int. Rev. Cytol.* 39:89–190, 1974.

Nicolson, G. L.; Smith, J. R.; and Poste, G. Effects of local anesthetics on cell morphology and membrane-associated cytoskeletal organization in Balb/3T3 cells. *J. Cell Biol.* 68:395–402, 1976.

Nilsson, S. K. Skin temperature over an artificial heat source implanted in man. *Phys. Med. Biol.* 35:366–383, 1975.

Nolan, R. J., and Faiman, M. D. Brain energetics in oxygen induced convulsions. *J. Neurochem.* 22:645–650, 1974.

Oliphant, W. D. High frequency therapy. III. Electrode theory and design. *Electronic Engineering* 16:252–255, 1943.

Oliphant, W. D. High frequency therapy. IV. The electric field in a dielectric. *Electronic Engineering* 16:296–299, 1943.

Ossovski, L., and Sachs, L. Temperature sensitivity of polyoma virus, induction of cellular DNA synthesis, and multiplication of transformed cells at high temperature. *Proc. Natl. Acad. Sci.* 58:1938–1943, 1967.

Ott, T. Heat center in the brain. *J. Nerv. Ment. Dis.* 14:152, 1887.

Overgaard, J. Combined adriamycin and hyperthermia treatment of a murine mammary carcinoma in vivo. *Cancer Res.* 36:3077–3081, 1976.

Overgaard, J. Influence of extracellular pH on the viability and morphology of tumor cells exposed to hyperthermia. *J. Natl. Cancer Inst.* 56:1243–1250, 1976.

Overgaard, J. Ultrastructure of a murine mammary carcinoma exposed to hyperthermia. *Cancer Res.* 36:983–987, 1976.

Overgaard, J. Effect of hyperthermia on malignant cells in vivo. A review and a hypothesis. *Cancer* 37:1–28, 1977.

Overgaard, J., and Heyden, G. Histological and histochemical studies on combined heat-roentgen treatment of a C3H mouse mammary carcinoma. *Histochemistry* 42:47–59, 1974.

Overgaard, K. Uber warmetherapie bosartiger tumoren. *Acta Radiologica* 15:89–100, 1934.

Overgaard, K., and Okkels, H. Uber den einfluss der warmebehandlung auf woods sarkom. *Strahlentherapie* 68:587–619, 1940.

Overgaard, K., and Overgaard, J. Investigations on the possibility of a thermic tumor therapy—II. Action of combined heat-roentgen treatment on a transplanted mouse mammary carcinoma. *Eur. J. Cancer* 8:573–575. 1972.

Overgaard, K., and Overgaard, J. Radiation sensitizing effect of heat. *Acta Radiologica: Therapy, Physics, Biology.* 13:501–511, 1974.

Palzer, R. J., and Heidelberger, C. Influence of drugs and synchrony on the hyperthermic killing of HeLa cells. *Cancer Res.* 33:422–427, 1973.

Palzer, R. J., and Heidelberger, C. Studies on the quantitative biology of hyperthermic killing of HeLa cells. *Cancer Res.* 33:415–421, 1973.

Papahadjopoulos, D. et al. Effects of local anesthetics on membrane properties. I. Changes in the fluidity of phospholipid bilayers. *Biochim. Biophys. Acta* 394:504–519, 1975.

Papahadjopoulos, D. et al. Effects of local anesthetics on membrane properties. II. Enhancement of the susceptibility of mammalian cells to agglutination by plant lectins. *Biochim. Biophys. Acta* 394:520–539, 1975.

Pautrizel, R. et al. Immunologie—stimulation, par des moyens physiques, des défenses de la souris et du rat contre la trypanosomose expérimentale. *C. R. Acad. Sci.* 268:1889–1892, 1969.

Pautrizel, R. et al. Immunologie—action de champs magnétiques combinés à des ondes électromagnétiques sur la trypanosomose expérimentale du lapin. *C. R. Acad. Sci.* 271:877–880, 1970.

Pautrizel, R.; Riviere, A. P.; and Berlureau, F. Immunologie—influence d'ondes électromagnétiques et de champs magnétiques associés sur l'immunité de la souris infestée par trypanosoma équiperdum. *C. R. Acad. Sci.* 263:579–582, 1966.

Payne, P. R., and Waterlow, J. C. Relative energy requirements for maintenance, growth, and physical activity. *Lancet* 2:210–211, 1971.

Percy, J. F. The results of the treatment of cancer of the uterus by the actual cautery, with a practical method for its application. *JAMA* 58:696–699, 1912.

Percy, J. F. A method of applying heat both to inhibit and destroy inoperable carcinoma of the uterus and vagina. *Surg. Gynecol. Obstet.* 17:371–376, 1913.

Percy, J. F. Heat in the treatment of carcinoma of the uterus. *Surg. Gynecol. Obstet.* 22:77–79, 1916.

Pettigrew, T. et al. Clinical effects of whole body hyperthermia in advanced malignancy. *Br. Med. J.* 4:679–682, 1974.

Pfeiffer, E. A. Electrical stimulation of sensory nerves with skin electrodes for research, diagnosis, communication and behavioral conditioning. A survey. *Medical and Biological Engineering* 6:637–651, 1968.

Pizzo, P. A.; Lovejoy, F. H.; and Smith, D. H. Prolonged fever in children: review of 100 cases. *Pediatrics* 55:468–473, 1975.

Pohl, H. A., and Crane, J. S. Theoretical models of cellular dielectrophoresis. *J. Theor. Biol.* 37:15–41, 1972.

Pomp, H.; Ipach, R.; and Jung, K. Veranderungen des saure-basen-haushalts unter der anwendung von hyperthermie. *Klin. Wochenschr.* 50:383–385, 1972.

Popovic, P.; Sybers, H. A.; and Popovic, V. P. Regression of spontaneous mammary tumors in C3H mice after 5-fluorouracil and differential hypothermia. *Panminerva Med.* 13:523–526, 1971.

Poste, G.; Papahadjopoulos, D.; and Nicolson, G. L. Local anesthetics affect transmembrane cytoskeletal control of mobility and distribution of cell surface receptors. *Proc. Natl. Acad. Sci. USA* 72:4430–4434, 1975.

Quinn, T. J., and Compton, J. P. The foundations of thermometry. *Rep. Prog. in Physics* 38:151–239, 1975.

Radigan, L. R., and Robinson, S. Effects of environmental heat stress and exercise on renal blood flow and filtration rate. *J. Appl. Physiol.* 2:185–191, 1949.

Rao, P. N., and Engelberg, J. HeLa cells: effects of temperature on the life cycle. *Science* 148:1092–1094, 1965.

Rawson, R. O., and Hardy, J. D. Visceral tissue vascularization: an adaptive response to high temperature. *Science* 158:1203–1204, 1967.

Reeves, O. R. Mechanisms of acquired resistance to acute heat shock in cultured mammalian cells. *J. Cell. Physiol.* 79:157–169, 1972.

Reichman, M., and Penman, S. Stimulation of polypeptide initiation in vitro after protein synthesis inhibition in vivo in HeLa cells. *Proc. Natl. Acad. Sci. USA* 70:2678–2682, 1973.

Reinhold, H. S. Improved microcirculation in irradiated tumors. *Eur. J. Cancer* 7:273–280, 1971.

Reiter, T. Uber spezifische wirkungen der ultra-kurzwellen. *Dtsch. Med. Wochenschr.* 59:160–166, 1933.

Repacholi, M. H. et al. Interaction of low intensity ultrasound and ionizing radiation with the tumor cell surface. *Phys. Med. Biol.* 16:221–227, 1971.

Riggle, G. C.; Bagley, D. H.; and Beazley, R. M. Microthermocouple probe for gradient temperature measurements. *Cryobiology* 10:345–346, 1973.

Rinaldini, L. M. An improved method for the isolation and quantitative cultivation of embryonic cells. *Exp. Cell Res.* 16:477–505, 1959.

Rivière M.-R. et al. Cancérologie—action de champs électromagnétiques sur les greffes de la tumeur T8 chez le rat. *C. R. Acad. Sci* 259:4895–4897, 1964.

Rivière, M.-R. et al. Cancérologie—phénomènes de régression observés sur les greffes d'un lymphosarcome chez des souris exposées à des champs électromagnétiques. *C. R. Acad. Sci.* 260:2639–2643, 1965.

Rivière, M.-R., and Guerin, M. Cancérologie—nouvelles réchérches éffectuées chez des rats porteurs d'un lymphosarcome lymphoblastique soumis a l'action d'ondes électromagnétiques associées à des champs magnétiques. *C. R. Acad. Sci.* 262:2669–2672, 1966.

Roberts, R. J.; Klaassen, C. D.; and Plaa, G. L. Maximum biliary excretion of bilirubin and sulfobromophthalein during anesthesia-induced alteration of rectal temperature. *Proc. Soc. Exp. Biol. Med.* 125:313–316, 1967.

Robinson, J. E. et al. Radiation and hyperthermal response of normal tissue in situ—3. Fraction studies. *Radiat. Res.* 59:141, 1973.

Robinson, J. E. et al. Tumor response to a three-fraction regimen combining hyperthermia and x-radiation. *Radiat. Res.* 59:185, 1973.

Robinson, J. E.; McCulloch, D.; and Edelsack, E. A. Microwave heating of malignant mouse tumors and tissue equivalent phantom systems. *J. Microwave Power* 11:87–98, 1976.

Robinson, J. E.; Wizenberg, M. J.; McCready, W. A. Combined hyperthermia and radiation suggest an alternative to heavy particle therapy for reduced oxygen enhancement ratios. *Nature* 251:521–522, 1974.

Robinson, J. E.; Wizenberg, M. J.; and McCready, W. A. Radiation and hyperthermal response of normal tissue in situ. *Radiology* 113:195–198, 1974.

Roe, C. F. Temperature regulation and energy metabolism in surgical patients. *Prog. Surg.* 12:96–127, 1973.

Rohdenburg, G. L., and Prime, F. The effect of combined radiation and heat on neoplasms. *Arch. Surg.* 2:116–129, 1921.

Rossi-Fanelli, A. et al. *Selective heat sensitivity of cancer cells.* Berlin: Springer-Verlag, 1977, pp. 1–189.

Roti-Roti, J. L., and Winward, R. T. The effects of hyperthermia on the protein-to-DNA ratio of isolated HeLa cell chromatin. *Radiat. Res.* 74:159–169, 1978.

Rotkovska, D., and Vacek, A. Modification of repair of x-irradiation damage of hemopoietic system of mice by microwaves. *J. Mirowave Power* 12:119–123, 1977.

Rowell, L. B.; Brengelmann, G. L.; and Murray, J. A. Cardiovascular responses to sustained high skin temperature in resting man. *J. Appl. Physiol.* 27:673–679, 1969.

Rubin, P., and Casarett, G. Microcirculation of tumors. Part 1: anatomy, function, and necrosis. *Clin. Radiol.* 17:220–229, 1966.

Ryan, G. B.; Unanue, E. R.; and Karnovsky, M. J. Inhibition of surface capping of macromolecules by local anaesthetics and tranquilizers. *Nature* 250:56–57, 1974.

Sapareto, S. A. et al. Effects of hyperthermia on survival and progression of Chinese hamster ovary cells. *Cancer Res.* 38:393–400, 1978.

Scheel-Kruger, J., and Hasselager, E. Studies of various amphetamines, apomorphine and clonidine on body temperature and brain 5-hydroxytryptamine metabolism in rats. *Psychopharmacologia* 36:189–202, 1974.

Schenberg-Frascino, A., and Moustacchi, E. Lethal and mutagenic effects of elevated temperature on haploid yeast. I. Variations in sensitivity during the cell cycle. *Molecular General Genetics* 115:243–257, 1972.

Schereschewsky, J. W. The action of currents of very high frequency upon tissue cells. A. Upon a transplantable mouse sarcoma. *Public Health Rep.* 43:927–939, 1928.

Schloerb, P. R. et al. Intracellular pH and buffering capacity of the Walker-256 carcinoma. *Surgery* 58:5–11, 1965.

Schmidt, K. L. Fever, a defense mechanism? *Dtsch. Med. Wochenschr.* 100:1805–1808, 1975.

Schmidt, V. D.; Roesner, D.; and Schuh, D. Untersuchungen uber den einfluss von hyperthermie und in vitro—thermosensibilisatoren auf das yoshida- und jensen-sarkom der ratte. *Archiv Geschwulstforsch.* 33:123–131, 1969.

Schochetman, G., and Perry, R. P. Characterization of the messenger RNA released from L cell polyribosomes as a result of temperature shock. *J. Mol. Biol.* 63:577–590, 1972.

Schramm, T., and Bender, E. In vitro-untersuchungen zur frage einer selektiven thermosensibilitat von tumorzellen. 1, 2. *Acta Biol. Med. Ger.* 16:17–21, 1966.

Schrek, R. Effect of heat and dimethyl sulphoxide on normal and leukaemic lymphocytes. *Lancet* 2:1020–1021, 1965.

Schrek, R. Sensitivity of normal and leukemic lymphocytes and leukemic myeloblasts to heat. *J. Natl. Cancer Inst.* 37:649–654, 1966.

Schulman, N., and Hall, E. J. Hyperthermia: its effects on proliferative and plateau phase cell cultures. *Radiology* 113:207–209, 1974.

Schutte, M. et al. Effects of anesthesia, surgery, and inflammation upon host defense mechanisms. I. Effects upon the complement systems. *Int. Arch. Allergy Appl. Immunol.* 48:706–720, 1975.

Schwan, H. P. Electrical properties of tissue and cell suspensions. *Adv. Biol. Med. Phys.* 5:147–209, 1954.

Schwan, H. P. Microwave radiation: biophysical considerations and standards criteria. *IEEE Trans. Biomed. Eng.* BME-19:304–312, 1972.

Schwan, H. P. et al. On the low-frequency dielectric dispersion of colloidal particles in electrolyte solution. *J. Phys. Chem.* 66:2626–2635, 1962.

Schwan, H. P., and Carstensen, E. L. Acoustic properties of hemoglobin solutions. *J. Acoust. Soc. Am.* 31:305–311, 1959.

Schwan, H. P., and Carstensen, E. L. Absorption of sound arising from the presence of intact cells in blood. *J. Acoust. Soc. Am.* 31:185–189, 1959.

Schwan, H. P., and Maczuk, J. Simple technique to control the stray field of electrolytic cells. *Rev. Sci. Instrum.* 31:59–62, 1960.

Schwan, H. P., and Piersol, G. M. The absorption of electromagnetic energy in body tissues. Part I. Biophysical aspects. *Am. J. Phys. Med.* 33:371–404, 1954.

Schwan, H. P., and Piersol, G. M. The absorption of electromagnetic energy in body tissues. Part II. Physiological and clinical aspects. *Am. J. Phys. Med.* 34:425–558, 1955.

Schwan, H. P.; Schwarz, G.; and Saito, M. On the orientation of nonspherical particles in an alternating electrical field. *J. Chem. Phys.* 43:3562–3569, 1965.

Selawry, O. Zur rolle erhohter korpertemperatur bei spontanremissionen menschlicher tumoren. *Int. Rundschau fur Physikalische Medizin* 10:151–153, 1957.

Sharp, J. C., and Paperiello, C. J. The effects of microwave exposure on thymidine-3H uptake in albino rats. *Radiat. Res.* 45:434–439, 1971.

Shibata, H. R., and MacLean, L. D. Blood flow to tumors. *Prog. Clin. Cancer* 2:33–47, 1966.

Shingleton, W. W. et al. Selective heating and cooling of tissue in cancer chemotherapy. *Ann. Surg.* 156:408–416, 1962.

Shingleton, W. W.; Parker, R. T.; and Mahaley, S. Abdominal perfusion for cancer chemotherapy with hypothermia and hyperthermia. *Surgery* 50:260–265, 1961.

Shoulders, H. S.; Turner, E. L.; and Scott, L. D. Observations on the results of combined fever and x-ray therapy in the treatment of malignancy. *South. Med. J.* 35:966–970, 1942.

Silver, S.; Poroto, P.; and Crohn, E. B. Hypermetabolic states without hyperthyroidism (nonthyrogenous hypermetabolism). *Arch. Intern. Med.* 85:479–482, 1950.

Simard, R.; Amalric, F.; and Zalta, J. P. Effet de la température supra-optimale sur les ribonucléoproteines et le RNA nucléolaire. *Exp. Cell Res.* 55:359–369, 1969.

Simard, R., and Bernhard, W. A heat-sensitive cellular function located in the nucleolus. *J. Cell Biol.* 34:61–76, 1967.

Sinclair, W. K., and Morton, R. A. X-ray and ultraviolet sensitivity of synchronized Chinese hamster cells at various stages of the cell cycle. *Biophys. J.* 5:1–25, 1965.

Singer, S. J., and Nicolson, G. L. The fluid mosaic model of the structure of cell membranes. *Science* 175:720–731, 1972.

Sisken, J. E.; Morasca, L.; and Kibby, S. Effects of temperature on the kinetics of the mitotic cycle of mammalian cells in culture. *Exp. Cell Res.* 39:103–116, 1965.

Skocpol, W. J., and Tinkham, M. Fluctuations near superconducting phase transitions. *Rep. Prog. Phys.* 38:1049–1097, 1975.

Smith, C. W. et al. Liquid-crystal optical activity for temperature sensing. *Appl. Phys. Lett.* 24:453–454, 1974.

Smith, J. H.; Robinson, S.; and Pearcy, M. Renal responses to exercise, heat and dehydration. *J. Appl. Physiol.* 4:659–665, 1952.

Song, C. W. et al. Vascularity and blood flow in x-irradiated Walker carcinoma 256 of rats. *Radiology* 104:693–697, 1972.

Sparks, F. C. et al. Complications of BCG immunotherapy in patients with cancer. *N. Engl. J. Med.* 289:827–830, 1973.

Sperelakis, N., and Lehmkuhl, D. Effects of temperature and metabolic poisons on membrane potentials of cultured heart cells. *Am. J. Physiol.* 213:719–724, 1967.

Sridhar, R., and Sutherland, R. Hyperthermic potentiation of cytotoxicity of Ro-07-0582 in multicell spheroids. *Int. J. Radiat. Oncol. Biol. Phys.* 2:531–535, 1977.

Stehlin, J. S. Hyperthermic perfusion with chemotherapy for cancers of the extremities. *Surg. Gynecol. Obstet.* 129:305–308, 1969.

Stehlin, J. S. et al. Perfusion for malignant melanoma of the extremities. *Am. J. Surg.* 105:607–614, 1963.

Stehlin, J. S. et al. Results of hyperthermic perfusion for melanoma of the extremities. *Surg. Gynecol. Obstet.* 140:338–348, 1975.

Stevenson, H. N. The effect of heat upon tumor tissue. *J. Cancer Res.* 4:54–56, 1919.

Stoyka, W. W., and Garvey, M. B. Haematological and cardiorespiratory responses to induced hyperpyrexia. *Can. Anaesth. Soc. J.* 21:325–334, 1974.

Straley, J. P., and Stephen, M. J. Physics of liquid crystals. *Rev. Mod. Phys.* 46:617–704, 1974.

Stratford, I. J., and Adams, G. E. Effect of hyperthermia on differential cytotoxicity of a hypoxic cell radiosensitizer, Ro-07-0582, on mammalian cells in vitro. *Br. J. Cancer* 35:307–313, 1977.

Strauss, A. A. et al. Surgical diathermy of carcinoma of the rectum. (Its clinical end results.) *JAMA* 104:1480–1484, 1935.

Strauss, A. A. et al. Immunologic resistance to carcinoma produced by electrocoagulation. *Surg. Gynecol. Obstet.* 212:989–996, 1965.

Strauss, A. A.; Appel, M.; and Saphir, O. Electrocoagulation of malignant tumors. *Am. J. Surg.* 104:37–45, 1962.

Strom, R. et al. The biochemical mechanism of selective heat sensitivity of cancer cells. IV. Inhibition of RNA synthesis. *Eur. J. Cancer* 9:103–112, 1973.

Strom, R. et al. Inhibition by elevated temperatures of ribosomal RNA maturation in Ehrlich ascites cells. *Cancer Biochem. Biophys.* 1:57–62, 1975.

Suit, H. D., Hyperthermic effects on animal tissues. *Radiology* 123:483–487, 1977.

Suit, H. D., and Shwayder, M. Hyperthemia: potential as an anti-tumor agent. *Cancer* 34:122–129, 1974.

Suryanarayan, C. R. Certain interesting observations during and after hydro-hyperthermic chemotherapy in advanced malignancies. (Preliminary survey.) *Indian J. Cancer* 3:176:181, 1966.

Suzuki, K. Application of heat to cancer chemotherapy—experimental studies. *Nagoya J. Med. Sci.* 30:1–21, 1967.

Swabb, E. A. et al. Diffusion and convection in normal and neoplastic tissues. *Cancer Res.* 34:2814–2322, 1974.

Swartz, R. D.; Sidell, F. R.; and Cucinell, S. A. Effects of physical stress on the disposition of drugs eliminated by the liver in man. *J. Pharmacol. Exp. Ther.* 188:1–7, 1974.

Szent-Gyorgyi, A. Bioelectronics and cancer. *Journal of Bioenergetics* 4:533–562, 1973.

Szmigielski, S. et al. Effect of microwaves combined with interferon and/or interferon inducers. (Poly 1–Poly C) on development of sarcoma 180 in mice. *J. Microwave Power* 11:174–177, 1976.

Tagashira, Y. et al. Continuous pH measuring by means of microglass electrode inserted in living normal and tumor tissues (2nd report), with an additional report on interaction of SH group of animal protein with carcinogenic agent in the carcinogenic mechanism. *Gan.* 45:99–101, 1954.

Tanaka, R., and Teruya, A. Lipid dependence of activity-temperature relationship of (Na^+, K^+) activated ATPase. *Biochim. Biophys. Acta* 323:584–591, 1973.

Tannock, I. F., and Steel, G. G. Quantitative techniques for study of the anatomy and function of small blood vessels in tumors. *J. Natl. Cancer Inst.* 42:771–782, 1969.

Tegtmeyer, P. et al. Regulation of tumor antigen synthesis by simian virus 40 gene A. *J. Virol.* 16:168–178, 1975.

Templeton, G. H. et al. Influence of temperature on the mechanical properties of cardiac muscle. *Circ. Res.* 34:624–634, 1974.

Thrall, D. E. et al. Response of cells in vitro and tissues in vivo to hyperthermia and X-irradiation. *Adv. Radiat. Biol.* 6:211–227, 1976.

Thrall, D. E., and Gillette, E. L. Combination of heat and ionizing radiation on C3H mouse skin and the C3H mouse mammary adenocarcinoma in vivo: significance of the order of application and quantitation of the heat effect (abstr.). *Radiat. Res.* 59:186, 1974.

Thrall, D. E.; Gillette, E. L.; and Bauman, C. L. Effect of heat on the C3H mouse mammary adenocarcinoma evaluated in terms of tumor growth. *Eur. J. Cancer* 9: 871–875, 1973.

Thrall, D. E.; Gillette, E. L.; and Dewey, W. C. Effect of heat and ionizing radiation on normal and neoplastic tissue of the C3H mouse. *Radiat. Res.* 63:363–377, 1975.

Turano, C. et al. The biochemical mechanism of selective heat sensitivity of cancer cells. III. Studies on lysosomes. *Eur. J. Cancer* 6:67–72, 1970.

Twentyman, P. R., and Bleehen, N. M. Changes in sensitivity to cytotoxic agents occurring during the life history of monolayer cultures of a mouse tumor cell line. *Br. J. Cancer* 31:417–423, 1975.

Twentyman, P. R.; Morgan, J. E.; and Donaldson, J. Enhancement by hyperthermia of the effect of BCNU against the EMT6 mouse tumor. *Cancer Treat. Rep.* 62: 439–443, 1978.

Umsawaski, T. Fever, infection and cancer patients. *J. Med. Assoc. Thai.* 58:183–185, 1975.

Unanue, E. R., and Karnovsky, M. J. Redistribution and fate of Ig complexes on surface of B lymphocytes: functional implications and mechanisms. *Transplant. Rev.* 114:185–210, 1973.

Van Cauwenberge, H., and Focan, C. Les états fébriles prolongés. *Révue Medicale de Liège* 29:305–309, 1974.

Vanwijck, R. R. et al. Stimulation or suppression of metastases with graded doses of tumor cells. *Cancer Res.* 31:1559–1563, 1971.

Vatner, S. F., and Braunwald, E. Cardiovascular control mechanisms in the conscious state. *N. Engl. J. Med.* 293:970–976, 1975.

Vermel, E. M., and Kuznetsova, L. B. Hyperthermia in the treatment of malignant diseases. *Vopr. Onkol.* 16:96–102, 1970.

Verney, G. F. The stimulation of skin temperature distributions by means of a relaxation method. *Phys. Med. Biol.* 20:384–394, 1975.

Vollmar, H. Uber den einfluss der temperatur auf normales gewebe und auf tumorgewebe. *Zeitschrift fur Kreislaufforschung* 51:71–99, 1941.

Von Ardenne, M. Selective multiphase cancer therapy: conceptual aspects and experimental basis. *Adv. Pharmacol. Chemother.* 10:339–380, 1972.

Von Schmahl, D.; Hollenriegel, K.; and Mundt, D. Tierexperimentelle untersuchungen zur krebs-mehrschritttherapie (nach M. v. Ardenne). *Archiv Geschwulstforsch.* 43:205, 1974.

Von Schmahl, D., and Noring, L. Extrem-hyperthermieversuche bei vier verschieden tumoren von ratten. *Zeitschrift fur Krebsforschung* 69:335–340, 1967.

Waddell, W. J., and Bates, R. G. Intracellular pH. *Physiol. Rev.* 49:285–329, 1969.

Walsh, W. M.; Shulman, R. G.; and Heidenreich, R. D. Ferromagnetic inclusions in nucleic acid samples. *Nature* 192:1041–1043, 1961.

Warburg, O.; Posener, K.; and Negelein, E. Uber den stoffwechsel der carcinomzelle 1. *Biochemische Zeitschrift* 152:309–344, 1924.

Warocquier, R., and Scherrer, K. RNA metabolism in mammalian cells at elevated temperature. *Eur. J. Biochem.* 10:362–370, 1969.

Warren, S. L. Preliminary study of the effect of artificial fever upon hopeless tumor cases. *Am. J. Roentgenol. Radium Ther. Nucl. Med.* 33:75–87, 1935.

Waterhouse, C. How tumors affect host metabolism. *Ann. NY Acad. Sci.* 230:86–93, 1974.

Waterhouse, C.; Fenniger, L. D.; and Keutmann, E. H. Nitrogen exchange and caloric expenditure in patients with malignant neoplasms. *Cancer* 4:500–514, 1951.

Waterhouse, C., and Kemperman, J. H. Carbohydrate metabolism in subjects with cancer. *Cancer Res.* 31:1273–1278, 1971.

Watkin, D. M., and Steinfeld, J. L. Nutrient and energy metabolism in patients with and without cancer during hyperalimentation with fat administered intravenously. *Am. J. Clin. Nutr.* 16:182–212, 1965.

Webb, P. Pain limited heat exposures. *Temperature—Its Measurement and Control in Science and Industry* 3:245–250, 1963.

Webb, P.; Annis, J. F.; and Troutman, S. J. Human calorimetry with a water-cooled garment. *J. Appl. Physiol.* 32:412–418, 1972.

Webb, P.; Garlington, L. N.; and Schwarz, M. J. Insensible weight loss at high skin temperatures. *J. Appl. Physiol.* 11:41–43, 1957.

Webb, S. J.; Stoneham, M. E.; and Frohlich, H. Evidence for non-thermal excitation of energy levels in active biological systems. *Physics Letters* 63:407–408, 1977.

Westermark, F. Uber die behandlung des ulcerirenden cervixcarcinoma mittels konstanter warme. *Centralblatt fur Gynakologie* 22:1335–1339, 1898.

Westermark, N. The effect of heat upon rat tumors. *Skandinavisches Archiv fur Physiologie* 52:257–321, 1927.

Westra, A., and Dewey, W. C. Variation in sensitivity to heat shock during the cell-cycle of Chinese hamster cells in vitro. *Int. J. Radiat. Biol.* 19:467–477, 1971.

Whitby, J. D., and Dunkin, L. J. Cerebral, oesophageal and nasopharyngeal temperatures. *Br. J. Anaesth.* 43:673–674, 1971.

Williams, N.; Atkinson, G. W.; and Patchefsky, A. S. Polymer-fume fever: not so benign. *J. Occup. Med.* 16:92–94, 1974.

Williams, R. E. et al. Utilization of fatty acid supplements by cultured animal cells. *Biochemistry* 13:1969–1977, 1974.

Wills, E. J.; Findlay, J. M.; McManus, J. P. A. Effects of hyperthermia therapy on the liver. II. Morphological observations. *J. Clin. Pathol.* 29:1–10, 1976

Wilmore, D. W. et al. Catecholamines: mediator of the hypermetabolic response to thermal injury. *Ann. Surg.* 180:653–669, 1974.

Wissler, E. H. Steady state temperature distribution in man. *J. Appl. Physiol.* 16:734–740, 1961.

Witz, I. P. et al. Tumor-bound immunoglobulins: their possible role in circumventing antitumor immunity. *Johns Hopkins Med. J.* 3:289–300, 1974.

Woodhall, B. et al. Effect of hyperthermia upon cancer chemotherapy; application to external cancers of head and face structures. *Ann. Surg.* 151:750–758, 1960.

Wright, G. L. Critical thermal maximum in mice. *J. Appl. Physiol.* 40:683–687, 1976.

Wuest, G. P. et al. In vitro effect of hyperthermia on the incorporation rate of nucleic acid precursors in tumors and normal tissues. *Zeitschrift fur Kreislaufforschung* 79:193–203, 1973.

Wybran, J. et al. Rosette-forming cells, immunologic deficiency diseases and transfer factor. *N. Engl. J. Med.* 288:710–713, 1973.

Wybran, J., and Fudenberg, H. H. Thymus-derived rosette-forming cells in various human disease states: cancer, lymphoma, bacterial and viral infections, and other diseases. *J. Clin. Invest.* 52:1026–1032, 1973.

Yasuhira, U. Acquired immunity to Ehrlich ascites tumor in mice after treatment with fresh or sonized tumor cells at various ages. *Acta Tuberculosea Japonica* 16:78–84, 1966.

Yatvin, M. B. Influence of membrane lipid composition on hyperthermic and radiation killing of *E. coli* cells. *Radiat. Res.* 70:610, 1977.

Yerushalmi, A. Cure of a solid tumor by simultaneous administration of microwaves and X-ray irradiation. *Radiat. Res.* 64:602–610, 1975.

Yerushalmi, A. Influence of metastatic spread of whole-body or local tumor hyperthermia. *Eur. J. Cancer* 12:455–464, 1976.

Yerushalmi, A., and Har-Kedar, I. Enhancement of radiation effects by heating of the tumor. *Isr. J. Med. Sci.* 10:772–776, 1974.

Young, G. F., and Lele, P. P. Focal lesions in the brain of growing rabbits produced by focused ultrasound. *Exp. Neurol.* 9:502–511, 1964.

Zimmer, R. P.; Ecker, H. A.; and Popovic, V. P. Selective electromagnetic heating of tumors in animals in deep hypothermia. *IEEE Trans. MTT* 19:238–245, 1971.

CHAPTER 12
Physical Models (Phantoms) in Thermal Dosimetry

Thomas C. Cetas

Introduction
Static Phantoms
Dynamic Phantoms
Conclusions

INTRODUCTION

Much of basic science is concerned with the selection, design, and elucidation of specialized models that represent one elementary property or characteristic. On the other hand, the applied sciences and arts, including medicine, are concerned directly with the complex, interactive realistic system. Much effort is devoted to understanding how knowledge gained from the idealistic models can help solve immediate problems, such as identification of the best available technique for heating a particular tumor. Physical models, or phantoms, can be of help in such problem solving. Nevertheless, it must be understood that phantoms are designed usually to model only a specific property of a complex system; it is seldom that all relevant parameters of the real system are modeled. Consequently, measurements on phantoms, or on any model, usually cannot be taken as quantitatively representative of the real system (Mortimer and Osborne 1935); only qualitative results or trends can be applied. In the field of hyperthermia phantoms are used most commonly to study complex electromagnetic power deposition patterns in heterogeneous tissues produced by the near field of an antenna.

The justification for this rather long philosophical perspective is to inject a note of caution. One of the most successful uses of phantoms is in radiation dosimetry, where phantoms model the relevant properties of the tissues so well that measurements with them can be applied quantitatively to patients. Variations of only a few percentage points results in changes in the radiation dose distribution. Consequently, therapeutic radiologists and dosimetrists rely heavily on their use. This reliance is not warranted in hyperthermia phantoms where tissue equivalence of all relevant parameters is not yet attainable. Alternatively, physicians and engineers from other backgrounds frequently dismiss phantom studies too readily, failing to appreciate their benefits in helping to understand the nature and pattern of a heating modality and in providing a standard means of comparing the physical (as opposed to clinical) effectiveness of various heating modalities or applicators.

This chapter discusses static (unperfused) phantoms in some detail, with some examples of uses. Three dynamic, or liquid perfused, phantoms will be described, although these are even more specialized to particular studies than are the static models. Occasion-

The author gratefully acknowledges the significant contributions of Dr. J. R. Oleson, Dr. M. R. Manning, P. R. Stauffer, and A. M. Fletcher. Supported in part by U. S. Public Health Service grants CA-17343 and CA-29653 from the National Institutes of Health.

ally, phantom materials are used to form a bolus to modify the heating patterns in the tissue or to help couple energy into the tissues. Perhaps, in the future, the greatest benefit of physical modeling will be to provide a means of experimentally evaluating the accuracy of the numerical models that are necessary for quantitative thermal dosimetry.

STATIC PHANTOMS

Static phantoms can be used only for power deposition studies since the thermal dissipation of tissues is not modeled by the phantoms. This is best illustrated by considering the bio-heat equation (Bowman, Cravalho, and Woods 1975; Chen and Holmes 1980; Shitzer 1975), which describes the energy balance in an infinitesimal tissue region as

$$\rho c \frac{dT}{dt} = k\nabla^2 T + \dot{q}_p + \dot{q}_b + \dot{q}_m , \qquad (1)$$

where ρ is tissue density, c is heat capacity, k is thermal conductivity, T is temperature, t is time, $\dot{q}$ is energy/unit time/unit volume, and the subscripts p,b,m, refer to external power deposition, blood flow, and metabolism, respectively. The term on the left represents the rate at which energy is accumulated in the volume leading to a rise in temperature. The first term on the right represents the diffusion of heat from the volume by thermal conduction processes. The second term represents the rate at which heat is deposited. The third term $\dot{q}_b$ represents heat dissipation by convection by the blood system. The last term represents heat generation from metabolism. These last two terms, especially the blood flow, are very important for living tissues, but are not present in static phantoms. The proper expression for phantoms is, in fact, the more conventional heat equation (Carslaw and Jaeger 1959),

$$\rho c \frac{dT}{dt} = k\nabla^2 T + \rho (SAR). \qquad (2)$$

Here, ρ,c,k now refer to the respective phantom properties. The power deposited per unit volume has been replaced by the product of density times the specific absorption rate (SAR) in W/kg which is becoming accepted as the dosimetric unit in biologically related electromagnetic studies (Guy 1975).

If the phantom is constructed so that it has a very small thermal conductivity, that is, $k\nabla^2 T$ is small com-

pared to the term on the left, then the effects of thermal conduction can be ignored and the expression reduces to the simple form, and

$$SAR = c \frac{dT}{dt} . \qquad (3)$$

Thus if one uses a burst of high power for a short time, then the power deposition pattern, that is, the SAR as a function of position, can be determined by measuring the rate of temperature increase in a specified volume. It was for these purposes that Guy (1971) developed this phantom system and introduced the use of a thermographic camera for measuring the resulting temperature distributions. If the heating time is long and the temperature gradients are large, then significant heat conduction will occur, and the temperature pattern no longer will be representative of the power deposition pattern.

Materials and Techniques

Various cuts of beef and saline solutions of differing concentrations are used frequently as phantoms for qualitative testing of heating applicators and as dummy loads for heating equipment. For such purposes, these materials are quite acceptable. They are not truly representative of real tissue, however. Excised tissue, especially after it has been bled and hung, no longer has the same electrical conductivity and dielectric constant as in vivo tissues. Guy, in 1971 and with Johnson in 1972, developed two basic types of phantom materials. The first represents high water content tissues, such as muscle or brain, and the second represents low water content tissues, such as fat or bone. The dielectric properties were based on measurements by Schwan, Carstenson, and Li (1956) and by Schwan and Piersol (1954). The muscle phantom material consists of salt, polyethylene, water, and a gelling material similar to that used in "Super-Stuff" (Bolus Stock TX−150, Oil Center Research, Lafayette, Louisiana), a child's toy. The bone and fat phantom materials are based on mixtures of aluminum powder, acetylene black, and a plasticizer; the latter are hard plastics and can be used as molds to hold the muscle phantom material. By appropriately varying the percentages of the constituents, the dielectric properties can be adjusted to represent tissues of differing water contents and at various electromagnetic frequencies. Recently Stuchley and Stuchley (1980) tabulated some of the dielectric properties of various mixtures along with dielectric properties of biological substances as a function of frequency. Schwan and Foster (1980) recently reviewed the electrical properties of tissues in a discussion of biophysical mechanisms. In our work, we use Guy's basic recipe but use an impedance bridge (HP 4815A RF Vector Impedance Meter, Hewlett Packard Corporation, Palo Alto, California) and probe (Hahn et al. 1980) to establish that dielectric parameters match those of the tissue to be modeled. Ho and colleagues (1971) have shown that the intrinsic thermal conductivity and heat capacity of phantom materials used by Guy are representative of tissues as well. We have confirmed this on similar type phantoms (Stauffer 1979; Stauffer, Cetas, and Jones 1981). Since the thermal conductivity is small, little error from thermal diffusion occurs during power deposition studies if the total time interval from the beginning of heating to the reading of the final temperatures occurs within 15 to 30 seconds.

The technique is to construct a phantom with dielectric properties that approximate those of the system to be modeled, such as a planar phantom in which muscle is packed into a mold made of fat and bone material. The phantom is constructed such that two parts can be separated to expose a cross-sectional plane for viewing by a thermographic camera (see figure 12.1) (Kantor and Cetas 1977). At microwave frequencies, polyethylene film can be used to separate the two halves of the phantom, and little error in the heating pattern will occur (Kantor and Cetas 1977). This is because the principal heating mechanism at

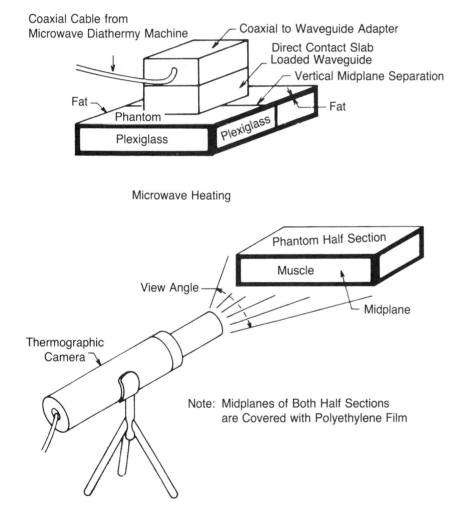

Figure 12.1. Diagram illustrating the technique of split-phantom power deposition studies. The phantom is assembled and heated rapidly with a short burst (several seconds) of high power. The heating applicator is removed, the phantom separated in the mid-plane, and the exposed surface is viewed with a thermographic camera. (Kantor and Cetas 1977. Reprinted by permission.)

these frequencies is dielectric heating arising from energy absorption by the polar water molecules. At lower frequencies, near 30 MHz, significant joule heating results from induced eddy currents flowing through large regions. Consequently, an insulating film in the current path would change the heating pattern. Under those circumstances, the use of cheesecloth to separate the section (Guy, Webb, and McDougal 1975) is more appropriate. Care must be taken, however, to avoid artifacts in temperature measurement arising from evaporative cooling of the exposed surface (Kantor and Cetas 1977). Figure 12.2 shows the heating pattern induced in a planar phantom by a loaded WR30 wave guide (Kantor and Cetas 1977) operating at 2.45 GHz. The bright areas in the thermogram represent warmer areas. Figure 12.2A is a diagrammatic sketch of the experiment; the shaded area refers to the area that became heated. The white rectangle with a vertical line on top and four lines at the bottom represents a prototype thermometer designed for use in electromagnetic fields (Cetas 1976). Figure 12.2B shows the heating pattern as represented on the gray scale on the thermographic camera. Figure 12.2C is a temperature profile taken along the white fiducial line shown in figure 12.2B. Figure 12.2D shows the temperature profile taken normal to the surface (i.e., vertical in fig. 12.2B). Figure 12.2E gives the temperature versus time curve as monitored by the thermometer and shows that cooling effects are small if the data are recorded within a few tens of seconds after the heating is completed. A thorough discussion of the use of a thermographic camera for thermometry, using phantoms as examples, has been reviewed by Cetas (1978).

In some cases it is not necessary to heat and separate two mated sections of phantom as just described. Rather, adequate information can be obtained from thermograms of an exposed surface, especially if the phantom surface has been covered with polyethylene to prevent evaporative cooling. An example is qualitative studies of the perturbation of the electromagnetic field by a bone embedded in musclelike material. Similarly, artifactual heating of electrically conductive thermometer probes can be demonstrated adequately.

Astrahan (1979a, 1979b) has described phantom materials made with clear gelatin and saline solution for use at frequencies around 1 MHz. It is not usually necessary at these frequencies to precisely model the inductive components of the impedance, especially if the heating technique involves resistive contact of the electrodes to the phantom or tissues. One interesting property of this phantom is that the melting temperature of the gelatin, which is a few degrees above room temperature, can be controlled as well. As the phantom is heated to its melting point, a liquid region forms. By injecting dye into the liquid region, it is possible to observe an isotherm at the liquid solvent interface through the transparent phantom. Astrahan does not indicate whether the electrical properties of the melted phase differ from those of the solid phase; presumably, they do not. Kopecky (1980) describes a dual chamber phantom that uses liquid dielectrics. The outer chamber is a temperature-controlled water bath, while the inner chamber contains a mixture of

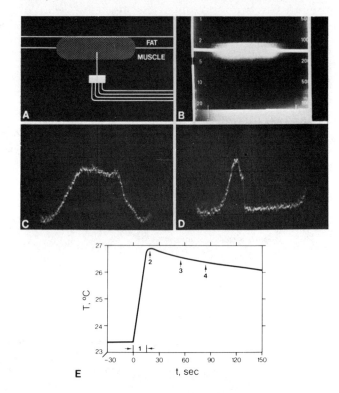

Figure 12.2. Temperature data for heating of a planar phantom with a loaded WR 430 waveguide applicator. Polyethylene film (0.002 in thick) was used to cover the mid-plane. *A*, Diagram of experimental geometry indicating heated region *(shaded)* and the placement of the prototype nonperturbing thermometer probe (tip is 6 mm below fat-muscle interface). *B*, Thermogram showing heated region *(white)*. *C*, Temperature profile of thermogram of phantom at depth indicated by the white fiducial marker in figure 12.3B. *D*, Temperature profile normal to fat-muscle interface showing depth of heating; taken by setting phantom on end. Unheated muscle is on the *left*, ambient background on the *right*. *E*, Copy of strip chart recorder trace that monitored probe temperatures. *1*, heating period (100 watts, 15 sec); *2*, the instant the phantom halves were separated; *3*, the instant the thermograph in figures 12.3B and 12.3C were taken; and *4*, the instant the thermogram in figure 12.3D was taken. (Kantor and Cetas 1977. Reprinted by permission.)

80% physiologically normal saline and 20% alcohol.[1] Microwave energy is absorbed in the inner saline bath, while the outer bath maintains the surface temperature of the phantoms. The heating pattern is determined by moving thermistors throughout the inner bath. Kopecky does not discuss thermometer probe orientation with respect to the electric field polarization, although presumably they were orthogonal. Also, he does not mention if thermally induced convection currents in the dielectric bath affected his results. He does show agreement between results using the liquid phantom and a Guy-type phantom study using a thermographic camera.

A variation on the idea of phantoms for power deposition studies is the use of bolus materials to shape the heating field in tissues so as to get a predictable power deposition pattern. The tissue surface temperature is controlled independently by the temperature of the bolus material. The bolus helps to focus the heating field onto the desired region and improves the coupling of power into the tissue. In particular, the impedance mismatch between air and tissue can be avoided by coupling directly from a 50 Ω antenna through bolus into tissues that at microwave frequencies have a characteristic impedance ($\sqrt{\varepsilon/\mu}$) on the order of 50 ohms (Johnson and Guy 1972; Schwann, Carstenson, and Li 1956; Schwan and Foster 1980; Schwan and Piersol 1954; Stuchley and Stuchley 1980). Hand and colleagues (1979) have described a liquid bolus material that they use to improve the heating pattern in rat tumors heated with 2.45-GHz microwave radiation. Their material is physiologically compatible so that it does not irritate immersed tissues. Cheung's mixture was mentioned above. For coupling microwave energy into tissues, Turner[2] uses deionized water bags between his applicator and the tissue. We have used KY-Jelly™ for similar purposes on some microwave heat treatments (Cetas, Connor, and Manning 1980). Similarly, solid bolus materials, such as silica gels packed in bean bags, can be used to couple the microwaves from the applicator into the tissues. If the bean bags are sufficiently porous, then forced air can be used for controlling skin temperatures.

Uses in Applicator Development, Evaluation, and Standardization

The first use (Guy 1971; Johnson and Guy 1972) of the phantom materials that Guy developed was to determine the power deposition patterns resulting from conventional heating modalities. This has led to the design of new antenna systems (Guy et al. 1978; Johnson and Guy 1972; Kantor and Cetas 1977; Kantor, Witters, and Greiser 1978; Kantor and Witters 1980), which have improved power deposition patterns. There are two principal advantages to using phantoms in this way. First, the characteristics of the applicator can be tested with engineering tools and techniques without introducing the complications of trials on animals. Second, the antenna characteristics of the applicator are determined independently of a cooling system, which is a property of the subject and not the applicator. From such studies, Guy (Guy 1971; Johnson and Guy 1972) proposed 915 MHz as an optimum frequency for diathermy, since at this frequency the best compromise was reached between the penetration depth of the electromagnetic wave in tissues and the dispersion of the wave in the near field of the antenna.

The characteristics of phantom materials can be well characterized and standardized. Thus, phantoms provide a means through which applicators can be compared. Kantor, at the Bureau of Radiological Health, has proposed the use of specific well-defined phantoms to demonstrate efficacy of heating of commercial diathermy equipment. His proposal has been incorporated into the Food and Drug Administration microwave diathermy standard (*Federal Register* 1980).

Thermographic studies with phantoms also have been useful in demonstrating the perturbing effects of metal probes in electromagnetic fields and the effects of sharp discontinuities in dielectric properties. Johnson and Guy (1972) demonstrated these effects with thermocouple probes placed in the head of a terminated cat. The head was subjected to microwave heating. This technique was used subsequently to demonstrate that specially designed thermometer probes (Cetas 1975; Larsen et al. 1979) do not perturb electromagnetic fields. Figure 12.3, taken from Cetas (1976), shows how ordinary thermistor probes perturb electric fields as well as self-heat directly in the field. Figure 12.4 is a thermogram showing the results of the heating of three specially designed thermometers. One of them uses high resistance leads; the second one is a fiberoptic probe; and the third is another style of high-resistance lead probe. All were inserted together into a nonlossy media, silicone oil at a close angle to the surface, and were heated in a strong electromagnetic field. The two high-resistance lead thermistor probes (top and bottom) still self-heat slightly, whereas the fiber optic probe (center) does not heat. Larson and co-workers (1979), however, again using thermography and phantoms, demonstrate that the insertion of a nonlossy dielectric thermometer into an energy absorbing medium, such as tissues, in fact causes a per-

---

[1] A. Cheung, University of Maryland, personal communication.
[2] P. F. Turner, BSD Corporation, Salt Lake City, Utah, personal communication.

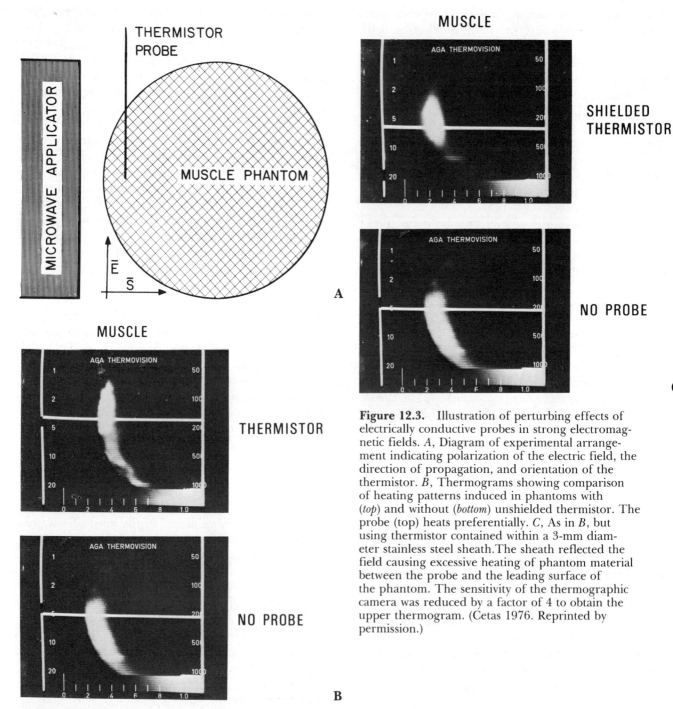

Figure 12.3. Illustration of perturbing effects of electrically conductive probes in strong electromagnetic fields. *A*, Diagram of experimental arrangement indicating polarization of the electric field, the direction of propagation, and orientation of the thermistor. *B*, Thermograms showing comparison of heating patterns induced in phantoms with (*top*) and without (*bottom*) unshielded thermistor. The probe (top) heats preferentially. *C*, As in *B*, but using thermistor contained within a 3-mm diameter stainless steel sheath. The sheath reflected the field causing excessive heating of phantom material between the probe and the leading surface of the phantom. The sensitivity of the thermographic camera was reduced by a factor of 4 to obtain the upper thermogram. (Cetas 1976. Reprinted by permission.)

turbation in the opposite sense: it acts like a cold heat sink. A true nonperturbing probe would have the same average energy absorption as the tissue it replaces.

At frequencies between 0.5 and 13.56 MHz we have studied the effects of placing electrically conducting needle thermometers in tissues being heated with radiofrequency (RF) currents. If a needle can be placed perpendicular to the electric field lines or along equipotential surfaces, then the needle will cause little perturbation to the electrical currents or to the heating pattern. If, however, the needle crosses equipotential surfaces, currents will flow along the needle, causing the tissue near the needle to heat. This is difficult to avoid since thermometers usually enter the tissue near the edge of a plate electrode. If the needle is inserted through dielectric tubing (e.g., Angio-Cath™), the currents will couple to the needle shaft only capacitively,

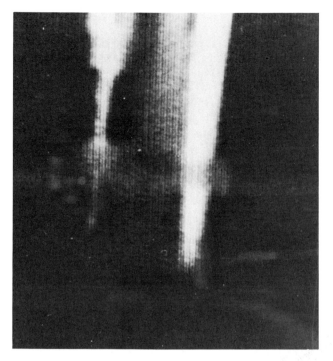

Figure 12.4. Thermogram after microwave irradiation of three minimally perturbing thermometer probes protruding into a silicone oil bath at a very close angle to the surface. On the *right* the probes are in air; in the *center*, they are just below the surface. The *upper* and *lower* probes are thermistors using very high resistance leads. The *middle* is an optical fiber-based thermometer.

and the needle heats only slightly. The effect is manifested clinically through patient complaints of hot spots in the vicinity of the needles and can be observed with a thermographic camera during treatment. The use of plastic catheters is a poor substitute for the use of nonperturbing fiber optic probes (Cetas 1976, 1982; Cetas and Connor 1978; Cetas, Connor, and Boone 1978; Christensen 1979). Nevertheless, such techniques can serve as stopgap measures until optical thermometers become available commercially.

Applications in Thermal Dosimetry

Thermal dosimetry for hyperthermic therapy implies the physical determination of the temperature versus time relationship for every point within the heated volume. This is still impossible, although much effort is being devoted to establishing measurement and numerical techniques that will be useful in determining the thermal dose. Phantom studies help considerably in this task. First, they help us to understand the power deposition pattern; second, they help to determine the effects of perturbing influences such as heterogeneous tissue properties or the insertion of electro-

magnetically perturbing thermometer probes. What they do not do is allow us to make measurements in phantoms and apply the results quantitatively to the heating of tumors in patients. Nevertheless, we will use examples to illustrate the use and limitations as aids to dosimetry in microwave and RF heating.

Microwave Heating of Surface Lesions

The first case involves the heating of a surface lesion of malignant melanoma. The patient had a large, fast-growing lesion in his right parotid salivary gland, involving also the tip of his earlobe. It was surgically resected, along with the inferior half of his ear, and skin was grafted over the defect. Melanoma subsequently invaded the skin graft as it healed and resulted in a new tumor mass about 4 cm x 6 cm in extent and 2 to 3 cm deep at the time the patient presented for hyperthermia. From the heating pattern in figure 12.2, we would expect exponential decay of the temperature distribution for 2.45-GHz microwaves propagating into the tumor. Figure 12.5A is a thermogram of the patient taken prior to heating, showing that the surface of the lesion is cold. It is also possible to locate the thermistor probes in the thermogram. The microwave applicator was oriented such that the electric field was perpendicular to the shaft of the probes. Figure 12.5B is a thermogram of the surface during heating, and figure 12.5C is a record of the temperature of the two probes. Surface temperatures, as determined thermographically, are indicated along the top of figure 12.5C. Blood flow and tissue properties changed the heating pattern dramatically from the expected dependence on $\frac{1}{e^2}$. Since the tumor involved the skin graft, the surface was permitted to become hot in order to allow temperatures at depth to rise to therapeutic levels. Following four hyperthermia treatments and concomitant radiotherapy (3200 rad in eight fractions), the tumor disappeared and the lesion healed.

The second case is shown in figure 12.6 (Manning et al. 1982). On this patient a large melanoma grew within the occipital region of the scalp. It was a contiguous mass adjacent to the skull and approximately 1.5 cm thick. From phantom studies we recognized that the strong dielectric discontinuity between the tumor and the skull would reflect much of the microwave energy back into the tumor, so little energy was expected to propagate into the brain. Figures 12.6A and 12.6B show thermograms taken during a heating. Figure 12.6C is a photograph of the tumor showing the location of a thermistor probe used to monitor subcu-

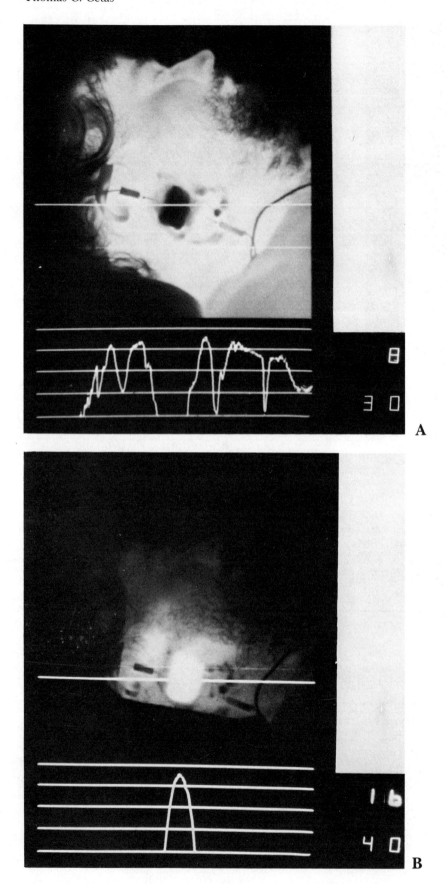

A

B

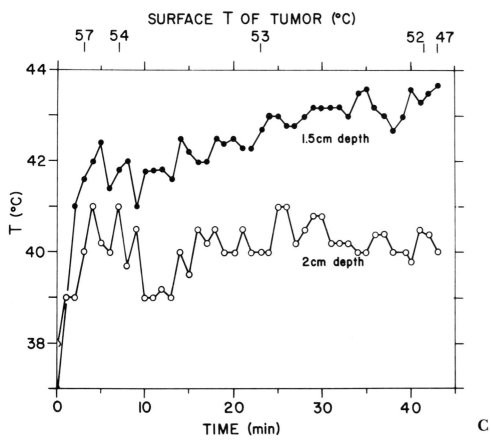

Figure 12.5. Thermal dosimetry data on patient with large melanoma over right parotid region. *A,* Thermogram of surface prior to heating with microwaves; *B,* typical thermogram taken during treatment; *C,* temperature versus time plots for two subcutaneous thermistor probes. The hottest temperature of the surface taken thermographically is indicated also. (Cetas, in press. Reprinted by permission.)

taneous temperatures. We found agreement to within 0.5°C between the surface measurement and probe reading throughout the several treatments. Phantom studies also showed that the thickness of the tumor is crucial to the success of the heating technique. If the lesions were very thin within the skin only, and a thick mass did not overlay the skull, then the energy would have propagated into the brain. Regions inaccessible to thermometers might have been heated. The thickness of this particular melanoma lesion was a significant fraction of the wavelength in tissue (about 1.5 cm). Again, the patient went on to a complete response for this lesion following heat and radiotherapy. He later died from distant untreated metastatic lung lesions that were present before this tumor was treated.

The next example demonstrates the power deposition pattern that results when a large area covering heterogeneous structures is heated by a relatively uniform microwave field. The patient had experienced a recurrence of breast cancer in her chest wall following a bilateral mastectomy. She was to be treated with hyperthermia and electrons. As for the patient with the occipital lesion discussed above, we anticipated that reflections from the rib cage would improve the homogeneity of the heating. The phantom shown in figure 12.7A was constructed to test this hypothesis. Rods of bone phantom represented the ribs, and a flat oblong disk represented the sternum. These were embedded in muscle phantom material as shown in the diagram. Overlaying this structure was a 1-cm-thick layer of muscle-equivalent material that represented the external chest wall and skin. Microwaves (2.45 GHz, WR430 waveguide antenna) were radiated onto the patient phantom from a height of 8 cm. The applicator was swept over broad arcs in order to illuminate the entire area. Three thermal studies were performed. Figure 12.7B shows a thermogram taken of the phantom skin surface following 2.5 minutes of heating. Note that the structure of the ribs is apparent, with the hottest areas occurring between the ribs. Nevertheless, the temperature pattern is smoothed because of thermal conduction and the diffuse reflection of the microwaves from the ribs. Figure 12.7C and 12.7D are thermograms taken from beneath the phan-

Thomas C. Cetas

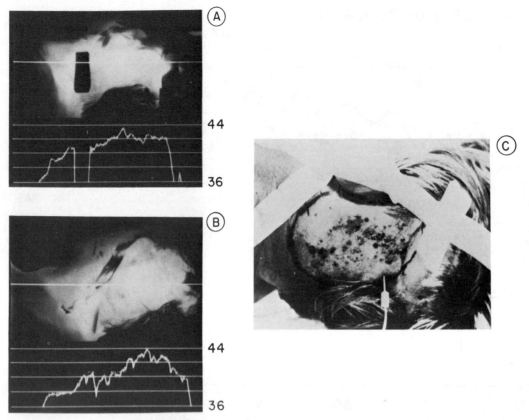

Figure 12.6. Typical thermograms (*A* and *B*) and photograph (*C*) of patient with melanoma over occipital region of skull. See text for discussion. (Manning et al. 1982. Reprinted by permission.)

tom and viewing a surface that corresponds to the internal surface of the rib cage. The heating field is still external. The thermogram in figure 12.7C was taken at 18 seconds into a 30-second heating. Areas directly beneath the ribs do not heat at all. Areas between the ribs heat rather dramatically because of superposition of energy reflected from the bone-muscle interface and the incident wave. It is interesting to note that along each of the heated strips nodes and antinodes can be observed. These correspond to the half-wavelength of the 2.45-GHz microwave field in muscle tissues ($\lambda = 1.75$ cm). The thermogram in figure 12.7D was taken about 30 seconds after that in figure 12.7C and 18 seconds after the heating. Thermal diffusion has already significantly altered the thermal pattern, as indicated by the rise in temperature of the ribs. In addition, a fringing field at the edges of the sternum has become apparent (note the bright horns at the top of the thermogram). Finally, we compared temperatures indicated by properly oriented thermocouples located in various places within the phantom. Thermocouples beneath the ribs or the sternum showed no temperature rise, in agreement with the thermograms. Thermocouples placed between and just above or just below the plane of the ribs showed definite hot spots when compared to probes placed just beneath or just above a particular rib. Nevertheless, the temperature rise at these hot spots was found to be approximately half that of the temperature rise on the surface as measured thermographically. Thus, the greatest power deposition is in the region anterior to the rib cage. If temperatures are monitored there, we need not be concerned with excessive temperatures occurring deep within the body where they cannot be measured. It should be emphasized that this study applies only to this particular wavelength of microwaves; longer wavelength fields, which tend to penetrate deeper, would have a quantitatively different heating pattern. Nevertheless, the observation that the bones markedly perturb the field would remain true for nearly all wavelengths.

Figure 12.8 is a series of thermograms taken of a patient who was treated with this technique. The temperature of the surface was reasonably uniform and does show some variations relating both to the vascular pattern and to the underlying structures. The ratio of forward and reflected power of the microwaves changed radically as the antenna was moved over the sternum and then to areas with thicker soft tissue masses. The black bars on the first thermogram taken

Physical Models (Phantoms) in Thermal Dosimetry

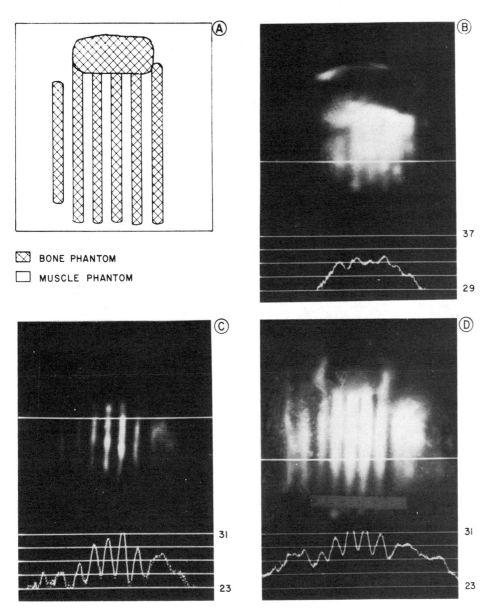

Figure 12.7. Phantom study representing microwave heating of a chest wall recurrence following mastectomy. See text for discussion.

at zero time are small markers placed on the patient's chest wall to help identify anatomically the region of heating. One thermistor probe was placed at an inferior lateral point on her chest wall about 5 mm subcutaneously (lower righthand corner in the thermogram). The probe indicated a gradual warm-up to 41.5°C at 18 minutes into the heat treatment and was stable for the remainder of the treatment. The surface warmed more quickly as measured thermographically and was at therapeutic temperatures by eight minutes into the treatment. Again, while the phantom work was instructive in terms of the power deposition, it was only of qualitative assistance in terms of the actual treatment. The patient experienced some cooling of the skin surface from convection and perspiration. The net gradient between the chest wall and the surface was less than that observed on the phantom.

Radiofrequency Current Heating

Electromagnetic currents in the frequency range of 0.1 to 10 MHz are used frequently for producing localized hyperthermia (Cetas and Connor 1978; Doss 1976; Doss and McCabe 1976; Hutson 1976; Oleson and Gerner 1982). This technique has certain advantages in that electromagnetic penetration depth is not

Thomas C. Cetas

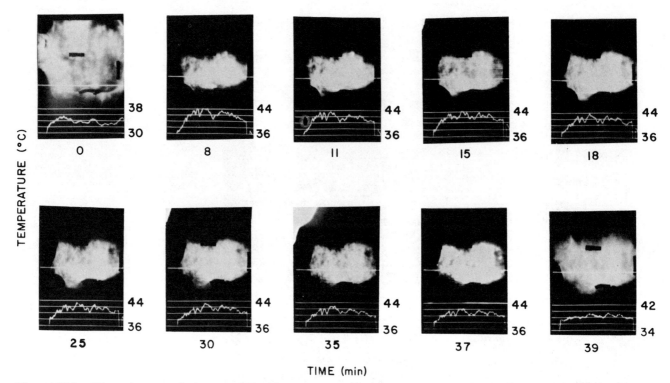

Figure 12.8. Thermograms of patient undergoing microwave heating of chest wall for recurrence following a bilateral mastectomy. The numbers to the *right* of each thermogram represent the temperature of the top and bottom scale lines of the thermogram. The number at the *bottom* is time in minutes from the start of heating.

a problem and that the heating field can be shaped to some extent by the configuration of the electrodes. Tissue impedances at these frequencies are predominantly resistive (Schwan, Carstenson, and Li 1956; Schwan and Foster 1980; Schwan and Piersol 1954; Stuchley and Stuchley 1980), and so the heating current can be understood in terms of analogies with low frequency or even direct current circuits. Finally, relatively high-powered equipment is available.

Some of the basic principles of the method can be illustrated with the phantom set-up in figure 12.9 (Oleson and Gerner 1982). Two plate electrodes are placed on either side of a rectangular phantom of muscle-type material. The cross-hatched regions represent a fat layer and three ribs constructed from fat and bonelike material. RF currents (about 1 MHz) pass from one plate through regions of muscle, fat, and muscle again, but around the ribs. The fat and muscle regions on the left side of the phantom are electrically in series so the high resistivity fat heats more. On the right, the bone and muscle are electrically in parallel, so the current flows through the lower resistivity muscle, which heats more. The ribs themselves do not heat although they have the same resistive properties as the fat. This rather simple phantom illustrates clearly that the nature of the heating pattern depends both on tissue properties and geometries that cannot necessarily be controlled by external means. This observation is true, of course, for all other heating modalities as well.

Figure 12.10 (Cetas, Connor, and Boone 1978) illustrates the effect of the size of the plate electrodes. The phantom (figure 12.10A) consists of a cylinder of muscle material with plate electrodes with diameters half that of the phantom and areas one-quarter that of the phantom cross-section. The phantom was heated with a pulse of RF power, was separated in the midplane, and both sections were exposed to the thermographic camera. Because the current was dispersed throughout the entire cross-sectional area of the phantom center but was confined to the cross-sectional areas of the plate at the ends, relatively superficial heating occurred near the plate electrodes. The power per unit volume is proportional to the square of the current density (A/m^2), so the heating per unit volume in the center is only one-sixteenth of that next to the plates. In figure 12.10B the plate area was increased to that of the phantom and the heating is uniform. Figure 12.10C is the same phantom as in figure 12.10B, except that the plates are at top and bottom, and a circular bone-type phantom has been placed in the center. As in figure 12.9, the current flows around the high resistivity bone material, causing cool shadows above and below the bone and slight hot spots on either side of the bone.

Physical Models (Phantoms) in Thermal Dosimetry

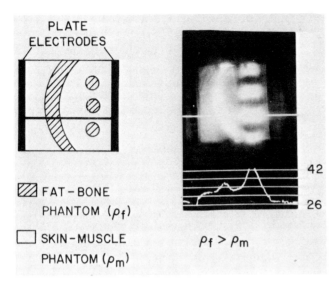

Figure 12.9. A phantom model of heating by RF capacitive coupling is shown. Resistivities ρ_f and ρ_m of fat and muscle, respectively, lead to greater ohmic heating not only in the fat layer, but also in muscle when currents are channeled around poorly conductive bone. (Oleson and Gerner 1982. Reprinted by permission.)

Figure 12.11A is a diagram of a phantom (Manning et al. 1982; Oleson and Gerner 1982) that was constructed to represent a large neck tumor growing anterior to the hyoid bone and below the chin. The tumor was approximately 8 cm in diameter. Figures 12.11B, 12.11C, and 12.11D are thermograms taken following heating of this planar phantom. The region in front of the hyoid was heated uniformly and the region behind this bone was protected from any heat. Figure 12.12 shows the temperature versus time plot of two thermistors located on either side of the tumor and several centimeters deep during a patient heat treatment. The temperatures at these two points are within a degree of one another again indicating relatively uniform heating for this very large tumor. The tumor completely regressed following a course of heat and radiation.

Figure 12.13 is a phantom study that represents heating of the entire neck cross-section. A bolus has been applied on either side of the neck to provide a rectangular cross-section. The shaded circle (fig. 12.13A) is a hollow plastic tube that represents the trachea. The bone structure representing the vertebral column is also apparent. Figure 12.13B is a surface thermogram made after heating, and figures 12.13C and 12.13D are computer generated relief plots of the thermogram showing the temperature (i.e., power deposition) pattern in two dimensions. The locations of the various lines on the relief plots are

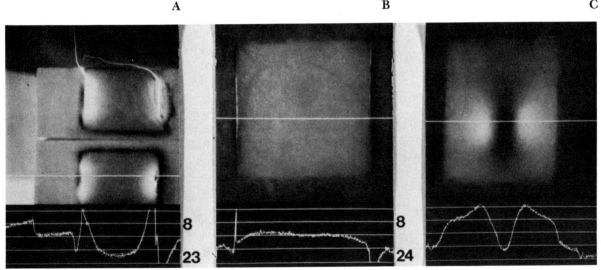

Figure 12.10. Phantom study of RF current heating using plate electrodes. Thermograms are taken with *white* as hot. A, A cylindrical muscle phantom was heated with a short current pulse (few seconds) at relatively high power and then was separated to expose the midplane to the thermographic camera. The *upper* and *lower* portions of the thermogram show the adjoining surfaces of the two halves of the phantom. The plate electrodes are circular with a diameter approximately half that of the phantom. B, Thermogram taken for the case where the plate electrodes cover the complete cross-section of a rectangular muscle phantom. The electrodes are again on each side. C, Thermogram showing the effects of a high resistivity bone phantom placed with the low resistivity muscle phantom. In this case, the electrodes are placed at the *top* and *bottom*, in contrast to the geometry of B. (Cetas, Connor, and Boone 1978. Reprinted by permission.)

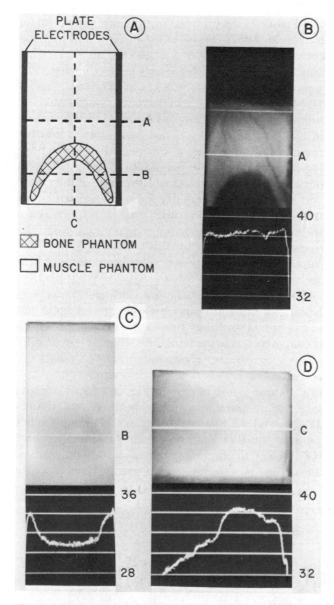

Figure 12.11. Phantom study of power deposition patterns using capacitively coupled RF at 3 MHz. Panel *A* is a schematic of phantom showing plate electrodes, anterior neck mass, and hyoid bone. *Dashed lines* correspond to white scan lines on thermograms. Panels *B*, *C*, and *D* are thermograms of surface temperatures. (Manning et al. 1982. Reprinted by permission.)

indicated by the marks in figure 12.13A. Again, extra heating occurs where the current is forced to pass through a smaller cross-sectional area.

Interstitial Heating

A form of heating that is very closely related to RF heating involves the use of interstitial electrodes. We have chosen to separate interstitial heating from RF heating with plates because interstitial heating is conveniently combined with brachyradiotherapy. The technique of interstitial thermoradiotherapy is described in chapter 21 and elsewhere (Cetas, Connor, and Manning 1980; Cetas et al. 1982; Manning, Cetas, and Gerner 1982; Manning et al. 1982). Further discussion of the dosimetry, including radiologic aspects, is given by Cetas and associates (1982). A description of another, but similar, interstitial heating system is given by Joseph and colleagues (1981).

Magnetic Induction Heating

As with other forms of inducing hyperthermia, magnetic induction has been used for many years, originating with D'Arsonval in 1893 (Süsskind 1979). It became a standard technique in physical therapy. One of the earliest phantom studies of the power deposition pattern was in 1936 by Pätzold and Wenk. More recently Storm and co-workers (1979, 1980) introduced the technique as a means of inducing deep-regional hyperthermia (see chapter 14). In this device the field is induced by a single turn current sheet with the appropriate tuning capacitance incorporated. A compact, highly efficient, low-loss heating coil results. To a first approximation, the magnetic field in the central plane of an induction coil is uniform. If a homogeneous dielectric, nonmagnetic conductive specimen is placed in this uniform magnetic field, the power deposited per unit volume will have the form

$$P = 2\pi f^2 \mu_0^2 \, \sigma \, r^2 H^2, \qquad (4)$$

where f is the frequency, H is the magnetic field, r is the radius from the center of the dielectric medium to the point of measurement, σ is the conductivity of the material, and μ_0 is the magnetic permeability, which is equal to that of free space (Stauffer 1979; Turner 1982). The heating pattern thus is parabolic ($\sim r^2$), with no power deposited at r = 0 and a maximum occurring at the outer radius of the body. It should be noted that the origin for this expression is the center of the dielectric body, not that of the coil.

Nevertheless, the human body is not homogeneous either electrically or thermally. Beneath the highly conductive skin frequently lies a layer of fat (low conductivity). Beneath this lies high conductivity muscle tissue and the internal organs. The total resistance of a current loop around the skin is high because the skin is thin, and hence relatively modest currents flow. In the fat layer, the low conductance implies high resistance for a current loop, and again little current will flow. In the high conductivity muscle, however, relatively substantial currents will be induced by the RF magnetic field. It is in this region that the predomi-

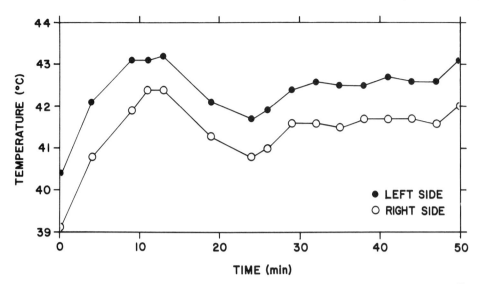

Figure 12.12. Thermistor probes at 2-cm depth from each lateral neck surface into the anterior tumor recorded these temperatures during the actual patient treatment corresponding to the phantom model in figure 12.11. (Manning et al. 1982. Reprinted by permission.)

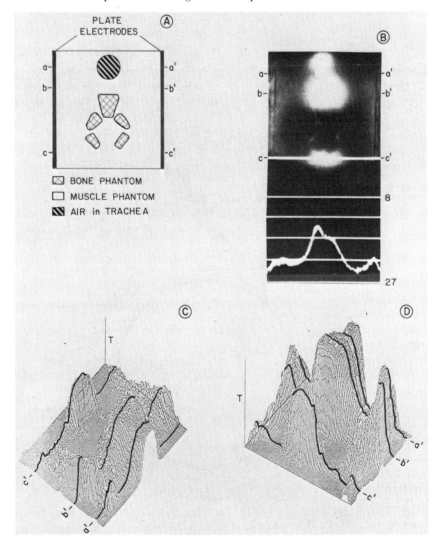

Figure 12.13. Phantom study of heating of the entire neck cross-section. See text for discussion.

nant heating will occur. Although even here, heterogeneities such as the vertebral column will perturb the currents, much as in the phantom trials illustrated in figures 12.9 and 12.10. Practically, and in agreement with Storm[3], we have found in clinical trials that when heating the thorax, current flow is constricted in the xiphoid region; hence, preferential heating occurs there. This can be prevented by applying a saline-soaked sponge as a current bolus directly on the xiphoid region. From this model also one would not expect energy to be deposited near the core of the patient. Heating at deep locations would occur from peripherally heated blood that is brought into the core by the circulation.

There is, however, another mechanism that may lead to local power deposition at depth in the body. If, for example, a large solid mass of high conductivity exists within the lungs, and it is lying within a region of low conductivity (normal lung), then eddy currents may be induced within this solid mass and cause this mass to heat preferentially. A similar situation could occur for an abdominal tumor surrounded by high resistivity tissue such as fat. The significance of this effect is being investigated as a means of understanding the results of others (Storm et al. 1979, 1980). The phantom studies required to test this hypothesis are difficult because of the problem of adequately modeling the heterogeneous properties of the tissues. In vivo studies are complicated by the effects of blood flow and by the continuing problem of measuring temperatures in the presence of the electromagnetic field. Non-field-perturbing optical thermometer probes will be available for these studies in the near future.

Magnetic Induction Heating of Ferromagnetic Implants

One final example of heating will be given to illustrate the use of static phantom materials in thermal dosimetry. This is the technique of using ferromagnetic implants (Stauffer 1979) to preferentially absorb energy from an imposed RF magnetic field. The tissue in the vicinity of the ferromagnetic implant is heated by thermal conduction from the hot seed rather than by direct absorption of energy from the magnetic field. Among the features of this technique is that the heating depends primarily upon the properties of the implants and their placement rather than those of the tissues. This technique leads to an additional way of performing interstitial thermoradiotherapy that has been shown (Cetas, Connor, and Manning 1980; Manning, Cetas, and Gerner 1982; Manning et al. 1982) to be

[3]F. K. Storm, personal communication.

particularly efficacious. Figure 12.14 is an idealized diagram of how such implants might be used in treating a glioblastoma. Stauffer (1979) has shown that the energy absorbed in one cubic centimeter of tissue that has been implanted by a long ferromagnetic seed is given by

$$P = \pi a \, (4\pi f \mu / \sigma)^{1/2} H^2. \quad (5)$$

Note that the power absorbed by the seed depends on the square root of the frequency f, whereas the energy absorbed from the magnetic field directly (equation 4) depends on the square of the frequency. In our laboratory we have determined that at frequencies near 13.56 MHz the tissue will absorb energy directly, whereas at lower frequencies near 1 MHz the ferromagnetic seeds will heat preferentially. Figure 12.15 is a computer reconstruction of a thermogram taken following a heating by magnetic induction of an array of 16 ferromagnetic seeds arranged in a muscle-equivalent phantom material. In order to understand this quantitatively, an analysis was performed in which the temperatures between the implants (figure 12.16, points b and c) was compared to the temperature of the implants (point a). Measured temperatures as a function of time during heating are shown in figure 12.17. In region I, the seeds heat rapidly. In region II, thermal conduction becomes important, and points b and c begin to warm steadily. In region III, a quasi-steady state has been established in which the seed temperature and the phantom material are warming at the same rate. That the seed temperatures are always significantly hotter than the surrounding medium causes no concern since the tissues next to seeds

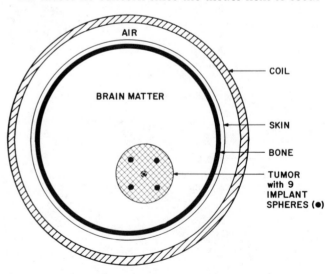

Figure 12.14. Diagram illustrating the technique of magnetic induction heating of ferromagnetic seeds in order to locally heat deep-seated tumors.

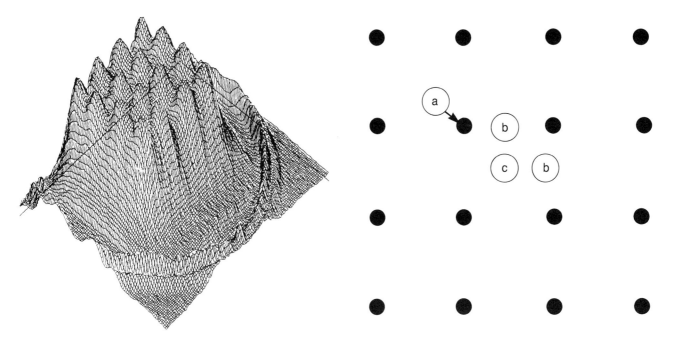

Figure 12.15. Computer-generated relief plot of thermogram of an array of 16 ferromagnetic seeds (2.4-mm diameter Ni rods, spacing 1.5 cm) imbedded in muscle phantom material and heated inductively.

Figure 12.16. Sketch showing points in an array to be used for thermal analysis. a, seed temperature; b, between nearest neighbor seeds; c, in center of a four-seed array.

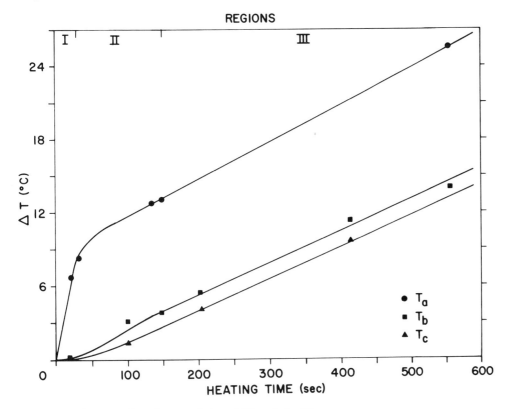

Figure 12.17. Heating characteristics of Ni rods in 16-seed array (see figs. 12.15 and 12.16) in a static muscle phantom.

would have been traumatized by the very act of insertion. Similar trauma occurs for a conventional radiation implant and heals quickly. It is interesting to note (fig. 12.17) that for the muscle-equivalent phantom material thermal conduction becomes important about one minute after the initiation of heating, as noted in our previous comments on other phantom systems.

The authors are developing metal alloys that pass from a ferromagnetic state at low temperatures through a Curie transition near therapeutic temperatures to a nonmagnetic state. Such seeds would no longer require any thermometer but rather would function as thermostats to establish the absolute temperature within the implanted region. Figure 12.18 shows the results of some preliminary materials produced by Demer and Demento of the University of Arizona, Department of Metallurgical Engineering. The 430-stainless-steel material remains ferromagnetic at all temperatures and was the substance used in the phantom trials described above. The nickel-4% silicon alloy passes through a broad Curie transition between 40°C to 50°C. Whether the transition can be made sharper and what constructions are required for making the seeds radioactive and biocompatible are under investigation. Clinical trials of the ferromagnetic seeds with ^{192}Ir sources and animal trials with the Curie point seeds are expected to begin soon.

DYNAMIC PHANTOMS

While the static phantoms discussed above are useful in studies of *power* deposition patterns in tissues by various modalities, they do not give true temperature distributions that occur in vivo because they do not account for the heat exchange owing to circulatory perfusion. In order to understand better the effects of blood flow upon the temperature distributions that result in living systems, the development of dynamic phantoms has been proposed for some time. The nature of such phantoms must be extremely complex, however, if they are to model realistically the complexities of in vivo thermal regulatory processes. Nevertheless, simple dynamic phantoms, while not necessarily realistic or quantitative, may help us to understand the contributions from various convective flow processes. An additional reason for constructing a dynamic phantom is to provide a true steady state temperature distribution, rather than a quasi-steady state distribution, such as that shown in figure 12.17. If one is testing the ability of a temperature controller, for example, to maintain a fixed temperature, then heat must escape from the system at a rate equal to the deposition rate.

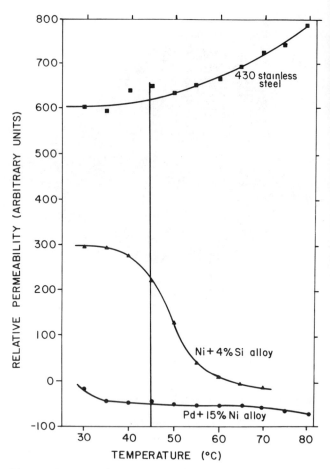

Figure 12.18. Relative permeability of three ferromagnetic alloys as a function of temperature. Note that the Ni–4%Si alloy passes through a Curie transition from the magnetic to nonmagnetic state between 40°C and 50°C.

Three simple dynamic phantom systems that were designed to look at certain aspects of convective heating transfer are discussed.

Theoretical Considerations

The theoretical considerations involved in designing a realistic thermodynamic phantom relate predominantly to the nature of the heat exchange caused by blood flow. Intuitively (Cetas and Connor 1978), we assume that the microcirculation can be modeled by a sponge with saline solution flowing slowly through it. The specific details of the microcirculation should not be important. The heat exchange by major vessels should be modeled adequately by plastic tubing of various diameters and fluid flow rates embedded in the sponge. In other words, the major vessels would appear as hot or cold pipes flowing through a homogeneous porous media that in turn has a slow mass flow through it. What is not at all clear from this intuitive approach is the size of vessels that represent

Physical Models (Phantoms) in Thermal Dosimetry

mined that the predominant heat exchange occurs in vessels of the order of 50 to 500 microns in diameter. Larger vessels tend to pass through the tissues as "cold pipes." Blood in capillaries is in equilibrium with the surrounding tissue mass so that no effective heat exchange occurs.

Examples of Dynamic Phantoms

Stauffer in 1979, in this laboratory (figure 12.19), has employed a simple dynamic phantom in which a sponge is perfused with saline solution at a known rate. Saline solution, which is forced vertically through a sponge, flows across the top surface and over the side to a collection and return reservoir. The temperature of the saline is maintained with a separate temperature controller. The flow is set such that the effective thermal conductivity as monitored by the Bowman thermal conductivity probe (Balasubramaniam and Bowman 1974, 1977) is equal to that recorded in animal brain tissue. This system was used to observe thermal patterns resulting from magnetic induction heating of ferromagnetic seeds as just described in the previous section. Here, the top of the ferromagnetic rods were just flush with the top of the sponge. The temperature of the seeds and the surrounding medium were monitored with the thermographic camera. The temperature of the seeds and the points between the seeds (a,b, and c, fig. 12.16) were plotted as before. The tempera-

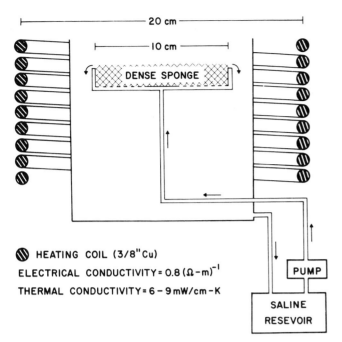

Figure 12.19. Diagram of dynamic phantom model.

major thermal perturbations and what size can be handled statistically. Chen and Holmes (1980) recently addressed this question using the techniques describing heat exchange that have been developed by mechanical engineers over several years. Assuming parameter values representative of tissues, they deter-

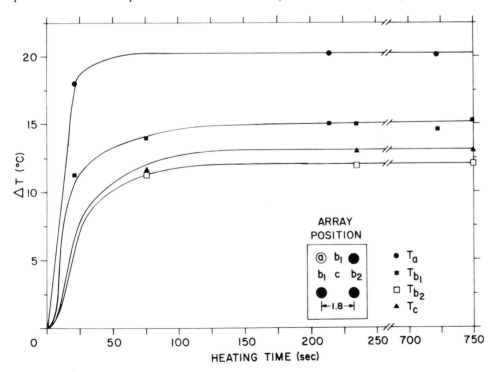

Figure 12.20. Heating characteristics of an array of four implants (430SS, 2.0-mm diameter, spaced 1.8 cm apart) in the dynamic phantom system (fig. 12.19).

275

ture pattern was smoother than in figure 12.15 but was skewed towards the edge of the sponge since the fluid flow on the surface was radially outward from the center. This skewing is characteristic of the phantom and may not be true of heated regions in vivo. It is clear from the temperature versus time plots (fig. 12.20) that a true steady state was reached. For this phantom, point b_2 was chosen so that it would not be influenced by fluids that had been heated beneath the surface, whereas point b_1 clearly was influenced by fluids heated below. A benefit of this phantom is that one can visualize how mass flow tends to make the heated region more uniform. Another benefit is that we are able to study systems where temperature regulation is involved, for example, using Curie point transitions in the ferromagnetic seeds (see figure 12.18) to regulate the temperatures within the volume. The application of these results to in vivo circumstances can be qualitative at best, however.

Sandhu, Bicher, and Hetzel (1982) have described a similar phantom in order to study the temperature distributions resulting from microwave heating (fig. 12.21A). Here again, saline solution flowing through a porous medium represents blood flow through the microcirculation. The geometric design is slightly different from ours. Fluid flow is driven by gravity and is controlled by the stopcock at the bottom. The bath at the top was maintained at a temperature of 37°C. Temperatures monitored at various points with thermocouples during heating are shown in figure 12.21B. The temperature pattern is skewed for this case of unidirectional flow, but the authors note that if blood flow in the microcirculation is omnidirectional then a more realistic pattern would be obtained by adding the values of temperature rise at contralateral symmetry points along the axis. Other observations are that when compared to the static case, maximum temperatures for a given power level in the center are lowered, and the heating pattern is broadened because of distribution of energy by convection. Finally, the dependence of the power deposition pattern on depth is comparable in terms of percentage of temperature rise.

Bowman (1980) has described a phantom system to study the influence of a large, thermally significant vessel. He selected glycerol for the primary medium because it has thermal properties similar to that of muscle. Through the glycerol media he passed a thin-walled glass tubing to simulate a large blood vessel. Temperature controlled water passed through the tubing at a controlled rate. By gradually bringing the thermal conductivity probe close to the vessel, it was possible to sense the diameter of thermal influence of the large vessel. Later, he confirmed the effect upon

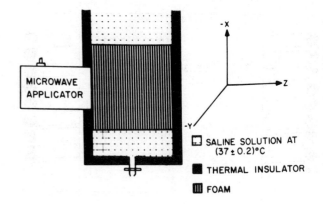

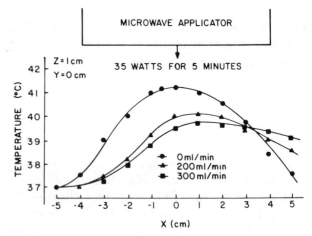

Figure 12.21. *A*, Schematic representation of dynamic phantom chamber of Sandhu, Bicher, and Hetzel (1982); *B*, steady state temperatures along x-direction for three flow conditions.

the probe by measurements in vivo using thermally significant vessels in isolated dog limbs.

CONCLUSIONS

Dynamic phantoms are more specialized than static phantoms in that they are designed to look at specific phenomena for purposes of better understanding a measurement technique or perhaps a means of heating. While they are designed to model the thermal dissipation properties of tissues, they are, nevertheless, restricted to systems that can be mechanically constructed, and so are not necessarily representative of living tissues. The goal of developing a realistic dynamic phantom system still appears to be valid. It would help us to understand not only power deposition in tissues but also the ultimate temperature patterns which result in tissues. Phantoms, including dynamic phantoms, are useful in that they provide a means to test mathematical models. This controlled arrangement is seldom possible in vivo, especially in human studies. The ultimate goal of thermal dosimetry is to predict the temperature distributions that will occur during therapy of patients with cancer when a particular form of heating is used. This implies numerical techniques involving large computers and a thorough knowledge of tissue properties. Studies with phantoms will be helpful in attaining that goal. Whether they will eventually be phased out as dosimetric tools as numerical techniques improve remains to be seen. Nevertheless, they always will be useful as a means of teaching and as a way of verifying that equipment indeed does work in the anticipated fashion.

References

Astrahan, M. A. Letter, hyperthermia phantoms. *Med. Phys.* 6:72, 1979a.

Astrahan, M. A. Concerning hyperthermia phantom. *Med. Phys.* 63:235, 1979b.

Balasubramaniam, T. A., and Bowman, H. F. Temperature field due to a time dependent heat source of spherical geometry in an infinite medium. *Journal of Heat Transfer Transactions ASME* 296:296–299, 1974.

Balasubramaniam, T. A., and Bowman, H. F. Thermal conductivity and thermal diffusivity of biomaterials: a simultaneous measurement technique. *Journal of Biomechanical Engineering Transactions ASME:* 99 (series K, no. 3):148–154, 1977.

Bowman, H. F. The bioheat transfer equation and discrimination of thermally significant vessels. *Ann. NY Acad. Sci.* 335:155–160, 1980.

Bowman, H. F.; Cravalho, E. G.; and Woods, M. Theory, measurement and application of thermal properties of biomaterials. *Annual Review of Biophysics and Bioengineering* 4:43–180, 1975.

Carslaw, H. S., and Jaeger, J. C. *Conduction of heat in solids*, 2nd edition. Oxford: Clarendon Press, 1959.

Cetas, T. C. Temperature measurement in microwave diathermy fields: principles and probes. In *Proceedings of the First International Symposium on Cancer Therapy by Hyperthermia and Radiation*, eds. M. J. Wizenberg and J. E. Robinson. Chicago: American College of Radiology, 1976, pp. 193–203.

Cetas, T. C. Practical thermometry with a thermographic camera: calibration, transmittance and emittance measurements. *Reviews of Scientific Instruments* 49:245–254, 1978.

Cetas, T. C. Thermometry. In *Therapeutic heat and cold*, 3rd edition, ed. J. F. Lehmann. Baltimore: Williams & Wilkins, 1982, pp. 35–69.

Cetas, T. C. Thermal dosimetry during hyperthermia. In *Proceedings of the International Symposium on Biomedical Thermology*, June 1981, Strasbourg, France, eds. M. Gauterie and E. Albert. New York: Alan R. Liss, in press.

Cetas, T. C. et al. Dosimetry of interstitial thermoradiotherapy, monograph 61. Bethesda, Md.: National Cancer Institute, 1982.

Cetas, T. C., and Connor, W. G. Thermometry considerations in localized hyperthermia. *Med. Phys.* 5:79–91, 1978.

Cetas, T. C.; Connor, W. G.; and Boone, M. L. M. Thermal dosimetry: some biophysical considerations. In *Cancer therapy by hyperthermia and radiation*, eds. C. Streffer et al. Munich and Baltimore: Urban & Schwarzenberg, 1978, pp. 3–12.

Cetas, T. C.; Connor, W. G.; and Manning, M. R. Monitoring of tissue temperature during hyperthermia therapy. *Ann. NY Acad. Sci.* 335:281–297, 1980.

Chen, M. M., and Holmes, K. R. Microvascular contributions in tissue heat transfer. *Ann. NY Acad. Sci.* 335:137–150, 1980.

Christensen, D. A. Thermal dosimetry and temperature measurements. *Cancer Res.* 39:2325–2327, 1979.

Doss, J. D. Use of RF fields to produce hyperthermia in animal tumors. In *Proceedings of the First International Symposium on Cancer Therapy by Hyperthermia and Radiation*, eds. M. J. Wizenberg and J. E. Robinson. Chicago: American College of Radiology, 1976, pp. 226–227.

Doss, J. D., and McCabe, C. W. A technique for localized heating in tissue: an adjunct to tumor therapy. *Med. Instrum.* 10:16–21, 1976.

Guy, A. W. Analysis of electromagnetic fields induced in biological tissues by thermodynamic studies on equivalent phantom models. *IEEE Trans. MTT* 19:205–214, 1971.

Guy, A. W. Letter. *J. Microwave Power* 10:358, 1975.

Guy, A. W. et al. Development of a 915-MHz direct contact applicator for therapeutic heating of tissues. *IEEE Trans. MTT* 26:550–556, 1978.

Guy, A. W.; Webb, M. D.; and McDougal, J. A. A new technique for measuring power deposition in phantoms exposed to EM fields of arbitrary polarization: example of the microwave oven. *IMPI Microwave Power Symposium*, University of Waterloo, Ontario, Canada, 1975, pp. 36–47.

Hahn, G. M. et al. Some heat transfer problems associated with heating by ultrasound, microwaves or radiofrequency. *Ann. NY Acad. Sci.* 335:327–346, 1980.

Hand, J. W. et al. A physiologically compatible tissue equivalent liquid bolus for microwave heating of tissues. *Phys. Med. Biol.* 24:426–431, 1979.

Ho, H. S. et al. Microwave heating of simulated human limbs by aperature sources. *IEEE Trans. MTT* 19 2:224–231, 1971.

Hutson, R. L. Calculations of two dimensional electric fields and potential distributions around electrodes in conducting media. *Med. Phys.* 3:256–258, 1976.

Johnson, C. C., and Guy, A. W. Nonionizing electromagnetic wave effects in biological materials and systems. *Proc. IEEE* 60:692–718, 1972.

Joseph, C. D. et al. Interstitial hyperthermia and interstitial iridium-192 implantation: a technique and preliminary results. *Int. J. Radiat. Oncol. Biol. Phys.* 7:827–833, 1981.

Kantor, G., and Cetas, T. C. A comparative heating pattern study of direct contact applicators in microwave diathermy. *Radio Science* 12:111–120, 1977.

Kantor, G., and Witters, D. M. A 2450 MHz slabloaded direct contact applicator with choke. *IEEE Trans. MTT* 28:1418–1422, 1980.

Kantor, G.; Witters, D. M.; and Greiser, J. W. The performance of a new direct contact applicator for microwave diathermy. *IEEE Trans. MTT* 26:563–568, 1978.

Kopecky, W. J. Using liquid dielectrics to obtain spatial thermal distributions. *Med. Phys.* 7:566–570, 1980.

Larsen, L. E. et al. A microwave compatible MIC temperature electrode for use in biological dielectrics. *IEEE Trans. MTT* 27:673–679, 1979.

Manning, M. R. et al. Clinical hyperthermia: results of a phase I trial employing hyperthermia alone or in combination with external beam or interstitial radiotherapy. *Cancer* 49:205–216, 1982.

Manning, M. R.; Cetas, T. C.; and Gerner, E. W. Interstitial thermoradiotherapy, monograph 61. Bethesda, Md.: National Cancer Institute, 1982.

Microwave diathermy products: performance standard. Department of Health and Human Services, FDA, BRH *Federal Register* 45 (147):50359–50368, July 29, 1980.

Mortimer, B., and Osborne, S. L. Tissue heating by shortwave diathermy. *JAMA* 104:1413–1419, 1935.

Oleson, J. R., and Gerner E. W. Hyperthermia in the treatment of cancer. In *Therapeutic heat and cold,* 3rd edition, ed. J. F. Lehmann. Baltimore: Williams & Wilkins, 1982.

Patzöld, J., and Wenk, P. Zur wirkungsweise des spulenfelds in der kurzwellentherapie: wärmessungen an geschieteten electrolyten in hoch frequenten spulenfeld. *Strahlentherapie* 55:692–707, 1936.

Sandhu, T. S.; Bicher, H.; and Hetzel, F. W. A realistic thermal dosimetry system, monograph 61. Bethesda, Md.: National Cancer Institute, 1982.

Schwan, H. P.; Carstenson, E.; and Li, K. The biophysical basis of physical medicine. *JAMA* 160:3, 191–197, 1956.

Schwan, H. P., and Foster, K. R. RF-field interactions with biological systems: electrical properties and biophysical mechanisms. *Proc. IEEE* 68:104–112, 1980.

Schwan, H. P., and Piersol, G. M. The absorption of electromagnetic energy in body tissues: part 1, biophysical aspects. *Am. J. Phys. Med.* 33:370–404, 1954.

Shitzer, A. Studies of bioheat transfer in mammals. In *Topics in transport phenomena,* ed. C. Gutfinger. New York: Halsted Press, 1975, p. 211.

Stauffer, P. R. A magnetic induction system for inducing localized hyperthermia in brain tumors. Master's thesis, University of Arizona, 1979.

Storm, F. K. et al. Normal tissue and solid tumor effects of hyperthermia in animal models and clinical trials. *Cancer Res.* 39:2245–2251, 1979.

Storm, F. K. et al. Hyperthermic therapy for human neoplasms: thermal death time. *Cancer* 4:1849–1854, 1980.

Stuchly, M. A., and Stuchly, S. S. Dielectric properties of biological substances—tabulated. *J. Microwave Power* 15:19–26, 1980.

Süsskind, C. The 'story' of nonionizing radiation research. *Bull. NY Acad. Med.* 55:1152–1163, 1979.

Turner, P. F. Deep heating of cylindrical or elliptical tissue mass, monograph 61. Bethesda, Md.: National Cancer Institute, 1982.

CHAPTER 13

Physical Aspects of Localized Heating by Radiowaves and Microwaves

Arthur W. Guy
Chung-Kwang Chou

Introduction
Biophysical Basis of Electromagnetic Heating
Applicators for Tissue Heating
Measurement of the Specific Absorption Rate (SAR)
Heating Patterns

INTRODUCTION

As early as 1909, de Keating-Hart used a combination of heat produced by high-frequency electrical currents and ionizing radiation for the treatment of cancer. These combined modalities have been used in the treatment of cancer in humans and laboratory animals over the past seven decades. Work prior to 1920 was summarized by Rohdenburg and Prime (1921); work between 1920 and 1937 was reviewed by Arons and Sokoloff (1937). The greatest reported success to date has resulted from the treatment of tumors by combined hyperthermia and ionizing radiation. Schereschewsky (1928) inhibited growth in most cases of transplanted mouse carcinoma, and in some cases completely eliminated the tumors, by treating them with 100-MHz ultra-shortwaves. For the next decade there was an average of one scientific paper per year reporting on the subject (Arons and Sokoloff 1937; Eidinow 1934; Fuchs 1936; Hasche and Collier 1934; Hill 1934; Johnson 1940; Mortimer and Osborne 1935; Pflomm 1931; Reiter 1932; Roffo 1934; Schliephake 1935).

Between 1931 and 1941 there were many basic problems in using shortwaves for the effective therapeutic heating of tissues. Most of these problems were related to the inability of investigators to quantify the actual power absorbed in tissues during treatment. The results of therapeutic treatment were left entirely to chance, and many quantitatively uncontrolled experiments resulted in contradictory statements in the medical literature.

Gessler, McCarty, and Parkinson (1950) appear to be the first group that used microwave frequencies in the experimental treatment of cancer. They were able to eradicate spontaneous mammary carcinoma in C3H mice with microwave exposure alone. Five years later, Allen (1955) cured Crocker sarcoma-39 in rats by exposing the animals to radiation and 2450-MHz microwaves for 10 to 20 minutes. Crile (1962) reported that growth of tumors in dogs and human beings was controlled by microwave diathermy and radiation combined. Crile concluded that prolonged elevation of temperature in certain cancers at levels between 42°C and 50°C selectively destroyed the tumors without damaging normal tissues. He felt that it was a secondary inflammatory reaction rather than the primary elevation of temperature that destroyed the tumors. Two years later, Cater, Silver, and Watkinson (1964) reported that combined therapy of radiation and subsequent 3-GHz microwave irradiation of the tumor

This work was supported by the Rehabilitation Service Administration grant 16−P−56818.

(47°C for 8 to 10 minutes) cured some rats with hepatoma-223 transplanted in the leg. The investigators noted that there were no long-term survivors treated by radiation alone or by hyperthermia alone, and that the tumor size was greater, and the mean survival time was significantly shorter, than in rats treated by the combined therapies. In the same year, Moressi (1964) found that mortality patterns were essentially identical in mouse sarcoma-180 cells exposed to 2450-MHz microwave radiation as in control cells held at the same temperature, ranging from 43°C to 48°C. He found that cell decay was highly temperature dependent, and undetected temperature deviations of no greater than 1°C could lead to erroneous interpretations.

Although the use of microwaves for therapeutic heating gained in popularity in the 1950s and early 1960s, interest in the use of shortwaves remained. Birkner and Wachsmann (1949) reported regressions and cures in skin carcinoma in patients exposed to shortwaves and ionizing radiation. Exposures of 82 patients for a period of 2.5 hours to 50-MHz shortwaves (tumor temperatures of 42°C to 44°C) alone produced regressions but no cures. When the shortwave exposure was combined with radiation, however, some cures were observed. Fuchs (1952) reported good clinical results when 50-MHz shortwave exposures of 10 to 20 minutes were followed by radiation exposure. He claimed that the good clinical result was an increased radiosensitivity owing to hyperemia and acceleration of metabolism as a result of the shortwave exposure.

In addition to the use of combined hyperthermia and x-ray therapy, interest developed in the use of microwaves for selectively heating tumors to provide for a more effective therapy with injected radioactive materials and chemotherapy. Copeland and Michaelson (1970) reported that the heating of Walker-256 carcinoma by selective radiation with 2800 MHz, 260 mW/cm^2 microwaves for five minutes induced a substantial increase in the amount of intravenously injected ^{131}I fibrinogen by 400%. Zimmer, Ecker, and Popovic (1971) reported the use of selective electromagnetic heating in tumors in hypothermic animals to enhance the action of chemotherapy.

Overgaard and Overgaard (1972a) provided an excellent review and reported on extensive and well-conducted experiments with 1200 mice in their laboratory, where transplanted tumors were permanently cured without damage to surrounding normal tissues by treatment with 27.12-MHz shortwave diathermy. They used a special high-frequency (HF) neutral thermocouple embedded in the tumor to provide automatic regulation of the shortwave output. Thus, it was possible to maintain any desired temperature continuously with a variation of about 0.1°C. With carefully controlled elevations in the range of 41.5°C to 43.5°C, they were able to work out a quantitative relationship between temperature and exposure time for curing the transplanted tumors. An analysis showed that the treatment induced histologic changes in the tumor cells without damaging the stromal and vascular cells in the tumor or in surrounding normal tissue. Immediately after treatment, definite changes were revealed in the mitochondria and lysosomes in the tumor cells. The intensity of these changes was directly related to the elevation of temperature, and became more pronounced within a few hours or days. They noted changes in the nuclei of the tumor cells and the chromosomal and nucleolar chromatin within the first few hours after exposure. They observed severe injury in all tumor cells 24 hours after exposure to a curative dose. Through histologic and biochemical observations, they obtained clues that allowed them to assume that the direct effect of the heat was due to a selective activation of the acid hydrolases located in the lysosomes of the tumor cells. In later work, Overgaard and Overgaard (1972b) found that the addition of a small dose of local radiation produced a highly significant intensification of the tumor-deleting effect. They found that successive application of heat and ionizing radiation doses, both substantially smaller than required in themselves to produce cures, produced a large number of total cures. They also noted that intervals of up to 24 hours between applications did not appreciably alter the curative effect. (See also chapter 7.)

From the mid-1970s on, interest in using electromagnetic fields, either alone or in combination with radiation, increased substantially, and a large number of favorable reports on the use of the therapy appeared in several symposia proceedings and publications. Among these are the proceedings of the First and Second International Symposia on Cancer Therapy by Hyperthermia and Radiation (Wizenberg and Robinson 1976; Streffer et al. 1977) and a special issue of *IEEE Transactions Microwave Theory and Techniques* on microwaves and medicine, with emphasis on the application of electromagnetic energy to cancer treatment (Guy 1978).

Some examples of more recent shortwave applications are the work of von Ardenne (1978); Overgaard (1978); Kim, Hahn, and Tokita (1978); and LeVeen and co-workers (1976). Continuing success is reported in the use of microwave hyperthermia as an adjunct in the treatment of tumors. Szmigielski and colleagues (1978) reported prolongation of the survival of mice bearing sarcoma-180 tumors when irradiated by 3000-MHz microwaves such that rectal temperatures increased by 3°C to 4°C. The inhibitory effect of microwave hyperthermia was enhanced by simultaneous treatment of the mice with interferon and interferon inducers. Mendecki and associates (1978) completely eradicated transplanted mammary adenocarcinoma in C3H mice and, in several cases, obtained favorable results in the treatment of basal cell carcinoma, malignant melanoma, and skin recurrence of carcinoma of the breast, using both 2450-MHz and 915-MHz microwave radiation. In these studies, the temperatures of the tumors were raised to the hyperthermic range of 42.5°C to 43.0°C. Nelson and Holt (1978) and Hornback and co-workers (1977, 1979) successfully treated cancers in patients with the UHF frequency of 433 MHz and ionizing radiation.

Recent papers on this subject presented at the Third International Symposium on Cancer Therapy by Hyperthermia, Drugs and Radiation are soon to be published as proceedings in the *Journal of the National Cancer Institute* (Dethlefsen and Dewey 1982).

The material covered in this chapter is a review of the biophysical basis of electromagnetic heating and the various shortwave and microwave applicators and their heating patterns. The application of an RF magnetrode for hyperthermia is discussed in chapter 14.

BIOPHYSICAL BASIS OF ELECTROMAGNETIC HEATING

Electrical Properties of Tissues

In order to comprehend some of the characteristics of shortwave and microwave interactions with biological materials, an understanding of the dielectric properties of biological tissues is necessary.

The dielectric behavior of the biological tissues has been evaluated thoroughly by Schwan and Piersol (1954, 1955), by Schwan alone (1957), and by other researchers, including Cook (1951a, 1951b, 1952) and Cole and Cole (1941). A tabulation of some important wave parameters is given in tables 13.1 and 13.2 (Johnson and Guy 1972). The first column lists selected frequencies between 1 MHz and 10 GHz. The frequencies of 27.12, 40.68, 433, 915, 2450, and 5800 are significant since they are used for industrial, scientific, and medical heating processes. The frequencies of 27.12, 915, and 2450 are used for diathermy purposes in the United States, whereas 433 MHz is authorized only in Europe for these purposes. The second column tabulates the corresponding wavelength in air, and the remaining columns pertain to the wave properties of a tissue group. Table 13.1 gives data for muscle, skin, or tissue of high water content, while table 13.2 is for fat, bone, and tissues of low water content. Other tissues containing intermediate amounts of water such as the brain, lung, bone marrow, and so forth, will have properties that lie between the tabulated values for the two listed groups. The table lists the dielectric properties and the depth of penetration of various tissues exposed to electromagnetic waves as a function of frequency. The interaction of electromagnetic waves with biological tissues is associated with these dielectric characteristics.

The action of electromagnetic fields on tissues produces two types of effects that control the dielectric behavior: (1) the oscillation of free charges or ions and (2) the rotation of polar molecules at the wave frequency. Free charge motion gives rise to conduction currents with an associated energy loss caused by electrical resistance of the medium. The rotation of polar molecules generates a displacement current in the medium with an associated dielectric loss caused by viscosity. These effects control the behavior of the complex dielectric constant. The effective conductivity accounts for both the conduction current and the dielectric losses of the medium.

The complex dielectric constant is dispersive (varies with frequency) because of the various relaxation processes associated with polarization phenomena. This may be illustrated by noting the dielectric properties given in tables 13.1 and 13.2. The decrease in the dielectric constant for tissues of high water content with increasing frequency is due to interfacial polarization across the cell membranes. The cell membranes, with a capacitance of approximately 1 $\mu F/cm^2$, act as insulating layers at low frequencies so that electromagnetic, field-induced currents are forced to flow around the membranes through the extracellular medium between the cells. The smaller cross-sectional pathways for the current result in a lower bulk conductivity of the tissues. At sufficiently low frequencies, the charging time constant is small enough to completely charge and discharge the membrane during a single cycle, resulting in a high tissue capacitance and therefore a high dielectric constant. When the frequency is increased, the capacitive susceptance of the cell increases, resulting in increasing currents in the intracellular medium with a resulting increase in total conductivity of the tissue. The increase in frequency also will prevent the cell walls from becoming totally charged during a complete cycle, resulting in a decrease in the dielectric constant. At a frequency of approximately 100 MHz, the cell membrane capacitive susceptance becomes sufficiently high that the cells can be assumed to be short-circuited. In the frequency range of 100 MHz to 1 GHz, the ion content of the electrolyte medium has no effect on the dispersion of the dielectric

Table 13.1
Properties of Microwaves in Biological Media

	Muscle, Skin, and Tissues with High Water Content				
Frequency (MHz)	Wavelength in Air (cm)	Dielectric Constant ϵ_H	Conductivity σ_H (S/meter)	Wavelength λ_H (cm)	Depth of Penetration (cm)
1	30,000	2000	0.400	436	91.3
10	3000	160	0.625	118	21.6
27.12	1106	113	0.602	68.1	14.3
40.68	738	97.3	0.680	51.3	11.2
100	300	71.7	0.885	27	6.66
200	150	56.5	1.00	16.6	4.79
300	100	54	1.15	11.9	3.89
433	69.3	53	1.18	8.76	3.57
750	40	52	1.25	5.34	3.18
915	32.8	51	1.28	4.46	3.04
1500	20	49	1.56	2.81	2.42
2450	12.2	47	2.17	1.76	1.70
3000	10	46	2.27	1.45	1.61
5000	6	44	4.55	0.89	0.788
5800	5.17	43.3	4.93	0.775	0.720
8000	3.75	40	8.33	0.578	0.413
10,000	3	39.9	10.00	0.464	0.343

SOURCE: Johnson and Guy 1972.

Table 13.2
Properties of Microwaves in Biological Media

	Fat, Bone, and Tissues with Low Water Content			
Frequency (MHz)	Dielectric Constant ϵ_L	Conductivity σ_L (mS/m)	Wavelength λ_L (cm)	Depth of Penetration (cm)
1	—	—	—	—
10	—	—	—	—
27.12	20	10.9– 43.2	241	159
40.68	14.6	12.6– 52.8	187	118
100	7.45	19.1– 75.9	106	60.4
200	5.95	25.8– 94.2	59.7	39.2
300	5.7	31.6–107	41	32.1
433	5.6	37.9–118	28.8	26.2
750	5.6	49.8–138	16.8	23
915	5.6	55.6–147	13.7	17.7
1500	5.6	70.8–171	8.41	13.9
2450	5.5	96.4–213	5.21	11.2
3000	5.5	110 –234	4.25	9.74
5000	5.5	162 –309	2.63	6.67
5800	5.05	186 –338	2.29	5.24
8000	4.7	255 –431	1.73	4.61
10,000	4.5	324 –549	1.41	3.39

SOURCE: Johnson and Guy 1972.

Table 13.3
Temperature Coefficient of Dielectric Constant and Specific Resistance of Body Tissues in Percent per Degree Celsius

	50 MHz	200 MHz	1000 MHz
$\Delta\epsilon/\epsilon$			
Tissues with high H$_2$O content	0.5	0.2	−0.4
Fatty Tissue	—	1.3	1.1
0.9% NaCl	−0.4	−0.4	−0.4
$\Delta\rho/\rho$			
Tissues with high H$_2$O content	−2	−1.8	−1.3
Fatty tissue	−(1.7−4.3)	−4.9	−4.2
0.9% NaCl	−2.0	−1.7	−1.3

SOURCE: Schwan 1965.
NOTE: The temperature dependence of resistance is comparable with that of saline solutions for all tissues with high water content. The dependence of the dielectric constant approaches that of saline solutions only at very high frequencies.

constant, so the values of ϵ_H and σ_H are relatively independent of frequency. Schwan and Piersol (1954; Schwan 1959) suggested, however, that suspended protein molecules with lower dielectric constants act as "dielectric cavities" in the electrolyte, thereby lowering the dielectric constant of the tissue. They attribute the slight dispersion of ϵ_H to the variation of the effective dielectric constant of the protein molecules with frequency. The final decline of ϵ_H and increase of σ_H at frequencies above 1 GHz can be attributed to the polar properties of water molecules, which have a relaxation frequency near 22 GHz.

The dielectric behavior of tissues with low water content is quantitatively similar to tissues with high water content, but the values of the dielectric constant ϵ_L and conductivity σ_L are an order of magnitude lower and are not as well understood quantitatively. This results from the unknown ratio of free to various types of bound water. There is also a large variation of dielectric constant in fat tissues of low water content. Since water has a high dielectric constant and conductivity, the electrical properties of fat will change significantly with small changes in water content.

The values of ϵ and σ also vary with temperature. This variation, discussed by Schwan (1965), is tabulated in table 13.3.

The dielectric properties of the tissues play an important part in determining the distribution of currents and associated heating patterns resulting from the application of electromagnetic fields. They also determine the reflected and transmitted power at interfaces between different tissue media and the amount of total power a given biological specimen will absorb when it is exposed to electromagnetic radiation.

Electromagnetic Heating Quantitation

The quantitation of electromagnetic heating can be illustrated by a simple analysis of the energy absorption caused by a high-frequency current applied to tissue through direct-contact electrodes, as shown in figure 13.1A. Here we consider a slab of tissue of thickness d sandwiched between conducting metal electrodes with surface A. An alternating current source with rms voltage V is applied to the electrodes. The resulting rms current I that will flow through the tissue is given by

$$I = YV, \quad (1)$$

where Y is the admittance measured at the metal electrodes. If we assume a uniform electric field and current flow, the admittance may be expressed as

$$Y = j\omega C^*, \quad (2)$$

where $\omega = 2\pi f$, f is the frequency of the applied voltage, and C^* represents a complex or "lossy" capacitance:

$$C^* = \epsilon_m^* A d^{-1}, \quad (3)$$

where ϵ_m^* is the complex permittivity of the tissue given by

$$\epsilon_m^* = \epsilon_o \epsilon_m - j\sigma_m \omega^{-1}. \quad (4)$$

$\epsilon_o = 8.85 \times 10^{-12}$ Farads/m is the permittivity of free space, ϵ_m is the dielectric constant of the tissue, and σ_m is the electrical conductivity of the tissue.

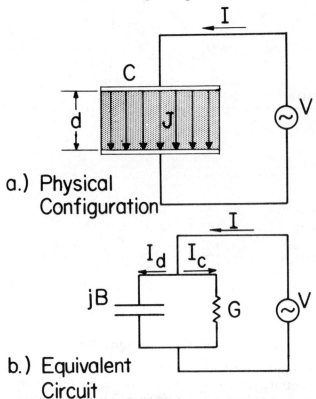

Figure 13.1. A, Physical configuration and B, equivalent circuit of heating homogeneous tissue with direct contact conducting plates.

The admittance may also be expressed in the standard form

$$Y = G + jB, \qquad (5)$$

where G is the conductance and B is the susceptance.
From the above equations

$$G = \sigma_m A d^{-1}, \qquad (6)$$

and
$$B = \omega C = \omega \epsilon_o \epsilon_m A d^{-1}, \qquad (7)$$

where $C = \epsilon_o \epsilon_m A d^{-1}$ is the capacitance for the electrodes, resulting from the real part of the tissue dielectric constant ϵ_m. By Joule's law, the power loss on the transfer of electrical energy into heat in the tissue is simply

$$P = GV^2. \qquad (8)$$

The tissue-filled capacitor may be represented by the equivalent circuit, shown in figure 13.1B, where the conduction current $I_c = VG$ passing through G is responsible for the tissue heating, and the current $I_d = VB$ flowing through the capacitor C is the dis-

placement current. The current density responsible for the heating is $J = I_c A^{-1}$. Since the electric field strength in the tissue is $E = V d^{-1}$, we may express the current density as

$$J = \sigma_m E \qquad (9)$$

and the power loss per unit volume in the tissue as

$$PA^{-1}d^{-1} = GV^2 A^{-1} d^{-1} = \sigma_m E^2. \qquad (10)$$

The National Council on Radiation Protection and Measurements (1981) recommends the use of the quantity *specific absorption rate* (SAR), given in units of watts per kilogram (W/kg) to quantify the rate of energy absorbed per unit mass in tissues exposed to electromagnetic fields. Thus, if we define W as the SAR, equation (10) may be written for any tissue with conductivity as

$$W = 10^{-3} \rho^{-1} \sigma E^2, \qquad (11)$$

where ρ is the density of the tissue in g/cm³.

Although the discussion above applies to relatively simple plane layer tissue geometries, it introduces some important dosimetric quantities and units that are used for the more complex exposure conditions at frequencies covering both the shortwave and microwave bands. The goal in the use of electromagnetic fields for therapeutic heating is to produce electric fields or currents in the region of tissue where the heating is desired, with minimal generation of fields in other regions of tissue. The fields must be maintained at sufficiently high levels to produce an SAR that will bring the tissue temperature to the proper level for the desired therapy. In cases where cooling mechanisms such as blood flow are minimal, the required SAR will be lower than that needed in other regions where cooling is greater. The SAR may be calculated from known field distributions by means of equation (11), or it may be measured directly from the rate of temperature rise using the equation

$$W = \frac{4186\, c \Delta T}{t} \text{ (W/kg)}, \qquad (12)$$

where c is the specific heat in kcal/kg·°C, T is the temperature rise in °C, and t is the exposure time in seconds.

Heating Characteristics in Tissue

The energy equation in the tissue heated with electromagnetic (RF) energy (Guy, Lehmann, and Stonebridge 1974) can be expressed as

$$\frac{d(\Delta T)}{dt} = \frac{1}{4186c}[W_a + W_m - W_c - W_b], \quad (13)$$

where W_a is the SAR of RF energy, W_m is the metabolic heating rate, W_c is the power dissipated by thermal conduction, and W_b is the power dissipated by blood flow, all expressed in W/kg, and $\Delta T = T - T_0$ is the difference between the tissue temperature T and the initial tissue temperature T_0 prior to treatment.

Within the therapeutic temperature range, the metabolic heating rate may be expressed as

$$W_m = W_o(1.1)^{\Delta T}, \quad (14)$$

where W_o is the initial metabolic heating rate.

The thermal conduction rate may be expressed as

$$W_c = \frac{k_c}{\rho}\nabla^2 T. \quad (15)$$

There k_c is the thermal conductivity of the tissue in mW/cm°C, and ∇^2 is the gradient operator.

If it is assumed that blood enters the tissue at arterial temperature T_a and leaves at tissue temperature T, the heat removed by blood flow is

$$W_b = k_2 m c_b \rho_b \Delta T^1, \quad (16)$$

where $\Delta T^1 = T - T_a$, c_b is the specific heat of blood, ρ_b is the density of blood in g/cm^3, m is the blood flow rate in milliliters per 100 g per minute, and the constant $k_2 = 0.698$. Typical physical and thermal properties of tissues are given in table 13.4 (Bluestein, Harvey, and Robinson 1968; Lehmann et al. 1965; Minard 1970; Nevins and Darwish 1970). When a therapeutic level of electromagnetic power, $50 < W_a < 170$ W/kg, is absorbed, ΔT will increase as shown in figure 13.2, with an initial linear transient period typically lasting about three minutes, with a time rate of increase,

$$\frac{d(\Delta T)}{dt} = \frac{W_a}{4186c}. \quad (17)$$

This period is followed by a nonlinear transient period usually lasting another 7 to 10 minutes, where ΔT becomes sufficiently large that blood flow and thermal conduction becomes important in dissipating the applied energy. In tissues with negligible or insufficient blood flow, the temperature will monotonically approach a steady-state value dictated by the magnitude of W_a as shown on the upper curve, where equilibrium is reached when $W_a = W_c$. For vascularized tissues, however, blood flow plays a significant part in heat dissipation, limiting the slope of the $d(\Delta T)/dt$ curve. In addition, for vascularized tissues a marked increase in blood flow will occur, caused by vasodilation when the temperature passes through the range of 42°C to 44°C. As a result, the temperature will drop and approach a steady-state value at a somewhat lower level,

Table 13.4
Thermal and Physical Properties of Human Tissues

Tissue	Specific Heat c(kcal/kg·°C)	Density ρ(g/cm3)	Metabolic Rate Wo (W/kg)	Blood Flow Rate m(ml/100 gm·min)	Thermal Conduction kc (mW/cm·°C)
Skel. muscle (excised)		1.07			4.4
Skel. muscle (living)	0.83		0.7	2.7	6.42
Fat	0.54	0.937			2.1
Bone (cortical)	0.3	1.79			14.6
Bone (spongy)	0.71	1.25			
Blood	0.93	1.06			5.06
Heart muscle			33	84	
Brain (excised)					5.0
Brain (living)			11	54	8.05
Kidney			20	420	
Liver			6.7	57.7	
Skin (excised)					2.5
Skin (living)			1	12.8	4.42
Whole body			1.3	8.6	

SOURCE: Guy, Lehmann, and Stonebridge 1974.

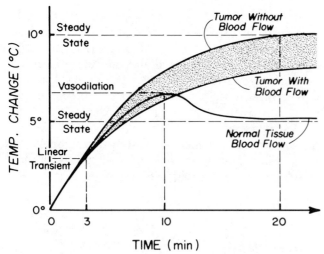

Figure 13.2. Schematic representation of temperature change in tumor and normal tissues under hyperthermia. (Modified from Guy, Lehmann, and Stonebridge 1974. Reprinted by permission. © 1974 by the IEEE.)

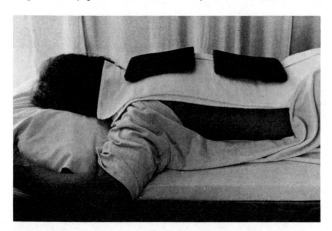

Figure 13.3. Capacitor electrodes for shortwave diathermy. (Guy, Lehmann, and Stonebridge 1974. Reprinted by permission. © 1974 by the IEEE.)

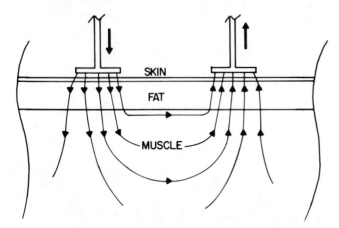

Figure 13.4. Field distribution in layered tissues exposed to capacitor-type electrode diathermy. (Guy, Lehmann, and Stonebridge 1974. Reprinted by permission. © 1974 by the IEEE.)

between 40°C and 41°C, as shown in figure 13.2, when $W_a = W_c + W_b$. In tumors, the blood flow is in general vigorous at the periphery and sluggish in the center (see chapter 8). Since the blood vessels in tumors are, in general, fully open during ordinary conditions, there is no vasodilation during the hyperthermia. After steady state conditions are reached, the final temperature of the tumor is higher than that of the surrounding normal tissue. The shaded area in figure 13.2 indicates the range of temperature rise in tumors. The lower boundary is for the periphery of the tumor, and the upper curve is for the center of the tumor where no blood flow exists. Clinical diathermy experience has shown that when vascularized normal tissue is exposed to hyperthermia, pain will be noted by the patient before any tissue damage can occur. In fact, the pain may be used as a guide to indicate that the tissue temperature has reached the required 43°C to 45°C for blood vessel dilation.

The SAR W_a must be sufficiently high so that the therapeutic level of temperature can be maintained over the major portion of the treatment period. If too little power is applied, the period of elevated temperature will be too short for any benefits. If too high a level is applied, the temperature can overshoot the safe level before the vasodilation can take effect. The pain sensors are a reliable and sensitive means for detecting this temperature range, however, and if the applied power level is set so that only mild pain or discomfort are first experienced by the patient, the vasodilation will be sufficient to limit or even lower the temperature in normal tissues to a level that is both tolerable and therapeutically effective. If the effective temperature is reached at the surface, it is felt as a mild burning sensation. On the other hand, if it is reached in the deeper tissues, it is felt as a dull aching type of pain.

APPLICATORS FOR TISSUE HEATING

Electromagnetic heating of tissues may be accomplished by four methods: (1) direct electric heating of muscle tissue by alternating currents through the use of interstitial or direct-contact electrodes, where the frequency is sufficiently high (greater than 100 kHz) to prevent the excitation of nerve action potentials; (2) noncontacting capacitive plates for application of currents at frequencies above 13 MHz; (3) solenoidal, or "pancake," magnetic coils, or loops that induce eddy currents in the tissue by alternating magnetic fields; and (4) radiation at ultra-high (UHF) or microwave frequencies. The latter three methods are discussed here.

Shortwave Applicators

Capacitor Electrodes

The earliest diathermy equipment consisted of a high-frequency generator, from which currents were applied directly to the tissues by contacting electrodes. As a result of uneven or poor contact, one of the greatest hazards was the production of burns, localized at the electrode-tissue interface. As the operating frequency of diathermy equipment increased, the electrodes were redesigned so that they did not have to make direct contact with the tissues, since displacement currents between the electrode plates and the anatomic surfaces were sufficient to couple energy to the tissue. Although capacitor electrode arrangements such as those shown in figure 13.3 are still used to treat patients with present-day 13.56- and 27.12-MHz diathermy equipment, there are many fundamental problems. Figure 13.4 illustrates how induced conduction currents in the tissue produce much greater power absorption in the subcutaneous fat than in the skin and muscle tissue, and how the divergence of the current tends to concentrate the power absorption in the superficial tissue adjacent to the electrodes. There is an order-of-magnitude greater heating in subcutaneous fat than in muscle or skin. Additional selective heating occurs in the fat owing to spreading of the fields as a function of distance from the electrodes. This, along with the lower specific heat in fat, results in a rate of heating more than 17 times greater in fat than in muscle. In addition, the blood-cooling rate is significantly less in fat, so the final steady-state temperature would be considerably higher than that in muscle.

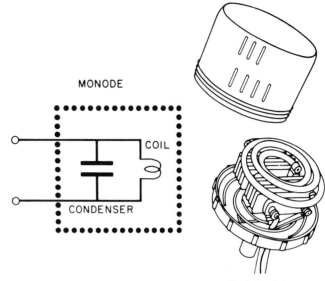

Figure 13.6. Compact-type induction coil with wiring arrangement (courtesy of Siemens-Reinger Werke Ag, Erlangen, Germany). (Guy, Lehmann, and Stonebridge 1974. Reprinted by permission. © 1974 by the IEEE.)

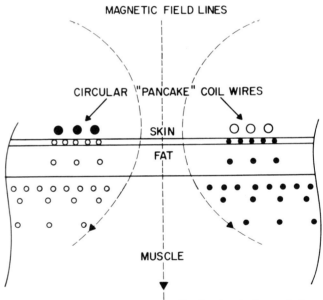

Figure 13.7. Induced current distribution in layered tissues exposed to induction pancake coil diathermy. (Guy, Lehmann, and Stonebridge 1974. Reprinted by permission. © 1974 by the IEEE.)

Induction Coils

Figures 13.5 and 13.6 illustrate various inductor coil configurations that induce circular eddy currents in the tissues by magnetic induction. Figure 13.5 illustrates the use of a large coil of insulated cable separated from the patient by toweling. The applicator shown in figure 13.6, a *monode*, is a more compact coil

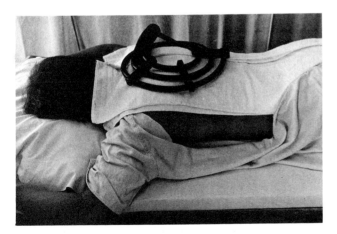

Figure 13.5. Induction coil (pancake coil) for shortwave diathermy. (Guy, Lehmann, and Stonebridge 1974. Reprinted by permission. © 1974 by the IEEE.)

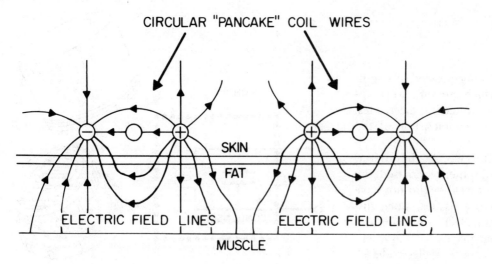

Figure 13.8. Field distribution in tissues caused by intercoil potentials of induction pancake coil shortwave diathermy. (Guy, Lehmann, and Stonebridge 1974. Reprinted by permission. © 1974 by the IEEE.)

and capacitor combination that may be spaced at various distances from the patient by an adjustable supporting arm (not shown). A cross-sectional view of the induced currents (figure 13.7) illustrates the superiority of the inductive coil over the capacitor electrode type. For this case, the induced electric fields are parallel to the tissue interfaces, and therefore are not greatly modified by the tissue boundaries. Ideally, the current density and heating will be higher in the muscle tissue where the conductivity is maximum, as shown schematically in figure 13.7. Under certain conditions where the diameter and spacing of the coil turns are excessive, or when the coil is placed too close to the tissue, more energy may be coupled to the subcutaneous fat than to the deeper tissues. This is caused by the sharp increase in magnetic field strength near the coils and the high electric field between the coil turns. This latter coupling is illustrated in figure 13.8.

Microwave Applicators

When microwave diathermy was first introduced in 1946, there was great hope that it would provide significantly improved heating patterns over those of shortwave diathermy. The shorter wavelength provided the capability to direct and focus the power and to couple it to the patient by direct radiation from a small, compact applicator. This was originally believed to be a distinct improvement over the quasi-static and induction-field coupling provided by a cumbersome capacitor and coil-type "pancake" applicators. The cross-sectional area of the directed power could be made smaller and used to provide much more flexibility in controlling the size of the area treated.

Radiating Type

The commercially available 2450-MHz diathermy

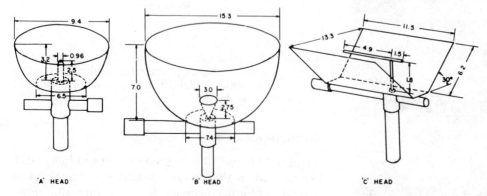

Figure 13.9. Radiation-type applicators used with 2450 MHz diathermy apparatus. (Guy, Lehmann, and Stonebridge 1974. Reprinted by permission. © 1974 by the IEEE.)

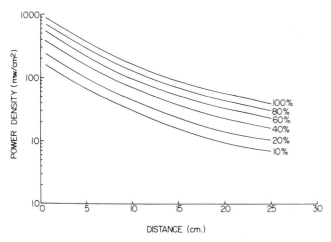

Figure 13.10. Power density along axis of maximum intensity of a Burdick C director at various power settings and distances. (Guy, Lehmann, and Stonebridge 1974. Reprinted by permission. © 1974 by the IEEE.)

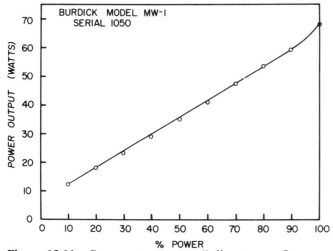

Figure 13.11. Power output to the C director as a function of the power setting on the Burdick diathermy unit. (Guy, Lehmann, and Stonebridge 1974. Reprinted by permission. © 1974 by the IEEE.)

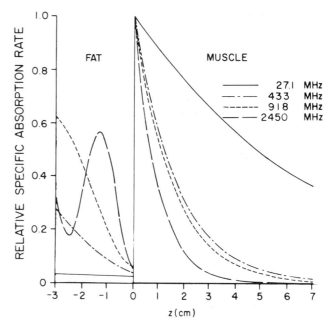

Figure 13.12. Relative SAR in plane layers of fat and muscle exposed to plane waves at different frequencies. (Johnson and Guy 1972. Reprinted by permission. © 1972 by the IEEE.)

equipment consists of a 100-W magnetron generator with a variable-power control calibrated in percentage of total output power. Various types of standard dipole and monopole applicators used with this generator are illustrated in figure 13.9. The radiation power density of the most widely used C director is shown in figure 13.10 as a function of the distance from the applicator and percentage of power output from the generator. The relationship between the percentage of power and actual power delivered to the antenna is shown in figure 13.11.

Direct Contact Type

Figure 13.12 shows clearly the superiority of the use of frequencies lower than 2450 MHz in terms of desirable therapeutic heating characteristics. Plane-wave or radiating-type sources become impractical to use at these lower frequencies, since the energy is impossible to focus into a beam with reasonably sized applicators, and the near-zone fields of the applicators extend to greater distances. Under these conditions, a pure radiation or far-zone field can be maintained only by placing the applicator at distances where large areas of the body would be exposed, and excessive power levels would be required. Thus, in order to obtain selective heating with reasonable input power levels (50 to 100 W), one must necessarily expose the tissues to the near-zone fields of the source. The induced fields in the tissue are then highly dependent on the source field distribution and frequency and may be considerably different from those produced by plane wave or radiation fields. The aperture source provides a reasonable model for studying the effect of source distribution, size, and frequency on the induced fields in the tissues. With such a source, the size and distribution of the induced fields may be controlled, resulting in improved diathermy applicators.

Guy and associates (1978a) have developed a 915-MHz direct-contact applicator (figure 13.13). This 13 × 13 cm cavity applicator is composed of two separate TE_{10} mode waveguides. A light weight dielectric foam (Emerson & Cuming, Inc., Canton, Maine) with $\epsilon' = 4$ is used to load the applicator to allow 915-MHz

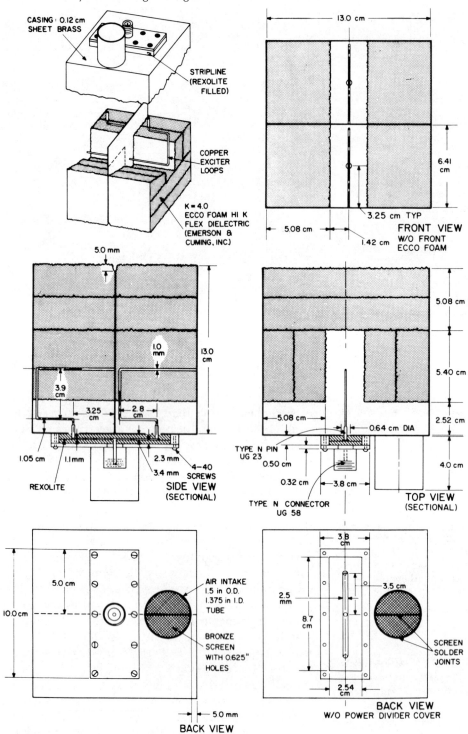

Figure 13.13. Design of the 915 MHz direct-contact applicator. (Guy et al. 1978a. Reprinted by permission. © 1978 by the IEEE.)

Figure 13.14. 2450 MHz circular polarized direct-contact applicator. (Kantor, Witters, and Greiser 1978. Reprinted by permission. © 1978 by the IEEE.)

wave propagation in the small sized waveguide feed system. The foam is also porous, which allows air to be blown through the applicator and onto the surface of the tissue being treated. To prevent hot spots, an acrylic radome is placed at the aperture to separate the metal edges from the tissue.

Kantor, Witters, and Greiser (1978) described a 2450-MHz circularly polarized applicator (figure 13.14). This applicator has a single feed and a pair of phase shifters for producing a circularly polarized TE_{11} mode. There is also an annular choke to control leakage.

A slab-loaded rectangular waveguide applicator also was reported by Kantor and Witters (1980). This is a standard WR-430 waveguide loaded with Teflon™ slabs with a dielectric constant of 2 (figure 13.15). These slabs provide more uniform heating patterns by forcing the fields to be uniform in the air space

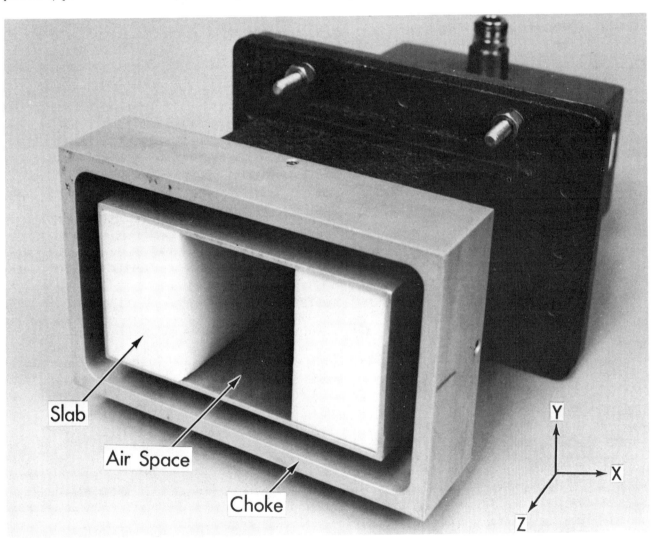

Figure 13.15. 2450 MHz slab-loaded direct contact applicator. (Kantor and Witters 1980. Reprinted by permission. © 1980 by the IEEE.)

between the slabs. The aperture is surrounded also by a choke.

Phase Array Contact Applicator

The SAR in deep tissues may be significantly increased over that produced with a single applicator by the use of phase array applicators. For example, if two identical applicators are placed on opposite sides of a body member, such as the leg or arm, and the applicators are fed with the same source so that the aperture field distributions are in electrical phase, the electric fields at the center of the cylinder, because of the two applicators, will directly add. Since the SAR is proportional to the square of the electric field, the SAR level and the temperature will be increased by a factor of 4. If two additional applicators are placed in the same way, with the center line of each perpendicular to the center line of the first two applicators, and they also are fed and phased, the total SAR from four applicators will be a factor of 8 above a single applicator. The fields of the two pairs of applicators do not add directly since they are perpendicular with each other. If we considered a spherical object being exposed the same way with four applicators, and another two are added to the remaining two open sides of the sphere, the SAR could be increased by a factor of 12 over that of a single applicator as compared with the surface SAR. An experimental demonstration of the phase array was done using the 915-MHz rectangular aperture applicators for exposing a cylindrical phantom model of the human arm shown in figure 13.16. The applicators were set up in phase using a system of hybrids (Guy and Chou 1977).

MEASUREMENT OF THE SPECIFIC ABSORPTION RATE (SAR)

The heating pattern in the tissue is synonymous with the SAR pattern, while the temperature distribution is a function of thermal diffusion, blood circulation, and time. Since the initial rate of change of temperature in the tissue owing to electromagnetic heating is equal to the SAR, it is more convenient to measure the SAR by temperature probes than through the use of larger and more complicated electric field sensors. The measurement must be done, however, before the temperature of the tissue is significantly altered by heat diffusion and blood flow. The application of high power of short duration is necessary for this measurement. Since the power need not be applied any longer than necessary to measure a change of a few degrees, the measurement can be done with no ther-

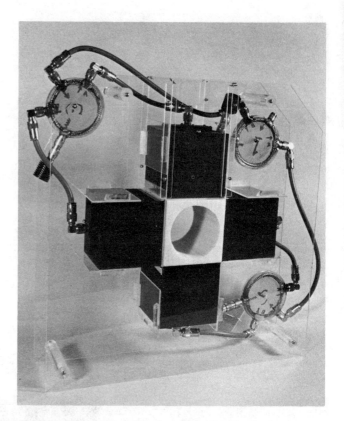

Figure 13.16. Phase array arrangement of four 915 MHz direct contact applicators for heating of the human arm.

mal damage to the tissue regardless of the applied power level.

The temperature probes used for determining the true rate-of-change of temperature must have a fast time response and be nonperturbing to electromagnetic fields. Recently, several nonperturbing temperature probes have been described and are commercially available (Bowman 1976; Rozzell et al. 1975; Wickersheim and Alves 1979)(see chapter 10). Since the heating pattern may vary considerably over a three-dimensional volume, a single point measurement will not define the pattern nor quantify the maximum SAR or temperature. Multiple probes, such as used on BSD-1000™ (BSD Corp., Salt Lake City, Utah) hyperthermia machine, can provide better information about the heating pattern, but to a limited extent.

Guy (1971b) has described a thermographic method for rapid measurement of SAR (not steady state temperature) in live tissue through the use of phantom models of the real tissues. This method uses a thermographic camera for recording RF-induced temperature changes over an internal surface of the exposed object. The phantoms are composed of materials with dielectric and geometric properties similar to the tissue structures that they represent. Phantom materials have been developed that simulate human fat, muscle, brain, and bone, in the absence of a functioning circulatory system. Models are designed to separate along planes perpendicular to the tissue interfaces after exposure to electromagnetic fields so that cross-sectional relative heating patterns can be measured with a thermograph. A silk screen is placed over the precut surface on each half of the model to hold the phantom tissue and to provide electrical contact between the two halves. In using the model, it is first exposed to the same EM source that will be used to expose actual tissue. After a short exposure to a power level much higher than normally used for therapy, the model is quickly disassembled and the temperature pattern over the surface of separation measured and recorded by means of a thermograph. Since the thermal conductivity of the tissue model is low, the rate of the change in measured temperature distribution after heating per time interval will closely approximate the heating distribution over the flat surface, except in regions of high temperature gradient where errors may occur because of appreciable heat diffusion. The thermograph techniques described for use with phantom models can be used in a similar way to quantify the SAR patterns in terminated animals exposed to electromagnetic fields (Chou and Guy 1977; Johnson and Guy 1972) and in live animals measured at their skin surface (Guy et al. 1978b).

HEATING PATTERNS

The heating (SAR) patterns obtained theoretically or experimentally from phantoms and human subjects are discussed in this section.

Shortwave Applicators

Guy, Lehmann, and Stonebridge (1974) have analyzed the SAR pattern in plane layers of tissues exposed to 27-MHz shortwaves produced by an induction coil (figure 13.17). The calculated patterns are shown in figures 13.18, 13.19, and 13.20 for different fat thicknesses and applicator spacings with variables defined in figure 13.17. The predicted heating patterns agree very well with the measurements using pig tissues (Lehmann et al. 1968). Figure 13.21 illustrates temperature measurements made in the thigh of human subjects exposed to shortwave diathermy. The inductive shortwave diathermy is more effective in elevating the temperature of deep tissue while maintaining cooler surface temperature than that possible with capacitive heating. The major disadvantage of the inductive applicator is the toroidal heating pattern, which is difficult to use for heating small tumors. The use of a magnetic-loop applicator (see chapter 14) has gained in popularity, however, in treating deep tumors in the torso.

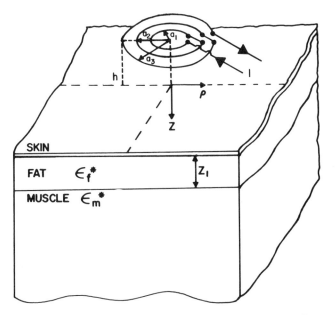

Figure 13.17. Geometry and coordinates for layers of skin, fat, and muscle tissue exposed to a flat "pancake" diathermy induction coil. (Guy, Lehmann, and Stonebridge 1974. Reprinted by permission. © 1974 by the IEEE.)

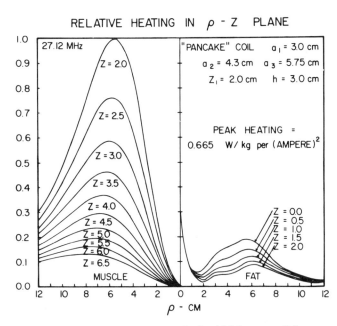

Figure 13.18. Calculated SAR in fat (thickness = 2.0 cm) at different depths exposed to 27.12-MHz induction coil diathermy. (Guy, Lehmann, and Stonebridge 1974. Reprinted by permission. © 1974 by the IEEE.)

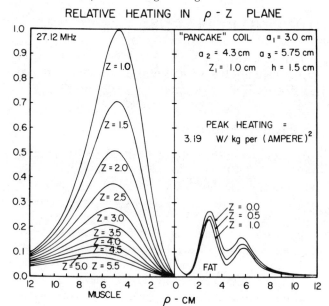
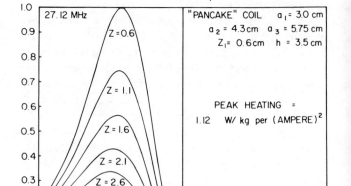

Figure 13.19. Calculated SAR in fat (thickness = 1.0 cm) at different depths exposed to 27.12-MHz induction coil diathermy. (Guy, Lehmann, and Stonebridge 1974. Reprinted by permission. © 1974 by the IEEE.)

Figure 13.20. Calculated SAR in fat (thickness = 0.6 cm) at different depths exposed to 27.12-MHz induction coil diathermy. (Guy, Lehmann, and Stonebridge 1974. Reprinted by permission. © 1974 by the IEEE.)

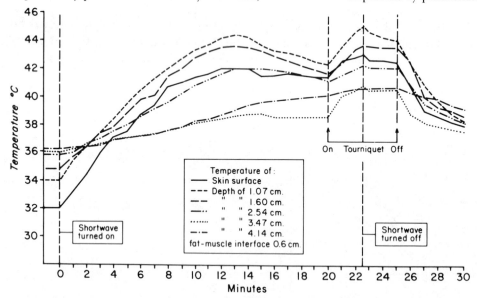

Figure 13.21. Temperature rise at different depths in human thigh ($Z_1 = 0.6$ cm and $h = 3.5$ cm, as defined in fig. 13.17) exposed to induction coil. (Lehmann et al. 1968. Reprinted by permission.)

Microwave Applicators

If other than a plane wave source is used to expose biological tissues, the SAR patterns are highly dependent on source size and field distribution. Many applications of microwave power in medicine require a quantitative understanding of the possible SAR or heating patterns in tissues obtained from the use of aperture and waveguide sources. Guy (1971a) has analyzed the case where a bilayered fat (thickness z) and muscle tissue model is exposed to a direct-contact aperture source of width, a, and height, b, as shown in figure 13.22. Figure 13.23 illustrates the results of this work, where the relative SAR levels in the x-z plane are plotted for $a = 12$ cm and $b = 2, 4, 12,$ and 26 cm. For comparison purposes, the SAR at the fat surface caused by a plane wave exposure is denoted by a dashed line in each figure. With small aperture heights, the applicator produces considerable super-

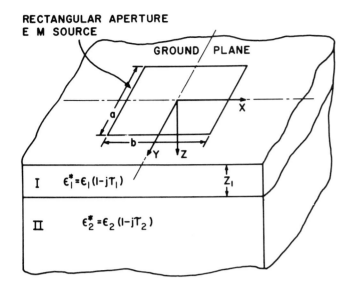

Figure 13.22. Direct contact aperture source on tissue layers. (Guy 1971a. Reprinted by permission. © 1971 by the IEEE.)

ficial heating, in excess of that produced by plane wave exposure, and minimal heating in the underlying musculature. As the aperture height increased, the relative heating in the deeper musculature increases greatly, compared to the superficial heating in the fat.

The SAR patterns in multilayered cylindrical tissues exposed to an aperture source can be determined also by representing the fields in the region as a summation of three-dimensional cylindrical waves, expressing the aperture field as a two-dimensional Fourier series and matching the boundary conditions. Ho and colleagues (1970, 1971) have calculated the SAR patterns for a number of different aperture and cylinder sizes. Typical results are shown in figure 13.24 for a human-arm-sized cylinder exposed to a surface aperture source 12 cm long with the electric field polarized in the direction of the axis. The patterns are plotted as a function of the radial distance from the center of the cylinder, for various circumferential angles, ϕ, measured with respect to the reference at the center of the aperture. The patterns are normal-

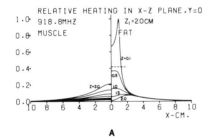

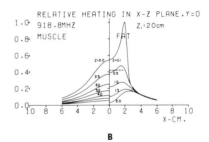

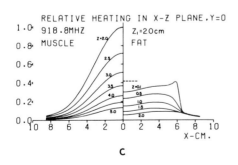

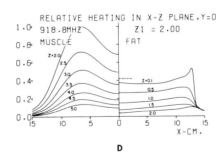

Figure 13.23. Relative SAR patterns in plane layers of fat and muscle exposed to TE_{10} mode waveguide aperture source with $a = 12$ cm, $f = 918.8$ MHz, and $Z_1 = 2$ cm for various aperture heights. For A–D, $b = 2$, 4.12, and 26 cm, respectively. (Guy 1971a. Reprinted by permission. © 1971 by the IEEE.)

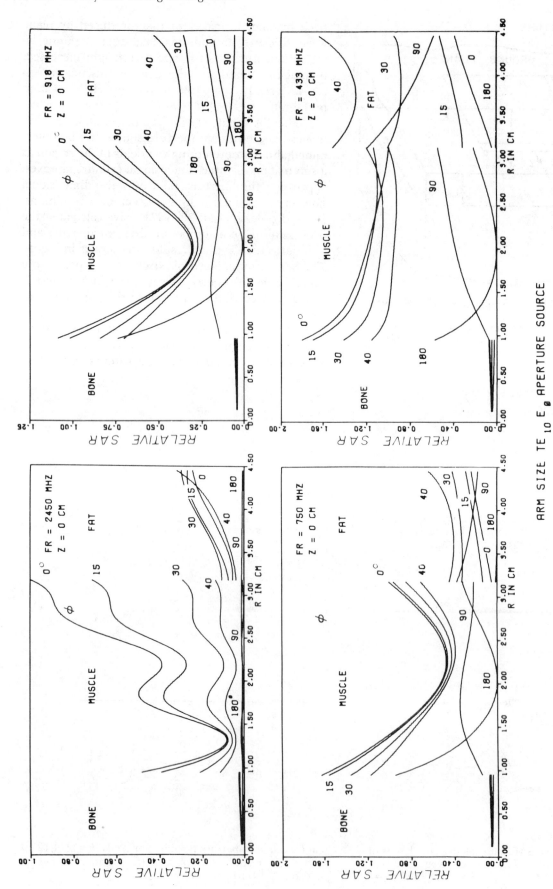

Figure 13.24. Relative SAR patterns in a cylindrical model of the human arm exposed to direct contact aperture source at four different frequencies. (Ho et al. 1971. Reprinted by permission. © 1971 by the IEEE.)

ized to the values calculated at the reference ($\phi = 0$) on the fat-muscle interface. The difference between the heating patterns calculated for cylindrical tissues and plane layered tissues demonstrates the significant role of tissue curvature in assessing the effectiveness and safety of devices designed for the medical applications of microwave energy.

All of the theoretical results discussed in this section point strongly to the ineffectiveness of 2450 MHz for diathermy, as noted in earlier reports by Schwan and Piersol (1954, 1955), Lehmann (1971), Lehmann and co-workers (1962a, 1962b, 1965), Guy (1971a, 1971b), and Guy and Lehmann (1966). Although the lower frequencies of 915 MHz authorized in the United States or 433 MHz authorized in Europe appear to be better choices, the theoretical data show that 750 MHz would be the best choice for a convenient clinical size applicator for heating 100- to 200-cm^2 regions.

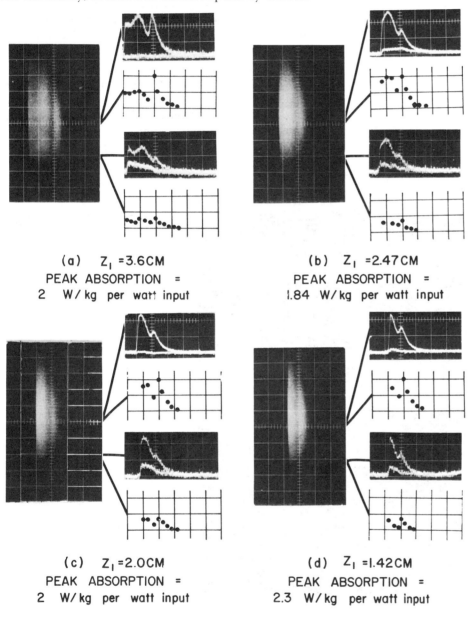

(a) Z_1 = 3.6 CM
PEAK ABSORPTION =
2 W/kg per watt input

(b) Z_1 = 2.47 CM
PEAK ABSORPTION =
1.84 W/kg per watt input

(c) Z_1 = 2.0 CM
PEAK ABSORPTION =
2 W/kg per watt input

(d) Z_1 = 1.42 CM
PEAK ABSORPTION =
2.3 W/kg per watt input

••• CALCULATED POWER ABSORPTION DENSITY

Figure 13.25. Thermograms and relative SAR patterns in plane tissue model exposed to 2450 MHz (director at 5-cm spacing). B scans taken at midline and 4 cm below. Dotted lines represent calculated SAR. (Guy, Lehmann, and Stonebridge 1974. Reprinted by permission. © 1974 by the IEEE.)

Radiation Type Applicators

Figure 13.25 illustrates the absorption patterns in plane fat-muscle tissue layers exposed to the diathermy C director as measured by the thermographic method for different fat thicknesses. The spacing between the applicator (plastic cover) and tissue surface was set to the clinically recommended value of 5 cm. The phantom models used for these studies were assembled by first constructing a 30 cm × 30 cm × 14 cm box with one-quarter-inch thick Plexiglas® sides, top, and bottom surfaces consisting of solid synthetic muscle. The thermograph camera was set to obtain a C scan (intensity proportional to temperature) over a two-dimensional area as shown by the large photographs in figure 13.25. The B scans consisted of two one-dimensional scans, one before and one after exposure, with the same horizontal scale as the C scans and a vertical deflection proportional to temperature (scale is given under each figure). The B scans were taken along the

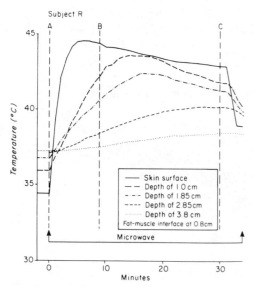

Figure 13.26. Temperature record in human thigh during exposure to microwaves applied at 2456 MHz using C director. (Lehmann et al. 1965. Reprinted by permission.)

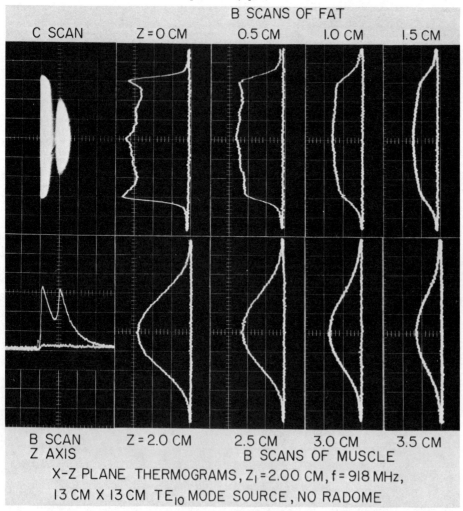

Figure 13.27. Thermograms of plane fat and muscle phantom exposed to 918-MHz direct contact applicator. Maximum SAR in muscle is 3.27 W/kg per watt input. (de Lateur et al. 1970. Reprinted by permission.)

horizontal midline and also along a parallel line 4 cm below the midline of the model.

Lehmann and associates (1965) have reported temperature measurements made in the thighs of human volunteers (with fat thicknesses of 0.8 cm and applicator spacings of 2 cm) exposed to the C director (figure 13.26). For these cases, the applied power was adjusted to the point where discomfort or mild pain was felt temporarily at the surface of the skin. The heating pattern of B director was shown to be toroidal like that of a shortwave pancake inductive applicator (Kantor 1977).

The depth of the heating patterns obtainable with the radiation type of microwave applicators is somewhat lower than that obtainable with shortwave inductive applicators. When all aspects are considered, the heating characteristics of the shortwave inductive diathermy applicator discussed in the previous section appear to be superior in terms of therapeutic value to those obtained with 2450-MHz modalities.

Direct Contact Type

The thermograms illustrated in figure 13.27 show SAR patterns in a muscle phantom model exposed to the 915-MHz direct contact applicator. The results show that the maximum absorption in the muscle is 3.27 W/kg per watt input to the applicator. Thus, an input power of approximately 50 W will produce an SAR of approximately 164 W/kg in the tissue, more than adequate for typical clinical applications. Taking density and specific heat into account, the maximum heating in the fat with the 915-MHz applicator was approximately 40% of that in the muscle. The maximum SAR per unit of incident power density was greater than that for plane wave exposure. This was expected since the applicator was designed to couple all of the applied energy to the tissue, whereas, with a plane wave source, a considerable amount of energy is reflected from the surface. The penetration and minimal fat heating characteristics in phantom models compare favorably with those of shortwave diathermy, with the additional advantage that the heating pattern is reasonably uniform, in contrast with the undesirable toroidal pattern of shortwave diathermy.

The 915-MHz UHF applicator was tested under clinical conditions by Lehmann and co-workers (1978) and de Lateur and colleagues (1970) by exposing the thighs of human volunteers. The experiments were conducted both with skin surface cooling and without cooling. Figure 13.28 shows the results of a typical experiment for a subject with 2.1 cm of subcutaneous fat. The initial five minutes of cooling lowered the skin

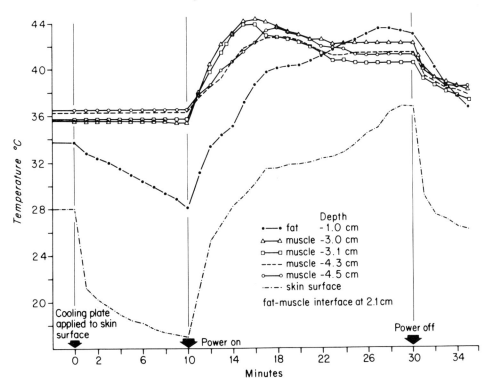

Figure 13.28. Temperature recorded in a human thigh at various depths of tissue exposed to 915 MHz direct contact applicator with surface cooling. (de Lateur et al. 1970. Reprinted by permission.)

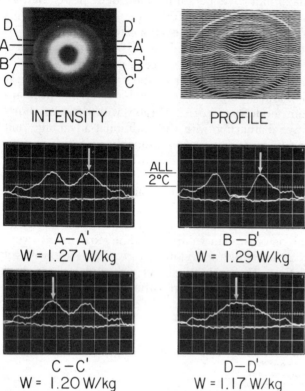

Figure 13.29. Thermograms of human arm model exposed to 915 MHz microwaves with a single direct contact applicator. (Guy 1971b. Reprinted by permission. © 1971 by the IEEE.)

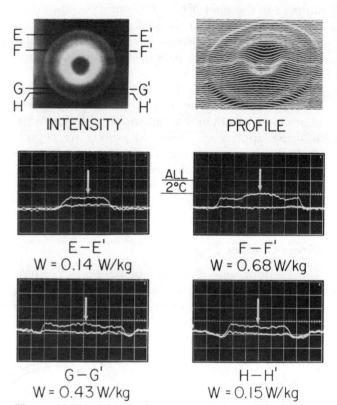

Figure 13.30. Thermograms in cross-sectional surface of human arm exposed to 915 MHz microwaves with phase array direct contact applicators shown in figure 13.16.

temperature below 15°C and, to a lesser degree, the fat temperature to 28°C. The muscle tissue was unaffected. When power was applied, the usual linear transient was observed, from which an SAR of 87 W/kg and 54 W/kg was calculated in the muscle at depths of 1.4 cm and 2.1 cm, respectively. An SAR of 56 W/kg at the center of the fat layer was calculated. During the 20-minute heating period, the muscle temperature as a function of time followed the characteristic trend illustrated in figure 13.2. The temperature reached a maximum of 45°C at a location 0.5 to 0.6 cm below the fat-muscle interface after a typical six-minute exposure period. At the time the maximum temperature was reached, an increase in blood flow occurred, resulting in a subsequent decrease in temperature in the normal tissue.

The experiments by Lehmann and associates (1968) carried out on human volunteers demonstrated that the 13-cm square radome-cooled direct-contact microwave applicator selectively heated musculature to a maximum level between 43°C and 45°C at a depth of 1 to 2 cm in the muscle. By increasing power levels sufficiently to compensate for blood cooling of tissue, it would be possible to return the musculature temperature to 43°C to 45°C and keep it to this temperature range for longer periods of treatment.

Kantor and Cetas (1977) thermographically measured the widths and depths of heating patterns of four 2450-MHz direct contact applicators, including a circular applicator and a slab-loaded waveguide applicator. The depth of heating was the same (2.2 cm) for all four applicators. The widths of the heating patterns are related to the aperture size and field distribution. When the choke was added to the slab-

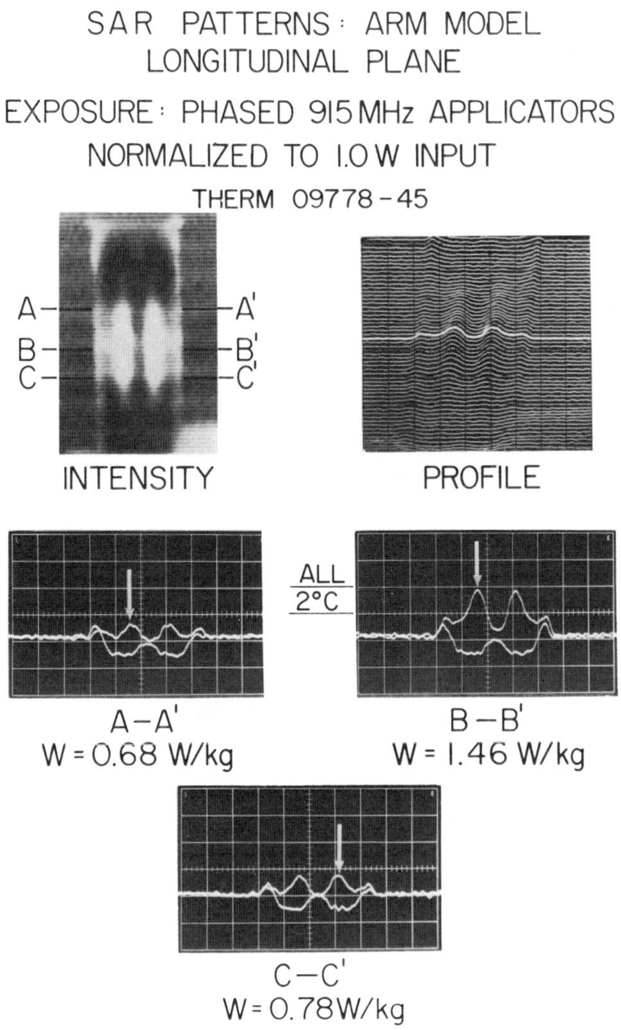

Figure 13.31. Thermograms in longitudinal surface of human arm exposed to 915 MHz microwaves with phase array direct contact applicators.

loaded waveguide, the width of heating pattern was reduced slightly compared with that of the original flange applicator (Kantor and Witters 1980). No data have been reported on the clinical use of these applicators.

The direct contact applicators not only provide better energy coupling into the tissue, but also reduce considerably the stray radiation (Guy, Lehmann, and Stonebridge 1974; Kantor and Cetas 1977; Lehmann, Stonebridge, and Guy 1979). This reduction in the stray radiation is an important safety consideration for both the patient and the operator.

Phase Array Applicators

The thermographically determined SAR distribution for a single applicator is shown in figure 13.29. When the same model was exposed with the four phase array applicators, the results shown in figures 13.30 and 13.31 were obtained. The results show a marked change of the SAR pattern, with maximum heating occurring at the bone-muscle interface and with uniform heating around the periphery of the bone. This would appear to be an ideal modality for treating bone tumors since the interface of maximum heating always would occur at the interface between healthy bone and the higher-water-content tumor tissue.

References

Allen, F. M. Biological modification of effects of roentgen rays. II. High temperature and related factors. *Am. J. Roentgenol.* 73:836–848, 1955.

Arons, I., and Sokoloff, B. Combined roentgenotherapy and ultra-short wave. *Am. J. Surg.* 36:533–543, 1937.

Birkner, R., and Wachsmann, F. Uber die kombination von rontgenstrahlen und kurzwellen. *Strahlentherapie* 79:93, 1949.

Bluestein, M.; Harvey, R. J.; and Robinson, T. C. Heat-transfer studies of blood-cooled heat exchangers. In *Thermal problems in biotechnology*. New York: American Society of Mechanical Engineers, 1968, pp. 46–81.

Bowman, R. R. A probe for measuring temperature in radio frequency heated material. *IEEE Trans. MTT* 24:43–45, 1976.

Cater, D. B.; Silver, I. A.; and Watkinson, D. A. Combined therapy with 220 kV roentgen and 10 cm microwave heating in rat hepatoma. *Acta Radiol.* 2:321–336, 1964.

Chou, C. K., and Guy, A. W. Microwave and RF dosimetry. Paper read at the workshop on the Physical Basis of Electromagnetic Interactions with Biological Systems. University of Maryland, College Park, Md., June, 1977, pp. 165–216.

Cole, K., and Cole, R. Dispersion and absorption in dielectrics. *J. Chem. Phys.* 9:34, 1941.

Cook, H. The dielectric behavior of some types of human tissues at microwave frequencies. *J. Appl. Phys.* 2:295, 1951a.

Cook, H. Dielectric behavior of human blood at microwave frequencies. *Nature* 168:247, 1951b.

Cook, H. A comparison of the dielectric behavior of pure water and human blood at microwave frequencies. *Br. J. Appl. Phys.* 3:249, 1952.

Copeland, E. S., and Michaelson, S. M. Effect of selective tumor heating on the localization of I^{131} fibrinogen in the Walker carcinoma 256. II. Heating with microwaves. *Acta Radiol.* 9:323–336, 1970.

Crile, G., Jr. Selective destruction of cancers after exposure to heat. *Ann. Surg.* 156:404–407, 1962.

de Keating-Hart. La fulguration et ses résultats dans le traitement du cancer, d'ares une statistique personnelle de 247 cas. Paris, 1909.

de Lateur, B. J. et al. Muscle heating in human subjects with 915 MHz microwave contact applicator. *Arch. Phys. Med.* 51:147–151, 1970.

Dethlefsen, L. A., and Dewey, W. C., eds. *Proceedings of the Third International Symposium on Cancer Therapy by Hyperthermia, Drugs and Radiation.* Bethesda, Md.: National Cancer Institute, 1982, in press.

Eidinow, A. Action of ultra-shortwaves on tumors. *Br. Med. J.* 2: 332, 1934.

Fuchs, G. Zur sensibilisierung rontgenrefractaer neoplasmen durch kurzwellen. *Strahlentherapie* 55: 3, 473–480, 1952.

Gessler, A. E.; McCarty, K. S.; and Parkinson, M. C. Eradication of spontaneous mouse tumours by high frequency radiation. *Exp. Med. Surg.* 8: 143, 1950.

Guy, A. W., and Lehmann, J. F. On the determination of an optimum microwave diathermy frequency for a direct contact applicator. *IEEE Trans. BME* 13:76–87, 1966.

Guy, A. W. Electromagnetic fields and relative heating patterns due to a rectangular aperture source in direct contact with bilayered biological tissue. *IEEE Trans. MTT* 19:214–223, 1971a.

Guy, A. W. Analyses of electromagnetic fields induced in biological tissues by thermographic studies on equivalent phantom models. *IEEE Trans. MTT* 19:205–214, 1971b.

Guy, A. W., ed. Special issue on microwaves in medicine, with accent on the application of electromagnetics to cancer treatment. *IEEE Trans. MTT* 26, no. 8, 1978.

Guy, A. W. et al. Development of a 915 MHz direct contact applicator for therapeutic heating of tissues. *IEEE Trans. MTT* 26:550–556, 1978a.

Guy, A. W. et al. Measurement of power distribution at resonant and non-resonant frequencies in experimental animals and models. *Scientific Report No. 11*, Bioelectromagnetics Research Laboratory, University of Washington. Final report for USAF/SAM Contract #41609-76-C-0032, 1978b.

Guy, A. W., and Chou, C. K. System for quantitative chronic exposure of a population of rodents to UHF fields. In *Biological effects of electromagnetic waves*, eds. C. C. Johnson and M. L. Shore. HEW Publication (FDA) 77-8011, vol. 2, pp. 389–410, 1977.

Guy, A. W.; Lehmann, J. F.; and Stonebridge, J. B. Therapeutic applications of electromagnetic power. *Proc. IEEE* 62:55–75, 1974.

Hasche, E., and Collier, W. A. Uber die beeinflussung bosartiger geschivulste durch ultrakurzwellen. *Strahlentherapie* 51:309–311, 1934.

Hill, L. Actions of ultra short waves on tumours. *Br. Med. J.* 2:370–371, 1934.

Ho, H. S. et al. Electromagnetic heating of simulated human limbs by aperture sources. In *Proceedings of the Twenty-third Annual Conference on Engineering in Medicine and Biology*, Washington, D.C., 1970, p. 159.

Ho, H. S. et al. Microwave heating of simulated human limbs by aperture sources. *IEEE Trans. MTT* 19:224–231, 1971.

Hornback, N. B. et al. Preliminary clinical results of combined 433 megahertz microwave therapy and radiation therapy on patients with advanced cancer. *Cancer* 40: 2854–2863, 1977.

Hornback, N. B. et al. Radiation and microwave therapy in the treatment of advanced cancer. *Radiology* 130:459–464, 1979.

Johnson, C. C., and Guy, A. W. Nonionizing electromagnetic wave effects in biological materials and systems. *Proc. IEEE* 60:692–718, 1972.

Johnson, H. J. The action of short radio waves on tissues. III. A comparison of the thermal sensitivities of transplantable tumours in vivo and in vitro. *Am. J. Cancer* 38: 533–550, 1940.

Kantor, G. New types of microwave diathermy applicators—comparison of performance with conventional types. *Proceedings of Symposium on Biological Effects and Measurement of Radio Frequency/Microwaves*, ed. D. G. Hazzard. Washington, D.C.: HEW publication (FDA) 77-8026, 1977, pp. 230–249.

Kantor, G., and Cetas, T. C. A comparative heating patterns study of direct contact applicators in microwave diathermy. *Radio Science* 12:111–120, 1977.

Kantor, G., and Witters, D. M., Jr. A 2450 MHz slab-loaded direct contact applicator with choke. *IEEE Trans. MTT* 28:1418–1422, 1980.

Kantor, G.; Witters, D. M., Jr.; and Greiser, J. W. The performance of a new direct applicator for microwave diathermy. *IEEE Trans. MTT* 26:563–568, 1978.

Kim, J. H.; Hahn, E. W.; and Tokita, N. Combination hyperthermia and radiation therapy for cutaneous malignant melanoma. *Cancer* 41:2143–2148, 1978.

Lehmann, J. F. Diathermy. In *Handbook of Physical Medicine and Rehabilitation*, eds. Krusen, Kottke, Elwood. Philadelphia: W. B. Saunders, 1971, pp. 273–345.

Lehmann, J. F. et al. Comparison of relative heating patterns produced in tissues by exposure to microwave energy at frequencies of 245 and 900 megacycles. *Arch. Phys. Med.* 43:69–76, 1962a.

Lehmann, J. F. et al. A comparative evaluation of temperature distributions produced by microwaves at 2456 and 900 megacycles in geometrically complex specimens. *Arch. Phys. Med.* 43:502–507, 1962b.

Lehmann, J. F. et al. Comparison of deep heating by microwaves at frequencies 2456 and 900 megacycles. *Arch. Phys. Med.* 46:307–314, 1965.

Lehmann, J. F. et al. Heating patterns produced by shortwave diathermy using helical induction coil applicators. *Arch. Phys. Med.* 49:193–198, 1968.

Lehmann, J. F. et al. Evaluation of a therapeutic direct-contact 915-MHz microwave applicator for effective deep-tissue heating in humans. *IEEE Trans. MTT* 26:556–563, 1978.

Lehmann, J. F.; Stonebridge, J. B.; and Guy, A. W. A comparison of patterns of stray radiation from therapeutic microwave applicators measured near tissue-substitute models and human subjects. *Radio Science* 14(6S):271–283, 1979.

LeVeen, H. H. et al. Tumor eradication by radiofrequency therapy: response in 21 patients. *JAMA* 235:2198–2200, 1976.

Mendecki, J. et al. Microwave-induced hyperthermia in cancer treatment: apparatus and preliminary results. *Int. J. Radiat. Oncol. Biol. Phys.* 4:1095–1103, 1978.

Minard, D. Body heat content. In *Physiological and behavioral temperature regulation*, eds. J. Hardy, A. Gagge, and J. Stolwijh. Springfield, Ill.: Charles C. Thomas, 1970, pp. 345–357.

Moressi, W. J. Mortality patterns of mouse sarcomas 180 cells resulting from direct heating and chronic microwave irradiation. *Exp. Cell Res.* 33:240–253, 1964.

Mortimer, B., and Osborne, S. L. Short wave diathermy—some biologic considerations. *JAMA* 153:1404–1413, 1935.

National Council on Radiation Protection and Measurements. Radiofrequency electromagnetic fields—properties, quantities and units, biophysical interaction, and measurements. In NCRP report no. 67, 1981, pp. 1–134.

Nelson, A. J. M., and Holt, J. A. G. Combined microwave therapy. *Med. J. Aust.* 2:88–90, 1978.

Nevins, R. G., and Darwish, M. A. Heat transfer through subcutaneous tissue as heat generating porous material. In *Physiological and behavioral temperature regulation*, eds. J. Hardy, A. Gagge, and J. Stolwijh. Springfield, Ill.: Charles C. Thomas, 1970, pp. 281–301.

Overgaard, J. Biological effect of 27.12-MHz short-wave diathermic heating in experimental tumors. *IEEE Trans. MTT* 26:523–529, 1978.

Overgaard, K., and Overgaard, J. Investigations on the possibility of a thermic tumour therapy. I. Short-wave treatment of a transplanted isologous mouse mammary carcinoma. *Eur. J. Cancer* 8:65–78, 1972a.

Overgaard, K., and Overgaard, J. Investigations on the possibility of a thermic tumour therapy. II. Action of combined heat-roentgen treatment on a transplanted mouse mammary carcinoma. *Eur. J. Cancer* 8:573–575, 1972b.

Pflomm, A. Experimentelle u. klinische untersuchungen uber die wirkung ultrakurzer elektrischen wellen. *Arch. Klin. Chir.* 166:251, 1931.

Roffo, A. E., Jr. Relation entre les ondes électriques et la multiplication cellulaire dans les cultures de tissus in vitro. *Arch. Diélectric Med.* 42:466–475, 1934.

Reiter, T. Researches sur les ondes utra-courtes. *Ann. d'Inst. Actinologie* 7:195–198, 1932.

Rohdenburg, G. L., and Prime, F. The effect of combined radiation and heat on neoplasms. *Arch. Surg.* 2:116–129, 1921.

Rozzell, T. C. et al. A nonperturbing temperature sensor for measurements in electromagnetic fields. *J. Microwave Power* 9:241–248, 1975.

Schereschewsky, J. W. The action of very high frequency upon a transplanted mouse sarcoma. *Public Health Report* 43:937, 1928.

Schliephake, E. *Short wave therapy*. London: Actinic Press, 1935, p. 181.

Schwan, H. P. Electrical properties of tissues and cells. *Adv. Biol. Med. Phys.* 5:147–209, 1957.

Schwan, H. P. Alternating current spectroscopy of biological substances. *Proc. IRE* 47:1841–1855, 1959.

Schwan, H. P. Biophysics of diathermy. In *Therapeutic heat and cold*, ed. S. Licht. New Haven, Conn.: Licht, 1965, pp. 63–125.

Schwan, H. P., and Piersol, G. M. The absorption of electromagnetic energy in body tissues, part I. *Am. J. Phys. Med.* 33:371–404, 1954.

Schwan, H. P., and Piersol, G. M. The absorption of electromagnetic energy in body tissues, part II. *Am. J. Phys. Med.* 34:425–448, 1955.

Streffer, C. et al., eds. *Proceedings of the Second International Symposium on Cancer Therapy by Hyperthermia and Radiation*. Baltimore and Munich: Urban & Schwarzenberg, 1977.

Szmigielski, S. et al. Inhibition of tumor growth in mice by microwave hyperthermia, polyriboinosinic-polyribocytidylic, and mouse interferon. *IEEE Trans. MTT* 26:520–522, 1978.

von Ardenne, M. On a new physical principle for selective local hyperthermia of tumor tissues. In *Proceedings of the Second International Symposium on Cancer Therapy by Hyperthermia and Radiation*, eds. C. Streffer et al. Baltimore and Munich: Urban & Schwarzenberg, 1978, pp. 96–104.

Wickersheim, K. A., and Alves, R. B. Recent advances in optical temperature measurement. *Industrial Research Development* 21(12):82–89, 1979.

Wizenberg, M. J., and Robinson, J. E., eds. *Proceedings of the First International Symposium on Cancer Therapy by Hyperthermia and Radiation*. Chicago: American College of Radiology, 1976.

Zimmer, R. P.; Ecker, H. A.; and Popovic, V. P. Selective electromagnetic heating of tumors in animals in deep hypothermia. *IEEE Trans. MTT* 19:232–238, 1971.

CHAPTER 14

Physical Aspects of Localized Heating by Magnetic-Loop Induction

F. Kristian Storm
William H. Harrison
Robert S. Elliott
Donald L. Morton

Introduction
Applicator Design
Field Distribution of a Magnetrode
Thermal Distribution of a Magnetrode
  Phantom Models
  Living Animal Models
Conclusions

INTRODUCTION

Regardless of advancing technology, it is unlikely that a single modality will be able to provide safe and effective hyperthermia for all tumors at all locations. Thus, instrumentation should be tailored precisely to those tumors and tumor sites being evaluated and treated. As outlined in the accompanying chapters, microwaves, radiowaves, and ultrasonic waves may be effectively used for surface or near-surface tumors, regional perfusion for specific diseases limited to the extremity, and systemic hyperthermia for diffuse metastases. Standard methods of localized heating by electromagnetic and acoustic waves, however, have proved so far to be ineffective or dangerous for high-temperature ($>42°C$) applications in deep-seated visceral human tumors; such responses have been due to preferential energy absorption within the superficial tissues (muscle or subcutis), reflection from tissue interfaces and air spaces, or overall host tolerance (Guy 1971; Storm et al. 1979a, 1981a, 1982a, 1982b, 1982c; Storm and Morton 1982).

The recent introduction of magnetic-loop induction hyperthermia utilizing a magnetrode may provide the safety and efficiency necessary to evaluate scientifically the role of localized heat therapy in internal human tumors (Elliott, Harrison, and Storm 1982; Storm et al. 1982b).

APPLICATOR DESIGN

Magnetrode magnetic-loop applicators (Henry Medical Electronics, Inc., Los Angeles, California) are self-resonant, noncontact circular structures with built-in impedance-matching circuitry that operate at 13.56 MHz (fig. 14.1). The element parameters were selected to produce very large circulating currents in the structure. These currents create a strong electromagnetic field in which the body or limb is immersed. Since the body is nonmagnetic, interaction is solely with the electric field, which consists of concentric circular flux lines.

The applicator is a coil comprised of a single turn of a rolled conducting sheet that overlaps itself in a nonconducting manner (fig. 14.2a). To rectify the problems of impedance transformation to a level appropriate to interface with the radiofrequency (RF) generator and the impedance adjustment created by the introduction of a limb or torso to be heated, a matching network (fig. 14.2b) is placed between the magnetrode and the generator. Simple adjustments of its controls reduce the reflected power to an insignifi-

Figure 14.1. A 20-inch magnetrode magnetic-loop applicator for transthoracic and transabdominal deep visceral hyperthermia.

cant level for a wide range of loads placed within the magnetrode.

FIELD DISTRIBUTION OF A MAGNETRODE

If the current distribution in the magnetrode is assumed to be entirely ϕ-directed, then the magnetic field can be expressed in cylindrical coordinates in the form

$$\mathbf{H} = [\mathbf{l}_r H_r(r,z) + \mathbf{l}_z H_z(r,z)]e^{j\omega t}. \quad (1)$$

Maxwell's equations reveal that in this case the electric field has only a ϕ-component, represented by $E_\phi(r,z)e^{j\omega t}$. Thus, the electric field lines are concentric circles.

The field distribution in the central transverse plane of the magnetrode is intermediate between what would be deduced for an infinitely long solenoid and for a single wire turn. For the latter, it is known that

$$H_z = \frac{2I}{a}\left\{\frac{K(k)}{[1 + (r/a)]} + \frac{E(k)}{[1 - (r/a)]}\right\}, \quad (2)$$

where a = the loop radius, $0 \le r \le a$, and K(k) and E(k) are elliptic integrals, wherein $k^2 = 4ar/(a + r)^2$. Since these elliptic integrals are tabulated functions, it is a simple matter to construct $H_z(r,0)$.

For a concentric circular contour C of radius r in the plane of the loop,

$$\oint_C \mathbf{E} \cdot d\mathbf{l} = 2\pi r E_\phi(r,0) = -j\omega\mu_0 \int_S \mathbf{H} \cdot d\mathbf{S}$$
$$= -j\omega\mu_0 \cdot \int_0^r H_z(r',0) 2\pi r' \, dr'. \quad (3)$$

When E_ϕ is computed from this relation, the result is as shown in figure 14.3. The central plane field distribution for a long solenoid is shown for comparison. The central plane E_ϕ distribution for a magnetrode is expected to lie between these two curves, with the intermediacy governed by the ratio L/a, with L the length of the magnetrode and a its radius.

If a nonmagnetic specimen of low conductivity is placed inside the magnetrode, one would expect only minor changes in the field distributions. Thus, the magnetrode appears to be a useful device for establishing an E field that is parallel to layer boundaries. The experimental evidence supports the thesis that this avoids the severe skin and subcutaneous fat burning often found when capacitor-type applicators are used for hyperthermia (Elliott, Harrison, and Storm 1982).

It also can be observed from the curves in figure 14.3 that there is a "dead spot" in the E field on the axis

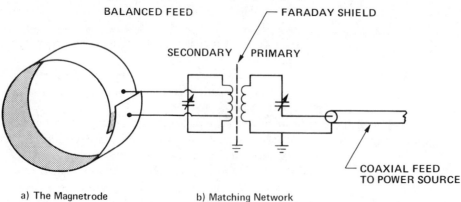

Figure 14.2. Diagram of magnetrode applicator, a, and matching network, b. (Elliott, Harrison, and Storm 1982.

Reprinted by permission. © 1982 by the IEEE.)

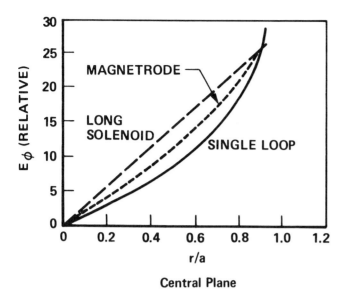

Figure 14.3. E_ϕ field distribution for a typical magnetrode contrasted to a long solenoid and a loop. (Elliott, Harrison, and Storm 1982. Reprinted by permission. © 1982 by the IEEE.)

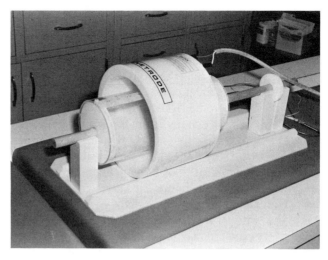

Figure 14.4. Phantom carrier, 12 cm in diameter and 30 cm in length, with dielectric support fixture. (Storm et al. 1982b. Reprinted by permission.)

of the magnetrode (Elliott, Harrison, and Storm 1982). This raises concern that there should be no heat generation in the portion of the specimen coinciding with the applicator axis, or a central cold spot. It must be remembered, however, that energy deposition is not the total process involved in eventual heat distribution, particularly in living specimens (see chapter 12).

THERMAL DISTRIBUTION OF A MAGNETRODE

Phantom Models

Phantom Carrier Design

The phantom carrier (fig. 14.4) consisted of a thin-walled Plexiglas® cylinder, 12 cm in diameter and 30 cm long, supported by a dielectric fixture that did not intrude upon the contents of the cylinder. The supporting fixture allowed concentric or acentric placement of the phantom carrier within a magnetrode applicator of 21-cm diameter and 13-cm width. In each experiment the magnetrode was centered at the long axis of the carrier. The long axis of the Plexiglas® carrier had a series of holes provided at 1-cm intervals for the introduction of needle thermometers to various depths, thus providing the means for two-dimensional temperature sampling within the phantom.

Tissue Equivalent Phantom

Standard homogeneous muscle-equivalent phantoms (Guy 1971) were prepared by combining by weight 76.5% saline solution (1.54 oz sodium chloride in 1 gal water), 15.2% powdered polyethylene (19.2 oz), and 8.4% Super-Stuff[1] (10.75 oz). This solution was poured into the phantom carrier and allowed to gel at room temperature before testing. In one experiment the phantom was placed concentrically within the magnetrode, and power at 300 W was applied for three minutes. There were significant radial and longitudinal thermal gradients observed, with decreasing temperatures with increasing depths or distances from the axial center of the electrode. This result was in accord with energy deposition predicted by electrodynamic theory and field strength measurements in air employing magnetic-loop induction hyperthermia (Storm et al. 1982b).

Animal Tissue Phantoms

Homogeneous animal tissue phantoms were prepared by packing the carrier with lean ground beefsteak. In one experiment, the phantom was placed concentrically within the magnetrode, and power at 500 W was applied for five minutes. In another experiment, the phantom was positioned acentrically within the magnetrode, and power at 400 W was applied for two and four minutes.

[1]Super-Stuff is the brand name of Bolus Stock No. TX 150, Oil Center Research, Lafayette, Louisiana.

One type of heterogeneous animal tissue phantom was prepared by introducing a 3-cm diameter thin-walled polyethylene cylinder as a peripheral void into the previously described ground-beef phantom. In one experiment, this phantom was placed concentrically within the magnetrode, and power at 400 W was applied for two and four minutes.

Other heterogeneous animal tissue phantoms were prepared by loosely packing the carrier with intact frankfurters in their skins (in effect, providing multiple contacting units of homogeneous animal tissue encased in a thin tissue capsule surrounded by multiple small air voids). Large voids were achieved by removing units or introducing plastic cylinders.

The thermal patterns produced in the various animal tissue phantoms were remarkably similar (Storm et al. 1982b). In each case, as predicted by electrodynamic theory (Elliott, Harrison, and Storm 1982), significant preferential peripheral heating was observed with minimal central heating. When these phantoms were placed concentrically within the magnetrode, there was a gradual radial reduction in temperature with increasing depth.

Placement of a homogeneous phantom off the central axis of the magnetrode resulted in some alteration of the thermal gradients, with preferential heating of the specimen at the site nearest the applicator. This suggested that the position of the test object within the field had some bearing on energy distribution.

It is more important, perhaps, that altered thermal gradients were observed in heterogeneous animal tissue phantoms. While preferential superficial tissue heating remained dominant, an irregular thermal distribution adjacent to voids and interfaces of varying types suggested that heterogeneous phantoms may be somewhat more apt representatives of the human situation, as has been suggested by Oleson and coworkers (in press).

Dead Animal Extremity Phantom

The dead animal extremity model consisted of a 12 × 19 × 25 cm leg of lamb with 2 cm subcutaneous tissue, at an ambient baseline temperature of 20°C. The extremity was placed concentrically within the magnetrode, and power at 300 W was applied for eight minutes. As shown in table 14.1, the dead animal extremity also demonstrated preferential peripheral heating, similar to that observed in tissue equivalent and artificial homogeneous and heterogeneous animal tissue phantoms, although altered heat distribution was noted in many areas of this complex tissue.

Living Animal Models

Heat distribution patterns were evaluated in large male dogs. The animals were shaved, anesthetized, and intubated without mechanical ventilatory support. The animals were suspended supine between dielectric supports to lie centrally within a 35-cm diameter magnetrode. A midline abdominal opening was made from the xyphoid to the pubis to allow temperature measurements, with particular care for thorough hemostasis. Upon opening the abdomen, an anterior air space occurred, and the intestines floated posteriorly. The dog spleen, being on a long and redundant mesentery, was easily displaced to a midline and central internal location without compromise of its blood supply. The exact size and location of each organ relative to the body wall were determined by caliper measurements. The dog then was repositioned such that the central spleen was lying in the center of the magnetrode. This provided a living model to assess thermal distribution patterns within surface tissues as well as peripheral (liver, kidney) and central (spleen) internal organs.

Temperature measurements were performed at room temperature (25°C to 27°C) with a Bailey™ microprobe thermometer (model BAT-12), autocorrelated for accuracy to 0.1°C at 0°C to 50°C, and Bailey™ 23-gauge microthermocouples of 0.15-sec time constant (Bailey Instruments, Inc., Saddle Brook, New Jersey). Temperature measurements were performed only during brief periods of RF wave cessation. Ambient temperatures prior to hyperthermia were obtained, and then the abdominal access opening was closed surgically to reproduce the normal condition. At the conclusion of each 10-minute hyperthermia application period, the surgical closure stitches were rapidly cut, and the abdomen was reentered. Serial temperature measurements were taken from surface tissues (skin, subcutaneous tissue, muscle, peritoneum), peripheral visceral organs (kidney, liver), and the central visceral organ (spleen). On each occasion of temperature measurement in these living tissues, care was taken to place the needle temperature probe into virgin tissue whose blood supply had not been potentially altered or disrupted from prior needle trauma. Fourteen serial tissue temperature measurements were possible during each monitoring period within two minutes. The actual temperatures obtained during serial monitoring were recorded as *measured °C*. A *corrected °C* was derived from the temperature decay curve and the timed temperature measurement interval for each tissue extrapolated back to time zero.

Table 14.1
Thermal Distribution Patterns in Dead Animal Extremity

Axis	Depth (cm)	Probe Location on Axis (cm)						
		−6	−4	−2	0	2	4	6
Y	1	27.3	29.7	36.0	32.4	30.0	26.2	25.5
	3	25.3	24.2	24.2	25.6	25.1	23.8	23.0
	5	22.0	21.5	22.1	23.4	23.2	22.9	23.1
	−5	Bone	25.8	28.2	29.5	27.1	26.5	26.6
	−3	Bone	35.4	34.2	32.5	Bone	Bone	Bone
X	1	28.2	28.4	29.5	27.5	26.2	27.0	25.0
	3	Bone	28.2	30.1	29.7	27.3	26.3	24.0
	5	Bone	25.2	25.8	25.3	24.0	24.9	25.5
	7	Bone	23.4	25.3	26.0	23.6	29.1	28.0
	9	Bone	23.3	25.3	29.5	25.8	32.2	31.2
		Measured °C						

NOTE: Measurements taken on leg of lamb, with Y axis 12 cm in diameter, X axis 19 cm in diameter, and 25 cm long. On each plane, axis and depth temperature measurements were performed. The extremity was placed centrally in a 21-cm diameter × 13-cm wide magnetrode at 20°C ambient. Hyperthermia application was 300 W for eight minutes. Complex thermal gradients were observed although peripheral heating predominated.

Table 14.2
Effect of Heat Dose on Thermal Distribution in Live Dog

Tissue Location	Organ	Moderate-Dose (250 W for 10 min) Measured °C (Corrected °C)		High-Dose (300 W for 10 min) Measured °C (Corrected °C)	
Surface	Skin	40.7	(40.7)	40.6	(40.6)
	Subcutis	43.0	(43.2)	47.6	(47.8)
	Muscle	43.0	(43.3)	45.5	(45.8)
	Peritoneum	41.8	—	42.3	—
Peripheral internal	Kidney surface	39.6	(39.7)	39.1	(39.2)
	Kidney central	40.0	(40.9)	38.7	(38.8)
Central internal	Spleen surface	40.3	(40.5)	39.6	(39.7)
	Spleen central	39.9	(40.1)	39.1	(39.2)
	Spleen 1 cm	40.0	(40.2)	38.8	(38.9)
	Spleen 2 cm	40.1	(40.3)	39.3	(39.4)

NOTE: Magnetrode hyperthermia at 250 W (moderate-dose) or 300 W (high-dose) was applied once for 10 minutes and the thermal gradients compared. Moderate-dose heat produced nearly equivalent peripheral and central internal heating with surface tissues at a physiologically tolerable temperature. High-dose heat produced preferential and injurious peripheral heating with a central "cold spot."

Effects of Heat Dose on Thermal Gradients

To first determine the effects of heat dose on thermal distribution patterns, a dog was subjected to high-dose (300 W for 10 minutes) and moderate-dose (250 W for 10 minutes) hyperthermia and the temperature responses compared (table 14.2). High-dose hyperthermia universally resulted in preferential and injurious peripheral tissue heating with a central cold spot similar to that observed in phantom models. Moderate-dose hyperthermia (as recommended in human therapy) for the same time interval resulted in virtually equivalent heating of peripheral and central visceral organs, with sparing of surface tissue. It appeared, therefore, that the amount of heat deposition per unit time (viz. the velocity of heating) had significant influence in determining thermal distribution patterns (Storm et al. 1982b). It is surmised that a "flash burn" effect from rapid high-dose hyperthermia does not allow sufficient time for effective heat transfer. By comparison, the safe and potentially tumoricidal uniform internal temperatures observed with moderate-dose hyperthermia might have been due to the allowance of adequate time for the recruitment of accessory vessels and increased blood flow to provide thermal adaptation, as in humans (Storm et al. 1979a).

Thermal Distribution Patterns in the Live Dog

A typical thermal distribution profile for the live dog is shown in table 14.3. During three sequential 10-minute hyperthermia treatments at 250 W (moderate-dose hyperthermia), all surface tissues remained within a physiologically tolerable temperature range and showed no signs of injury (fig. 14.5), similar to the results in human clinical trials (Baker et al. 1982; Storm et al. 1979a, 1979b, 1981a, 1981b, 1982a, 1982c, 1982d; Storm and Morton 1982). Virtually uniform internal heating occurred. In contradistinction to the observations in phantoms, potentially tumoricidal temperatures of 42°C or above occurred in both peripheral and central organs to a nearly equivalent degree. Measured and corrected intraparenchymal temperature samples of all visceral organs evaluated showed a temperature gradient of less than 1°C.

These findings suggest that the thermal distribution in living systems is not the simple product of energy deposition per se as predicted by electrody-

Table 14.3
Thermal Distribution Patterns in Live Dog

Tissue Location	Organ	Ambient °C	Tx. No. 1 Measured °C (Corrected °C)	Tx. No. 2 Measured °C (Corrected °C)	Tx. No. 3 Measured °C (Corrected °C)
Surface	Skin	34.5	40.7 (40.7)	44.6 (44.6)	44.2 (44.2)
	Subcutis	35.7	43.0 (43.2)	44.7 (44.9)	44.6 (44.8)
	Muscle	36.8	43.0 (43.4)	44.4 (44.7)	43.8 (44.1)
	Peritoneum	—	41.8 —	42.8 —	42.6 —
Peripheral internal	Liver surface	—	40.1 (40.3)	41.3 (41.5)	42.4 (42.6)
	Liver central	37.3	40.2 (40.4)	41.4 (41.6)	42.4 (42.6)
Peripheral internal	Kidney surface	—	39.6 (39.7)	40.7 (40.8)	41.4 (41.5)
	Kidney peripheral	—	39.9 (40.1)	41.2 (41.4)	42.2 (42.4)
	Kidney central	37.0	40.0 (40.9)	41.0 (41.1)	42.3 (42.4)
Central internal	Spleen surface	—	40.3 (40.5)	41.1 (41.3)	42.7 (42.9)
	Spleen central	37.2	39.9 (40.1)	41.4 (41.7)	42.4 (42.6)
	Spleen 1 cm	—	40.0 (40.2)	41.3 (41.5)	42.1 (42.3)
	Spleen 2 cm	—	40.1 (40.3)	41.4 (41.6)	42.0 (42.2)
	Spleen 3 cm	—	— —	41.5 (41.7)	42.2 (42.4)

NOTE: A 19.5-kg anesthetized male dog was subjected to three sequential magnetrode hyperthermia applications at 250 W for 10 minutes, with 7 to 10 minutes between applications. Nearly uniform internal organ heating occurred at safe surface temperatures. These results were typical of those obtained in other dogs.

Table 14.4
Thermal Distribution Patterns of Live versus Dead Dog

Tissue Location	Organ	Alive Measured °C (Corrected °C)		Dead Measured °C (Corrected °C)	
Surface	Skin	42.6	(42.9)	39.5	(39.8)
	Subcutis	43.7	(44.2)	42.7	(43.2)
	Muscle	44.2	(44.6)	46.0	(46.6)
	Peritoneum	44.6	—	48.4	—
Peripheral internal	Liver surface	42.4	(42.8)	52.1	(52.5)
	Liver central	42.4	(42.8)	48.1	(48.5)
Peripheral internal	Kidney surface	41.4	(41.5)	41.2	(41.4)
	Kidney peripheral	41.4	(41.6)	42.3	(42.5)
	Kidney central	41.5	(41.6)	43.1	(43.3)
Central internal	Spleen surface	42.1	(42.3)	39.0	(39.2)
	Spleen central	42.1	(42.3)	39.9	(40.1)
	Spleen 1 cm	42.6	(42.8)	39.2	(39.2)
	Spleen 2 cm	41.5	(41.7)	39.3	(39.6)

NOTE: One 26-Kg anesthetized male dog was subjected to three sequential magnetrode hyperthermia applications at 250 W for 10 minutes. The dog was terminated, allowed to cool to room temperature, then subjected to an identical series of hyperthermia applications. Comparison of thermal gradients in the live dog showed uniformity of deep-heating as observed in other live animals, while the same dead animal showed preferential peripheral heating and a central cold spot, as in the phantom models.

namic theory (Elliott, Harrison, and Storm 1982) and as manifested in phantom models. Other mechanisms, probably blood flow and possibly differences in tissue conductivity, as well as unknown factors, must ultimately influence the overall thermodynamic patterns achieved in living animals and humans (Storm et al. 1982b).

Thermal Distribution Patterns in Living versus Dead Dog

The thermal distribution profiles of the same dog subjected to equivalent hyperthermia while alive and dead are shown in table 14.4. While alive, nearly uniform internal organ heating occurred as previously observed. The dead animal, however, showed significant and injurious peripheral organ heating with a central cold spot similar to that observed in phantom models. This finding illustrates the importance of thermoregulatory systems in determining the final heat distribution in living animals.

CONCLUSIONS

The recently introduced magnetrode RF magnetic-loop induction applicator is being evaluated at many centers, and initial results in deep visceral human tumors have been encouraging (see chapter 15).

Temperature measurements on homogeneous and heterogeneous tissue equivalents, on animal tissue phantoms, a dead animal extremity, a dead dog, and on a live dog subjected to rapid high-dose heating confirmed electrodynamic theory and showed preferential and potentially injurious peripheral heating with a central cold spot. Sequential moderate-dose heating (as advised in humans) in live dogs showed virtually equivalent (<1°C) and potentially tumorici-

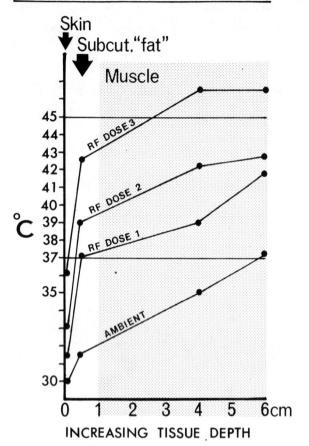

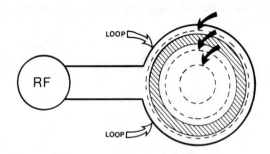

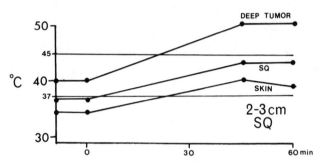

Figure 14.5. Performance of the magnetic-loop applicator in human thigh with 1 cm of subcutaneous tissue at three RF doses. No surface cooling was employed. Note the uniformity of deep heating at the higher dose levels with minimal heating of surface tissues. (Storm et al. 1982a. Reprinted by permission.)

dal temperatures of 42°C or more in both peripheral and central internal organs, with no injury to surface tissue; such findings are in accord with the thermodynamic theory that predicts heat redistribution in the presence of a functioning vascular system. These results confirm the findings in the human normal extremity (fig. 14.5) and those observed in clinical cancer trials (fig. 14.6) (Baker et al. 1982; Storm et al. 1979a, 1979b, 1980, 1981b, 1982a, 1982d; Storm and Morton 1982).

These data indicate that magnetic-loop induction can produce potentially effective deep internal heat in living animals which can be useful in human hyperthermia trials involving deep-seated tumors (see chapter 15).

Figure 14.6. The magnetic-loop applicator forms concentric circular flux lines in which the body or limb is immersed. This allows effective hyperthermia of deep-seated tumors (intraabdominal sarcoma, 10 × 10 cm) with surface tissue sparing in patients with 1 cm or more of subcutaneous tissue (*SQ*). (Storm et al. 1981a. Reprinted by permission.)

References

Baker, H. W. et al. Regional hyperthermia for cancer. *Am. J. Surg.* 143:586–590, 1982.

Elliott, R. S.; Harrison, W. H.; and Storm, F. K. Hyperthermia: electromagnetic heating of deep-seated tumors. *IEEE Trans.* BME 29:61–64, 1982.

Guy, A. W. Analysis of electromagnetic fields induced in biological tissues by thermodynamic studies on equivalent phantom models. *IEEE Trans.* MTT 19:205–210, 1971.

Oleson, J. R. et al. Magnetic induction heating in heterogeneous material. In *Proceedings of the Third International Symposium on Cancer Therapy by Hyperthermia, Drugs and Radiation*, eds. L. A. Dethlefsen and W. C. Dewey. Bethesda, Md.: *J. Natl. Cancer Inst.* (special journal supplement), in press, 1982.

Storm, F. K. et al. Normal tissue and solid tumor effects of hyperthermia in animal models and clinical trials. *Cancer Res.* 39:2245–2251, 1979a.

Storm, F. K. et al. Relationship between tumor type and capacity to induce hyperthermia by radio frequency. *Am. J. Surg.* 138:170–174, 1979b.

Storm, F. K. et al. Hyperthermia therapy for human neoplasms: thermal death time. *Cancer* 46:1849–1854, 1980.

Storm, F. K. et al. Clinical radio frequency hyperthermia by magnetic-loop induction. *J. Microwave Power* 16:179–184, 1981a.

Storm, F. K. et al. Radio frequency hyperthermia of advanced human sarcomas. *J. Surg. Oncol.* 17:91–98, 1981b.

Storm, F. K. et al. Clinical radio frequency hyperthermia: a review. In *Proceedings of the Third International Symposium by Hyperthermia, Drugs and Radiation*, eds. L. A. Dethlefsen and W. C. Dewey. Bethesda, Md.: *J. Natl. Cancer Inst.* (special journal supplement), in press, 1982a.

Storm, F. K. et al. Thermal distribution of magnetic-loop induction heating: effect of the living state and velocity of heating. *Int. J. Radiat. Oncol. Biol. Phys.* 8:865–871, 1982b.

Storm, F. K. et al. Clinical RF hyperthermia by magnetic-loop induction: a new approach to human cancer therapy. *IEEE Trans. MTT*, in press, 1982c.

Storm, F. K. et al. Thermochemotherapy for melanoma metastases in liver. *Cancer* 49:1243–1248, 1982d.

Storm, F. K., and Morton, D. L. Localized hyperthermia in the treatment of cancer. In *International advances in surgical oncology*, vol. 5, ed. G. P. Murphy. New York: A. R. Liss, 1982, pp. 261–275.

CHAPTER 15

Animal and Clinical Studies with Microwave and Radiowave Hyperthermia

F. Kristian Storm
Donald L. Morton

Introduction
Clinical Methods
Thermal Effects on Normal Tissue
Animal Tumor Investigations
Clinical Cancer Trials
Conclusions

INTRODUCTION

The physiologic effects of localized hyperthermia were known to the ancients, but it was not until this century and the discovery of electromagnetic radiation that effective deep heating was possible. Theilhaber, in 1919, was one of the first to treat inoperable human cancers by diathermy and believed he could demonstrate an objective tumor response. Bishop, Horton, and Warren in 1932, and Warren in 1935, applied shortwave heating to several cancer patients and also observed apparent tumor regression. Almost 20 years later, Gessler, McCarthy, and Parkinson (1950) found that microwave heating of murine tumors would cause complete destruction of tumor without serious injury to the host animal. By 1962, Shingleton had developed a specialized shortwave apparatus and confirmed that tumors in rabbits and dogs could be effectively heated without damage to normal tissues. At that time, Crile (1961, 1962) reviewed progress to that time and proceeded to apply microwave diathermy to tumors in mice, dogs, and man. He, too, found that some tumors could be selectively destroyed by heat without damage to surrounding tissues, but felt that safe hyperthermia was not possible unless tumors were superficially located. In 1971, Goldenberg and Langner investigated various time intervals of shortwave hyperthermia on transplanted human tumors and found that inhibition of tumor growth was proportional to the duration of heating. In that year, Zimmer and Ecker attempted to achieve differential heating of normal tissues versus tumor in S-band and X-band microwave fields by first applying deep hypothermia before heat therapy. In 1972, the Overgaards studied murine tumors heated in a shortwave diathermy field. They were able to heat subcutaneous tumors to 47°C without apparent damage to skin or underlying tissues and formulated extensive heat dose-time treatment schedules.

Extensive in vitro data were accumulating at that time to suggest that tumors may be slightly more thermosensitive than normal cells at 41°C to 44°C, temperatures readily achieved by electromagnetic radiation. These observations, together with the realization that some tumors could be selectively heated to 45°C and above without injury to adjacent normal tissues, suggested that hyperthermia might play a role in clinical cancer therapy.

The authors gratefully acknowledge the significant contributions of Beverly Drury, R.N.; Mitzi Benz, R.N.; Larry R. Kaiser, M.D.; Thomas H. Weisenburger, M.D.; Robert G. Parker, M.D.; and Charles M. Haskell, M.D. This work was supported by U.S. Public Health Service grant CA-24883 from the National Institutes of Health.

CLINICAL METHODS

Local Hyperthermia

Local hyperthermia is produced by transmission of electromagnetic energy to a limited area. Several innovative methods have been used to concentrate this energy directly within tumor. The earliest attempts used electrocautery or electrofulguration, although these methods required direct visualization of the tumor, and extensive adjacent normal tissue damage was caused by the extremely high temperatures produced.

Interstitial hyperthermia was achieved by passing a localized current field (LCF) of 500 kHz between electrodes implanted directly in tumor (Doss 1976; Sternhagen et al. 1978). While this technique is invasive, it appears to have a significant effect on small surface tumors where the extent of tumor is known (e.g., oropharynx, vagina, rectum). A subminiature rigid coaxial waveguide, the *microwave syringe*, also has been used to "inject" microwave fields into deep tissues (Taylor and Robinson 1978).

Noninvasive localized hyperthermia usually has been achieved by specially constructed *microwave waveguides* (UHF, L-band, S-band), which deposit externally applied energy into a defined area of tissue. This approach has proved to be quite useful in superficial cancers, although penetration has been limited to several centimeters. Multiple guides have been organized in a phase array in an attempt to "focus" greater energy within the deeper tumors. Testing of this approach is taking place and results are preliminary (Samaras 1981; Gibbs 1981). Parallel opposed *capacitive electrodes* (HF) have been placed in intimate contact with tumors to create localized hyperthermia within intervening tissues (Dickson et al. 1977; Dickson and Shah 1977; Goldenberg and Langner 1971; LeVeen et al. 1976; LeVeen, Ahmed, and Piccone 1980; Marmor, Hahn, and Hahn 1977; Overgaard and Overgaard 1972; Shingleton 1962; Storm et al. 1979a, 1979b, 1980a, 1980b, 1981b, in press; Storm and Morton 1979; Sugaar and LeVeen 1979; Westermark 1927). The field may be "shaped" by varying the contour and placement, as well as by the size of the two electrodes.

These techniques of localized hyperthermia generally have been reserved for exposed or accessible tumors. Relatively high intratumor temperatures have been achieved by many of these methods; however, their safe application has depended on preferential concentration of more electromagnetic energy within tumor than in surrounding tissues.

Regional Hyperthermia and Selective Tumor Heating

At temperatures from 41°C to 44°C tumor cells may be slightly more thermosensitive than normal cells (Bender and Schramm 1966; Cavaliere et al. 1967; Crile 1961, 1962; Giovanella et al. 1973; Overgaard and Overgaard 1972). These temperatures have been readily achieved, even at considerable depth, using HF, UHF, L-band, and S-band frequencies. Since the electromagnetic energy is not specifically localized within the tumor, the term *regional hyperthermia* best describes the condition where both tumor and adjacent tissues are exposed to similar amounts of incident heat.

At temperatures of 45°C and above, host tolerance becomes the prime consideration (Hardy et al. 1965; Cunningham 1970). During early studies of regional electromagnetic hyperthermia, several investigators observed a rise of 5°C to 10°C in intratumor temperature over that of the surrounding normal tissues, even though both received similar doses of heat. This phenomenon was termed *selective tumor heating* and appears to be the result of some unique physiologic trait(s) common to many solid tumors, such that they act as a heat reservoir. This natural ability of tumors to differentially retain heat is distinguished from localized hyperthermia, where energy is focused or concentrated directly within the tumor.

Selective tumor heating probably was discovered by Nils Westermark in 1927. He placed rodents bearing sarcomas and carcinomas between lead electrodes covered by compresses soaked in a common salt solution and passed high frequency currents through a portion of the animal. He observed that prior to heating, the temperature of the tumor was 0.5°C to 5°C lower than that of the surrounding tissues, but upon heating, tumor temperatures from 44°C to 48°C could be attained without destruction of adjacent normal tissue. He postulated that differences in vascularity might account for the heat variations and stated, "It is a well-known fact that these rat-tumors are poorly supplied with vessels. With regard to their temperature, therefore, they cannot derive any great influence through their circulation, their thermal regulation, therefore, being apparently bad." Since these pioneering experiments, several investigators have confirmed that ambient tumor blood flow is generally less than that of normal tissues (Gullino and Grantham 1961; Mantyla 1979). LeVeen and associates (1976) postulated that as a tumor grows it does not generate the integrated ramification of vessels that is present in normal tissues. The absence of such a system produces a high resistance to blood flow and a reduced ability to efficiently exchange incident heat. Alternately, Storm and colleagues (1979b, 1980b) have suggested that many tumors act as a heat reservoir because their neovascularity is physiologically unresponsive and cannot augment blood flow to adapt to thermal stress, as does normal vasculature (fig. 15.1). While these are attractive explanations for the observed phenomenon, both theories remain unproved, and other factors still unknown may play a significant role in selective tumor heating (Oleson et al., in press).

LeVeen and co-workers (1976) applied regional hyperthermia to humans and confirmed selective tumor heating in three patients. Using capacitance electrodes at 13.56 MHz and 1 to 4 W/cm^2 for 30 minutes, they found intratumor temperatures 8°C to 10°C higher than those of adjacent normal tissue, with minimal destruction of normal tissue. As found by previous investigators using these standard methods, they concluded that to avoid undesirable heating of the skin and subcutaneous tissue, which occasionally resulted in burns, energy was transmitted best to a surgically exposed tumor. In an attempt to target radiofrequency (RF) energy directly on tumor in order to avoid surface tissue injury, LeVeen, Ahmed, and Piccone (1980) subsequently developed equipment (Triport-222™, Life Extension Technology, Inc., Westport, Connecticut) that employed multiple portals of entry with three sequentially activated electrode pairs. Each pair of electrodes was of opposite potential and was activated with a short burst (0.1 sec) of power, after which the power was instantaneously switched by electronic means to the next electrode pair.

Cooled-capacitance electrodes have penetrated safely the tissues of dogs, sheep, and pigs (Storm et al. 1979b), but effective cooling (even at 3°C) in humans has been confined to depths of less than 1 cm of subcutaneous tissue. Use of this method has therefore been limited to surface or near-surface tumors.

Recently, Storm, Harrison, Elliott, and Morton have developed a fundamentally new RF device that heats by *magnetic-loop induction*, which has been significantly more effective for producing deep internal hyperthermia than standard methods (fig. 15.2) (see chapter 14). This approach, using a Magnetrode™, permits effective heating of the deep tissues without dangerous power consumption in the surface layers and is effective in patients with more than 1 cm of overlaying tissues. Nearly all of the information about the effects of deep internal hyperthermia in animals and humans has been generated using this device (Storm et al. 1982).

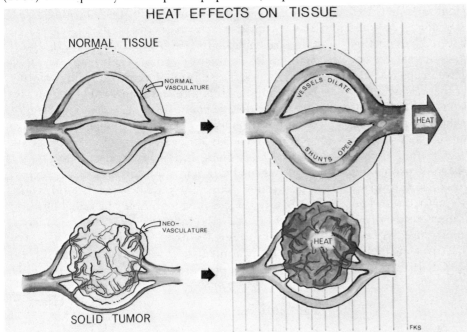

Figure 15.1. Postulated mechanism of selective solid tumor heating. Compared to tumors, normal tissues have a relatively high ambient blood flow, which increases in response to thermal stress, thereby dissipating heat. Tumors, with relatively poor blood flow and unresponsive neovasculature, are incapable of augmenting flow and act as a heat reservoir. (Storm et al. 1979b. Reprinted by permission.)

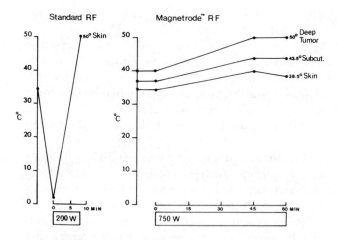

Figure 15.2. Comparison of standard versus Magnetrode™ radiofrequency (13.56 MHz) hyperthermia of the human abdomen. Standard heating with 225-sq-cm paired capacitively coupled contact electrodes in a patient with 2-cm subcutaneous tissue resulted in a 50°C *skin* temperature within 10 minutes at 200 W (approx. 1 W/cm²), even with surface cooling at 3°C. Magnetrode™ heating at 750 W with a 20-in circular noncontact electrode in a patient with a 3-cm subcutaneous tissue and a 15 × 15 × 18-cm retroperitoneal sarcoma resulted in a 50°C *tumor* temperature, with surface tissues remaining within a physiologic temperature range.

The term "selective tumor heating" actually is a misnomer. More precisely, tumors that respond to this method are relatively incapable of cooling as compared to normal tissues. This may seem like semantic game playing, but acceptance of this simple concept leads to the suggestion that vasoactive drugs may enhance temperature differentials, or that raising ambient tissue temperatures before treatment (theoretically possible with adjuvant regional thermal perfusion or whole-body hyperthermia) might further augment tumor heating capacity.

Regional hyperthermia is capable of producing potentially tumoricidal temperatures of 41°C to 44°C, temperatures well tolerated by normal tissues. The exact location and extent of the tumor need not be known in order for therapy to be effective; in fact, many such tumors may be safely selectively heated at or above 45°C (Storm et al. 1979b, 1981b, 1982; Storm and Morton 1979).

THERMAL EFFECTS ON NORMAL TISSUE

Skin and Subcutaneous Tissue

Rat skin tolerates heat to 45°C for 1.5 hours and to 46°C for 1 hour. Necrosis has been observed after 46°C for 1.5 hours. Pig skin apparently adapts to 50°C for a four-minute duration, although necrosis results at eight minutes (Westermark 1927). Graded RF doses for 1 to 10 w/cm² applied for three minutes to canine skin produced occasional first-degree burns (erythema) at 42°C to 44°C (corresponding to 1 to 2 W/cm²), second-degree burns (edema and bullae formation) at 45°C to 46°C (3 to 4 W/cm²), and third- and fourth-degree burns (full-thickness skin and muscle necrosis) at and above 50°C (5 to 10 W/cm²) (Storm et al. 1979b).

In humans, the upper limit of thermal tolerance appears to be about 45°C, imposed simultaneously by the threshold of pain at this temperature and irreversible tissue damage that begins at this temperature (Hardy et al. 1965). When temperatures above 39°C are applied to human skin, local thermal conductance increases rapidly because of increased local blood flow and convective heat transfer to the central blood volume. The consequence of this increased normal tissue perfusion is progressive difficulty in raising local tissue temperature above physiologic temperature (Stolwijk 1976; Stolwijk and Hardy 1965).

In our experience with sustained RF hyperthermia, normal human skin tolerated temperatures below 45°C for 5 to 15 minutes, the time generally required before maximum local heat dissipation occurs. We observed erythema, blister formation, and abrupt burning pain within minutes of peak heating above 45°C during the early phases of our investigations with standard capacitance electrodes. We found that normal skin would tolerate without injury only 0.5 to 1.5 W/cm², an increase incapable of producing internal heating. In patients with less than 1 cm of subcutaneous tissue, dosage could be increased safely to 3 to 4 W/cm² by the addition of surface cooling from 15°C to 25°C using cooled surface contact electrodes (Henry Medical Electronics, Inc., Los Angeles, California).

Abnormal skin and subcutaneous tissue, such as split-thickness skin grafts, pedicle grafts, and tissues previously subjected to high doses of radiation, with resultant fibrosis and brawny edema, sustained burns at temperatures as low as 40°C. This finding suggested that noncancerous tissues with an inadequate or impaired blood supply are also thermosensitive.

Extremities

In general, standard methods of electromagnetic heating have poor tissue penetration with decreasing thermal gradient with increasing depth. Thus, there are few historical studies on the effects of deep hyperthermia in extremities. During our initial investigations to determine muscle, nerve, and vessel thermal tolerance, dog, sheep, and pig extremities were heated to 41°C to 44°C at 3- to 5-cm depth for 15 to 30 minutes using water-cooled capacitance electrodes to protect surface tissues (Storm et al. 1979b). At these temperatures, histologic examination failed to reveal abnormalities, and in no instance was motor or vascular injury apparent over an observation period of one to six weeks.

Interestingly, sequential temperature recordings in animal extremities subjected to a constant heat dose showed that when normal muscle reached 43°C to 44°C, there was an abrupt temperature reduction that could not be overcome without a substantial increase in applied heat (fig. 15.3) (Storm et al. 1979b). This observation was made in humans treated with microwave diathermy, and the heat plateau could be overcome by an arterial occlusive tourniquet (Lehmann, DeLateur, and Stonebridge 1969). Guy, Lehmann, and Stonebridge (1974) suggested that at about 44°C there appeared to be a triggering of muscle blood flow, coincident with the perception of pain. Teleologically, it seems reasonable to assume that the evolving human systems would have developed such a mechanism to dissipate heat to prevent the tissue necrosis that begins at slightly higher temperatures. In our experience, this threshold of thermal adaptation appears to vary anywhere from 43°C to 45°C in human normal muscle, and is generally perceived as a sudden deep ache. Continued hyperthermia in the presence of such pain invites a rapid exhaustion of compensatory host mechanisms and potential normal tissue injury.

Viscera

There have been few studies on internal organ hyperthermia produced by electromagnetic waves. Using a frequency of 13.56 MHz and water-cooled capacitance electrodes at 15°C to 25°C, or the Magnetrode™, we subjected dogs to hyperthermia equivalent to 42°C to 45°C internal heat dosage for 15 minutes over the upper chest, lower chest/upper abdomen, and lower abdomen/spinal cord and then observed the results for two to three weeks. During this interval, there was no clinical evidence of internal organ injury. Posttreatment cardiac enzymes were transiently elevated, but no abnormalities were apparent by electrocardiogram. Hepatic transaminases and amylase were transiently elevated, but there were no clinical signs of compromise to internal organs. Unfortunately, prolonged continuous heating at these temperatures universally resulted in fatal cardiac tachyarrhythmias (Storm et al. 1979b).

In order to determine whether any particular normal organ was subject to "selective heating," we also subjected dogs to graded heat doses from 37°C to 49°C and took direct temperature measurements of heart, lung, esophagus, liver, stomach wall, stomach contents (solid and liquid), gallbladder, bile, spleen, pancreas, kidney, small bowel wall and contents, colon, and stool. Heating was remarkably uniform in all normal organs evaluated, and it was noteworthy that no preferential heating or hot spots occurred in any normal viscera (Storm et al. 1979b, 1982).

During human cancer trials of visceral heating with the Magnetrode™ there was no clinical evidence of normal tissue damage as a result of deep heating. Direct temperature measurements of normal lung, esophagus, liver, or rectum failed to reveal selective heating of these organs at incident heat doses sufficient to cause effective tumor heating (Storm et al. 1979a, 1979b, 1980b). Whether selective heating of any normal organ occurs with other methods or at other frequencies is undetermined.

ANIMAL TUMOR INVESTIGATIONS

The older literature contains several examples of tumor regression and apparent permanent cures with hyperthermia produced by high-frequency currents. While the validity of the methods of temperature recording used in this era are highly suspect, the results of these early investigations are of historical interest. In 1927, Westermark studied 109 rats with Flexner-Jobling carcinoma or Jensen's sarcoma. Tumors were inoculated into subcutaneous tissue and treated by interposition between contoured lead electrodes when they attained a size of 0.3 to 2.4 cm. The response was similar in both tumors: complete necrosis occurred at

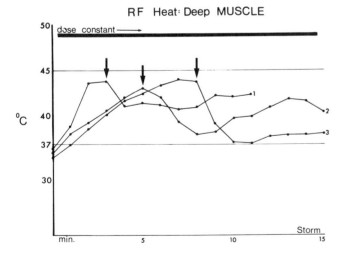

Figure 15.3. Hyperthermia response of normal muscle (dog). Spontaneous cooling in the presence of constant heat dose is consistent with physiologic thermal adaptation. (Storm et al. 1979b. Reprinted by permission.)

44°C after 180 minutes, at 45°C after 90 minutes, at 46°C after 50 minutes, at 47°C after 30 minutes, and at 48°C after 20 minutes. No histologic changes in the tumors were found immediately after treatment, although progressive necrosis was observed several days after heating, with total necrosis by day six. Westermark found that heating to 48°C for 20 minutes was injurious both to rat tumors and to the normal germinal epithelium of the testis, but that at temperatures from 44°C to 47°C, tumor necrosis occurred at shorter intervals while germinal epithelium tolerated more prolonged heating. In 1928 and again in 1933, Schereschewsky reported on treatment of mice bearing transplanted sarcoma CR-180 with pairs of insulated copper plates when tumors reached 4 to 12 mm in size. Of 403 animals, 100 recovered from their tumors and remained disease-free. After a single two- to four-minute treatment at 48°C to 49°C, tumors immediately became softer and smaller, with complete regression by 10 to 14 days. Tumor regrowth, if it occurred, became manifest within 10 days. Similar results were found in rats inoculated with sarcoma CR-10 and in chickens with Rous fowl sarcoma. In 1937, Dickens, Evans, and Weil-Malherbe applied shortwaves at 88 MHz to mice with Walker-256 carcinoma and rats with Jensen's sarcoma. Tumors were treated once for four minutes, between paired brass disks insulated with rubber. Of 286 surviving mice, 4% had tumor cure at temperatures below 48°C, and 63% had cure at temperatures above 48°C. Of 106 surviving rats, 26% had cure occurring below 50°C, and 88% were cured at temperatures above 50°C. Effectively treated tumors again showed little histologic change immediately after heating, but developed progressive pyknosis and necrosis after several days. Using this same experimental model in 1940, Johnson found that the exposure time required to produce 50% cures were 6 hours, 1.5 hours, and 45 minutes for the Walker-256 rat carcinoma at 43.5°C, 45°C, and 47°C, respectively, whereas times for the Jensen's rat sarcoma for the same range of temperatures were 3 hours, 1 hour, and 15 minutes. The immediate result of treatment was a marked swelling of tumor—an increase often by as much as 30% of its diameter—that persisted for up to 48 hours, followed by softening and progressive shrinkage of the mass.

Carter, Silver, and Watkinson (1964) were among the first to adequately monitor temperatures in an electromagnetic field. Using hepatoma-223 transplanted into the leg muscles of August-strain rats, microwave heating was applied with a magnetron of the continuous wave type with a power output of 500 W, operating at a frequency of 3000 MHz, and a waveguide 7×3.5 cm in cross-section. They found no demonstrable tumor effect at 45°C to 47°C in 33 rats, with exposure times of 5 to 10 minutes. In 1971, Goldenberg and Langner evaluated growth inhibition of GW-77 human colonic tumors growing in bilateral hamster cheek pouches exposed to 4°C elevation over ambient temperature (39°C maximum). Heat was administered to a unilateral tumor by shortwave diathermy (Ultratherm 525™, Siemens-Reiniger, Erlangen, West Germany) with parallel opposed noncontact electrodes. They found that the degree of tumor growth inhibition was proportional to the duration of heating, with a maximum effect of about 35% ($\pm 5\%$) growth inhibition after daily 30-minute treatments on seven consecutive days. Interestingly, contralateral nontreated and presumably normothermic cheek pouch tumors also displayed some growth inhibition, suggesting possible abscopal antitumor hyperthermia effects. In carefully controlled investigations in 1972, Overgaard and Overgaard heated HB mammary carcinomas in C3H mice by a modified shortwave diathermy apparatus at 27.12 MHz with an output of 1 W and insulated electrodes of 2 to 4 cm^2. By automatic regulation of the generator output, it was possible to maintain any desired temperature continuously with a variation of about 0.1°C. They treated 1200 mice with tumors measuring $5 \times 5 \times 6$ to 8 mm and showed that thermal effect was related to heat dose and length of exposure, such that similar cure rates (22% to 25%) were obtained at 42°C after 120 minutes, 42.5°C after 90 minutes, 43°C after 60 minutes, and at 43.5°C after 45 minutes. Extensive serial histologic examination of treated tumors revealed an effect on both rapidly and slowly proliferating tumor tissue, but virtually no alteration of stroma or vascular cells occurred at these temperatures. There was subsequent rapid new growth of connective tissue and simultaneous resorption of the tumor cells. In 1976, Mendecki, Friedenthal, and Botstein analyzed the effects of microwave heating at 2450 MHz, employing a specially built (RCA Laboratories™, Princeton, New Jersey) 10-W generator and an applicator designed to produce heat penetration limited to hemispherical volume of approximately 1.5 cm in diameter. They reported 100% cure in 54 C3H mice bearing a 6-mm diameter mammary carcinoma transplanted in the subcutaneous tissue after four treatments with tumor surface temperature at 43°C for 45 minutes and an intratumor temperature of 43.5°C. In 1977, Marmor, Hahn, and Hahn treated EMT-6 sarcomas and KHJJ carcinomas in BALB/cka mice at 43°C, 43.5°C, and 44°C, for 5, 10, 20, 30, and 40 minutes with a prototype generator at 13.56 MHz (Critical Systems, Inc., Palo Alto, California). Temperatures were measured by in situ thermistors placed at right angles to the lines of current flow. The EMT-6 tumor was highly sensitive to cure: a five-minute exposure at 44°C cured nearly 50% of the tumors. Cure rate was a function of temperature and exposure time. The KHJJ carcinoma was more resistant, although a majority of the animals treated at 43.5°C were cured of their tumors. Tumor cell survival studies showed that cell kill was similar to that for water bath heating. Direct cell kill could not account for the observed cures, however, and additive mechanisms appeared to be implicated in the tumor eradication.

These investigations concentrated on the tumoricidal effects of temperatures between 39°C and 45°C. In general, the results were in agreement with Arrhenius's formula for the rate of a chemical reaction with temperature (see chapter 3). More recently, the evaluations of higher temperatures have shown that above 42°C, heating time for tumor destruction may be halved. In 1977, Dickson and Shah treated New Zealand white rabbits with transplants of VX-2 carcinoma in extremity muscle. Shortwave-induced hyperthermia was provided by a 30-W generator at 13.56 MHz (Critical Systems, Inc.) and parallel-opposed contact flat-paddle electrodes of various dimensions. Fifteen tumors were selectively heated to 47°C to 50°C for 30 minutes, and 10 tumors completely regressed. In a similar study, Dickson and colleagues (1977) later reported that the temperature of skin and normal muscle remained 3°C to 4°C below the minimal temperature in the tumor. Seven of ten VX-2 carcinomas regressed completely after a single treatment at 47°C for 30 minutes. The host was cured and there was no damage to normal tissues. Subsequently, Storm and co-workers (1979b) selectively heated spontaneously arising canine tumors (fig. 15.4) using water-cooled capacitance electrodes at 13.56 MHz, a prototype impedance matching network, and a generator of 1000-W capacity. Tumors heated at and above 50°C cooled very slowly compared to normal muscle (fig. 15.5), consistent with the theory that many solid tumors have a relatively poor ability to dissipate incident heat. It should be noted that tumors heated at these extremely high temperatures showed no evidence of liquifactive necrosis, and adjacent normal tissues were undamaged.

The cumulative results of these animal tumor investigations indicate that heat alone may be an effective tumoricidal agent. At temperatures between 42°C and 45°C, tumors may be slightly more thermosensitive than normal adjacent tissues and may be killed after prolonged heating. Selective tumor heating above 45°C has been achieved also in many solid ani-

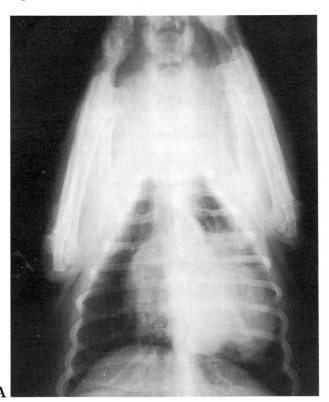
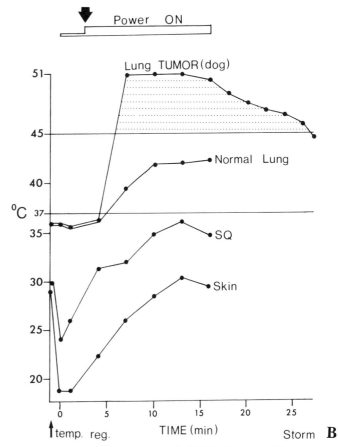

Figure 15.4. Selective hyperthermia of intrathoracic sarcoma (dog). *A*, Radiograph showing tumor of hemithorax; *B*, thermal profile. (Storm et al. 1979b. Reprinted by permission.)

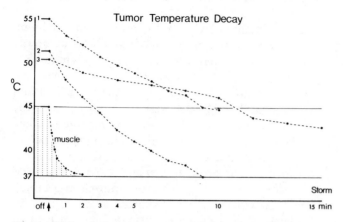

Figure 15.5. Delayed heat dissipation of tumors (*1, 2, 3*) versus normal muscle (dog). (Storm et al. 1979b. Reprinted by permission.)

mal tumors. High rates of tumor kill have been reported after relatively short periods of heating.

CLINICAL CANCER TRIALS

Electrocoagulation

Electrocoagulation or electrofulguration has been part of the surgeon's armamentarium against exposed tumors for nearly 75 years, especially for palliative management of inoperable carcinoma of the rectum. Recently, however, this method has been employed as a curative treatment for this disease (Madden and Kandalaft 1971). The depth of heat penetration is dependent upon the intensity of the current, the size of the electrode (usually a wire loop), the resistance of the tissues, and the duration of exposure. The procedure is performed through a large proctoscope, with the patient under spinal anesthesia, and requires 15 to 45 minutes in the operating room. The tumor is partially resected over several weeks and frequently requires multiple procedures.

The effectiveness of this therapy depends upon the skill and experience of the operator. Most surgeons have reserved electrocoagulation for palliative management of low-lying obstructive tumors to avoid a colostomy in patients who have extensive metastatic disease beyond the pelvis, and for selected patients with small superficial nonulcerating rectal cancers who are poor surgical risks for general anesthesia and abdominal-perineal resection (Wanebo and Quan 1974).

Microwave Thermal Therapy

In 1968, Hartman and Crile treated four children with osteogenic sarcoma of the long bones by first surgically exposing the tumor to isolate it from adjacent normal tissues, and then directly heating the tumor with microwave diathermy (frequency not specified) for 15 to 24 minutes at 50°C to 60°C. Temperatures were monitored by thermocouples placed directly into the bone tumors. Two patients suffered third-degree burns. One patient survived five years, and three died of local tumor recurrence and metastases at 6 to 17 months. These authors suggested that temperatures above 50°C for 30 minutes would be required for total tumor destruction.

More recently, Luk, Phillips, and Holse (1980) conducted a clinical trial using 915-MHz or 2450-MHz microwaves at 40 to 100 W on superficial tumors, including locally recurrent breast, nasopharyngeal, and anal carcinoma, and other metastatic tumors in the body surface. Eleven patients were treated at a mean temperature of 42.5°C for one hour, three times per week for two to three weeks. Temperatures were measured serially with a thermistor during five-second intervals of microwave cessation. Two patients had a complete response to treatment (defined as total tumor disappearance for one month), two had a partial response, and seven had no response.

Joines and Raymond U (1977) reported the effect of 2450-MHz microwave heating, using a direct-contact applicator with a variable aperture, on seven superficial cancers, including metastatic carcinoma, recurrent breast carcinoma, and melanoma. They found 40% tumor regression within 4.5 weeks after five treatments of 43.5°C for 30 minutes administered every other day. Temperatures were measured by tissue-implantable thermistor probes at right angles to the microwave field. After several pilot studies, they concluded that optimal microwave frequencies for heating cancerous tissue is in a range between 500 and 1000 MHz.

Because most investigations have combined microwave heating with other modalities, usually radiation therapy, there has been a paucity of evaluable clinical information on microwave hyperthermia as a single agent.

Radiofrequency Thermal Therapy

In early 1975, Kim, Hahn, and Tokita (1978) initiated investigations on human tumors using RF inductive hyperthermia at 27 MHz (International Medical Electronics, Ltd., Kansas City, Missouri). They studied 24 cutaneous tumors in 12 patients, including melanoma (9), chondrosarcoma (1), adenocarcinoma (1), and unknown primary (1). Temperature profiles were made with fine wire thermocouple probes inserted into tumors and adjacent normal tissues at the same depth (approximately 1.5 cm) at sites where maximum power would be generated. Temperatures then were recorded within a few seconds of power shut-off. At the initiation of treatment, tumor and normal tissue temperatures averaged 35°C; there was a rapid rise to about 40.5°C within 7 to 10 minutes. Normal tissue plateaued at this temperature, but the average tumor temperature continued to rise to 42°C or more. This maximum temperature was maintained throughout a 30-minute observation period. There was selective tumor heating in 20 of 24 patients (83%) and in 17 patients (71%) temperatures at or above 42°C were achieved. Later experiments (Kim and Hahn 1979) were with temperatures from 41°C to 43.5°C for 30 to 40 minutes applied two to nine times in 19 patients with cutaneous cancers, including mycosis fungoides (6), melanoma (6), lymphoma cutis (2), Kaposi's sarcoma (2), chondrosarcoma (1), thyroid cancer (1), and unknown primary (1). The investigations achieved complete response (disappearance of palpable nodules within the treated field in four patients (21%) and partial response (more than 50% reduction in tumor volume) in six patients (32%). They concluded that heat alone at these temperatures caused partial tumor response, but that the duration of the response was transitory.

In 1976, LeVeen and associates reported their experience with RF heating of both superficial and deep internal cancers in 21 patients. Neoplasms treated included squamous cell carcinoma of the head and neck (7), carcinoma of lung (7), adenocarcinoma of colon (4), hypernephroma (1), and metastatic carcinoma to bone (2). Hyperthermia was produced by a 1-kW generator at 13.56 MHz, which transmitted energy to the patient through a pair of insulated capacitance electrodes of varying types. Most patients evidently were given heat doses that they could tolerate under sedation without pain so as to avoid burns and damage to normal tissue, and received doses from 1 to 4 W/cm^2 over a period of 30 minutes. In seven patients, intratumor temperatures were recorded (method not specified) and were always over 46°C, with a mean temperature of 48.4°C. In three patients the temperature of the tissue adjacent to the tumor also was determined, and tumor temperatures were 8°C to 10°C higher than in the normal tissues. After one to nine treatments, substantial tumor necrosis or regression was observed in all patients. These investigators concluded that human cancers could be selectively heated, and that RF therapy consistently resulted in death of cancerous tissue with minimal destruction of normal tissue.

Subsequently, LeVeen, Ahmed, and Piccone (1980), applied capacitance electrodes with multiple portals of entry (Triport-222™). Using this approach, 32 patients with advanced lung cancer (squamous cell, 26; adenocarcinoma, 6) were treated (schedule and additional therapies not specified) and followed from one to five years. Tumor temperatures were measured with thermocouples, where possible, and were maintained above 45°C. Results showed seven patients (22%) alive after one year, with six apparently tumor-free at one year and two patients tumor-free at three years. Using this methodology, Sugaar and LeVeen (1979) reported on the histologic response of three lung tumors preoperatively treated with RF hyperthermia (dose, schedule, and temperature not specified) compared to 15 resected control tumors. They believed they could identify three phases of thermal injury that were interrelated with the immune system. They postulated that after an initial plasmocytic response (stage I), there is a lymphocytic effector response in the tumor stroma (stage II) before a generalized breakdown of stromal vasculature and connective tissue occurs, to allow masses of small lymphocytes to gain direct access to heat-damaged malignant cells (stage III). These observations have not been confirmed.

In 1977, after initial investigations in phantoms and spontaneously arising animal tumors, Storm and co-workers instituted a phase I RF therapy trial in cancer patients with advanced disease who had failed on all standard methods of therapy, including surgery, radiation therapy, chemotherapy, or combination therapy (Storm et al. 1979a, 1979b, 1980a, 1980b, 1981a, 1981b; Storm and Morton 1979). Hyperthermia was produced by RF waves at 13.56 MHz at 50 to 1000 W of absorbed power. Superficial tumors and those with less than 1 cm of overlying normal tissues were treated with paired surface contact electrodes of 100 to 225-cm^2 area, with or without surface cooling (Henry Medical Electronics, Inc.). Deep subcutaneous and internal tumors were treated by a Magnetrode™ (Henry Medical Electronics, Inc.) through a section of the body from 10 to 20 inches wide without preferential surface tissue heating. All patients underwent pretreatment and posttreatment physical and neurologic examination. Laboratory evaluation included complete blood count, protime, partial thromboplastin time, platelet count, SMA-12, and urinalysis. Patients undergoing transthoracic heating also had electrocardiographic and cardiac isoenzyme determinations. Patients who received transabdominal heat also had hepatic transaminase and amylase evaluations. During hyperthermia, the patients' serial core temperature and vital signs were recorded. Intratumor temperatures and temperatures of the skin, subcutaneous tissue, and adjacent normal tissue were recorded by needle thermistors at three- to five-minute intervals. Needle thermometers were inserted into the tissue via plastic locating tubes during brief periods of wave cessation. This method was comparable with temperature measurements by an alcohol thermometer in situ in phantoms. This trial provided much of the current knowledge of the effect of deep hyperthermia on advanced malignancy.

Tumor Heating Capacity

Of 89 tumors evaluated in skin, subcutaneous tissue or muscle, intraabdominal viscera, intrathoracic viscera or bone, temperatures at and above 42°C were possible in 69 (78%) tumors, at and above 45°C in 32 (36%) tumors, and at and above 50°C in 22 (25%) tumors (Storm et al. 1982; Baker et al. 1982).

Tumor Histology

Of 52 tumors evaluated, intratumor temperatures of 42°C or greater in 42 tumors (81%) and of 45°C or greater in 23 tumors (44%) appeared to be independent of the histologic type of tumor (table 15.1). Temperatures of 42°C or greater were observed in 14 of 18 melanomas (78%), 13 of 13 sarcomas (100%), five of nine adenocarcinomas (56%), three of three teratocarcinomas, one of two epidermoid carcinomas, and six of seven less common tumors. Temperatures of 45°C or greater were obtained in 9 of 18 melanomas (50%), 6 of 13 sarcomas (46%), two of nine adenocarinomas (22%), three of three teratocarcinomas, one of two epidermoid carcinomas, and two of seven other cancers, with normal tissues remaining within physiologic temperature range (Storm et al. 1979a). In 29 of 52 tumors, temperatures of 45°C or greater could not be achieved without injury to normal tissues. No correlation with tumor type could be established for this finding, however. Since vascularity and blood flow vary significantly among tumors (Gullino and Grantham 1961; Mantyla 1979), it has been postulated that potentially effective hyperthermia may not be possible in many types of cancer. These data suggest that potentially tumoricidal hyperthermia at and above 42°C may be achieved in the treatment of most primary and metastatic solid human tumors, regardless of histologic type. Since many tumors could not be selectively heated to temperatures greater than 45°C, factors other than histologic type of malignancy ultimately must determine the overall ability to achieve high-temperature hyperthermia. Since normal tissues appear to augment blood flow in response to thermal stress, it is possible that some tumors retain this ability.

Tumor Size

Thermal response by tumor size also was evaluated in 89 tumors (table 15.2). Thirty-six of the tumors were less than 5 cm in least dimension, and 53 were 5 cm or larger. Sixty-one percent of small tumors and 89% of large tumors could be heated at or above 42°C. Interestingly, large tumors had a greater potential for maintaining temperatures at and above 45°C (43%) than did small tumors (25%). Most of the tumors that could not be heated to 45°C displayed physiologic adaptation to heat similar to that of adjacent normal tissues. Thermal adaptation was most often observed in lesions less than 5 cm in size, and rarely in lesions of 5 cm or greater, probably as a result of diminished vascular integrity in the larger tumors. These results (Storm et al. 1982) indicate that capacity for tumor heating may be influenced by tumor size. It is well known that standard methods of cancer therapy, including surgery, radiation therapy, chemotherapy, and immunotherapy are most effective against smaller cancers. The trend of these data suggests that localized hyperthermia may be uniquely effective against large cancers, for which little effective therapy currently exists.

Superficial Tumors

Nineteen tumors arising in skin or subcutaneous tissue were evaluated, and selective heating at 45°C and above was observed in 14 tumors (74%). Tempera-

Table 15.1
Tumor Thermal Response by Histologic Type

Histology	No.	≥42°C (%)	≥45°C (%)	≥50°C (%)
Melanoma	18	14 (78)	9 (50)	7 (39)
Sarcoma	13	13 (100)	6 (46)	6 (46)
Adeno Ca	16	11 (69)	4 (25)	2 (13)
Terato Ca	3	3	3	3
Epidermoid	2	1	1	1
Totals	52	42 (81)	23 (44)	19 (36)

SOURCE: Storm et al. 1979a.

Table 15.2
Tumor Thermal Response by Size

Size	No.	42°C–44°C (%)	45°C–49°C (%)	≥50°C (%)
<5 cm	36	22 (61)	9 (25)	7 (19)
≥5 cm	53	47 (89)	23 (43)	15 (28)

SOURCE: Storm et al. 1982.

Animal and Clinical Studies with Microwave and Radiowave Hyperthermia

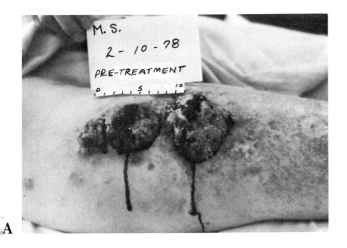

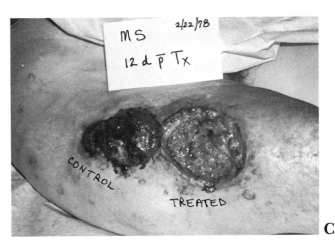

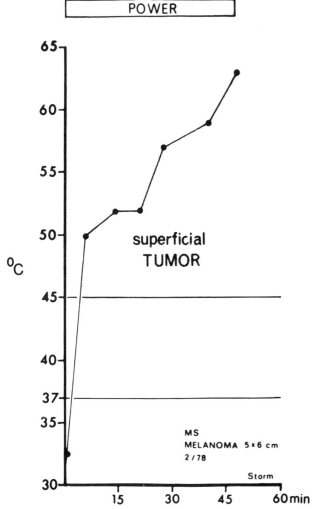

Figure 15.6. Hyperthermia of surface melanoma (human). *A*, Pretreatment, showing two large exophytic extremity tumors; *B*, thermal profile of treated tumor; *C*, posttreatment, showing slough of treated tumor versus nontreated tumor. (Storm et al. 1979b. Reprinted by permission.)

tures as high as 70°C were achieved, with up to 57°C at 10-cm tumor depth. Surface tumors treated at or above 50°C for 15 minutes on one or more occasions generally showed histologic evidence of coagulative necrosis and would slough within 10 to 14 days (figs. 15.6A, B, C).

Visceral Tumors

Seventeen intrathoracic or intraabdominal tumors were evaluated initially. Selective heating to 45°C and above was possible in six tumors (35%), and all were 5 cm or larger in size. Figure 15.7 illustrates the temperature profile in a patient who had a recurrent 10 × 15 × 30-cm primary intraabdominal sarcoma that progressed despite surgical resection, radiation, and chemotherapy. Figure 15.8 shows the temperature profile of a 10 × 10-cm hepatic metastasis in a patient with colon carcinoma. While intratumor temperatures frequently were not uniform, tumors heated to 50°C and above for 15 minutes on one or more occasions generally showed coagulative necrosis and intravascular thrombosis. In contrast to superficially exposed tumors, effectively heated visceral tumors would remain intact with little change in size and with no evidence of systemic tumor breakdown products by serum creatinine, urate, or urinary protein determinations. Serial biopsies of these internal tumors revealed few functional vessels and progressive tumor replacement by scar (figs. 15.9A, B, C), as others have noted (Overgaard and Overgaard 1972). We have postulated that the attendant vascular necrosis and thrombosis that occurred at these high temperatures prevented absorption of the tumor and allowed only fibrous replacement over a prolonged interval, similar to a healing infarction. Therefore, direct biopsy rather than size measurement appeared to be necessary to assess therapeutic benefit in visceral cancers treated by hyperthermia.

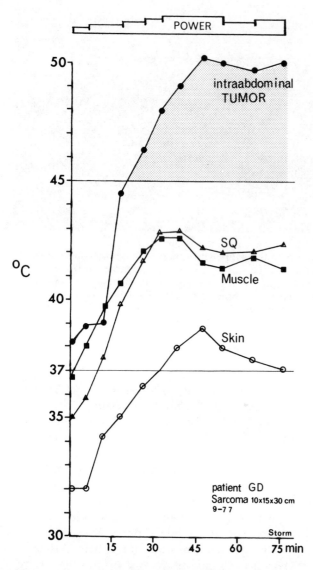

Figure 15.7. Thermal profile of recurrent primary intraabdominal sarcoma, $10 \times 15 \times 30$ cm, showing selective tumor heating (human). (Storm et al. 1979b. Reprinted by permission.)

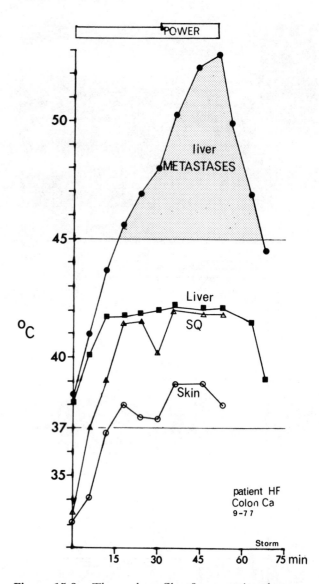

Figure 15.8. Thermal profile of metastatic colon carcinoma in liver, 10×10 cm, showing selective tumor heating (human). (Storm et al. 1979b. Reprinted by permission.)

Thermal Death Time

The effects of localized RF hyperthermia by temperature and exposure time were evaluated for 44 tumors in 38 patients with histologically proved cancer (fig. 15.10) (Storm et al. 1980b). At least half of the tumors studied were 5 cm or larger in their least dimension. Tumors were located in skin, subcutaneous tissue, or muscle (18), the intraabdominal cavity (19), the intrathoracic cavity (4), and neck (3). Tumor types included sarcoma (15), melanoma (13), adenocarcinoma (7), epidermoid carcinoma (3), and other less common tumors (6). Treatments were administered once, once weekly for two to five weeks, or daily for 10 working days, from 13 to 600 minutes total treatment time, at tumor temperatures from 40°C to more than 50°C. Effects of hyperthermia were assessed by multiple random needle (Tru-Cut®) biopsies taken before, during, and serially after therapy. Since previous research had shown that effectively heated deep tumors were slowly replaced by scar with little change in size, biopsies, rather than other criteria of tumor response, were necessary to assess the short-term effects of therapy. Gross morphologic and histopathologic evaluations of tumor changes also were performed after surgical resection or at autopsy. The tumoricidal effects of hyperthermia by temperature (°C) and total treatment time (minutes) were evaluated by various pathologists who had no knowledge of the specific

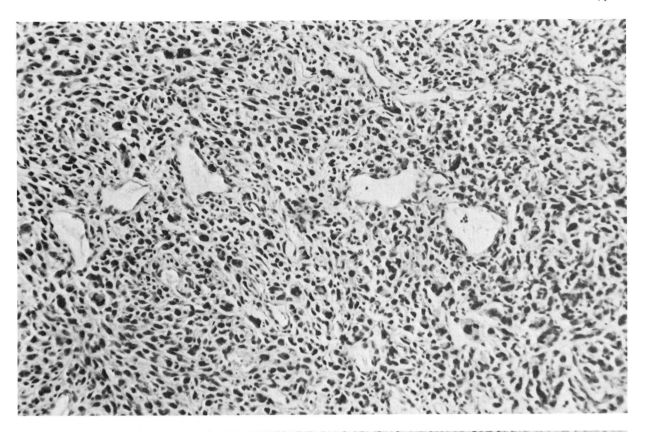

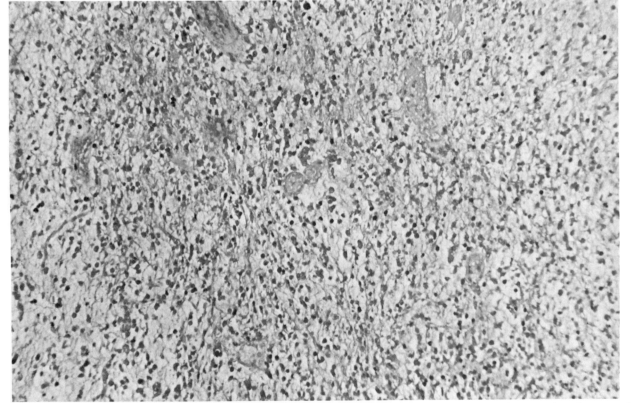

Figure 15.9. Photomicrographs of human intraabdominal sarcoma treated five times at 50°C or more for 15 to 30 minutes. A, Pretreatment, showing pleomorphic, hyperchromatic nuclei, prominent nucleoli, moderate amounts of cytoplasm, scant amounts of supporting tissue, and intact vessels; B, at two weeks, showing coagulation necrosis and vascular thrombosis;

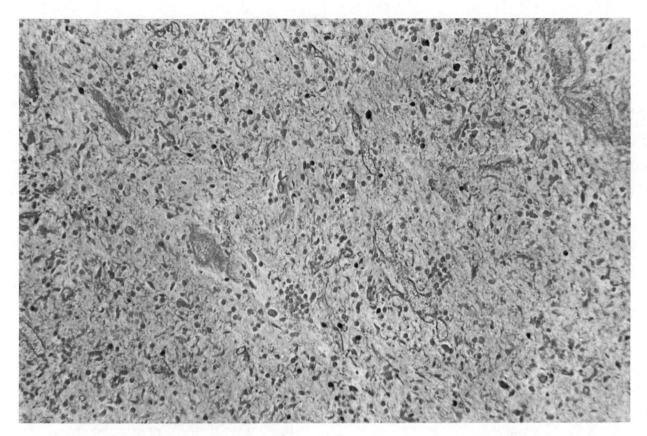

Figure 15.9 continued *C*, at three months, showing marked fibrosis, tumor cell ghosts, and obliterated vascular channels. (×100) (Storm et al. 1981a. Reprinted by permission.)

treatment regimen given. Overall tumor necrosis was defined as the percentage of tumor cells showing total absence of nuclei or replacement by fibrous tissue, compared to pretreatment. Changes in tumor size also were recorded serially. The results showed that a single treatment for 17 to 45 minutes at 50°C or greater resulted in 20% to 100% tumor necrosis, whereas lower temperatures had no apparent effect. Two or three weekly treatments for 30 to 72 minutes total time at 45°C to 50°C produced a 70% to 100% necrosis, whereas 40°C to 45°C temperatures produced almost as much necrosis but required more than twice the time. Five weekly and 10 daily treatments at 40°C to 45°C for 135 to 600 minutes total time produced tumor necrosis; however, for the same amount of time, temperatures greater than 45°C were most effective. Significant coagulative necrosis and vascular thrombosis again occurred in most effectively heated tumors. Surface tumors generally would slough, and even though deep internal tumors appeared to remain the same size in those patients for whom long-term follow-up was possible, these tumors were replaced with varying amounts of fibrous tissue, as observed previously. These clinical results suggested that the necrosis was related to both temperature and treatment time, and that higher temperatures and/or longer durations of therapy were most beneficial. Even at very high temperatures, however, total tumor necrosis by our criteria was rarely possible. This result was ascribed to the cooling influence of the microscopic blood vessels that perfuse and pervade spontaneously arising human malignancies, particularly at the tumor periphery. It appeared that both superficial and internal human cancers that arose in skin and subcutaneous tissue with a responsive sympathetic nervous system, or in intraparenchymal organs, had thermal responses different from those predicted by tumor models or induced tumors in animals.

Thermal Toxicity

Hyperthermia was well tolerated in the mildly sedated, awake patient. Normal tissues in the treatment field adjacent to tumors, including skin, subcutaneous tissue, abdominal and thoracic musculature, esophagus,

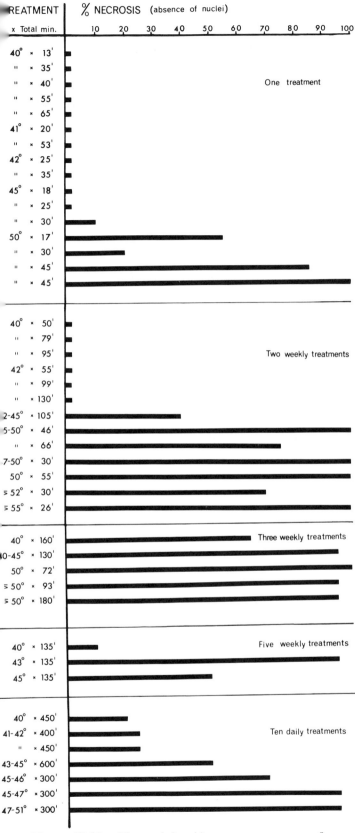

Figure 15.10. Thermal dose/time tumor response of one to five weekly or 10 daily treatments (human). Percentage of necrosis was estimated from multiple random biopsies before and after therapy, and from surgical resection or later autopsy. (Storm et al. 1980b. Reprinted by permission.)

lung, liver, and gut ordinarily could be maintained at a physiologic temperature of less than 45°C with appropriate RF application, since all of these tissues were observed to be capable of thermal adaptation. In patients undergoing deep visceral hyperthermia at 500 to 1000 W absorbed power density, we frequently observed diaphoresis, skin flushing, a modest rise in core temperature (0.5°C to 1°C), respiratory rate (2 to 12 rpm), systolic blood pressure (20 to 40 mm Hg), and a moderate to marked rise in pulse (20 to 60 beats per minute) (fig. 15.11). Thus, it appeared that patients dissipated localized heat by evaporation and increased peripheral blood flow, rather than by increased ventilation. Since these initial investigations by Storm and colleagues, more than 500 patients have received more than 5000 individual treatments at multiple centers. More than 75% of these patients have undergone intrathoracic or intraabdominal hyperthermia with no clinical or laboratory evidence of internal organ injury. No significant arrhythmias, cardiac isoenzyme changes, or abnormal serum parameters of internal organ function have been reported at heat sufficient to produce effective tumor heating. No rise in serum creatinine or urate levels was observed in patients with large tumors heated to 45°C to 50°C; therefore, prophylactic intravenous hydration and uricosuric agents were discontinued after initial trials. Three patients with superficial tumors heated to 50°C to 70°C had slough of immediately overlying skin; otherwise, injury to normal skin has not been observed. Two obese patients (with 2- to 3-cm subcutaneous tissue) had small localized areas of subcutaneous fibrosis as a result of temperatures of 42°C to 45°C; no other adverse reactions in normal surface tissues were observed after deep hyperthermia. The patients with surface tumors heated at 50°C to 70°C probably had overlying skin slough caused by direct heat conduction or obliteration of dependent vasculature. The localized areas of subcutaneous fibrosis in the very obese patients probably were the result of the extremely poor vascularity in the tissue. Second degree burns occurred at temperatures under 45°C in a patient with a pedicle graft and one with fibrous tissue and brawny edema. This result suggests that noncancerous tissues with inadequate blood supply also may be thermosensitive.

CONCLUSIONS

Hyperthermia has been shown to be a potentially effective tumoricidal agent in animal models. Preliminary investigations in humans indicate that with proper microwave and RF instrumentation, superficial and deep internal hyperthermia are possible with

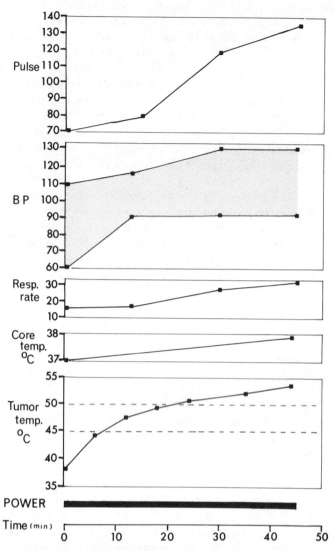

Figure 15.11. Alteration of vital signs in an awake patient during visceral hyperthermic tumor therapy. (Storm et al. 1979b. Reprinted by permission.)

minimal host toxicity. Clinical human trials suggest that higher temperatures and longer durations of treatment are most effective, although the optimum dose/treatment regimen has yet to be established. It is unknown whether continuous or fractionated schedules are most effective. Hyperthermia appears to be possible in most varieties of human solid tumors, even though it is impossible to predict individual tumor heating capacity at this time. Since many effectively heated visceral tumors may not regress in size, serial histopathologic examinations (Storm et al. 1979a, 1979b, 1980b) and/or assessment of tumor doubling time (Storm et al. 1979b) appear necessary to evaluate therapeutic results. Whenever possible, each patient should serve as his own control. Prospective randomized clinical trials are warranted.

References

Baker, H. W. et al. Regional hyperthermia for cancer. *Am. J. Surg.* 143:586–590, 1982.

Bender, E., and Shramm, T. Untersuchungen zur thermosensibilitat von tumor und normalzellen in vitro. *Acta Biol. Med. Germ.* 17:527–543, 1966.

Bishop, F. W.; Horton, R.; and Warren, S. L. A clinical study of artificial hyperthermia induced by high frequency currents. *Am. J. Med. Sci.* 184:512–532, 1932.

Carter, D. B.; Silver, I. A.; and Watkinson, D. A. Combined therapy with 220 kr roentgen and 10 cm. microwave heating in rat hepatoma. *Acta Radiologica Therapy Physics Biology* 2:321–336, 1964.

Cavaliere, R. et al. Selective heat sensitivity of cancer cells: biochemical and clinical studies. *Cancer* 20:1351–1381, 1967.

Crile, G., Jr. Heat as an adjunct to the treatment of cancer. Experimental studies. *Cleve. Clin. Q.* 28:75–89, 1961.

Crile, G., Jr. Selective destruction of cancers after exposure to heat. *Ann. Surg.* 156:104, 1962.

Cunningham, D. J. An evaluation of heat transfer through the skin in the human extremity. In *Physiological and behavioral temperature regulation*, eds. J. D. Hardy, A. P. Gagge, and J. A. J. Stolwijk. Springfield, Ill.: Charles C Thomas, 1970, pp. 100–105.

Dickens, F.; Evans, S. F.; and Weil-Malherbe, H. The action of short radiowaves on tissues. II. Treatment of animal tumors in vivo. *Am. J. Cancer* 30:341–354, 1937.

Dickson, J. A. et al. Tumor eradication in the rabbit by radio frequency heating. *Cancer Res.* 37:2162–2169, 1977.

Dickson, J. A., and Shah, S. A. Technology for the hyperthermic treatment of large solid tumors at 50°C. *Clin. Oncol.* 3:301–318, 1977.

Doss, J. D. Use of RF fields to produce hyperthermia in animal tumors. In *Proceedings of the First International Symposium on Cancer Therapy by Hyperthermia and Radiation*, eds. M. J. Wizenberg and J. E. Robinson. Chicago: American College of Radiology, 1976, pp. 226–227.

Gessler, A. E.; McCarthy, K. S.; and Parkinson, M. D. Eradication of spontaneous mouse tumors by high frequency radiation. *Exp. Med. Surg.* 8:143–148, 1950.

Gibbs, F. A. Clinical evaluation of a microwave/radiofrequency system (BSD Corporation) for induction of local and regional hyperthermia. *J. Microwave Power* 16:185–192, 1981.

Giovanella, B. C. et al. Selective lethal effect of supranormal temperatures on mouse sarcoma cells. *Cancer Res.* 33:2568–2578, 1973.

Goldenberg, D. M., and Langner, M. Direct and abscopal antitumor action of local hyperthermia. *Z. Maturforsch.* 26:359–361, 1971.

Gullino, P. M., and Grantham, F. H. Studies on the exchange of fluids between host and tumor. II. The blood flow of hepatomas and other tumors in rats and mice. *J. Natl. Cancer Inst.* 27:1465–1491, 1961.

Guy, A. W.; Lehmann, J. F.; and Stonebridge, J. B. Therapeutic applications of electromagnetic power. *Proc. IEEE* 62:55–75, 1974.

Hardy, J. D. et al. Skin temperature and cutaneous pain during warm water immersion. *J. Appl. Physiol.* 20:1014–1021, 1965.

Hartman, J. T., and Crile, G., Jr. Heat treatment of osteogenic sarcoma. *Clin. Orthop.* 61:269–276, 1968.

Johnson, H. J. The action of short radiowaves on tissues. III. A comparison of the thermal sensitivities of transplantable tumours in vivo and in vitro. *Am. J. Cancer* 38:533–550, 1940.

Joines, W. T. et al. Techniques and results of using microwaves and x-rays for the treatment of tumors in man. Paper presented at the International Symposium on the Biological Effects of Electromagnetic Waves, November 1977, Airlie, Va.

Kim, J. H., and Hahn, E. W. Clinical and biological studies of localized hyperthermia. *Cancer Res.* 39:2258–2261, 1979.

Kim, J. H.; Hahn, E. W.; and Tokita, N. Combination hyperthermia and radiation therapy for cutaneous malignant melanoma. *Cancer* 41:2143–2148, 1978.

Lehmann, J. F.; DeLateur, B. J.; and Stonebridge, J. B. Selective muscle heating by shortwave diathermy with a helical coil. *Arch. Phys. Med.* 50:117–132, 1969.

LeVeen, H. H. et al. Tumor eradication by radio frequency therapy. Response in 21 patients. *JAMA* 235:2198–2200, 1976.

LeVeen, H. H.; Ahmed, N.; and Piccone, V. A. RF therapy: clinical experience. The thermal characteristics of tumors. *Ann. N.Y. Acad. Sci.* 335:362–371, 1980.

Luk, K. H.; Phillips, T. L.; and Holse, R. M. Hyperthermia in cancer therapy. *West. J. Med.* 132:179–185, 1980.

Madden, J. L., and Kandalaft, S. Electrocoagulation in the treatment of cancer of the rectum; continuing study. *Ann. Surg.* 174:530, 1971.

Mantyla, M. J. Regional blood flow in human tumors. *Cancer Res.* 39:2304–2306, 1979.

Marmor, J. B.; Hahn, N.; and Hahn, G. M. Tumor cure and cell survival after localized radio frequency heating. *Cancer Res.* 37:879–883, 1977.

Mendecki, J.; Friedenthal, E.; and Botstein, C. Effects of microwave-induced local hyperthermia on mammary adenocarcinoma in C3H mice. *Cancer Res.* 36:2113–2114, 1976.

Oleson, J. R. et al. Magnetic induction heating in heterogenous material. *J. Natl. Cancer Inst.*, in press.

Overgaard, K., and Overgaard, J. Investigations on the possibility of a thermic tumour therapy—I. *Eur. J. Cancer* 8:65–79, 1972.

Samaras, G. M. et al. Clinical hyperthermia systems engineering. *J. Microwave Power* 16:161–169, 1981.

Schereschewsky, J. W. The action of currents of very high frequency upon tissue cells. Mouse and fowl sarcoma. *Pub. Health Rep.* 43:927–945, 1928.

Schereschewsky, J. W. Biological effects of very high frequency electromagnetic radiation. *Radiology* 20:246–253, 1933.

Sternhagen, C. J. et al. Clinical use of radio frequency current in oral cavity carcinomas and metastatic malignancies with continuous temperature control and monitoring. In *Proceedings of the Second International Symposium on Cancer Therapy by Hyperthermia and Radiation*, eds. C. Streffer et al. Baltimore and Munich: Urban & Schwarzenberg, 1978, pp. 331–334.

Shingleton, W. W. Selective heating and cooling of tissue in cancer chemotherapy. *Ann. Surg.* 156:408–416, 1962.

Stolwijk, J. A. J. Physiological response to whole-body and regional hyperthermia. *Proceedings of the First International Symposium on Cancer Therapy by Hyperthermia and Radiation*. Chicago: American College of Radiology, 1976, pp. 163–167.

Stolwijk, J. A. J., and Hardy, J. D. Skin and subcutaneous temperature changes during exposure to intense thermal radiation. *J. Appl. Physiol.* 20:1006–1013, 1965.

Storm, F. K. et al. Human hyperthermic therapy: relationship between tumor type and capacity to induce hyperthermia by radio frequency. *Am. J. Surg.* 138:170–174, 1979a.

Storm, F. K. et al. Normal tissue and solid tumor effects of hyperthermia in animal models and clinical trials. *Cancer Res.* 39:2245–2251, 1979b.

Storm, F. K. et al. Hyperthermia in cancer treatment: potential and progress. In *Practical oncology for the primary care physician*, vol. 1, ed. G. Sarna. Boston: Houghton Mifflin, 1980a, pp. 42–52.

Storm, F. K. et al. Hyperthermia therapy for human neoplasms: thermal death time. *Cancer* 46:1849–1854, 1980b.

Storm, F. K. et al. Radio frequency hyperthermia of advanced human sarcomas. *J. Surg. Oncol.* 17:91–98, 1981a.

Storm, F. K. et al. Clinical radio frequency hyperthermia by magnetic-loop induction. *J. Microwave Power* 16:179–184, 1981b.

Storm, F. K. et al. Clinical radio frequency hyperthermia: a review. *J. Natl. Cancer Inst.*, in press.

Storm, F. K., and Morton, D. L. Localized hyperthermia in the treatment of cancer. *Int. Adv. Surg. Oncol.* 5:261–275, 1982.

Sugaar, S., and LeVeen, H. H. A histopathologic study on the effects of radio frequency thermotherapy on malignant tumors of the lung. *Cancer* 43:767–783, 1979.

Taylor, L. S., and Robinson, J. E. Devices for microwave hyperthermia. In *Proceedings of the Second International Symposium on Cancer Therapy by Hyperthermia and Radiation*, eds. C. Streffer et al. Baltimore and Munich: Urban & Schwarzenberg, 1978, pp. 115–117.

Theilhaber, A. *Munch. Med. Wochenschr* 71:126, 1919.

Wanebo, H. J., and Quan, S. H. Failures of electrocoagulation of primary carcinoma of the rectum. *Surg. Gynecol. Obstet.* 138:174–176, 1974.

Warren, S. L. Preliminary study of the effect of artificial fever upon hopeless tumor cases. *Am. J. Roentgenol.* 33:75–87, 1935.

Westermark, N. The effect of heat upon rat tumors. *Skand. Arch. of Physiol.* 52:257–322, 1927.

Zimmer, R. P., and Ecker, H. A. Selective electromagnetic heating of tumors in animals in deep hyperthermia. *IEEE Trans. BME* 19:238–245, 1971.

CHAPTER 16

Physical Aspects and Clinical Studies with Ultrasonic Hyperthermia

Padmakar P. Lele

Introduction
Rationale for Use of Ultrasound, Focusing, Beam-Steering, and Intensity Modulation
System Instrumentation and Technique
Measured Temperature Distributions Produced by Stationary or Steered Beams of Unfocused or Focused Ultrasound in Experimental Studies In Vitro and In Vivo
Preliminary Studies on Tumor Temperature Distributions and Response in Various Tumors

INTRODUCTION

Although the interest in hyperthermia as a modality in treatment of cancer has a long but checkered history, the current resurgence of interest (Wizenberg and Robinson 1976; Streffer et al. 1978; American Cancer Society 1979; Jain and Gullino 1980; Dewey and Dethlefsen, in press) is likely to establish hyperthermia as a primary mode of cancer therapy, as important as surgery, radiation, or drugs. This optimism stems from its emphatic demonstration of significant antitumor effects in patients in whom all conventional modes of therapy had failed. The current cycle of interest has generated a scientific rationale for its use based on a demonstrably greater heat sensitivity of malignant cells as compared with normal cells, and has led to the realization of the importance of (and difficulties in) controllable heat delivery to the treatment volume.

RATIONALE FOR USE OF ULTRASOUND, FOCUSING, BEAM-STEERING, AND INTENSITY MODULATION

Performance Specifications for Equipment for Local Thermotherapy

In order to define the performance specifications for equipment for thermotherapy on solid, malignant tumors, it may be useful to summarize here the currently available, noncontroversial information or beliefs on the effects of heat on tumors and normal tissues presented elsewhere in this book.

1. Malignant cells and tumors appear to be more sensitive to heat-induced effects than normal cells and tissues, but the difference in their heat sensitivities, though not known with any precision, is small (Pettigrew et al. 1974; Larkin 1979). *Therefore, the technique should enable induction of hyperthermia localized to the tumor or a predetermined target volume, thus sparing normal tissues in the path or surrounding the tumor. The spatial fall-off of temperature beyond the target volume should be steep.*

2. Studies on cell cultures indicate that a small difference in heat dose (temperature-duration history) produces a large difference in cell kill (Westra and Dewey 1971). *Therefore, the level of hyperthermia should be precisely controllable.*

3. The possibility that a subtherapeutic heat dose may enhance tumor growth cannot be excluded, although there is no firm evidence to indicate

This work was supported in part by U. S. Public Health Service grants CA–16111, CA–26232, CA–30944, and RR–00088 from the National Institutes of Health.

that it does so. *To be on the safe side, temperature distribution within the hyperthermia region should be uniform and at the therapeutic level. More importantly, uniform treatment fields are essential for establishing temperature-duration relationships with any accuracy, for hyperthermia alone or with radiation or chemotherapy.*

4. Tumor vasculature, and consequently the heat dissipation capacity, may vary with the tumor type, size, and location from tumor to tumor and within different regions of the same tumor (Mäntylä 1979; Peterson 1980). *Therefore, it should be possible to control heat deposition in different regions of the target volume to meet these varying requirements. It would be preferable to be able to measure heat dissipation in different regions, and to be able to predict heat deposition requirements as a part of therapy planning.*

5. Large tumors have an ischemic or necrotic core, which heats to a high temperature when the tumor is subjected to hyperthermia and leads to breakdown and regression of the tumor. The high temperature and its duration, as well as the histologic appearance of the tumor after treatment, is consistent with heat-induced coagulation necrosis of the tumor (Storm et al. 1980). The temperatures in the peripheral regions of the tumor are, however, lower. *The technique should enable the therapist, not the tumor, to control the temperature and the occurrence or nonoccurrence of coagulation and its location and volume. The hyperthermia technique should enable production of the preselected level of temperature either below or at that required for coagulation, at the desired location, and permit its maintenance for the preselected duration. It would appear also that it is more important to be able to heat smaller tumors with "normal" blood perfusion and the well perfused, growing edge of larger tumors to a therapeutically adequate temperature, than to generate high temperatures in the necrotic core.*

To meet these objectives, the technique must permit complete control of heat generation in different regions of the target volume.

Rationale for Noninvasiveness

In production of hyperthermia for cancer therapy, noninvasiveness of the technique is of great importance since disruption of blood vessels and mechanical probing of the tumor might increase metastasis. Ultrasonic and electromagnetic (EM) radiations are the two principal modalities potentially useful for this purpose.

Plane Wave Fields

The amplitude and distribution of the hyperthermia, that is, its volume extent and uniformity, are governed by the patterns of heat generation and of heat diffusion from conduction and blood flow. Since heat diffusion cannot effectively be controlled, the optimization of the technique for production of uniform, controllable hyperthermia relies almost entirely on the control over the pattern of heat generation in the tissues by the modality used.

Let us consider the pattern of heat generation in tissues by EM fields and ultrasound. If a uniform intensity field of any form of radiant energy were to impinge upon a medium with uniform attenuation and thermal characteristics (fig. 16.1A), the intensity would decay exponentially with increasing depth in the medium (fig. 16.1B). Since, for a given absorption coefficient, the rate of heat generation is proportional to the local intensity, it would also decay exponentially with depth. Given uniform thermal properties—specific heat, heat conduction, and heat diffusion by blood flow—the pattern of temperature distribution would also be similar, being highest at the surface and decaying exponentially with depth (fig. 16.1C). In the case of biological tissues in vivo, it is not possible to compensate for this gradient of temperature, to any useful extent, by cooling of the skin. Cooling may, on the other hand, cause elevation of subcutaneous temperatures from the resulting vasoconstriction. Attenuation of ultrasonic or EM energy in transit through an attenuating medium is inversely related to the wavelength in the medium. Therefore, for a given intensity at the surface, less energy is lost in transit to a given depth in the medium, and higher local intensity is obtained at the longer wavelengths at lower frequencies than at higher frequencies. Figure 16.2 shows the local intensity relative to that at the surface, at different depths in typical mammalian tissues, for ultrasonic frequencies of 0.1 to 5.0 MHz. Note that higher local intensities are obtained at a given depth at lower frequencies. Note also that energy not attenuated in the target volume and overlying tissues will be transmitted into deeper tissues and may undergo multiple reflections, refraction, and scattering until it is absorbed completely, since it cannot leave the body. Thus, for instance, if a frequency of 0.3 MHz is used to attempt to heat a tumor 10 cm below the surface, almost 80% of the incident energy will be transmitted deeper and lead to heating of deeper tissue, especially of soft tissue-bone interfaces. As discussed later, the use of such low frequencies could lead also to the occurrence of cavitation. Note that it is not possible to produce a preferentially greater temperature rise at depth by using energy at two or more different wavelengths (or frequencies), since for each of the wavelengths (or frequencies) the intensity is highest at the surface and decays exponentially, albeit over different path lengths (fig. 16.3).

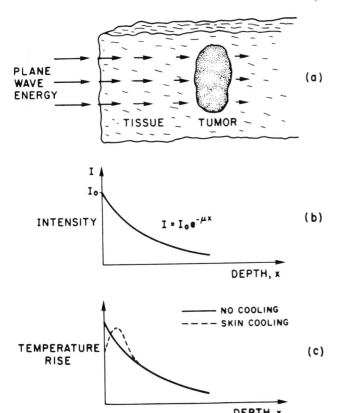

Figure 16.1. Intensity and temperature distribution patterns in a homogeneous medium with plane wave radiation. (Lele 1980. Reprinted by permission.)

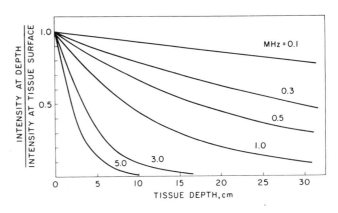

Figure 16.2. Decrease in plane wave intensity with increasing depth in soft tissues for ultrasound at different frequencies. Absorption coefficient is assumed to be 0.35 dB/cm/MHz, typical of most soft tissues.

The same considerations apply to the situations in which the source, a radiofrequency (RF) or microwave antenna or a transducer, is placed intraluminally within a cavity (fig. 16.4). The intensity and temperature rise will be highest at the tissue surface and exponentially lower at depth. Cooling of the mucosa may protect it from heat damage, but the intratumoral temperature will decay exponentially with increasing distance from the source, as in figure 16.1C. Similarly, a microwave antenna inserted into the tumor will produce highest temperatures near the tip of the inserted probe, as shown in figure 16.5, despite the invasiveness of the procedure.

Multiple-Beam Superposition

One of the ways in which higher intensities could be obtained at depth, compared to those at the surface, is by superposition of two or more beams entering the medium through different surface portals, but superimposed on the target at depth. Rotation of the source of energy in an arc centered at the deep target is commonly practiced in radiation therapy to deliver a

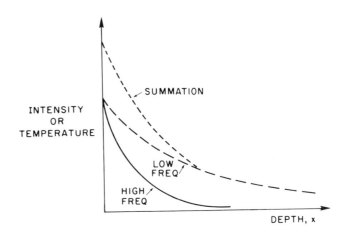

Figure 16.3. Intensity and temperature distribution patterns for simultaneous irradiation with energy at two different frequencies. (Lele 1980. Reprinted by permission.)

335

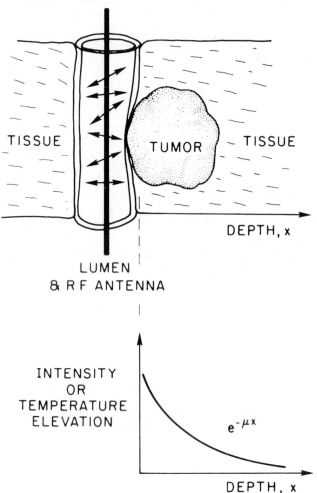

Figure 16.4. Intensity and temperature distribution patterns for an intraluminal RF antenna. (Lele 1980. Reprinted by permission.)

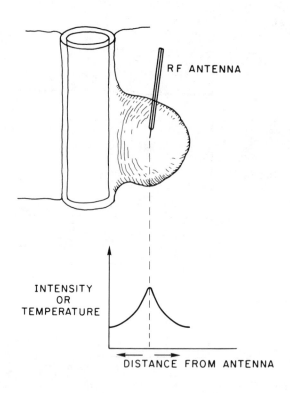

Figure 16.5. Intensity and temperature distribution patterns for an intratumoral RF needle antenna. (Lele 1980. Reprinted by permission.)

higher total dose to a deep target volume than to the intervening tissues. Superposition of multiple beams of electromagnetic or ultrasonic energy, however, poses serious problems of phasing because of their relatively longer wavelengths compared to x-rays. If two beams are out of phase where they overlap, they would interfere destructively, and the local intensity (and the consequent heat generation) may be lower in the overlap than in the individual beams. The phase at any point in the medium is of course governed both by the path length (in fractions of a wavelength) and the velocity in the medium. It is also well to bear in mind that the intensity in each beam will decay exponentially from the surface inward. The maximum depth at which a dose higher than that at the surface could be achieved would depend on both the attenuation characteristics of the intervening tissues and the size of the available portal. Furthermore, with multiple overlapping, but stationary, beams, maximal heating would occur at the point(s) of their intersection. This is borne out in measurements of temperature distributions in experimental animals in vivo described later.

Inhomogeneous Media and Nonuniform Fields

In reality, the tissue media are not uniform in their attenuation or heat transfer characteristics, nor do the EM or ultrasound sources have a uniform intensity distribution either in near or far fields. These facts complicate matters further, as far as heat generation rates are concerned. Local heat transfer, by conduction and blood perfusion, would tend to smooth out the temperatures, and under steady state conditions the differences would not be as large as under transient conditions. Nevertheless, so long as heat generation is uncontrollable, so will be the temperatures. With EM fields (Guy, Lehmann, and Stonebridge 1974; Schwan 1980), the differences between the attenuation coefficients of skin, fat, and muscle are rather remarkable; for ultrasound they are comparatively slight. Bone presents a problem to both modalities, and will be discussed later. Avascular tissues, such as the lens oculi, also need special consideration because of their limited heat dissipation capacity. The occurrence of cataracts with exposure to EM waves at intensity levels very much lower than those used for therapeutic hyperthermia is well known.

Focused Fields

The use of a single convergent (or focused) beam greatly alleviates the problems of field inhomogeneity and of phasing (fig. 16.6A). The path length from the radiating element to the focus is fixed, and there is little possibility of destructive interference in the tissues at the focus, unless the propagation velocities and path lengths in different intervening tissues are significantly different. Higher intensities can be achieved at the desired target at depth (fig. 16.6B) with correspondingly higher rates of heat generation and correspondingly higher temperatures (fig. 16.6C). Specifically, the temperature rise at the surface over the skin, the mucosa, or granulation tissue, and the relatively less vascular subcutaneous tissues, can be held to a negligible value. The intensity decays rapidly in tissues beyond the focus, by beam divergence as well as by attenuation, thus minimizing the temperature elevation in tissues, such as in bone, underlying the tumor.

Aperture

The gain in intensity with progressive convergence of the field toward the focus is reduced by the loss from attenuation in the tissues as the path length increases. Therefore, with a beam of a given angle of convergence and for a given attenuation in the tissue, there is no effective gain in intensity beyond a certain depth (fig. 16.7). This depth can be increased by increasing the angle of convergence (use of a larger aperture) as seen in figure 16.7, or by use of a lower frequency to reduce attenuation.

Not only the attenuation, but also the absorption coefficients of tissues, increase with frequency: that is, the shorter the wavelength, the higher both the attenuation and absorption. In a plane progressive ultrasonic wave, attenuation in soft tissues is only slightly greater than absorption (Lele and Senapati 1977; Wells 1977); most of the energy that is attenuated is absorbed locally. The energy attenuated in tissues superficial to the tumor is thus not only wasteful, but since it generates heat it is also detrimental to the goal of sharply localized heating of the tumor. Lowering the attenuation by use of longer wavelength (lower frequency) would reduce the loss of energy and heat generation in transit to the focus in the tumor. Similarly, it would reduce the heat generation within the tumor, requiring proportionately higher intensities to

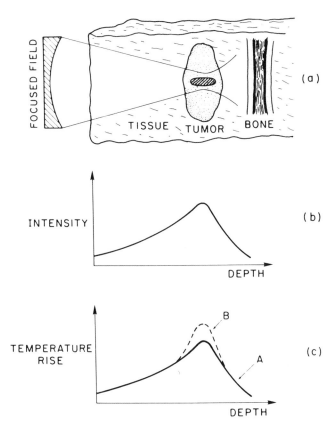

Figure 16.6. Intensity and temperature distribution patterns with a focused field. In *a*, the experimental design is shown; *b* shows the plot of intensity versus depth. In *c*, the curve labeled *A* represents the pattern of temperature distribution, if the ultrasonic and thermophysical properties of the tumor were the same as those of the tissues. Curve *B* represents the pattern of temperature distribution, if the ultrasonic absorption coefficient in the tumor was higher than that of tissues and/or the heat diffusivity in the tumor was lower than that in tissues. (Lele 1980. Reprinted by permission.)

achieve therapeutic temperature levels. The choice of the proper wavelength (frequency), which is crucial to localization of hyperthermia, and of the aperture is thus an exercise in optimization of these conflicting requirements for size and depth of the specific tumor and the absorption/attenuation properties of the tumor and the overlying tissues. These computations, followed by calculation of heat generation in the tumor and heat loss by conductivity and blood flow, can be performed easily on a computer during therapy planning to determine the wavelength (frequency), aperture, and power requirements.

Choice of Wavelength

The wavelength also determines the size and shape of the focus and, therefore, the minimum size of the

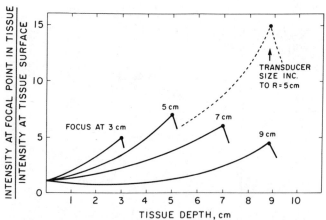

Figure 16.7. Variation of intensity gain with the depth of placement of the focus of a transducer, 3 cm in radius and 10 cm focal length, in a tissue with an absorption coefficient of 0.3 dB/cm (*solid lines*). Note the increase in gain with an increase in the aperture of the transducer (*dotted line*).

target volume that can be selectively heated. In practicable focusing systems with apical angles of 45° to 60°, the focal regions are elongated axially, the ratio of their half-power beam length to the half-power beam width being approximately 6 : 1 to 4 : 1. For production of localized hyperthermia in deep tumors, the longest wavelength that can be used is thus governed not by the lateral (or radial) dimension of the tumor, but by its axial length or thickness and should be approximately one-fifth of that dimension (fig. 16.8). Thus, for heating a spherical tumor 3 cm in diameter, for example, the wavelength probably should not be more than about 5 or 6 mm, and for a 6-cm diameter tumor, not more than about 1 cm. For EM waves, the data available in the literature (Guy, Lehmann, and Stonebridge 1974; Schwan 1980) indicate that for these wavelengths in tissues, the half-power penetration depths are only a few millimeters. It would appear, therefore, that some very ingenious approaches would be required to accomplish localized heating of tumors situated at depths greater than approximately 15 mm from the skin or the mucous membranes by EM radiation. Ultrasound, at these wavelengths, can penetrate to the deepest tumors, but needs to be coupled to the body through a column of fluid, such as degassed water.

EM energy, although restricted in its penetration depth in tissues, has an unquestionable advantage over ultrasound in that it can traverse air. It is thus potentially useful for heating tumors surrounded by air, such as those in the lung. In addition, its transmissibility through air makes EM energy easier to couple to large or irregular contours or ulcerated tumor surfaces. In such cases, EM heating would be preferable.

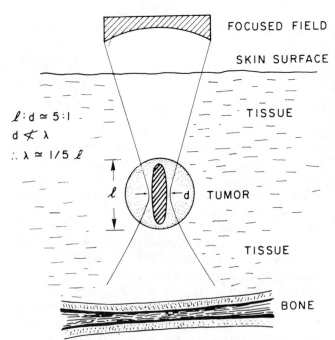

Figure 16.8. The relationship between the maximal permissible wavelength for production of localized hyperthermia and tumor dimensions. (Lele 1980. Reprinted by permission.)

On the other hand, the necessity of positive coupling in the case of ultrasound confers a measure of safety to the therapy personnel and the patient since there is no "leakage" of energy or inadvertent radiation of nontargeted tissues, as is likely to happen with EM energy. Other potential problems in local hyperthermia in different target organs by ultrasound and EM energy are discussed in a previous publication (Lele 1980). The usefulness of the two modalities in production of local hyperthermia is discussed succinctly by Hunt (in press).

Beam Steering and Intensity Modulation

Focusing concentrates the energy emanating from the transducer into a small focal region that serves as a noninvasive heat source within the tissue. The position and the intensity of the source can be controlled externally to generate the amount of heat required to produce and sustain hyperthermia in each region of the tumor. The rate of deposition of energy at any region must be adjusted to the regional heat removal capacity and needs to be higher during induction of hyperthermia than for its maintenance. Considering the time constants of temperature decay in tissues in vivo, the energy deposition must be refreshed every 3 to 10 seconds. The focus thus can be moved about in different regions of the target volume over this period. It is also necessary to modulate the intensity at the transducer to compensate for differences in attenuation in the tissues overlying different regions of the target volume or differences in their thermal needs. Obviously, the intensity modulation has to be tied to the transducer location for the specific region of the tumor.

It is not necessary to distribute the energy evenly throughout the entire target volume for induction of uniform temperature elevation. In fact, such uniform energy deposition, even in a homogeneous medium, would lead to nonhomogeneous temperature distribution, as shown in figure 16.9A. The temperatures at the surface of the target volume, through which heat is predominantly lost to the surrounding "normothermic" tissues, would necessarily be lower than those in the central regions. In order to maintain temperature uniformity throughout the target volume, more energy needs to be delivered at the periphery than in the inner regions (fig. 16.9B). As shown later, it is indeed possible to induce uniform hyperthermia in target volumes up to approximately 20 mm in diameter by deposition of heat at the periphery alone. In larger volumes, a small proportion of energy needs to be deposited in the central region. Such control over the pattern and magnitude of energy deposition is presently possible only by the use of steered, intensity-modulated, focused ultrasound (or its limited use version, the annular focus lens, described later), which was developed and evaluated extensively in animal tissues in vitro and in vivo, in transplanted murine tumors, and in spontaneous tumors in dog and human patients in this laboratory. This system, at present, is unique in its capability to deliver uniform dose distributions in tissues and tumors in which it has been evaluated. It is also unique in its capability to selectively heat preselected target volumes located at depth, noninvasively and without unacceptable or significant temperature elevation in tissues overlying or surrounding the target volume.

Distribution of intensity to achieve a homoge-

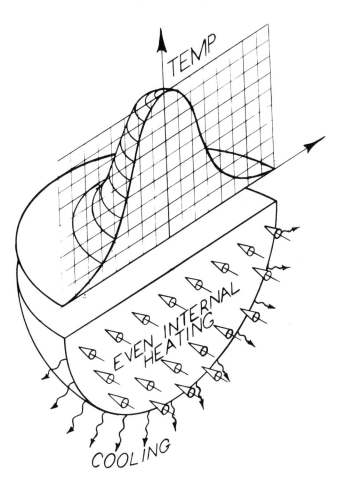
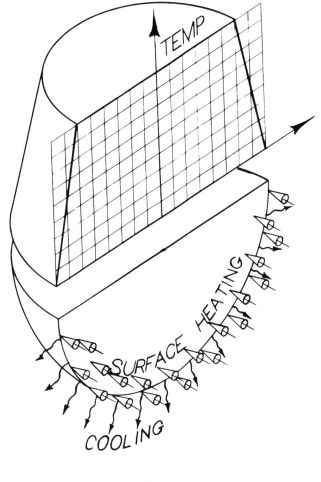

Figure 16.9. Temperature distributions in a tissue volume resulting from deposition of energy *(A)* uniformly through its volume and *(B)* preferentially on its surface.

neous distribution of temperature in the critical treatment volume is planned for each individual tumor. In addition, the heat dissipation capacity of the target tissue is measured by using an ultrasonic heat pulse technique during therapy planning. This value is then used in a modified bio-heat transfer model to calculate the pattern of heat deposition necessary to maintain isothermic temperature distributions under steady state conditions. For deep seated tumors, the thermal dose to transmission tissues or surrounding normal structures is minimized in order to achieve a satisfactory therapeutic ratio.

Energy Requirements

An approximate estimate of the energy requirements for heating a given target volume may be obtained by assuming it to have the same thermophysical properties as water and a volume larger than the target by a factor proportional to the average (or estimated) blood flow for the target tissue (Altman and Ditmer 1971; Bowman, Cravalho, and Woods 1971). For example, the power required to raise the temperature of a tumor 6 cm in diameter through 7°C in 10 minutes can be calculated as follows.

Let us assume that the heat capacity of the tumor is equal to that of water, and that blood flow adds an equivalent of 70% volume. To estimate the heat requirements of the perfused tumor, let us then consider its effective diameter to be 8 cm.

Specific heat, $C = 1$ cal/gm.°C. Since 1 cal = 4.2 watt sec, and $\rho = 1$ gm/cm^3, where ρ = density, $\rho C = 4.2$ watt. sec/cm^3.°C.

Therefore, power required,

$$Q = (\text{volume}) \cdot (\rho C) \cdot (\Delta T)/\Delta t \quad (1)$$

where, ΔT = temperature rise, °C and, Δt = time in sec over which the temperature is to be raised, or

$$Q = (4/3 \, \pi \, R^3) \cdot (\rho C) \cdot (\Delta T)/\Delta t, \quad (2)$$

where R = radius of tumor.

For a tumor with $R = 4$ cm (8-cm diameter), $\Delta T = 7$°C, $\rho C = 4.2$ watt. sec/cm^3 · °C, and $\Delta t = 10 \times 60$ sec.

$Q = 13$ watts for 10 min to raise the temperature of a well-perfused tumor 6 cm in diameter by 7°C in 10 minutes.

The acoustic power needed to generate 13 watts of heat in a tumor with an intensity absorption coefficient of 0.2 dB/cm, using a 1-MHz transducer, 6 cm in diameter, and focused at 12 cm can easily be calculated to have an average intensity of 55 W/cm^2. If the focusing lens was plano-concave, the peak intensity would be approximately 165 W/cm^2. Note that the attenuation in the path to the target volume has not been included in this calculation. This would increase the intensity needed at the transducer.

The steady state requirements to maintain the hyperthermia of 7°C can be estimated using Fourier's law of heat conduction, which over surface of a sphere simplifies to:

$$Q = 4 \cdot \pi \cdot R \cdot T_{max} \cdot K, \quad (3)$$

where Q = power in watts, R = radius of the tumor, and K = thermal conductivity of the tissue.

Assuming $K = 20 \times 10^{-3}$ W/cm°C for highly perfused tissue in vivo (Bowman, Cravalho, and Woods 1971), to sustain the 7° hyperthermia in the above tumor the steady state power requirement would be 7 watts, or about half of that needed to raise the temperature by 7°C.

Heat Transfer Modeling

Note that we have assumed the target to be highly perfused, with perfusion equal to that of normal kidney or liver, to estimate the maximum power requirements that the hyperthermia system should be able to deliver to produce hyperthermia in tumors of a clinically relevant size. Calculations based on the general bio-heat transfer equation (BHTE) (Pennes 1948; Jaeger 1952; Wulff 1974), assuming that capillary blood flow represents an isotropic heat sink, indicate that blood flow can contribute a heat sink of about 50% of that from heat conduction. This calculation has been borne out by comparison of power requirements to produce hyperthermia in tumors and several tissues in vivo with perfusion, to those required post mortem in the same tissues without perfusion.

Further analyses, modeling, and experimental verification have been conducted for prediction of the pattern of temperature rise or hyperthermia distribution resulting from different insonation patterns in tumors with different thermophysical properties in vivo and post mortem (Parker and Lele 1980). Interesting data have been obtained on the effects of blood perfusion on tissue temperature distributions during insonation. Using a circular transducer trajectory for energy deposition at the surface of a cylinder of tissue, such as the skeletal muscle or a well-perfused tumor, isothermal distribution can be obtained in tissue volumes up to approximately 20 mm in diameter. With larger trajectory diameters, or in tissues with larger blood flow (such as the brain), there is a dip in the temperature in the center (fig. 16.10) under in vivo conditions, but isothermal distributions are obtained on occlusion of the blood flow or post mortem. A relatively simple heat transfer model adequately predicts the relationship between the magnitude of the temperature dip and the rate of tissue blood perfusion. It should be emphasized, however, that although predictive modeling is presently possible, with assumption of the blood flow being an isotropic heat sink, there is an urgent and compelling need to develop models that would incorporate realistic situations, such as anisotropy, the presence of larger vessels, and vascular imperfections like arteriovenous anastomoses or shunts (Peterson 1980).

Avoidance of Collapse Cavitation

The comparatively high acoustic power requirements for induction of hyperthermia may lead to cavitation damage because of the high peak intensities if sharp focusing were used. Specially designed "smeared focus" lenses, in which peak intensities are lowered without compromising the angle of beam convergence, are therefore used. The peak intensities employed for production of hyperthermia are always well below the threshold for cavitation-induced damage in organized mammalian tissues. Extensive theoretical and experimental research on ultrasonically induced cavitation in mammalian tissues in vitro and in vivo was conducted in this laboratory (Lele, Senapati, and Hsu 1973; Lele 1977). It was found that although bubble-oscillation type of cavitation occurs even at intensity levels associated with diagnostic ultrasound, collapse cavitation does not occur in organized tissues until peak focal intensity is higher than 1450 W/cm^2 at 2- to 3-MHz frequency. Tissue damage by cavitation is characterized by the occurrence of hemorrhage within the tissue. In well over a hundred experiments in vivo in which acoustic emission was monitored during repeated insonations, and the tissues were subsequently subjected to histologic examination as serial sections, in no instance was hemorrhage detected unless collapse cavitation (indicated by wide band or anharmonic acoustic emission) was present during insonation. The threshold for cavitation, as well as for occurrence of hemorrhage, could be raised by insonation under hyperbaric conditions (42 atmospheres). Some of these data are given by Lele (1977). The threshold is lower at lower frequencies. This, along with the low level of ultrasonic absorption (and thus the heat gen-

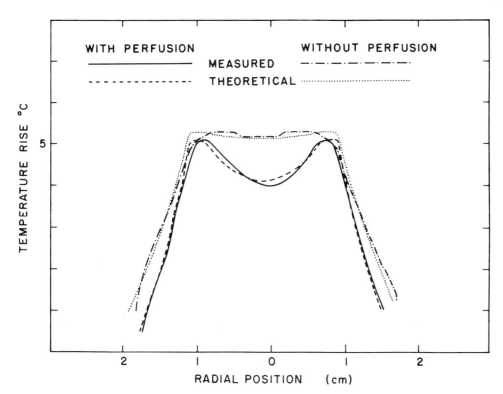

Figure 16.10. Measured and calculated steady-state temperature distributions in the brain of the cat, with and without blood perfusion, during heating by a circular insonation pattern 2 cm in diameter. (Adapted from Parker and Lele 1980.)

eration) at low frequencies, militates against the use of ultrasound below a frequency of about 500 kHz. In production of hyperthermia, the permissible peak intensity is pegged to the frequency and is always held well below the threshold for collapse cavitation. Histologic examination of the tissues of dogs subjected to repeated hyperthermia sessions at 40°C over a 19-month period during the development of the hyperthermia system failed to reveal any pathology, except for fibrosis along the tracks of the thermocouples inserted for measurement of temperature distributions during hyperthermia.

Permissible Transient Peak Temperature

The peak temperature rise in the tissue at the focus can be very high, even at subcavitation intensities and in spite of the use of smeared-focus lenses. Since the transducer is either moving in a predetermined trajectory or is pulsed for a short time, the duration of such temperature rise is brief. Previous studies (Lele 1977) have shown that heat-induced damage to biological tissues is dependent on both the temperature and its duration. The temperature threshold for damage rises as the duration of exposure is shortened. This relationship, which can be explained by the theory of reaction kinetics, is reproduced in figure 16.11. It is clear from these data that there is no likelihood of damage to normal mammalian brain, liver, or muscle tissues by temperatures up to 65°C for 0.1 seconds (the maximum duration of exposure of any site with the lowest translocation velocity used). Since no data are available for tumors, however, the permissible instantaneous peak temperature is restricted to 2°C above the desired steady-state hyperthermia temperature. The temperature excursion is therefore greatest when the target tissue is cool, and is reduced to a maximum of 2°C as the desired hyperthermia level is approached. Thus, the acoustic power delivered to the tissue is maximum at the initiation of hyperthermia and diminishes progressively as the desired hyperthermia temperature is approached.

Need for Multiple Transducers

With these restrictions on the peak insonation intensity, it often may not be possible to deliver sufficient power to induce hyperthermia in large or very vascular tumors using a single transducer translocated with a trajectory period of 3 to 10 seconds. Since heat generation at any given point in the medium is a function both of the local intensity and duration of insonation, higher local intensity can be used, if the duration of local insonation is reduced, by increasing the translocation velocity. The focal trajectory also can be made longer to deliver the energy over a larger volume within the target, based on heat transfer considerations. For larger tumors, or when the portal available is

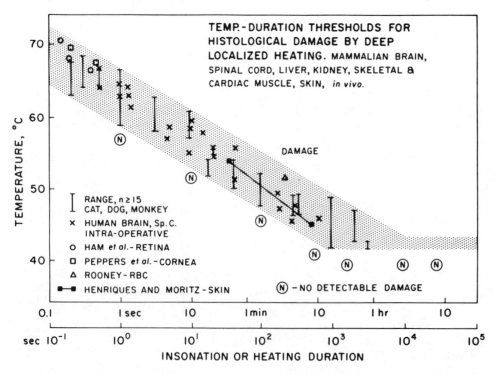

Figure 16.11. Temperature-duration thresholds for histologic damage by localized heating at depth of mammalian tissues in vivo.

sufficiently large, multiple transducers can be used to increase the total delivered power.

To prevent the possibility of local damage to tissue at the focus, the present computer-controlled hyperthermia system is programmed so that operation in a continuous wave mode is disabled unless the transducer(s) is (are) moving at a preselected velocity. Pulsed operation is possible only below preset limits of power and repetition rates.

Problems at the Soft Tissue-Bone Interface

The other area of concern is the possibility of partial reflection and refraction of the ultrasonic beam at tissue interfaces within the body, which might cause unacceptable interfacial heating and/or displacement of hyperthermia distribution. The problem of bone interface has been discussed earlier. It should, however, be added that under in vivo conditions, there is little or no specular reflection, but only diffuse backscattering owing to the roughness of the periosteal surface. Little ultrasonic energy is transmitted to the bone except when the angle of incidence is large, and mode conversion takes place. Excessive heating of the bone or attached tissues can be avoided by making sure that it lies in the diverging field beyond the focus, and the angle of the central ray or axis of ultrasound beam is no larger than about 33°. In a previous study in the cat, related to neurosurgical application of focused ultrasound, lesions were placed in the maxillary division of the trigeminal nerve, pituitary body, and mammillary body—structures that lie directly on the base of the skull—in 114 animals (Basauri and Lele 1962). The targets were ablated, but the bone underlying the lesion (and the structures surrounding the target) showed no evidence of damage. A comparable study was conducted in the spinal cord of the squirrel monkey to develop the technique of ultrasonic commissural myelotomy for relief of intractable pain. No damage to the vertebral bodies underlying the spinal cord was detected in any of the 38 monkeys studied (Lele 1967). For coupling energy into the bone, such as for hyperthermia of osteogenic sarcoma, the focal region needs to be placed directly on the bone, and the angle of incidence needs to be larger than about 45°. Similar consideration applies to interfaces with air-filled tissues, such as the lung or the intestines.

Reflection, Refraction, and Scattering in Soft Tissues

Reflectivity is proportional to the degree of mismatch of characteristic acoustic impedances of contiguous tissues. For soft tissues, these values do not vary greatly. The maximum reflectivity at soft tissue interfaces, such as those between muscle, kidney, spleen, or liver, is less than 38 dB below that of a perfect reflector (Wells 1969). This means that the maximum reflected energy (under normal incidence and large, flat interfaces) can be approximately 10^{-4} of the incident power, and under normal conditions in vivo, it would be even less. The reflectivity of the fatty tissue-muscle interface, such as in postmenopausal breast, is somewhat higher and is expressed in its better echographic visibility. There is little heating in fat because of its low absorption coefficient.

Refraction of acoustic wave, like that of EM waves, is related to the ratio in the velocities of its propagation in the two contiguous media, which for ultrasound depend on their elastic moduli. The velocities in the soft tissues of interest vary within the range of 1581 meters/sec for muscle to 1549, 1561, and 1566 meters/sec for liver, kidney, and spleen, respectively (Wells 1969), yielding an average refractive index of 1.014 for the muscle/organ interface. This may shift the focus, at deep targets, axially by 1%, which is negligible compared to cardiorespiratory motions of the body tissues. The absence of any significant refraction of acoustic waves within the body is borne out also by the close depiction by cross-sectional ultrasonograms of the cross-sectional anatomy, except in the eye, in which the relatively high propagation velocity in the lens occuli may introduce distortions.

The magnitude of ultrasonic energy scattered from soft tissue interfaces is even lower—by orders of magnitude—than that reflected, particularly at wavelengths useful for hyperthermia (Lele and Senapati 1977; Wells 1977). As stated previously, in a plane progressive ultrasonic wave, attenuation in soft tissues in vivo is only slightly greater than absorption (Lele and Senapati 1977; Wells 1977). This implies that the combined magnitude of beam refraction, diffraction, and scatter to distant regions is negligible. It is thus obvious that internal reflections, refraction, and scattering are not likely to impose significant restrictions on the predictability and reproducibility of the distribution of hyperthermia by ultrasound in intact animals in vivo. No aberrant or unpredicted temperature distributions were found in any of the several hundred hyperthermia experiments using focused ultrasound in which the temperatures were measured at 400 to 800 points in the regions of interest. Some of the data are presented later. Furthermore, the procedure for induction of hyperthermia by focused ultrasound currently requires placement in the tumor of a fine calibrated thermocouple and the localization of the thermojunction by using low-intensity pulsed ultrasound and a computer-controlled scanning procedure. The location of the thermojunction is also checked by ultrasound scans (or by A-mode) and/or radiographically. This procedure would compensate automatically for any refraction. Interfacial heating has been observed only in experiments involving dissection and separation of tissue planes for placement of thermocouples leading to entrapment of air between different fascial planes.

Applicability to Intraabdominal Targets

Lastly, because of presence of gas in the intestines, some concern exists regarding the applicability of ultrasonic hyperthermia to intraabdominal targets. It would be reasonable to assume that well-defined hyperthermia could be produced in any solid mass that can be well visualized by ultrasonic scanning; this is, therefore, preferable to radiographic visualization for this application.

SYSTEM INSTRUMENTATION AND TECHNIQUE

From the above discussion it is clear that the use of a focused beam of radiant energy is essential for production of hyperthermia localized to a deep target volume. The distribution of energy within the target volume needs to be tailored to the specific heat transfer characteristics of the particular target in order to obtain uniformity of temperature elevation throughout the target volume. This entails beam steering and intensity modulation. Since tumors are generally of irregular shapes and at varying depths, both the beam steering and intensity modulation can best be achieved with the use of a computer. The computer can be used also for optimization of the ultrasonic wavelength (frequency) and aperture of the focused beam for the depth of the target and for the determination of the focal trajectory and velocity needed for uniformity of temperature elevation throughout the target volume.

Based on these considerations, the steered, intensity modulated, focused ultrasound hyperthermia system consists of the following subsystems:

1. Insonation.
2. Temperature measurement.
3. Translation and rotation.
4. Computer for insonation beam steering, intensity modulation, and control. It is also used for gathering and displaying temperature information and physiologic parameters and for therapy planning.

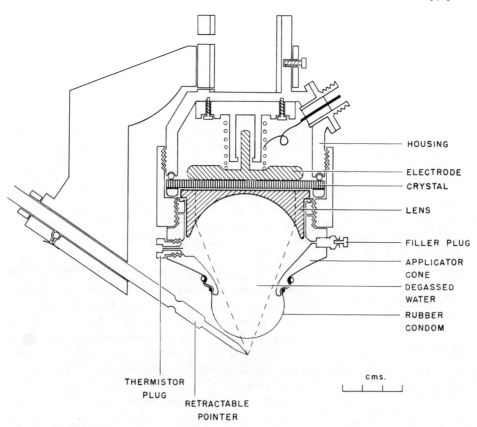

Figure 16.12. Schematic of an insonation head. (Lele 1962a. Reprinted by permission.)

Insonation

Insonation Head

The insonation head holds the transducer, focusing lens, and applicator cone. Its design permits interchangeability of these components (of the same diameter), conferring flexibility in choice of the frequency of ultrasound, plane wave or focused field generation, and choice of focal length (or aperture of the system), with an appropriate applicator cone for coupling. The design specifications, which are unchanged since 1959, are shown in figure 16.12 and detailed by Lele (1962a). Three different sizes of insonation heads enable use of transducers 8, 12, and 16 cm in diameter, yielding power ratios of approximately 1:2:4 and permitting optimization of focal length and aperture to meet the insonation field requirements of tumors at different depths. Adjustable, retractable, and detachable pointers indicate the location of the focus to aid in positioning the insonation head and to delineate the treatment portal. Optical projectors to simulate the cone of ultrasound are under development.

Transducers

Cross-cut quartz crystals 8, 12, and 16 cm in diameter are used in spite of their higher initial cost as compared to ceramics, because of their stability at the high output power levels and the prolonged periods of continuous wave operation required in hyperthermia studies. Most of the 8-cm-diameter transducers have been in use since 1959 and cover the range of 0.6 to 6.0 MHz by operation at the fundamental and harmonic frequencies. Larger-diameter transducers are used for deep tumors and therefore operate only at the lower frequencies, 0.6 and 0.9 MHz.

Focusing Lenses

These are made from polystyrene or Rexolite™ for impedance matching and low attenuation loss properties. For the three transducer sizes, the focal lengths vary from 6 cm to 30 cm, yielding solid angles of 45° to 22½°. Normal lenses have a smeared spot focus; line-focus lenses are used for insonation through the intercostal spaces for subpleural lesions; annular focus lenses, with or without a central "leak", are used in stationary insonation systems. Sharply focused, multi-focus, and Fresnel lenses, as well as zone plates, are used for special applications. Computer programs for lens design have been developed for computing lens curvatures for specified focal characteristics and lens materials, and a lens machining program has been developed to generate machinist tables for lathe operation for fabrication of lenses. For isonation of deep targets through restricted access, such as for hyperthermia of the posterior pharynx or tonsil through the mouth, combinations of exponentially tapered horns and ultrasonic wave guides (fig. 16.13) were developed. A convergent beam of ultrasound is projected at the exit aperture of the waveguide and has been successfully used in dog patients. An example of the intensity distribution pattern is shown in figure 16.14. Further development of these is necessary, however.

Applicator Cones

As shown in figure 16.12, applicator cones hold degassed water for acoustical coupling of the lens to the body portal. Different configuration is needed for each diameter and focal length for unimpeded transmission of ultrasonic energy while minimizing size. Temperature of the degassed water in the cone can be held at 37°C by a thermistor and heating element incorporated in the cone walls. Provision is made for circulation of chilled degassed water.

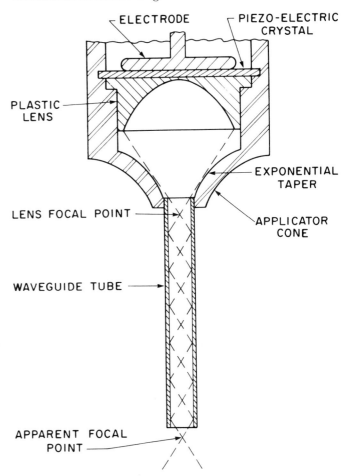

Figure 16.13. Exponentially tapered horn and waveguide for insonation of deep targets with limited access.

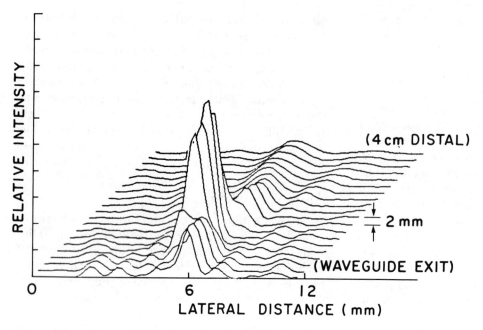

Figure 16.14. Intensity distribution at the exit aperture of the waveguide shown in figure 16.13.

Radiation Pressure Gauge

The calibration of the acoustical output and the insonation system checkout are performed by radiation pressure measurements under actual conditions of use, at three or more power levels (Lele 1962a).

Degassing

Water or saline solutions used for acoustical coupling are degassed to prevent enucleation of gaseous bubbles during insonation. Either boiling or mechanical deaeration under vacuum are adequate.

Several other items, such as sensitized microthermocouples for determination of field patterns, piezoelectric microprobes, suspended ball radiation pressure detector, Schlieren visualization, plastic bar lesion visualization (Lele 1962b), spectrum analyzers, and so forth, are used in development of the hyperthermia system but are not necessary for routine use.

Temperature Measurement

Fine gauge chromel-constantan thermocouples are used with an ice reference-bath in a conventional manner. Current-limited input cables (less than 3 microamp per lead) are used to connect these to the strip chart recorder for electrical safety of the patient (fig. 16.15).

Transducer-driving electronics and the configuration in the hyperthermia system is shown schematically in figure 16.15. Continuous monitoring of the forward and reverse power separately by the computer enables cessation of insonation if the latter increases, since it denotes loss of coupling in the acoustic system by inclusion of air or other similar events.

Translation and Rotation Subsystem

This subsystem is used for precision positioning of the insonator and execution of the trajectory as determined in the computer in therapy planning and is mounted on a heavy, movable base with adjustable height (fig. 16.16). The major components are shown schematically in figure 16.17. The X-Y-Z translator tables move in orthogonal axes. The rotatable support, which carries the insonator (one or more insonation heads), can be manually adjusted up to 45° in two planes to permit proper positioning for coupling to tumors in different locations. Fine control required for correct positioning and dynamic control necessary for treatment of many of the tumors, however, are presently lacking in these two motions. The insonation head generally is mounted at an angle pointing inward to the center of the focal trajectory to ensure that the energy is distributed over as large an area as possible in the tissues above the target volume. The rotation of the insonation head, therefore, is coupled to the X-Y-Z translators through software.

Two stepper motors are used to retract thermocouple probes (or arrays of thermocouple probes) in precision increments of 0.5 mm or more through the region of interest, allowing sufficient time after each step for thermal equilibration of the probe. The temperature measurement at each point is digitized after

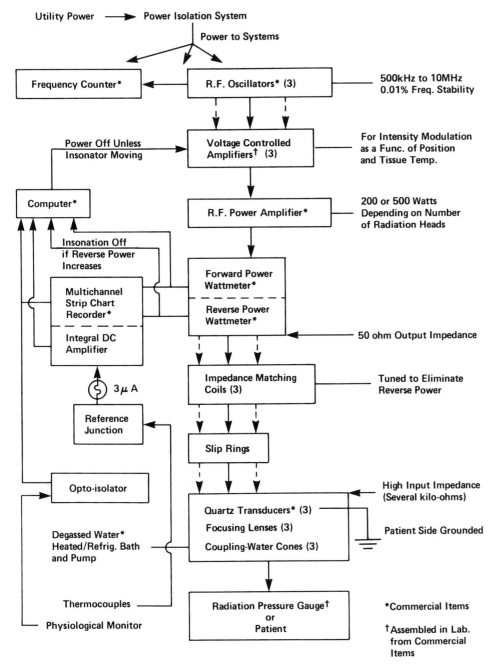

Figure 16.15. Schematic of the ultrasonic and temperature measurement subsystems.

suitable amplification and stored in the computer (as well as recorded on the strip-chart recorder), together with positional information. The computer also sends out a pulse to the event marker of the strip-chart recorder after every step. Profiles of temperature distribution can then be plotted from the stored or recorded information.

Computer System

The computer hardware is shown schematically in figure 16.18. It should be pointed out that the write bits, through the bits interface, can control up to a maximum of six stepping motors, whether for driving axes of the positioning table or for sampling temperature distributions.

Software and System Functions

The PDP 11/04 minicomputer operates under a DEC RT-11 operating system with an ESI PASCAL compiler and support package. A special software system,

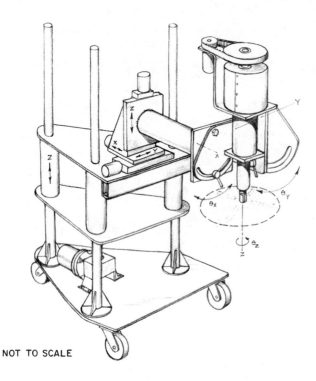

Figure 16.16. Sketch of the mechanical and electrical components of the translation and rotation subsystem.

built in-house and known as "the interpreter," is used to run all experiments and patient therapy. The interpreter treats all functions, such as transducer movement or temperature measurement, as subroutines and prompts the therapist for commands or information in a conversational style. Other features of the interpreter include task priority assignment, scheduling and handling, real time interrupt handling capabilities, high and low level language compatibility, and on line experiment modification capabilities. Complex procedures are executed by grouping functions, or subroutines, under a name known as a *macro*. When, in response to a prompt from the interpreter, the user types in a valid macro name, all the subroutines defined by that macro will be executed.

As an example, the macro name, SCAN, defines a group of subroutines that build up a profile of an ultrasonic intensity distribution. Individual routines ask the user questions concerning scaling and direction of the scan and pass these parameters on to other functional blocks that pulse the ultrasonic transducer, measure the output of a sensing probe, store the result, and move the probe to a new location where the process is repeated. An example of these intensity scans, plotted by a system graphics routine, is shown in figure 16.14. By regrouping these individual subroutines under other names, new procedures can be defined. The interpreter allows for quick, on-line defining or regrouping of these macros, or procedures. Rather than produce an exhaustive list of subroutines or functions, an example of a therapy session will be given, along with descriptions of relevant features.

Technique

As in radiation therapy, a hyperthermia therapy session begins with treatment planning, in which a thermal-acoustic model of the tumor is used to guide the choice of ultrasonic frequency, beam shape, and trajectory. Specifically, the planning program asks for information on the tumor size and dimensions, the expected absorption coefficient and thermal diffusivity, the presence and location of any possible ultrasonic reflectors such as bone or air cavities, and the desired temperature increase. The program responds by calculating the optimum frequency, focal length, intensity distribution along the Z axis, expected power levels, and time required to arrive at steady state temperature elevations. The program then asks for any corrections and will reoptimize the parameters using the new information.

The actual treatment session begins with the positioning of the insonator over the target volume. The coordinates, in the case of superficial tumors, are determined visually. In the case of deep tumors, they are read from ultrasound scans, CAT scans, or x-rays, with reference to fixed accessible reference points. ECG leads are hooked up, and a hypodermic needle thermometer probe is inserted into the tumor. The probe contains two microthermocouples, one bare, and the other coated with a material with a known acoustical absorption coefficient. The placement of the needle probe can be checked by A-mode or B-scan data. A real-time clock routine (at highest software priority) is then started. This routine, activated every five milliseconds, monitors heart rate data each period and samples temperatures, which are recorded continuously on the strip-chart recorder every fortieth period, or five times per second. These data are stored for later analysis and display. Next, the macro FIND will move the insonation head that will be used in treatment in a search pattern around the tumor region while pulsing it at very low power levels. When the focal region is positioned near a thermocouple, a sharp temperature rise (on the order of 1°C) will be recorded. The routine will then perform a fine step search to locate the intensity maxima in X, Y, and Z coordinates. Thus, the transducer coordinate system can be aligned with the absolute location of an inserted thermocouple. Once aligned, a single short burst of ultrasound at a precalibrated intensity creates a tem-

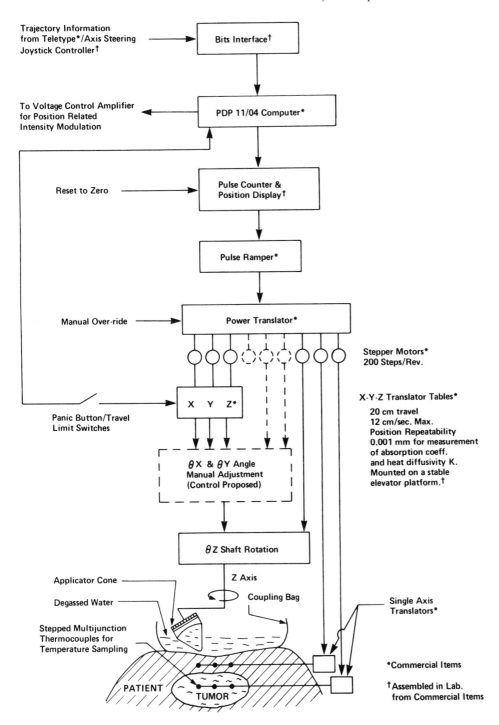

Figure 16.17. Schematic of the translation and rotation subsystem. Two stepper motors are used for retraction of the thermocouple probe(s) in steps of 0.5 mm or more for measurement and mapping of temperature distributions through the target and surrounding tissue volumes.

perature rise and fall, which is recorded. The second thermocouple is similarly found, the ultrasonic burst repeated, and the temperature rise and fall is again recorded. These two sets of data yield information on the total attenuation in the path to the target, and the absorption coefficient and thermal diffusivity of the tumor. This, however, takes time, especially since the computer in use is a relatively slow machine. A quick estimate of the total heat diffusivity in the tumor is obtained by measuring the interburst interval between successive ultrasonic bursts necessary to maintain the local temperature at a steady elevated tem-

Padmakar P. Lele

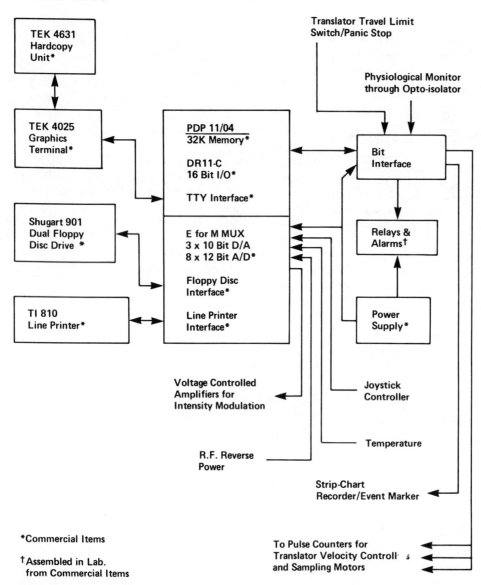

Figure 16.18. Schematic of the computer hardware of the hyperthermia system.

perature (about 1°C above baseline). This information is used to determine the minimum velocity of the focal trajectory. The desired transducer movement pattern is then generated. The entire trajectory, with power levels specified as a function of position, can be produced by software alone or by moving the transducer in a test pattern using a joystick control. The software pattern generation routines allow the therapist to specify a variety of geometric shapes based on tissue heat transfer considerations, which have been found to be useful in heating tissue. These include concentric circles and ellipses, and arbitrary straight line segment shapes. The joystick control allows the therapist to drive the transducer into desired locations, and build up a pattern that can be repeated by the computer. The patterns are broken up into discrete steps, which are stored in motion tables. During treatment, the real-time clock routine reads the motion table every five milliseconds and sends out the appropriate commands to each axis. A power level table is constructed with the motion table; thus, for each position a corresponding RF power output to the transducer can be specified. As the treatment begins, the baseline RF power levels can be adjusted manually or by keyboard commands to the interpreter. Once the desired temperature elevation has been reached, a feedback routine can make adjustments to the RF power to maintain temperatures within 0.1°C of the preset value. During treatments, alarms are sounded and action taken for a variety of error conditions, including the detection of out-of-bounds values of temperature, RF power levels, or transducer position. Positioning adjustments and changes in trajectory and power levels can be executed on line through the interpreter. Sampling motors used with micropositioners (single axis translators) are used to retract the thermocouples through the tissue, thus building up a spatial profile of temperatures. After the desired time-temperature dose has been reached, the ultrasonic insonation and movement are stopped. The thermocouples are left momentarily in place to monitor the cooling rate and to record optional posttreatment thermal pulse measurements. Finally, the treatment records are placed on disk for permanent storage, and are used for display and analysis. Figure 16.19 shows a time-temperature histogram from a therapy session on a dog patient. Note the relatively short time spent at subtherapeutic temperatures during the entire session, including the induction of hyperthermia and after its termination.

Safety Features

A number of features built into the system hardware and software serve to protect the patient from overexposure to ultrasonic energy, or from any electrical leakage from the RF or AC power circuits. These are summarized below for reference. Refer to figures 16.15, 16.17, and 16.18.

RF Overpower

The reverse and forward power levels are continuously monitored through the computer A/D circuits. The reverse power level is a sensitive indicator of loss of transducer coupling or change in tuning and is, therefore, an essential parameter to monitor. The software checks for absolute levels and sudden changes in reverse and forward power levels. If these occur, the RF power is reduced to zero through the voltage-controlled amplifier (VCA). To guard against the possibility of failure of both the RF system and the computer, a relay circuit cuts the power to the RF amplifier if the reverse RF power increases beyond a preset-limit for five milliseconds or more.

Motion Stop-RF Stop

Software will zero the RF input to the power amplifier within five milliseconds if the heating pattern movement is interrupted by the panic switch, by limit-of-motion detectors, or by keyboard commands. Power output cannot be initiated at any time without motion, or without special software override commands that allow pulsing or low level output from a stationary transducer.

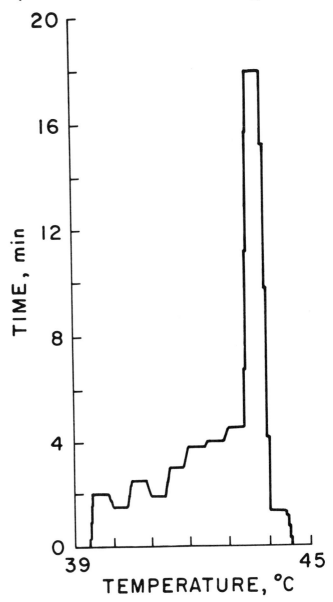

Figure 16.19. Time-temperature histogram of a hyperthermia therapy session in a dog patient.

Out-of-Bounds Temperature

Temperatures are monitored through the strip-chart recorder amplifier continuously and by computer A/D circuits. In addition to the continuous records on the strip chart recorder, absolute temperatures are calculated five times per second and are stored for later presentation and analysis. (Higher sampling rates will require a faster processor, since the associated calculations must compete against the other tasks required of the computer, such as driving the stepper motors. The slow 11/04 processor presently employed is a limiting factor on the number of tasks that can be carried out simultaneously.) An audible alarm alerts the therapist to out-of-bounds values.

ECG Monitoring

The arrival of a QRS complex, as detected by a physiologic monitor, is checked for every five milliseconds. (This can be updated more frequently than the temperature tasks since the former involves simply checking a single bit.) The interarrival times are converted to heart rate, and an alarm is activated if the rate is out of preset bounds.

Electrical Faults

Patients are protected from electrical failure of any device connected to them, such as ECG and thermocouples, by optical isolation, microcurrent limiting cables, and isolation of power supplies to equipment connected to the patient, as appropriate.

MEASURED TEMPERATURE DISTRIBUTIONS PRODUCED BY STATIONARY OR STEERED BEAMS OF UNFOCUSED OR FOCUSED ULTRASOUND IN EXPERIMENTAL STUDIES IN VITRO AND IN VIVO

As discussed previously (Beam Steering and Intensity Modulation), deposition of energy in patterns calculated from the measured heat diffusivity characteristics of the target tissue is necessary for generation of uniform temperature distributions in the treatment field. For optimization of these procedures, detailed, point-to-point measurement of temperature distributions in a variety of biological tissues and tumors, especially in vivo, under different conditions of insonation, was conducted. Not only was this information necessary for the development, evaluation, and optimization of the hyperthermia system, but also a detailed knowledge of actual temperature distributions in tissues of different acoustical characteristics, different patterns and rates of blood perfusion (Peterson 1980), and of different vascular responsivities to heat was necessary for purposes of developing a predictive heat generation/heat transfer model. Perhaps more importantly, it was realized that detailed measurement of temperature distributions would be impossible under clinical conditions in most of the patients—both human and canine—because of the necessity of using invasive methods to measure local temperatures with meaningful spatial resolution. In most patients, this would limit temperature measurement to a few points along one or two tracks through the tumor. Therefore, extensive studies were continued in transplanted tumors in animals (and in spontaneous tumors in dogs before euthanasia) to establish the range of variability in the distribution of hyperthermia and predictability of the hyperthermia distribution from measurements made along one or two tracks. Results in the tumors studied indicate that the dimensions of hyperthermia volume can be controlled rather precisely, and the temperature distributions in this volume are uniform. The uniformity of temperature distribution within the target zone and the ability to control the spatial extent of the hyperthermia field make it possible to undertake quantitative evaluations of graded heat doses on tumors in vivo and to attempt to correlate them with cellular effects on the constituent cells.

Temperature distributions resulting from insonation with stationary and steered beam(s) of unfocused or focused ultrasound were measured in tissue-equivalent phantom, beef muscle in vitro, in dog muscle mass, and in several types of transplanted murine tumors in vivo. In some experiments in vivo, the live, anesthetized rat bearing a tumor 10 x 10 x 10 cm was placed inside the abdomen of an anesthetized dog to simulate a deep-seated tumor. Arrays of four to six thermocouples, stepped through the volume of interest under computer control, were used to measure the steady-state temperatures at 600 to 800 locations in both in vitro and in vivo experiments. Confirmation of results was sought in spontaneous tumors in dog patients in vivo, using fewer multithermocouple probes. Salient features are summarized below.

Thermometry

Electrically insulated Chromel-constantan thermocouples, 25 to 125 microns in diameter, unsheathed or sheathed in 25- to 23-gauge insulated hypodermic tubing (for insertion into muscle and tumors in dogs in vivo), were used for temperature measurement. Because of their small size compared to the ultrasonic wavelength in the tissues (2.5 to 0.5 mm), they were essentially nonperturbing. Time constants (time required to reach 63.2% of the value of a new temperature, after a step-change) were negligible for steady-state temperature measurements. Heat conduction errors also were insignificant for unsheathed thermocouples; for thermocouples sheathed in insulated stainless steel hypodermic tubing, they were significant (10% to 20%) for temperature gradients of 10° per mm or more. The magnitude of the error diminished considerably (to approximately 5%) with smaller temperature gradients. Explicit correction algorithms (Singh and Dybbs 1975) were applied to data obtained with sheathed thermocouples in regions of steep temperature gradients. Thermocouple calibration was periodically checked at three or more temperatures

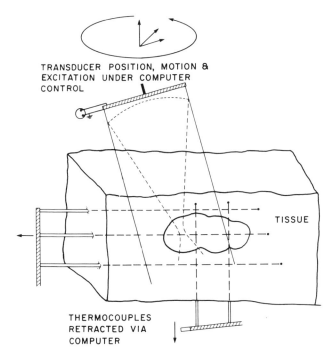

Figure 16.20. Experimental set up for generation of hyperthermia by plane or focused ultrasonic fields and measurement of tissue temperature distributions by thermocouple arrays stepped under computer control. (Lele and Parker 1982. Reprinted by permission.)

against a secondary reference thermometer, the calibration of which was traceable to the National Bureau of Standards.

The unsheathed thermocouples were threaded through the specimen, and both their ends were secured to a yoke. The sheathed thermocouples, calibrated in millimeters, also were inserted all the way through the specimen to ascertain their position, and then retracted into the tissue and secured to a yoke. For measurement of temperature distributions in the specimen, four to six thermocouples were placed in two or more planes through the volume of interest. All the thermocouples in a given plane were secured to a common yoke (fig. 16.20). After induction of hyperthermia of approximately 6°C, when the temperature had stabilized, the thermocouples were retracted in 0.5- or 1.0-mm steps using the computer, stepping motors, and translators. A five-second interval was allowed for thermal equilibration at each new location before taking measurements. Usually, one additional thermocouple was placed in the center of the region and held stationary to monitor any changes in the local temperature over the duration of the experiment.

For every thermocouple, the temperature and the location of the thermojunction at each step were recorded on a multichannel strip-chart recorder and simultaneously digitized and stored in the computer.

These data, from 600 to 800 locations within the sample, were plotted on graph paper, and points at the same temperature were connected manually by a line. The results presented are thus actual temperature distributions measured in the medium and *not* computer simulations, which they resemble (specially when steered, focused ultrasound was used for hyperthermia).

Results

The results of these efforts have been presented in detail elsewhere (Lele and Parker 1982), and are summarized below.

Phantoms

A phantom that is truly tissue equivalent in acoustic absorption/attenuation (including frequency dependence) and heat-diffusion characteristics is not yet available. The use of simple phantoms (Astrahan 1979a, 1979b) yields misleading results. In view of the complexity and flexibility of the heat diffusion characteristics that would be required in a phantom for its use as a predictive model, development of a computer phantom with heat generation and heat transfer models is presently considered to be more appropriate than that of a physical phantom.

Plane Wave Ultrasound

In normal tissues (fig. 16.21), as well as in tumors (fig. 16.22), insonation with plane wave ultrasound, 0.6 to 3.0 MHz in frequency, results in spatially nonuniform hyperthermia characterized by the existence of a small, almost punctate, region of maximum temperature rise, the depth of which appears to depend on the absorption coefficient and heat transfer properties of the medium and cannot be altered by spatial manipulation of the plane wave ultrasonic source. The existence of an ultrasound reflecting structure below the target may lead to the generation of a second region of hyperthermia at depth. The region of maximum temperature elevation is eccentric relative to the transducer and, owing to the smallness of its size, it cannot easily be located except by thorough scanning of the region in all three orthogonal planes. It is thus likely to be missed by temperature measuring procedures practicable under clinical conditions. This may explain the occurrence of burns observed by various investigators in animal (Marmor et al. 1978) and human (Marmor et al. 1979) tumor treatments using plane wave ultrasound. Furthermore, the resultant nonuniformity of temperature distributions within the target volume (fig. 16.21B) renders impossible any precise correla-

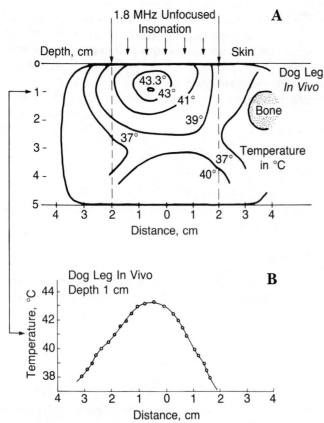

Figure 16.21. Steady state temperature distribution in the gluteal muscle mass of dog in vivo from plane wave 1.8-MHz ultrasound. *A*, isotherms plotted from temperatures measured at 1-mm spacing along 9 tracks; *B*, the temperature distribution along the track at 1-cm depth. (Lele and Parker 1982. Reprinted by permission.)

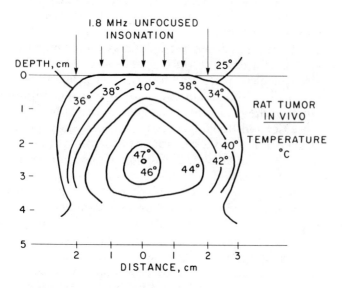

Figure 16.22. Isotherm plot of hyperthermia in rat tumor in vivo using unfocused 1.8-MHz ultrasound. Note presence of small "hot spot" at 3-cm depth. (Lele and Parker 1982. Reprinted by permission.)

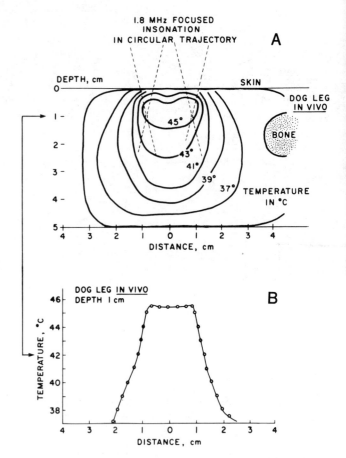

Figure 16.23. Steady-state temperature distribution in the gluteal muscle mass of dog in vivo from steered, focused 1.8-MHz ultrasound. *A*, isotherms plotted from temperatures measured at 1-mm spacing along 9 tracks; *B*, the temperature distribution along the track at 1-cm depth. Note the uniformity of temperature distribution within and the sharp fall-off of temperature outside the hyperthermia field. In contrast to hyperthermia by plane wave ultrasound in the same preparation, shown in figure 16.21, note the absence of any reflective heating at the depth of 3.5 cm. (Lele and Parker 1982. Reprinted by permission.)

tion of the temperature and duration of hyperthermia (either alone or in combination with radiation or chemotherapy), with resultant effects on tumors. Therefore, plane wave ultrasound is not the best treatment choice, even for therapy of superficial lesions.

Steered, Focused Ultrasound

All the above problems and difficulties are obviated by use of steered, focused ultrasound, which enables precise tailoring of the heat dose to individual tumors. Spatially uniform levels of hyperthermia, restricted to the target volume, located superficially (fig. 16.23) or at depth (figs. 16.24, 16.25, 16.26), can be achieved with equipment presently available.

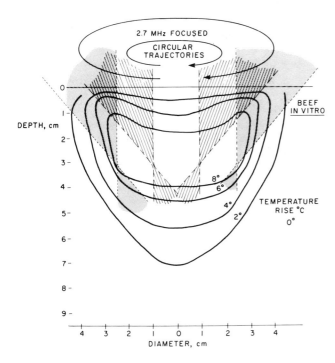

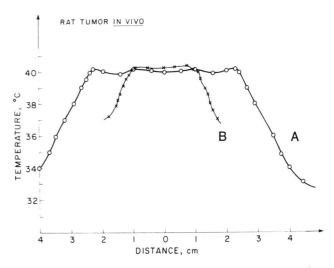

Figure 16.24. Temperature distribution in beef muscle mass in vitro using beams focused at 3-cm depth in circular trajectories. The vertical limits of the 8° isotherm can be raised or lowered by changing frequency (thus changing the attenuation rate) or, to a lesser extent, by changing the angle of incidence of the focused beams. (Lele and Parker 1982. Reprinted by permission.)

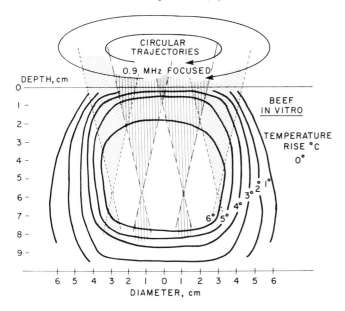

Figure 16.25. Temperature distribution in beef muscle mass in vitro using beams focused at 6-cm depth in circular trajectories. Note broad region of uniform hyperthermia to a depth of 7.5 cm. Note the use of a lower frequency than that for hyperthermia at depth of 1 to 4 cm shown in figure 16.24. (Lele and Parker 1982. Reprinted by permission.)

Figure 16.26. Temperature distributions in a large rat tumor, in vivo, using 2.7-MHz beams focused at 3-cm tissue depth, with circular trajectories. Data taken at 3-cm depth. A, 4.5-cm-diameter circle of insonation (with fill-in at 2.0-cm diameter); B, 2.0-cm diameter circle of insonation. Note that the temperature across the tumor is 40.1°C ± 0.1°C and the rapid spatial temperature decay away from heated region in both cases. (Lele and Parker 1982. Reprinted by permission.)

The ability of the system to produce a discrete volume of uniform hyperthermia in deep tumors in vivo was evaluated in anesthetized rats bearing large nonnecrotic transplanted tumors (such as shown in fig. 16.27) placed deep in the peritoneal cavity of anesthetized dogs (fig. 16.28). Thus, the circulation in both the tumor and the overlying tissues was intact. The results obtained in one such experiment are shown in figure 16.29. Note that while the intratumoral temperature was raised by 11°C, the temperature rise in the overlying tissues of the dog was only 2°C. The irregularity in the temperature distribution across the tumor was due to intermittent loss of contact between the tumor surface and the overlying tissues of the dog, with respiratory excursions from displacement of coupling gel. In experiments of short duration this can be avoided by flooding the peritoneal cavity with a normal saline solution, which, however, often leads to the development of pulmonary edema.

Focusing, beam steering, and intensity modulation are necessary to achieve uniformity of control over resultant temperature distributions. Cross-fired beams superimposed at the target volume (fig. 16.30) or multiple stationary transducers focused at adjacent points within the target tissue (fig. 16.31) yield results (fig. 16.32) that are similar to those with plane wave ultrasound.

The capability of being able to heat a discrete

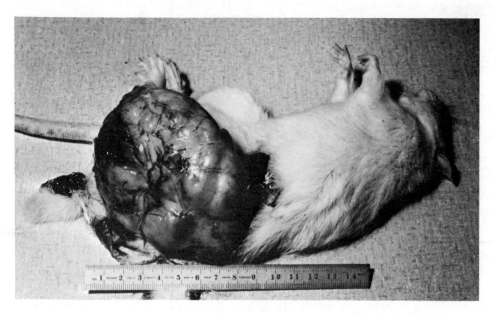

Figure 16.27. Transplanted renal adenocarcinoma grown to a large size in the flank of a rat. The tumor for transplant was kindly supplied by Dr. Ralph deVere White.

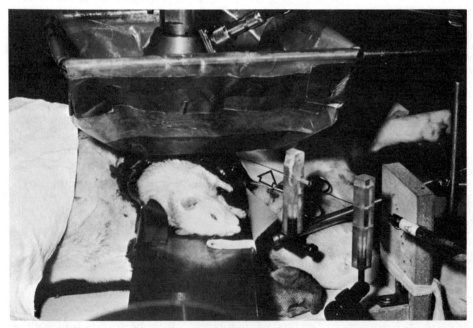

Figure 16.28. An anesthetized tumor-bearing rat, with a tumor similar to that shown in figure 16.27, placed in the peritoneal cavity of an anesthetized dog, as a model for deep tumors for hyperthermia temperature distribution studies in vivo. The *arrow* points to the intratumoral, stepped thermocouple probe.

tumor volume at depth to a uniform hyperthermia temperature has the advantages of

1. Sparing of overlying and adjacent normal tissues.
2. Uniform treatment field dosage, which is essential for establishing temperature/duration relationships with any accuracy, whether for tumor response or underlying biological mechanisms, for hyperthermia alone or with radiation or chemotherapy.
3. Ease of temperature measurement, since temperature at one or a few locations is representative of temperature distribution throughout the treatment volume, the diameter of which can be controlled easily without affecting the spatial fall-off of temperature in the surrounding tissue (fig. 16.26). Simple, inexpensive thermocouples are adequate for temperature measurement in ultrasonic fields.

Physical Aspects and Clinical Studies with Ultrasonic Hyperthermia

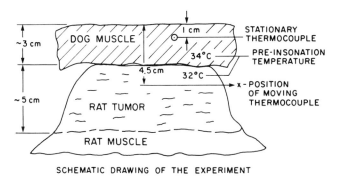

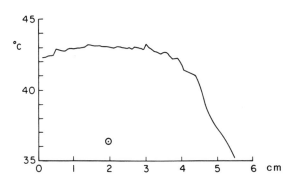

Figure 16.29. Temperature distributions in a simulated, deep tumor, in vivo, shown in figure 16.28. See text. (Lele and Parker 1982. Reprinted by permission.)

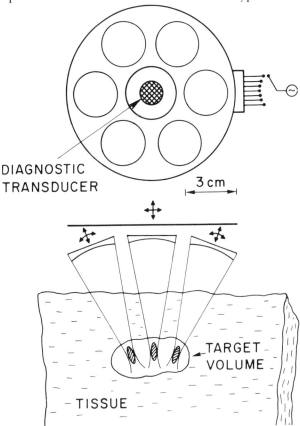

Figure 16.31. Multiple, stationary, targeted, focused, or unfocused transducer system. The orientation of each transducer is adjustable.

PRELIMINARY STUDIES ON TUMOR TEMPERATURE DISTRIBUTIONS AND RESPONSE IN VARIOUS TUMORS

Transplanted Murine Tumors

Tumor-bearing rodents were used in large numbers for comparison of different ultrasonic methods for production of hyperthermia and for optimization of a steered, intensity-modulated, focused ultrasound system. Although tumor regression was observed in almost all of the animals exposed to plane wave ultrasound, it was not possible to correlate the magnitude of tumor regressions with the temperature and duration (heat-dose), since the stepped thermocouple technique revealed that different parts of the tumor were subjected to different levels of hyperthermia. It was observed also that many of the tumors thus treated appeared to develop the shape of a crater (occasionally with central ulceration), suggesting that tumor regression in the central area was accompanied by continued tumor growth at the periphery. In contrast, with

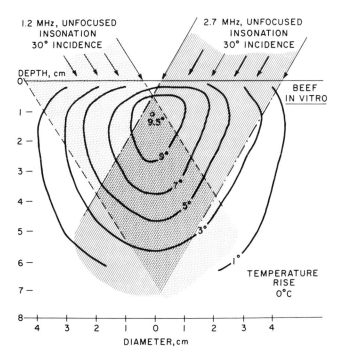

Figure 16.30. Temperature distribution resulting from two cross-fired plane wave ultrasound beams, superimposed at the target volume. (Lele and Parker 1982. Reprinted by permission.)

357

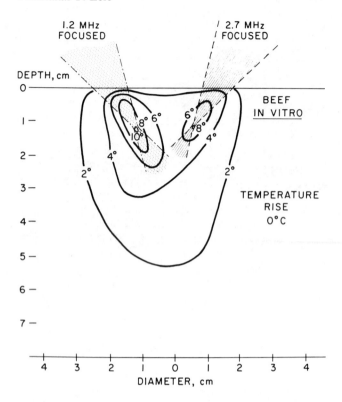

Figure 16.32. Temperature distribution from stationary, targeted, transducers shown in figure 16.31. (Lele and Parker 1982. Reprinted by permission.)

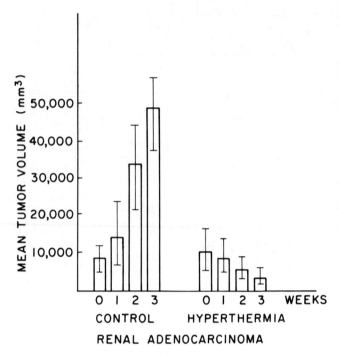

Figure 16.33. Effect of local hyperthermia at 42.5°C for 20 minutes, once a week, on the growth of renal adenocarcinoma in the rat. Studies were started four weeks after implantation of tumor. (Preliminary data from a study by Drs. Babayan, Lele, and Krane.)

steered, focused ultrasound, tumor regression was found to occur with two or three sessions at 43°C and appeared to be uniform across the entire treated area. Data from all these animals, used primarily for studies on distribution of hyperthermia, have not been used for deducing quantitative dose-effect relationships because of the large numbers of thermocouple probes that had repeatedly been inserted and retracted through the tumor. Results of dose-effect studies conducted after standardization of the technique and use of one or two multithermocouple probes are shown in figures 16.33 and 16.34.

Spontaneous Tumors in Dog Patients

Thirteen dogs with inoperable or recurrent tumors were treated using steered, focused ultrasound (Lele et al. 1980a). Most of the dogs were old and had concomitant diseases.

Prior to initiation of therapy, biopsy of the tumor and, if indicated, of the regional lymph nodes for pathologic diagnosis and staging, regional x-ray films for detection of local bone invasion, and chest films for detection of metastasis were obtained. Tumor was observed for an adequate period to establish the tumor growth rate before therapy.

Of 13 dogs, 12 were referred to the study because their tumors were inoperable, and no other therapeutic avenues were available. Of the 12, 10 had tumors in the oral cavity or posterior pharynx. In 2 of the 12, the lesion involved the face. In 1 of 13, the tumor was an osteosarcoma at the lower end of the tibia (in the hock), but amputation was refused. Of 13 lesions, 12 were within 5 mm of a bone—the mandible, maxilla, tibia, or the frontal bone, and 10 of 13 showed bone erosion on roentgenograms. Of 13, 1 was in the tonsillar fossa; 3 of 13 lesions were single, and in 10 of 13 the disease was locally disseminated.

The histologic diagnosis was

Fibrosarcoma	4
Melanoma	2
Melanosarcoma	1
Undifferentiated sarcoma	1
Mastosarcoma	1
Osteosarcoma	1
Malignant odontoma	1
Squamous cell carcinoma	1
Tonsillar carcinoma	1

The therapy protocol is shown schematically in figure 16.35. Treatments were given at weekly intervals with measurement of tumor temperature at three or more locations and stepped retraction before end of session.

1. Hyperthermia alone was applied, starting with 42.5°C for 20 minutes for three sessions, and escalating by 0.5°C every fourth session if there was no response.
2. If the tumor showed regression or stopped growing in volume, treatment was continued at that temperature for six weeks and then discontinued for six weeks, and a second biopsy was obtained.
3. If the tumor showed no regression, or continued to grow after three sessions at 44°C for 20 minutes (end of 12 weeks), ionizing radiation, 600 or 660 rad (depending on tumor type) with a 140 Kvp Keleket™ superficial x-ray therapy unit operated at 100 Kvp, 8 mA was added, being delivered to the tumor immediately after hyperthermia at 44°C for 20 minutes for six weeks.
4. If the tumor recurred, hyperthermia at 44°C for 20 minutes was combined with radiation and continued for six weeks as in 2.

In large, rapidly growing, mucosal or submucosal tumors with a natural drainage portal in the oral cavity or the aerodigestive tract, causing difficulty in eating or swallowing or respiration, in situ cytoreduction of 25% to 33% of tumor volume was carried out. The procedure was repeated in other regions of the tumor with an interval of one to two weeks, followed by tumor-bed hyperthermia as in 1 through 4 above. Regional films for local bone damage, chest films for metastasis, and tumor biopsy six to eight weeks following last treatment were obtained if the animal was surviving, or at autopsy immediately following death.

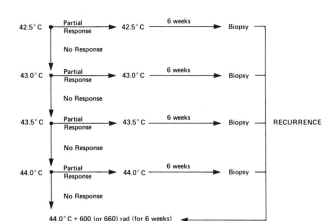

Figure 16.35. Protocol for the treatment of spontaneous tumors in dogs with hyperthermia alone or, with hyperthermia and ionizing radiation.

Procedure

The dogs were anesthetized with nitrous oxide-Halothane anesthesia delivered intratracheally for the treatments. Vital signs and ECG were monitored routinely. Tumor measurements were made, and the tumor was photographed in color from two or three angles. One or two multithermocouple probes placed into the tumor and two unsheathed probes placed on the surface were used to monitor temperatures during hyperthermia. One of the thermocouples was placed in contact with the underlying bone, when present. In oral tumors, hyperthermia was induced by using the exponential horn wave-guide insonator shown in figure

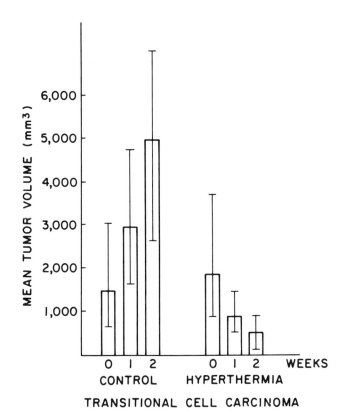

Figure 16.34. Effect of local hyperthermia at 42.5°C for 20 minutes, once a week, on the growth of transitional cell carcinoma. Studies were started three weeks after implantation of tumor. (Preliminary data from a study by Drs. Babayan, Lele, and Krane.)

16.13. The field pattern is shown in figure 16.14. Therapy planning was accomplished following the procedures described earlier. Intensity modulation to compensate for regional differences in the tumor was under computer control at all times. But the reference level of intensity was controlled manually, until the intratumoral temperature reached the predetermined hyperthermia temperature. Maintenance of this temperature for the duration of hyperthermia was under computer control. Before the end of the prescribed duration of hyperthermia, one of the intratumoral thermocouple probes was retracted in steps to ascertain the intratumoral temperature distribution.

For treatment of tumors by cytoreduction in situ, the combination of temperature and duration was selected with reference to the extensive data on heat tolerance of normal tissues, shown in figure 16.11. Different temperature-duration combinations were selected along the lower edge of the plotted data to test their equivalence in causing coagulation necrosis of the tumor tissue leading to nonhemorrhagic cytoreduction.

The tumor was again photographed in color at the end of each treatment. The temperature history of the tumor from the beginning to the end of each therapy session was plotted out from the data in the computer to determine the total duration to which the tumor was subjected at each temperature, during induction and maintenance of hyperthermia, and cooling to normothermic levels. An example of such a plot is given in figure 16.19.

Results

If the treatment resulted in coagulation necrosis, a change in the color of the tumor was always immediately evident. This was followed by sloughing off of the tumor coagulum in six to eight days. With temperatures and durations used in protocols 1 through 4, there was little discernible color change, if any. The entire series of treatments, involving more than 113 insonation sessions, was remarkable in the absence of any unintended toxicity, except in two instances of skin injury resulting from the difficulties in positioning of the insonation head using the mechanical focus-indicating pointer system. No changes in the radiographs of the bone attributable to the treatment were seen in any animals, except in the osteosarcoma. Analysis of the strip-chart records of the temperature rise and the insonation power showed that in most, if not all, instances, following three to five minutes of intratumoral temperature of 43°C there was a sudden drop in temperature, and more insonation power was needed for restitution of the temperature to 43°C.

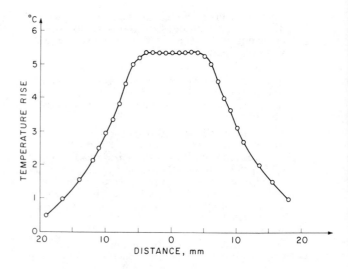

Figure 16.36. Temperature distribution in a mandibular odontoma in a dog. See text.

Three to five minutes after restitution of this temperature, the power requirements were lower. This phenomenon was similar to that observed in human patients, as discussed later in this chapter and shown in figure 16.40, and possibly denotes a thermovascular response.

The temperature distribution across the tumor plotted from data obtained by retracting the thermocouple probe, before termination of hyperthermia, showed uniform temperature distribution, with steep fall-off beyond the hyperthermia field in almost all instances. A sample is shown in figure 16.36. The presence of bone underlying the tumor is seen not to perturb the temperature distribution in the tumor. The temperatures measured at the bone surface were the same as those in overlying soft tissues when the tumor was on or within 5 mm of the bone. With tumors separated from the bone by 1 cm or more, the surface temperature at the bone was lower than the intratumoral temperature. Analysis of the computer-generated histograms of cumulated duration at various temperatures (fig. 16.19) shows that little time was spent at subtherapeutic temperatures during the induction of hyperthermia and during cooling after termination of insonation. The duration at temperatures of 42°C to 43°C was higher than those at lower temperatures because of the thermovascular response, causing a dip in the temperature, referred to above.

The temperature-duration combination for coagulation necrosis of tumor tissue did not differ greatly from that of normal tissues (fig. 16.11), although in most instances it was approximately 1°C lower at any given duration.

The oncologic results in this series of patients, small as it was, were as follows:

All four fibrosarcomas and the odontoma showed

total regression with hyperthermia alone, but recurred in six weeks. After hyperthermia combined with ionizing radiation, the patients were disease free for periods ranging from 10 to 18 months. Biopsies showed no tumor cells in the treatment field.

Melanomas, melanosarcoma, undifferentiated sarcoma, osteosarcoma, and tonsillar carcinoma responded to hyperthermia alone with enduring local control. All these tumors, except osteosarcoma, were sensitive to hyperthermia at 42.5°C for 30 minutes. There were no metastases to lungs in patients with clean chest films at the initiation of therapy. This fact is especially noteworthy in the case of osteosarcoma, which in dogs metastasizes early to lungs (at approximately 12 weeks). With successful treatment of the primary tumor, the dog remained free of lung metastases for 9½ months. As of December 1980, three of the patients are disease free 9 to 14 months after therapy. Others died from nononcologic disease or remote spread of malignancy. No disease was found at autopsy at the site of the primary lesion.

Spontaneous Tumors in Human Patients

Introduction

Feasibility and toxicity studies of local tumor hyperthermia by focused ultrasound in human patients were conducted (October, 1979 to April, 1980) collaboratively with Dr. Kaiser of the Pondville Hospital, Walpole, Massachusetts; Dr. Feldman of Boston University Medical Center, Boston, Massachusetts; Drs. Frei and Ervin of the Sidney Farber Cancer Institute, Boston, Massachusetts; and Drs. Bertino and Kowal of Yale-New Haven Hospital, New Haven, Connecticut, under protocols approved by the human subjects committees of the collaborating institutions, the Massachusetts Institute of Technology, and the MIT Clinical Research Center (Lele et al. 1980b, 1980c).

Two special semi-portable systems of simplified insonation equipment, not requiring the use of the computer, were developed for production of uniform distribution of hyperthermia in superficial tumors and were evaluated in experimental animals in vitro and in vivo before use in human patients. A simple, mechanical system, which enables generation of circular focal trajectories up to 20 mm in diameter, and a stationary system were used. One was a motor-driven system that rotated an insonation head in a circular trajectory of adjustable diameter. The insonation head was pointed inward to the center of rotation, so that the portal was always larger than the focal trajectory. Optical motion detectors were incorporated in the design to turn the insonation off if the rotational speed fell below a preset limit, to prevent any possibility of focal damage in the target tissue. The other, a stationary insonator, consisted of a focusing system that projected a pattern of intensity distribution similar to that of a focused transducer rotating in a circular trajectory (Lele 1981). The transducer could be operated either in continuous wave or pulsed modes. To increase the diameter of region of uniformity from approximately 20 mm to about 30 mm, a controllable "central leak" was added to the annular focus design of the lens. A diagrammatic cross section through the lens and the intensity distribution are shown in figure 16.37. Note that the focused beam projected is circular, and therefore only regularly shaped volumes of a given tumor could be heated. The clinical results must be viewed in this context.

Objectives

The objectives of focused ultrasound treatments of superficial human tumors were:

1. To determine the feasibility of production of uniform distribution of hyperthermia in regularly shaped volumes up to 3 x 3 x 4 cm

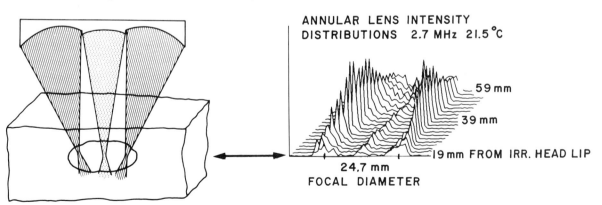

Figure 16.37. Annular-focus Lens with a central "leak": diagrammatic cross section and intensity distribution. (Adapted from Lele 1981.)

deep, in superficial or readily accessible tumors in diverse anatomic locations.
2. To determine the occurrence of any resultant toxicity.
3. To estimate the effect of this procedure on local control or tumor growth delay.

Patient Eligibility

Eligibility for treatment by focused ultasound was determined according to the following criteria:

1. Patients must have histologically verified malignant tumor.
2. The disease must be advanced, recurrent, or metastatic, and refractory to other anticancer modalities.
3. The patient's general condition must be satisfactory to allow transportation to the treatment facility.
4. The patient must give informed consent to the procedure and an explicit indication that (s)he is aware of the investigational character of the procedure, the need for insertion of hypodermic needles containing thermocouples for temperature measurement, and of the possibility of pain or burning of the skin during treatment.

Pretherapy Assessment

The pretherapy assessment consisted of:

1. Measurement or estimation of three dimensions, independently by two of the investigators.
2. When accessible, a photographic record.
3. When inaccessible to physical examination, an ultrasound scan, CAT scan, or radiograph.
4. Measurement of temperature sensibility of skin overlying tumor.

Treatment

Treatment plan was as follows:

Pretreatment trial session consisted of acoustical measurements and hyperthermia to 42°C for 20 minutes, with measurement and recording of temperature at two locations in hyperthermia target volume, surrounding regions, and especially at bone surface, if applicable. Thermocouple probe was retracted to determine temperature distribution when feasible. In subsequent treatments, temperatures were measured and recorded at a minimum of three sites—one within the hyperthermia target volume, one outside, and one on the skin/mucosa. Treatments were given once a week for three sessions at 42.5°C for 20 minutes. Temperature was escalated in steps of 0.5°C, if no response in three sessions at the previous temperature. Maximum intratumoral temperature was not to exceed 45°C for 20 minutes. Temperature was to be lowered if the patient complained of pain. Treatment was to be interrupted if pain continued. Treatment was to be discontinued if any significant toxicity, burn, or hemorrhage was evident. Reports of pain or discomfort were entered on strip-chart records. If patient has more than one tumor of comparable size, one was treated and the other was used as a control.

Assessment of Hyperthermia Reactions and Tumor Response

Evaluation of patient response was based on the following observations:

1. Skin and/or mucosal reactions was observed and recorded immediately after each hyperthermia session.
2. The subjective assessment of the patient on the local sensory effects of hyperthermia treatment, if any, was obtained and recorded immediately after hyperthermia.
3. Tumor size was measured at weekly intervals during therapy by the same two investigators as performed pretherapy assessment.
4. Tumor size was measured at regular intervals of two to four weeks following cessation of therapy, for 16 to 24 weeks, if possible.
5. Tumor biopsy was obtained between 12 and 24 weeks posttherapy.
6. The following scoring system was used for grading skin reactions:

   0.0 No visible reaction
   1.0 Slight but definite erythema
   2.0 Moderate erythema
   3.0 Severe erythema (deep red)
   4.0 First sign of breakdown (bulla or vesicle formation)
   5.0 Definite moist desquamation over less than 50% of field
   6.0 Moist desquamation over more than 50% of field
   7.0 Complete breakdown

Procedure

The procedure for treatment was as follows:

1. Warmth sensibility, that is, the thresholds (the elevation of temperature in degrees centigrade

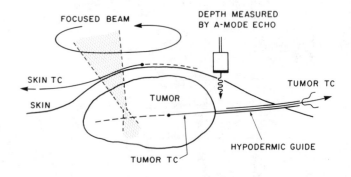

Figure 16.38. Thermometry procedure in human patients. See text. Annular focus lens was substituted for steered focused beam in part of the study. Coupling water bath is not shown.

over the resting skin temperature) for the perception of "warmth" and "hot" in the skin over the treatment field over the tumor, was measured using an electrothermal stimulator (Lele 1962c). This threshold was measured with the skin temperature at rest at the normal ambient temperature of 20°C, and after skin cooling (if this was necessary during hyperthermia treatment), since the thresholds are known to vary with the skin temperature (Lele 1954; Lele, Weddell, and Williams 1954). These data were obtained in an attempt to study the relationship between the occurrence of heat injury to the skin with the sensation of heat-pain. In normal skin, acclimated to normal ambient temperature, the sensation of heat-pain is evoked at a lower threshold than (and serves as a warning of impending) heat injury. Patients with head and neck tumors were given a push button to sound a buzzer or were asked to raise a finger if they felt any discomfort during hyperthermia and to indicate the sensation felt on a chart (pain, hot, stinging, burning, pressure) rather than make verbal reports, to avoid any movement of the target tissue.

2. Intratumoral thermocouple probe(s) was (were) inserted under local anesthesia at a site remote from the treatment volume (fig. 16.38). This was to prevent (1) reduced heat sensibility of the skin in the hyperthermia field, and (2) modification of tumor response to local hyperthermia by the anesthetic agent. The depth of the intratumoral thermocouple was determined by A-mode ultrasound at a frequency of 15 MHz ($\lambda = 0.15$ mm), yielding axial resolution of 0.25 mm. Skin temperature was measured by an unsheathed thermocouple, carefully located in the treatment field.

3. A wide-mouthed open bag made of thin film of soft polyethylene was used to hold degassed water for coupling the ultrasonic source to the skin when steered, focused ultrasound was used. With annular-focus lens, the outlet of the conical applicator was coupled directly to the skin by ultrasonic coupling gel. The temperature of water in the applicator was controlled.

4. Hyperthermia was induced under manual control of intensity. The hyperthermia level was maintained through a feedback control.

5. Problems of motion of the region under treatment, such as respiratory excursions or movements of the throat with swallowing, were solved by development of a spring-loaded transducer support system.

6. In a sample of tumors located at sites considered to be difficult to heat satisfactorily with ultrasound, such as those overlying bone or lungs, and with any changes in ultrasonic frequency, transducer, or lens, before the end of the treatment session both the skin and intratumoral thermocouple probes were retracted manually in small increments, to determine spatial temperature distributions.

7. The hyperthermia field was examined and the patient was questioned on local sensations.

Results

The results were as follows:

1. Heat sensibility of the skin over the tumor was frequently diminished. Thirty of the 44 tumors treated were situated directly over a bone, such as mandible, sternum, ribs, or humerus and/or an air cavity, such as lung or aerodigestive cavity.

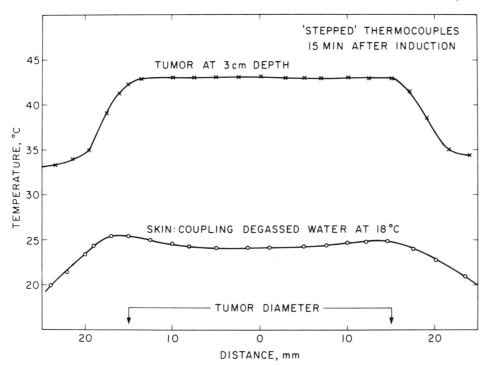

Figure 16.39. Cutaneous and intratumoral temperature distributions with the use of the annular focus lens.

In 27 of 203 hyperthermia sessions in these 30 patients, a thermocouple was placed in contact with the bone and its position verified by A-mode pulse-echo ultrasound. An additional thermocouple monitored intratumoral temperature. Owing to optimization of ultrasonic frequency, focusing characteristics, and beam angulation, presence of the bone was found not to distort the intratumoral temperature distribution appreciably, nor was the temperature at the periosteum found not to differ by more than ±0.3°C from the intratumoral temperature. None of the patients reported occurrence of deep-seated pain during hyperthermia. No symptoms or signs of bone damage were reported subsequently; nor was any evidence of bone damage seen in three patients examined radiographically 10 to 12 weeks after hyperthermia. Temperature measurements at tissue-air interfaces in the thorax were not made because of the danger of inducing pneumothorax. No unexpected temperatures were found at the mucosa of buccal or aerodigestive tracts during hyperthermia of overlying tumors.

2. In thin cutaneous and/or subcutaneous tumors, the skin was cooled by cooling the coupling degassed water during insonation in order to prevent skin burn or pain. No instances of skin burn or pain were encountered in any of the patients. Sensations of "hot" or "pain" could be elicited at skin temperatures as low as 23°C by suddenly raising it to 24°C. Skin temperature could be maintained as much as 20°C below intratumoral temperature at a depth of 3 cm (figs. 16.39, 16.40).

3. Tumor temperature could be raised to the level desired in all tumors studied, except perhaps one. This was a pulsatile, frontal osteosarcoma in which measurement of temperature at depth was infeasible because of absence of information on the location of the meninges and of the tumor vasculature. The highest temperature used was 43°C, except in one patient with a schwannoma, in whom it was 45°C at one session. Temperature distributions with the annular focus lens were uniform to± 0.25°C across the hyperthermia volume (fig. 16.39). With the use of steered, focused beam with intensity modulation, the variation was ± 0.1°C. Intratumoral temperature of 43°C or higher sustained for three to five minutes evoked a vasomotor response leading to a drop in the temperature. Increased power was needed to restore the temperature to the previous level (fig. 16.40). The power required to sustain this temperature became slightly reduced within three to five minutes after restoration of the temperature.

4. At termination of hyperthermia, temperatures at depth decayed slowly as compared to a quick drop in skin temperature (fig. 16.40).

5. Seven of the patients had diffuse tumors with

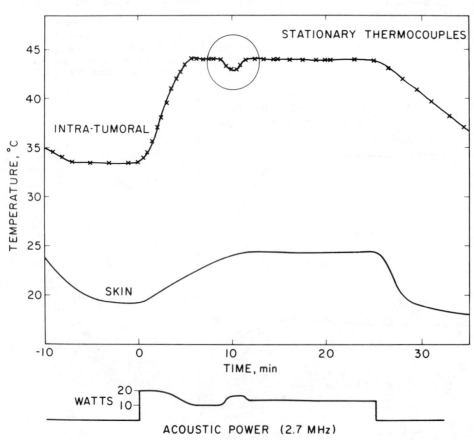

Figure 16.40. Temporal course of intratumoral and cutaneous temperatures and the acoustical power. The thermovascular response is encircled.

Table 16.1
Objective Responses of Spontaneous Human Tumors to Steered, Focused Ultrasound

Tumor	No.	Regression 0%	<50%	>50%	100%
Inflammatory carcinoma, breast	1	1			
Schwannoma	1	1*			
Renal cell carcinoma	1	1			
Acinar cell carcinoma	2			2	
Melanoma	2				2†
Undifferentiated sarcoma	3			1	2
Chest-wall metastases of breast carcinoma	3			3	
Squamous cell carcinoma	23	1	6‡	14	2
Totals	36	4	6	20	6

*Tumor located in amputation stump and invading skin. Patient reported feeling of "tenseness" within the region lasting approximately 30 hours following hyperthermia. Increase in pre-existing redness of skin following 45°C for 20 minutes. There was no change in tumor measurements. Biopsy results awaited (12 June 1980).
†Both tumors disappeared after two sessions at 42.5°C for 20 minutes. Control tumors continued to grow even with concurrent chemotherapy. In one patient there was hemorrhage from multiple sites in control tumors three weeks following last hyperthermia.
‡In one patient with previous neck dissection there was referred pain (bearable) along the scar, which was remote from the treatment area.

undefined boundaries. Hyperthermia resulted in subjective and objective improvements in the status of the tumors in five of these seven patients. Improvements included partial relief of pain when previously present, decrease in amount of drugs required for control of pain, reduced trismus, and flattening and shrinkage of tumor. These improvements, however, could not be quantified with confidence. Adequate temperature measurements could not be made in one case (osteosarcoma). These eight patients are excluded from assessment of response to hyperthermia. The objective response in the remaining 36 patients determined from measured cross-diameters is presented in table 16.1.

6. Maximum response in each patient occurred early in the course of treatments. Partial response (> 50% regression) was obtained in 72% (26 of 36) of tumors, with complete regression in 17% (6 of 36) of tumors. It must be understood that because of the fixed diameter of the field, in some tumors only a part was exposed to hyperthermia. In eight of the tumors in which the tumor regression was less than 50%, biopsy 12 to 24 weeks after end of hyperthermia treatments showed absence of viable tumor cells. Thus, even though there was no shrinkage in tumor volume, the malignancy was controlled in these cases. In one tumor, which did not respond, biopsy showed presence of viable tumor cells.

7. Eleven of the patients with multiple tumors were on chemotherapy (cisplatin and bleomycin, cyclophosphamide, methotrexate, and cisplatin). In these, two comparable tumors were studied; one was treated with hyperthermia, the other served as control. In nine patients, the control tumors showed no response to chemotherapy alone, but the tumors treated additionally with hyperthermia showed partial regression. In two patients, the control tumors showed partial regression, while the tumors treated additionally with hyperthermia showed complete regression.

8. There was no evidence of subjective or objective dose limiting toxicity. Some of the patients tended to fall asleep during hyperthermia and had to be kept awake by being talked to to prevent any sudden motions on awakening from sleep. The absence of toxicity with the use of this technique is in marked contrast to the pain, blistering, and burning of the skin, or "shelling out" of the tumor reported by investigators using plane wave ultrasound (Marmor et al. 1979).

References

Altman, P. L., and Ditmer, D. S., eds. *Respiration and circulation: biological handbooks.* Bethesda, Md.: Federation of American Societies for Experimental Biology, 1971.

American Cancer Society. Conference on Hyperthermia in Cancer Treatment, San Diego, Calif., September 1978, ed. J. W. Milder. *Cancer Res.* (special supplement) 39: 2231–2340, 1979.

Astrahan, M. A. Letter on hyperthermia phantom. *Med. Phys.* 6:72, 1979a.

Astrahan, M. A. Concerning hyperthermia phantom. *Med. Phys.* 6:235, 1979b.

Basauri, L., and Lele, P. P. A simple method for production of trackless focal lesions with focused ultrasound: statistical evaluation of the effects of irradiation on the central nervous system of the cat. *J. Physiol.* 160: 513–534, 1962.

Bowman, H. F.; Cravalho, E. G.; and Woods, M. Theory, measurement, and application of thermal properties of biomaterials. In *Annual review of biophysics and bioengineering*, ed. L. J. Mullins. 4:43–80, 1971.

Dewey, W. C., and Dethlefsen, L. A., eds. *Proceedings of the Third International Symposium on Cancer Therapy by Hyperthermia, Drugs and Radiation.* Bethesda, Md.: *J. Natl. Cancer Inst.* (special journal supplement), in press.

Guy, A. W.; Lehmann, J. F.; and Stonebridge, J. B. Therapeutic applications of electromagnetic power. *Proc. IEEE* 62:55–75, 1974.

Hunt, J. W. Applications of microwave, ultrasound, and radiofrequency heating in vivo. In *Proceedings of the Third International Symposium on Cancer Therapy by Hyperthermia, Drugs and Radiation*, eds. L. A. Dethlefsen and W. C. Dewey. Bethesda, Md.: *J. Natl. Cancer Inst.* (special journal supplement), in press.

Jaeger, J. C. Conduction of heat in tissues supplied with blood. *Br. J. Appl. Physiol.* 3:221–222, 1952.

Jain, R. K., and Gullino, P. M., eds. *Thermal characteristics of tumors: applications in detection and treatment. Ann. NY Acad. Sci.* 335:1–542, 1980.

Larkin, J. M. A clinical investigation of total-body hyperthermia as cancer therapy. *Cancer Res.* 39:2252–2254, 1979.

Lele, P. P. Relationship between cutaneous thermal thresholds, skin temperature and cross-sectional area of the stimulus. *J. Physiol.* 126:191–205, 1954.

Lele, P. P. A simple method for production of trackless focal lesions with focused ultrasound: physical factors. *J. Physiol.* 160:494–512, 1962a.

Lele, P. P. Irradiation of plastic with focused ultrasound: a simple method for evaluation of dosage factors for neurological applications. *J. Acoust. Soc. Am.* 34:412–420, 1962b.

Lele, P. P. An electrothermal stimulator for sensory tests. *J. Neurol. Neurosurg. Psychiatry* 25:329–331, 1962c.

Lele, P. P. Production of deep focal lesions by focused ultrasound—current status. *Ultrasonics* 5:105–112, 1967.

Lele, P. P. Thresholds and mechanisms of ultrasonic damage to 'organized' animal tissues. In *Proceedings of a Symposium of Biological Effects and Characterizations of Ultrasound Sources*, eds. D. G. Hazzard and M. L. Litz. DHEW Pub. (FDA) 78-8048, 1977.

Lele, P. P. Induction of deep, local hyperthermia by ultrasound and electromagnetic fields. *Radiat. Environ. Biophys.* 17:205–217, 1980.

Lele, P. P. An annular-focus ultrasonic lens for production of uniform hyperthermia in cancer therapy. *Ultrasound Med. Biol.* 7:191–193, 1981.

Lele, P. P. et al. Local hyperthermia by focused ultrasound: technique and results in spontaneous tumors in dogs (abstr.). In *Proceedings of the Third International Symposium on Cancer Therapy by Hyperthermia, Drugs and Radiation*, eds. L. A. Dethlefsen and W. C. Dewey. Bethesda, Md.: National Cancer Institute, 1980a.

Lele, P. P. et al. Clinical evaluation of local hyperthermia by focused ultrasound (abstr.). In *Proceedings of the Third International Symposium on Cancer Therapy by Hyperthermia, Drugs, and Radiation*, eds. L. A. Dethlefsen and W. C. Dewey. Bethesda, Md.: National Cancer Institute, 1980b.

Lele, P. P. et al. Treatment of advanced squamous cell cancers of the head and neck region by focused ultrasound. Presented at the International Head and Neck Oncology Research Conference, Rosslyn, Va., 1980c.

Lele, P. P. and Parker, K. J. Temperature distributions in tissues during local hyperthermia by stationary or steered beams of unfocused or focused ultrasound. *Br. J. Cancer* 45(suppl. V):108–121, 1982.

Lele, P. P., and Senapati, N. The frequency spectra of energy back-scattered and attenuated by normal and abnormal tissue. In *Recent advances in ultrasound in biomedicine* vol. 1, ed. D. N. White. Forest Grove, Oreg.: Research Studies Press, 1977, pp. 55–84.

Lele, P. P.; Senapati, N.; and Hsu, W. L. Mechanisms of tissue-ultrasound interaction. In *Proceedings of the Second World Congress on Ultrasonics in Medicine*, eds. M. de Vlieger, D. N. White, and V. R. McCready. Amsterdam: Excerpta Medica, 1973, pp. 345–352.

Lele, P. P.; Weddell, G.; and Williams, C. M. The relationship between heat transfer, skin temperature and cutaneous sensibility. *J. Physiol.* 126:206–234, 1954.

Mäntylä, M. J. Regional blood flow in human tumors. *Cancer Res.* 39:2304–2306, 1979.

Marmor, J. B. et al. Treating spontaneous tumors in dogs and cats by ultrasound-induced hyperthermia. *Int. J. Radiat. Oncol. Biol. Phys.* 4:967–973, 1978.

Marmor, J. B. et al. Treatment of superficial human neoplasms by local hyperthermia induced by ultrasound. *Cancer* 43:188–197, 1979.

Parker, K. J., and Lele, P. P. The effect of blood flow on temperature distributions during localized hyperthermia. *Ann. NY Acad. Sci.* 335:64–65, 1980.

Pennes, H. H. Analysis of tissue and arterial blood temperature in the resting human forearm. *J. Appl. Physiol.* 1:93–122, 1948.

Peterson, H. I. *Tumor blood circulation.* Boca Raton, Fla.: CRC Press, 1980.

Pettigrew, T. R. et al. Clinical effects of whole-body hyperthermia in advanced malignancy. *Br. Med. J.* 4:679–682, 1974.

Schwan, H. P. Electromagnetic and ultrasonic induction of hyperthermia in tissue-like substances. *Radiat. Environ. Biophys.* 17:189–203, 1980.

Singh, B. S., and Dybbs, A. Error in temperature measurements due to conduction along the sensors. *ASME* 75-WA/HT-92, 1975.

Storm, F. K. et al. Hyperthermic therapy for human neoplasms: thermal death time. *Cancer* 46:1849–1854, 1980.

Streffer, C. et al., eds. *Proceedings of the Second International Symposium on Cancer Therapy by Hyperthermia and Radiation.* Baltimore and Munich: Urban & Schwarzenberg, 1978.

Wells, P. N. T. *Physical principles of ultrasonic diagnosis.* London and New York: Academic Press, 1969.

Wells, P. N. T. *Biomedical ultrasonics.* London and New York: Academic Press, 1977.

Westra, A., and Dewey, W. C. Variation in sensitivity to heat shock during the cell-cycle of Chinese hamster cells in vitro. *Int. J. Radiat. Biol.* 19:467, 1971.

Wizenberg, M. J., and Robinson, J. E., eds. *Proceedings of the International Symposium on Cancer Therapy by Hyperthermia and Radiation.* Chicago: American College of Radiology, 1976.

Wulff, W. The energy conservation equation for living tissue. *IEEE Trans. BME* 21:494–495, 1974.

CHAPTER 17

Regional Perfusion Hyperthermia

Renato Cavaliere
Bruno Mondovi
Guido Moricca
Giorgio Monticelli
Pier Giorgio Natali
Francesco Saverio Santori
Franco Di Filippo
Antonio Varanese
Luigi Aloe
Alessandro Rossi-Fanelli

Introduction
   Effect of Hyperthermia on Experimental Tumors
   Potentiation of Hyperthermic Effect
   Elevated Temperature and Immunity
Experimental Animal Models
   Local and Regional Hyperthermia
   Whole-Body Perfusion Hyperthermia
Clinical Applications
   Hyperthermic Perfusion
   Hyperthermic Chemotherapeutic Perfusion
   Sympathectomy and Hyperthermic Perfusion
Clinical Results
Discussion and Conclusions

INTRODUCTION

In the last 10 years there has been a tremendous increase in research on elevated temperature therapy as a possible cancer treatment. Only a few of these experiments will be discussed here. Unless otherwise specified, the terms *elevated temperature* and *hyperthermia* are defined as exposure of tissues to temperatures slightly above the physiologic level, or between 41.5°C and 42.5°C.

The use of heat as a cancer treatment started more than a century ago, when Bush (1866) noticed that a sarcoma of the face disappeared after a high fever caused by erysipelas. Since then, the idea of some sort of hyperthermic cancer treatment has periodically been reinvestigated. Beginning in 1967 (Cavaliere et al. 1967), we have reported long-term regression of malignant tumors in humans by hyperthermic limb perfusion and have studied the possible mechanisms of action on a biochemical level.

Elevated temperatures similarly affect both in vitro and in vivo tumors. In the latter case, however, other factors are also active, among which immunologic reactions and tumor vascularization appear to be significant. An important, but sometimes forgotten, factor in cancer research is the significance of the differences between the cancer cell and its normal counterpart. If an effective anticancer therapy is to be developed, these differences must be recognized and exploited. This is not a simple problem, however; there are many variables that must be taken into consideration. For example, the rate of cellular reproduction is extremely important and has many consequences. The evaluation of damage to a cancer cell by a particular therapy is also difficult. There may be histologic, ultrastructural, or biochemical damage, or there may be damage that is not apparent in any of these areas, especially immediately following the treatment. Transplantation of the tumor into experimental animals is a problem also, as many tumors have a low rate of establishment. In addition, there may be difficulties with the normal control cell; it is not always possible to assume that human tissues will respond in the same way as experimental animal tissue. Evaluation of cell viability can be difficult to determine, as there has been considerable criticism of the dye-exclusion test (Giovanella, Lohman, and Heidelberger 1970). We prefer to determine viability on the basis of biochemical changes and growth of the cells in culture.

Under the same conditions of temperature and

The authors gratefully acknowledge the significant support of the Italian National Research Council, grants 800159596 and 800151096.

time of exposure, is the cancer cell more damaged by heat than is the normal cell? The answer is affirmative according to the following data:

1. Fibrosarcomas, as compared to mouse embryonic fibroblasts, and human cancer cells, as compared with the normal surrounding tissue (with equal rates of cellular reproduction), have exhibited higher thermal sensitivity (Giovanella et al. 1973; Giovanella, Stehlin, and Morgan 1976; Giovanella and Mondovi 1977; Stehlin et al. 1975).
2. The biochemical alterations studied in the cases of Novikoff's hepatoma and minimum-deviation Morris-5123 hepatoma show that they are less resistant to elevated temperatures than rat regenerating liver, which is used as a fast-growing normal control tissue (Mondovi 1969a, 1969b, 1976; Cavaliere et al. 1967, 1978b, 1979a; Strom et al. 1977).
3. The reduction or disappearance of malignant tumors that have been heated has been observed in vivo; the normal surrounding tissues have not been damaged (Mondovi 1976). The effectiveness of hyperthermia as a new weapon against human malignant tumors has been demonstrated by numerous investigators (Cavaliere et al. 1967). These authors have supplied substantial evidence about the thermal sensitivity of malignant tumors both in vitro and in vivo; the first patients treated by this group in 1964 are alive and well 16 years later, having been treated with hyperthermia alone.

This chapter is divided into two parts: the effect of hyperthermia on experimental tumors and clinical hyperthermic perfusion studies.

The Effect of Hyperthermia on Experimental Tumors

Rohdenburg and Prime, in 1921, noticed that pieces of sarcoma-180 did not form tumors when injected into mice when they had been incubated at 42°C for 90 minutes. The first definitive study indicating some biochemical alterations caused by elevated temperatures was performed in 1927 by Westermark. He demonstrated that following exposure to supranormal temperatures (42°C to 44°C) for various lengths of time, the Flexner-Jobling carcinoma and Jensen's sarcoma exhibited marked inhibition in cellular respiration. Subsequently, oxygen uptake of Novikoff's hepatoma cells incubated at 42°C for 60 minutes was found to be remarkably lower than at 38°C (Cavaliere et al 1967; Mondovi et al. 1969a). This inhibition did not occur in rat regenerating liver cells, used as a control tissue (fig. 17.1) under the same conditions. Thus, it appears that different tumors have different levels of thermosensitivity and require different temperatures

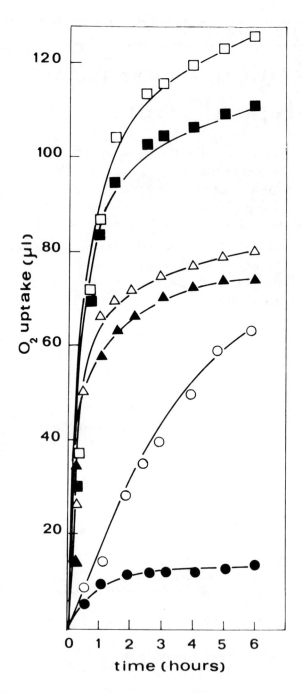

Figure 17.1. Oxygen uptake of Novikoff hepatoma, rat liver, and regenerating rat liver cells, incubated at 38°C and 42°C in the presence of 0.015 M D-glucose and 0.013 M succinate. ●, Novikoff hepatoma cells at 42°C; ○, Novikoff hepatoma cells at 38°C; ▲, rat liver cells at 38°C; △, rat liver cells at 42°C; ■, regenerating liver cells at 42°C; □, regenerating liver cells at 38°C. (Cavaliere et al. 1967. Reprinted by permission.)

Table 17.1
Thermosensitivity of Tumor and Normal Cells In Vitro

Type of Cells	Parameter Observed	Temperature and Duration of Exposure			
		42°C 2 hr	43°C 30 min	43°C 2 hr	43°C 3.5 hr
Novikoff hepatoma	O₂ uptake	15	90–40*	30	15–20
	DNA synthesis	—	28	4	—
	RNA synthesis	—	30–40	15–30	—
	Protein synthesis	—	—	5–10	—
Morris-5123 hepatoma	O₂ uptake	100	—	90–95	70–75
	DNA synthesis	61	—	37	—
	RNA synthesis	—	—	83	—
	Protein synthesis	55	—	40	—
Regenerating liver	O₂ uptake	100	—	100	100–105
	DNA synthesis	—	—	105–115	—
	RNA synthesis	—	—	95	—
	Protein synthesis	—	—	105–120	—

SOURCE: Moricca et al. 1977.
NOTE: Comparative effects of preincubation at 42°C or 43°C on rates of oxygen consumption, nucleic acid synthesis, and protein synthesis at 38°C by Novikoff hepatoma, Morris-5123 minimal deviation hepatoma, and regenerating liver cells. The values are expressed as percentage of the activity of analogous preparations kept at 38°C for the same time and under the same conditions.
* A progressive increase in the extent of inhibition.

Table 17.2
Effect of Temperature on ^{3}H Thymidine Incorporation into Novikoff Hepatoma Cells

	Temperature of Preincubation				
	38°C	40°C	41°C	42°C	43°C
Disint/min/μg DNA	776	326	68	40	36
Inhibition (%)	0	52	90	94	95

SOURCE: Mandovi et al. 1969b.
NOTE: Cell suspensions were preincubated for two hours in Krebs-Ringer solution at 38°C, 40°C, 42°C, and 43°C. Incorporation of the labeled compound was carried out by adding 0.6 μC/mg dry weight of ^{3}H thymidine to the washed cells resuspended in Eagle's Basal medium and by incubating for one hour at 38°C.

and times of incubation for maximum effect. For instance, in the minimum-deviation Morris-5123 hepatoma, the damage becomes clear only after 3.5 hours of incubation at 43°C (table 17.1). No general pattern has been observed in the inhibition of anaerobic glycolysis by elevated temperatures; some tumors exhibit this effect and others do not (Strom et al. 1977).

Following an elevated temperature exposure, the inhibition of biosynthesis of nucleic acids and proteins is very high and is evident very early: in Novikoff's hepatoma the incorporation of tritiated thymidine in the DNA is 94% inhibited after two hours of incubation at 42°C, and some inhibition is already evident after a 30-minute incubation at 42°C (tables 17.1 and 17.2). Biosynthesis of RNA and proteins is inhibited also to a greater or lesser degree in different tumors (Mondovi 1976; Mondovi et al. 1969b) (table 17.1).

The role of plasma membranes as possible targets of hyperthermia is not established. The membranes seemed at first to be a primary target (Strom et al. 1969); however, subsequent results indicate that the irreversible damage to the tumor membrane that is caused by hyperthermia is a secondary effect (Strom et al. 1977, 1973; Mondovi 1979). This does not mean that membrane damage is not a significant part of hyperthermia's effectiveness. Tumor membranes are also more sensitive to chemical compounds like polyenic antibiotics (Mondovi et al. 1971). The decrease in RNA production can be partially restored by increasing the concentration of nucleoside precursors (Strom et al. 1973). This effect could be linked with the loss of nucleosides or nucleotides to the medium, since the permeability of the mouse or rat ascites tumor increases between 24°C and 46°C. There is also an increase in the energy of activation of about 20 kcal/mole, as predicted by the theory of Arrhenius. Subsequent results have indicated the direct involvement of some enzymes that are necessary for the biosynthesis and completion of nucleic acids, particularly those involved in the synthesis of pre-RNA (Strom et al.1975).

Morphologically, the nucleolus seems to be hyperthermia's preferred target (Love, Soriano, and Walsh 1970). Immediately following treatment, however, the tumor cells do not appear to be more damaged than the normal cells (Mondovi 1975). It is noteworthy that tumor cells killed by exposure to 42.5°C for two hours do not show any morphologic alteration when observed under the optical or electron microscope, if fixed immediately after treatment.[1]

Recently, Fajardo and co-workers (1980) described the structural damage in hyperthermically treated cells observed under light and electron microscopy. EMT-6 tumor was implanted in BALB/CKA mice, and hyperthermia was performed by radiofrequency (RF) at 44°C for 30 minutes. Focal cytoplasmic swelling, rupture of the plasma membrane, and peripheral migration of heterochromatin were observed five minutes after the initiation of therapy. Fragmentation of the cytoplasm occurred after 30 minutes, and most of the cells became necrotic two to six hours later. The cells were completely destroyed after 48 hours. The authors suggested that the physical changes in the plasma membrane should be attributed to the necrosis induced by heat. Unfortunately, this research was performed on implanted tumors, and there is no data available on directly heated tumor cells. This could explain the discrepancies between these results and those observed by Giovanella.

Mondovi (1975) has reported that the activity of the mitochondria are not compromised by elevated temperatures. Conflicting reports have been released on whether lysosome function is altered, and there are no conclusive results (Turano et al. 1970) concerning their postulated decomposition by elevated temperatures (Overgaard 1977). A lowering of the intracellular pH during the beginning of the incubation periods probably causes some damage, thereby enhancing the hyperthermic effect (Mondovi 1979). A similar condition is indeed observed in the biosynthesis of DNA, which is strongly inhibited by the lower pH caused by lactic acid formation (table 17.3).

It is highly probable that the repair systems of tumor cells are involved in the mechanism of biological damage. Thrall, Gillette, and Bauman (1973) have reported that the original heat-induced damage to tumor cells could be repaired if the heat was supplied in fractionated doses. Giovanella, Lohman, and Heidelberger (1970) have reported on a very significant discovery: normal cells are greatly damaged 24 hours after exposure to elevated temperatures, but the damage is quickly repaired. The tumor cells also are greatly damaged after 24 hours, but only a few of these cells survive because most are incapable of repair. Moreover, different stages of the cell cycle appear to show different sensitivities to elevated temperatures. It seems that the most labile phase is the one immediately following DNA synthesis (Giovanella, Mosti, and Heidelberger 1969; Palzer and Heidelberger 1973).

Dewey and colleagues (1971) and Westra and Dewey (1971) observed a specific denaturing action on the spindle proteins of tumor cells exposed to elevated temperatures during the M phase. An increase in the protein/DNA ratio in HeLa cells was noticed after 30-minute heating at 45°C, without any manifest alteration of DNA. This has led to the postulation that proteins in the heated tumor cells were denatured and bound to chromatin more efficiently than in the control cells (Roti-Roti and Winward 1978). According to Tomasovic, Turner, and Dewey (1978), an increased amount of nonhistonic proteins were found associated with DNA isolated from CHO cells that were heated at 45.5°C. This increase was directly proportional to the amount of time of exposure to elevated temperature. These studies have led to the supposition that chromosomal proteins are involved in the mechanism of cell death by elevated temperature (Dewey et al. 1971).

The Potentiation of Hyperthermic Effect

The effectiveness of hyperthermia may be potentiated by chemical compounds, ionizing radiation, or both. Ben-Hur, Prager, and Riklis (1978) noted that the addition of polyamines to the culture medium caused a synergistic effect, which resulted in the death of Chinese hamster V-79 fibroblasts subjected to elevated temperature. The synergistic effects of the following compounds decreased in this order: spermine, spermidine, cadaverine, putrescine. The highest inhibition was obtained by treating simultaneously with polyamines and heat, and the synergistic effect decreased as the length of time between the two treatments in-

---

[1] B. C. Giovanella, personal communication.

Table 17.3
Effect of D-Glucose and pH on ^{3}H Thymidine Incorporation into Novikoff Hepatoma Cells

Buffer Used	Temperature of Preincubation (°C)	D-Glucose Added to 0.015 M	Initial pH	pH After Incubation	pH After Incorporation	Disint/min/ μg DNA
A	38	No	7.1	6.3	6.1	380
A	38	Yes	7.1	6.1	5.7	110
A	43	No	7.1	6.3	5.9	53
A	43	Yes	7.1	6.3	5.8	15
B	38	No	7.5	8.4	7.8	11,000
B	38	Yes	7.5	7.4	7.7	2550
B	43	No	7.5	8.4	7.6	0
B	43	Yes	7.5	7.2	7.8	0

SOURCE: Mondovi et al. 1969b.
NOTE: Cell suspension was preincubated for two hours in the following solutions at 38°C and 43°C in the presence or absence of 0.015 M D-glucose:
A: Eagle's Basal medium;
B: Krebs-Ringer bicarbonate phosphate solution: in the usual Krebs-Ringer-phosphate solution, 42 volumes (out of 100) NaCl were substituted by isotonic NaHCO (gas with air + 5% CO_2).
Incorporation of ^{3}H thymidine (0.6 μC/mg dry weight) was performed by incubating the washed cells at 38°C for one hour: A, in Eagle's medium; B, in a solution containing equal parts of Eagle's medium and Krebs-Ringer solution.

creased. The order in which the treatments were given had little influence on this effect. The spermine appears to be the major factor in determining the thermal sensitivity of these cells, and the heat-induced modifications occurred at the intracellular level.

Ben-Hur and Riklis (1978) observed that elevated temperatures caused a reversible inhibition of ornithine decarboxylase activity. Therefore, it appeared that the heat acted only on the translation of the enzyme and did not inactivate it. The latter observation correlates well with the hyperthermic inhibition of protein biosynthesis (Ben-Hur and Riklis 1978; Mondovi et al. 1969b). The heat probably damages the biosynthetic machine, for example the polysomes, and inactivates the biosynthetic enzymes. When cells are treated with hyperthermia in the presence of spermine, there is a delay of about three hours in the recovery of the ornithine-decarboxylase activity after transferring the cells to a 37°C medium. The interaction between the polyamines and heat could occur either at the level of the nucleic acids (Ben-Hur and Riklis 1978, 1979a) or the membranes (Ben-Hur and Riklis 1979a; Gerner et al. 1980). Mondovi (1979) recently suggested that the mechanism of amine enhancement of tumor damage also should be due to aldehydes formed from amines by amine oxidase. In fact, exogenous purified bovine plasma amine oxidase, in the presence of spermine, causes a dramatic drop of ^{3}H thymidine incorporation into DNA of Ehrlich ascites cells. It is interesting to note that when the tumor cells are preincubated in the presence of bovine plasma amine oxidase, without addition of substrate or inhibitor, the inhibition of DNA synthesis is much more pronounced at 43°C than at 37°C. This phenomenon should be due to enhanced permeability of Ehrlich ascites cells at 43°C, thus facilitating the exit of intracellular polyamines, which in turn are oxidized by the added enzyme in the incubation mixture.

Encouraging clinical results have generated a tremendous interest in the combination of hyperthermia with antiblastic (chemotherapeutic) compounds and radiotherapy. The elevated temperature seems to increase the sensitivity of some tumor cells to different chemical compounds. This sensitivity is particularly prominent when thiotepa, Adriamycin, or bleomycin are used (Hahn, Braun, and Har-Kedar 1975). The increase in chemosensitivity of the cell culture has depended upon the absolute temperature, the duration of hyperthermic exposure, the drug concentration, and the length of its use. In the case of bleomycin, it was noticed that the sensitivity was highest when the elevated temperature was applied simultaneously, but there was a substantial sensitization even when the drug was applied before or after the heat treatment. The radiosensitizing drugs (such as misonidazole) also have increased hyperthermic effect in some investigations. Further information on these topics are given in the clinical correlation section of this chapter.

A combination of hyperthermia and ionizing radiation for the treatment of tumors has been employed more frequently (Bronk 1976; Dietzel 1975; Kim, Kim, and Hahn 1974). The first investigations go back to 1913 by Muller (1913), and the research in this field has increased enormously since then, particularly in the last few years. Crile (1963) stated that the association of 30 minutes of heat at 44°C and 2000 rad applied to sarcoma-bearing mice sterilized 50% to 80% of the tumors but also caused the destruction of the feet of the mice. The use of only one agent did not cause damage to healthy tissues but sterilized only 25% to 40% of the tumors. If we compare the doses of radiation alone required to kill the same number of cells as radiation and hyperthermia combined, we arrive at a quantitative expression of the potential synergistic effect of these two modalities. The concept is the same as that of *relative biological effectiveness* (RBE) or the *oxygen enhancement ratio* (OER). Robinson, Wizenberg, and McCready (1974) proposed the term *thermal enhancement ratio* (TER). The synergism between elevated temperature and radiation probably is due to the different ways in which these two treatments act on the cell cycles and processes (Ben-Hur, Bronk, and Elkind 1972; Robinson and Wizenberg 1974). Westra and Dewey (1971) have shown that the radioresistant S phase is heat sensitive in both HeLa cells and hamster cells. It was suggested that the repair processes for sublethal radiation damage are hindered at 41°C and 42°C (Ben-Hur, Elkind, and Bronk 1974; Gerner and Leith 1977). During the survival trials at temperatures higher than 41.5°C, the following effects have been observed: inhibition of the repair processes and potentiation of the damage caused by radiation. The sequence of treatments, the temperature used, and the quality of radiation employed are the parameters necessary to consider in order to find the optimum condition for obtaining the best results on the tumor without attendant damage to the normal tissues.

Gillette and Ensley (1979) studied mouse adenocarcinoma in vivo in order to compare damage to the tumor with that to the skin. They observed a therapeutic gain factor (TER tumor:TER skin) of 1.3 when radiation was applied during heating, and a therapeutic gain factor (TGF) of 0.9 when radiation was used before heat. Their best TGF was obtained by heating for one hour at 42.5°C. Heat treatments with brief intervals appear to have produced the best therapeutic effects, especially when radiation is applied during the period of elevated temperature or immediately following it.

Giovanella, Lohman, and Heidelberger (1970) and Miyakoshi and colleagues (1979) have independently observed that by heating V-79 cells for two hours at 42°C, cell sensitivity to another exposure at 44°C decreases, while a first exposure at 44°C increases the response of the cells to a second exposure at 42°C. When these cells are irradiated and then exposed to 44°C for 15 minutes followed by 42°C for 60 minutes, the "radio-stimulating" effect is greater than with 42°C first, followed by 44°C. The authors suggest that the most effective combination is heating for a short time at 44°C, then for a longer time at 42°C, with simultaneous radiotherapy. The exposure at 44°C would have a destructive effect on the repair system for damage produced by the exposure at 42°C. By inverting the phases, that is, heating at 42°C first, and 44°C later, a certain amount of thermoresistance might be produced. The combined treatment of 44°C to 42°C plus x-rays (300 rad at 30 rad per minute) probably inhibits the repair process for damage by radiation. The factors of greatest significance in clinical treatment appear to be the absolute temperatures employed and the timing and sequence of applications.

Joshi and Barendsen (1978) have investigated the ureter transplantable carcinoma in a strain of BN "inbred" (RUC-2) rats. Heating for one hour at 41°C and applying ionizing radiation 30 minutes later produced an increased response to the x-rays and a degree of sensitization not very different from that observed when the heat treatment started 30 minutes after irradiation. When the cells were heated and irradiated simultaneously, the damage was greater. The authors suggest that when it is necessary to give the hyperthermic treatment shortly before or after radiation, the treatment at 41°C is more effective after the irradiation. Interestingly, they felt that the effects caused by heat persisted for a much longer time than the damage created by radiation.

The mechanism of synergistic action of thermoradiotherapy was studied by Ben-Hur, Elkind, and Bronk (1974) by using double radiation doses on hamster cells. The cells were incubated at supranormal temperatures for two hours after the first radiation exposure. The survival curve after the second radiation dose shows that the ability to repair sublethal damage from the first radiation dose progressively vanishes when the temperature increases from 37°C to 41°C. At temperatures higher than 41°C another phenomenon of cell death appears to add to the inhibition of repair and apparently continues for some time after the cells are brought back to 37°C. This effect remains steady for the first five hours, and is not entirely abolished after 21 hours. A normal response to radiation comes only after some cellular generation (190 hours). It is therefore presumed that heat irreversibly damages the repair enzymes, and the reinstatement of a normal radiosensitivity requires either the synthesis of an enzymatic pool, or the reestablishment of structures that would guarantee this synthesis.

Ben-Hur and Riklis (1979b) have shown, moreover, that spermine enlarges the synergistic interaction of heat and ionizing radiation; the effect of the radiation is increased, and the inhibition of repair of the sublethal damage caused by the radiation also is increased. Incubating the cells for one hour at 42°C in the presence of 0.15 mM spermine gives the same potentiation of radiation damage as when the cells are kept at 42°C for 2.5 hours. In addition, the spermine increases the ability of heat to produce a transient thermal resistance.

In practice, most investigators have used the heat treatment first and the radiation immediately afterward. Kim, Hahn, and Tokita (1978) have treated human malignant melanomas with hyperthermia (30 to 45 minutes with a 27.12-MHz source) and radiation (300 rad) for a total of 10 times in four weeks, and have obtained good results. Meyer, Hopwood, and Gillette (1979) have carefully studied the relationship among cell damage, time lapse, and sequence of treatments with hyperthermia and radiation, using a mouse (MADCA P-37) carcinoma heated to 42.5°C and irradiated with 400 to 500 rad. They also found that the greatest decrease in cell survival was obtained by the simultaneous use of radiation and heat. The combined effect decreased as the time interval between heat and radiation increased. The sensitivity of tumor cells was 10 times greater when they were irradiated during the 120-minute elevated temperature treatment than it was when they were irradiated three hours before or after the heat treatment. When tumor cells were heated from 20 to 60 minutes, the survival rate was lower when heat followed irradiation; longer treatments (120 minutes) with the reverse sequence did not exhibit a significant difference in effectiveness. This probably is due to the damage that the cells sustained during the time they were exposed to heat. Biochemically and ultrastructurally, it can be speculated that the damage by radiation can be repaired within two hours, while that caused by the hyperthermic treatment can be directly correlated with the time of exposure. These cells are nonsynchronized; therefore those in the S phase, which are sensitive to heat and radioresistant, undergo a gradual elimination by heat. In consequence, more cells will survive the short hyperthermic treatments than the long ones.

The best results in vivo (maximum damage to the tumor and minimum damage to the healthy tissues [skin]) have been obtained by employing radiation during short periods of elevated temperature (TER). The synergistic effect between elevated temperature and radiation is particularly noticeable in hypoxic conditions. Even in highly hypoxic tumors, the thermal cell death for 50% of tumors (TCD_{50}) was increased by elevated temperatures, and the remaining cells were considerably more sensitized (Robinson, Wizenberg, and McCready 1974; Robinson et al. 1974). The OER of the hyperthermia-radiation combination is comparable to that of radiation with a high TER. In other words, whereas radiation alone requires a high oxygen pressure to be effective, it appears that ionizing radiation in combination with hyperthermia is very effective for hypoxic tumors. Kim, Kim, and Hahn (1974) observed that the OER value for ^{60}Co irradiation in HeLa cells was 2.9 at 37°C. By keeping the same cells at 42°C, the OER value decreased to 1.58. Since hypoxic tumor cells are relatively radioresistant and survive radiotherapeutic treatment, the inability of hypoxia to protect tumors from hyperthermic damage may be exploited in therapeutics (Robinson et al. 1974; Robinson, Wizenberg, and McCready 1974). Thrall, Gillette, and Dewey (1975) observed that radiation lessens the LD_{50} values in conditions of hyperbaric oxygenation with 724 rad, under normal conditions with 1210 rad, and in hypoxic conditions with 1656 rad. Probably, hyperthermia causes an increase in tissue oxygenation. The best results were obtained by irradiating in the presence of hyperbaric oxygen, followed immediately by elevated temperature. The repair of heat-induced damage was evaluated by submitting the hypoxic tissue to radiation at progressively greater time intervals from a 15-minute hyperthermic treatment at 44.5°C. After about 12 hours an almost total repair of hyperthermic damage was observed.

Elevated Temperature and Immunity

Since the beginning of the century it has been suspected that the therapeutic effect of elevated temperature on malignant tumors is linked to immune phenomena. After inoculating advanced cancer patients with bacterial toxins, Coley (Nauts, Fowler, and Bogatko 1953) noticed that their tumors regressed. He thought that an immune mechanism was responsible for this regression. Recent research has shown that immunity may be an important component of the host reaction to elevated temperatures.

Mondovi and co-workers (1972) reported that Ehrlich ascites cells that were treated with hyperthermia had increased immunogenicity and inhibited the growth of tumors in mice inoculated with live cancer cells. In addition, it was found that some heated cells have a higher immunogenicity than cells that have been treated with ionizing radiation. The length of the treatment and the choice of temperature were again extremely important. Heating the cells at 42.5°C for three hours produced a remarkable effect, but after heating at the same temperature for six hours, the immunogenicity was nearly absent (fig. 17.2). These observations have led to the supposition that on the

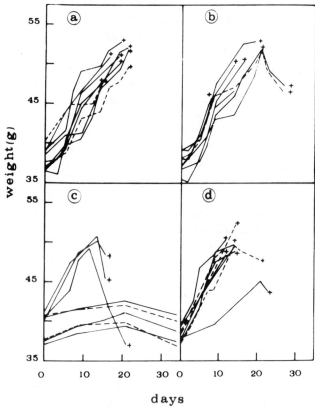

Figure 17.2. Tumor growth after inoculation of 10^7 viable tumor cells to intact mice *(a)* or to mice immunized with cells that have been exposed to radiation *(b)*, to 42.5°C for three hours *(c)*, to 42.5°C for six hours *(d)*. (Mondovi et al. 1972. Reprinted by permission.)

cell surface there is an antigenic determinant that is stimulated by a moderate amount of heat, leading to an increase in immunogenicity; however, an excessive prolongation of elevated temperature damages these supposed surface determinants.

It is probable that different tumors in different hosts will respond differently to the same treatment. Discrepancies sometimes can be ascribed to differences in experimental conditions. Suit, Sedlacek, and Wiggins (1977), for instance, observed that the immunization from radiation-killed cells was greater than that induced by hyperthermically killed cells. This could be due to the use of only syngeneic tumors and a temperature of 43.5°C. Mondovi and colleagues (1972) and Schechter, Stowe, and Moroson (1978), using 42.5°C, showed an indirect stimulating effect on the host immune response owing to the increased antigenicity of the heated tumor cells. They observed a decrease in the growth rate of both the primary tumor and the metastases when the primary tumor was heated for two periods of 90 minutes each. These authors suggested that elevated temperatures indirectly affect the immune system, because when heat was applied to the healthy limb (opposite that bearing the tumor), no apparent effect on the growth of the primary tumor or the metastases was observed. Schechter, Stowe, and Moroson (1978) believe that the increase in metastases sometimes observed after heating is caused by overheating, rather than by inadequate elevated temperature. It has been demonstrated that exposure of tumor cells to sublethal elevated temperatures causes a slight increase in their susceptibility to lysis by antiserum plus complement. In addition, strains of immunoresistant tumor cells are sometimes more thermoresistant than their immunosensitive counterparts (Strom 1970). Goldenberg and Langner (1971) have observed the disappearance of tumors on the opposite side to that being heated, and Strauss (1969) has observed a similar effect on metastases of tumors that have been necrotized by electrocoagulation. Shah and Dickson (1978) also have described an augmentation of the host immune response by local elevated temperature. They contend that generalized (total-body) hyperthermia could damage lymphoid tissue and thus result in a net negative effect on the immune state. They have also stated that the B-lymphocytes of the host probably are more susceptible to heat than are T cells. Yerushalmi (1976) also believes that local hyperthermic treatment of the tumor is more effective than a total-body treatment, which could give rise to a greater number of metastases. Harris (1976), on the other hand, has remarked that T lymphocytes are extremely sensitive to elevated temperatures. He observed that the T lymphocytes were losing lytic action on the tumor cells after 45 minutes at 43°C. Yerushalmi and Weinstein (1979), however, believe that the T lymphocytes are not involved in the body's defense against tumor growth, since the combination of localized hyperthermia and radiation prolonged the life of nude athymic mice injected with Lewis carcinoma.

Szmigielski and Jamak (1978) suggest that local treatment of tumors with microwaves (2450 MHz) produces either a specific stimulation, or stimulation of nonspecific cell-mediated immune mechanism. Hyperthermia may influence host immune defenses apart from its action on the tumor. Yerushalmi (1978) has demonstrated that when healthy animals are treated with combined hyperthermia-radiation therapy, they have a much lower rate of "takes" to transplantation of experimental tumors. Liburdy (1979), on the other hand, observed that when mice were exposed to RF (27 MHz)-induced hyperthermia, they exhibited significant changes in the distribution and functioning of lymphocytes. He feels that the action of the radiowaves on the immune system is not direct, but may be mediated through the action of steroids.

The possible immunologic aspects of treatment may be potentiated if the tumor cells are first heated and subsequently injected with complete Freund adjuvant (Bourden and Halpern 1976). Stehlin and colleagues (1975) reported that patients with melanomas of the extremities treated with hyperthermic perfusion demonstrated increased cytotoxicity of both lymphocytes and sera against melanoma cells.

EXPERIMENTAL ANIMAL MODELS

Local and Regional Hyperthermia

When considering the use of hyperthermia experimentally and clinically, one is compelled to examine the two means of thermal application: whole-body hyperthermia and local-regional hyperthermia. The first method is more challenging—the "new frontier"—but its application is potentially more hazardous because of our limited understanding of the pathophysiologic components of hyperthermia. Local-regional hyperthermia is less complicated technically and, presumably, offers greater comfort and safety to the patient.

The intent of local-regional hyperthermic treatment is to expose the tumor and surrounding tissues to above-normal temperatures of 42°C to 42.5°C; the possibilities of practical application at first seemed numerous. A series of preliminary trials, however, showed that the hyperthermia provoked by external sources of heat is neither constant nor uniform, whereas the goal of any form of hyperthermic treatment lies precisely in exposing the entire mass of neoplastic cells to the constant action of high temperature maintained at a given level throughout the area over a given period of time (i.e. two to four hours). Furthermore, the thermal gradients that occur in a part of the body that is exposed to historic external sources of heat (infrared rays, diathermy, or other) are perfectly normal when two factors are considered: (1) the dissimilarity of the tissues that make up the area being treated, which do not behave like physically inert materials; and (2) the temperature-dispersing action of the blood circulating through the district that is exposed to the heating agent. These factors suggested to us the possibility of using circulating blood as a heat transfer mechanism—that is, as the vehicle to transport and distribute the heat to the tumor and the tumor-bearing district by extracorporeal circulation technique, to allow evaluation of the biochemical and ultrastructural effects of heat, as well as clinical efficacy.

The first experiments carried out on the hindquarters of dogs showed that regional extracorporeal circulation could raise the temperature of the perfused limb, and that the increase in extremity temperature varied according to the temperature of the blood circulating in the limb. It was also observed that after a certain period of time the distribution of above-normal temperatures was uniform throughout the perfused area. Finally, after an initial period characterized by the presence of thermal gradients, limb temperature was stabilized at a steady state and the ratio between the temperature applied to the perfused limb, and the temperature dispersed by the limb no longer varied (fig. 17.3).

Although it was established that an effective hyperthermic treatment can be obtained through perfusion, many other problems were unsolved. These involved principally the unknown technical factors and doubts about relying on hyperthermic perfusion to ensure the selective action of heat on the cancer cells. Solutions to these problems were sought through experiments performed on laboratory animals. Dogs were used to assess the technique and the means of perfusion and to standardize them for subsequent clinical applications; it was also necessary to see how the perfused areas reacted, that is, to identify possible local circulatory side effects or functional alterations and to find out what type of heat-induced lesions, either immediate or delayed, could eventually contribute to the failure of the hyperthermic treatment.

The experimental models on rats were designed to help evaluate the response of the cancer cells to hyperthermic perfusion. Rats were chosen because of the facility with which experimental tumors could be

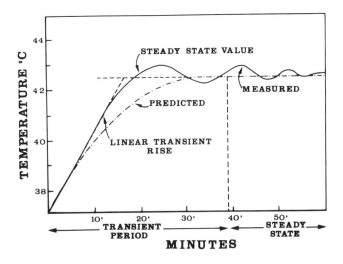

Figure 17.3. The behavior of temperature during hyperthermic perfusion. After a transient period during which the temperature rises, a steady-state period is achieved where the oscillations of temperature must be known and maintained above the level of efficacy, 42°C. (Cavaliere et al. 1980. Reprinted by permission.)

obtained and transplanted and because much experimenting and testing could be carried out within a reasonably short period of time.

Perfusion of the Limbs in the Dog

Regional hyperthermic perfusion was first performed on the hindquarters of adult male dogs weighing about 20 kg. The perfusion technique described below remained unchanged until the first clinical applications were carried out.

The surgical procedure involved the preparation and cannulation of the iliac vessels: the common iliac artery and the external iliac vein were connected to a Sigmamotor™ heart-lung machine equipped with a digital pump; the hypogastric vessels were temporarily closed off and to obtain maximum "isolation" of the limb, a tourniquet was applied to the root of the limb and secured in place with a Steinmann pin. At the beginning, perfusion was performed with whole blood, but after our first experience, it was diluted to 50% with an isotonic solution. This was siphoned from the iliac vein and pumped into the iliac artery after being oxygenated in a disk oxygenator and warmed in a heat exchanger in circulating water, thermostatically regulated at 43.5°C. Dilution of the blood soon became indispensable owing to a high incidence of hemolysis and sludging. Thus, after our first experiments, the blood was diluted at a 1:1 ratio using saline solution.

The first significant conclusion drawn was that the normal tissues of the dog can tolerate temperatures of 42°C for two hours without damage. Beyond this temperature, some normal tissue lesions may appear, probably owing to microcirculatory failure caused by injury to vessel walls. Furthermore, comparison of the experiments conducted with and without the tourniquet at the root of the limb, indicated that central venous pressure (CVP) and both rectal and esophageal temperatures should be monitored, in addition to all the other parameters that normally are recorded during major surgical operations. In fact, continuous CVP monitoring provided valuable information on possible shunts between the perfusion circuit and systemic circulation. This information was particularly useful when the tourniquet was not in use.

To verify the effectiveness of hyperthermic perfusion for the treatment of cancer, further experiments were conducted on seven dogs with spontaneous osteosarcoma or soft-tissue sarcoma of the limbs, according to the described technique and maintenance of temperatures at maximum tolerable values. After cannulation, the artery and vein (axillary or iliac) were connected to a heart-lung machine. The perfused limb was maintained at a temperature of 43°C.

In these extreme conditions, pronounced edema of the perfused limb was observed and, in some cases, the skin took on a bronze color. In one experiment, 12 hours after treatment the temperature of the perfused limb was lower than that in the rest of the body; the limb continued to cool until dry gangrene set in on the fifth day. In another five animals and in spite of pronounced edema, the condition of the limb gradually improved within one to two weeks, presenting no further complications. In one of these animals, the tumor mass was considerably reduced in size, until 12 months later when the tumor again began to grow. Follow-up was possible on the four other animals for at least 12 months. Biopsies of the treated tumors, performed serially, showed the presence of rare neoplastic cells, always associated with irreversible degenerative phenomena (Taglia 1971).

Perfusion of the Limbs in the Rat

Neoplastic cells (2.5×10^7) of Yoshida solid sarcoma and Novikoff hepatoma were transplanted into the thighs of adult male Wistar rats, weighing approximately 250 g and given a standard diet. When the tumor had reached a volume of 2 cm^3 seven to ten days after the inoculation, hyperthermic perfusion was performed as follows: the iliac artery and vein were cannulated, and the limb was perfused using a small peristaltic pump especially designed for this purpose. The pump was equipped with a disk-oxygenator and a heat exchanger maintained at 43.5°C to 44°C, using an independent water circulating unit. Several needle thermistors were inserted into the muscles of the limb and in multiple areas of the tumor to monitor the effectiveness of perfusion as a heat transfer. A tourniquet was placed at the root of the limb. Hyperthermic treatment was considered initiated only when the temperature of the entire tumor had reached 42.0°C (fig. 17.4).

The following observations were made: oxygen consumption of the tumor perfused at 38°C was the same as that of an identical unperfused limb, whereas the oxygen consumption of a tumor that had been perfused at 43°C was almost completely inhibited after 45 minutes of perfusion (fig. 17.5). Similarly, the incorporation of labeled uridine and thymidine in the nucleic acids of the tumors that had been perfused at 42°C was strongly inhibited in comparison to that observed in the control groups of rats whose limb tumors had been perfused at 38°C or not at all (table 17.2).

Transplantability tests showed that the perfused Yoshida solid sarcoma and Novikoff hepatomas stopped growing after perfusion at 42°C for 30 minutes; on the contrary, tumors perfused at 38°C continued to grow.

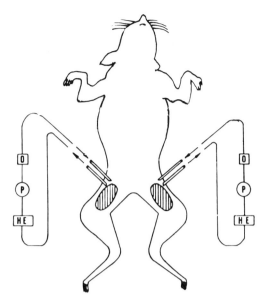

Figure 17.4. Hyperthermic perfusion of the limbs of a rat. O = oxygenator, P = pump, HE = heat exchanger.

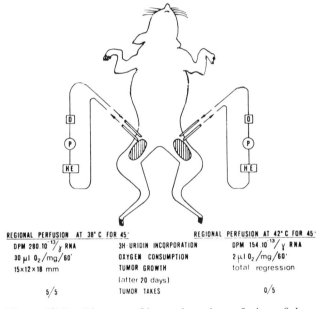

Figure 17.5. Diagram of hyperthermic perfusion of the limbs of a rat, showing comparative values of 3H-uridine incorporation, oxygen consumption, and tumor growth after treatments at 38°C and 42°C.

Regression was complete in Yoshida solid sarcomas that had been perfused at 42°C for 45 minutes and left in situ. Tissue samples removed from the tumor site 10 days after treatment were histologically examined, and no viable neoplastic cells were observed.

Perfusion of Internal Organs

Encouraged by the results of the limb perfusion models that had provided extremely interesting data confirming both the therapeutic action of heat and the reliability of perfusion as a means of achieving effective hyperthermic treatment, investigations were conducted on the kidney and the liver. The kidney was selected because it is fairly easy to perfuse—even in animals as small as the rat—and because it is a double organ, allowing selective observation of the results of hyperthermic perfusion, even in the event of partial or total failure of the treatment.

Our objectives were more ambitious in the case of the liver, as a valid treatment by hyperthermic perfusion could resolve one of the most complex problems of clinical oncology.

Kidneys of adult male Wistar rats, weighing about 250 g, were treated. Novikoff's hepatoma cells (2×10^7) were injected under the capsule of the upper pole of the left kidney in one group and Yoshida sarcoma cells were used in another group. Seven days later an extensive surgical approach was employed for the perfusion of the kidney, using the same equipment as previously described for the perfusion of the limbs of rats. Two techniques were adopted: (1) perfusion carried out through the renal vein and the main trunk of the renal artery with direct suture of these vessels at the end of perfusion, or (2) for a tumor implanted at the upper pole of the kidney, perfusion through the upper branch of the renal artery with ligature of the branch at the end of perfusion. The latter procedure avoids the complications inherent in a difficult direct suture of a small artery or in a permanent internal bypass. Otherwise, ligature of the upper branch of the renal artery may only partially affect the oxygen supply to the organ with the tumor and does modify the response of the tumor cells to heat treatment. Perfusion blood flows of about 2 ml/m were employed; the temperature was controlled by thermistors inserted into the tumor and into the subcapsular area of the normal part of the kidney. Before starting the treatment, the oxygenator was primed with blood diluted at a 1:1 ratio with a heparinized saline solution, and the temperature of the heat exchanger was set at 44°C. In the control group of animals, the same system of perfusion was adopted, but the temperature was set at 38°C.

In spite of complexities of a technical nature that necessarily limited the number of valid experiments, we obtained interesting results. Tumors perfused for 30 minutes at 42°C were not transplantable, but of tumors that had been perfused at 38°C, 100% tumor "takes" were obtained. Tumors of 20% of the animals had regressed after hyperthermic perfusion and there was no evidence of any significant distant negative effects on the function of the host organ.

The experimental models that involved isolated perfusion of the liver were aimed at evaluating two important factors: (1) the functional behavior of hepatic parenchyma when heated and (2) the behavior of the perfused tumors (viz., transplantability and evolution following treatment).

Positive results of such testing in the rat would have been very beneficial, perhaps determinant, bearing in mind the facility with which experimental tumors may be obtained and the large number of animals on which the investigations may be carried out. Unfortunately, however, this was not possible since perfusion via the hepatic artery in the rat does not heat the liver, since the portal blood supply cannot be deviated or interrupted. The technical difficulty of achieving an effective portacaval anastomosis and the extremely rapid onset of irreversible intestinal damage preclude such manipulation. Therefore, we tried to use the portal circle itself by cannulating the portal vein at its central part. Certain (by means of a sky-blue tracer) that the liver had been perfused, the cannula was hooked to the femoral artery of another rat via the heat exchanger. Thus, portal blood from the intestinal area of the first animal was carried via a reversed T cannula inserted in the portal vein of the second, guaranteeing the survival of the first rat whose liver was to be treated by perfusion. The experiments were carried out on male Wistar rats weighing 250 g and fed a standard diet. Four days before treatment, cells of Walker-256 carcinoma were inoculated into the median lobe of the liver in a suspension with a volume of 0.3 ml containing 2×10^7 viable cells. Unfortunately, only 1 of 20 animals in both groups was evaluable. It was noted, however, that the size of the hepatic tumor in the treated group had decreased significantly in comparison to those in the control group. This experimental technique was found to be too complicated to be reliable, however.

Two fairly satisfactory results were subsequently obtained from larger animals, particularly from the dog. Hyperthermic perfusion via the portal blood was successfully performed by taking blood from the superior mesenteric vein and passing it through a pump and a heat exchanger before perfusing the liver. A second model involved the performance of a true circuit of isolation perfusion: side-to-side portacaval anastomosis was performed, and blood of the suprahepatic vein was deviated from the inferior vena cava, then diluted, heated, and pumped into the hepatic artery. During this procedure the inferior vena cava was temporarily occluded upstream of the anastomosis and downstream of the outlet of the suprahepatic vein; the flow of portal blood to the liver was also blocked (fig. 17.6). The caval circulation was restored

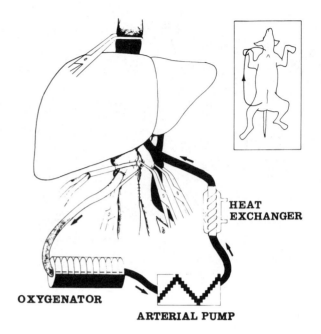

Figure 17.6. Model of isolated perfusion of the liver through hepatic artery.

by means of a femoral-jugular bypass (fig. 17.6). This model, which is technically very complicated, was performed on only a limited number of animals, but it was possible to study the behavior of certain highly important parameters. The aim was to determine, in sufficiently reliable terms, the possible types of damage that could be caused to the function of the normal hepatic cell by hyperthermic perfusion.

At present we are studying the modifications of the soluble liver protein before and after perfusion as a possible parameter to evaluate the hepatic damage caused by hyperthermia. Significant conclusions have not been reached.

Whole-Body Perfusion Hyperthermia

At the beginning of our investigation, we considered the possibility of a hyperthermic treatment extended over the entire body to have intriguing clinical implications. The principal difficulties to be confronted at that time were:

1. The lack of reliable knowledge regarding the physiopathologic aspects of hyperthermia
2. A complete absence of reliable techniques to reach and then maintain body temperature at a given level for a specific period of time

The technical problem was approached first, since the definition of an adequate experimental model would allow us to obtain the necessary information on the reaction of the whole body when totally subjected to hyperthermia. It had to be determined whether organs or parenchyma would be damaged to the point of making any concrete program unfeasible or excessively dangerous.

The first trials were carried out on healthy adult dogs of average size, weighing 18 to 20 kg. Hyperthermia was induced using a Siemens (Erlangen, West Germany) machine, the Ultratherm 608™, at a frequency of 27.12 MHz by means of a cable placed above the animal. Thermodispersion was limited by placing both the dog and the coil in a cabinet where forced ventilation could be maintained at controlled temperature and humidity (fig 17.7). Rectal and esophageal temperatures were recorded continuously, and the muscle temperatures were measured by thermistor needle probes every 10 minutes after switching off the radiofrequency (RF) machine.

The application of temperatures of up to 42°C did not cause significant lesions to the cardiovascular system. Changes were limited to sinus tachycardia, but neither rhythm nor coronary circulation were disturbed. At a temperature of 42.5°C or over, irreversible ventricular fibrillation took place. Concurrent with the increase in cardiac rate, pronounced tachypnea developed. In the dog, this is attributed to the body's effort to counteract hyperthermic stress, since heat dispersion through the skin is possible only in small areas (such as arm pits and groin). The onset of pronounced tachypnea, which strongly alters the efficiency of breathing, made respiratory support necessary. The animal was given curare, intubated, and assisted with intermittent positive pressure. The only alteration observed regarding arterial blood pressure was a lowering of the systolic pressure, although it remained sufficiently strong to ensure adequate tissue perfusion. In fact, diuresis remained spontaneously active. Central venous pressure (CVP) was reduced initially (to 3 cm H_2O), then settling to normal levels, rising dangerously only coincidentally with the disturbances in cardiac rhythm.

The effects of hyperthermia on the central nervous system (CNS) merits special attention. Recordings in the two cerebral hemispheres showed that temperature increase is higher in the brain than in the rest of the body; intracerebral temperatures were always 0.4°C to 0.6°C higher than the temperatures recorded in the esophagus, rectum, pleura, and liver. When hyperthermia is continued indefinitely at the esophageal temperature of 42°C, the animal dies, and although no causes of death are observed in the other organs or structures, post mortem examination has always revealed pronounced cerebral edema and an extremely high number of petechial hemorrhages scattered over the surface of the cerebral hemispheres.

Our next experiment involved cross-circulation between two dogs. We connected the common carotid artery of a heated dog to that of another dog whose temperature was kept at a normal level. We observed that between the two animals, the auxiliary animal died more rapidly (when the cerebral hemispheres had been perfused by warmed blood to 42.5°C) than the dog exposed to diathermy. A single trial of this kind obviously cannot be considered significant, but it is worth noting that this same phenomenon arises when hyperthermia is applied for long periods, or beyond three to four hours, the time foreseen for clinical applications. The use of mechanical, positive-pressure respirators during long-running hyperthermia on animals is able to provoke CNS lesions quite similar to those described.

Finally, evaluations of the functional activity of other organs and systems did not present any significant or easily reversible changes that could be attributed to the effect of hyperthermia. This was demonstrated in the lungs, liver, and kidneys, the only

Figure 17.7. Cabinet used for whole-body hyperthermia in which forced ventilation is maintained at both a controlled temperature and humidity. (Cavaliere et al. 1977. Reprinted by permission.)

condition being that the temperatures not exceed 42°C for a maximum time of two to four hours.

With the collaboration of the Veterinary Hospital directed by Dr. L. Taglia in Rome, we performed whole-body hyperthermia on a nine-year-old female German shepherd with breast carcinoma. General anesthesia was given using short-acting barbiturates and curare. After intubation maintained by oxygen and halothane, the dog was then placed in a cabinet, under the radiowave heating coil. After 40 minutes, rectal temperature was 43.6°C, which compared favorably with the temperature of 44°C in the tumor. Hyperthermia had to be interrupted after about one hour, however, owing to the onset of rhythm disturbances appearing as bursts of polymorphic ventricular extrasystolic beats that did not respond to pharmacologic treatment. Wakening was satisfactory, however, and a complete return to normal conditions was observed after eight hours. Two weeks later a biopsy of the tumor revealed vast areas of necrosis and degeneration of the neoplastic cells. During the 18 months that followed there were no signs of further tumor growth (the stop-effect of hyperthermia), even though the dog's owner refused to permit exploratory surgery that would have permitted us to verify what had happened within the tumor mass, which was objectively prominent but no longer growing. A second whole-body hyperthermia treatment was performed on a 13-year-old male toy Schnauzer suffering from a rectal carcinoma. Although this animal presented an azotemia of only 0.40%, its urine had a very low specific gravity and contained albumin, erythrocytes, and numerous hyaline casts (chronic interstitial nephritis). After 30 minutes of hyperthermia, there was marked tachycardia and a very intense mitral thrill: hyperthermia was interrupted. When his condition seemed to have returned to normal, the animal unexpectedly died. Histologic analysis of a sample of the tumor revealed no significant modifications.

We then treated six bitches between the ages of 8 and 13 years, all suffering from malignant neoplasms of the breast. Hyperthermia was induced and maintained for 120 to 150 minutes (never more or less) at esophageal and rectal temperatures of about 41.8°C. Temperatures in the tumor ranged from 42°C to 42.3°C. All the animals survived the treatment without difficulty, and prolonged follow-up was possible on five of them. One died of the disease within the year. The tumor of the second continued to grow, and a mastectomy was performed three months after the hyperthermic treatment. In the three other cases, no pathologic phenomena of any type attributable to neoplastic disease were observed.

CLINICAL APPLICATIONS

Introduction

The results obtained through systematic research, both basic and experimental, demonstrate the following points:

1. Heat can and does provoke damage to neoplastic cells both in vivo and in vitro.
2. This action is selective, since temperatures of 42°C to 42.5°C do not provoke irreversible damages to normal cells or parenchyma.
3. The effect of the heat is manifested, at least in experimental models, in whatever way the temperature is obtained, even though the perfusion technique seems to have a more rapid effect.

At this stage, we applied hyperthermia to clinical practice. Perfusion of the isolated extremity offered certain advantages compared to external sources of heat. Many aspects of the technique had, in fact, already been developed by Creech and co-workers (1961) for the treatment of limb tumors where high doses of antiblastic drugs[2] were transported to the neoplastic mass, while minimizing systemic toxic effects. In theory, this concept was valid also for hyperthermia, for if irreparable damage occurred from the hyperthermic perfusion, or if no therapeutic effects were obtained, amputation was always possible. In fact, the only conventional treatment still possible for some tumors of the limbs is amputation. Consequently, it is for precisely this type of tumor that we tried hyperthermic perfusion. The applicability of hyperthermic perfusion was, therefore, restricted at the beginning and for a certain number of years thereafter. The first results, reported by Cavaliere and colleagues in 1967, had demonstrated, even in humans, the therapeutic effect of hyperthermia and the selective heat sensitivity of cancer cells. Nevertheless, the incidence of postperfusion complications was unacceptably high. As it was later possible to demonstrate, this was due in part to the inadequacy of the equipment (digital pump, heat exchanger, temperature-recording systems, external applications of infrared rays, excessive thermodispersion, etc.) and in part to an incomplete understanding of the pathophysiologic aspects of hyperthermia (systemic reaction to the aggression of regional hyperthermia, reabsorption of the by-products of tumor destruction, acute secondary renal failure, etc.).

As our experience progressed and our understanding of these pathophysiologic problems grew, a satisfactory standardization of the techniques was reached and is described in the following paragraphs. Extend-

[2]The term *antiblastic drug* is synonymous to chemotherapy agent.

Table 17.4
Classification According to the M. D. Anderson Hospital and Tumor Institute

Stage I	....	No metastasis
	A:	primary removed
	B:	primary intact
Stage II	....	Local metastasis
Stage III	....	Regional metastasis
	A,	in transit
	B,	lymph node(s)
	AB,	skin and lymph node(s)
Stage IV	....	Distant metastasis

SOURCE: Sugarbaker and McBride 1976.

ing the applicability of this technique, perfusion became routine treatment, since the incidence of death and morbidity during treatment was almost the same as that of a major surgical operation routinely performed. Hyperthermic and hyperthermic antiblastic perfusion have been used for many years now to treat those patients with limb tumors in the clinical stage in which treatment is considered curative. More explicitly, the criteria for hyperthermia include stage I melanoma, when radial growth of lesions corresponds to Clark's level III or higher (Clark 1969), or when thickness is more than 1.5 mm according to Breslow (1970), and limb melanoma at stages II, IIIA, IIIB, and IIIAB[3] (table 17.4). For stage IV melanoma, hyperthermic perfusion is clearly palliative. Where osteosarcoma and soft tissue sarcomas are concerned, it is clear that when there are distant metastases, hyperthermic perfusion cannot be considered a curative treatment. In primary and/or locally recurrent tumors, however, perfusion alone or in association with other therapies has proved effective in obtaining complete local-regional control of the disease. At any rate, the use of hyperthermic perfusion appears to be justified. Even in patients with advanced tumors of the limbs, hyperthermic perfusion has proved useful, provoking regression without complications. Furthermore, there is evidence suggestive of positive effects on the immune system.

Methods

Hyperthermic Perfusion

The patient is placed under general anesthesia. In treating the upper limb, the patient is placed in a supine position with the head turned to the opposite side and the arm in an abducted position with the palm downward. The tip of the shoulder is slightly raised

---
[3]The classification adopted is that of the M. D. Anderson Hospital and Tumor Institute, Houston, Texas.

over the level of the operating table. To perform this operation (which also involves a sympathectomy, described later), an incision is made 4 cm above the clavicle, on the midportion of the sternocleidomastoid muscles, extending downward and laterally along the deltopectoral groove. At this point, the incision is deepened. The cephalic vein is left untouched but the major and minor pectoralis tendons are cut. The axillary artery and vein of the first and second portion are isolated. Their branches are interrupted or temporarily ligated; the artery and vein are cannulated using U.S.C.I. cannula (16- and 18-gauge), inserted at the division line between the first and second portion of the vessels. The cannula are connected to a heart-lung machine. The machine is equipped with a roller pump, disk oxygenator, previously primed with an isotonic solution containing heparin (25 mg), and heat exchanger; the heat exchanger is operated by an independent water-circulating unit that maintains the water temperature at a constant given level. An electromagnetic flow meter and polygraph, for the recording of the perfusion flows and pressures, are inserted into the arterial line. A thermistor is inserted just next to the arterial cannula to monitor the actual temperature of the blood entering the artery. Electrocardiogram, arterial pressure, CVP, and respiration are monitored; all are extremely important to secure a true balance between systemic and perfusion circulations. A tourniquet at the root of the limb is not used since compression at high temperatures and over long periods of time could provoke serious damage to the nerve trunks, with possible impairment to the mobility of the perfused limb. Needle thermocouple probes are inserted into the muscles of the arm and forearm and into different parts of the tumor(s); other probes are inserted into the rectum and midthoracic esophagus. All the thermometers are connected with a Data Logger-Fluke™ Model 2200 control system, equipped with a recorder and alarm. This shows temperature variations and signals any temperature drop out of the given range. It is a highly sensitive system, with an accuracy of two- to three-tenths of a degree. Once the thermometers have been inserted, the limb is wrapped in water-circulating rubber blankets thermostatically regulated to limit heat dispersion from the skin.

Perfusion then begins. When starting up the arterial pump, heparin (50 mg) and a beta-blocking agent (0.40 mg pindolol) are injected into the artery. The starting temperature of 38°C in the artery is gradually raised to 42.5°C and even to 43°C. Perfusion pressure, which is kept slightly higher than the systemic pressure, is constantly regulated according to the changes in the CVP. Flows are predetermined but subject to adjustment pending the amount of return venous flow, which takes place by siphoning, to avoid use of a second pump and possible additional damage to the blood cells. After 45 to 60 minutes of perfusion, the temperature of the tumor rises to 42°C. At this point, the temperature of the heat exchanger can be reduced by 0.5°C, and the hyperthermic treatment is under way. Perfusion is continued for a minimum of two hours to a maximum of four hours, depending on the size and histologic type of the tumor and on the condition of the patient. In general, both rectal and esophageal temperatures increase by 1.0°C or slightly more.

When the treatment has been completed, the tourniquet is removed, and the circuit is flushed with an isotonic solution containing 1 million IU of antikallikrein, and the vessel incisions are sutured. Only the pectoralis major muscle is reconstructed. During the time of the perfusion, and for a few days thereafter, controlled osmotic diuresis and low-level metabolic alkalosis are maintained in order to prevent the onset of acute renal failure. The lost blood is reintegrated; the amount of water and electrolyte administered is the amount necessary to maintain a slightly positive balance even with a highly active diuresis. Hematologic and blood gas parameters are checked every 30 minutes, both in the limb and systemically. An infusion of 1.4% sodium bicarbonate is administered for a few days.

The procedure for treating the lower limbs is essentially the same, although the surgical approach is different. A transverse abdominal incision is made approximately 2 cm below the umbilicus, ipsilateral to the limb to be perfused, providing simultaneous easy access to the vessels and the lower lumbar sympathetic chain. The rectus muscle and the external oblique and internal oblique muscles are incised in the direction of the skin incision. The transversus abdominis may be incised in the direction of its fibers. The peritoneum is reflected. The end of the aorta on the left or the beginning of the lower vena cava on the right is exposed, as well as the common iliac vessels and their junction. At this stage, a sympathectomy of the lower lumbar chain (the last three lumbar ganglia) is performed; the iliolumbar, circumflex, and lower epigastric vessels are temporarily ligated and the internal iliac artery and vein are temporarily occluded with gauze strips; the common iliac artery and the external iliac vein are cannulated (U.S.C.I. 18- and 20-gauge, respectively) via a transverse incision in the anterior vessel wall. Thermocouples are inserted in the same way as for the upper limbs, into the thigh and leg muscles and into the tumor. The limb is wrapped in thermal dispersion-control blankets, and perfusion is performed according to the procedure already described.

Hyperthermic perfusion may be performed via other vessels but we chose the iliac and axillary vessels for three reasons: (1) the possibility of using the same approach for sympathectomy, (2) the facility with which other lower vessel portions can be used in a second treatment, and (3) the possibility of observing, during the first operation, the condition of the lymph nodes, located immediately above the regional ones, to help determine the precise staging of the tumor. Frozen sections of the interpectoral and supraclavicular lymph nodes for the upper limbs, and of the retrocrural, iliac, and obturator lymph nodes for the lower limbs, are always made. A regional lymphadenectomy is never routinely performed except in cases where the nodes are grossly involved with tumor.

Hyperthermic Chemotherapeutic Perfusion

The technical aspects of hyperthermic antiblastic perfusion do not differ greatly from hyperthermic perfusion. Hyperthermic antiblastic perfusion is less complicated, requires less time, and is less costly.

The surgical approach is identical, but a tourniquet is applied at the root of the limb after cannulating the artery and vein and before perfusion is started. This is done to prevent leakage of the antiblastic drugs into the systemic circulation. The tourniquet is maintained by Steinmann pins inserted into the trapezius muscle for perfusion of the upper limbs and into the anterior-superior iliac spine for perfusion of the lower limbs. The drugs are injected in fractionated doses, at 10-minute intervals, into the arterial cannula as soon as perfusion begins, but not before the temperature of 42°C has been reached in the perfusion circuit. In using melphalan (L-phenyl alanine mustard), temperature must reach no more than 41.8°C, as the drug then becomes ineffective. The antiblastic drugs used are melphalan, 1 to 1.2 mg/kg of body weight for the treatment of melanoma, and 0.8 mg/kg followed by 0.015 mg/kg of actinomycin D for the treatment of osteosarcoma and soft tissue sarcoma. After 90 to 120 minutes of perfusion, the circuit is washed, and the incisions in the vessel walls are sutured.

Hyperthermic antiblastic perfusion, therefore, requires less time and the monitoring of perfusion flows or temperatures is unnecessary. Furthermore, it is not absolutely essential to record rectal and esophageal temperatures. This type of perfusion is performed routinely in four hours from the time the patient is anesthetized to the time of waking. CVP, ECG, and diuresis must be checked regularly. Diuresis, in particular, is always osmotically active during the operation itself and for at least 24 hours thereafter.

Several variations of this procedure also are employed (Storm, Sparks, and Morton 1979; Dyson, Storm, and Emerson, in press).

Sympathectomy and Hyperthermic Perfusion

After having standardized the perfusion technique described in the preceding paragraphs, complications that had previously occurred (reported by Cavaliere and associates in 1967) disappeared almost entirely. Mortality was reduced to 3%. A decisive advance in reducing morbidity was made by associating sympathectomy with hyperthermic perfusion, both with and without the use of antiblastic drugs.

The decision to perform a sympathectomy evolved gradually. It was observed that many patients complained of limb pain during the follow-up period. For a long time the problem of postperfusion limb pain was not considered important; it was thought to be a transient problem caused by edema and venous stasis (and was at times associated with failure of the external popliteal sciatic nerves and the characteristic phenomenon of foot drop). Having overcome this last problem through routine anterolateral fasciotomy, attention was directed to the pain developing after perfusion. Although the perfused limb appeared normal, it was observed that patients often complained of long-lasting diffuse pain. There was also a significant discrepancy between the intensity of this persistent pain and the seriousness of any circulatory disturbances that may have occurred after perfusion. Pain disappeared almost completely, however, when sypathetic blocks were made. The sooner the blocks were made following the operation, the sooner the pain diminished. All this pointed to the existence of a pathogenesis that probably was mediated by the autonomic nervous system. In the light of data reported in the literature (Challenger 1974; Fontaine 1977; Litwin 1962), it was considered that this type of pain might be due to a form of reflex-sympathetic dystrophy. On the other hand, these investigators agreed with Gros (1974) that "syndromes from irritation of the autonomous nervous system are caused either by irritation of the blood vessels or by irritation of the sympathetic nerve trunks." Hyperthermic perfusion can, therefore, be considered a sufficient stimulus to produce painful symptomatology of the perfused limb. Thus, we associated sympathectomy with perfusion, as a sympathectomy does not add greatly to the surgical difficulties.

Exposing the vessels, as in the hyperthermic perfusion procedure, a sympathectomy is easily performed. For the treatment of the upper limbs (Hershey and Calman 1967), the clavicular head of the sternocleidomastoid muscle, the anterior scalene muscle, and the thyrocervical arterial trunk are sectioned above the clavicle to obtain access to and identify the stellate ganglion. Having depressed the pleural dome, a nerve hook is placed under the sympathetic chain, and the various rami are clipped with silver ligatures; the chain is divided between the distal part of the stellate ganglion and the third dorsal ganglion. In order to avoid causing the Bernard-Horner syndrome, the proximal part of the stellate ganglion is left in situ. The axillary vessels are then cannulated and perfusion is performed (fig. 17.8).

For lower limb tumors, once the retroperitoneal structures are exposed as described, the lumbar chain can be palpated laterally against the vertebrae. By gently lifting the chain with silk threads, it is possible to identify the last three lumbar ganglia and their rami (fig. 17.9). A ligature is made between the second lumbar ganglion and the terminal division rami; clips are placed on the collateral rami and on the terminal branches; at this point, the sympathectomy has been completed (Simeone 1977).

The results obtained were better than expected. The limb pain disappeared, and the sympathectomy considerably reduced edema and postperfusional venous stasis (Cavaliere et al. 1979b).

CLINICAL RESULTS

Melanoma of the Limbs

From October, 1964 to December 1979, 91 patients were treated by perfusion for melanoma. We have adopted the classification of the M. D. Anderson Hospital and Tumor Institute (Sugarbaker and McBride 1976) (table 17.4).

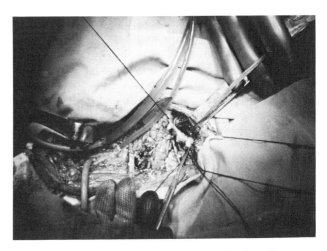

Figure 17.8. Axillary vessels are cannulated, and at the same time the distal part of the stellate ganglion is removed.

Figure 17.9. The iliac artery and vein are cannulated and at the same time the last three lumbar ganglions are removed.

Patients were grouped according to the stage and the type of perfusion (table 17.5). Perfusion is now being performed for the treatment of stage I melanoma as well, as long as the Clark's histologic level is III or more (Clark et al. 1969) and/or when the tumor is at least 1.5 mm thick, according to Breslow (1970).

In each case, several biopsies of the lymph nodes were made. A lymphadenectomy, as stated, was performed only in cases where lymph nodes were involved. The results have been evaluated in terms of survival rates, taking into account patients who were included in the first clinical trials. Some of these early patients demonstrated completely negative responses to treatment: four patients treated by hyperthermic perfusion and perfusion immediately followed by excision died less than five years posttreatment (Cavaliere 1976). Such results decrease the survival rate when the overall data are examined; however, these trials are of extreme importance to a correct application of hyperthermic perfusion. For example, in three of our patients with primary melanoma, only an incisional biopsy of the melanoma was made, and the patients were perfused later because of the degree of invasion that existed.

The five-year survival rate of patients treated for melanoma, excluding stage IV melanoma, was 62% in those subjected to hyperthermic perfusion and 59% in those treated with hyperthermic antiblastic perfusion (table 17.6). Therefore, a close similarity exists between the two groups.

It should be emphasized that between the two techniques, hyperthermic perfusion is technically more difficult and more costly. It should be established, therefore, whether the two techniques have equal therapeutic effects. In stage I and II melanoma, no great differences existed in the final results (after perfusion), but in the case of regionally spreading melanoma, that is, with in-transit metastasis and/or regional involvement of the lymph nodes (stages IIIA, IIIB, and IIIAB), it is doubtful that the two techniques provide similar results.

We tried to clarify this aspect of hyperthermic treatment by means of an evaluation of acquired data. From this, the five-year survival rate, limited to local-regional melanoma, is higher for patients treated with hyperthermic perfusion that for those treated with hyperthermic antiblastic perfusion. Examining the survival curves (Kaplan-Meier actuarial method) obtained with stages IIIA, IIIB, and IIIAB, no differences exist for the first three years. After this period, the survival

Table 17.5
Breakdown of Evaluated Patients According to Stage of Melanoma

Stage of Melanoma	No. of Patients	Hyperthermic Perfusion Alone	Hyperthermic Antiblastic Perfusion
I	10	3	7
II	13	3	10
IIIA	20	6	14
IIIB	15	6	9
IIIAB	18	6	12
IV	15	6	9
Total	91	30	61

SOURCE: Cavaliere et al. 1980.

Table 17.6
Five-Year Survival Rates Calculated Using the Kaplan-Meier Method

Stages of Disease	Hyperthermic Perfusion Alone (%)	Hyperthermic Antiblastic Perfusion (%)
I, II, IIIA, IIIB, IIIAB	62	59
II, IIIA, IIIB, IIIAB	57	47
IIIA, IIIB	79	65

SOURCE: Cavaliere et al. 1980.

rate is higher in favor of hyperthermic perfusion; this difference after five years (fig. 17.10) becomes more than 10%. Variations in this order, however, are not of great significance when applied to a small number of patients; furthermore this was not a randomized study. Figures 17.11 and 17.12 illustrate the results that were obtained with treatment by hyperthermic perfusion of patients with diffused local-regional melanoma.

Osteosarcoma of the Limbs

Between October, 1964 and December, 1979, 57 patients with osteosarcoma underwent treatment. The first nine patients were subjected to hyperthermic perfusion, and three of them can today be considered free of disease—8, 10, and 11 years after treatment. No other form of therapy was necessary. The other patients died from metastases. Metastasis might have been due to incomplete exposure of the bone tumor to heat, owing to the circulatory situation of the bone. Considering this notion and attempting to improve our results, we modified the treatment procedure by combining hyperthermic perfusion and/or hyperthermic antiblastic perfusion with systematic amputation of the limb, performed a few weeks after the perfusion. The time lapse between perfusion and amputation was set arbitrarily at four weeks, based on previous results that suggested that the presence of perfused tumor cells helped significantly in prevent-

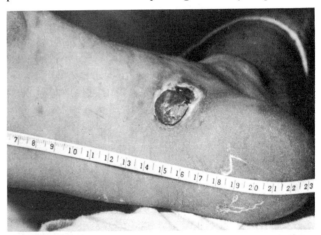

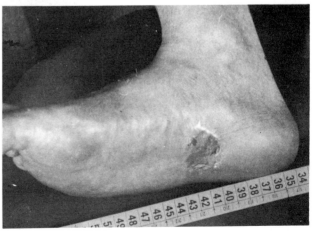

Figure 17.11. Results obtained with hyperthermic perfusion for treatment of stage II melanoma of the foot A, before and B, one year after treatment.

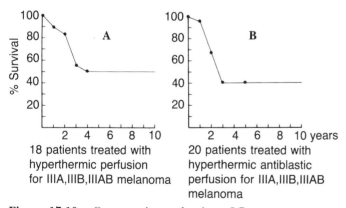

Figure 17.10. Comparative evaluation of five-year survival rates obtained in patients with stage IIIA, IIIB, and IIIAB melanoma, treated with (A) hyperthermic perfusion and (B) hyperthermic antiblastic perfusion. (Cavaliere et al. 1980. Reprinted by permission.)

Table 17.7
Osteosarcoma (Human)

Treatment	Disease-Free Survival*		
	No./Patients	(%)	Years
Hyperthermic perfusion followed by amputation at 4 wks	7/12	60	7–10
Hyperthermic antiblastic perfusion followed by amputation at 4 wks	13/17	75	1–5
	9	55	>2

*Kaplan-Meier actuarial scale.

ing further metastasis. This was found also to be true for treatment of melanomas.

Perfusion followed by amputation was performed on 29 patients with a success rate of 60%, according to Kaplan-Meier actuarial method (table 17.7). This is far greater than the rates obtained with conventional treatments, including those where major ablative surgery is performed (Moricca et al. 1977). Furthermore, analyses of tumors that were excised during amputation showed that hyperthermic perfusion had caused extensive necrosis and that the tumors were surrounded by a zone of bone sclerosis (stop effect); the appearance was that of sclerotic border of a benign tumor. Although amputation was preferred to disarticulation, recurrence was never observed in the stump.

These observations, and the predilection of osteosarcoma to afflict the young (40% of whom refuse ablative surgery), led us in 1975 to begin new studies aimed at verifying whether salvage of the limb was possible without reducing the survival rates. The treatment involves multiple therapeutic steps, from hyperthermic antiblastic perfusion to en bloc resection (when possible), bone reconstruction, and adjuvant chemotherapy. Amputation or resection is performed, depending on the state of the tumor and limb. Figures 17.13 and 17.14 present the various treatment schedules that have been defined.

The results obtained in patients treated between 1975 and 1979 (minimum elapsed time: 15 months) can be summarized as follows: of a total of 20 patients, 11 (55%) are free of disease; 8 (40%) have died (7 of lung metastases and 1 of diffused metastases); the presence of lung metastases has been recently detected in one patient. The average elapsed time between treatment and death was 15.5 months. Nine of the patients were subjected to resection, three partially and six totally, with reconstruction of the resected bone. In two patients, reconstruction involved the use of specially constructed articulated prostheses (fig. 17.15A). In one patient, reconstruction required a prosthesis that we designed and built (fig. 17.15B). In three patients, the bone was reconstructed using a portion of the contralateral bone and bone chips taken from the iliac crest; the arthrodesis thus obtained was stabilized by means of an intramedullary device that was pressure-fixed to the adjacent healthy bone (Cavaliere et al. 1978a)(fig. 17.16). In three other patients, a partial local resection was made. Of the nine patients

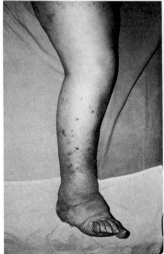
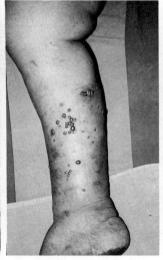
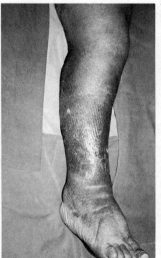
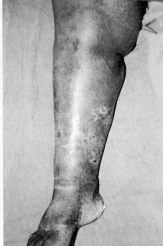

Figure 17.12. Results obtained with hyperthermic perfusion for the treatment of advanced limb melanoma. A and B, stage IIIA melanoma of the leg before perfusion; C and D, one year after perfusion.

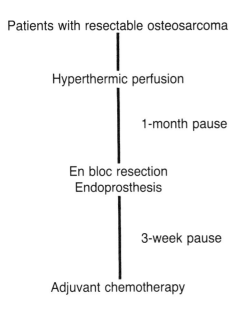

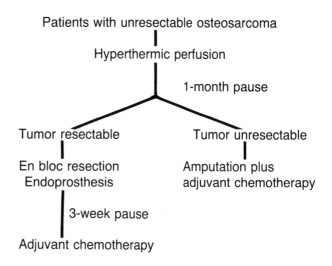

Figure 17.13. Diagram of "limb salvage" protocol of treatment of resectable osteogenic sarcoma of the limb. (Cavaliere et al. 1980. Reprinted by permission.)

Figure 17.14. Diagram of "limb salvage" protocol of treatment of unresectable osteogenic sarcoma of the limb. Cavaliere et al. 1980. Reprinted by permission.)

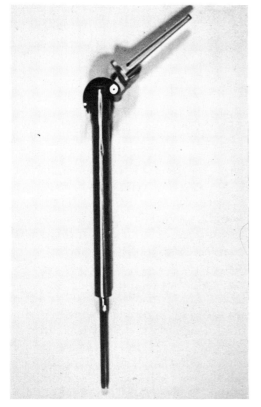
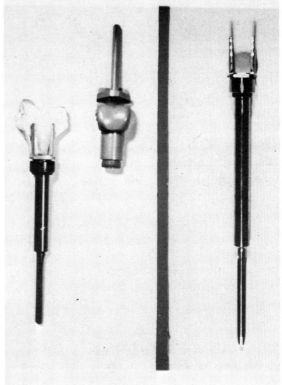

Figure 17.15. *A*, Articulated metallic endoprosthesis; *B*, specially designed and constructed metallic prostheses.

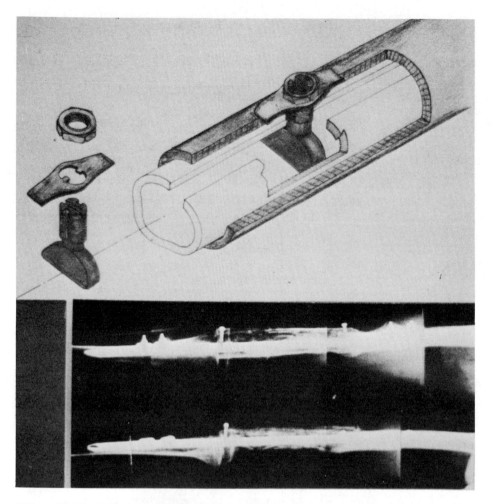

Figure 17.16. Drawing of the technique used for arthrodesis; x-ray shows distant clinical results.

that were subjected to resection, five (55%) are free of disease, while four (45%) died of metastases. No local recurrences were observed (fig. 17.17). Of 14 patients with osteogenic sarcoma that were treated, nine are alive (58%), five died (35%), and one now has lung metastases.

Soft Tissue Sarcoma

From October, 1964 to December, 1979, 50 patients with soft tissue sarcoma were treated, either by hyperthermic perfusion or by hyperthermic antiblastic perfusion, followed by excision of the tumor or amputation of the limb, depending on the histologic type of cancer, its size, and location. The five-year survival rates were 53% and 56%, respectively, for hyperthermic perfusion and hyperthermic antiblastic perfusion (table 17.8).

In the treatment of melanoma and osteosarcoma, the results have improved over the past few years—since we began performing tumor excision a few weeks after perfusion rather than immediately following treatment.

In considering our results, it must be emphasized that almost all were cases of recurrent soft tissue sarcoma that had previously been treated with radiation therapy and chemotherapy. The results cannot be considered entirely satisfactory, however, for there have been recurrences, confirming the difficulty of completely controlling these tumors, either locally or regionally, by any method of therapy.

With these considerations in mind, we modified the treatment by the random addition, between perfusion and excision, of either intraarterial drug infusion (Adriamycin) or radiotherapy (fig. 17.18). The tumor was then excised and adjuvant chemotherapy administered. Our main objectives were to obtain local-regional control of the tumor without resorting to radical surgery and distant control of the disease through (1) the potential enhancement of the immune system (which a

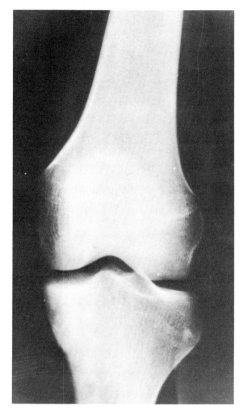 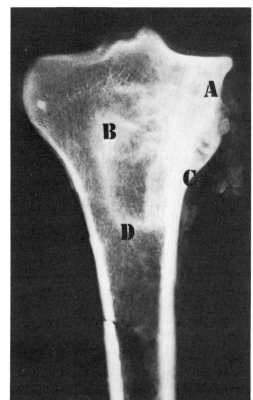 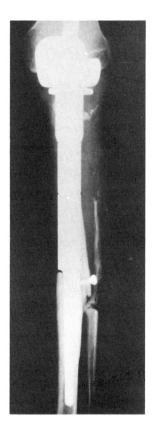

Figure 17.17. *A*, Osteosarcoma of proximal end of left tibia. *B*, X-ray of a section of the resected bone. The periosteal reaction is well structured with a continuous bone layer. There is sclerosis of the spongy bone in the demarcation zone between healthy and neoplastic bone. *Letters* indicate points where biopsy was performed with negative results. *C*, En bloc resection of the knee and proximal end of the tibia and substitution with total knee prosthesis; results 25 months later. (Cavaliere et al. 1980. Reprinted by permission.)

Table 17.8
Soft Tissue Sarcoma (Human)

Treatment	Survival at 5 Years* (%)
Hyperthermic perfusion followed by surgery	53
Hyperthermic antiblastic perfusion followed by surgery	56

*Survival at 5 yrs. Based upon Kaplan-Meier actuarial scale.

perfused tumor left in situ seems able to provoke), and (2) adjuvant chemotherapy.

Preliminary results of three-year trials following this multistep treatment confirm that local-regional disease control can be attained. No recurrences have been observed, and the histologic patterns of the excised tumors confirm this. Of the ten patients treated, only two died (12 and 19 months after the treatment), both from lung metastases. In terms of limb function (fig. 17.19), the results are very satisfactory, better than the results obtained with conservative or very radical surgery (Cavaliere et al. 1980). These preliminary results allow consideration of this multistep treatment as the most effective currently available against

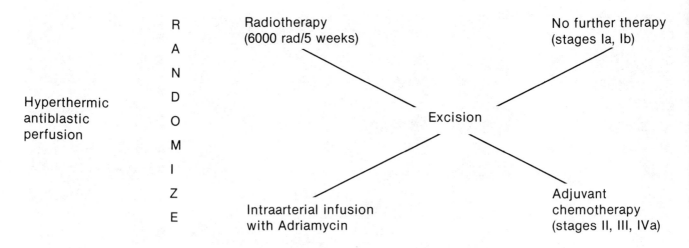

Figure 17.18. Diagram of treatment protocol for osteosarcoma. Adriamycin was administered in two cycles in one month, with dosages of 10 mg per day for 10 days, at 10-day intervals. Staging by American Joint Committee Classification.

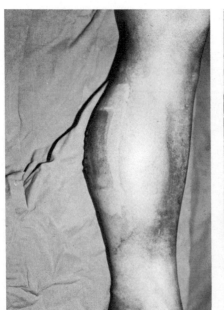
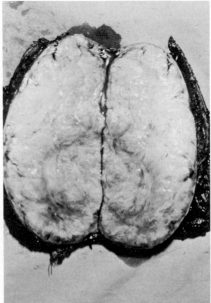
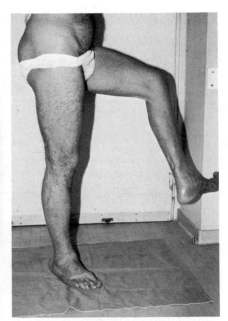

Figure 17.19. *A*, A large recurrent fibrosarcoma of the calf before treatment. *B*, The completely necrotic resected tumor mass. *C*, Distant clinical results after conservative surgical procedure. (Cavaliere et al. 1980. Reprinted by permission.)

recurrent soft tissue sarcoma. At the same time, it emerges from a cooperative evaluation with other clinical trials in which radical surgery is associated with radiotherapy and/or infusional chemotherapy. Hyperthermic perfusion is undoubtedly the most important phase in obtaining both local and distant control of the disease in a significant number of patients.

Whole-Body Hyperthermia

No description of this technique is given because clinically no technique has yet been developed that is universally satisfactory (see chapters 18 and 19).

In treating two patients several years ago, we used the technique previously described in the section on experimental models. One of the patients had disseminated melanoma and was paralyzed from cerebral metastases. After 18 hours of hyperthermia at 41.8°C, there was an improvement in the paralysis and regression of the cutaneous metastases of the tumor. The duration of the response was very short, however, and paralysis recurred after 30 days. The patient underwent a second whole-body treatment but died shortly after; an autopsy was not possible. In the second case, a 13-year-old child with disseminated rhabdomyosarcoma was treated. Again we obtained a very short-duration of improvement of symptoms. Post mortem examination revealed that metastases (found in almost all organs) appeared to be regressing.

These partially successful experiments convinced us that there is a need to develop more "pilotable" techniques that can be repeated as often as necessary, according to the evolution of the patient's disease. Convinced that heated circulating blood as a heat transfer remains, for reasons not fully understood, the best way to perform reliable hyperthermia (Cavaliere et al. 1980), we concentrated our efforts on transporting and maintaining heat through hyperthermic perfusion.

We developed an initial technique, in which the aorta, cannulated via the hypogastric artery, is used as inflow, or entry, while the inferior vena cava, which is cannulated via the external iliac vein, is used as outflow or exit. Blood is drawn from the inferior vena cava and passed through an oxygenator and a heat exchanger at 43°C before being reintroduced into the body via the aorta. The patient is placed under general anaesthesia and wrapped in an overalls-type garment made of an insulating material containing circulating water. Monitoring is continuous and complete. The water circulation is computer controlled so that flows and temperatures can be maintained according to the patient's skin temperature. The aim is to eliminate thermo-dispersion from the skin. If this dispersion is not abolished, it is not possible to raise the esophageal temperature above 41.3°C, unless the temperature of the heat exchanger is raised to a dangerously high level. The overalls in which the patient is wrapped contains the necessary openings to permit the surgical approaches.

Only slightly optimistic conclusions can be drawn from this initial testing. Perhaps the technique can be performed with far less complicated methods by resorting to the very same principles of shunting by hemodialysis as reported by Parks (see chapter 19), ensuring repeatability. This is undoubtedly an area of clinical oncology in need of much further study.

DISCUSSION AND CONCLUSIONS

Without doubt, hyperthermia is able to provoke the regression of a tumor. The mechanism of tumor cell damage by hyperthermia is not completely understood, however.

Experimental results, both personal and those of other authors, seem to confirm that the effects of heat are expressed by means of multiple mechanisms that produce a rather well-known damaging effect.

It is essential to focus on each component of the heating process so that the correct technique for administering heat and identifying the relationship between intensity and duration can be applied. These goals and the realization of maximum patient benefit are elusive. It is likely that the action of the heat involves many mechanisms and, if we accept this point of view, we must direct our efforts to identifying and characterizing them.

The mechanism of selective heat damage to tumor cells can be summarized as follows:

1) Biochemical damage from heat appears first in nucleic acids and protein synthesis machinery. There also is some damage to cell membranes.
2) Increased immunogenicity of heated tumor cells.
3) Vascular damage attributable to hyperthermia.

In particular, the study of vascular components by means of electromagnetic waves has demonstrated that the temperature of the tumor is higher than that of the surrounding tissues (Jain, Grantham, and Gullino 1979; Song 1978; Storm et al. 1979), and selective damage of newly formed capillaries occurs in solid human malignant tumors (Sugaar and LeVeen 1979). Thus, the best effects should be obtained by means of electromagnetic waves. Nevertheless, this does not correspond to clinical data in which the best results have been obtained, and are still obtainable, with perfusion—using circulating blood as a heat transfer throughout the neoplastic tissues.

The use of hyperthermic perfusion avoids the pitfall of thermo-dispersion that is related to circulating blood and permits the maximum homogeneity of heating. It is also true that the clinical effects of hyperthermic perfusion are closely related to microcirculatory modifications, caused by perfusion itself and the consequent onset of inflammatory phenomena (Moricca et al. 1977). On the contrary, during perfusion we were never able to observe the effects described by Sugaar and LeVeen, even though temperature of the tumor and the normal tissues in the perfused limb were monitored constantly.

Even the interpretation of these phenomena is far from clear, although it appears that the vascular component of the action mechanism of hyperthermia cannot easily be reconducted to a problem of thermal gradients between normal tissues and neoplastic populations. Until this aspect of the problem has been resolved and until it is possible to "pilot" the clinical applications of hyperthermia, even under its vascular mediation aspect, the supremacy of perfusion as a means of transferring heat probably will continue.

The current preference for perfusion as a treatment for limb tumors is likely to continue, even though the results are not rigorously reproducible. At this stage, however, less predictable factors must be considered, such as the interactions between hyperthermia and the immunologic host response. We must mention our experimental results which agree with those of Goldenberg and Langner (1971) on the regression of normothermic tumors, contralateral to those subjected to hyperthermia. Furthermore, there are clinical observations concerning the disappearance, often only temporarily but nevertheless definitely, of distant untreated metastases after isolated limb perfusion of melanoma and osteosarcoma.

The results obtained for osteosarcomas are extremely demonstrative as far as the modification of the immunologic response of the host is concerned. A 60% disease-free survival rate, of up to five years for patients treated with perfusion and amputation (amputation always being performed at a later date), cannot be explained if we do not hypothesize that in some way hyperthermia modifies the immunogenicity of the neoplastic cells to a significant degree. This is demonstrated clearly by the percentages of disease-free survivals that do not appear to have been altered if an en bloc resection of the perfused tumor is performed, instead of amputation a few weeks after the hyperthermic perfusion. Whether these modifications of the immune response of the host are due, as seems likely, to a clarification by means of hyperthermia of antigenic factors of the cells' surface, is a question that still must be answered, even if clinical observations seem to point to this phenomenon. For example, the hypothesis agrees that prolonged excessive heating eliminates the increase of immunogenicity (Mondovi et al. 1972) and agrees with the opinion of Schechter, Stowe, and Moroson (1978) on increased antigenicity of the heated tumor cells. Indirectly, Dickson's (1979) observations that local-regional hyperthermia is capable of evoking, from an immunologic point of view, a greater reaction than whole-body hyperthermia, confirm this, although Dickson attributes this phenomenon to the damage produced by whole-body hyperthermia to the lymphoid system. In fact, if our opinion is correct, the absence of absolute reproductivity of the results in clinical practice should be related to the way the heat is applied, as well as to the inevitable differences in the reactions of each patient.

In light of these considerations, we must stress that we have established the duration of hyperthermic perfusion in human beings in a totally arbitrary way, and it was possible only later, on the basis of our results, to establish the most beneficial time for heat exposure of a minimum of two hours to a maximum of four hours.

Only after obtaining a better understanding of the action mechanism of the heat and all its components will it be possible to perfect the techniques of application, using to the utmost the selective heat sensitivity of the neoplastic cells. Until then, especially in cases of osteosarcomas and melanomas, our results must be very carefully considered.

The most interesting aspect of hyperthermic treatment of osteosarcoma involve the increase in the percentage of survival rates and the possibility, in our most recent trial, of saving the limb every time that the tumor extent has enabled us to perform an en bloc resection. It is extremely significant that we have never observed, in our multistep treatment, local-regional recurrences that must, undoubtedly, be attributed to hyperthermia.

In the cases of limb melanoma, our results force us to conclude that treatment by hyperthermic perfusion is the best treatment of all possible therapies.

Aside from the initial stages of the disease, even when there is local-regional spreading, especially in the case of in-transit metastasis, the percentage of positive results with hyperthermia is significant. This has been verified by other authors, and especially by Stehlin (1975, 1976) and Creech (1961) who obtained significant results with antiblastic perfusion in normothermia, and then improved their results by using hyperthermic antiblastic perfusion. This introduced a new question into the field of hyperthermic perfusional treatment: Which is more advantageous, hyperthermic perfusion or hyperthermic antiblastic perfusion?

On the basis of our personal experience, we find no major differences between the two types of perfusion in treating osteosarcoma and soft tissue sarcoma; hyperthermic antiblastic perfusion accomplishes efficiently the first phase of the multistep treatment established for these tumors. On the contrary, hyperthermic perfusion usually is used in regionally spreading melanoma. Positive results are 10% higher in the actuarial evaluation over a five-year survival period. Obviously, in referring to a certain number of patients, an increase of 10% is only an indicative value. Consequently, we began a randomized trial comparing the two methods integrated within the project of the Consiglio Nazionale delle Ricerche, "Controllo della Crescita Neoplastica," Sottoprogetto Terapie Associate. A certain period of time will be necessary for a given number of patients to enter into this trial; therefore, reliable results will not be available for some time. This study is highly appropriate, since from a technical point of view, hyperthermic antiblastic perfusion is a simpler technique not requiring special equipment, except for an unsophisticated heart-lung machine and a system for monitoring temperatures. Complications are extremely rare, and only a few hours of time are required.

Hyperthermic perfusion, on the other hand, is technically more complex, requires more complicated equipment, and more time. There is a greater likelihood of complications in the perfused limb—even though today there are no serious complications at a systemic or local-regional level.

We now perform hyperthermic perfusion on only a part of those patients with local-regional spread of melanoma, both in the randomized study and in patients with isolated or distant metastases, regardless of the type of primary tumor of the limb. This choice is based on the theory that the enhancement of the immune response is greater when antiblastic drugs are not employed, even though many authors do not agree with this opinion (Stehlin 1976).

When considering whole-body hyperthermia, it becomes obvious that if it is performed in accordance with the informative principles of hyperthermia, such treatment could constitute a great step in our fight against advanced cancer and perhaps in the prevention of metastasis. There are serious difficulties, however, with treatment by whole-body hyperthermia: those of a technical nature and those caused by the body's reaction to heat. It appears that the best technique—that which gives the best results—involves a perfusional system, but always with a possible risk involved.

Among the six patients with diffuse skin and visceral metastases of melanoma that have been treated so far, we obtained regression of visible neoplastic lesions in only about 30%; however, the temperature did not reach more than 41.6°C in a period ranging between 45 and 90 minutes. Only recently have we been able to perfect the overall unit that regulates thermo-dispersion at the skin level, permitting maintenance of higher temperatures. The perfusional technique used by Parks (see chapter 19) seems to be extremely important, allowing repeated applications of whole-body hyperthermia and minimizing technical problems and the surgical approach. Investigators must continue, however, to seek eventual modifications of the immune response on the part of the individual subjected to whole-body hyperthermia, bearing in mind Dickson's (1979) caution: even if the experimental results cannot always be applied to human pathology, it is clear that a lowering of the body's defenses could constitute a valid reason to terminate the routine use of whole-body hyperthermic perfusion.

Another consideration is the association of the administration of antiblastic drugs to whole-body hyperthermia, as is presently being done for local-regional perfusion. Obviously, the problem of establishing the correct doses of these antiblastic drugs will be resolved eventually by treatment by whole-body antiblastic hyperthermia. In our view, these problems are a part of the research program already in progress, and our opinions will be expressed at its conclusion. Our research deals also with perfusion of the organs (particularly the liver), which we have studied extensively on an experimental basis, but have not studied in a clinical setting.

Hyperthermic treatment for tumors merits the great attention and significant effort required to understand the mechanisms involving the selective action of supranormal temperatures in cancer cells.

References

Ben-Hur, E.; Bronk, B. V.; and Elkind, M. M. Thermally enhanced radiosensitivity of cultured Chinese hamster cells. *Nature New Biology* 238:209–211, 1972.

Ben-Hur, E.; Elkind, M. M.; and Bronk, B. V. Thermally enhanced radioresponse of cultured Chinese hamster cells: inhibition of repair of sublethal damage and enhancement of lethal damage. *Radiat. Res.* 58:38–51, 1974.

Ben-Hur, E.; Prager, A.; and Riklis, E. Enhancement of thermal killing by polyamines. I. Survival of Chinese hamster cells. *Int. J. Cancer* 22:602–606, 1978.

Ben-Hur, E., and Rilkis, E. Enhancement of thermal killing by polyamines. II. Uptake and metabolism of exogenous polyamines in hyperthermic Chinese hamster cells. *Int. J. Cancer* 22:607–610, 1978.

Ben-Hur, E., and Riklis, E. Enhancement of thermal killing by polyamines. IV. Effect of heat and spermine on protein synthesis and ornithine decarboxylase activity. *Cancer Biochem. Biophys.* 4:25–31, 1979.

Ben-Hur, E., and Riklis, E. Enhancement of thermal killing by polyamines III. Synergism between spermine and radiation in hyperthermic Chinese hamster cells. *Radiation Res.* 78:321–328, 1979.

Bourdon, G., and Halpern, M. B. Immunité antitumorale induite par l'administration de cellules tumorales homologues et isologues traitées par chauffage ménagé. *C.R. Acad. Sc. Paris.* 282 (Série D):1571–1574, 1976.

Breslow, A. Thickness, cross-sectional areas and depth of invasion in the prognosis of cutaneous melanoma. *Ann. Surg.* 172:902–908, 1970.

Bronk, B. V. Thermal potentiation of mammalian cell killing: clues for understanding and potential for tumor therapy. *Adv. Radiat. Biol.* 6:267–324, 1976.

Bush, W. Uber den einfluss, welchem heptigere erysipeln zuweilen auf organisierte neubildungen austuben. *Verhandl. Naturh. Preuss. Rhein. Westphal.* 23:28–30, 1866.

Cavaliere, R. Regional hyperthermia by perfusion. In *Proceedings of the Second International Symposium on Cancer Therapy by Hyperthermia and Radiation*, eds. M. J. Wizenberg and J. E. Robinson. Chicago: American College of Radiology, 1976, pp. 251–265.

Cavaliere, R. et al. Selective heat sensitivity of cancer cells. *Cancer* 20:1351–1381, 1967.

Cavaliere, R. et al. Cabinet used for whole-body hyperthermia. *Recent Results Cancer Res.* 59:112–151, 1977.

Cavaliere, R. et al. Proposal for a new multidisciplinary approach for the treatment of osteogenic sarcoma. *Chemioterapia Oncologica* 4(suppl.):366–373, 1978a.

Cavaliere, R. et al. Infusione e perfusione dei tumori del fegato: osservazioni cliniche e sperimentali. Atti II Congresso della Società Italiana di Chirurgia e Oncologia, Milano, Novembre 1978b.

Cavaliere, R. et al. Fattori di rischio nelle terapie ipertermiche delle neoplasie degli arti. Atti del III Congresso della Società Italiana di Chirurgia Oncologica, Roma, Novembre 1979a.

Cavaliere, R. et al. Hyperthermic perfusion with sympathectomy for the treatment of advanced painful limb tumors. In *Advances in pain research and therapy*, vol. 2, eds. J. J. Bonica and V. Ventafridda. New York: Raven Press, 1979b.

Cavaliere, R. et al. Heat transfer problems during local perfusion in cancer treatment. *Ann. NY Acad. Sci.* 335:311–325, 1980.

Challenger, J. M. Sympathectomy nervous system blocking in pain relief. In *Relief of intractable pain*, vol. 1, ed. M. S. Swerdlow. Amsterdam: Excerpta Medica, 1974.

Clark, W. H., Jr. et al. The histogenesis and biological behaviors of primary human malignant melanomas of the skin. *Cancer Res.* 29:705–726, 1969.

Creech, D. et al. Cancer chemotherapy by perfusion. *Adv. Cancer Res.* 6:111–147, 1961.

Crile, G., Jr. The effect of heat and radiation on cancer transplanted in the feet of mice. *Cancer Res.* 23:372–380, 1963.

Dewey, W. C. et al. Heat induced lethality and chromosomal damage in synchronized Chinese hamster cells treated with 5-bromodeoxyuridine. *Int. J. Radiat. Biol.* 20:505, 1971.

Dickson, J. A. The effects of hyperthermia in animal tumor systems. *Recent Results Cancer Res.* 59:43–111, 1979.

Dietzel, F. Tumor and temperatur, aktuelle probleme bei der anwendung thermischer verfahren in onkologie und strahlentherapie. Baltimore and Munich: Urban & Schwarzenberg, 1975.

Dyson, C. W.; Storm, F. K.; and Emerson, R. C. A simplified technique for hyperthermic limb perfusion. *J. Extracorporeal Technol.*, in press.

Fajardo, L. F. et al. Effects of hyperthermia in a malignant tumor. *Cancer* 45:613–623, 1980.

Fontaine, R. Histoire de la sympathectomie lombaire de sa naissance à ce jour. *Acta Chir. Belg.* 1:3–16, 1977.

Gerner E. W. et al. Enhancement of hyperthermia-induced cytotoxicity by polyamines. *Cancer Res.* 40:432–438, 1980.

Gerner E. W., and Leith, J. T. Interaction of hyperthermia with radiation on different linear energy transfer. *Int. J. Radiat. Biol.* 31:283–288, 1977.

Gillette, E. L., and Ensley, B. A. Effect of heating order on radiation response of mouse tumor and skin. *Int. J. Radiat. Oncol. Biol. Phys.* 5:209–213, 1979.

Giovanella, B. C. et al. Selective lethal effect of supranormal temperatures on mouse sarcoma cells. *Cancer Res.* 33:2568–2578, 1973.

Giovanella, B. C.; Lohman, W. A.; and Heidelberger, C. Effects of elevated temperatures and drugs on the viability of L 12110 leukemia cells. *Cancer Res.* 30:1623–1631, 1970.

Giovanella, B. C., and Mondovi, B. Selective heat sensitivity of cancer cells. In *Recent Results in Cancer Res.*, eds. A. Rossi-Fanelli et al. Berlin and New York: Springer-Verlag, 1977, pp. 1–6.

Giovanella, B. C.; Mosti, R.; and Heidelberger, C. Biochemical and biological effects of heat on normal and neoplastic cells. *Proc. Am. Assoc. Cancer Res.* 10:28, 1969.

Giovanella, B. C.; Stehlin, J. S., Jr.; and Morgan, C. A. Selective lethal effect of supranormal temperatures on human neoplastic cells. *Cancer Res.* 36:3944–3950, 1976.

Goldenberg, D. M., and Langner, M. Direct and abscopal antitumor action of local hyperthermia. *Z. Naturforsch* 26b:359–361, 1971.

Gros, D. Pain and autonomic nervous system. In *Advances in neurology*, vol. 4: *International Symposium on Pain*, ed. J. J. Bonica, New York: Raven Press, 1974.

Hahn, G. M.; Braun, J.; and Har-Kedar, I. I. Thermochemotherapy synergism between hyperthermia (42°C–43°C) and adriamycin (or bleomycin) in mammalian cell inactivation. *Proc. Natl. Acad. Sci. USA* 72:937–940, 1975.

Harris, J. W. Effect of tumor-like assay conditions, ionizing radiation and hyperthermia on immune lysis of tumor cells by cytotoxic T-lymphocytes. *Cancer Res.* 36: 2733, 1976.

Hershey, F. B., and Calman, C. H. Sympathectomy. In *Atlas of vascular surgery*. St. Louis: C. V. Mosby, 1967.

Jain, R. K.; Grantham, F. H.; and Gullino, P. M. Blood flow and heat transfer in Walker 256 mammary carcinoma. *J. Natl. Cancer Inst.* 62:927–933, 1979.

Joshi, D. S., and Barendsen, G. W. Thermal enhancement of the effectiveness of gamma radiation for induction of reproductive death in cultured mammalian cells. *Int. Radiat. Biol.* 34:233–243, 1978.

Kim, J. H.; Hahn, E. W.; and Tokita, N. Combination of hyperthermia and radiation therapy for cutaneous malignant melanoma. *Cancer* 41:2143–2148, 1978.

Kim, J. H.; Kim, S. H.; and Hahn, E. W. The radiosensitization of hypoxic tumor cells by hyperthermia. *Radiat. Res.* 59:140–145, 1974.

Liburdy, R. P. Radiofrequency radiation alters the immune system: modulation of T- and B-lymphocyte levels and cell-mediated immunocompetence by hyperthermic radiation. *Radiat. Res.* 77:34–46, 1979.

Litwin, M. S. Postsympathectomy neuralgi. *Arch. Surg.* 84:121, 1962.

Love, R.; Soriano, R. Z.; and Walsh, R. J. Effect of hyperthermia on normal and neoplastic cells "in vitro." *Cancer Res.* 30:1525–1533, 1970.

Meyer, K. R.; Hopwood, L. E.; and Gillette, E. L. The response of mouse adenocarcinoma cells to hyperthermia and irradiation. *Radiat. Res.* 78:98–107, 1979.

Miyakoshi, J. et al. Combined effect of X-irradiation and hyperthermia (42 and 44°C) on Chinese hamster V-79 cells "in vitro." *Radiat. Res.* 79:77–88, 1979.

Mondovi, B. Biochemical and ultrastructural lesions. In *Proceedings of the First International Symposium on Cancer Therapy by Hyperthermia and Radiation*, eds. M. J. Wizenberg and J. E. Robinson. Chicago: American College of Radiology, 1976, pp. 3–15.

Mondovi, B. Temperature range and sensitivity (discussion). *Ann. N.Y. Acad. Sci.* 335:202–203, 1979.

Mondovi, B. et al. The biochemical mechanism of selective heat sensitivity of cancer cells. I. Studies on cellular respiration. *Eur. J. Cancer* 5:129–136, 1969.

Mondovi, B. et al. The biochemical mechanism of selective heat sensitivity of cancer cells. II. Studies on nucleic acids and protein synthesis. *Eur. J. Cancer* 5:137–146, 1969b.

Mondovi, B. et al. Effect of polyenic antibiotics on Ehrlich ascites and Novikoff hepatoma cells. *Cancer Res.* 31:505–509, 1971.

Mondovi, B. et al. Increased immunogenecity of Ehrlich ascites cells after heat treatment. *Cancer* 30:885–888, 1972.

Moricca, G. et al. Hyperthermic treatment of tumors: experimental and clinical applications. *Recent Results Cancer Res.* 59:112–151, 1977.

Muller, C. Die Krebskrankheit und ihre behandlung mit rontgenstrahlen und hoch frequenter electrizitat, dzw. diathermie. *Strahlentherapie* 2:170–191, 1913.

Nauts, H. C.; Fowler, G. A.; and Bogatko, F. H. A review of the influence of bacterial infection and of bacterial products (Coley's Toxins) on malignant tumours in man. *Acta. Med. Scand.* 145(suppl. 276):1–103, 1953.

Overgaard, J. The effect of local hyperthermia alone, and in combination with radiation, on solid tumors. In *Proceedings of the Second International Symposium on Cancer Therapy by Hyperthermia and Radiation*, eds. C. Streffer et al. Baltimore and Munich: Urban & Schwarzenberg, 1978, pp. 49–61.

Overgaard, J. Effect of local hyperthermia on the acute toxicity of misonidazole in mice. *Br. J. Cancer* 39:96–98, 1979.

Palzer, R. J., and Heidelberger, C. Influence of drugs and synchrony on the hyperthermic killing of HeLa cells. *Cancer Res.* 33:422–427, 1973.

Robinson, J. E. et al. Radiation and hyperthermia response of normal tissues "in situ." 3. Fraction studies. *Radiat. Res.* 59:141, 1974.

Robinson, J. E., and Wizenberg, M. J. Thermal sensitivity and the effect of elevated temperatures on the radiation sensitivity of Chinese hamster cells. *Acta Radiol. Ther. Phys. Biol.* 13:241–248, 1974.

Robinson, J. E.; Wizenberg, M. J.; and McCready, W. A. Radiation and hyperthermal response of normal tissues "in situ." *Radiology* 113:195–198, 1974.

Rohdenburg, G. L., and Prime F. Effect of combined radiation and heat on neoplasms. *Arch. Surg.* 2:116–129, 1921.

Roti-Roti, J. L., and Winward, R. T. The effects of hyperthermia on the protein-to-DNA ratio of isolated HeLa chromatin. *Radiat. Res.* 74:159–169, 1978.

Schechter, M.; Stowe, S. M.; and Moroson, H. Effects of hyperthermia on primary and metastatic tumor growth and host immune response in rats. *Cancer Res.* 38:498–502, 1978.

Shah, S. A., and Dickson, J. Effect of hyperthermia on the immune response of normal rabbits. *Cancer Res.* 38:3518–3522, 1978.

Simeone, F. A. The lumbar sympathetic-anatomy and surgical implications. *Acta Chir. Belg.* 76:17–26, 1977.

Song, C. W. Effect of hyperthermia on vascular functions of normal tissues and experimental tumors: brief communication. *J. Natl. Cancer Inst.* 60:711–713, 1978.

Stehlin, J. S. Hyperthermic perfusion with chemotherapy by perfusion. In *Proceedings of the First International Symposium on Cancer Therapy by Hyperthermia and Radiation*, eds. M. J. Wizenberg and J. E. Robinson. Chicago: American College of Radiology, 1976, pp. 266–271.

Stehlin, J. S. et al. Results of hyperthermic perfusion for melanoma of the extremities. *Surg. Gynecol. Obstet.* 140:338–348, 1975.

Storm, F. K.; Sparks, F. C.; and Morton, D. L. Treatment of melanoma of the lower extremity with intra-lesional BCG and hyperthermia perfusion. *Surg. Gynecol. Obstet.* 149:17–21, 1979.

Storm, F. K. et al. Normal tissue and solid tumor effects of hyperthermia in animal models and clinical trials. *Cancer Res.* 39:2245–2251, 1979.

Strauss, A. A. *Immunologic resistance to carcinoma produced by electrocoagulation*. Springfield, Ill.: Charles C Thomas, 1969.

Strom, R. et al. Effect of temperature on potassium-dependent stimulation of transcellular migration in normal and neoplastic cells. *FEBS Lett.* 3:343–347, 1969.

Strom, R. Ricerche sul meccanismo d'azione del calore sui tumori. *Atti Soc. It. Cancer* VII, pf. 2:49–60, 1970.

Strom, R. et al. The biochemical mechanism of selective heat sensitivity of cancer cells. IV. Inhibition of RNA synthesis. *Eur. J. Cancer* 9:103–112, 1973.

Strom, R. et al. Inhibition by elevated temperatures of ribosomal RNA maturation in Ehrlich ascites cells. *Cancer Biochem. Biophys.* 1:57–62, 1975.

Strom, R. et al. Selective heat sensitivity of cancer cells. Biochemical aspects of heat sensitivity of tumor cells. *Recent Results Cancer Res.* 59:7–35, 1977.

Sugaar, S., and LeVeen, H. H. A histopathologic study on the effects of radiofrequency thermotherapy on malignant tumors of the lung. *Cancer* 43:767–783, 1979.

Sugarbaker, E. V., and McBride, C. M. Survival and regional disease control after isolation-perfusion for invasive stage I melanoma of the extremities. *Cancer* 37:188–194, 1976.

Suit, H. D.; Sedlacek, R. S.; and Wiggins, S. Immunogenicity of tumor cells inactivated by heat. *Cancer Res.* 37:3836–3837, 1977.

Sutherland, R., and Macfarlan, W. Cytotoxicity of radiosensitisers in multicell spheroids: combination treatment with hyperthermia. *Br. J. Cancer* 37(suppl. III):168–171, 1978.

Szmigielski, S., and Janiak, M. Reaction of cell-mediated immunity to local hyperthermia of tumors and its potentiation by immunostimulation, a review. In *Cancer therapy by hyperthermia and radiation*, eds. C. Streffer et al. Baltimore and Munich: Urban & Schwarzenberg, 1978, pp. 80–88.

Taglia, L. Hyperthermia in the therapy of neoplasms in the dog. Paper read at Congresso Mondiale di Veterinaria, Citta del Messico, August 1971.

Thrall, D. E.; Gillette, E. C.; and Bauman, C. L. Effect of heat on the C_3H mouse mammary adenocarcinoma evaluated in terms of tumor growth. *Eur. J. Cancer* 9:871–875, 1973.

Thrall, D. E.; Gillette, E. L.; and Dewey, W. C. Effect of heat and ionizing radiation on normal and neoplastic tissue of the C_3H mouse. *Radiat. Res.* 63:363–377, 1975.

Tomasovic, S. P.; Turner, G. N.; and Dewey, W. C. Effect of hyperthermia on nonhistone proteins isolated with DNA. *Radiat. Res.* 73:535–552, 1978.

Turano, C. et al. The biochemical mechanism of selective heat sensitivity of cancer cells. III. Studies on lysosomes. *Eur. J. Cancer* 6:67–72, 1970.

Westermark, H. The effect of heat upon rat tumors. *Skand. Arch. Physiol.* 52:257–322, 1927.

Westra, A., and Dewey, W. C. Variation in sensitivity to heat shock during the cell cycle of Chinese hamster cells "in vitro." *Int. J. Radiat. Biol.* 19:467–477, 1971.

White, J. C. Sympathectomy for relief of pain. In *Advances in neurology*, vol. 4: *International Symposium on Pain*, ed. J. J. Bonica. New York: Raven Press, 1974, pp. 629–638.

Yerushalmi, A. Influence of metastatic spread of whole body on local tumor hyperthermia. *Eur. J. Cancer* 12:455–463, 1976.

Yerushalmi, A. Stimulation of resistance against local tumor growth, of hosts pretreated by combined local hyperthermia and X-irradiation. *Bull. Cancer (Paris)* 65:475–478, 1978.

Yerushalmi, A., and Weinstein, Y. Stimulation of resistence to tumor growth of athymic nude mice pre-treated by combined local hyperthermia and X-irradiation. *Cancer Res.* 39:1126–1128, 1979.

CHAPTER 18
Systemic Hyperthermia: Background and Principles

Joan M. Bull

Background
Methodology
Cancer Responses

BACKGROUND

Nature's systemic hyperthermia, fever, is one of the body's defenses against infectious challenge. In 1870, Busch, a German physician, observed that several cancer patients experienced spontaneous tumor regression after incurring erysipelas with concomitant high fever. Coley (1893) made similar observations and actually induced erysipelas in patients with cancer and observed their tumor regression. Using antibacterial toxins, Coley's toxins, he began to treat cancer patients. The patients treated with early nonpurified toxins developed high fevers. At least three of these patients had tumor regressions, which were thoroughly documented by Nauts, Swift, and Coley (1946). Their report included histopathology of the tumors and measurements of the tumor regression following the toxin therapy. Interestingly, subsequent purification of the toxin was accompanied by less fever, and while purified toxins did not induce tumor regression, it is evident that in the late nineteenth century the role of fever as an antineoplastic agent was being considered.

During the 1930s, Stafford Warren (1935) pioneered the treatment of patients with combined radiation therapy and systemic hyperthermia induced by "heat boxes." Warren described treatment sessions lasting up to 21 hours using this technique. His observations were particularly remarkable because he accomplished the treatments before intravenous fluid and electrolyte support was available. With burgeoning developments in ionizing-radiation therapy, however, interest in systemic fever therapy waned.

Heat was not systematically used as a form of cancer treatment again until 1967, when Cavaliere and associates successfully treated melanoma and sarcoma by hyperthermic limb perfusion (see chapter 17). Their report stimulated a study of the effect of heat on tumor cells in the laboratory. Palzer and Heidelberger (1973) and others (Giovanella, Tohman, and Heidelberger 1970) found in vivo and in vitro that tumor cells were sensitive to heat (see chapter 4). Moreover, selective heat killing of tumor cells in comparison to normal cells was observed (Overgaard and Overgaard 1972). They demonstrated that in vitro cell kill was directly proportional to both the length of time of heat exposure and the degree of temperature to which the cells were exposed. Specifically, either higher temperature or longer time exposure increased cytotoxicity. There was minimal cytotoxicity below 41.5°C, and cytotoxicity was not selective above 45°C.

The efficiency of heat-induced cytotoxicity increases dramatically between 41.5°C and 45°C. A time efficient temperature range of 43°C to 45°C is achievable, however, only by using local hyperthermia techniques. Such methods are required because of the physiologic intolerance of whole animals to systemic temperatures above 42°C for time durations longer than minutes (Pettigrew et al. 1974a). The interest in lesser temperatures, and therefore less efficient whole-body hyperthermia techniques, derives from indications that temperatures of 41.5°C to 42°C are associated with anecdotal tumor regression (Bull et al. 1978; Pettigrew et al. 1974b). Furthermore, treatment of metastatic disease requires a systemic, not local, approach. Also, current preclinical work suggests a synergism of the 41°C to 42°C-temperature range with both radiation and some chemotherapeutic agents. The preliminary clinical work that is the foundation of derivative attempts to combine the modality of heat with chemotherapy and radiation is described here.

METHODOLOGY

There are basically three methods of inducing systemic fever. The first method employs bacterial, or other exogenous protein, or chemical compounds injected intramuscularly or intravenously to cause systemic temperature elevation. Individual patient responses to various bacterial by-products (endotoxins) are quite variable, however, and this historical method is not currently popular owing to the difficulty in predicting and controlling both the degree and the time of temperature elevation.

A second means of inducing systemic hyperthermia employs the epidermal organ (the skin). This method of elevating whole-body temperature is accomplished by several techniques. In the current decade, whole-body hyperthermia was first demonstrated to be safe, reliable, and reproducible by Robert Pettigrew and associates, of Edinburgh, Scotland. They use warm paraffin (wax) and warm anesthetic gases to induce systemic temperature elevation. The patient, monitored with esophageal, rectal, and tympanic membrane thermistor temperature probes, is immersed in warm paraffin. The fusion point of the wax is 43°C to 45°C; the wax therefore, acts as an insulating barrier to the skin at the interface of the skin and wax. The melting point of the wax is 55°C. Thus, the liquid wax contributes more than 300 k calories per minute to a patient whose skin is protected from thermal injury by the solid phase wax adjacent to the skin. The wax also insulates the subject against loss of endogenous metabolic heat. Pettigrew and co-workers elevate whole-body temperature 3°C to 6°C per hour by this technique. They have treated patients for as long as 20 consecutive hours, although the usual time of treatment varies between two and six hours. The patients are anesthetized using spinal anesthesia and endotracheally administered respiratory anesthetic agents. This method has been shown to be safe and efficient, and is currently in use by Pettigrew and colleagues (1974a) and by Blair and Levin (1978).

A variation of this technique also uses the skin as a heat-inducing organ but does so by enclosing the body in a warm-water circulating suit. The patient's own metabolic heat is contained by a vapor-impermeable barrier. Esophageal and rectal temperature monitoring is done with calibrated thermistor temperature probes. Heat is applied to the skin with a high-flow, low-pressure water suit. Induction temperature can be hand controlled with careful monitoring (Barlogie et al. 1949; Larkin 1977), or regulated by a microprocessor-controlled feedback-loop based on esophageal and rectal temperatures and keyed to keep the core temperature (esophageal and rectal temperatures) always below 42°C (Bull et al. 1979a). The time to temperature induction with both the wax and water suit techniques is two to three and one-half hours. Most commonly, this form of hyperthermia application is induced using endotracheally administered respiratory gas general anesthesia. The Seimans water-cabinet (Seimans Corporation, Erlangen, West Germany) is a variation of this technique (Neumann et al., in press).

The National Cancer Institute (NCI) team used a feedback loop microprocessor-controlled warm-water suit but employed intravenous sedation with spontaneous room air respiration. The difference in anesthesia is important because the type of anesthesia affects a patient's systemic pH and the vascular flow to visceral organs (Lees et al., in press). Using spontaneous respiration, the intravenous anesthesia favors a systemic alkalotic pH. The endotracheal respiratory gas anesthesia technique of Pettigrew and co-workers, of Blair and Levin, and of Larkin and colleagues allows patients to be at physiologic or somewhat acidic pH. This may affect the response to hyperthermia of both the tumor and normal tissue (Nielsen and Overgaard 1979).

Elevation in core temperature results from a combination of heat delivered added to the endogenous metabolic heat of the subject. Cooling of the patient after hyperthermia with paraffin or water suit is done by opening the insulating layer or garment, lowering the water temperature, and cooling the patient by rapid air convection. The time required to cool patients is quite brief compared with that of the heating phase, and patients return to 37°C core temperature usually by 20 to 45 minutes after therapy.

Heating the body through the skin surface causes profound cardiovascular stress. In studies done by

Bull and co-workers (1979a) using a flow direction triple lumen thermodilution pulmonary artery catheter to monitor right atrial pulmonary artery, pulmonary capillary wedge pressure, and cardiac output, the systemic arterial pressure and pulmonary wedge pressure fell, and the cardiac index rose dramatically from a median value of 3.3 ± 2 l/min/M^2 to a high of 7.2 ± 0.3 l/min/M^2, or nearly twofold, in 13 patients treated to a systemic temperature of 41.8°C for two hours. Cardiac rate increased with each degree rise of temperature, with a maximum cardiac rate reaching 160 ($\pm$ 9) beats per minute at 41.8°C. These findings indicate a high output cardiac state during temperature induction and maintenance phase on whole-body hyperthermia. It should be noted that no cardiac damage was documented at any time with heat alone at these temperatures.

Patients lose large amounts of fluid in perspiration with all induction whole-body heating methods. The fluid loss is replaced with a half to one liter per hour of dextrose and saline. Changes in serum electrolytes do not occur with adequate intravenous fluid replacement.

The serum CPK values at 24 hours are elevated. The elevation is caused by an increase in the CPK isoenzyme, indicating skeletal muscle as the tissue causing the general enzyme elevation (Bull et al. 1979a). Use of the skin induction technique causes an acute decrease in creatinine clearance during the actual heating period. The creatinine clearance rapidly returns to normal after cooling, however, probably because of a preferential shunting of blood from the visceral organs to the skin (Lees et al., in press).

Pettigrew and colleagues (1974a) have documented that when the systemic temperature is controlled below 41.8°C, there is little elevation in the liver enzymes LDH, SGOT, SGPT, or bilirubin. When the temperature rises above 41.8°C, however, an increase in these values indicates hepatic toxicity. Bull and co-workers (1979a), Parks and colleagues (1979), and Barlogie and associates (1949) have documented a fall in serum phosphate and magnesium during whole-body heating. It had been considered that hypophosphatemia and hypomagnesemia occur as a response to the respiratory alkalosis that occurs during the procedure while using spontaneous respiratory support, for no urinary or stool loss of phosphate or magnesium were documented. Since the patients that were maintained at normal pH also developed hypophosphatemia, it seems plausible that an intracellular shift of the serum phosphate and magnesium occurs as a direct response to heat. Further elucidation of the actual cellular physiology with intracellular pH probes is required.

Toxicities observed with the surface heating method include nausea and vomiting in approximately half of the patients as reported by all investigators. The nausea occurs in the first 12 hours after heating. Approximately 40% of the patients experienced diarrhea. Twenty percent of patients experience a posthyperthermia fever as high as 41°C from 6 to 24 hours following the cooling phase. This fever gradually subsides over two to five days. The fevers occur in the absence of a documented infectious source, and appear to have no prognostic value as far as tumor response. The water blanket technique occasionally causes superficial pressure burns on heel and buttock pressure areas.

While seizure or coma occur with heat stroke, there has been no seizure or other central nervous system dysfunction observed in the series of Pettigrew and colleagues (1974a) or in the NCI series (Bull et al. 1979a), although seizures have been described by Barlogie and associates (1949). All investigators have reported instances of reversible peripheral neuropathy that resemble the peripheral neuropathy that is caused by the vinca alkaloid and platinum compound chemotherapeutic agents.

The third means of inducing whole-body hyperthermia involves a surgical procedure developed by Parks and Frazier (see chapter 19). The core temperature is elevated internally from the blood vessels by employing an arteriovenous (AV) shunt and an extracorporeal heat exchanger. This method is time efficient in that patients can be elevated to 41.5°C to 42°C within 30 to 90 minutes; therefore, the time to temperature induction is much shorter than with surface heating methods. The cardiac output was documented in four cases to be elevated to 10.8 l/min/M^2. The pulmonary artery and pulmonary artery wedge pressures were minimally affected by the hyperthermia. In contrast to the observations made using the water suit method, Parks (chapter 19) describes an increase in urine phosphate to 38 to 104 mEq/hr during treatment and not returning to baseline values for 96 hours. No incidence of hemolysis was noted.

It is critical to define and compare differences in organ or tumor vascular response in patients treated by the AV shunt technique and by the skin heating methods.

CANCER RESPONSES

Heat Alone

The results of studies documenting cancer responses to systemic heat alone should be considered anecdotal in nature because of the small patient numbers. Pettigrew and co-workers (1974b) reported 38 patients with advanced previously treated tumors who were treated with hyperthermia alone in 188 treatment sessions. The treatment dose was four hours at 41°C followed by two treatments of four hours each at 41.8°C. They observed objective tumor responses in previously treated unresponsive sarcoma, gastric carcinoma, colon carcinoma, melanoma, carcinoma of the lung, and neuroblastoma. The responses to heat alone are interesting because the responses were in unresponsive tumor histologies that had failed other therapies. Of note, these investigators observed one instance of extremely rapid and dramatic tumor lysis that proved lethal because of the associated hepatic and renal decompensation.

Larkin and colleagues (1977) described frequent objective responses in tumors and also reported rare rapid, dramatic, but morbid responses in large tumors involving the liver. Bull and associates (1978) reported a series of 14 patients treated for four hours at 41.8°C. Of these 14 patients, four had objective tumor responses (<50% of tumor regression). Two responses were of colon carcinoma metastatic to the liver, one of a melanoma, and one of an adrenal carcinoma.

There is no information, however, to suggest at what interval the heat treatments should be given or the optimum duration of therapy for any patient or any type of tumor. In order to gain information about this important topic a large series of patients with single histologies is necessary.

Heat and Chemotherapy

The question of timing sequence is even more complicated when one begins to look at the interactions of heat plus chemotherapy. There have been several phase I studies done examining the effects of heat plus Adriamycin (Bull et al. 1979b), heat plus cyclophosphamide (Parks et al. 1979), and heat plus the nitrosoureas, methyl-CCNU and BCNU (Bull et al. 1980). These studies indicate a synergism of heat with Adriamycin and heat with the nitrosoureas. Whether this interaction of heat and drug is differentially cytotoxic to tumor compared to normal tissue is not yet clear. All series combining heat and drug have pragmatically administered drug simultaneously with heat without attempt to sequence the heat and drug. The question of time and sequence is extremely important and should be more completely explored with a preclinical animal model; however, the absence at present of a suitable animal model makes it necessary to look at the question in phase I clinical trials.

The whole question of the vascular supply of the tumor as altered by heat supplied by the skin, by AV shunt, or by alteration by vasoactive drugs on the effect of chemotherapy is open. These effects on the metabolism and activation of drugs and final effect on tumor are not understood. The effect on the hepatic, renal, and cellular metabolism of pharmacologic agents has not yet been approached in systemic hyperthermia. The effect of whole-body heat on the immune system is also unknown.

Whole-body hyperthermia has only recently been reported in combination with radiation (see chapter 19). In vitro data suggest that radiation effect is synergistically increased by temperatures between 41°C and 42°C. In addition to the current investigations of local hyperthermia with radiation, a means of increasing tumor kill to large abdominal or thoracic tumors using systemic hyperthermia should be explored.

The field of systemic hyperthermia is in its infancy. The investigation of the optimal sequencing of heat alone, the optimal time sequencing of heat plus chemotherapy, the optimal effect of changes in blood flow to the chemotherapy pharmacology, as well as exploring the effect of whole-body heat with radiation therapy as a means of augmenting treatment to deep tumors, should be the future task of investigators.

References

Barlogie, B. et al. Total body hyperthermia with and without chemotherapy for advanced human neoplasms. *Cancer Res.* 39:1481–1489, 1949.

Blair, R. M., and Levin, W. Clinical experience of induction and maintenance of whole body hyperthermia. In *Proceedings of the Second International Symposium on Cancer Therapy by Hyperthermia and Radiation*, eds. C. Streffer et al. Baltimore and Munich: Urban & Schwarzenberg, 1978, pp. 318–321.

Bull, J. M. et al. Whole body hyperthermia—now a feasible addition to cancer treatment. *Proc. Am. Soc. Clin. Oncol.* 19:405, 1978.

Bull, J. M. et al. Whole body hyperthermia: a phase-I trial of a potential adjuvant to chemotherapy *Ann. Intern. Med.* 90:317–322, 1979a.

Bull, J. M. et al. A phase-I trial of systemic heat and Adriamycin. *Proc. Am. Soc. Clin. Oncol.* 20:398, 1979b.

Bull, J. M. et al. Whole body hyperthermia combined with nitrosourea in malignant melanoma. *Proc. Am. Soc. Clin. Oncol.* 21:481, 1980.

Busch, W. Ueber den eingluss, welchen heftigere erysepeln zuweilen auf organisirte neubeldungers ausüben. *Verhanal. d. Naturh. Verd. Pruess*, Rheinl. u. Westphal, Bonn 23:28–30, 1866.

Cavaliere, R. et al. Selective heat sensitivity of cancer cells. Biochemical and clinical studies. *Cancer* 20:1351–1381, 1967.

Coley, W. B. The treatment of malignant tumors by repeated inoculations of erysipelas; with a report of ten original cases. *Am. J. Med Sci.* 105:487–511, 1893.

Giovanella, B. C.; Tohman, W. A.; and Heidelberger, C. Effects of elevated temperatures and drugs on the viability of L1210 leukemia cells. *Cancer Res.* 30:1623–1631, 1970.

Larkin, J. M. et al. Systemic thermotherapy: description of a method and physiologic tolerance in clinical subjects. *Cancer* 40:3155–3159, 1977.

Lees, D. E. et al. Internal organ hypoxia during hyperthermia cancer therapy in humans. In *Proceedings of the Third International Symposium on Cancer Therapy by Hyperthermia, Drugs and Radiation*, eds. L. A. Dethlefsen and W. C. Dewey. Bethesda, Md.: *J. Natl. Cancer Inst.* (special journal supplement), in press.

Nauts, H. C.; Swift, W. E.; and Coley, B. L. The treatment of malignant tumors by bacterial toxins as developed by the late William B. Coley, M.D., reviewed in the light of modern research. *Cancer Res.* 6:205–216, 1946.

Neumann, H. et al. Moderate whole body hyperthermia in treatment of small cell carcinoma of the lung—a pilot study. In *Proceedings of the Third International Symposium on Cancer Therapy by Hyperthermia, Drugs and Radiation*, eds. L. A. Dethlefsen and W. C. Dewey. Bethesda, Md.: *J. Natl. Cancer Inst.* (special journal supplement), in press.

Nielson, O. S., and Overgaard, J. Effect of extracellular pH on thermotolerance and recovery of hyperthermic damage in vitro. *Cancer Res.* 39:2772–2778, 1979.

Overgaard, K., and Overgaard, J. Investigations on the possibility of a thermic tumour therapy. I. Short-wave treatment of a transplanted isologous mouse mammary carcinoma. *Eur. J. Cancer* 8:65–78, 1972.

Palzer, R. J., and Heidelberger, C. Studies on the quantitative biology of hyperthermic killing of HeLa cells. *Cancer Res.* 33:415–421, 1973.

Parks, L. C. et al. Treatment of far advanced bronchogenic carcinoma by extracorporeally induced systemic hyperthermia. *J. Thorac. Cardiovasc. Surg.* 78:883–897, 1979.

Pettigrew, R. T. et al. Circulatory and biochemical effects of whole body hyperthermia. *Br. J. Surg.* 61:727–730, 1974a.

Pettigrew, R. T. et al. Clinical effects of whole body hyperthermia in advanced malignancy. *Br. Med. J.* 4:679-682, 1974b.

Warren, S. L. Preliminary study of the effect of artificial fever upon hopeless tumor cases. *Am. J. Roentgenol. Radium Ther.* 33:75–87, 1935.

CHAPTER 19
Systemic Hyperthermia by Extracorporeal Induction: Techniques and Results

Leon C. Parks
George V. Smith

Historical Considerations
Eligibility of Patients and Disease
The Prehyperthermia Evaluation
Technique
Management During Hyperthermia
Measurement of Temperature
Dosage and Effectiveness
Consideration for Increasing Dosage
Combined Radiation and Hyperthermia
Tumor Pathology after Hyperthermia
Immunologic Effects
Anticancer Effects: Clinical Trials
Role of the Surgeon
Special Problems
Conclusions

HISTORICAL CONSIDERATIONS

The recent advent of safe and efficient methods for production of hyperthermia, temperatures greater than 41.5°C (106.7°F), has reawakened interest in this century-old approach to cancer treatment. Early observations of tumor regression following happenstance infection led Coley (1893) and others (Fowler 1969; Miller and Nicholson 1971; Nauts, Fowler, and Bogatko 1953), to deliberate injections of bacterial substances, but the potential therapeutic effects of infection were forgotten in the rush to participate in the rapid advances of surgery, radiotherapy, and pharmacology. Those who continued investigation of the effects of infection on cancer may have retarded for decades the development of this modality by attributing the anticancer effects observed to a stimulation of immunity rather than a possible response to fever per se. This uncertainty contributed to the origins of cancer immunotherapy, whose practitioners 80 years after Coley's report still would inject bacteria (Hersh et al. 1976)—a complex solution to the perhaps simple need for a method by which to induce artificial fever.

Recognized again as "nature's instrument" (Wizenberg and Robinson 1976), fever (local, regional, and systemic) is being induced in cancer patients by a variety of physical methods, whose advantages and limitations are poorly characterized but important to ascertain. The principal issue is a recognition that most significant malignancy is a systemic disease at diagnosis, if not at onset whose effective therapy will require whole-body hyperthermia for optimal effectiveness.

Proponents of local and regional heating suggest that the destruction of some portion of a cancer by heat can evoke an all-encompassing, presumably immunologic, whole-body anticancer response, but data indicative of such an effect remain scant. In some animal tumor models local heating of the primary tumor will suppress unheated implants (Dickson and Shah 1978), and cauterization of human colorectal tumors results in the cure of some patients presumed to have distal disease (Madden and Kandalaff 1971). It has not been shown, however, that these effects are generally applicable to management of advanced cancer. The development of various forms of electromagnetic (microwave, radar, radio, differing but in

---

The authors gratefully acknowledge the significant contributions of Deanna Minaberry, R.N.; Ronald J. Jackson; Bernard Hickman, M.D.; Patricia Roane, C.R.N.A.; Pat Wigley; Doyle P. Smith, M.D.; Scott McCraw, C.R.N.A.; M. Don Turner, Ph.D.; William A. Neely, M.D.; James B. Grogan, Ph.D.; Richard McMillan, R.N.; Patricia Etheridge, R.N.; Sharron Poole; Elene Griffin Thomas; James M. Goodman; Tomiko Mita; and William DeVeer.

wavelength) inductive heating of tumor tissue should soon enable adequate testing of this hypothesis, but early results suggest that most disseminated tumors are not curable by this approach. Therefore, it may be fairly proposed that whole-body cancer requires whole-body hyperthermia.

A variety of methods for induction of whole-body hyperthermia by surface heating have been used in recent clinical trials. Pettigrew and colleagues (1974) showed that by immersion in molten wax, the core body temperature could be elevated to 41.8°C with an acceptable incidence of adverse effects and the regression of advanced solid tumor malignancies. Another technique of surface heating was proposed by Larkin (1979), who, by wrapping patients in heating blankets and insulating materials, induced hyperthermia in patients with preterminal malignancy and recently reported a 43% objective response rate with modest complications. A similar approach was employed by Barlogie and co-workers (1979), who reported that of eleven evaluable patients with advanced metastatic disease, seven achieved stable disease status, and four manifested objective signs of tumor regression after serial treatments at 41.9°C to 42.0°C for four-hour durations. Bull and colleagues, after development of a high flow, warm-water perfusion suit, exposed 14 patients to several sessions each of four hours at 41.8°C, demonstrating that whole-body hyperthermia was feasible. They reported that antitumor activity was suggested by development of three objective responses among the 14 patients. While these studies demonstrate that whole-body hyperthermia of a magnitude active against cancer may be produced by surface heating, the methodology is cumbersome and has had limited acceptance. Thus, the lack of a means for ready induction of precision hyperthermia applicable to large numbers of patients appears to have deterred investigation of this modality.

In 1977, Parks, Frazier, and Mountain proposed the following criteria for a desirable system of whole-body hyperthermia:

1. Induce systemic temperatures of 41.5°C to 43.0°C.
2. Within one hour elevate a patient's temperature to that selected for treatment.
3. Be capable of maintaining hyperthermia for 24 to 72 hours.
4. Provide fine temperature control and autoregulation of temperature.
5. Readily enable a rapid cooling as well as a rapid heating effect.
6. Do not require general anesthesia or, if necessary, enable it to be of minimal depth.
7. Allow patients undergoing hyperthermia to be maintained in a setting conducive to conventional care.
8. Provide hyperthermia in diverse settings, such as radiation therapy or isolation areas.
9. Enable the repetitious induction of sessions of hyperthermia over intervals of several days or weeks.
10. Induce hyperthermia with reasonable safety in neutropenic or thrombocytopenic patients.

Surface heating by application of warm water suits, blankets, or wax appeared to satisfy some of the listed criteria, while others appeared best fulfilled by use of an extracorporeal circuit for temperature control. Some advantages and disadvantages of these nonexclusive approaches are shown in table 19.1. On balance, the extracorporeal approach was deemed advantageous, and laboratory and clinical studies of the method were undertaken.

The extracorporeal induction of systemic hyperthermia is derived in part by relating concepts developed for regional hyperthermic perfusion (Krementz et al. 1977), open heart surgery, and hemodialysis. This methodology has been employed at The University of Mississippi Medical Center (UMC) to the extent that by January 1, 1980, after 30 months of effort, 371 treatments had been administered to 102 patients. This chapter is a synthesis of that experience, and although derived in the setting of a unique methodology, much of what was learned appears generally applicable to the practice of whole-body hyperthermia.

ELIGIBILITY OF PATIENTS AND DISEASE

It should be borne in mind that the following reflects the authors' views at a time when a substantial experience with whole-body hyperthermia has been gained. The opinions expressed are those currently held, which are less conservative than those existing at the beginning of the UMC series.

Whole-body hyperthermia in our hands is now sufficiently safe to be considered for use as adjuvant therapy in patients not proved to have recurrent malignancy. Such a view would not have been responsible during the early stages of development of our procedure. Thus, physicians proposing to use whole-body hyperthermia for cancer treatment should not necessarily consider the following as guidelines for immediate implementation, but as that which may be possible with some experience with the modality and with its effects upon normal and malignant tissues in humans.

Most patients in average condition for their years are capable of tolerating whole-body hyperthermia of 41.5°C to 41.8°C for eight hours of treatment, and such treatments can be repeated at several-day inter-

Table 19.1
Comparison of Effects of Induction of Whole-Body Hyperthermia by Surface and Extracorporeal Heating

Surface Heating	Extracorporeal Heating
Simpler mechanism for short-duration HT	Uniquely suited to long-duration HT
Slow rate of induction and change	Rapid induction and temperature alternation
Minimal patient manipulation	Surgical implantation of shunt required
Risk of skin burn/maceration	Negligible risk of skin injury
Reversal of normal temperature gradients, maximum heat at surface	Normal direction of temperature gradients, from core to skin
Maximum heat on pain/temperature skin receptors?, increase catechol release	Minimal surface heating, may reduce catechol response
Requires patient wrapping; difficult to manipulate extremities	Minimal insulation necessary, normal sweat evaporation; easy patient manipulation
Slow response makes precise temperature control difficult	Rapid response allows tight temperature control
Difficult access to ECG leads, IV, and arterial lines, wounds, ostomies	Normal access to patient
No risk of thromboembolic problem	Some risk of thromboembolism

NOTE: HT = hyperthermia.

vals to a total of at least four treatments without undue difficulty. The only specific contraindication to hyperthermia now recognized is symptomatic congestive heart failure in a patient who is receiving optimal medical support with digoxin and diuretics. Patients having modest symptoms of congestive heart failure, who are not receiving these medications, often will tolerate hyperthermia after digitalization and reduction of total body water. A second general requirement is adequate pulmonary function, with the single best criteria appearing to be an FEV_1 exceeding 1.2 liters. This need for adequate cardiopulmonary function appears to be the principle determinant for patient acceptance or exclusion.

Less important factors to consider are the presence of significant ascites, which generally must be resolved prehyperthermia either by diuresis or a LeVeen shunt, and a requirement for renal function adequate to maintain a serum creatinine of approximately 1.5 mg%. While it is possible to treat patients with moderate ascites, or with poor renal function, the difficulties with fluid and electrolyte management become substantial, and the patient risk is considerably higher than usual.

Jaundice, central nervous sytem (CNS) metastasis, history of myocardial infarction or alcohol abuse, significant peripheral vascular disease, inferior vena cava obstruction, large tumor mass, correctable infection, and a host of similar factors have all been found not to significantly impair a patient's ability to tolerate whole-body hyperthermia. The oldest patient in our series is 72 years of age, and the youngest is 18. Because we have easily managed 10- to 15-kg laboratory animals undergoing extracorporeal hyperthermia, a lower age limit may be considered.

There has been little evidence that one type of tumor is more sensitive to hyperthermia than another, although there is some indication that hypernephroma and large cell and small cell bronchogenic carcinoma are unusually susceptible to hyperthermia. Conversely, there is a suggestion that breast cancer is less responsive to hyperthermia than most of the other malignancies treated; however, the breast cancers in this series have all been very far advanced failures of multimodality therapy, and conclusions based on these resistant malignancies may not be valid for earlier disease. In general, it is felt at this time that the probability of obtaining a satisfactory anticancer response is relatively independent of the tumor site and histology.

The decision as to the stage at which a tumor becomes a suitable candidate for hyperthermia is related to the probability of effectively treating the malignancy with other modalities. Unresectable cancer of the pancreas and adenocarcinoma of the lung are such resistant lesions that hyperthermia may be recommended upon their diagnosis, and patients need not be required to have received treatment with other modalities such as radiation or chemotherapy. Instead, it may be advantageous to use these modalities in conjunction with hyperthermia as a combined therapeutic approach. In contrast, at this time, breast cancer should not be treated with hyperthermia until the feasibility of hormonal therapy has been evaluated, and at least one treatment course of combined chemotherapy has been given. A modification of this view is beginning to be considered, however, for it may be advantageous to apply hyperthermia in a setting where initial therapy with hormones, chemotherapy, or radiation has damaged a tumor and produced a "crippled cell" population. Hyperthermia has a substantial potential to interfere with cell repair processes and its application to tumors after initiation of an initial tumor regression may be highly advantageous. Thus, in the months ahead, the inclusion of hyperthermia in initial treatment plans for recurrent breast cancer should be considered.

It is considered appropriate at present to employ hyperthermia for treatment of recurrent bronchogenic cancer of every type except oat cell, of all unresectable adenocarcinomas of the gut and unresectable esophageal cancer regardless of cell type, for advanced hypernephroma, for breast cancer that has not responded to initial palliative therapy, and for many recurrent or metastatic sarcomas of adults. Other difficult-to-treat advanced malignancies may be candidates for hyperthermia, but there is not yet sufficient evidence on which to base an objective opinion.

It is becoming appropriate to consider the use of hyperthermia as an adjuvant to surgery for malignancies that are resectable but that have a high probability of recurrence. Possible candidates for this approach are patients with bronchogenic cancers, which manifest vascular or lymphatic involvement; most patients having esophageal, gastric, or pancreatic neoplasms; patients having colon adenocarcinoma with vascular involvement or extensive lymphatic permeation; and patients with other neoplasms whose disease is aggressive. The acceptable safety level of hyperthermia and its effect on these diseases, even when they are far advanced, warrant a consideration of use of this modality against minimal or microscopic disease. The use of hyperthermia as an adjuvant must be developed in carefully controlled circumstances, and controlled clinical trials of this approach may now be in order.

It must be noted, again, that the authors' opinions regarding the application of hyperthermia is the result of two and one-half years of clinical experience with hundreds of hyperthermia treatments. Those who are initiating work in this area might at first seek to treat only relatively young and healthy patients with malignancies clearly unamenable to any other form of therapy. As experience is gained, the indications for treatment may become less rigid and the clinical base more broad. Hyperthermia is a treatment modality that becomes simple and safe only with experience. Its use should be initiated with caution.

THE PREHYPERTHERMIA EVALUATION

Specialty evaluations, which experience suggests are routinely desirable, are pulmonary function testing and a cardiology consult with echocardiogram. Hyperthermia is a state of high cardiopulmonary stress during which pulmonary gas exchange and cardiac output increase greatly. Patients with a major abnormality of any of these functions tolerate hyperthermia poorly. Modest congestive failure, if untreated at the time of evaluation, is not necessarily a contraindication to hyperthermia, as the administration of cardioglycosides and diuretics may enable those patients to tolerate treatment adequately. Consideration should be given to a history of previous irradiation of the heart or lungs, or the previous administration of Adriamycin, as the adverse effects of these agents may be exacerbated by hyperthermia.

Although abnormalities of both ventricular function and conduction may be unmasked by the increased cardiac work load of hyperthermia, a number of patients with a past history of myocardial infarction and three patients with mild-to-moderate aortic stenosis have undergone hyperthermia without difficulty. One patient with mitral regurgitation, not detected during initial evaluations, developed severe postshunt cardiac failure, which required shunt ligation and prohibited treatment. While this suggests that the shunt requisite to the extracorporeal approach may impose a disadvantageous cardiac burden, in reality it does not seem to; hearts unable to withstand the modest increase in cardiac output produced by the AV shunt likely will not tolerate the much greater demands of hyperthermia.

Pulmonary function appears adequate for hyperthermia when patients have an FEV_1 greater than 1200 ml and a pO_2 greater than 60 mm Hg on room air. The numerous other characteristics of pulmonary functions that may be obtained provide little additional useful information. In general, a patient who can unhesitatingly walk the length of a hospital corridor or climb a single flight of stairs has adequate pulmonary reserve for treatment with hyperthermia.

Patients having severe pleural or pericardial effusions or airway obstruction often respond to corrective measures sufficiently to be able to tolerate treatment. These problems should be corrected if possible, even when modest, for the increased circulatory dynamics and fluid exchanges of hyperthermia often will worsen such conditions. The routine use of tube thoracostomy is advised for pleural collections, and surgical removal or windowing of the pericardium provides better control of pericardial effusion than simple aspiration.

Bronchoscopy or tomography of major airways adjacent to a tumor mass may reveal an unsuspected degree of airway compromise. Some obstructions may be relieved by endoscopic resection or cauterization. Occasionally, a situation may warrent tracheostomy and placement of long, cuffed endotracheal tube through a tumor-narrowed airway. Such tubes are uncomfortable and predispose to pulmonary infection, but nonetheless may be required if patients are to be treated, as hyperthermia routinely will exacerbate airway obstruction by causing edema and hemorrhage within the airway containing the compromising mass. As truly effective regimens of hyperthermia are developed, the surgical resection of tumor mass compromising an airway may become the preferred approach.

Evaluation of the abdomen frequently requires the use of ultrasound or isotope scans for the detection of ascites and the quantitation of tumor mass. Ascites, if present in more than modest amounts, may best be managed by peritovenous shunting via a LeVeen shunt or similar prosthetic device. This is recommended because the fluid and electrolyte shifts of hyperthermia are difficult to control when ascites exist and because the increased circulatory dynamics of hyperthemia will exacerbate acites, which may interfere with diet, mobilization, and the pulmonary function of these seriously ill patients.

Smoldering intraabdominal infections such as the undeclared postsurgical or peritumor abscess, or that produced by ulceration of tumor within a bowel lumen, may cause severe sepsis if organisms or endotoxin are circulated by the hyperdynamic blood flow of hyperthermia. Surgical resection or drainage of these lesions can prevent lethal endotoxemia, if they are identified by contrast radiologic studies or sonography before treatment. Too often, recurrent intraluminal tumor is poorly suited to resection, but in such cases the sterilization of the gastrointestinal tract by administration of "bowel prep" antibiotics may reduce the risk of sepsis. It is also helpful to administer broad spectrum antibiotics to these patients for 24 to 48 hours before hyperthermia treatment.

Partial or complete bowel obstruction generally will not soon be relieved by hyperthermia, as tumor regression does not occur until some weeks after completion of treatment. During that interval the obstruction may become worse. Therefore, it is often appropriate to provide surgical relief of intestinal obstruction before beginning hyperthermia, particularly if oral alimentation will be desired or required. Otherwise, a significant period of hyperalimentation and bowel decompression by an indwelling tube will be required, both of which pose discomforts and hazards for the patient.

Bile duct obstruction, identifiable by chemical and x-ray contrast studies, may also be slow to abate and can be a source of serious sepsis if organisms reach the entrapped bile. Surgical decompression of biliary obstruction is rarely simple in these patients and at present is appropriate only for a selected minority of individuals. Nonetheless, the forceful insertion of T tubes through obstructing tumor or the Longmire operation for hepatoenteric diversion (Longmire and Sanford 1948) may provide critical hepatic decompression, without which hyperthermia may be poorly tolerated. Four of the 102 UMC patients have required such intervention. The hazard posed by biliary obstruction is not the abnormal liver functions per se, as a number of the patients treated in this series tolerated hyperthermia well despite jaundice, but rather is the potential for sepsis.

Renal function is adequate for hyperthermia if the serum creatinine level is 1.5 mg/dl or less, and a number of patients having only one kidney have been treated successfully. Urinary infection should be searched for by culture of a specimen prehyperthermia and, if present, eliminated before beginning treatment. It must be strongly emphasized that tumor mass may enlarge after hyperthermia and abate slowly, so that if an IVP performed for physical or function abnormalities reveals any suggestion of bilateral ureteral obstruction, then urinary diversion, temporary or permanent, must be considered before beginning hyperthermia.

The possibility of impending pathologic fractures should be kept in mind when dealing with patients with advanced malignancy, particularly those with a predilection for bone involvement, such as with bronchogenic, breast, and renal carcinomas. Bone pain or tenderness should be sought by history and physical exam, and, if present, should be thoroughly characterized by plain films of the skeleton. The bone scan is less useful as it shows early bone involvement of little significance in patients with known disseminated disease. When significant bone destruction, particularly of the femur, humerus, or vertebral column is detected, however, orthopedic evaluation should be undertaken before proceeding with hyperthermia treatment. Since hyperthermia-induced tumor regression is slow in onset, and significant bone repair even slower, we suggest that orthopedic repair of badly damaged bones should be undertaken before beginning hyperthermic treatments. Otherwise, the manipulation and mobilization of patients may be impaired, or a fracture may occur after starting treatments when its operative repair may be less safe because of impending neutropenia. Therefore, early correction of significant orthopedic problems is advised.

Other problems analogous to these exist, but the ones discussed exemplify the principle that compromised organ function should be restored to a reasonable degree when practical before undertaking hyperthermia. There is no evidence that hyperthermia administered within several weeks of surgical procedures interferes with wound healing. In fact, clinical experience suggests the contrary. A total of 113 surgical procedures have been performed prehyperthermia in our series, and despite the advanced disease of the patients and their debilitated circumstances, there has not been a single instance of wound infection or failure of primary healing.

TECHNIQUE

Extracorporeal induction of hyperthermia is based on integration of three components: (1) an arteriovenous shunt for vascular access, (2) a heat exchanger incorporating extracorporeal circuit (ECC), and (3) a temperature regulation device (TRD), which monitors temperature at a selected site and autoregulates temperature. Preclinical evaluation of a prototype of this system was conducted in spring of 1977, and the concepts developed were first applied to humans on August 31, 1977. As of December 31, 1979, this methodology has been used to provide 371 patient treatments (fig. 19.1).

The shunt must be of a physical size such as to accommodate a flow of 500 ml or more per minute. The flow of ordinary hemodialysis shunts is much less than this, but uremic patients are hypocoagulable in contrast to patients with advanced cancer, many of whom have sticky, hypercoagulable blood (Elias, Shukla, and Mink 1975), which will clot dialysis shunts. This requirement for high flow originates not only from a consideration of prevention of thrombosis, but also because the physical act of patient heating requires a caloric input into a patient, and the increased rate of heat transfer possible with high flow results in more rapid patient heating (fig. 19.2). The increased flow and the physical size of the requisite large shunt would greatly increase the risks of hemorrhage and infection if ordinary shunt approaches, with devices that transverse the skin, were used. The problem of infection in these patients is of particular concern because their immunocompetence is often attentuated by their malignancy, previous chemotherapy, radiotherapy, and other factors. Therefore, to obtain a high-flow shunt, with minimal risk of infection and hemorrhage, a new approach to vascular access was developed (fig. 19.3).

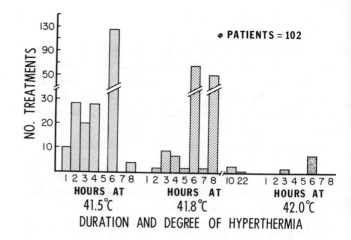

Figure 19.1. Magnitude and duration of patient treatments given from August 1977 to December 1979.

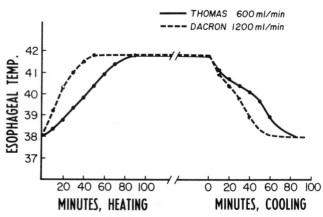

Figure 19.2. Comparison of the rates of heating and cooling obtained with moderate- or high-flow arteriovenous shunts.

Under general or spinal anesthesia, a 100-cm length of 8-mm diameter woven Dacron[1] tubing is sewn end-to-side to the common femoral vein and placed as a long, tight "U" under the skin and subcutaneous tissue of the anterior thigh. This wound is allowed to heal for several days before beginning treatments. Hyperthermia is begun by inducing general endotracheal anesthesia and surgically prepping and draping the skin over the shunt. A small incision is made over the end of the shunt and a short length of tubing brought through the skin. Heparin 150 units/kg is administered and the exposed shunt tubing clamped and divided for connection to an extracorporeal circuit.

[1] Meadox Medicals, Oakland, New Jersey 07436.

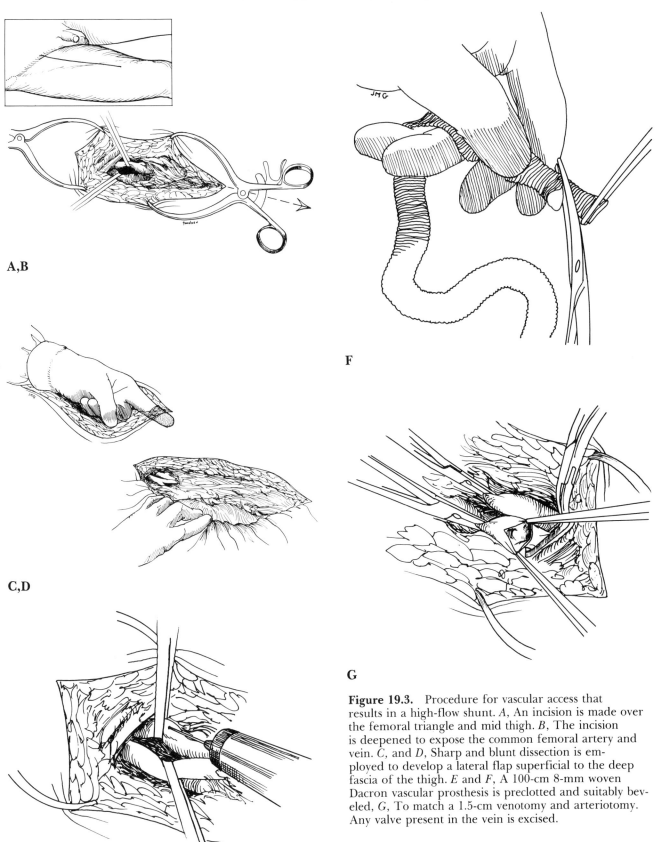

Figure 19.3. Procedure for vascular access that results in a high-flow shunt. *A*, An incision is made over the femoral triangle and mid thigh. *B*, The incision is deepened to expose the common femoral artery and vein. *C*, and *D*, Sharp and blunt dissection is employed to develop a lateral flap superficial to the deep fascia of the thigh. *E* and *F*, A 100-cm 8-mm woven Dacron vascular prosthesis is preclotted and suitably beveled, *G*, To match a 1.5-cm venotomy and arteriotomy. Any valve present in the vein is excised.

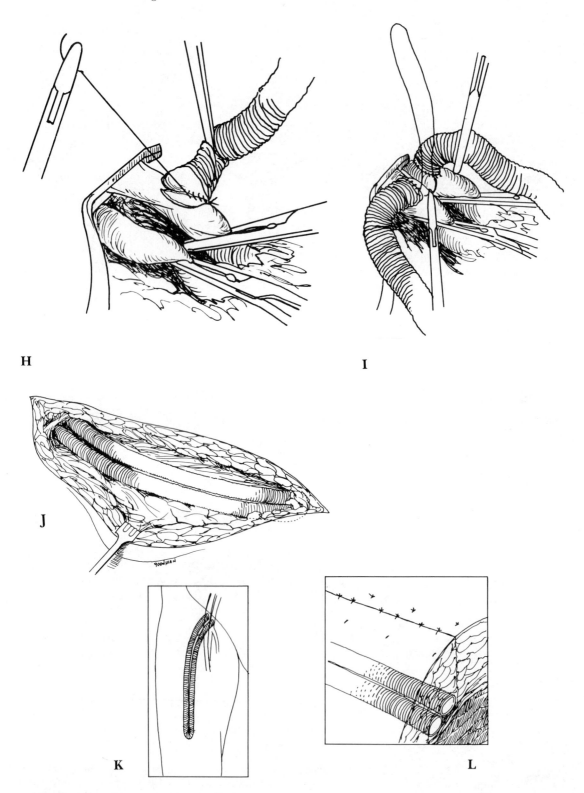

Figure 19.3. continued *H* and *I*, End to side anastomoses of the prosthesis to the artery and vein are created with a single simple running suture of 5-0 Prolene™. *J, K,* and *L,* The anastamosed prosthesis is placed under the lateral flap and the wound closed with meticulous technique.

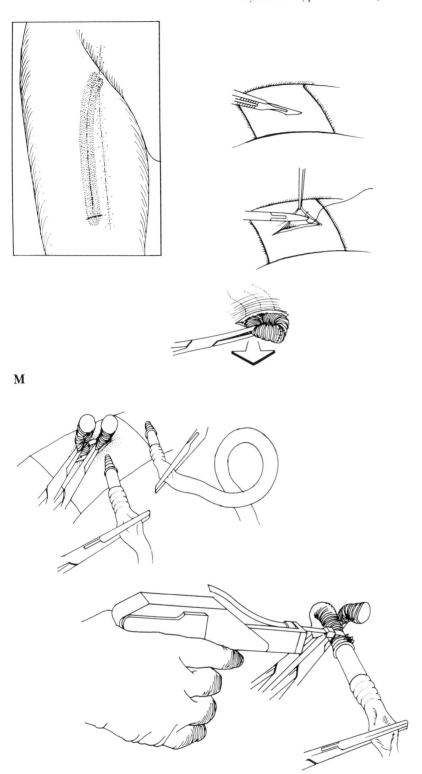

M, After the implant incision heals, treatment is begun of an anesthetized patient by placing a small incision over the distal end of the shunt. Subdermal vessels are suture ligated, after which a portion of the shunt is exposed. N, After heparinization, the shunt is clamped, divided, and connected to a primed, heat exchanger incorporating, extracorporeal circuit.

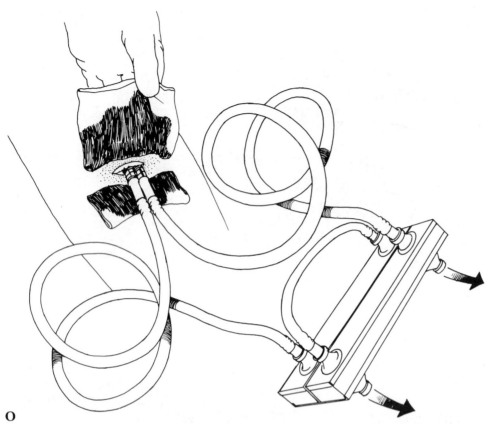

Figure 19.3. continued *O*, The shunt and ECC tubing are covered with povidone-iodine soaked gauze and secured in position with tape and bandages. The heat exchangers are then placed on the TRD and heating begun. After cooling, the circuit is severed from the prosthesis, which is reanastamosed and the incision closed.

The extracorporeal circuit (ECC)[2] is constructed of a ten-foot and a four-foot length of quarter-inch diameter Tygon™ tubing joined by plastic connectors to a single or twin heat exchanger (HE). All surfaces of the ECC, which will be in contact with blood, are coated with a TDMAC-heparin complex to reduce their thrombogenicity. The ECC is primed with approximately 100 ml of heparinized saline and connected to the tubing of the shunt by plastic or metal cannulas, which are secured in place by nylon snap bands. The removal of clamps from the shunt allows blood to flow through the ECC at approximately 2000 ml per minute. The exposed shunt tubing is drenched with providone-iodine solution, covered with bandages, and secured with tape.

The ECC tubing to the arterial side of the heat exchanger is led through a roller pump to reduce and regulate flow in the ECC to 900 to 1500 ml per minute. The heat exchangers of the circuit are attached to the TRD[3], which upon activation supplies temperature control fluid to the exchangers. Within the TRD are two reservoirs of water, one maintained at 30°C by a refrigeration unit and the other maintained at 49°C by a heating coil. By means of proportioning valves, water from either reservoir, or any blend from both, is pumped through the heat exchangers at 10 liters per minute. A thermistor probe placed at any desired site in the patient's body is connected to the TRD, and the signal from it controls the position of the proportioning devices. At the beginning of hyperthermia, 49°C water is applied to the heat exchangers, but as the patient's temperature approaches that programmed for treatment, the TRD automatically provides progressively cooler fluid to the heat exchangers until the patient's temperature stabilizes at that desired. Thereafter, the patient's temperature is continuously autoregulated at the level preset into the TRD, and little physician or nurse attention is required (fig 19.4).

[2]Research Against Cancer, Inc., 2648 Crane Ridge Drive, Jackson, Mississippi 39216.

[3]IBID.

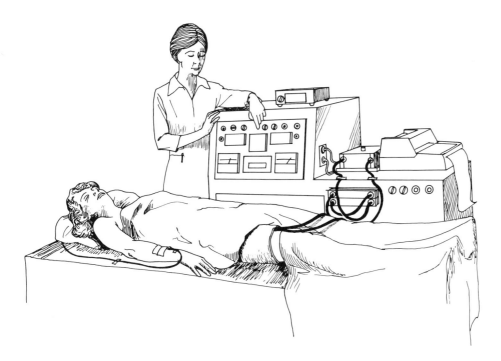

Figure 19.4. Overview of extracorporeal induction of hyperthermia showing the shunt area, the circuit, and its Temperature Regulation Device (TRD).

A hyperthermia treatment is concluded by resetting the TRD, which then provides progressively cooler fluid to the heat exchanger until the patient is cooled to the desired temperature, usually 38.0°C. After cooling, the shunt is exposed and again surgically prepped and draped. The shunt limbs are clamped, and the Dacron exposed to the skin is excised, disconnecting the circuit. This excision of a portion of the shunt on completion of each treatment is the reason for implanting a long "U" loop of Dacron, which can be used for five or six treatments. The clean shunt ends are anastomosed with a simple running stitch of 5-0 Prolene™ and the skin incision is closed with silk sutures and lightly dressed. Anesthesia is reversed, and the patient usually is immediately extubated and returned to his room on an ordinary medical-surgical floor. It has been uncommon for the patients to require recovery room or intensive care unit facilities, but it is useful for them to have a private nurse for the first eight hours posthyperthermia.

The shunt, extracorporeal circuit, and TRD rapidly induce and precisely control systemic hyperthermia. The esophageal temperature of a 70-kg individual may be elevated to 41.8°C within 35 minutes of beginning heating and a uniform core body temperature, as manifested by urinary bladder temperature, will be obtained in another 20 to 30 minutes. A given level of hyperthermia once obtained is held within rigid limits, as the TRD will always autoregulate temperature within 0.1°C accuracy and will maintain temperature to within 0.05C of set-point for more than 80% of the time. There has never been an instance of temperature overshoot with this method.

Cooling is also efficacious, and upon completion of treatment the patient temperature may be reduced as rapidly as cardiovascular stability will allow. The system is capable of cooling the patient so rapidly as to cause modest high output, low peripheral resistance, and hypotension. Thus, the current approach is to reduce patient temperature 0.5°C during the first 15 minutes of cooling, 1.0°C during the second 15 minutes, and as rapidly as possible thereafter. Using such an approach, hypotension during cooling occurs infrequently.

ECC-related complications have been uncommon, and there has been no detectable patient injury caused by this method in the 371 applications of the procedure. Small deposits of fibrin upon the surfaces of the heat exchangers occasionally develop when systemic heparinization is conducted at a 100 unit/kg level, but have never been seen with a 150 unit/kg heparin dosage. There has been no instance of air embolism or harmful circuit disruption. Surprisingly, there has never been a single instance of gross hemolysis detectable by examination of the serum or urine. A transient rise in serum bilirubin frequently is noted posttreatment, however, and may be contributed to by a decreased red blood cell survival.

In general, the ECC methodology is simple, precise, and applicable to large numbers of patients. While the use of the Dacron shunt requires the assistance of a surgeon at the beginning and conclusion of the treatment, it is expected that a totally implantable large flow shunt well suited for use by registered nurses and technicians will soon be available. This device will enable hyperthermic treatment of large numbers of patients by paramedical personnel, which is seen as a significant advance in the use of this modality. An alternative approach is use of the Thomas shunt (Thomas 1970), a large transcutaneous device whose Silastic™ and Dacron tube is also sutured to the femoral vessels and provides flow of approximately 500 ml per minute. While the connection and disconnection of the ECC to this device is simpler than with use of the Dacron shunt, the low flow of this shunt results in both slow heating and a high incidence of thrombosis. The propensity for infection of Thomas shunts also severely limits their application.

The approach described has proved satisfactory and appears to provide a practical means for induction of whole-body hyperthermia. The techniques involved are quite comparable to those of hemodialysis.

MANAGEMENT DURING HYPERTHERMIA

Design of Treatment Facility

The hyperthermia treatment area should be modeled after hemodialysis or intensive care facilities rather than an operating room. Once connection of a patient to the extracorporeal circuit and TRD has been accomplished and hyperthermia induced, the patients are surprisingly stable, and their treatment becomes routine.

The current practice at UMC is to assign one nurse anesthetist (C.R.N.A.) to each patient and one registered nurse (R.N.) to every two patients. The anesthetist is responsible for proper anesthesia, ventilation, and volume support. The R.N. oversees the control of patient temperature, and is responsible for the administration of crystalloid solutions and drugs, the function of the extracorporeal circuit, and the obtaining and recording of hematologic and metabolic data. Both the C.R.N.A. and the R.N. share the responsibility for protecting the patient from pressure injuries to skin, the monitoring of fluid input and output, and the care of the various tubes and wounds common to these patients.

One surgeon supervises these activities and can easily manage the simultaneous treatment of two patients. It appears likely that one surgeon and one anesthesiologist can conveniently manage as many as 8 to 12 patients simultaneously undergoing hyperthermia, and there has been discussion as to whether it might be appropriate to assign two patients to one anesthetist. Much of the cost of hyperthermia is anesthesia-related, and the efficient use of personnel is critical to cost control.

This concept of a number of individuals simultaneously undergoing treatment under the care of relatively few physicians can best be developed if patient treatments are done in large open areas rather than in individual rooms, and is a principal reason for developing hyperthermia along ICU practices rather than those of the operating room. The UMC experience has been so conducted, and the low incidence of infection observed indicates that extracorporeal methodology can safely be practiced in such units, as has been the practice with hemodialysis for years.

A criterion of note for the treatment area is that temperature control within the room may require more cooling capacity than usual. Any device, such as the TRD, that incorporates both heating and cooling units that strive against one another generates substantial heat. Several of these units operating in proximity may cause room temperature to become elevated with discomfort to personnel, and worse, instability of temperature monitoring and readout devices. Most telethermometer systems incorporate circuits for minimizing drift caused by changes in environmental temperature, but practice has indicated that unless room temperature is controlled at a stable level, errors will develop in the instrumentation, which will jeopardize the safety of the patient.

In addition to space and temperature requirements, the usual ICU requirements for electrical safety, oxygen and vacuum outlets, and equipment for monitoring EKG and arterial blood pressures are requisite. There is a question as to whether the use of nitrous oxide as part of the anesthetic for hyperthermia is desirable and, if so, provision for its supply is appropriate. Also of issue is whether ventilation of the patient can best be accomplished with anesthesia machines or by a respirator, and a preferred usage has not been established. A convenient laboratory for determination of arterial blood gases, electrolytes, and calcium and phosphate levels is required and must be staffed at all times. Other laboratory determinations are required on a less urgent basis and can be readily handled by the usual clinical laboratory.

Table 19.2
Routine Pretreatment Orders for Hyperthermia in A.M.

1. Full liquid supper
2. NPO after midnight
3. Medications:
   Decadron™ 2 mg IM at midnight and 6 A.M.
   Kefzol™ 1 gm IM at 6 A.M.
   _____
   _____
4. Send to O.R. with patient:
   Hyperalimentation 1000 ml
   Intralipid™ 500 ml
   Keflin™ 1 gm × 4 doses
   K_2PO_4 15 mM vials #8
   Solu-Cortef™ 250 mg amps #3
5. On call to O.R. Give:
   Valium™ _____
   Scopolamine _____
6. Place sponge cushion and lapidus mattress on bed
7. Tap water enema this evening
8. Check chart for CBC, SMAC within last 48 hours; send samples if incomplete data
9. Type and cross match two units whole blood, "for O.R. in A.M."
10. MUST HAVE BEFORE SENDING TO O.R. (check off):
    _____ CBC   _____ Chest x-ray
    _____ SMAC   _____ ECG   _____ Surgical consent
    _____ PT, PTT   _____ Support stockings removed
    _____ Urinalysis
    _____ Informed consent for hyperthermia
11. Record: Height _____ Weight _____

General Patient Care

Standard prehyperthermia orders are given the night prior to hyperthermia treatment (table 19.2). Patients are brought to the treatment area in their hospital bed on which an alternating pressure air mattress has been placed. One moderate caliber peripheral IV and one central venous line are inserted, and the administration of blood begun immediately. A standard blood pressure cuff and EKG electrodes are attached. General endotracheal anesthesia is induced, and the patient placed on a ventilating device. Using the technique previously described, the surgeon, with the assistance of the R.N., connects the extracorporeal circuit to the patient and also to the TRD. A thermistor probe is placed in the esophagus, and a thermistor-tipped catheter[4] is placed in the urinary bladder. During induction of hyperthermia, the TRD monitors and controls esophageal temperature while the bladder probe is connected to an independent monitoring device. After stabilization of core temperature, the leads are switched, and subsequently the TRD is controlled by bladder temperature.

The patient is covered with an ordinary sheet and a light hospital blanket while the high-flow Dacron shunt is in use. If the Thomas shunt is being employed, an additional blanket is generally placed on the patient to reduce heat loss. Foam pads are placed under the ankles, extremities, back, and at other pressure points. Great care is taken to protect the peroneal and the ulnar nerves. Throughout treatment the position of the patient's extremities is constantly changed and all major joints are repeatedly flexed. Such practices prevent both nerve injuries and muscle stiffness, which otherwise frequently occur. A liquid-filled pillow is placed under the back of the patient's head to prevent pressure lesions, which result in hair loss. Ointment is applied to the eyes to prevent the cornea from drying during treatment. No effort is made to control loss of perspiration, and surface heating is not employed unless there is a particular problem with cutaneous disease, such as inflammatory breast cancer or melanomatosis. For those conditions, a thermistor is applied to the skin and a small, water circulating, heating blanket is applied over the area of tumor involvement to ensure adequate heating of the tumor-involved surface.

At completion of treatment, the patient is cooled, the circuit disconnected, and anesthesia terminated. The patient's clothing and bed are changed, and a standard posthyperthermia regimen ordered (table 19.3). Patients generally are stable and awake within one hour of completing treatment. Patients with special problems, such as a comprised airway, are admitted to the recovery room for overnight surveillance.

Anesthesia Management

General endotracheal anesthesia is induced with thiopental sodium (200 to 350 mg) and succinylcholine (60 to 100 mg). A standard low-pressure cuffed endotracheal tube is passed transorally, positioned in the

---
[4]Research Against Cancer, Inc., 2648 Crane Ridge Drive, Jackson, Mississippi 39216.

Table 19.3
Routine Posthyperthermia Orders

1. VS qh × 4, q2h × 4, then q4h. Call physician if BP below 100, P above 130, T above 102°F.
2. NPO until A.M., full liquid breakfast, regular diet thereafter. May have sips of water and ice chips.
3. 60% O_2/face mask at 10L/min until A.M., then D/C.
4. IV fluids via central line.
   1000 ml D5½NS + K_2PO_4 (K^+ 22mEq, $PO_4^=$ 15mM) at 80 ml/hr.
5. MEDICATIONS:
   Morphine _____
   Keflin™ 2.0 gm. IV q4h × 6 doses only.
   _____
   _____
   _____
6. CBC, SMAC in A.M. and q.d.
7. Stat lytes at 4, 8, and 12 hrs post HT.
8. If K 3.0 to 3.5 give KCl 15 mEq qh × 2, 10 mEq qh × 4. If K 3.6 to 4.0 give KCl 10 mEq qh × 4.
9. Change leg dressing q shift; check for bruit.
10. Ankle-to-knee stockings.
11. D/C Foley and K_2PO_4 solution in A.M.; then KVO IV with D5 0.2NS at 40 ml/hr.

NOTE: HT = hyperthermia.

usual manner, and connected to an anesthesia machine for assisted ventilation. One hundred percent oxygen is administered, since a low arterial pO_2 has occurred in some patients when anesthetic levels of nitrous oxide were administered. Positive end expiratory pressure (PEEP) of 5 cm is applied, as failure to take this precaution will result in alveolar-capillary dysfunction and hypoxemia during treatment. Anesthesia is maintained by periodic administration of morphine to a total of 1 to 2 mg/kg and Valium™ to 10 to 20 mg, with relaxation being maintained by intermittent administration of tubocurarine to 2 to 6 mg. Recently, the inhalation fluorocarbon anesthetic, Ethrane™, has been administered at 0.25% without evidence of hepatic injury. This approach appreciably lessens the narcotic dosage required, and appears to be particularly useful for extended duration treatments of 10 hours or more. Ethrane anesthesia has now been used for more than 75 treatments.

It is a preferred technique to administer the majority of an anesthetic agent during the induction of hyperthermia and thereafter reduce the amounts administered. The anesthetic requirement decreases significantly as hyperthermia develops, as it appears to exert an anesthetic and amnesic effect itself. Nonetheless, the average patient will be fully alert and uncomfortable at 41.8°C if significant doses of anesthetics are not administered. As the termination of treatment is approached, the short-acting narcotic Fentanyl™ is substituted for morphine and is administered in 2 to 5 mg doses as necessary. Continuation of the administration of long-action narcotics into the hour preceding patient cooling will result in an unresponsive patient at completion of treatment and is undesirable.

There has been little indication that hyperthermia appreciably alters the metabolism of the common narcotics, but there is some suggestion that their effects increase in the posttreatment period, either because of a release phenomenon or because metabolism and excretion of these agents decreases. For this reason it is important to administer the majority of the drugs given during the early part of hyperthermia rather than evenly throughout its course, and it is particularly desirable to avoid the administration of long-acting anesthetics at the end of treatment.

The reaction of hyperthermic patients to barbiturates is of special note. It appears that hyperthermia greatly impairs the metabolism of barbiturates, and the administration of pentobarbital at the ordinary dosage of 30 mg/kg will result in an anesthetic state for as long as 96 hours. Peculiarly, the dose of barbiturate required to induce anesthesia is not appreciably lessened by hyperthermia, but the duration of the anesthetic effect is increased such that barbiturates should not be given to hyperthermic patients without great caution.

Volume, Fluid, and Electrolyte Support

A significant advance in patient management at UMC has been the subsitution of the tranfusion of whole blood and fresh frozen plasma (FFP) for the administration of large volumes of crystalloid solutions. Hyperthermia induces a high cardiac output, a low peripheral arterial resistance, and an increased venous capacitance. These responses require that the intravascular volume be augmented. It has been the practice of some investigators to administer crystalloid at 700 to 900 ml per hour to offset the relative hypovolemia of hyperthermia, as well as the fluid losses of respiration and the excretion of sweat and urine. If this practice is continued for four to six hours, however, the body tissues become edematous, resulting in the development of scleral edema, laryngeal swelling, "wet lung," and, in some instances, intracerebral edema. Not only do these complications pose specific organ-related problems, but the greatly increased total body water contributes to the usual posthyperthermia diuresis, and dangerous hypokalemia and hypophosphatemia ensue. These problems are signifcantly lessened by the following regimen for volume, fluid, and electrolyte support.

Two units of whole blood are administered at the very beginning of treatment, and often before anes-

Table 19.4
Protocol for Fluid/Electrolyte/Volume Management During Hyperthermia

Time	Solutions		Crystalloid	Colloid
Induction				
0–90 min	Normal saline	1000 ml	1000	
	Whole blood	2 unit		1000
		Subtotal	1000 ml	1000 ml
Treatment				
91–450 min (or longer)	D5½NS 1000 ml + K$_2$PO$_4$ (K$^+$ 66mEq) + Heparin 100 u/kg + Solu-Cortef™ 250 mg at 250 ml/hr		1500	
	Fresh frozen plasma 6–8 units (add heparin 1000 u to each unit)			2000
	Hyperalimentation fluid at 40 ml/hr		240	
	NSS for meds (antibiotics, etc.)			500
		Subtotal	1740 ml	2500 ml
Cooling				
451–560 min	D5 ½NS 1000 ml + K$_2$PO$_4$ (K$^+$ 44mEq) + KCl 56 mEq at 100 ml/hr		150	
	Hyperalimentation fluid at 40 ml/hr		60	
		Subtotal	210 ml	
		Total for Treatment	3250 ml	3500 ml

thesia is induced. Additional volume expansion, either with FFP or whole blood, depending on the hematocrit, is administered as necessary during hyperthermia to maintain a satisfactory central venous pressure and urine output. The average patient will receive four to eight units of FFP during an eight-hour treatment at 41.8°C. The anesthetist will also infuse 500 to 1000 ml of saline over an eight-hour treatment for the administration of pharmacologic agents. If urine output tapers, he also may administer a fluid challenge of 200 to 500 ml of saline, but an effort is made to avoid this practice. The majority of crystalloid support is given by the hyperthermia nurse as follows:

D5½NS (dextrose 5%; 0.45% sodium chloride) 1000 ml + K$_2$PO$_4$ (66mKq K$^+$, 45 mMole PO$_4^=$) + heparin 100 units/kg + Solu-Cortef™ 250 mg is administered into the outflow port of the extracorporeal heat exchanger at 250 ml per hour. Intralipid™ 500-ml total is administered through this same line. A line for blood sampling is placed into the inflow port of the heat exchanger and small amounts of heparinized saline continuously administered through it. Conceptually, the administration of heparinized saline into the beginning of the extracorporeal circuit near the arterial inflow could by regional heparinization reduce any tendency to clotting within the circuit, but as this has not been a problem, a more convenient site of infusion is employed.

A standard 25% dextrose, 10% protein hyperalimentation fluid containing the usual mineral and vitamin supplements is administered via the central venous line at a rate sufficient to increase the blood glucose to 400 to 600 mg/dl. This usually requires a rate of 40 to 60 ml per hour. Thus, the crystalloid replacement approximates 300 ml per hour, as summarized in table 19.4, which, together with the blood and plasma also administered, will maintain a stable central venous pressure of 4 to 10 ml H$_2$O and a urine output of 100 to 200 ml per hour indefinitely in most patients. Even patients with marginal cardiopulmonary function tolerate this regimen well, and intrapulmonary and peripheral arterial pressures are well maintained (fig. 19.5).

Metabolic Considerations

All major organ systems are stressed by systemic hyperthermia. Specific measures must be taken to prevent central nervous system injury and to offset kidney dysfunction, which can be caused by hyperthermia (fig. 19.6). Other organs that are vulnerable to hyperthermia are the liver, the intestinal mucosa, the bone

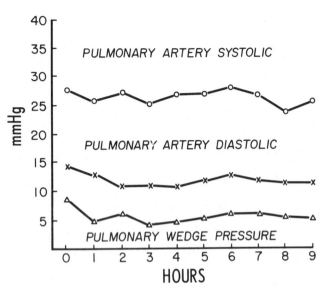

Figure 19.5. Effects of hyperthermia and the UMC fluid/volume replacement schedule on intrapulmonary vascular pressures. (Parks et al. 1979. Reprinted by permission.)

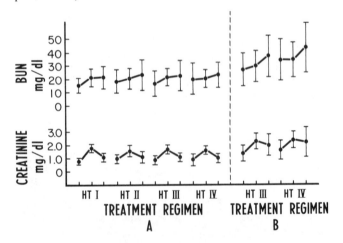

Figure 19.6. Effects of hyperthermia on renal function. A transient renal dysfunction develops as the intensity of treatment is increased. Treatment A is 4 × 6 hours at 41.5°C; Treatment B is 1 × 6 hours at 41.5°C, 1 × 6 hours at 41.8°C, and 2 × 8 hours at 41.8°C.

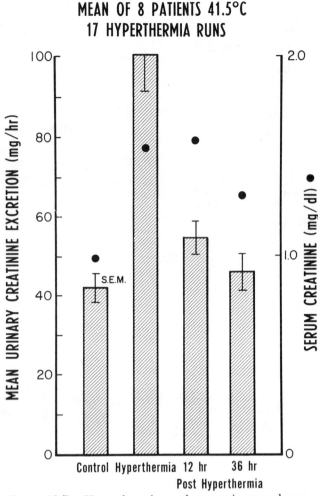

Figure 19.7. Hyperthermia produces an increased creatinine excretion, presumably by increased turnover rather than rhabdomyolysis, as CPK levels remain low.

marrow, the peripheral nerves, the skeletal muscle (fig. 19.7), and the skin. All require protective measures, many of which are metabolic in nature.

Hyperthermia greatly increases energy production by the body, with a concomitant increase in oxygen consumption and carbon dioxide production, while the respiratory quotient rapidly alters to a value consistent with metabolic utilization of lipids rather than carbohydrates (figs. 19.8 and 19.9). These findings appear significant, since an important tissue that ordinarily derives little energy from lipid consumption is the brain (White, Handler, and Smith 1973), and perhaps its extensions, the spinal cord and peripheral nerves. This may partly explain the vulnerability of the central nervous system to fever and heat stroke. Many cancer patients have depleted carbohydrate and fat stores, and it seems cogent to administer these substances to ensure the availability of adequate precursors to marginally competent metabolic pathways. Furthermore, our preliminary laboratory data suggest metabolic manipulation may improve brain metabolism during hyperthermia, and the treatment of patients with higher temperatures for longer durations may be facilitated by development of these measures.

Serum levels of calcium and phosphate tend to fall precipitously but can be maintained by infusion of adequate amounts of these ions (fig. 19.10). This is important as severe hypophosphatemia is a reported cause of central nervous system injury and rhabdo-

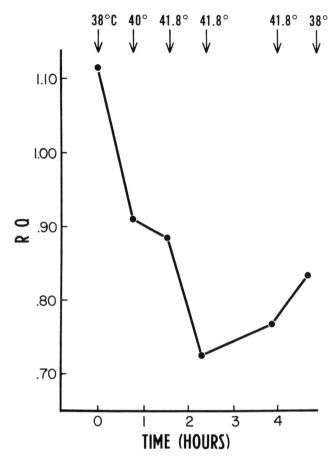

Figure 19.8. The declining respiratory quotient as body temperature becomes elevated. A rapid metabolic shift to fat utilization occurs despite the presence of high blood glucose levels.

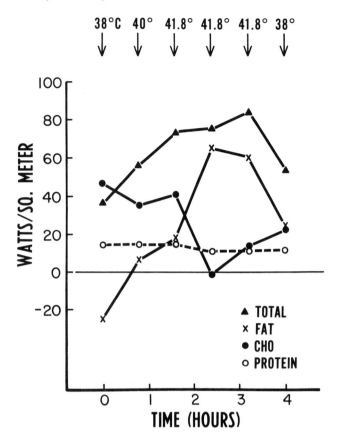

Figure 19.9. Effects of hyperthermia on energy production from different foodstuffs. The shift from carbohydrate to fat utilization is clearly seen.

myolysis (Knochel 1977; Knochel et al. 1978). The administration of substantial amounts of phosphate appears to prevent neurologic and muscle injury, however, as neither has been seen since beginning this practice but were observed before its use. We have not conducted proper balance studies to determine the cause of the hypophosphatemia of hyperthermia, but renal excretion of phosphate is greatly increased during treatment (fig. 19.11). Whether the low serum phosphate levels that develop during hyperthermia simply reflect increased excretion or some more specific phenomenon occurring within the patient is not known.

Calcium excretion is increased during hyperthermia and the administration of phosphate also contributes to hypocalcemia, which may reach dangerous levels. Accordingly, calcium gluconate 1.0 gm is infused over 30 minutes at the beginning, midpoint, and conclusion of an eight-hour hyperthermia treatment; the serum calcium level is carefully monitored during the posttreatment interval. NOTE: it cannot be too strongly stressed that the mixing of calcium and phosphate during their administration may produce an insoluble precipitate having an unusually lethal effect upon the pulmonary microcirculation. The production of this hazardous mixture must be carefully avoided.

The increased oxygen consumption and energy production caused by hyperthermia suggests a need to maintain greater than normal blood and tissue tensions of oxygen. This requirement is increased further in that the dissociation of oxygen from hemoglobin is hindered by high temperatures. The maintenance of arterial pO_2 of 60 to 80 mm Hg has been associated with neurologic injury in three patients: spinal cord irradiation may have been a contributing factor in one patient. Therefore, it has become the established practice to maintain arterial oxygen tensions as high as possible during hyperthermia by the use of 100% oxygen for ventilation.

The impression of a need for maximal oxygenation has resulted in discontinuance of nitrous oxide as an anesthetic agent. While this may be an inaccurate conclusion based on limited data which interferes with

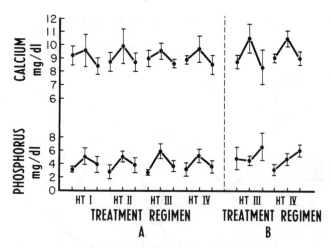

Figure 19.10. Effects of hyperthermia and the UMC replacement schedule upon serum calcium and phosphate levels. Severe hypocalcemia and hypophosphatemia will develop unless these ions are aggressively replaced. Treatment regimens are identical to those in figure 19.6.

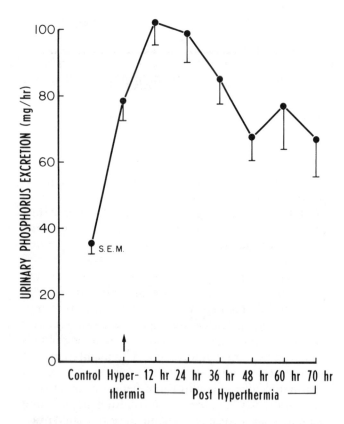

Figure 19.11. The marked and prolonged increase in urinary excretion of phosphorus associated with hyperthermia.

the utilization of an otherwise useful anesthetic agent, the injuries observed have sufficed to discourage the maintenance of any but the highest possible oxygen tension during treatment. There has been no indication that hyperthermia increases the pulmonary toxicity of high alveolar oxygen tensions, as both microscopic examination of pulmonary tissue and sequential assessment of pulmonary function have been normal.

A controversy does exist as to the relationship between the anticancer effects of hyperthermia and the oxygen tension existing in malignant tissue under treatment. Until there is a clear demonstration that high arterial pO_2 significantly reduces the antitumor effect of hyperthermia, it is recommended that the highest possible oxygen tension be maintained in patients undergoing hyperthermia.

Effects of Pharmocologic Agents During Hyperthermia

There is a general belief that drug metabolism and excretion are increased by fever and that febrile patients may require increased drug dosages, but such has not been the case for the patients in the UMC series. A variety of pharmacologic agents have been given to patients undergoing hyperthermia, and their effects are little changed, or substantially increased, rather than diminished (table 19.5). This may reflect an impairment of metabolic and excretory mechanisms, but for whatever reasons, the effects of common drugs upon hyperthermia patients are either unchanged or greater than those ordinarily expected. These findings suggest that the administration of drugs whose effects during hyperthermia are unknown should be be approached with caution.

Laboratory Assessment

Careful and frequent laboratory monitoring of the patients is critical to their safety. Samples may be obtained from indwelling arterial lines and central venous catheters, or may be obtained with particular ease from the ECC circuit. The UMC approach to blood sampling is to insert a #19-gauge scalp vein needle through the Silastic™ of the heat exchanger into the blood path. Typically, pressures in the circuit are low, and a slow drip of heparinized saline through the needle flushes the system continually when samples are not being withdrawn. Care should be taken to place the needle upstream of any infusions and downstream from a roller pump, if used, to obtain uncontaminated samples without the problems of infusion against arterial pressure.

The ECC circuit poses the possibility of continuous automated online blood sampling and analysis, an attractive concept not yet realized. One possible monitoring approach would be the direct insertion of electrodes into the ECC circuit for measurement of pH, pCO_2, pO_2, and certain specific ions such as K^+. These techniques have not yet been tested, but the ability to employ such devices for continuous monitoring may become a significant advantage of the extracorporeal method.

Arterial blood gases are drawn from the extracorporeal circuit at hourly or more frequent intervals. The values obtained must be corrected to patient temperature, and particular care must be taken to rapidly cool blood samples with ice, lest erroneous values be obtained. Ventilation must be increased during hyperthermia to maintain a normal pCO_2, but metabolic acidosis is unusual (Parks et al. 1979), and its appearance should initiate a search for a specific cause, such as hypoperfusion secondary to inadequate intravascular volume. Sepsis from an unknown or recognized abscess also may cause metabolic acidosis, and as sepsis may be difficult to recognize in the hyperthermic tachycardic patient, the appearance of metabolic acidosis should suggest its presence if another source for the acidosis cannot be identified.

The importance of monitoring and supporting serum calcium and phosphate levels has been indicated. An obstacle to this surveillance is that most clinical laboratories are not accustomed to providing rapid determinations of serum calcium and phosphate levels. The development of a stat laboratory principally devoted to the assessment of hyperthermia patients for these and other metabolites will facilitate safe treatment of patients.

Effects of Antineoplastic Drugs

There is growing evidence that hyperthermia potentiates the antineoplastic effects of many chemotherapeutic agents. Critical issues that must be resolved to effectively combine drug and heat therapy are (1) the choice of drug, (2) the dosage to be given, (3) the timing of administration, and (4) the number of repetitions. All of these possible combinations must be interrelated to the permutations for hyperthermia: (1) the magnitude of treatment temperature, (2) the duration of treatment, (3) the intertreatment intervals, and (4) the number of repetitions. Although investigation of these issues is becoming more active, there still is scant data on which to base clinical trials.

The cytotoxic effects of several common chemotherapeutic agents such as Bleomycin™ and 1,3-bis (2-chlorethyl)-1-nitrosourea (BCNU) clearly are potentiated by hyperthermia (Hahn 1979; Marmor 1979). There is also substantial laboratory evidence that hyperthermia potentiates radiation therapy, and the radiomimetic nature of alkylating agents suggests their effects may be similarly potentiated. The only large experience in the clinical application of hyperthermia and chemotherapy is that of hyperthermic regional perfusion of extremities, usually performed for recurrent or aggressive melanoma. In that setting better survival and regression rates appear to occur when L-phenylalanine mustard and heat are combined, as contrasted with the effects of either agent alone (Stehlin et al. 1975). Phase I studies of hyperthermia as an adjuvant to chemotherapy are only beginning to be reported. No randomized comparison of hyperthermia versus hyperthermia and chemotherapy has yet been reported. Therefore, the experience of UMC with the combination of hyperthermia and chemotherapy is of interest.

The initial UMC protocol provided for the randomization of patients with bronchogenic cancer to either 41.5°C for six hours of hyperthermia alone, or hyperthermia for six hours plus Cytoxan™ (CTX) 250 mg/M^2. As laboratory studies have indicated that chemotherapy is most effective if given in proximity to hyperthermia, the CTX was administered intravenously as a one-hour infusion during the midpoint of hyperthermia. These treatment regimens were given at one- to two-week intervals for four repetitions and had no significant marrow-suppressive effects. Alopecia did not occur, nor was cystitis observed in the CTX-treated patients.

The clinical assessment of 10 patients allocated to each group suggested the combination of hyperthermia and CTX was more effective than hyperthermia alone. This opinion was strengthened by two experiences with patients having large cell bronchogenic carcinomas, which appeared eradicated by treatment with hyperthermia and CTX; however, fatal central nervous system recurrences developed five months posthyperthermia and autopsies disclosed carcinomatous meningitis without evidence of any recurrence on the body side of the blood-brain barrier. Assuming that systemic hyperthermia induces temperatures in the central nervous system equivalent to those elsewhere in the body, these findings suggested that the hyperthermia-CTX combination was significantly more effective than hyperthermia alone, as CTX does not penetrate the blood-brain barrier. While other explanations for the behavior of these malignancies are possible, the findings were sufficiently striking to cause the addition of BCNU to the treatment regimen because of its ability to penetrate the blood-brain barrier. Randomization between hyperthermia alone and

Table 19.5
Effects of Hyperthermia upon Pharmacologic Actions of Common Drugs

Pharmacologic Agent	Dosage Evaluated	Effects in Hyperthermia Patients
Anesthetic Agents		
Atropine	0.4 mg IV q4hr	Unchanged. Note HT patients have a predilection to sinus tachycardia and dry airways with cast formation by bronchial mucus. Use should be minimized.
Barbiturates (Pentothal™ and pentobarbital)	Pentothal 200–350 mg pentobarbital 30 mg/kg × 1	Markedly increased. HT does not diminish the dose required for induction of anesthesia, but greatly prolongs the duration of the effect. Should be used with caution.
Fentanyl™	1–2 ml PRN	Unchanged. Good supplemental anesthetic agent.
Morphine	1–2 mg/kg	Unchanged. Preferred anesthetic agent.
Valium™	5–10 mg IV q1–2hr	Unchanged. Good supplemental anesthetic agent.
Anectine™	60–100 mg	Unchanged. Good for induction relaxing agent.
Pauvulon™	6–8 mg q2–4hr	Unchanged. Principal relaxing agent.
Antibiotics		
Keflin™	2 gm q4–6hr	No evidence of nephrotoxicity. Not one *Staph* infection in more than 300 patient treatments.
Tobramycin™	1 mg/kg q8hr	Apparently unchanged. No ototoxicity and no recognized increase in nephrotoxicity.
Mandole™	2 gm q6hr	Uncertain toxicity. Recent substitution of Mandole™ for Keflin™ has been associated with an increased instance of transient elevations of serum creatinine. A causal relationship is suspected but unproved. Caution is advised.

Other antibiotics, including Cleocin™, Amikacin™, Amphotericin™, and Chloramphenicol™ have been administered to patients undergoing hyperthermia. Customary dosages have been used, and no unusual toxicity has been observed, but caution is advised, particularly with Amphotericin™.

Pharmacologic Agent	Dosage Evaluated	Effects in Hyperthermia Patients
Cardiovascular Agents		
Digoxin™	Full digitalization patients have received as much as 0.75 mg during HT	Unchanged. There has been a notable lack of arrhythmias during extracorporeal induction of HT, regardless of digitalization.
Inderal™	0.5 mg–1 mg IV PRN	Increased. Has rarely been administered for supraventricular tachyarrhythmia, but usage has been associated with bronchospasm. Use with caution.
Dopamine™	5–20 mcg/kg/min	Unchanged. Agent of choice for hypotension of cooling and has been employed for endotoxin-related hypotension during treatment. It has not contributed to tachyarrhythmia.
Diuretics		
Lasix™	40–100 mg q1–4 hr at start of HT then PRN	Unchanged. Preferred agent for induction of diuresis at start of HT. No ototoxicity has been observed.
Ethacrynic acid	50 mg IV PRN	Unchanged. Alternate agent for induction of diuresis during HT. No toxicity noted.
Mannitol™	25 gm IV q2–4hr	Unchanged. Used for prevention of CNS edema in patients with brain metastasis.

Table 19.5 *Continued*

Pharmacologic Agent	Dosage Evaluated	Effects in Hyperthermia Patients
Antineoplastics		
Cytoxan™	250 mg/M² IV (for first two treatments)	Increased. Doses of Cytoxan™/BCNU normally well tolerated may produce severe marrow depression of HT patients. The listed dosage is safe in patients with normal reserves but should be reduced in those with previous radiation or chemotherapy.
BCNU	50 mg IV during midpoint of first two treatments (with CTX); 75 mg IV for four treatments as sole agent (maximum dose)	Increased. See above. Also concerned about increased incidence of "busulfan lung" related to this agent.
Adriamycin™	70 mg/M² IV 24 hr pre HT. None if 550 mg/M² within two months of HT	Unchanged. There has been a notable lack of cardiac dysfunction in HT patients previously receiving maximal treatment with Adriamycin. Occult cardiac failure or conduction abnormalities have not been made manifest by HT, although expected. Would beware.
Cisplatin™	100 mg/m² (treatment 1); 100 mg total (treatment 2)	Two patients initially receiving cisplatin manifested no increased effects. One patient previously treated with cisplatin was given additional dose intra-arterially during HT and developed a renal injury requiring dialysis. Significant tumor control occurred in all three patients. Administer with caution.
Other		
Heparin™	Initial 150 units/kg; continuous 25 units/kg per hr	Unchanged. Standard regimen for TDMAC-heparin coated ECC. Not known to be adequate for untreated circuits and not advised for such.

NOTE: HT = hyperthermia.

with CTX was stopped, and a total of 60 patients have now been treated with combined hyperthermia, CTX, and BCNU.

The initial dosages administered were six hours at at 41.5°C; CTX, 250 mg/M²; and BCNU, 50 mg, with the drugs administered over one hour at the midpoint of hyperthermia and treatments repeated at approximately four- to seven-day intervals. This was a toxic regimen, which caused moderate to severe marrow depression in patients presumed to have normal bone marrow reserves, and caused severe and prolonged marrow suppression in patients who had previously received chemotherapy or pelvic irradiation (fig. 19.12). While no deaths occurred from the combination, platelet transfusions and aggressive antibiotic coverage of neutropenia often were required, and the anticancer effects observed were not felt to warrant the toxicity produced. Therefore, the dosage of chemotherapy was reduced, and CTX 250 mg/M² and BCNU 50 mg were administered only during the first two hyperthermia treatments. This regimen did not cause severe marrow suppression even in patients treated with hyperthermia for eight hours at 41.8°C. Thrombocytopenia in the 40,000 to 60,000 range did develop in approximately 20% of these patients and required platelet transfusions, but suppression of the WBC to below 1000 was not seen in the patients treated with this protocol. Patients who had received previous chemotherapy or pelvic irradiation were given CTX 125 mg/M² and BCNU 25 mg during the first two treatments and tolerated treatment well.

CTX recently has been deleted from the UMC protocols, and patients now receive BCNU 75 mg during the midpoint of each of four treatments with hyperthermia. This change was effected because of uncertainty as to the activation of CTX in patients with liver functions altered by malignancy or hyperthermia and also because of continued concern regarding brain metastasis. This dosage is at the upper limits of that possible in patients with normal bone marrow reserves, and must be reduced if there is a history of previous chemotherapy, irradiation, or marrow com-

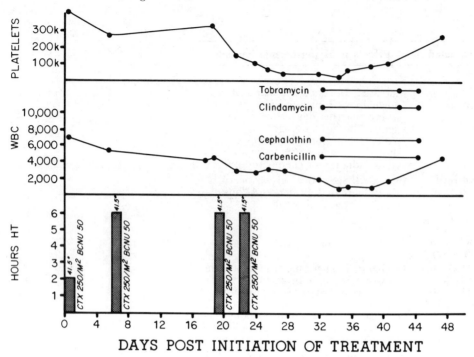

Figure 19.12. The toxicity of administering Cytoxan™ 250 mg/M² and BCNU mg during four consecutive hyperthermia treatments. The bone marrow reserve was normal pretreatment. *HT* = hyperthermia.

promise. The dosage may be excessive, as the platelet counts of most patients fall to 50,000 or less, and a number have required platelet transfusions. The neutropenia induced by this dosage has not been severe, as white counts have ranged from 2000 to 3000 throughout the posttreatment period (fig. 19.13). This change in the treatment protocol has been made too recently to ascertain the effects on overall tumor regression, but the initial impression is that a significant number of regressions are developing. A total of 87 patients have now received CTX and/or BCNU as an adjuvant to 24 to 30 hours of whole-body hyperthermia at 41.5°C or 41.8°C. The principal adverse effect has been the marrow suppression discussed previously, which occurs infrequently with the modified dosages noted.

There has been one other severe complication of combination chemotherapy and thermal therapy: the development of a fatal interstitial pulmonary fibrosis with histologic characteristics identical to the drug-induced pulmonary pathology sometimes referred to as "busulfan lung" (Rosenow 1972). This injury has been attributed to other aklylating agents, including CTX (Sostman, Matthay, and Putnam 1977), and most recently to BCNU (Richter et al. 1979). Only one case of this complication has been identified in the UMC series, and as its clinical characteristics are not subtle, it seems probable that only a single instance has occurred. The lack of toxic effects of repeated hyperthermia with CTX and BCNU on pulmonary function, as manifested by serial evaluation of arterial blood gases, has been reported elsewhere (Parks et al. 1979), and there was no evidence of progressive deterioration of pulmonary function. As the UMC incidence of bulsulfan lung is only 1%, and since the use of chemotherapy appears to increase the anticancer effects of hyperthermia, we have continued to employ these agents. Inasmuch as bulsufan lung may be reversible if diagnosed early (Richter et al. 1979), it is important to bear this entity in mind if pulmonary deterioration occurs in the patient treated with the combination of heat and alkylating agents.

Other side effects have been notable for their absence. There has been one instance of hematuria developing during treatment, which necessitated cessation of hyperthermia. Cystoscopy revealed a modest cystitis in an area of catheter balloon contact; the inflammation may have been potentiated by CTX. Anorexia, nausea, vomiting, or diarrhea have been infrequent, mild, and transient. Alopecia has never occurred. There have been no other adverse effects noted.

Two other regimens of chemotherapy have been administered to a few patients. Adriamycin 70 mg/M² has been given 18 hours *prior* to treatment to two patients with sarcoma and one with oat cell lung cancer. All patients had extensive pulmonary involvement and were not able to complete their planned treatment. These patients are of note only in that the patients did not experience any cardiotoxic effects,

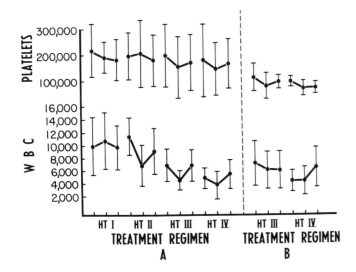

Figure 19.13. Marrow suppressive effects of administering BCNU 75 mg during the midporton of the first three hyperthermic treatments of patients with normal reserves. Treatment regimens are the same as those for figure 19.6.

which occurred when lower doses of Adriamycin were administered *during* hyperthermia (Kim et al. 1979).

It is possible that the liberation of epinephrine and norepinephrine, which occurs when Adriamycin is administered under hyperthermic conditions, may be avoided by administering this agent at least several hours prior to hyperthermia. Since there is evidence from clinical trials of regional perfusion to indicate hyperthermia is active against sarcoma (Krementz et al. 1977; Stehlin 1976; Stehlin et al. 1975), since Adriamycin is active against sarcoma and an agent potentiated by hyperthermia (Marmor 1979), there is a need to develop treatment approaches that avoid the cardiotoxic effects that may be induced by combined hyperthermia and Adriamycin therapy.

A second small group of patients with melanoma has received cisplatin concurrent with hyperthermia. One patient at high risk for postsurgical recurrence received cisplatin 100 mg/M^2 and cisplatin 100 mg only during the first two of four six- to eight-hour treatments of 41.8°C. There were no side effects and no evidence of recurrence six months posttreatment. A second patient with extensive axillary and pulmonary involvement by melanoma, who manifested progression after a standard regimen of 41.5°C hyperthermia, received three additional treatments at 41.8°C, during the first of which cisplatin 100 mg/M^2, and during the second cisplatin 100 mg only, were administered. There were no adverse effects and substantial regression developed. A third patient with extensive lower extremity and groin involvement by melanoma resistant to regional perfusion and systemic cisplatin, 650-mg total dose, received cisplatin 50 mg and cisplatin 100 mg as continuous intrafemoral arterial infusions during two hyperthermia treatments of 41.8°C for six hours. Profound tumor regression occurred, but was accompanied by a renal tubular injury that required dialysis. This nephrotoxicity did not completely resolve. These cases suggest that under some, but not all conditions, cisplatin and hyperthermia may safely be combined, and their combination may have a significant ability to cause regression of advanced melanoma.

Inpatient Posthyperthermia Care

Patients having an ordinary or better general physical condition tolerate hyperthermia well and generally do not require any unusual supportive measures in the intervals between treatment or immediately after completion of a treatment series. Electrolyte imbalances, particularly hypokalemia and hypophosphatemia, are the most common abnormalities and must be sought and corrected. Patients with major tumor burdens treated at 41.8°C frequently develop a one- to three-day depression of creatinine clearance, and the administration of potential nephrotoxic drugs, such as aminoglycosides, must be carefully controlled during these intervals. Thrombocytopenia, if seen, occurs only in the latter aspect of a treatment period but may require platelet transfusion. Anemia or neutropenia occur infrequently in patients with normal marrow reserves. Patients often manifest moderate fatigue and mild anorexia in the first week after completion of a four-treatment regimen and need encouragement to ambulate and eat. These feelings usually abate two to four weeks posthyperthermia, after which time most patients are active and often ravenously hungry.

Patients in poor condition because of extensive pulmonary or abdominal involvement need aggressive supportive measures to tolerate the stresses of hyperthermia. One seemingly important aspect of their care is to provide nutritional support by hyperalimentation throughout hyperthermic treatments and well into the recovery period. Complications attending this practice have been infrequent, but it is the practice not to push the fluid/calorie load to the degree sometimes recommended for depleted patients. Specifically, no more than 3000 ml of a standard solution is ever administered in a 24-hour period, and more commonly the infusion rate is 80 to 100 ml per hour. These patients are less tolerant of parenteral nutrition than many because of their disease, the periodic administration of steroids, and frequent infections common to advanced cancer. In such a setting, aggressive hyperalimentation may be more injurious than beneficial.

A particular issue not yet resolved is whether patients treated with hyperthermia for extensive disease should receive a prolonged course of steroids in the posttreatment period. Hyperthermia may produce pleural effusions, malaise, fever, leukocytosis, lethargy, and anorexia, most of which respond well to high-dose steroid treatment. Peritumor edema and intratumor congestion frequently develop in tumor masses, causing enlargement and possible compression of important adjacent structures; this effect also may be mitigated by steroid administration. The principal deterrent has been a concern that the administration of steroids may reduce the anticancer effects of hyperthermia either by stabilizing damaged cells or possibly by suppressing an immune response. Both concepts are theoretical objections for which there is little evidence. To date, the routine administration of steriods to hyperthermia patients has not been implemented but remains under consideration.

Outpatient Follow-up

Upon discharge, patients are returned to the care of a local physician experienced in oncology. It is recommended that a CBC and electrolyte/metabolite screen be obtained weekly for the first two weeks and monthly thereafter. Significant abnormalities are uncommon, but calcium and phosphate levels may need to be supplemented, and occasionally the administration of potassium is necessary. A transient anemia may develop because of the short half-life of transfused and heated blood, and perhaps one in fifteen patients will need transfusion in the postdischarge interval. Another relatively common occurrence is the development of urinary tract infections secondary to repeated catheterizations, and a urine culture should be obtained one month posttreatment, or sooner if symptoms occur. Routine measures of shunt care analogous to those used by dialysis patients are indicated and will prevent delayed shunt infection.

Oncologic evaluation for progression or regression of tumor should be conducted at monthly intervals by usual techniques. It is important to convey to the local physician that an enlargement of tumor mass in the first one or two months posttreatment may or may not be a sign of progression. This uncertainty as to the significance of apparent tumor enlargement poses considerable difficulty as decisions regarding additional therapy are contemplated. To simply watch a lesion enlarge without implementing any other available therapy, in the belief that the lesion will ultimately regress, requires an expression of faith in hyperthermia not yet warranted by the overall experience. However, most patients undergoing treatment with hyperthermia have malignancies resistant to all known therapy, and the physician has little recourse but to observe the disease course and hope for the best.

A decision as to whether an observed enlargement of tumor mass represents a "progression to regression" phenomenon or is, in fact, healthy growing tumor, can be made only with experience and difficulty. One possible approach is the transient administration of high-dose steroids, which will quickly diminish edematous masses, but theoretically may interfere with the anticancer effects of hyperthermia.

The following comments may help with the assessment of malignancy in the posthyperthermic period, but it should be kept in mind that the experience is limited and the impressions are preliminary ones.

The radiologic characteristics of an intrathoracic tumor mass may progress from a sharply defined lesion to a larger mass whose borders are somewhat fuzzy and indistinct. Over several weeks, this appearance may evolve toward a larger, still less well-defined lesion, which then may rapidly and totally regress (fig. 19.14). If such a phenomenon develops about intrahepatic tumor, there may be a progression of hepatomegaly, and the onset of jaundice, which oncologists ordinarily regard as an ominous event. Similarly, if there is tumor mass about hollow viscera, such as bowel or ureter, partial or complete obstruction may develop and require surgical measures for relief.

A sense of discouragement and inclination to curtail treatment frequently develops when a patient, family, and physician encounter these apparent indications of tumor growth. Further treatment may be deemed inappropriate, but the UMC experience suggests that life-maintaining measures should be aggressively employed until it is certain that a malignancy has progressed and become overwhelming. Several of the UMC patients allowed to expire because of apparent tumor progression have been shown by autopsy to have been developing massive necrosis of their malignancies and a regression whose benefits would likely have been significant if aggressive support had been pursued.

For such reasons, the best management approach to patients apparently developing progression after hyperthermia may be their early return to the treatment center. If progression in fact appears to be occurring, patients may be retreated with higher doses of hyperthermia and/or other chemotherapeutic agents. The important error to avoid is the cessation of support at a time when support may be most indicated.

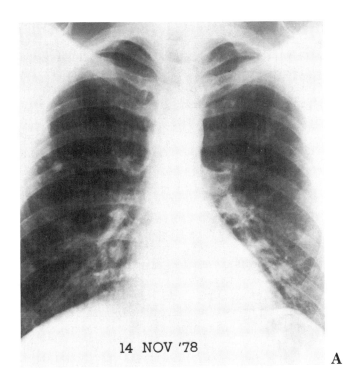

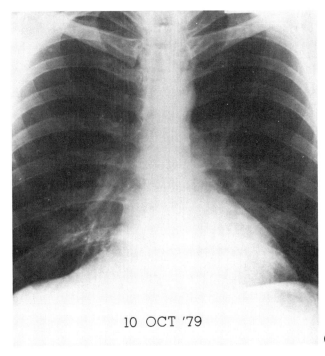

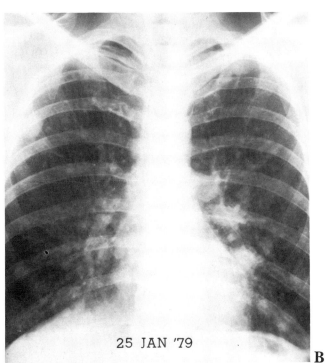

Figure 19.14. *A*, A 44-year-old patient with progressive metastatic hypernephroma despite nephrectomy and medroxyprogesterone therapy. *B*, Apparent progression two months posthyperthermia with increases in the size and number of lesions. The lesion changes began to clear one month later. *C*, Complete regression one year posttreatment.

MEASUREMENT OF TEMPERATURE

Although illogical, there has been less attention directed to the problem of assessing the temperature of hyperthermia patients than might be assumed. Both the patients and the malignancies undergoing whole-body hyperthermia will live or die by temperature differences of only a few tenths of a degree centigrade, as temperatures above 42.5°C are lethal to the host, and those below 41.5°C are ineffectual against tumor. Ordinary clinical and laboratory instrumentation will not reliably measure the narrow temperature differences between cancer kill and patient injury, and telethermometer devices specially designed for hyperthermia are required.

Even when accurate thermometers are available, it is uncertain where temperature should be measured because the temperatures of the different organs and tissues of hyperthermic mammals may differ widely (Dickson, MacKenzie, and McLeon 1979; Parks, Frazier, and Mountain 1977). Curious phenomena exist: arterial blood is a cool tissue even though people tend to think of themselves as hot-blooded creatures. The blood that flows in arteries is relatively cool because it has just passed through the lungs, where a cooling effect is exerted by the inspired gases and the loss of heat by evaporation. Thus, the temperature of arterial blood will depend on whether humidified or nonhumidified air is inspired, and also on the temperature of air entering the trachea, particularly if the warming and humidifying passageways or the upper airways are bypassed by an endotracheal tube.

The differences in blood temperature, which may be produced by manipulation of inspired gases, are sufficient for therapeutic consideration. For example, the relatively avascular metastases of intraabdominal malignancies will be preferentially heated and the brain cooled if cool, low-humidity gases are delivered to the lungs, because esophageal and tympanic temperatures will remain 0.2°C to as much as 0.6°C cooler than core body temperature. In contrast, a well-vascularized or intrapulmonary malignancy may be disadvantageously cooled, unless the patient is ventilated with heated and moisturized gases, which then become an important component of the therapeutic approach.

The UMC experience indicates that the preferential site for monitoring temperature of systemic hyperthermia is the bladder, with esophageal and tympanic temperatures also being monitored. The usefulness of tympanic temperatures is limited, however, by the inaccuracies of the only readily available instrumentation for their measurement. While a thermistor probe of acceptable accuracy that is suitable for insertion into the esophagus can be obtained, the reliability of esophageal temperatures for conducting hyperthermia is limited by the presence of at least three temperature zones within the esophagus: (1) that adjacent to the right mainstem bronchus, (2) that adjacent to the left atrium, and (3) that of the remaining esophagus. These differences may be readily observed by advancing a probe into the stomach and slowly withdrawing it, whereupon differences of as much as 0.6°C frequently will be found. Since even insertion of a measured amount of esophageal probe places its sensor blindly, esophageal temperature appears inherently unsatisfactory for the conduct of hyperthermia treatment.

The monitoring of esophageal temperature is advantageous in one regard: it appears to be the most dynamic of the sites that can be measured with precision. The temperatures at sites indicative of true core body temperature, such as the bladder, are significantly lower than those of the esophagus, and presumably of the brain, during rapid induction of systemic hyperthermia. If the temperature of only the slowly heating bladder is monitored when hyperthermia is induced, a significant and possibly lethal overshoot may occur, whose hazard is directly proportional to the rapidity with which hyperthermia is induced. The rapid induction afforded by extracorporeal circulation requires monitoring of esophageal temperature, but when steady state temperatures in the bladder and esophagus are reached, esophageal temperatures will be 0.2°C to as much as 0.5°C less than bladder temperature.

Although previously mentioned, it should be reemphasized that it is critical for the investigators of this new field to be able to share their findings with one another. Until the relationship between the temperatures of the bladder, esophagus, and other sites are characterized and understood, it must be accepted that the beneficial and adverse effects noted with one approach to treatment may bear little or no relationship to those of other regimens. For example, the NCI experience indicates that four-hour treatments at 41.8°C produce elevated CPK values at 24 hours; whereas the eight-hour treatments at 41.8°C bladder temperature performed at UMC are not associated with significant abnormalities (fig. 19.15). Only by understanding that the small differences that exist between temperatures of different organ sites are important can investigators begin to appreciate the criticality of the site of temperature measurement.

The telethermometer device used at UMC consists of a thermistor-tipped 16 Fr. Foley catheter[5] which is placed in the bladder in a routine manner.

---

[5]Research Against Cancer, Inc., 2648 Crane Ridge Drive, Jackson, Mississippi 39216.

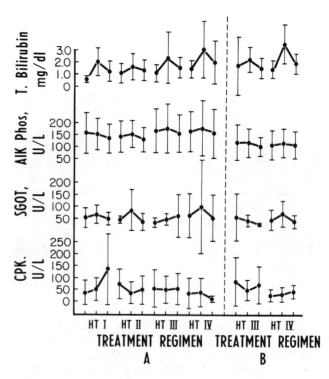

Figure 19.15. Effects of hyperthermia upon liver functions and serum CPK levels, none of which are significantly affected. Treatment regimen A is 4 × 6 hours at 41.5°C; B is 1 × 6 hours at 41.5°C, 1.6 hours at 41.8°C, and 2 × 8 hours at 41.8°C.

Each thermistor catheter is individually calibrated against a National Bureau of Standards reference thermometer and shown to be accurate to within 0.05°C at 41.5°C. This disposable catheter and its accompanying digital telethermometer have been simple to use and reliable. Rectal probes are not recommended as they may be expelled by peristalsis or become encased in stool, leading to erroneously low readouts and excessive patient heating if treatment is controlled from this site.

The measurement of intratumor temperatures by needle probes has been of interest, but neither the probes nor the ability to place them have been sufficiently consistent to facilitate conduct of clinical treatment. A limited number of determinations of temperatures made within the pulmonary artery suggest they correlate well with intrabladder temperature, but the approach poses difficulties, even with flow-directed catheterization.

In summary, the experience of UMC suggests that the measurement of urinary bladder temperature is advantageous for the conduct of whole-body hyperthermia.

DOSAGE AND EFFECTIVENESS

There is considerable evidence from small animal and tissue culture models that the magnitude of temperature induced during hyperthermia must be at least 41.5°C to produce an anticancer effect (Dickson 1976). Dickson's analysis of these data also suggests that at a temperature of 41.5°C a treatment duration of 30 to 40 hours is necessary if this magnitude of hyperthermia is to have significant antitumor effects. As the magnitude of hyperthermia is increased, the requisite treatment duration decreases. But even at 42.0°C, perhaps with current practices, the highest whole-body temperature safe for humans, a treatment duration of 20 or more hours appears necessary for any substantial anticancer effects.

There are few investigations reported that address the question of whether hyperthermia treatment is more effective when given as a single continuous treatment or when administered as repetitious short courses. There are data indicating that any thermal tolerance acquired by malignant cells disappears 24 to 48 hours posthyperthermia, suggesting that if repetitious treatments are employed, they should be administered at intervals of two or more days. Practical considerations that bear on these issues include the difficulties of organizing teams for around-the-clock treatments, the unavailability of laboratory support during late hours, and the general experience that the shorter the duration of stresses, such as hyperthermia, anesthesia, and fluid/electrolyte shifts, the better they are tolerated by patients.

The UMC experience suggests, however, that long-duration hyperthermia may be more practical than generally realized. Extracorporeal induction of hyperthermia is particularly well suited to long treatments, as the patient's body surfaces are accessible for manipulation to prevent pressure sores or stiff joints, and the skin remains dry and does not become macerated as with surface heating. It has been our observation that the repetitious administration of hyperthermia as six- or eight-hour treatments every two to three days induces a certain degree of physical and mental exhaustion in the patients. The periods of time without oral ingestion of nutrients before, during, and after treatments increase nutritional depletion, despite the use of hyperalimentation, resulting in general weakness and lack of a sense of well-being. The obtaining of multiple blood samples proves tiresome, as do the repetitious chest films and ECGs required by hospital policy between each administration of general anesthetic. Many of these expensive tests can be avoided if fewer and longer treatments prove practical and safe.

Almost all patients tolerate well a hyperthermia treatment of 41.8°C magnitude and eight-hour duration. There has been no indication of any trend toward progressive organ dysfunction during long treatments, and our experience suggests that treatments of 10- to 15-hour durations may be safe for most patients. The longest treatment administered at UMC to date is 22 hours at 41.8°C. This schedule was well tolerated by one patient with extensive breast cancer and marginal pulmonary function.

Indeed, patients with extensive disease, especially pulmonary involvement, may tolerate a lesser number of long-duration treatments better than the effects of repeated short treatments. The anticancer effects of hyperthermia may induce considerable edema about intrapulmonary lesions which, together with the effects of repeated sessions of mechanical ventilation, can induce pulmonary decompensation requiring long-term respirator support. Since long-term mechanical ventilation may predispose to life-threatening infection in these compromised patients, a preferable approach may be to administer a single long-duration treatment requiring only a single recovery effort.

There is, however, one clinical situation in which initial treatments of long duration may be hazardous: the treatment of large tumor burdens of fast-growing malignancies, such as melanoma and poorly differentiated bronchogenic carcinoma. Even short-duration treatments may cause, in these patients, a sustained posttreatment autogenous fever of 39.5°C to 41.0°C, a high cardiac output, and a progressive deterioration of alveolar gas exchange. This circumstance may be difficult to reverse in patients treated for only three hours, and it is unlikely that treatment of longer duration would be survivable with present means of support.

It is also unwise to proceed with an initial treatment of long duration in patients with compromised organ systems. Therefore, it is recommended that patients with marginal function of critical organs, particularly cardiopulmonary and renal systems, receive an initial treatment of three-hour duration. The response of the patient to that treatment maybe used to plan further treatments. If the short-duration "test" treatment is well tolerated, it is likely that treatments of eight-hour durations will be safe.

Finally, the UMC experience indicates that a total treatment duration of at least 12 hours at temperature, given in no less than six-hour sessions, is required for a significant antitumor effect. During early UMC protocols, patients received brief treatments of one- to three-hour durations, often administered at weekly intervals. Such treatments never produced significant tumor regression, and subsequent experience suggests they are unlikely to do so. A treatment regimen of 24 hours at 41.5°C hyperthermia administered as four six-hour treatment sessions appears to represent a minimum level at which an antitumor effect may be induced in 30% to 50% of the patients, even with combined chemotherapy. Furthermore, the regressions induced in a number of patients treated with this schedule were transient, with reactivation occurring at 4 to 12 months. If these findings prove the general rule, those undertaking cancer treatment with whole-body hyperthermia must consider the problems of anesthesia, laboratory support, and other hospital services required for long-duration hyperthermia treatment.

CONSIDERATION FOR INCREASING DOSAGE

Although adverse neurologic effects reported to occur from sunstroke and hyperpyrexia may be avoided by proper patient support during induced hyperthermia, our experience indicates that the tolerance of central nervous system tissue to hyperthermia still is a limiting factor. Generally, 42.0°C appears to be the upper limit of constant whole-body hyperthermia safe for most individuals, although Frazier has shown that this temperature may safely be exceeded for short periods of time (Parks et al. 1979). There is no indication in the literature, however, that temperatures much above 42.0°C are tolerable by humans for any significant period, and the UMC experience supports that contention.

Patients receiving treatment at 42.0°C recover much more slowly from anesthesia than those treated at 41.8°C and may manifest transient abnormal neurologic findings. The difference in the effects induced by increasing treatment temperatures from 41.8°C to 42.0°C is a very real argument for accurate temperature monitoring. It is a strong clinical impression that temperatures greater than 42.°C are dangerous with present means of support, and treatment exceeding this level should be approached with great caution.

There are, however, sound reasons for treating patients with temperatures higher than 42.0°C, if possible, as it seems likely that some tumors will prove resistant to lesser temperatures regardless of treatment duration; the length of treatment of other malignancies may be shortened significantly if higher magnitudes of hyperthermia are applied. Theoretically there are a number of ways by which treatment at temperatures higher than 42.0°C (supraHT) may be achieved, and several of these have been investigated in the UMC laboratories. The approaches include both physical and biochemical methodology, and there is a critical need for investigation of these areas.

A simple example of physically induced supraHT is the application of electrically or water-heated blankets to tumor-bearing skin areas of patients undergoing whole-body hyperthermia. A common indication for this approach is chest-wall involvement by recurrent breast cancer with concomitant distal metastasis. By use of carefully applied local heating, the skin and cutaneous implants may be heated to 42.5°C for at least eight hours without serious burn problems. If such local heating is not applied, skin temperatures are likely to be lower than those required for an antitumor effect. The transabdominal circulation of heated peritoneal dialysate also may be employed to create regional hyperthermia within the abdominal cavity (Smith, McMillan, and Parks, in press), both by itself and in conjunction with whole-body hyperthermia. Particular indications for such treatment may be disseminated ovarian cancer or the type of gut adenocarcinoma that produces massive numbers of peritoneal implants without metastasis to distant sites. The heating of the portal circulation by peritoneal "thermal dialysis" can induce regional hepatic hyperthermia and may prove an additional means for treatment of tumor involvement of that organ.

Since the extracorporeal method has proved to be an efficient means for induction of both systemic and regional extremity hyperthermia, it seems reasonable to surmise that other body regions may be superheated by combining systemic and regional extracorporeal hyperthermia. One approach contemplates the transfemoral insertion of an aortic catheter whose tip is positioned at the level of the diaphragm. The infusion of heated blood through this catheter may induce supraHT in all the abdominal viscera. Although there are a number of technical obstacles to this approach, they appear to be surmountable, and laboratory development and clinical evaluation of this approach is ongoing.[6]

Electromagnetic induction of regional hyperthermia has proved possible by a variety of techniques (see chapter 15), but it apparently has not been attempted in patients undergoing systemic hyperthermia. While it should be possible to obtain localized supraHT by such an approach, there are complicated issues involved. Selective tumor heating may be dependent on relative avascularity of tumor masses, and tumor perfusion may be increased by the hyperdynamic circulation of systemic hyperthermia. Normal tissues, already at hyperthermia temperatures, may become more susceptible to excessive localized heating, and the therapeutic differential between normal and malignant tissues may narrow. While such problems may or may not exist, it seems clear that combined methodologies of local or regional and systemic hyperthermia should be evaluated.

Another approach to obtaining whole-body supraHT, which is undergoing evaluation at M. D. Anderson Hospital, involves rapid alternation of patient temperature by the extracorporeal method. The extracorporeal technique can provide a very rapid heating and cooling effect, with temperature changes occurring as fast as 0.1°C every 15 seconds. With such rapid heating it has been shown that it is possible to take patients to a whole-body temperature of 43.0°C, as measured in the esophagus for short intervals, before rapidly returning to a baseline temperature of 41.8°C. This technique referred to as *pulse hyperthermia* by Dr. Howard O. Frazier of Houston (Parks et al. 1979), may be an important development, since it allows very high temperatures to be induced for short periods; the initial tumor regressions observed with this approach have been noteworthy. The process is suited to automation using microprocessor technology, and detailed clinical investigation of its effectiveness is forthcoming.

In addition to the physical approaches for induction of whole-body supraHT, there are at least two biochemical approaches to support of the central nervous system during hyperthermia that may allow treatment temperatures to be increased. The approaches under investigation at UMC are the intravenous administration of the ketone body d,l-β-hydroxybutyrate and the ATP precursor fructose diphosphate (FDP). Preliminary data suggest that both of these substances may be useful substrates for energy production by neural tissue during hyperthermia, and that their availability may exert protective effects in tissues attempting metabolic processes at high temperatures. No data is yet available to indicate whether the protective effect will be exerted preferentially in normal tissue or whether a degree of protection may also be conveyed to malignant cells. In this regard, there are fewer theoretical hazards to the use of ketone bodies than FDP, as the latter may indeed support the anaerobic metabolism of malignancy. Investigation of the use of these substances and others as methods of biochemical support of patients undergoing hyperthermia is warranted.

COMBINED RADIATION AND HYPERTHERMIA

Three international symposia on cancer therapy by hyperthermia and combined therapy have been held, and the proceedings from these are a valuable compilation of information on biological effects of hyperthermia and, most particularly, its combination with ionizing radiation. (Wizenberg and Robinson 1976; Streffer et al. 1977). The principal issues for those who subscribe to combination hyperthermia and radiation are the timing and dosage relationships of the two agents (see chapter 20). It is clear that hyperthermia potentiates the effects of radiation on malignant tissue and also, perhaps to a lesser degree, its effects on normal structures (Suit and Gerweck 1979). There is controversy as to whether radiation is most advantageous if given before, during, or after hyperthermia, and on the intervals between treatment. Much of the available data on the effects of these agents is derived either from in vitro studies or the effects of radiation on animal tumor models, which typically receive single-fraction hyperthermia of only short duration.

Twenty-seven patients have now received radiation/systemic hyperthermia combination therapy at the University of Mississippi. The treatment approach has been to use one-half of the radiation dosage customary for the cancer problem under treatment, and to administer as much of the radiation as possible immediately prior to hyperthermia. Patients having tumor recurrence in an area previously treated with a normal dose of radiation are given small "booster" doses, if thought tolerable, and normal organs are

---

[6] R. H. Bartlett, personal communication.

shielded when possible. All radiation is administered from a standard cobalt source. Adjuvant chemotherapy is administered in conjunction with hyperthermia and radiation, with the typical drug and dosage being BCNU 75 mg (total dose) given during the midpoint of the first two treatments. This dosage is reduced from those customarily used in patients not receiving radiation treatment.

In addition to the patients treated with radiation in immediate proximity to hyperthermia, a total of 57 patients had received radiation prior to hyperthermia. The importance of this consideration is illustrated by the following case, which is the second instance of adverse effects from radiation and hyperthermia we have experienced.

A patient with bronchogenic carcinoma and mediastinal involvement received 4500 rad administered in split increments of 2500 rad and 2000 rad, each applied over a 10-day course approximately one month prior to hyperthermia. The tracheal air column was shielded during the second course. Following hyperthermia, the patient developed paraplegia, which was at a level coincident with the upper margin of the radiation field. He died of pulmonary infection. Post mortem examination revealed neural cell necrosis in the irradiated portion of the spinal cord. It may be pertinent that this patient had low pO_2 measurements of 70 to 100 mm, maintained for 15-minute intervals during each hour of hyperthermia as part of his protocol treatment, and such relative hypoxemia is no longer permitted in patients previously irradiated. The development of paraplegia in this patient at the level of the upper margin of the radiation port is a clear warning to beware the potential of radiation and whole-body hyperthermia for damaging normal CNS tissues and possibly other organs. Two additional CNS casualties have been observed in recently radiated patients. No similar neurologic complication has been observed in patients receiving 1000 rad to 2500 rad immediately prior to hyperthermia or in patients receiving full doses of radiation two or more months before hyperthermia.

The anticancer effects produced by hyperthermia and radiation have been striking. To date, there has been objective evidence for tumor control in every patient receiving combined radiation and hyperthermia. This has occurred even though some of the lesions treated, such as adenocarcinoma of the lung, would be expected to be radioresistant. Tumor response has been quick and substantial or complete. With the exception of the case described above, and one similar to it but less devastating to the patient, there has been no instance of significant side effects,
even with the use of whole-abdominal irradiation, including the right hepatic lobe.

TUMOR PATHOLOGY AFTER HYPERTHERMIA

An effort has been made to characterize the effects of hyperthermia upon tumor tissue, not only to attempt to understand the effects of the treatment, but also to enable predictions as to what might constitute adequate treatment for a given lesion. If, for example, one could by biopsy obtain tissue that showed that a tumor was irreversibly injured by a given dose of hyperthermia, further treatments could be avoided. Conversely, if the tumor were still viable, further treatment could be initiated. In general, however, a correlation between posttreatment morphology and tumor regression or progression has been difficult to ascertain, and the following observations must be considered tentative. It is also quite possible that an entirely different set of observations and conclusions will be produced by study of lesions treated with different regimens than those used in this series.

Gross examination of tumor mass recently treated by hyperthermia frequently reveals a 20% to 30% enlargement of the mass. Typically, the tumor softens and vascular congestion and intra- and peritumor edema become manifest. Tumor visualized through a bronchoscope often is notably more friable, and its manipulation or biopsy may produce troublesome bleeding. Rarely are there signs of acute inflammation within or about tumor nodules. This state ordinarily continues to and throughout tumor regression. As regression becomes manifest, the tumor mass decreases in size, while a bland necrosis expands within it. Skin lesions typically will ulcerate and discharge their contents, leaving only a smooth ulcer that steadily heals over. Lesions involving visceral organs, as seen by x-ray, steadily decrease in size and finally disappear. If lesions in this state become available for gross examination, they are found to be soft and cystic, and on section, contain white, cheesy, necrotic tissue. The absorption of this product may be so complete that a chest film showing lesions several centimeters in diameter ultimately will clear with no residual scar or other abnormality.

The histologic effects of hyperthermia upon tumors have been even more difficult to assess with certainty because most advanced malignancies manifest considerable necrosis, and a variation of histology that is dependent on the site of sample. Therefore, it is necessary to be quite cautious before attributing to hyperthermia any acute cell death noted histologically. Ordinary tumor necrosis typically is a spectrum of phenomena, however, which range from the ghost cell of recent necrosis to the formation of calcification and fibrosis in long-dead tissue. Furthermore, ordinary necrosis tends to occur in central locations and does not extend to the periphery of tumor masses.

In contrast to these findings, the injury produced by hyperthermia typically displays hallmarks of a recent event, notably the presence of inter- and intracellular edema, which cause ballooning of both cytoplasm and nuclei to produce cellular characteristics similar to those of radiation damage. Dead cells are manifested by disrupted ghosts, which are distributed across the tumor and are present to the periphery of masses, even in well-vascularized areas. There is some suggestion that an ischemic event is associated with cell kill by hyperthermia, as the distribution of lesions often follows the wedge shape of an infarct pattern and acute thrombosis of small vessels may be identifiable. Whether this occurs as a cause or effect of the impact of hyperthermia upon malignant tissue is uncertain.

The histology of tumors treated with hyperthermia often is disturbingly normal until a sudden regression occurs, sometimes several months posttreatment. During the interval, the cells appear quite viable and frequently display normal features such as mitosis, mucin production, and similar features of tumor activity. Then, within days of such findings, an acute bland necrosis of the entire tumor mass may occur and lead to complete necrosis. The reasons for this phenomenon are unclear, but its existence seriously restricts the use of interval biopsy findings as a guide to the treatment of patients (see also chapter 15).

IMMUNOLOGIC EFFECTS

There have been periodic suggestions that destruction of a tumor by heat may produce a systemic anticancer response of an immune nature (Strauss et al. 1965; Szmigielski and Janiak 1977) (chapter 23). Such a concept has been used to explain the favored survival of patients having improved survival time following local or regional treatment of cancers by electrocautery (chapter 14), regional hyperthermic perfusions (chapter 17), or electromagnetic induction of localized hyperthermia (chapter 15). Although the beneficial effects of these treatments have been occasionally attributed to "immunostimulation," laboratory evidence for the induction of clinically relevant tumor specific immune responses by these approaches has been lacking. Furthermore, there is a growing suspicion that there may be little, if any, relationship between immunocompetence and advancement or regression of malignancy (Benjamini and Renick 1979). Thus, the hypothesis that hyperthermia evokes a significant, specific anticancer immune response remains questionable and warrants further investigation.

There is no doubt that the suppression of immunocompetence, caused by conventional forms of cancer therapy such as radiation and chemotherapy, plays a significant role in the adverse effects the patients experience. Significant neutropenia inevitably follows intensive chemotherapy, and the resulting deficiencies of phagocytosis, cell-mediated reactivity, and humoral immunity result in frequent severe infections, which are a major cause of morbidity and mortality in patients undergoing treatment for advanced malignancy (Bodey 1973). Similar deleterious effects may be produced by radiation; its marrow-suppressive effects frequently limit the treatment that may be administered to the neoplasm. Therefore, it is of interest to consider the effects of hyperthermia upon the immune system, as there are preliminary indications that the immunosuppressive effects of hyperthermia may be less than those of other available treatments for advanced malignancy (see chapter 23).

At UMC, efforts have been made to characterize certain immune functions of patients treated with systemic hyperthermia. The effects of hyperthermia on the ingestive and bactericidal capacity of polymorphonuclear leukocytes (PMN) for *Staphylococcus* were assessed by assays designed by Grogan and Miller (1973) who demonstrated that a patient's PMN ingestive capacity was not altered after systemic hyperthermia. The data also indicated that there was a tendency for the bactericidal capacity of the PMN to increase, particularly in patients having a depressed bactericidal capacity prior to hyperthermia treatment (fig. 19.16).

Patients also were evaluated for T- and B-cell responses by phytohemagglutinin (PHA), pokeweed mitogen (PWM), and rosette-forming cell (RFC) assays; these results are shown in table 19.6. These data are consistent with other studies, which suggest that T- and B-cell function of patients with advanced cancer are suppressed, as compared with normal function. The data indicate, however that while suppression of lymphocyte proliferation does occur during systemic hyperthermia, there is a recovery of this function within 48 hours of treatment. The suppressions observed are less than those generally reported for high-dose combination chemotherapy regimens, even though the patients reported here received adjuvant alkylating agents during hyperthermia.

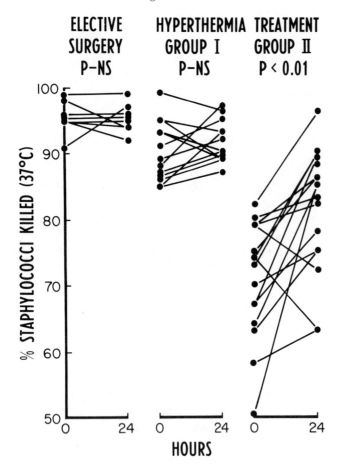

Figure 19.16. Bacterial activity of polymorphonuclear leukocytes 24 hours before and 24 hours after hyperthermia. Normal function is not augmented but depressed responses appear to be stimulated by treatment. (Grogan, Parks, and Minaberry 1980. Reprinted by permission.)

These laboratory findings are consistent with the clinical observation that serious infections occurred infrequently in the patients treated in this series with systemic hyperthermia. Although every patient had repetitious exposure to potential sources of serious sepsis, such as central venous lines, repeated insertion of bladder catheters, prolonged exteriorization of vascular prostheses, and prolonged endotracheal intubation for anesthesia and respiratory support, the overall incidence of infection was low, perhaps strikingly so. Infections, when present, responded to treatment in the ordinary fashion, and there was no single documented instance of the explosive bacterial or atypical infection that is seen in transplant or other suppressed patients.

ANTICANCER EFFECTS: CLINICAL TRIALS

It should be understood that the investigation at UMC has sequentially considered three issues: (1) the safety and efficacy of extracorporeal induction of hyperthermia, (2) the requirements for metabolic and physiologic support of patients undergoing hyperthermia, and (3) the anticancer effects of treatment. It was necessary to resolve the first two issues before the third could be seriously examined. The patients initially treated were in difficult straits and ill suited to long-term survival studies. As experience with hyperthermia has accumulated, the probability of survival has increased (fig. 19.17). As of December 31, 1979, a total of 102 patients, ranging in age from 18 to 72 years, have received 371 treatments. Most of the patients had

Table 19.6
Effects of Hyperthermia upon Immune Responses as Determined by Phytohemagglutinin (PHA), Pokeweed Mitogen (PWN), and Rosette-Forming Cell Assays (RFC)

	T- and B-Cell Responses		
	PHA	PWM	RFC
Normal	30.23 ± 9.2	15.84 ± 7.3	64.4 ± 11.1
Ca, pre HT	27.2 ± 9.4	11.0 ± 6.5	58.3 ± 8.4
Ca, mid HT	12.0 ± 7.3	5.2 ± 4.7	28.7 ± 8.7
Ca, end HT	12.8 ± 10.0	4.7 ± 2.4	28.4 ± 14.2
Ca, post HT	28.9 ± 11.6	6.5 ± 4.8	35.8 ± 12.2

NOTE: HT = hyperthermia.

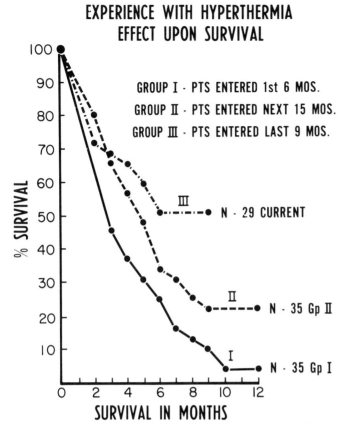

Figure 19.17. The influence of experience on survival after hyperthermia. This does not reflect differences in patient selection but may be attributable in part to better protocols.

Table 19.7
Patient Treatment prior to Hyperthermia

Treatment	No. of Patients
Surgery + radiation + chemo Rx/hormone Rx	28
Surgery + radiation	15
Surgery + hormone Rx	5
Surgery + chemo Rx	7
Surgery, resection	11
Radiation + chemo Rx	7
Radiation	12
Chemo Rx	3
Hormone Rx	1
None, exploratory, or biopsy	13

Table 19.8
Site and Histology of Patients Receiving Hyperthermia as Primary Treatment

Site	Type	No. of Patients
Lung	Adenocarcinoma	3
Pancreas	Adenocarcinoma	3
Stomach	Adenocarcinoma	2
Gallbladder	Adenocarcinoma	1
Liver	Hepatoma	1
Various sites	Epidermoid	3
Total		13

undergone previous treatment (table 19.7). The distribution of histologic types of malignancy and the sites of origin are given in tables 19.8 and 19.9.

Seventy-two patients (71%) have survived two months or more and have allowed evaluation of the effects of hyperthermia on their malignancies. Objective evidence of regression has developed in 43 patients, including 11 complete responses. Clear failure of treatment has occurred in 14%, and another 12% of patients have had responses of questionable benefit. Long-term (greater than six months) responses include posttreatment survivals of 28, 23, 19, 17, 16, 14, 14, 11, 11, 9, 8, 8, 8, 7, 7, 7, 7, and 7 months, representing 18% of the patients. A number of these patients are well, and will survive for still longer periods. Another 16 patients are alive and well at two to six months.

The overall mortality from all causes has been 72% for a 28-month period, and the current survival curve is illustrated in figure 19.18. Since 86% of the patients comprising this survival curve had inadequate response to treatment by radiation and/or chemotherapy, and all had far-advanced malignancy, often of a preterminal nature, such survivals are noteworthy. Further insight into the anticancer effects of hyperthermia may be gained by consideration of the following histories.

Patient 1
A 45-year-old man with extensive pulmonary involvement by hypernephroma progressive after nephrectomy and treatment with medroxyprogesterone. He received hyperthermia treatments for six hours at 42.0°C × 2 and at 41.8°C × 2, with adjuvant CTX 250 mg/M^2 and BCNU 50 mg being administered during the first two treatments. Complete regression developed and the chest film is normal 17 months posttreatment.

Patient 2
A 41-year-old woman with melanoma of the

Table 19.9
Site of Origin and Histology of All Malignancies Treated

Site	Type	No. of Patients
Lung	Squamous	19
	Large cell	9
	Small cell	3
	Adeno	6
	Alveolar	1
Total		38
Gastrointestinal		
Esophagus	Adenocarcinoma	2
Gastric	Adenocarcinoma	2
Pancreatic	Adenocarcinoma	4
Colon	Adenocarcinoma	11
Hepatoma	Adenocarcinoma	2
Gallbladder	Adenocarcinoma	2
Total		23
Kidney	Adenocarcinoma	7
Ovarian	Adenocarcinoma	2
Breast	Adenocarcinoma	11
Skin	Melanoma	9
Bone/soft tissue	Sarcoma	4
Total		33
Other	Squamous	5
	Adenocarcinoma	1
	Lymphoma	2
Total		8

right arm initially treated by wide surgical excision. Four months later, right breast and right axillary involvement became manifest, with 38 positive nodes being removed from the axilla. She received hyperthermia for eight hours at 41.8°C × 4, with adjuvant cisplatin 100 mg/M² and 100 mg administered during the first and second treatments. Since then, she has remained free of recurrence and is now 10 months posttreatment.

Patient 3
A 42-year-old woman with duodenal and biliary obstruction by pancreatic adenocarcinoma. She received 2000 rad of radiotherapy two weeks prehyperthermia, followed by four hyperthermia treatments of six- to eight-hour duration each at 41.8°C to 42.0°C. CTX 250 mg/M² and BCNU 50 mg were administered during the first two treatments. Complete regression developed, with relief of GI and biliary obstruction. This state has been maintained for 10 months.

Patient 4
A 28-year-old man with hepatoma resistant to hepatic arterial infusion chemotherapy, first seen with a serum bilirubin of 9 and severe ascites. Patient required a LeVeen shunt for the ascites and a mesocaval shunt for a variceal bleed before any treatment could be initiated. He subsequently received hyperthermia treatments of four-hour duration at 41.5°C × 4, with adjuvant CTX 250 mg/M² and BCNU 50 mg being administered during the first two treatments. Complete regression developed, which was maintained for 16 months before reactivation occurred. Retreatment with 41.8°C for eight-hour duration × 4, accompanied by adjuvant BCNU 75 mg × 4, again resulted in substantial regression now maintained for five months since the second treatment.

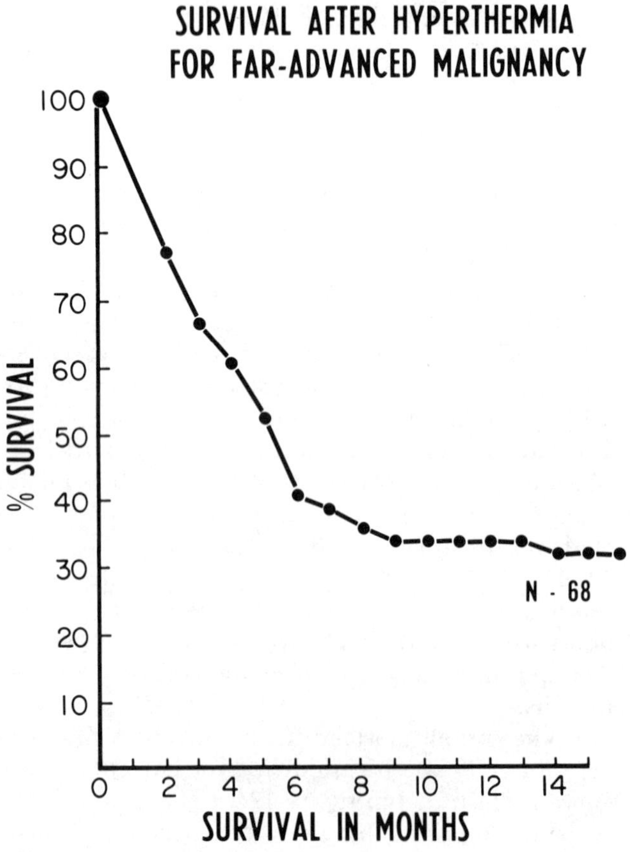

Figure 19.18. Overall survival in phase II patients. All have been at risk for at least three months and post-hyperthermia for six to twelve months.

Patient 5
A 45-year-old man with recurrent hypernephroma causing paraplegia, inferior vena cava obstruction, and intermittent small bowel obstruction. He received hyperthermia of 41.5°C for six-hour duration × 4 without chemotherapy, after which there was no progression for 15 months. Reactivation was treated by hyperthermia of 41.8°C for six-hour duration × 4 with adjuvant CTX 250 mg/M^2 and BCNU 50 mg × 2, after which progression has not been manifest for 12 months.

Patient 6
A 45-year-old man with unresectable epidermoid cancer of the bronchus-intermedius secondary to mediastinal involvement, including partial superior vena cava obstruction. He received hyperthermia of six-hour duration at 42.0°C × 2 with adjuvant CTX 500 mg/M^2 administered during each treatment. Substantial regression developed, which was maintained for 27 months before reactivation recurred. The therapy was repeated and has resulted in further regression now maintained for four months.

ROLE OF THE SURGEON

The anticancer effects of whole-body hyperthermia appear to be different from those observed when tumors respond to radiation or chemotherapy alone. As a rule, if those agents are effective, tumors will steadily decrease in mass after initiation of therapy, and the function of compromised structures will steadily improve. In contrast, the mass of a tumor does not regress soon after therapy when treated with the temperatures and duration of whole-body hyperthermia used in our series. On the contrary, the tumor will stabilize or even enlarge for a period of several months. The cause of this effect is uncertain, but it may be that hyperthermia causes a failure of tumor cell replication, and that mass reduction does not become apparent until cells not only fail to divide, but live out their life span and expire. Whatever the cause, persistent or enlarging tumor mass after treatment can cause several problems. In the UMC experience, the worst problems have been continued or progressive compromise, that may lead to obstruction, of airway, biliary, ureteral, and intestinal lumens. Another significant cause of problems has been infection and hemorrhage in large persistent tumor masses undergoing regression posthyperthermia.

For these reasons the use of surgery for debulking and bypass procedures prior to initiating hyperthermia has progressively increased. Where possible, compromised organs have been relieved by radiation before hyperthermia, but as many of the patients have had failures of primary radiotherapy, debulking often has had to be accomplished surgically if at all. There has always been some suggestion that such procedures be conducted prior to undertaking chemotherapy or, in particular, immunotherapy, since there then would be a lesser tumor burden to treat. This approach has received limited recognition, however, because its potential to increase the effectiveness of palliative therapy has not been considered sufficient to warrant the morbidity involved.

The experiences of this investigation suggest that the management of large tumor masses with hyperthermia will pose certain hazards. Clearly, the necrosis of large amounts of tumor tissue will foster development of infection or exsanguinating hemorrhage. The vascular congestion and edema that appear within tumors under treatment may further embarrass a compromised airway, worsen biliary or ureteral obstruction, or convert partial intestinal obstruction to a complete blockage. The development of edema or vascular congestion within brain metastasis may produce or exacerbate elevated intracranial pressure, with resultant neurologic deterioration. For these reasons, the debulking of symptomatic or potentially harmful tumor masses prior to undertaking systemic-hyperthermia may be a valid consideration when feasible.

The diversity of surgical procedures that were necessary to enable the UMC patients to undergo hyperthermia are listed in table 19.10. Many of these procedures were performed after considerable soul searching, and surgical colleagues felt that some of the procedures were controversial; however, the adverse effects of necrosis developing in large tumor masses are reflected in the 14 deaths in the UMC series that were caused by these events. The problems caused by necrosis have been sepsis, 6; hemorrhage, 5; hypercalcemia, 1; and platelet consumption, 2. In most instances, the responsible mass technically could have been removed and certainly could have been debulked if resection had not been deferred as "unconventional." Because of such findings, there has been a progressive and careful implementation of extended surgical resections prior to hyperthermia. The results of this approach have been generally gratifying and suggest a need for a surgical presence on the hyperthermia team (Streffer et al. 1977).

As yet, there is no significant experience with the application of hyperthermia to patients without

Table 19.10
Surgical Procedures Necessary Before Hyperthermia

Operation	No. of Procedures
Vascular Operations	
HT shunt	
Dacron™	79
Thomas	23
Reshunt	10
Revision	9
Extra anatomic bypass	3
Hepatic ligation	2
Mesocaval	2
Carotid ligation	1
Total	129
Thoracic Operations	
Tracheotomy, cervical	17
Tracheotomy, mediastinal	1
Tube thoracostomy	28
Open lung biopsy	3
Relief SVC obstruction	2
Pericardial window	2
Bronchpscopy, diagnostic	13
Bronchoscopy, therapeutic	56
Major	(8)
Minor	(114)
Total	122
Extended Operations	
Multiorgan	8
Bowel	4
Total	12
Urologic Operations	
Diversion	1
Reimplant	3
Nephrectomy	1
Total	5
General Surgical Operations	
Resection small bowel obstruction	7
LeVeen shunt	4
Biliary diversion	4
Total	15
Other Operations	
Biopsy, diagnostic	23

NOTE: HT = hyperthermia.

known residual disease and high risk of early recurrence. Examples of clinical circumstances where such adjuvant treatment may be appropriate include bronchogenic cancer found to exhibit vascular invasion after resection, intestinal malignancies found to involve several lymph nodes or to manifest vascular invasion, or any resected aggressive malignancy (such as pancreatic adenocarcinoma). Even though these lesions are "completely" removed, the ultimate risk of recurrence may exceed 90%, and the application of hyperthermia to the minimal disease that undoubtedly persists postresection is attractive. Since there is substantial evidence that hyperthermia potentiates the anticancer effects of radiation, the use of this modality in conjunction with surgical resection and adjuvant hyperthermia may be important if local recurrence is probable. A possibly ideal patient for a combined treatment approach would have a resectable bronchogenic carcinoma shown to involve mediastinal or supraclavicular nodes without extensive distant disease. Treatment of such an individual might include a limited resection of the primary lesion, mediastinal and supraclavicular irradiation, and then hyperthermic treatment during which adjuvant chemotherapy of a radiomimetic nature would be administered. In such a setting, resection of the primary lesion would greatly reduce problems of pulmonary hemorrhage and/or infection, which otherwise may prove troublesome during subsequent hyperthermia treatment.

SPECIAL PROBLEMS

Respiratory Insufficiency

Hyperthermia is a condition of considerable cardiopulmonary stress, since it causes both cardiac output and oxygen consumption approximately to double. The increased pulmonary blood flow is not well tolerated by patients with cor pulmonale, and those with significant restrictions of mechanical or diffusion function also require prolonged posttreatment support. These problems suggest that if patients with cardiopulmonary restrictions are to be treated, every effort should be made to improve or offset the deficient functioning of these organs prior to undertaking hyperthermia.

The most correctable situation frequently encountered is that of malignant effusion which, although relatively asymptomatic prehyperthermia, may be a cause of significant respiratory embarrassment in the posttreatment period. The frequency and rapidity with which these effusions reaccumulate after thoracentesis, despite the use of various sclerosing agents, has led to an abandonment of these measures. Currently, malignant effusions are treated by insertion of a chest tube, which is left for a long period of time, usually 10 to 14 days. Typically, this procedure is performed in an operating room to minimize the risk of infection, as the patients are unusually susceptible to any bacterial contamination of the pleural space. As a rule, this procedure is conducted under general anesthesia, since the pain threshold of many of these patients has become limited during their multiple hospitalizations. Also, it is frequently necessary to explore digitally the pleural space, and this cannot be done comfortably under local anesthesia. Postinsertion, chest tubes must be carefully managed by frequent dressing changes and the application of antiseptic ointment in a manner analogous to that used for care of central venous lines. By these measures, difficulty with infection has been minimized, and patients have tolerated the usual regimen of steroid and antineoplastic agents without difficulty.

There are also circumstances where the probability of tumor response is high, but because of mechanical restrictions the patient's pulmonary functions are marginally competent or inadequate to undertake hyperthermia. In such circumstances, the use of prolonged intubation and respiratory support, initiated prior to undertaking hyperthermia and continued for a significant portion of the recovery period, may be considered. Such circumstances can face the physician, patient, and family with difficult decisions as to the wisdom of undertaking such an endeavor. Only time and experience will provide an adequate basis for judgment under these circumstances.

Should respiratory support be elected, there are several practical problems to be dealt with. Although there is an increasing tendency to rely on transoral or transnasal endotracheal intubation for significant periods, the UMC experience suggests that this may not be the optimal approach. Because patients receiving this measure would typically have intrapulmonary malignancy and frequently endobronchial or intratracheal lesions, and because of the increased pulmonary vascularity and lymph flow during hyperthermia, the patient's problem with secretions will be greater than usual. Furthermore, the presence of abnormal secretions or bleeding within airways may persist for much longer periods than customarily encountered. These findings have led us to resort to elective tracheostomy earlier and more frequently than would be the case for patients with trauma or other acute conditions. Tracheostomy is not innocuous, and two UMC patients, whose mediastinums were heavily irradiated, have died of hemorrhage from innominate arteries. On balance, however, the use of tracheostomy has seemed advantageous over that of long-term nasotracheal intubation.

A particular hazard to be avoided is alveolar degeneration secondary to oxygen toxicity. While there is no evidence that hyperthermia increases the incidence of oxygen-induced alveolar membrane proliferation, such findings would not be unexpected, inasmuch as hyperthermia increases susceptibility of other organs to noxious agents. Furthermore, the alkylating agents CTX and BCNU, which appear to be potentiated by hyperthermia, have independent toxic effects on the pulmonary alveolus, which can result in deterioration of diffusion capacity. Therefore, it is recommended that patients undergoing respiratory support receive only the required FiO_2, and that continual efforts be made to limit the concentration of oxygen inspired and the duration of dependence on ventilatory support.

The high-flow shunts required for extracorporeal induction of hyperthermia may offer another approach to support of respiratory insufficiency encountered in these patients. The long-term use of membrane oxygenation has been proposed for such circumstances (Bartlett and Gazzaniga 1978), but has found limited application, particularly as an approach to support of adults with pulmonary insufficiency. Nonetheless, substantial technology for this purpose exists and may be more applicable to these patients than most. A traditional problem with the implementation of prolonged ECC respiration has been vascular access. Since this is already available in patients undergoing extracorporeal hyperthermia, the development of programs for membrane support of some of these patients may be expedient.

Jehovah's Witnesses

The treatment with hyperthermia of patients who refuse blood transfusion poses some difficulties, and it seems on first impression that they should not be accepted for such protocols. They are, however, as entitled as the rest of us to treatment of malignant disease, and they may have a special requirement for hyperthermia in that they are poor candidates for extensive surgical procedures. In addition, most of these individuals having malignancy will present with low serum hemoglobin values, which cannot be augmented. The treatment of such patients may provide useful information as to the relationship between hyperthermia, tissue oxygenation, and tumor control. If such patients are to be treated, the following difficulties should be recognized.

Since circulating hemoglobin ordinarily constitutes a major portion of the blood buffer system, anemic patients possess a relatively high probability of developing respiratory acidosis during the induction phase of hyperthermia. Careful attention must therefore be paid to blood gases until their stabilization occurs, and it is frequently necessary to conduct treatments at a relatively low pCO_2. In addition to this respiratory compensation, the periodic administration of sodium bicarbonate also is frequently required to prevent serious acidosis. The second problem is, of course, that ordinary schedules of volume expansion with blood and plasma solutions cannot be followed, and greatly increased amounts of crystalloid must be administered. The use of dextran solutions has been of value in these circumstances and has not been accompanied by any unusual adverse effect, particularly that of hemorrhagic tendency. Through these measures, patients have been successfully exposed to hyperthermia for treatments up to six-hour duration at 41.8°C, while their hematocrit was maintained in a range of 12 to 14. Two patients have been so managed, one with far advanced Hodgkin's disease recurrent after all usual measures, and the other with extensive hepatic involvement by colon carcinoma. Both patients tolerated hyperthermia adequately, and both manifested tumor regression.

This limited experience provides no real insight as to whether it may be advantageous to treat patients with relatively low hematocrits; however, by varying hematocrit and inspired FiO_2, differentials between the oxygen tension in normal tissue and in malignant tissue can be created, which may be of therapeutic value.

Myasthenia Gravis

Another interesting problem encountered in the UMC series has been the treatment of a patient with myasthenia gravis with recurrent malignant thymoma involving the mediastinum and left ventricle; her neoplastic disease had failed multiple conventional approaches to therapy. There is a known tendency for myasthenia to worsen during hot summer months or with febrile illness, so there was considerable speculation as to the effects of hyperthermia upon this patient. She received a total of 21 hours at 41.8°C, during which adjuvant CTX 250 mg/M^2 and BCNU 50 mg were administered. This limited treatment was provided because previous chemotherapy and radiation had greatly reduced her bone marrow reserve, prohibiting the delivery of a full course of hyperthermia.

The myasthenia transiently worsened during the immediate posttreatment period, with the steroid dosage required for its control increasing from 40 mg of prednisone every other day to 100 mg every other day. Within one month of treatment, however, it was possible to begin rapidly tapering the steroid dosage until it was reduced to 20 mg q.o.d., with the patient being less symptomatic than before treatment. Chest films reveal regression of her mediastinal masses and the left ventricle lesion is stable or perhaps in a phase of scarification. The patient is active full-time in her employment as a teacher and is feeling well nine months posttreatment.

This single case suggests that these patients may be subjected safely not only to hyperthermia, but also to the various anesthetics and other manipulations required for their treatment. As malignant thymoma and myasthenia are distressing afflictions that progress rapidly once they are found to be resistant to the usual therapeutic modalities, this experience suggests that it may be reasonable to consider them for inclusion in hyperthermia protocols.

CONCLUSIONS

A new methodology for hyperthermia, the use of extracorporeal technology for induction and precision regulation of these temperatures, has been tested in humans and found safe and effective. Methods of support by which seriously ill patients can be safely maintained through long periods of high fever have been developed. These techniques are developed to the extent that it has become routine to maintain cancer patients at whole-body temperatures of 41.8°C or higher for long periods of time.

Considerable evidence has accrued that indicates that whole-body hyperthermia has an anticancer effect against most, and perhaps all, solid tumor malignancies, even when far advanced. It is clear that hyperthermia potentiates at least some chemotherapeutic agents, and that it will potentiate radiotherapy to a marked degree. Some malignancies, ordinarily resistant to conventional therapy have been totally regressed by this modality, and there is no certainty that they will reappear.

This is encouraging in that hyperthermia is still in its infancy as a modality of cancer treatment, and it seems likely that the incidence and magnitude of its anticancer effects will increase with experience. There are a host of permutations of treatment plans to explore, and intensive investigation of this agent appears indicated.

It is now unequivocally clear that whole-body hyperthermia can be induced in cancer patients, and that some will benefit from this treatment. This is not enough, but it is a beginning.

References

Barlogie, B. et al. Total-body hyperthermia with and without chemotherapy for advanced human neoplasms. *Cancer Res.* 39:1481–1489, 1979.

Bartlett, R. H., and Gazzaniga, A. B. Extracorporeal circulation for cardiopulmonary failure. *Curr. Probl. Surg.* 25:5–65, 1978.

Benjamini, E., and Renick, D. M. Cancer immunotherapy: facts and fancy. *CA* 29:362–370, 1979.

Bodey, G. Infections in patients with cancer. In *Cancer medicine*, eds. J. F. Holland and E. Frei III. Philadelphia: Lea & Febiger, 1973, pp. 1135–1165.

Bull, J. M. et al. Whole body hyperthermia: a phase-I trial of a potential adjuvant to chemotherapy. *Ann. Intern. Med.* 90:317–323, 1979.

Coley, W. R. The treatment of malignant tumors by repeated inoculations of erysipelas: with a report of ten original cases. *Am. J. Sci.* 105:487–511, 1893.

Dickson, J. A. Hazards and potentiation of hyperthermia. In *Proceedings of the First International Symposium on Cancer Therapy by Hyperthermia and Radiation*, eds. M. J. Wizenberg and I. E. Robinson. Chicago: American College of Radiology, 1976.

Dickson, J. A.; MacKenzie, A.; and McLeon, K. Temperature gradients in pigs during whole body hyperthermia at 42°C. *J. Appl. Physiol.* 47:712–717, 1979.

Dickson, J. A., and Shah, S. A. Stimulation of an antitumor immune response in VX 2-bearing rabbits by curative hyperthermia. In *Proceedings of the Second International Symposium on Cancer Therapy by Hyperthermia and Radiation*, eds. C. Streffer et al. Baltimore and Munich: Urban & Schwarzenberg, 1978, pp. 294–296.

Elias, E. G.; Shukla, S. K.; and Mink, I. V. Heparin and chemotherapy in the management of inoperable lung cancer. *Cancer* 36:129–136, 1975.

Fowler, G. A. *Beneficial effects of acute bacterial infections or bacterial toxin therapy on cancer of the colon or rectum.* New York: New York Cancer Research Institute, monograph #10, 1969.

Grogan, J. B., and Miller, R. C. Impaired function of polymorphonuclear leukocytes in patients with burns and other trauma. *Surg. Gyncol. Obstet.* 137:784–788, 1973.

Grogan, J. B.; Parks, L. C.; and Minaberry, D. Polymorphonuclear leukocyte function in cancer patients treated with total body hyperthermia. *Cancer* 45:2611–2615, 1980.

Hahn, G. M. Potential for therapy of drugs and hyperthermia. *Cancer Res.* 39:2264–2268, 1979.

Hersh, E. et al. Induced stimulation of cellular and humoral immunity in patients with malignancy. In *Research report, 1976*. Houston, Tex.: The University of Texas Press, 1976.

Kim, Y. D. et al. Hyperthermia potentiates doxorubicin-related cardiotoxic effects. *JAMA* 241:1816–1817, 1979.

Knochel, J. P. The pathophysiology and clinical characteristics of severe hypophosphatemia. *Arch. Intern. Med.* 137:203–220, 1977.

Knochel, J. P. et al. Hypophosphatemia and acute rhabdomyolysis. *Clin. Res.* 26:565A, 1978.

Krementz, E. T. et al. Chemotherapy of sarcomas of the limbs by regional perfusion. *Ann. Surg.* 185:555–564, 1977.

Larkin, J. M. A clinical investigation of total body hyperthermia as cancer therapy. *Cancer Res.* 39:2252–2254, 1979.

Longmire, W. P., Jr., and Sanford, M. C. Intrahepatic cholangiojejunostomy with partial hepatectomy and biliary obstruction. *Surgery* 129:264–276, 1948.

Madden, J. L., and Kandalaff, S. The treatment of cancer of the rectum by electrocoagulation. *Ann. Surg.* 174:530, 1971.

Marmor, J. B. Interactions of hyperthermia and chemotherapy in animals. *Cancer Res.* 39:2269–2276, 1979.

Miller, T. R., and Nicholson, J. T. End results in recticulum cell sarcoma of bone treated by bacterial toxin therapy alone or combined with surgery and/or radiotherapy (47 cases) or with concurrent infection (5 cases). *Cancer* 27:524–548, 1971.

Nauts, H. C.; Fowler, G. E.; and Bogatko, F. A. A review of the influence of bacterial infections and bacterial products (Coley's toxins) on malignant tumors in man. *Acta Med. Scand.* 145(suppl. 276):4–19, 1953.

Parks, L. C. et al. Treatment of far-advanced bronchogenic carcinoma by extracorporeally induced systemic hyperthermia. *J. Thorac. Cardiovasc. Surg.* 78:883–892, 1979.

Parks, L. C.; Frazier, O. H.; and Mountain, C. F. Paper read at the Fourth Annual Conference of the Denton A. Cooley Society, June 1977, Houston, Tex.

Pettigrew, R. T. et al. Clinical effects of whole body hyperthermia in advanced malignancy. *Br. Med. J.* 4:679–682, 1974.

Richter, J. E. et al. Pulmonary toxicity of bischloronitrosourea. *Cancer* 43:1607–1612, 1979.

Rosenow, E. C. The spectrum of drug-induced pulmonary disease. *Ann. Intern. Med.* 77:977–991, 1972.

Smith, G. V.; McMillan, R.; and Parks, L. C. Intra-abdominal hyperthermia by peritoneal perfusion. *J. Natl. Cancer Inst.*, in press.

Sostman, H. D.; Matthay, R. A.; and Putnam, C. F. Cytotoxic drug induced lung disease. *Am. J. Med.* 62:608–615, 1977.

Stehlin, J. S. Regional hyperthermia and chemotherapy by perfusion. In *Proceedings of the Second International Symposium on Cancer Therapy by Hyperthermia and Radiation*, eds. M. J. Wizenberg and J. E. Robinson, Chicago: American College of Radiology, 1976, pp. 266–271.

Stehlin, J. S. et al. Results of hyperthermic perfusion for melanoma of the extremities. *Surg. Gynecol. Obstet.* 140:339–348, 1975.

Strauss, A. et al. Immunologic resistance to carcinoma produced by electrocoagulation. *Surg. Gynecol. Obstet.* 121:989–996, 1965.

Streffer, C. et al., eds. *Proceedings of the Second International Symposium on Cancer Therapy by Hyperthermia and Radiation.* Baltimore and Munich: Urban & Schwarzenberg, 1977.

Suit, H. D., and Gerweck, L. E. Potential for hyperthermia and radiation therapy. *Cancer Res.* 39:2290–2298, 1979.

Szmigielski, S., and Janiak, M. Reaction of cell-mediated immunity to local hyperthermia of tumors and its potentiation by immunostimulation—a review. In *Proceedings of the Second International Symposium on Cancer Therapy by Hyperthermia and Radiation*, eds. C. Streffer et al. Baltimore and Munich: Urban & Schwarzenberg, 1977, pp. 80–88.

Thomas, G. D. Large vessel arteriovenous shunt for hemodialysis: a new concept. *Am. J. Surg.* 120:244, 1970.

White, A.; Handler, P.; and Smith, E. L. *Principles of biochemistry.* New York: McGraw-Hill, 1973, p. 970.

Wizenberg, M. J., and Robinson, J. E., eds. *Proceedings of the First International Symposium on Cancer Therapy by Hyperthermia and Radiation.* Chicago: American College of Radiology, 1976.

CHAPTER 20

Thermoradiotherapy: Molecular and Cellular Kinetics

Leo E. Gerweck

Introduction
Independent Effects of Heat Combined with Ionizing Radiation
Synergistic Interaction of Heat Combined with Ionizing Radiation
Response of Cells to Fractionated Heating and Radiation
Summary

INTRODUCTION

The rationale for the adjunctive use of hyperthermia with ionizing radiation for the treatment of patients with cancer has been developed largely from quantitative in vitro studies. Studies at the molecular and cellular levels provide the framework within which tissue effects can be understood and may be suggestive of particular treatment protocols. Furthermore, by providing insight into the nature and mechanism of hyperthermic effects, these studies should suggest steps that could be taken to increase the therapeutic effectiveness of hyperthermia. Nevertheless, these cellular effects will be modified by tissue level phenomena and represent only part of the complex interactions between heat, radiation, and tissue that occur in the patient.

The study of the combined use of hyperthermia and radiation is warranted by their complementary cytotoxic effects at the cellular level. Interaction between heat and radiation may be independent, additive, or synergistic, depending on the magnitude of the hyperthermic treatment, the time interval between treatments, the dose rate of the radiation, and the order in which the two treatments are applied. Certain important microenvironmental factors, which may be without effect or antagonistic to one agent, may interact synergistically with another. It is not clear if the synergistic interaction of radiation with hyperthermia will be more pronounced in tumor than in normal tissue; the independent use of these agents may be more efficacious than applications that involve synergism. In addition to the cytotoxic and sensitizing effects of heat, hyperthermia also exerts pronounced effects on cell growth rate and progression.

While characteristics of the response of cells to single doses of heat, alone or in combination with radiation, are being elucidated, less is known of the lethal and radiosensitizing effects of multiple heat treatments. Since it is unlikely that tumor sterilization will be obtained with large single doses of heat or radiation, this topic should receive additional attention at the cellular and tissue levels.

The lethal, nonlethal, and radiosensitizing effects of heat on the cellular and molecular levels are discussed in this chapter. Emphasis is placed on topics that are relevant to the tissue and clinical responses to hyperthermia. Much of the material presented in this chapter has been recently reviewed by Dewey and colleagues (1980).

This work was supported by U.S. Public Health Service grant CA−22860-02 from the National Institutes of Health.

Leo E. Gerweck

INDEPENDENT EFFECTS OF HEAT COMBINED WITH IONIZING RADIATION

Characteristics of the Response of Cells to Single Doses of Heat or Radiation

When they are compared, mammalian cell heat-survival curves and ionizing radiation survival curves exhibit several similarities and differences. For example, at 43°C the initial cellular inactivation rate of Chinese hamster ovary (CHO) cells increases with heating time to a relatively constant rate of inactivation. This rate can be approximated by a straight line when surviving fraction is plotted logarithmically versus linear treatment time. These kinetics are similar to those observed for CHO cells exposed to a radiotherapeutic conventional dose rate of 200 rad per minute, as seen in figure 20.1. When hyperthermic treatment temperature increases or decreases only a few degrees, however, the kinetics of cell killing are markedly affected, whereas a two- to threefold decrease or 10^{10}-fold increase in dose rate is without significant effect on the response of cells to radiation (Gerweck et al. 1979). The illustrated cellular responses indicate that the time of exposure to hyperthermia or radiation, as well as the exact temperature, will have a substantial impact on fractional cell kill.

Substantial variations in the thermal sensitivity of various cell lines at a specific temperature also have been noted (Kano, Miyakoshi, and Sugahara 1978; Gerweck and Burlett 1978; Raaphorst et al. 1979), and these variations are more marked than normally observed in response to radiation. In a study of seven cell lines, Raaphorst and associates (1979) found that the range of treatment times at 45.5°C, which was required to reduce the surviving fraction to 0.01, varied by a factor of approximately 2.7 for six of the lines. The seventh line required a heat treatment nearly seven times longer than the most sensitive line. In contrast, the maximum variation in radiation dose to reduce survival to 0.01 was a factor of 1.6 for the same seven cell lines. At present, very few of the cell lines exposed to hyperthermia have been of human origin, and insufficient quantitative information is available to characterize the thermal sensitivities of normal and malignant human cells as determined by loss of reproductive capacity. It should be pointed out, however, that there is no established correlation between thermal and radiation sensitivity. This lack of correlation is apparent in figure 20.2, where the responses of two human glioblastoma cell lines (A2 and A7) and CHO cells are compared (Gerweck and Burlett 1978). Lines

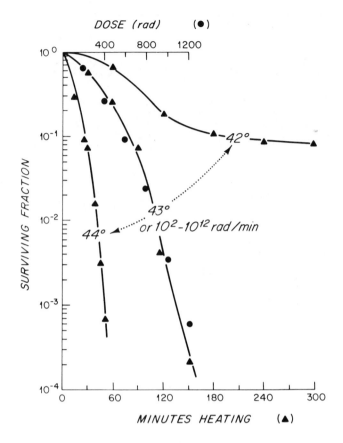

Figure 20.1. The lethal response of exponential-phase CHO cells heated at various temperatures is shown by *triangles*. Response of the same cells to radiation is indicated by *circles*. The data plotted show that the response of cells to radiation is dependent only on total dose over the dose rate examined, whereas the response of cells to hyperthermia is governed by both the time treated and precise temperature. Radiation data are from Gerweck and co-workers (1979) after adjusting for RBE differences between 50 kVp x-rays (2000 rad/min), 280 kVp x-rays (200 rad/min), and 600 kV electrons (10^{12} rad/min).

A2 and A7 are equally resistant to heat but are most sensitive (A2) or resistant (A7) to radiation; heat-sensitive CHO cells exhibit intermediate radiosensitivity. Similar results have been obtained by Raaphorst and co-workers (1979). These data suggest that cells resistant to one agent may or may not be resistant to the direct effects of the other agent.

Cell Cycle Effects of Heat

Cellular sensitivity to radiation or hyperthermia (Bhuyan et al. 1977; Palzer and Heidelberger 1973; Westra and Dewey 1971) varies throughout the cell cycle. Of most interest is the relative sensitivity of S-phase cells to heat compared to their relative resistance to radiation. This differential cell cycle effect is more prominent under treatment conditions that result in sub-

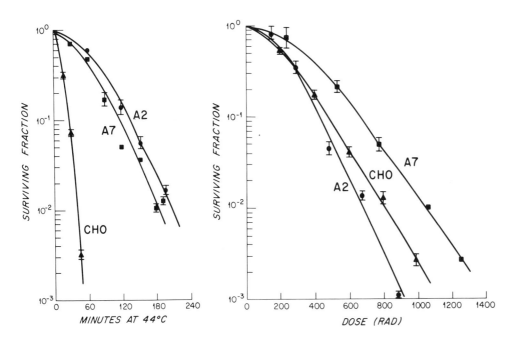

Figure 20.2. Two human glioblastoma cell lines, A2 and A7, and a CHO cell line were exposed to 44°C hyperthermia *(left panel)* or x-rays. The heat resistant line A2 is radiosensitive, and the heat sensitive CHO cells exhibit intermediate radiation resistance (Gerweck and Burlett 1978. Reprinted by permission.)

stantial cell killing, but may decrease at temperatures or treatment times that result in less cell killing (Bhuyan et al. 1977; Westra and Dewey 1971), as shown in figure 20.3. The relative thermal resistance of density-inhibited plateau-phase cells compared to exponential phase cells (Kase and Hahn 1975; Schlag and Lücke-Huhle 1976) (under equivalent nutrient conditions) probably is due in part to the large G_1 fraction of plateau-phase cells. Clinically, the greater thermal sensitivity of cycling S-phase cells may yield efficacious results if the fraction of cells cycling in tumors is substantially larger than in dose-limiting associated normal tissues. Moderate heat treatment would be selectively lethal to radiation-resistant S-phase cells.

Influence of the Microenvironment on the Lethal Effect of Heat or Radiation

Studies in rodent and human tissue using a variety of direct and indirect methods indicate that the concentrations of certain metabolites differ in normal and tumor interstitial fluid. These include a reduced concentration of oxygen and glucose and elevated levels of lactic acid, and possibly carbon dioxide, in tumor tissue. While hypoxic regions in tumors may increase the rate of glucose metabolism and lactic acid production, a high rate of lactate production also may occur in well-oxygenated tumors or regions of tumors. These findings and their impact on hyperthermic and radiation therapy have been reviewed by Suit and Gerweck (1979).

Under hypoxic conditions the dose of radiation required to yield a specific surviving fraction is increased by a factor of approximately 2.5 or greater, compared to cells heated under oxygenated conditions, as seen in figure 20.4. The presence of hypoxic foci in tumor tissue presumably inhibits the efficacy of radiotherapy. In contrast, under similar conditions, acutely hypoxic cells are at least as sensitive to hyperthermia as oxygenated cells (figure 20.4).

Similar results have been observed (Bass, Moore, and Coakley 1978; Gerweck, Nygaard, and Burlett 1979; Overgaard and Bichel 1977; Power and Harris 1977) in experiments where other factors, viz., pH and glucose content, were similar under oxygenated and hypoxic conditions. In addition to these observations, studies with CHO cells have shown that under prolonged hypoxic conditions, the sensitivity of cells to hyperthermia gradually increases (Gerweck, Nygaard, and Burlett 1979), as shown in figure 20.5. Cells maintained hypoxic for up to 10 hours and then heated at 42°C are as sensitive as oxygenated or acutely hypoxic cells. Sensitivity to hyperthermia gradually increases, however. After 30 hours of culturing time under hypoxia, survival is decreased by a factor of 5 compared to aerobic cells. The increasing sensitivity to hyperthermia as a function of culturing time under hypoxia contrasts with the unchanging radioresistance exhib-

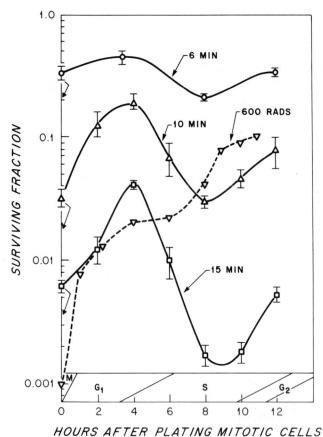

Figure 20.3. Variation in the fraction of CHO cells surviving heat or ionizing radiation delivered during various phases of the cell cycle. Cells were heated at 45.5°C. The position of the cells in the cycle at the time of treatment is indicated on the *abscissa*. Except for cells heated in mitosis, all survival points have been corrected for cellular multiplicity; a multiplicity correction for mitotic cells would lower the survival levels, as indicated by the *arrows*. (Westra and Dewey 1971. Reprinted by permission.)

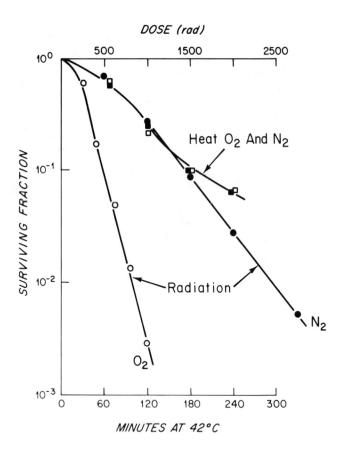

Figure 20.4. The absence of oxygen during treatment substantially reduces the sensitivity of CHO cells to ionizing radiation. In contrast, thermal sensitivity is not influenced by oxygen deprivation immediately before and during treatment. Radiated cells in oxygen (○) versus nitrogen (●); heated cells in oxygen (□) versus nitrogen (■).

ited by hypoxic cells. The similar thermal sensitivity of oxygenated and acutely hypoxic cells and the greater sensitivity of chronically hypoxic cells provides a strong rationale for the use of hyperthermia to decrease the probability of local tumor control failure caused by the presence of radioresistant hypoxic foci.

In addition to the reduced oxygen concentration, a number of in vivo pH measurements suggest that the interstitial fluid pH in tumors is low compared to that in normal tissue. These studies indicate the tumor fluid pH is 0.3 and 0.5 units below the normal tissue value of approximately 7.4 (owing to increased lactate and carbon dioxide); these findings have been previously summarized by Gerweck, Nygaard, and Burlett (1979). In none of the published studies has it been possible to ascertain the pH in various regions of the tumor tissue. It is likely, however, that the hydrogen ion concentration approaches normal values in the proximity of functioning blood vessels and decreases well below the average value in poorly vascularized areas.

The lethal response of cells at various hydrogen ion concentrations to elevated temperature has been examined in only a few cell lines (Freeman, Dewey, and Hopwood 1977; Meyer, Hopwood, and Gillette, in press; Overgaard and Bichel 1977), and only qualified statements regarding the quantitative relationship between these factors can be made. Studies with CHO cells at 42°C (Gerweck 1977) demonstrate that a significant increase in sensitivity to hyperthermia occurs when extracellular pH drops to 7.0 and below, as indicated in figure 20.6. It should be noted also that reduction in pH is without effect on radiation sensitivity (upper curve, fig 20.6).

Variation in the pH-sensitizing effect as a function of temperature has been examined in CHO (Gerweck 1977) and in cultured human glioblastoma cells. The ratio of the survival curve slopes of pH 7.4 com-

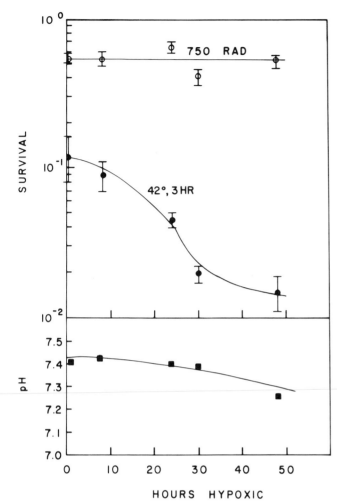

Figure 20.5. Culturing under hypoxic conditions for prolonged periods of time increases the sensitivity of cells to hyperthermia, but not radiation. For the data shown, CHO cells were cultured under low density conditions to minimize changes in sensitivity caused by medium depletion. The change in pH owing to cell metabolism (indicated in the *lower panel*) was insufficient to alter cell sensitivity to heat. (Gerweck, Nygaard, and Burlett 1979. Reprinted by permission.)

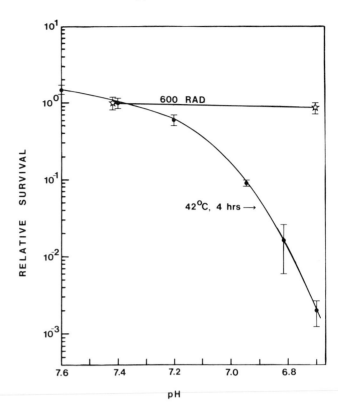

Figure 20.6. Asynchronous CHO cells were heated or irradiated under various extracellular pH conditions as indicated. All data is normalized to a surviving fraction of 1 at pH 7.4. The cells were maintained at the indicated pH for a total of five hours.

pared to pH 6.7 is plotted as a function of temperature in figure 20.7. For both cell lines, the enhancement ratio at 44°C and above was approximately 1.25 to 1.4. For the glial cells, the pH effect increases substantially at 43°C, whereas for the CHO cells the enhancement effect is more pronounced at temperatures below 43°C. These data indicate that temperatures that are only moderately lethal at normal pH for each cell line elicit the strongest pH sensitizing effect. Dewey and colleagues (1980) reported that reduction of pH after heat treatment enhanced hyperthermic lethality at temperatures of 43°C and above. This posttreatment enhancement was not substantial at survival levels above 0.01. Overgaard and Bichel (1977) reported that reduction of pH from 7.2 to 6.4 substantially increased the sensitivity of a murine plasmacytoma cell line to 42.5°C hyperthermia.

In addition to oxygen and pH, the interstitial fluid glucose content in tumors is reduced at sites distant from capillaries because of the high rate of glucose consumption and limited blood oxygen supply. Hahn (1974) investigated the importance of glucose availability on the heat sensitivity of oxygenated cells with an inhibitor of glucose transport (5-thio-D-glucose). The results indicated that glucose did not play a detectable role in cellular response to 43°C hyperthermia. Similar results were obtained by Kim, Kim, and Hahn (1978); however, when 5-thio-D-glucose was added to cells deprived of oxygen, sensitivity to 42°C hyperthermia was markedly increased (fig. 20.8). Presumably, the removal of both oxygen and glucose results in a pronounced drop in the ability of cells to support normal cellular metabolic processes and results in cell death. This would be expected to occur more rapidly at elevated temperatures. Detailed studies designed to determine which of the three nutritional effects (low O_2, low pH, and low glucose) would

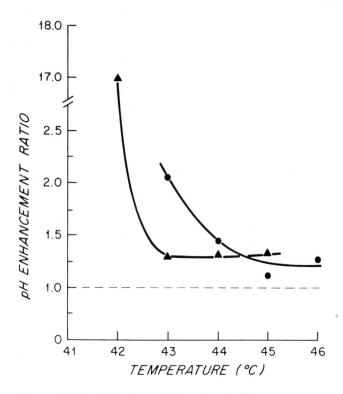

Figure 20.7. The slopes of heat inactivation curves determined at pH 7.4 were divided by heat inactivation curve slopes obtained at pH 6.7. This ratio is indicated as the *pH-enhancement ratio* on the *ordinate* and was obtained at the temperatures indicated on the *abscissa*. (Gerweck and Richards 1981. Reprinted by permission.)

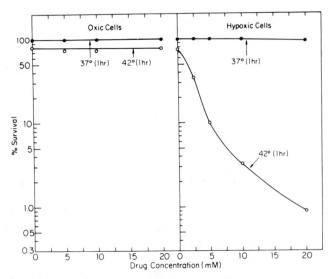

Figure 20.8. Survival of oxygenated and hypoxic HeLa cells exposed to 42°C for one hour with varying concentrations of 5-thio-D-glucose. The plating efficiency of both oxygenated and hypoxic cells was 60%. (Kim, Kim, and Hahn 1978. Reprinted by permission.)

have the greatest influence on tumor thermal sensitivity have not been performed. Under static in vitro conditions and beginning with physiologic concentrations of glucose, metabolic reduction of pH appears to play the dominant role in increased sensitivity to hyperthermia under oxygenated or hypoxic conditions (Gerweck 1977; Gerweck, Nygaard, and Burlett 1979).

In summary, oxygen, hydrogen ions, and glucose have been shown to modify the sensitivity of cells to hyperthermia or radiation. Low pH, low oxygen, and glucose concentrations are features of tumor interstitial fluid that are either without affect or decrease the effectiveness of ionizing radiation, but increase cellular sensitivity to hyperthermia.

Influence of Heat on Cell Cycle Progression

The kinetics of cell progression and division after heat treatment differ substantially from those following radiation treatment (1971). Six minutes of heat treatment of CHO cells to 45.5°C, which reduces survival to approximately 30%, delays the entry of cells into mitosis for 11 hours. The 11-hour delay was observed in cells treated at various stages of G_1, S, and S/G_2. Treatment during mitosis induced a substantially greater mitotic delay. Heated mitotic cells generally failed to complete cytokinesis, and 90% were tetraploid at the next division. In contrast to hyperthermia-induced mitotic delay, a dose of radiation that reduces cell survival to 30% induces division delays of only one to four hours and does not block cytokinesis of cells heated in mitosis.

Miyamoto, Rasmussen, and Zeuthen (1973) studied the effect of a 41.3°C, 41.6°C, or 41.9°C temperature shock of one hour on mouse fibroblast cells. Throughout the temperature range examined, division delay time increased substantially with increasing temperature. Furthermore, delay time increased with cell age and was maximal in late G_2. The effect of treatment on mitotic cells was dependent on the treatment temperature and apparently on the stage of mitosis. This study suggests that following moderate heat treatment, an asynchronous population of cells should become enriched in the G_2 or G_2 + M phase and depleted in G_1 in a transitory manner. This has been observed by Kal and Hahn (1976), Kase and Hahn (1975), and by Schlag and Lücke-Huhle (1976), and may influence the response of cells to fractionated heat treatment. Unfortunately, the magnitude of redistribution following single heat treatment in vitro, and especially in vivo (Kal and Hahn 1976), does not appear adequate to substantially affect cellular or tissue response to hyperthermia. Furthermore, attempts to enhance synchrony in mammalian cells with multiple heat shocks have not been successful (Miyamoto, Rasmussen, and Zeuthen 1973).

For studies of the potential use of hyperthermia in cancer therapy, the most acceptable index for evaluating the magnitude of a hyperthermic insult at the cellular level is loss of reproductive capacity. Following heating or irradiation, reproductively dead cells may exclude vital stains for hours or days. Leeper[1] studied posttreatment proliferation of CHO cells following heat or radiation treatments that reduced survival to approximately 2% (division delay times were approximately 90 and 13 hours, respectively). Although the rate of cell detachment and lysis varied with the treatments, it was observed that the majority of reproductively dead cells were capable of one or more cell divisions.

In summary, following single heat or radiation treatments, partial and transitory cell-cycle redistribution occurs. The importance of redistribution on thermal response either in vitro or in vivo has not been demonstrated. Most prominent is the relatively marked division delay observed following heat treatment compared to that following radiation treatment.

Molecular Effects of Hyperthermia and the Target of Cell Lethality

The lethal effects of ionizing radiation are thought to be related to DNA damage—principally strand breaks that are not completely or correctly rejoined. In comparison, current knowledge of the target and mechanism of heat killing is limited. A number of heat-induced alterations in cellular structure and metabolism have been described; however, none of them have been unequivocally linked with cell death. Morphologic changes visualized by light and electron microscopy following hyperthermia are numerous. Immediately following heat treatment in the 43°C to 45°C range, the surface morphology of cells is substantially altered. The surface membrane becomes undulated with blebs and ruffles and shows a substantial decrease in microvilli (Bass, Moore, and Coakley 1978; Lin, Wallach, and Tsai 1973). Intracellularly, changes in the granular component of the nucleolus have been noted (Simard and Bernhard 1967) and associated with inhibition of synthesis and processing of ribosomal RNA (Warocquier and Scherrer 1969). Other changes that have been described include shrinkage of the nucleus and condensation of the heterochromatin. In the cytoplasm, the Golgi complex hypertrophies, and the number of lysosomal vacuoles increases. Polyribosomes dissociate to monoribosomes, and mitochondrial membranes become disrupted, with a condensation or emptying of the mitochondrial matrix (Kwock et al. 1978; Overgaard 1976, 1979) (see chapter 7).

The plethora of heat-induced visible lesions at the subcellular level are associated with marked alteration in cellular metabolism. Virtually every aspect of cellular metabolism that has been examined is altered at temperatures above 42°C, including respiration and DNA, RNA, and protein synthesis. Studies of this nature are more quantitative than the lesions described by light or electron microscopy; however, hyperthermia-induced perturbations of biosynthetic activity do not exhibit a strong dependence on hyperthermic exposure. For example, Reeves (1971) found that the incorporation of acid-insoluble precursors into DNA, RNA, and protein was inhibited in a heat-sensitive pig kidney cell line and in a heat-sensitive clonal subline to a similar extent. In addition, heat treatments that were lethal to only 10% to 20% of the cells resulted in a three- to fivefold, or greater, reduction in the incorporation of these precursors.

The inhibition in DNA synthesis, at the replicon level, appears to be due primarily to a block in DNA synthesis initiation (Dewey et al. 1980). Hyperthermia also delays the ligation of small molecules ($\approx 1.5 \times 10^7$ daltons) into large molecules (4×10^8 daltons), as shown in figure 20.9 (Dewey, et al. 1980). The delay in rejoining may result in an increase in illegitimate exchanges between DNA molecules and possibly gives rise to the chromosome aberrations that are observed following heat treatment of cells during the DNA synthetic stage (but not in other stages of the cell cycle). It should be pointed out, however, that hyperthermia does not appear directly to induce DNA strand breaks (Corry, Robinson, and Getz 1977; Dikomey 1978). The quantitative relationship between cell survival and chromosome aberrations in heated S-phase cells is identical to that observed for ionizing radiation, which probably kills cells by damage to DNA. The chromosome aberration frequency is far lower in G_1- and G_2-phase cells, and aberrations do not play a significant role in cell killing in non-S-phase cells (Westra and Dewey 1971). While substantial direct effects of heat ($\leq 45°C$ to $46°C$) on DNA have not been observed, hyperthermia increases the amount of nonhistone protein isolated with DNA (Clark and Lett 1976; Lett and Clark 1978; Tomasovic, Turner, and Dewey 1978). Roti-Roti, Henle, and Winward (1979) found that the activation enthalpy for the initial rapid increase in chromatin protein content in HeLa cells was

---
[1] Dennis Leeper, 1979, personal communication.

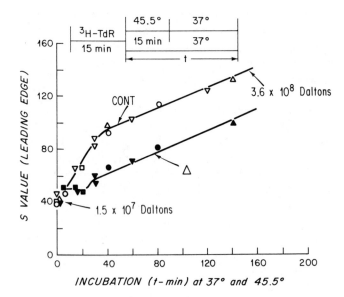

Figure 20.9. Effect of heat treatment (▲) on processing of pulse-labeled DNA into high molecular weight molecules. Monolayer cultures of CHO cells were pulse labeled with 2μCi/ml ^{3}H-thymidine (20 Ci/mM) for 15 minutes, washed twice with medium containing 10μg/ml stable thymidine, and then either incubated at 37°C (*Cont.*, ○, △, □, ▽) or heated (▲) in medium at 45.5°C for 15 minutes prior to incubation at 37°C for various periods (●, ▲, ■, ▼). Cells were trypsinized and prepared for alkaline sucrose gradient sedimentation analyses. The different symbols represent four different experiments. To obtain a curve that depicts the rate of processing of small DNA molecules synthesized during the ^{3}H pulse into large molecules, the unnormalized alkaline sucrose gradient sedimentation profiles were used to obtain S values of the leading edges of the main peaks (*ordinate*) after various chase times (*t-min, abscissa*). (Dewey et al. 1980. Reprinted by permission.)

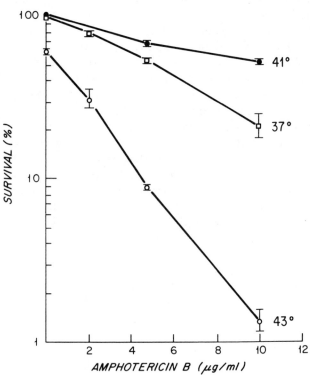

Figure 20.10. Dose response of HA1 cells to amphotericin B at different temperatures. The zero dose point represents the effect of heat alone during the one-hour exposure; 43°C alone causes a measurable reduction in plating efficiency. Estimates of survival were obtained under the assumption that the reduction in cell number in the treatment vessel is the result of disappearance of only those cells that have lost their reproductive capacity. The differences between the 37°C and 41°C data are the extremes seen over several experiments and were not always observed. (Hahn, Li, and Shiu 1977. Reprinted by permission.)

86.8 kcal/mole, whereas for cell lethality, an enthalpy of 153.8 kcal/mole was observed. The heat-induced increase in chromatin protein content likely represents a lesion to the normal chromatin replicative and repair processes; however, these relationships have not been established.

Several studies suggest that damage to the plasma membrane may be critically involved in cell killing. Hahn, Li, and Shiu (1977) (fig. 20.10), observed that the sterol-binding drug amphotericin B strongly sensitized Chinese hamster cells (Hal) to 43°C hyperthermia. The greatly enhanced killing would appear to reflect, directly or indirectly, interactions between the drug and hyperthermia at the level of the cell membrane. Yatvin (1977) was able to increase the thermal sensitivity of *Escherichia coli* K-1060 by increasing the ratio of saturated to unsaturated fatty acids in the cell membrane. Thermal sensitivity was increased also by heating *E. coli* in 10 mM procaine. In addition to these studies, the thermal sensitivity of several cell lines appears to correlate with membrane cholesterol content, according to Cress, Herman, and Gerner (1978). This suggests that plasma membrane may be involved in cell killing because membrane fluidity decreases with cholesterol content. Studies by Verma and Wallach (1976) have shown that pH-sensitive membrane transitions involving lipoprotein complex occur in the physiologic temperature range.

These studies indicate that factors, or drugs, that tend to destabilize membranes also increase sensitivity to hyperthermia. Several of the drugs used in these studies may also affect a wide variety of biomolecules in the cell. In addition, drug-induced changes in the cell membrane may cause changes in the intracellular concentration of simple ions and other metabolites, which could influence the intracellular target of hyperthermic lethality. For example, changes in culture medium tonicity have been shown to markedly influence thermal sensitivity (Raaphorst and Dewey 1978). In addition, immediate changes in transport and membrane permeability evaluated with fluorescein diacetate and eosin after heat treatment do not quantitatively correlate with loss of reproductive capacity. Reeves (1971) observed an enhanced "leakiness" of uridine but not leucine loaded heat sensitive cells compared to a heat-resistant line.

In summary, temperatures of 41°C and above are lethal to mammalian cells. For the wide variety of lesions induced, no single molecular lesion has been identified with cell lethality. Chromosome aberrations may represent critical lesions for cells heated during the DNA synthetic phase. Damage to the plasma membrane also may be directly or indirectly involved in the killing process. The mechanisms of heat and radiation lethality differ, their effects also differ on the cellular level in several complementary respects. Probably most important is the increased sensitivity of cells to hyperthermia under low, or chronic (but not acute), hypoxic conditions. Hyperthermia should be useful for eliminating hypoxic populations and also should be selectively damaging to tumors because of the low interstitial fluid pH of this tissue. In addition, radioresistant S-phase cells are relatively heat sensitive. These complementary effects suggest that hyperthermia could be effectively used with radiation for the treatment of patients with cancer. In addition to these independent effects, heat and radiation interact synergistically if time between treatment is on the order of a few hours or less, as described in the following section.

SYNERGISTIC INTERACTION OF HEAT COMBINED WITH IONIZING RADIATION

Several experiments combining heat and ionizing radiation in vitro have demonstrated that the interaction of these modalities depends on a large number of variables. The magnitude of interaction between heat and radiation depends on temperature, time at elevated temperature, sequence of heat and radiation treatment, and time interval between these treatments. Furthermore, response varies with the dose rate and quality of radiation. Depending on these variable treatment conditions, the interaction may be independent, additive, or synergistic. Several of these relationships have been established only over very narrow treatment conditions, and additional studies will be required before generalizations can be made. Nevertheless, a number of trends are discernible among the published results.

Influence of Sequence, Time Interval, and Temperatures on Heat and Radiation Response

One of the initial studies of the effects of various elevated temperatures and treatment sequences on radiation sensitivity was performed by Robinson, Wizenberg, and McCready (1974). Chinese hamster lung cells were exposed to x-rays (100 rad per minute) during the midportion of a two-hour exposure to 37.5°C, 40°C, 41°C, and 42°C. The results, which are shown in figure 20.11, demonstrate that sensitization was significant at 40°C. At 42°C, the cells were twice as sensitive as those irradiated at 37.5°C, as determined by a comparison of the slopes of the inactivation curves. These investigators also found that at 42°C, heating times as short as 15 minutes, either immediately prior to or following irradiation, produced significant and equivalent sensitization. Increasing the heating time to two hours resulted in only a moderate increase in sensitivity (fig. 20.12), indicating that a maximum radiation sensitizing effect may occur with increasing heating time. The development of a maximum radiosensitizing effect with increasing heating time at 42°C was reported also by Dewey and coworkers (1979), and at 45.5°C in synchronous CHO cells by Gerweck, Gillette, and Dewey (1975).

Although exceptions have been reported (Ross-Riveros and Leith 1979), it appears in general that the sensitizing efficacy of hyperthermia increases with temperature if the time of treatment at elevated temperature is constant (Loshek, Orr, and Solomonidis 1977; Robinson and Wizenberg 1974; Sapareto, Hopwood, and Dewey 1978). This is illustrated in figure 20.13 (Loshek, Orr, and Solomonidis 1977), in which the radiation sensitivity, $K'r$, is seen to increase with temperature for a constant heat-exposure time. Figure 20.14 (Loshek, Orr, and Solomonidis 1977) shows that at least for the particular protocol examined (radiation followed by heat), direct cell killing by heat (K_H) increases more rapidly with temperature than does radiosensitization by hyperthermia (K_I). Although sensitization tends to increase with temperature for a constant time of exposure to heat, if the time of exposure to various temperatures is adjusted so that the lethal effects of heat alone are equivalent, the sensitizing effect of hyperthermia appears reasonably temperature-independent (Sapareto, Hopwood, and Dewey 1978). Two recent reports indicate, however,

Leo E. Gerweck

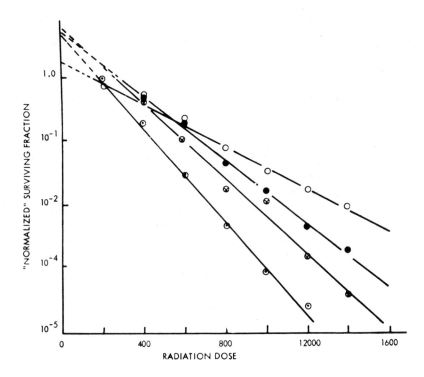

Figure 20.11. Radiation survival curves for Chinese hamster cells subjected to thermal treatment (two hours before, during, and after irradiation). The curves are normalized to a surviving fraction of 1.0 at zero radiation dose (rad); ○, 37.5; ●, 40.0; ⊗, 41.0; ⊙, 42.0°C. (Robinson and Wizenberg 1974. Reprinted by permission.)

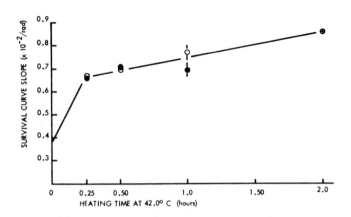

Figure 20.12. The effect of heating time and the heating-irradiation sequence on radiation sensitivity; treatment temperature of 42°C. ○, After irradiation; ●, before irradiation; ⊗, before, after, and during irradiation. (Robinson and Wizenberg 1974. Reprinted by permission.)

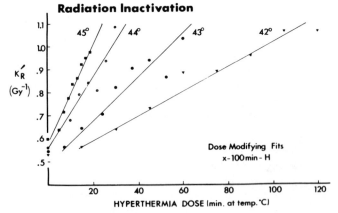

Figure 20.13. Radiosensitivity, K'_R, to 250 kV x-rays at room temperature, followed 100 minutes later by a hyperthermia exposure at the indicated temperature and duration. K'_R was calculated assuming a common radiation extrapolation number for each experiment. *Data points* represent the mean of two to five experiments. (Loshek, Orr, and Solomonidis 1977. Reprinted by permission.)

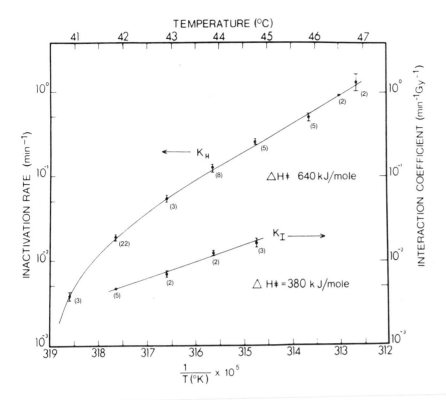

Figure 20.14. Arrhenius plot of the thermal sensitivity, K_H, and the interaction coefficient K_I. The values for K_H include additional data derived from experiments within which only the thermal sensitivity was assayed. *Error bars* represent the standard error of the mean for the number of observations indicated in *parentheses*. (Loshek, Orr, and Solomonidis 1977. Reprinted by permission.)

that sensitization is somewhat greater at 42.5°C than above or below this temperature (Ross-Riveros and Leith 1979; Sapareto, Raaphorst, and Dewey 1979). Additional studies with several cell lines and various levels of thermal damage will be required to determine the generality of these observations. Reported maximum enhancement ratios of approximately 2, determined by comparing survival curve slopes, have been reported at various temperatures (42°C to 45°C); however in most cases, experimental enhancement ratios of 1 to 1.6 have been reported (Dewey et al. 1977, 1980; Joshi, Barendsen, and van der Schueren 1978; Loshek, Orr, and Solomonidis 1977; Ross-Riveros and Leith 1979; Sapareto, Hopwood, and Dewey 1978; Robinson and Wizenberg 1974).

Robinson and Wizenberg (1974) observed equivalent thermal sensitization to radiation cell killing if heat was administered before or after irradiation. In contrast, Gerweck, Gillette, and Dewey (1975) observed in both G_1- and S-phase CHO cells that 45.5°C hyperthermia was more effective when delivered before irradiation than after irradiation. Studies by Joshi, Barendsen, and van der Schueren (1978) have shown that the relative efficacy of heat before or after

irradiation varies with the treatment temperature and/or magnitude of the heat treatment. Survival following heat treatment of RUC-2 cells to 41°C for one hour was less when treatment was given immediately after irradiation, compared to heat treatment immediately prior to irradiation, as shown in figure 20.15. At 43°C for 10 minutes, the treatment-surviving fractions were essentially equal and independent of treatment sequence, whereas at 45°C, heat treatment for 10 minutes prior to irradiation reduced survival more than heating after irradiation. At all temperatures, simultaneous heating and irradiation was at least as effective, or more effective in terms of cell killing, than heating after irradiation. The greater killing efficacy observed when RUC-2 cells were simultaneously heated and irradiated has also been observed in CHO cells (Sapareto, Hopwood, and Dewey 1978) and in a mammary carcinoma cell line, as seen in figure 20.16 (Meyer, Hopwood, and Gillette 1979). For the temperature (42.5°C) and duration of hyperthermia (20 to 120 minutes) used in the experiment shown in figure 20.16, Meyer, Hopwood, and Gillette (1979) observed very little sensitization or interaction between heat and radiation when three to four hours separated the

Leo E. Gerweck

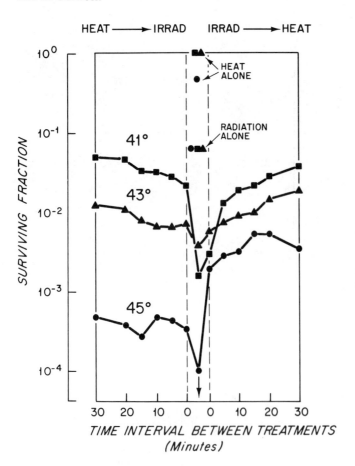

Figure 20.15. Influence of time interval between heat treatment and irradiation on rat ureter carcinoma cells (RUC-2) in culture. The response of cells to heat treatment alone at 41°C for one hour, 43°C for 10 minutes and 45°C for 10 minutes is indicated. The dose of radiation was 1000 rad in all cases. *Curves* connect data points resulting from the combined treatments. Curves have been redrawn and error bars omitted. (Joshi, Barendsen, and van der Schueren 1978. Reprinted by permission.)

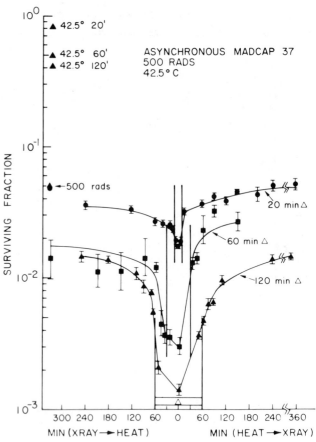

Figure 20.16. Effect of 500 rad before, during, or after 20, 60, or 120 minutes of heating at 42.5°C. The duration of the heat treatment is delineated by the *heavy lines* where the zero point represents the midpoint of heating. The *time scale* represents minutes before or minutes after the midpoint of heating. The surviving fraction after the heat treatment alone was 0.82, 0.50, and 0.41 for 20, 60, and 120 minutes of heating, respectively. The surviving fraction with 500 rad alone was 0.05. (Meyer, Hopwood, and Gillette 1979. Reprinted by permission.)

treatments. In studies performed at higher temperatures, where the lethal effects of heat alone were more pronounced (Gerweck, Gillette, and Dewey 1975), heat treatment sensitized cells to radiation for up to 15 hours after treatment; however, when cells were irradiated and then heated, survival increased relatively quickly with kinetics similar to those observed in figure 20.16. In a study of two cell lines, Li and Kal (1977) observed that the degree of sensitization varied between the lines when heat followed radiation by less than one hour.

In summary, as the time interval between heat and radiation decreases, the killing efficiency of the combined modalities increases. The lethal interaction of heat and radiation is most pronounced when cells are treated with both agents simultaneously. Sensitization may or may not increase to a maximum with increasing time of exposure to hyperthermia. The thermal sensitizing efficacy increases with increasing temperature if the heating time is constant. The synergistic interaction between heat and radiation appears to decay when more than a few hours separate the treatments. When the effects of heat alone are substantial, however, heated cells may remain sensitive to subsequent radiation for several hours. Studies have not been performed that show that the kinetics and magnitude of sensitization vary between normal and neoplastic cells.

Influence of the Cell Cycle on the Response of Cells to Heat and Radiation

The lethal response of cells to radiation varies throughout the cell cycle (fig. 20.3). The complementary effect of heat and radiation would flatten the curve of the age response of cells to combined treatment if their interaction were equivalent throughout the cell cycle; however, equivalent interaction does not occur throughout the cell cycle, as seen in figure 20.17 (Gerweck, Gillette, and Dewey 1975). Equivalent interaction would result in the response indicated in figure 20.17 by the dashed curve, which was obtained by multiplying the surviving fraction from heat alone by the surviving fraction from radiation alone. Cells entering S (indicated in lower right curve) are relatively sensitized compared to cells in G_1. Similar results have been reported by Kim, Kim, and Hahn (1976) following heat treatment of HeLa cells at 43°C for 0.5 hours. These observations suggest that tumors, which are comprised of a significantly larger fraction of cycling cells than are in the surrounding normal tissue, may be selectively damaged by the combination of heat and radiation.

Influence of Dose Quality and Dose Rate on the Interaction of Hyperthermia and Radiation

The lethal effect of a single specified dose of radiation decreases when delivered in two fractions separated by several minutes to a few hours. Studies by Ben-Hur, Elkind, and Bronk (1974) have shown that the recovery or repair process is abolished when cells are maintained at elevated temperatures between acute radiation exposures. Repair of sublethal radiation damage also can be blocked by sufficient hyperthermic treatment prior to fractionated radiation treatment (Gerweck, Gillette, and Dewey 1975). These studies suggest that the radiation enhancement effects of hyperthermia would be most pronounced under radiation exposure conditions where repair of sublethal radiation damage is prominent, as occurs with low-dose-rate irradiation. Similarly, the accumulation and repair of sublethal damage is less prominent in cells exposed to high linear energy transfer (LET) radiation and, therefore, the radiosensitizing effects of heat would be expected to be less prominent with radiation of this quality. Ben-Hur, Elkind, and Bronk (1974) irradiated cells at 3.3, 12, and 360 rad per minute while they were maintained at various temperatures. The results of studies at 3.3 rad per minute are shown in figure 20.18. As temperature increased from 37°C to 42°C, the survival curve slopes decreased by a factor

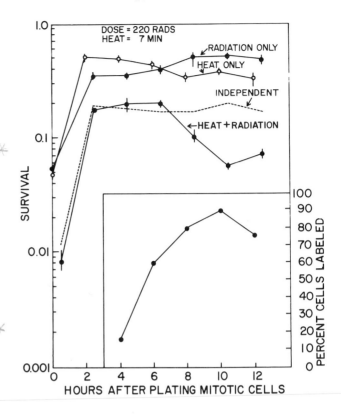

Figure 20.17. Synchronous mitotic cells were plated, and as they progressed through the cell cycle, they were irradiated or heated, or both. When the treatments were combined, the cells were irradiated 30 minutes after heat treatment. The curve labeled *independent* was constructed by multiplying the surviving fraction for radiation only by the surviving fraction for heat only. The labeling index of untreated cells is indicated in the *lower right* portion of the figure. Except for cells heated in mitosis, all points were corrected for multiplicity. (Gerweck, Gillette, and Dewey 1975. Reprinted by permission.)

of approximately 6. In contrast, at 360 rad per minute and 42°C, hyperthermia increased the rate of inactivation by a factor of 1.5.

Gerner and Leith (1977) compared the response of cells to x-rays and ^{12}C ions at 37°C and immediately following 43°C treatment for 60 minutes. The results (fig. 20.19) clearly indicate that heat most efficiently sensitized cells exposed to low LET radiation. Damage induced by the high LET events produced by ^{12}C was not noticeably modified by hyperthermic treatment.

Influence of pH and Oxygen Concentration on the Interaction Between Hyperthermia and Radiation

Prolonged oxygen deprivation and reduced pH sensitize cells to hyperthermia (Gerweck, Nygaard, and

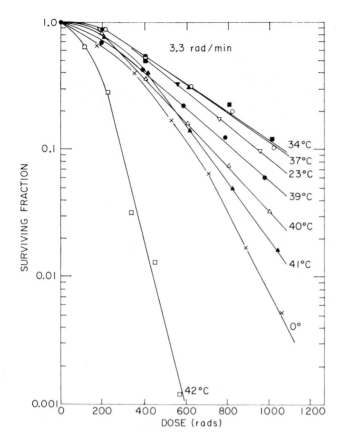

Figure 20.18. Survival of Chinese hamster cells irradiated (250 kVp 3.3 rad/min) at the indicated temperatures. All cells were suspended in the medium DuM-20 for five hours, which blocked cell division during treatment and reduced survival to 50% to 70%. The survival of irradiated cells was normalized to that of unirradiated cells that were given the same treatment. (Ben-Hur, Elkind, and Bronk 1974. Reprinted by permission.)

Burlett 1979). The available data do consistently demonstrate that these tumorlike microenvironmental conditions enhance the lethal interaction between heat and radiation. The influence of simultaneous heat and radiation on the radiation oxygen enhancement ratio (OER) has been examined by Robinson, Wizenberg, and McCready (1974). Mouse bone marrow cells were irradiated during a one-hour heat treatment at 37.5°C, 42.5°C, or 43°C. The resulting OERs were 2.47, 1.69, and 1.38, respectively. Similar results were observed by Kim, Kim, and Hahn (1975), who heated HeLa cells to 42°C for two hours following radiation treatment. In contrast, Power and Harris (1977) found no reduction in OER under controlled pH conditions. For these studies, V-79 and EMT-6 cells were heated to 43°C for 45 minutes immediately before or immediately after irradiation. No reduction in OER was observed in yeast cells irradiated at various above normal tempera-

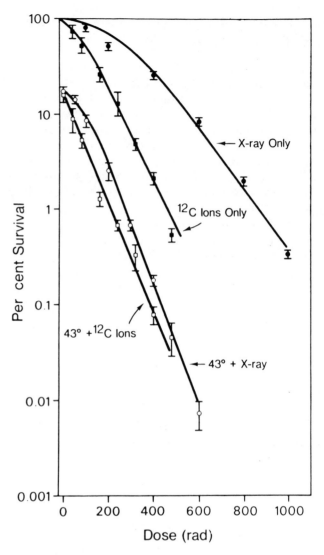

Figure 20.19. Survival response of asynchronous CHO cells. Cultures were treated at either 37°C (*closed circles*) or 43°C (*open circles*) for 60 minutes before irradiation with 4 MeV x-rays, or at 37°C (*closed squares*) or 43°C (*open squares*) for 60 minutes before irradiation with accelerated carbon ions (LET = 34.1 kev/μ). Standard errors of the mean are shown by the *bars* on the data points. (Gerner and Leith 1977. Reprinted by permission.)

tures (Kiefer, Kraft, and Hiawica 1976). Additional studies undertaken under controlled nutrient and pH conditions will be required to resolve these differences.

The response of cells to hyperthermia and radiation under low medium pH conditions has received scant attention. Dewey and colleagues (1980) reported that heat treatment for one hour at 42.5°C followed by radiation was more lethal at pH 6.6 than at pH 7.4. Most of the additional cell killing observed at low pH was due to the effect of heat alone. It is not clear whether cells heated to equivalent survival levels under dissimilar pH conditions would be more or less sensitive to the combined treatments of heat and radiation. Unless it is conclusively demonstrated that the radiation OER is reduced at hyperthermia, and that synergism is more pronounced under acidic conditions, the rationale for combining these two agents in a manner to elicit synergism is weakened.

The Mechanism of Thermal Radiosensitization

Although the inactivation mechanisms and lethal lesions of hyperthermia and ionizing radiation apparently differ, the synergistic response of cells to these agents suggests that their effects interact at a molecular level. Most studies have examined the influence of heat on the molecular processes thought to be responsible for radiation killing. Considerable evidence suggests a nuclear target location for radiation killing, and more specifically, damage to DNA. Radiation-produced DNA strand breaks, if not properly repaired, may lead eventually to chromosomal aberrations. It has been suggested that this chromosomal damage is the cause of radiation cell death (Dewey et al. 1971). Dewey, Sapareto, and Betten (1978) have shown that the variation in radiation sensitivity of mitotic, G_1-, and S-phase Chinese hamster cells closely correlates with the radiation-induced chromosomal aberration frequency. Heat treatment prior to irradiation increased both the radiation sensitivity and frequency of chromosomal aberrations of mitotic, G_1-, and S-phase cells by factors of 1.0, 1.2, and 1.5, respectively. A similar relationship was obtained, however, between aberration frequency versus log cell survival, regardless of the phase of the cell cycle treated and the treatments (radiation or heat plus radiation) evaluated (see fig. 20.20). This relationship, and the ability of approximately one aberration per cell to reduce survival to 0.37, suggests that heat enhances radiation cell killing by increasing the aberration frequency per unit dose.

The hyperthermia-induced increase in chromosome aberration frequency may be due to an increase in the amount of DNA damage produced by radiation or repair of this damage. No significant decrease in the molecular weight of DNA was observed by Lett and Clark (1978) when CHO cells receiving 1000 rad were heated to 45.5°C for 7 to 17 minutes prior to irradiation. Lett and Clark (1978) and Corry, Robinson, and Getz (1977), however, found heat treatment delayed the rejoining of radiation-induced single and double strand DNA breaks in a heat-dose-dependent manner.

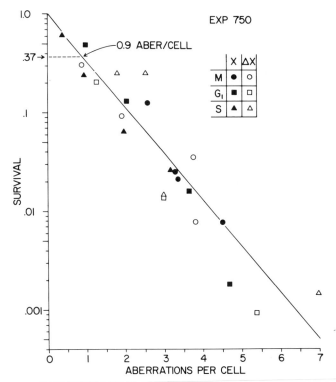

Figure 20.20. The data plotted shows survival of CHO cells as a function of aberrations per cell. The values for survival and aberration frequency were normalized to 1.0 and 0.0, respectively, for zero radiation dose. Note that for cells irradiated (X) or heated and then immediately irradiated ($\triangle X$) in either M-, G_1-, or S-phase, log survival is linearly related to aberration frequency, with about one aberration per cell corresponding to 37% survival. (Dewey, Sapareto, and Betten 1978. Reprinted by permission.)

Clark and Lett (1976) also reported that the eventual return of the rejoining kinetics to the normal rate correlated with the eventual loss of thermal sensitization. Heat-induced delay in rejoining of DNA strand breaks may result in an increase in improper rejoining between DNA molecules, which would result in chromosomal aberrations. A similar hypothesis has been advanced to account for heat-induced chromosomal aberrations during DNA synthesis.

In addition to the hyperthermic-induced delay in strand break rejoining, Warters and Roti-Roti (1979) have shown that hyperthermia delays excision of radiation-induced base damage. These investigators have determined that the excision delay was not due to thermal inactivation of excision enzymes. The proteins that become bound to DNA as a result of heat treatment may limit access of the enzyme to the damaged base. The lesions causing the delay in ligation of strand breaks have not been identified; however, studies by Dube, Seal, and Loeb (1977) and by Dewey and colleagues (1980) have shown that thermal inactivation of the repair enzyme β polymerase could play a role in delayed repair.

In summary, hyperthermia appears to potentiate the efficacy of radiation by inhibiting the repair of radiation-induced lesions in DNA. This delay may result in the observed increased frequency of chromosomal aberrations when heat treatments are combined with radiation. The lesions responsible for heat killing have not been identified, and an interaction between these lesions and radiation has not been demonstrated. The lethal interaction between heat and radiation becomes more pronounced as the time interval between the treatments decreases. Sensitization increases with the magnitude of the heat treatment, which also influences the relative efficacy of the treatment sequence. In general, radioresistant S-phase cells are sensitized to a greater degree than G_1-phase cells. The relative efficacy of heat plus radiation increases with decreasing dose rate, but synergistic interaction is not apparent when hyperthermia is combined with high LET radiation. Experimental evidence does not suggest that either the magnitude or kinetics of interaction differ between normal and cancer cells; however, this possibility has not been systematically investigated. The influence of pO_2 and pH on hyperthermic sensitization has not been adequately investigated. Presumably, pH, pO_2, and other metabolite concentrations could change during and following an initial heat and radiation treatment and would influence the response of tissue to subsequent treatments.

RESPONSE OF CELLS TO FRACTIONATED HEATING AND RADIATION

Complete local control of most tumors with single doses of heat, alone or in combination with other agents, does not appear likely without unacceptable normal tissue damage. A substantial decrease in the lethal response of cells to hyperthermia occurs when a specified heat dose is delivered in two or more fractions. This sparing effect of fractionation has been demonstrated at cell and tissue levels, and is discussed in chapter 6. The development of thermal resistance during fractionated heat treatments could eliminate or enhance the differential sensitivity to hyperthermia that may occur between normal and tumor tissue. This differential is expected, in part, because of the different microenvironmental conditions, such as pH, occurring in tumor and normal tissue. Of importance

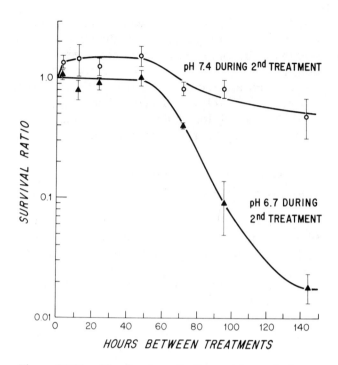

Figure 20.21. The fraction of cells surviving a single continuous heat treatment of six hours at pH 7.4 was divided into the surviving fraction of cells exposed to three hours of treatment at pH 7.4 followed by three hours of additional treatment at pH 7.4 or 6.7. The time interval between the split fractions is indicated on the *abscissa*. The mean and standard error of two to five experiments for each fractionation interval is indicated. (Gerweck, Jennings, and Richards 1980. Reprinted by permission.)

for therapeutic reasons is the possibility that the development of thermal tolerance may reduce or eliminate the pH-sensitizing effect or alter factors that influence the tumor and normal tissue responses. Thermal sensitivity at the time of the second and subsequent treatments would be governed primarily by the rate of decay of thermal tolerance. The response of cells to fractionated heat or heat plus radiation under normal and tumorlike nutritional conditions (such as low pH and pO_2) warrants investigation. Nielsen and Overgaard (1979) reported that reduction of pH from 7.2 to 6.5 reduced, but did not eliminate, the development of resistance in murine tumor L1A2 cells exposed to fractionated heat treatment at 42°C. Recent studies performed in this laboratory show that the viable progeny of CHO cells heated for three hours at 42°C (survival 10%) remain resistant to second heat treatments for two or more generations. Furthermore, as indicated in figure 20.21, previously heated cells are only slightly more sensitive to subsequent treatment at pH 6.7 compared to pH 7.4. Beginning 72 hours after the initial treatment, thermotolerance declines and is lost by 144 hours, regardless of pH level. These data imply that hyperthermia should be delivered over sufficiently long interfraction intervals so that thermotolerance has declined, and microenvironmental effects, such as low pH, may play a role in producing a differential effect between tumor and normal tissue. The rate of decay of thermotolerance may or may not be dependent on the proliferation kinetics of the tissue or on the temperature and duration of the heat treatment. Basic studies of the response of cells to multiple heat treatments are needed.

SUMMARY

Quantitative in vitro studies indicate that heat plus radiation is complementary and independent, or complementary and additive, or synergistic. Synergistic interaction is most readily demonstrated when the two treatments are applied simultaneously or within a few hours of each other. The magnitude and time interval over which synergism predominates is strongly dependent on the magnitude (temperature and time of heat treatment) of the thermal treatment. In vitro studies have not demonstrated that the interaction is more pronounced in tumor cell lines as compared to normal tissue cells. Nevertheless, the relatively greater effects of heat on S-phase cells, hypoxic cells, glucose-depleted cells, or cells existing in an acidic environment suggest that heat should be useful for supplementing the therapeutic effects of radiation.

It is important to note that heat may result in physiologic changes in locally treated tissue (as discussed by Song in chapter 8) that could influence response to subsequent radiation treatment. Studies in animal tumors have shown that heat damages tumor vascular tissue. The resultant tumor hypoxia would render these cells resistant to subsequent radiation for an unknown period of time. If heat treatment followed radiation by several hours, however, the selective effects of heat on the nutritionally deprived tissue would not be compromised.

The development and decay of thermal tolerance will have a substantial impact on the responses of both tumor and normal tissues to heat treatment. It is not unlikely that thermotolerance decreases, or possibly eliminates, the differential response of these tissues to single heat treatments. Additional studies should indicate more clearly whether heat fractionation schedules that allow thermal tolerance to decay between heat treatments should be employed. Nevertheless, the studies reviewed in this chapter indicate that the lethal effects of heat as influenced by the tumor microenvironment make it an attractive adjunctive therapy in combination with radiation for the treatment of patients with cancer.

References

Bass, H.; Moore, J. L.; and Coakley, W. T. Lethality in mammalian cells due to hyperthermia under oxic and hypoxic conditions. *Int. J. Radiat. Biol.* 33:57–67, 1978.

Ben-Hur, E.; Elkind, M. M.; and Bronk, B. V. Thermally enhanced radioresponse of cultured Chinese hamster cells: inhibition of repair of sublethal damage and enhancement of lethal damage. *Radiat. Res.* 58:38–51, 1974.

Bhuyan, B. K. et al. Sensitivity of different cell lines and of different phases in the cell cycle to hyperthermia. *Cancer Res.* 37:3780–3784, 1977.

Clark, E. P., and Lett, J. T. The effect of hyperthermia on the rejoining of DNA strand breaks in x-irradiated CHO cells (abstr.). *Radiat. Res.* 67:519, 1976.

Corry, P. M.; Robinson, S.; and Getz, S. Hyperthermic effects on DNA repair mechanisms. *Radiology* 123:475–482, 1977.

Cress, A. E.; Herman, T. S.; and Gerner, E. W. Cholesterol content and thermal sensitivity of mammalian cells *in vitro* (abstr.). *Radiat. Res.* 74:477, 1978.

Dewey, W. C. et al. Radiosensitization with 5-bromodeoxyuridine of Chinese hamster cells x-irradiated during different phases of the cell cycle. *Radiat. Res.* 47:672–688, 1971.

Dewey, W. C. et al. Cellular responses to combinations of hyperthermia and radiation. *Radiology* 123:463–474, 1977.

Dewey, W. C. et al. Cell biology of hyperthermia and radiation. In *Radiation biology in cancer research*, eds. R. E. Meyn and H. R. Withers. New York: Raven Press, 1980, pp. 589–621.

Dewey, W. C.; Sapareto, S. A.; and Betten, D. A. Hyperthermic radiosensitization of synchronous Chinese hamster cells: relationship between lethality and chromosomal aberrations. *Radiat. Res.* 76:48–59, 1978.

Dikomey, E. Repair of DNA strand breaks in Chinese hamster ovary cells at 37 or 42°C. *Proceedings of the Second International Symposium on Cancer Therapy by Hyperthermia and Radiation*, eds. C. Streffer et al. Baltimore and Munich: Urban & Schwarzenberg, 1978, pp. 146–148.

Dube, D. K.; Seal, G.; and Loeb, L. A. Differential heat sensitivity of mammalian DNA. *Biochem. Biophys. Res. Commun.* 76:483–487, 1977.

Freeman, M. L.; Dewey, W. C.; and Hopwood, L. E. Effect of pH on hyperthermic cell survival: brief communication. *J. Natl. Cancer Inst.* 58:1837–1839, 1977.

Gerner, E. W., and Leith, J. T. Interaction of hyperthermia with radiations of different linear energy transfer. *Int. J. Radiat. Biol.* 31:283–288, 1977.

Gerweck, L. E. Modification of cell lethality at elevated temperatures. The pH effect. *Radiat. Res.* 70:224–235, 1977.

Gerweck, L. E. et al. Repair of sublethal damage in mammalian cells irradiated at ultrahigh dose rates. *Radiat. Res.* 77:156–169, 1979.

Gerweck, L. E., and Burlett, P. The lack of correlation between heat and radiation sensitivity in mammalian cells. *Int. J. Radiat. Oncol. Biol. Phys.* 4:283–285, 1978.

Gerweck, L. E.; Gillette, E. L.; and Dewey, W. C. Effect of heat and radiation on synchronous Chinese hamster cells: killing and repair. *Radiat. Res.* 64:611–623, 1975.

Gerweck, L. E.; Jennings, M.; and Richards, B. Influence of pH on the response of cells to single and split doses of hyperthermia. *Cancer Res.* 40:4019–4024, 1980.

Gerweck, L. E.; Nygaard, T. G.; and Burlett, M. Response of cells to hyperthermia under acute and chronic hypoxic conditions. *Cancer Res.* 39:966–972, 1979.

Gerweck, L. E., and Richards, B. Influence of pH on the thermal sensitivity of cultured human glioblastoma cells. *Cancer Res.* 41:845–849, 1981.

Hahn, G. M. Metabolic aspects of the role of hyperthermia in mammalian cell inactivation and their possible relevance to cancer treatment. *Cancer Res.* 34:3117–3123, 1974.

Hahn, G. M.; Li, G. C.; and Shiu, E. Interaction of amphotericin B and 43° hyperthermia. *Cancer Res.* 37:761–764, 1977.

Joshi, D. S.; Barendsen, G. W.; and van der Schueren, E. Thermal enhancement of the effectiveness of gamma radiation for induction of reproductive death in cultured mammalian cells. *Int. J. Radiat. Biol.* 34:233–243, 1978.

Kal, H. B., and Hahn, G. M. Kinetic responses of murine sarcoma cells to radiation and hyperthermia *in vivo* and *in vitro*. *Cancer Res.* 36:1923–1929, 1976.

Kano, E.; Miyakoshi, J.; and Sugahara, T. Differences in sensitivities to hyperthermia and ionizing radiation of various mammalian cell strains *in vitro*. In *Proceedings of the Second International Symposium on Cancer Therapy by Hyperthermia and Radiation*, eds. C. Streffer et al. Baltimore and Munich: Urban & Schwarzenberg, 1978, pp. 188–190.

Kase, K., and Hahn, G. M. Differential heat response of normal and transformed human cells in tissue culture. *Nature* 255:228–230, 1975.

Kiefer, J.; Kraft, W. W.; and Hiawica, M. M. Cellular radiation effects and hyperthermia. Influence of exposure temperature on survival of diploid yeast irradiated under oxygenated and hypoxic conditions. *Int. J. Radiat. Biol.* 30:293–300, 1976.

Kim, S. H.; Kim, J. H.; and Hahn, E. W. The radiosensitization of hypoxic tumor cells by hyperthermia. *Radiology* 114:727–728, 1975.

Kim, S. H.; Kim, J. H.; and Hahn, E. W. The enhanced killing of irradiated HeLa cells in synchronous culture by hyperthermia. *Radiat. Res.* 66:337–345, 1976.

Kim, S. H.; Kim, J. H.; and Hahn, E. W. Selective potentiation of hyperthermic killing of hypoxic cells by 5-thio-D-glucose. *Cancer Res.* 38:2935–2938, 1978.

Kwock, L. et al. Impairment of Na^+-dependent amino acid transport in a cultured human T-cell line by hyperthermia and irradiation. *Cancer Res.* 38:83–87, 1978.

Lett, J. T., and Clark, E. P. Effects of hyperthermia on the rejoining of radiation-induced DNA strand breaks. In *Proceedings of the Second International Symposium on Cancer Therapy by Hyperthermia and Radiation*, eds. C. Streffer et al. Baltimore and Munich: Urban & Schwarzenberg, 1978, pp. 13–18.

Li, G. C., and Kal, H. B. Effect of hyperthermia on the radiation response of two mammalian cell lines. *Eur. J. Cancer* 13:65–69, 1977.

Lin, P. S.; Wallach, D. F. H.; and Tsai, S. Temperature-induced variations in the surface topology of cultured lymphocytes are revealed by scanning electron microscopy. *Proc. Natl. Acad. Sci. USA* 70:2492–2496, 1973.

Loshek, D. D.; Orr, J. S.; and Solomonidis, E. Interaction of hyperthermia and radiation: temperature coefficient of interaction. *Br. J. Radiol.* 50:902–907, 1977.

Meyer, K. R.; Hopwood, L. E.; and Gillette, E. L. The response of mouse adenocarcinoma cells to hyperthermia and irradiation. *Radiat. Res.* 78:98–107, 1979.

Meyer, K. R.; Hopwood, L. E.; and Gillette, E. L. The thermal response of mouse adenocarcinoma cells at low pH. *Eur. J. Cancer*, in press.

Miyamoto, H.; Rasmussen, L.; and Zeuthen, E. Studies of the effect of temperature shocks on preparation for cell division in mouse fibroblast cells (L cells). *J. Cell Sci.* 13:889–900, 1973.

Nielsen, O. S., and Overgaard, J. Effect of extracellular pH on thermotolerance and recovery of hyperthermic damage *in vitro*. *Cancer Res.* 39:2772–2778, 1979.

Overgaard, J. Influence of extracellular pH on the viability and morphology of tumor cells exposed to hyperthermia. *J. Natl. Cancer Inst.* 56:1243–1250, 1976.

Overgaard, J. Ultrastructure of a murine mammary carcinoma exposed to hyperthermia in vivo. *Cancer Res.* 36: 983–995, 1979.

Overgaard, J., and Bichel, P. The influence of hypoxia and acidity on the hyperthermic response of malignant cells in vitro. *Radiology* 123:511–514, 1977.

Palzer, R. J., and Heidelberger, C. Influence of drugs and synchrony on the hyperthermic killing of HeLa cells. *Cancer Res.* 33:422–427, 1973.

Power, J. A., and Harris, J. W. Response of extremely hypoxic cells to hyperthermia: survival and oxygen enhancement ratios. *Radiology* 123:767–770, 1977.

Raaphorst, G. P. et al. Intrinsic differences in heat and/or x-ray sensitivity of seven mammalian cell lines cultured and treated under identical conditions. *Cancer Res.* 39: 396–401, 1979.

Raaphorst, G. P., and Dewey, W. C. Enhancement of hyperthermic killing of cultured mammalian cells by treatment with anisotonic NaCl or medium solutions. *J. Thermal Biol.* 3:177–182, 1978.

Reeves, O. R. Mechanisms of acquired resistance to acute heat shock in cultured mammalian cells. *J. Cell. Physiol.* 70:157–170, 1971.

Robinson, J. E., and Wizenberg, M. J. Thermal sensitivity and the effect of elevated temperatures on the radiation sensitivity of Chinese hamster cells. *Acta Radiologica* 3: 241–248, 1974.

Robinson, J. E.; Wizenberg, M. J.; and McCready, W. A. Combined hyperthermia and radiation suggest an alternative to heavy particle therapy for reduced oxygen enhancement ratios. *Nature* 251:521–522, 1974.

Ross-Riveros, P., and Leith, J. T. Response of 9L tumor cells to hyperthermia and X-irradiation. *Radiat. Res.* 78: 296–311, 1979.

Roti-Roti, J. L.; Henle, K. J.; and Winward, R. T. The kinetics of increase in chromatin protein content in heated cells: a possible role in cell killing. *Radiat. Res.* 78:522–531, 1979.

Sapareto, S. A.; Hopwood, L. E.; and Dewey, W. C. Combined effects of X-irradiation and hyperthermia on CHO cells for various temperatures and orders of application. *Radiat. Res.* 73:221–233, 1978.

Sapareto, S. A.; Raaphorst, G. P.; and Dewey, W. C. Cell killing and the sequencing of hyperthermia and radiation. *Int. J. Radiat. Oncol. Biol. Phys.* 5:343–347, 1979.

Schlag, H., and Lücke-Huhle, C. Cytokinetic studies on the effect of hyperthermia on Chinese hamster lung cells. *Eur. J. Cancer* 12:827–831, 1976.

Simard, R., and Bernhard, W. A heat-sensitive cellular function located in the nucleolus. *J. Cell Biol.* 34: 61–76, 1967.

Strom, R. et al. The biochemical mechanism of selective heat sensitivity of cancer cells. IV. Inhibition of RNA synthesis. *Eur. J. Cancer* 9:103–112, 1973.

Suit, H. D., and Gerweck, L. E. Potential for hyperthermia and radiation therapy. *Cancer Res.* 39:2290–2298, 1979.

Tomasovic, S. P.; Turner, G. N.; and Dewey, W. C. Effect of hyperthermia on nonhistone proteins isolated with DNA. *Radiat. Res.* 73:535–552, 1978.

Verma, S. P., and Wallach, D. F. H. Erythrocyte membranes undergo cooperative, pH-sensitive state transitions in the physiological temperature range: evidence from Raman spectroscopy. *Proc. Natl. Acad. Sci. USA* 73: 3558–3561, 1976.

Warocquier, R., and Scherrer, K. RNA metabolism in mammalian cells at elevated temperatures. *Eur. J. Biochem.* 10:362–370, 1969.

Warters, R. L., and Roti-Roti, J. L. Excision of X-ray-induced thymine damage in chromatin from heated cells. *Radiat. Res.* 79:113–121, 1979.

Westra, A., and Dewey, W. C. Variation in sensitivity to heat shock during the cell-cycle of Chinese hamster cells in vitro. *Int. J. Radiat. Biol.* 19:467–477, 1971.

Yatvin, M. B. The influence of membrane lipid composition and procaine on hyperthermic death of cells. *Int. J. Radiat. Biol.* 32:513–521, 1977.

CHAPTER 21
Interstitial Thermoradiotherapy

Michael R. Manning
Eugene W. Gerner

Introduction
Rationale
Physics and Engineering Considerations
Thermal Dosimetry
Clinical Results
Summary and Perspectives

INTRODUCTION

Over the past decade, significant advances have been made in the clinical treatment of human cancers (Kaplan 1979). These developments have occurred, in part, because of refinements in surgical and radiation therapy techniques, the discovery and more rational use of effective chemotherapy agents, the use of drugs and radiation in combined modality therapies, and the introduction of cytotoxic agent sensitizers that act to improve the efficacy of either radiation or drugs. For radiation therapy, the search for clinically effective sensitizers has included studies using hyperbaric oxygen chambers or hypoxic cell radiosensitizers to combat the problem of tumor cell hypoxia. The use of high linear energy transfer (LET) radiations also is being studied by several groups.

Still, the failure of local control remains a significant problem for various tumor types and locations. Certain malignancies of the head and neck are examples. Suit (1969) has estimated that several thousand patients will succumb to cancer each year because their primary head and neck tumors have not been controlled locally. Glioblastoma multiforme is a disease that is almost uniformly fatal because current approaches have been unsuccessful in achieving local control. In this disease, Urtasun and colleagues (1976) have shown that the existence of hypoxic tumor cells contribute significantly to the resistance of this tumor to conventional radiotherapy techniques. Certain gynecologic tumors, including late stage ovarian and pelvic malignancies, also have very poor local response rates to current therapeutic techniques, possibly because of tumor cell hypoxia. In specific instances, malignant melanoma can present as a local control problem; it is thought that this may be due to its unique radiation response, as depicted by a significantly large shoulder on its radiation survival response curve (Barranco, Romsdahl, and Humphrey 1971).

These few examples are presented to emphasize that although it is clear human cancer is a systemic disease, and curative therapies must approach it as such, the failure of local control often proves fatal before systemic disease becomes life threatening. It is from this point of view that we, and others, initiated research programs to investigate the use of hyperthermia combined with radiation with the goal of achieving local tumor control in those malignancies where failure to do so currently results in high mortality rates.

This work was supported by U. S. Public Health Service grant CA-17343. The authors gratefully acknowledge the significant contributions of T. C. Cetas, Ph.D.; W. G. Connor, Ph.D.; R. C. Miller, M.D.; J. R. Oleson, M.D., Ph.D.; S. Aristizabal, M.D., and E. Surwitt, M.D.

Michael R. Manning and Eugene W. Gerner

RATIONALE

The biological rationale for the use of hyperthermia combined with radiation is discussed in chapter 20. It is worthwhile to note again the work of Ben-Hur, Bronk, and Elkind (1972), which demonstrated that hyperthermia was most effective at potentiating the cytotoxic effects of low-dose-rate radiation (Ben-Hur, Elkind, and Bronk 1974). This is due presumably to the ability of heat to inhibit sublethal radiation damage repair and to enhance lethal radiation damage. Subsequent to these cell culture studies, we pursued this approach in transplantable tumors in mice (Miller et al. 1978) and in spontaneously occurring tumors in dogs and cats (Miller et al. 1977). Consistent with studies by Ben-Hur, Elkind, and Bronk, we found that hyperthermia combined with low-dose-rate radiation was quite effective in terms of controlling tumor growth. Surprisingly, however, the results of these studies suggested that normal tissues surrounding the treated tumors tolerated the combined therapy remarkably well. While enhanced normal tissue reactions often were seen from the combined heat and radiation treatments (Miller et al. 1976; Robinson, Wizenberg, and McCready, 1974), the effects on tumors were more remarkable (Thrall, Gillette, and Dewey 1975). While these results are not clearly understood, it is our impression that factors such as hypoxia, low pH, and nutrient deprivation in tumors may be significant contributors to the differential response to this therapy observed in malignant tissues compared to normal tissues. The importance of these and other factors has been reviewed in this text by Gerweck and elsewhere by Dewey and co-workers (1980).

The biological rationale for hyperthermia combined with low-dose-rate radiation is supported firmly by these experimental studies; however, other considerations were equally, if not more, important in the clinical development of this combined approach. First, interstitial and intracavitary radiotherapy has long been used in specific situations in order to deliver superlethal doses to tumors while sparing normal tissues. Because the use of interstitial radiotherapy requires manual placement of the radiation sources, this technique generally has been confined to accessible tumors, such as gynecologic tumors of the pelvis, although Feder, Syed, and Neblett (1978) and Tak (1978) have pioneered a resurgence in the use of interstitial therapy to treat other sites. Thus, the goal of obtaining specific tumor localization of radiation dose weighed strongly in our interest in developing interstitial thermoradiotherapy, since many studies had demonstrated enhancement of normal tissue radioresponses by heat (see chapter 22 for review). The physics and engineering considerations of producing uniform heat fields in human tumors was the second important consideration in our decision to develop interstitial thermoradiotherapy. Prior to 1974, the clinical application of hyperthermia had involved the

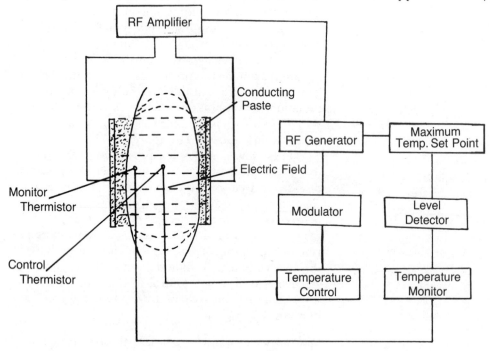

Figure 21.1. Low frequency RF generator and set-up. Block diagram of system using 500-kHz RF waves to produce and control local hyperthermia. (Gerner et al. 1975. Reprinted by permission.)

use of water baths (Hartman and Crile 1968) and conventional diathermy (Warren 1935) techniques to deliver heat doses to tumors. These approaches were fraught with numerous technical problems, of which nonuniform heating in tumors was significant. Also, while technical developments were being made in the use of externally applied radiofrequency (RF) and microwave sources, significant problems were apparent. These problems were highlighted by the work of Guy, alone (1971) and with Lehmann (1966), which demonstrated that depth of penetration was limited using single microwave applicators and that temperature uniformity was complicated by tissue interfaces and inhomogeneities. Thermal dosimetry also was a problem since conventional metallic probes were known to perturb temperature fields produced by RF and microwave applicators (Cetas 1978). A final concern was the cost of developing devices to induce heat fields for treating deep seated human tumors, using either RF, microwave, or ultrasound technology.

In 1974, J. D. Doss of the Los Alamos Scientific Laboratory in Los Alamos, New Mexico, constructed a 500-KHz RF generator that averted many of the technical problems discussed above. A schematic of this device used in our studies is shown in figure 21.1. First, the temperature fields are produced by passing the RF currents directly through the volume of interest. While this approach shared with interstitial radiotherapy the disadvantage of being an invasive technique, it also shared the advantage of being able to localize heat fields. Second, conventional thermometers, such as thermistors and thermocouples, could be used within the treatment fields because perturbation at these low frequencies was minimal (Cetas and Connor 1978). As shown in figure 21.1, the system could be controlled by using appropriate feed-back systems. This technique shared with interstitial radiotherapy the advantage of being able to construct desired dose distributions simply by changing the number of sources (electrodes) or configuration of sources, such that producing either regular or irregular heating patterns was conceptually and technically feasible. The cost of the unit was rather modest. Finally, since this interstitial thermotherapy approach required implanted conducting electrodes, it was obvious on technical grounds alone that it could be combined with interstitial radiotherapy by using the metallic covers and/or guides of the interstitial radiation sources. Thus, the rationale for interstitial thermoradiotherapy is derived from biological, clinical, and physical considerations relating to the application of heat and radiation in human therapy.

PHYSICS AND ENGINEERING CONSIDERATIONS

Details of the technology involved in producing hyperthermic fields in phantoms, animals, and human sub-

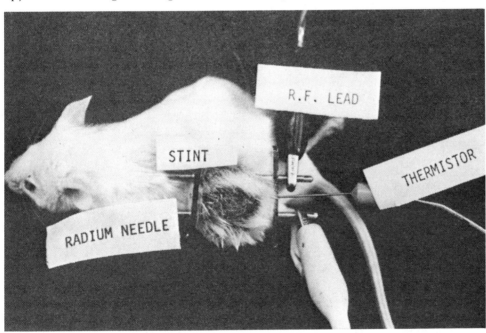

Figure 21.2. This transplanted mouse mammary tumor is being irradiated with *radium needles*. The needles are connected to the *RF generator* and are acting as heating electrodes. A *thermistor* is placed in the tumor to monitor temperature and to control the tumor temperature via a feedback mechanism. (Miller, Connor, and Boone 1977. Reprinted by permission.)

jects using interstitial thermotherapy techniques have been previously published (Cetas and Connor 1978; Cetas, Connor, and Manning 1980; Gerner 1975; Miller et al. 1977). Figure 21.2 shows a simple treatment set-up delivering a combined interstitial thermoradiotherapy treatment to a transplantable mouse mammary tumor (Miller, Connor, and Boone 1977). The radiation dose is delivered using radium needles, while the heat dose is delivered using the metal casings of the radium needles as electrodes to produce the localized RF current fields that generate the heating patterns. Figure 21.3 shows a phantom study of this treatment example. From this figure, it is apparent that heating is rapid and, for this small phantom, heating is uniform. As mentioned previously, larger volumes and a variety of fields can be produced with this technique by modifying the number and arrangement of needle electrodes. Figure 21.4 displays an example of a phantom study using two parallel rows of needles to heat a large volume. Again, heating is rapid. The volumes directly around the needles heat most, but with time, the treatment volume reaches a rather uniform temperature owing to thermal conduction. This figure also demonstrates that the excessive heating at the tips of the needles can be reduced by coating the sharp point of the needle with a dielectric material. Figures 21.5 and 21.6 show two examples of the application of this technique in dogs. Thermographic camera measurements were used to follow surface temperatures of the heated volumes. Figure 21.5 shows an example of rather uniform heating using parallel rows of implanted needles. While the dosimetry technique used here (thermographic camera) does suffer from the disadvantage of measuring only surface temperature, it has the advantage of being noninvasive and is able to provide a complete single plane temperature map. This figure again shows the ability to raise temperatures rapidly using this technique and that temperatures do remain elevated after the RF power is turned off. Figure 21.6 demonstrates that heating is not always uniform using this technique. In this case, a perianal adenoma in a dog was treated, again using parallel rows of needles. While the explicit reasons for the inhomogeneous heating patterns are unknown, presumably they relate to tissue inhomogeneities and physiologic factors, such as cooling by the blood.

In the examples of interstitial thermotherapy described here, heat fields are produced by passing an alternating electrical current through the treatment volume between the electrodes implanted in the tissue. The current has two separate components, which can be referred to as resistive and capacitive. The capacitive component causes no energy dissipation in the tissue and, therefore, does not contribute to heat pro-

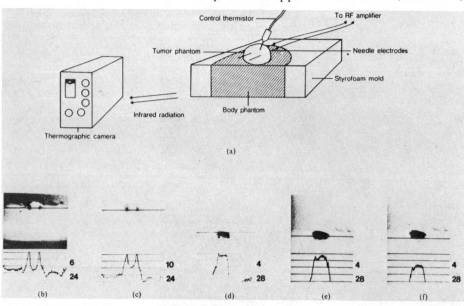

Figure 21.3. Phantom studies of solid, protruding tumor (2-cm diameter) heated by two interstitial electrodes. *a* Diagram of phantom configuration. *b–f* Thermograms taken during heat treatment. In the upper portion, warm areas are black. The graph at the bottom gives the temperature profile along the line indicated by the fiducial marker in the upper portion. The upper number on the right corresponds to the full-scale sensitivity in °C and the lower gives the temperature represented by the lower line on the graticule; *b* thermogram taken 10 seconds after heating at 5 W was begun; *c* thermogram taken a few seconds before the set temperature (5°C above initial temperature) was reached (at 65 sec); *d–f* thermograms taken several minutes after reaching the control temperature and after steady-state conditions were established. The thermograms differ in the position where the line temperature profile was taken. (Cetas and Connor 1978. Reprinted by permission.)

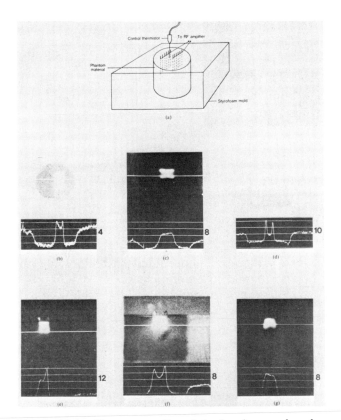

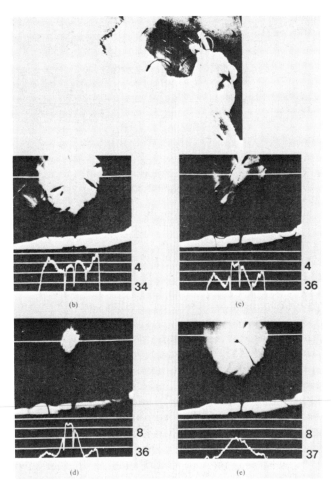

Figure 21.4. Phantom studies of the heating produced by two parallel rows of needle electrodes. The heated field is 3 cm long, 2 cm wide, and 2 cm deep. On the thermograms, white is hot. The number to the right indicates the full scale sensitivity in °C. a, Diagram of experimental configuration; b, thermogram taken through a covering of household plastic wrap during initial warm-up; c, thermogram taken immediately after warmup, with needles and plastic cover removed. In d–g the front section of the phantom was removed so that a vertical surface 1 mm in front of the rows of needles could be monitored thermographically: d, thermogram showing excessive heating owing to intense fields around needle points; e, thermogram showing that excessive heating on left was reduced by a dielectric coating on the sharp point of the needle; f, thermogram taken with all needles having dielectric-coated points. Heating along needle is uniform although still greater than in the region between the needles; g, thermogram of heating pattern after the set temperature is reached. (Cetas and Connor 1978. Reprinted by permission.)

Figure 21.5. Thermographic monitoring of RF current heating in a dog's shoulder. Heating method similar to that described in phantom in figure 21.2. a, Photograph of the experimental configuration; b, thermogram taken before heating; c, thermogram taken as heating begins; d, thermogram taken of steady-state conditions approximately midway through treatment; e, thermogram taken 10 seconds after treatment showing residual heat in treatment area. (Cetas and Connor 1978. Reprinted by permission.)

duction in the treatment volume. Thus, it is advantageous to minimize the capacitive component of the current. This can be accomplished by reducing the frequency, since impedance to the capacitive current is inversely related to frequency. There is a practical lower limit to the frequency used in this application, as nerve and muscle fibers may be depolarized at frequencies less than 100 kHz. We have employed interstitial thermotherapy primarily in the range of 0.1 to 3 MHz. The rate of energy dissipation per cubic centimeter in tissue, using these localized RF current fields, can be calculated using the relationship

$$P \text{ (watts/cm}^3\text{)} = J \text{ (amperes/cm}^2\text{)}^2 \cdot \sigma \text{ (ohm} \cdot \text{cm)} , \quad (1)$$

where P is the volume rate of energy dissipation, J is the resistive root-mean-square current density, and σ is the resistivity of the tissue in the treatment volume. The heating patterns can then be constructed by appropriately choosing electrode size and shape, numbers of electrodes and electrode configuration, as shown in several previous examples (figs. 21.3–21.6).

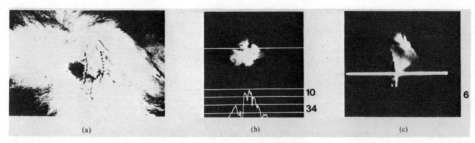

Figure 21.6. Treatment of a canine perianal adenoma using same heating configuration as in figures 21.2 and 21.3. *a*, Photograph of treatment configuration with control thermistor shown at the bottom, inserted into the tumor; *b*, thermogram taken during treatment. Note that the tumor is cooler than the region in upper part of field. An extra thermistor probe is inserted at the top. *c*, Close-up thermogram of the heated field. The region of the tumor is cooler (under fiducial line), and a hot spot exists in region of the extra thermistor probe. (Cetas and Connor 1978. Reprinted by permission.)

THERMAL DOSIMETRY

In our clinical studies of interstitial thermoradiotherapy, thermal dosimetric studies are done using combinations of phantom studies and direct temperature measurements during treatment. The phantom studies have the advantage of providing complete thermal distributions, using thermographic camera monitoring techniques. These studies are then valuable aids to the thermistor and thermocouple measurements that are obtained during the actual clinical treatments. In this section, the applicator, treatment, and dosimetry from one patient is discussed.

A 34-year-old woman presented with a leiomyosarcoma overlying the left vaginal wall, extending from the upper to the lower midline. This lesion was 4 to 5 cm in thickness and 8 cm in depth into the vagina. The patient had metastatic disease in the lungs in addition to this mass partially occluding the rectum. Figure 21.7A shows a schematic of the applicator used. This modified Syed-type applicator included a central copper obturator (hollow) and 19 hollow needle guides for radiation source placement. The obturator was 11 cm long, while the needle guides were 12 cm in length. Figure 21.7B shows another view of this applicator to depict the relationship between the obturator and needle guides and to show the placement of thermometer probes. Phantom studies of this treatment applicator have been previously discussed (Cetas, Connor, and Manning 1980). Briefly, the phantom studies demonstrated that heating between the outer row of needles and the central obturator would not provide a sufficient heat dose to the outer edge of the tumor. Therefore, it was decided to add a second heat dose, now using only the semiconcentric rows of needle guides. This strategy, shown in figure 21.8 along with the initial heating procedure, did result in adequate heating of the outer rim of the malignancy. Subsequent thermistor measurements made during the course of therapy demonstrated that temperatures between 44°C and 46°C were achieved for depths up to 8 cm along the obturator and for distances up to 3 cm from the obturator surface (fig. 21.9).

Thus, it is quite clear that static phantom studies cannot be used to predict exact temperatures in human patients. We have found them useful, however, as in the patient discussed here, to help identify prob-

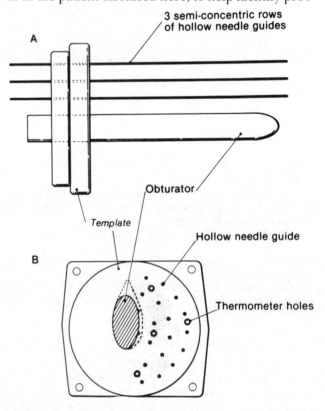

Figure 21.7. *A*, Line drawing of the interstitial applicator used to treat the patient with leiomyosarcoma of the vagina. *B*, Line drawing indicating placement and orientation of the applicator relative to the anatomic position and the tumor. (Cetas, Connor, and Manning 1980. Reprinted by permission.)

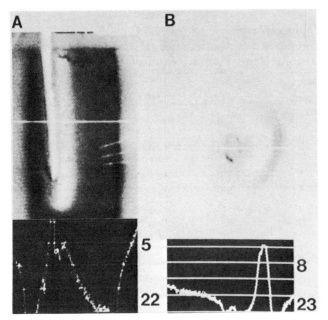

Figure 21.8. Thermograms of heating patterns induced in phantoms simulating treatment of the patient with vaginal leiomyosarcoma. *A*, First treatment with the copper obturator serving as one electrode and the outer row of needles connected together as the other; *B*, second treatment with all the inner needles connected together as one electrode and the outer needles as the other electrode as in *A*. (Cetas, Connor, and Manning 1980. Reprinted by permission.)

lems in applicator designs before patient treatments so that improved therapeutic application can be achieved. In addition, the ability to derive complete two-dimensional thermal distributions with the thermographic camera in phantom and with some clinical studies greatly augments our ability to interpret single- and multiple-point thermistor measurements obtained during the actual treatment.

CLINICAL RESULTS

The data presented here is an update of a larger study (Manning et al. 1982). To date, twenty-five patients have been accessioned into this phase I trial employing interstitial thermoradiotherapy. Twenty-seven lesions have been treated. One patient with multiple melanoma was treated with interstitial thermoradiotherapy for two lesions, while another patient with two epithelioid sarcomas received this treatment for both masses. Patient entry into this trial was based on failure to respond to conventional therapies, a life expectancy of three months, and accessibility of the lesion(s). A variety of tumor types and sites were treated. These are summarized in table 21.1 and included squamous cell carcinoma (6), adenocarcinoma (14), melanoma (3),

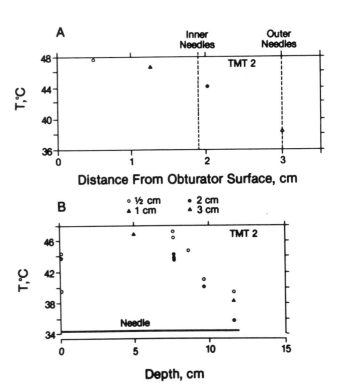

Figure 21.9. Measured temperatures at steady-state conditions during the treatment of the patient with vaginal leiomyosarcoma. *A*, Temperatures at the hottest point in each catheter as a function of radial distance from the plastic obturator; *B*, temperatures as a function of depth for each of the catheter positions in *A*. Penetration depth of the needles is indicated also. (Cetas, Connor, and Manning 1980. Reprinted by permission.)

sarcoma (3), and carcinosarcoma (1). The patients ranged in age from 24 to 81 years and included 6 males and 19 females.

Because this was a phase I trial, a range of radiation doses was administered. Many of the patients had received prior radiotherapy. In all cases, a dose thought to be maximally tolerable of interstitial radiation was prescribed. This range of doses is shown in table 21.2, along with histologies and tumor response for comparative purposes. Low-dose radiation ranged from 40 rad/hour to 80 rad/hour. In all these patients, only one heat dose was administered and always at the time of insertion of the interstitial radiation device, done under general anesthesia. Radiation doses were administered using either radium-226 or afterloaded iridium-192. The prescribed heat dose was always 43°C to 44°C (minimum tumor temperature) for 30 minutes. Patient 11 was the only exception to this protocol. She received two heat doses of 43°C for 30 minutes, one immediately preceding the radiation dose and one immediately following the end of the radiation dose.

Table 21.1.
Number of Patients Treated by Site and Histologic Type of Cancer

Histology	Site	Number of Patients
Squamous cell	Head and neck	3
	Vaginal cuff	1
	Cervix	2
Adenocarcinoma	Breast, skin recurrence	5
	Colon/rectum recurrence, pelvic floor, perineum, rectum	6
	Gall bladder umbilical node	1
	Endometrium	1
	Ovarian, pelvis recurrence, vagina	1
Melanoma	Skin recurrence	3
Sarcoma	Leiomyosarcoma, vaginal metastases, epithelioid	1 / 2
Carcinosarcoma	Ovary, pelvis recurrence, vagina	1

Table 21.2.
Comparison of Radiation Doses for Each Patient by Histology and Tumor Reponse

Patient Number	Total Rad Dose	Histology	Response
1	4500	Squamous cell carcinoma	CR
2	3990	Squamous cell carcinoma	CR
3	4000	Adenocarcinoma	PR
4	4000	Melanoma	CR
5	3960	Adenocarcinoma	CR
6	2000	Squamous cell carcinoma	CR
7 a	4500	Melanoma	CR
7 b	4000	Melanoma	CR
8	4000	Adenocarcinoma	CR
9	2160	Adenocarcinoma	PR
10	3000	Adenocarcinoma	CR
11	4560	Sarcoma	CR
12	1800	Adenocarcinoma	PR
13	3000	Adenocarcinoma	PR
14	2100	Adenocarcinoma	CR
15	2000	Squamous cell carcinoma	CR
16	3000	Adenocarcinoma	CR
17	3000	Adenocarcinoma	CR
18 a	4300	Sarcoma	CR
18 b	4500	Sarcoma	CR
19	2000	Squamous cell carcinoma	PR
20	2000	Carcinosarcoma	PR
21	2000	Adenocarcinoma	PR
22	2500	Squamous cell carcinoma	PR
23	2000	Adenocarcinoma	PR
24	1500	Adenocarcinoma	PR
25	4000	Adenocarcinoma	CR

NOTE: Tumor responses were evaluated over the course of several months. Responses were evaluated based on volume reductions following treatment. Less than a 50% decrease in tumor volume was scored as no response (NR). Greater than 50% but less than 100% reduction in tumor volume was scored as a partial response (PR). Complete tumor disappearance in the treatment volume was scored as a complete response (CR).

Table 21.3 summarizes these responses. Of note is that we have yet to observe less than a partial response in patients treated with interstitial thermoradiotherapy, even though most diseases treated have previously failed conventional radiotherapy procedures, and the doses used in this study are rather small. Currently, 10 of 27 treated lesions have shown at least a partial response. Of this group, five patients are still alive. Previous experience suggests that some of these patients may go on to complete responses. All have experienced subjective relief in addition to their quantitative tumor responses. Nearly two-thirds, or 63%, of the patients treated in this trial have shown complete responses to this therapy. Of special note is that in no case has tumor regrowth been observed in patients showing complete responses to interstitial thermoradiotherapy, suggesting complete tumor eradication. The level of response by tumor histology has been nearly uniform. Complete responses have been observed in four of six squamous cell carcinomas, seven of fourteen adenocarcinomas, three of three melanomas, and three of three sarcomas.

Complications have been minimal and include one vaginal-rectal fistula, surgically repaired, colostomy performed; one superficial focal soft tissue necrosis on vaginal wall; one fatty tissue liquefaction with subsequent healing; and one mild proctitis (with ^{226}Ra needles) with minor burning around entrance and exit sites and complete subsequent healing. Burns have been observed around the entry and exit points of the radiation sources (electrodes); however, this has been only a minor complication. Following therapy, these burns rapidly heal with good normal tissue regrowth.

The following case report is presented as an example of the types of patients and responses observed in this trial.

Case History
A 67-year-old female presented to the clinic with a recurrent squamous cell carcinoma on

Table 21.3
Summary of Responses to Interstitial Thermoradiotherapy in 25 Patients Treated for 27 Lesions

Response	Number (%)	Length of Observation
Complete	17 (63)	4–30 months*
Partial	10 (37)	4–18 months†
None	0	—

*No regrowth observed in treated site in any patient; two patients had recurrences outside treatment field; five patients alive.
†No disease progression observed during follow-up; five patients alive; one patient NED in treatment site after surgical removal of residual tumor.

the left lateral free margin of the tongue. She previously had a local excision and prompt recurrence. Surgery was attempted; however, the disease was not eradicated. She was referred for consideration of radiation therapy. It was elected to treat this recurrence with definitive radiation combining external treatment plus implant. She received an external dose of 5000 rad encompassing the lesion on the tongue through parallel portals, which also encompassed the upper echelon of lymph nodes. She then received an interstitial implant, using radium needles, and a further 3000 rad over 50 hours was delivered. This brought the total dose to the lesion to 8000 rad. Approximately two years after completion of this treatment, the patient was found to have recurrent squamous cell carcinoma at the primary site in the left lateral tongue, and chemotherapy was begun. Only a partial response was obtained in the tumor, which regrew, and the patient ultimately underwent a left hemiglossectomy and mandibulectomy. Less than six months later, she developed an obvious tumor on the remaining right side of the tongue. At this time, she was referred for consideration of experimental interstitial thermoradiotherapy.

She was treated with hyperthermia (43°C for 30 minutes plus interstitial radiation [iridium-192]). Six hollow needle guides were at the midplane of the mouth, and five guides were used in the lateral row. These were connected to the radiofrequency generator and the heat treatment was commenced. Several thermistors were placed in between the parallel needle rows to monitor temperatures. The lesions came to temperature quickly (within five minutes) and were maintained at 43°C for 30 minutes. Temperature was monitored in two different locations in the midline between the parallel rows.

The tumor itself was rather exophytic, somewhat ulcerative, and undoubtedly hypoxic in nature. During the heat treatment, there seemed to be apparent shrinkage of the tumor. It took on a darker, almost purplish color, became less moist, and obviously less vascular. By the end of the 30-minute treatment, the central, more ulcerative necrotic portion of this tumor sloughed without any significant bleeding. At this point, the hollow plastic tubes were then inserted through the needle guides and the needle guides removed. Subsequently, the iridium ribbons, each containing six seeds, were placed in the medial six rows and lateral five rows. Computer dosimetry revealed that the 60 rad per hour line totally encompassed the lesion. The iridium was left in place for 75 hours, thus delivering 4500 rad to the implant. After the iridium and hollow tubes were removed, there was no evidence of infection. At this time, the tumor appeared very necrotic, and a large portion of it could be removed easily, without bleeding, with a cotton-tip applicator. The patient was having no significant pain. The patient was seen one month after the procedure. The area of necrotic tumor slough and site of implant had healed. There was one small area of residual necrosis present. By two months, recurrent tumor was noted in the remaining tongue outside of the interstitial thermal radiotherapy zone. The patient was restarted on systemic chemotherapy, but expired at six months posttreatment of widespread metastatic neoplasm. Even though there was bulky recurrent tumor in the remainder of the tongue, the area treated with interstitial radiation and heat appeared free of tumor.

SUMMARY AND PERSPECTIVES

Our initial clinical studies employing interstitial thermoradiotherapy have been encouraging. Tumor responses have been significant, while normal tissue complications have not been severe. The efficacy observed to date for this modality undoubtedly involves a number of biological and physical factors. These include better dose localization, more effective heating, inhibition of sublethal damage repair, tumor hypoxia and low pH, and deficiencies in tumor versus normal tissue microcirculation. How these factors rank or interrelate is unknown. Clearly, however, the results obtained suggest that expanded trials of interstitial thermoradiotherapy in human cancer therapy are warranted where local control is a problem.

Future plans should involve the initiation of phase II clinical trials to establish efficacy of this modality in specific tumor types and sites. We are currently commencing trials of cutaneous and subcutaneous melanoma, cervix, and head and neck cancers. All these cancers have the advantage of being accessible to this technique, but this does not imply that interstitial radiotherapy is limited to these lesions. Using implanted electrodes, the problem of glioblastoma multiforme could be addressed in the near future, as these currently are unresponsive tumors with dismal treatment responses. Other deep seated tumors also may be amenable to this general approach when combined as an intraoperative technique, and certainly other approaches need to be pursued. Interstitial thermoradiotherapy offers many advantages now, in terms of cost, simplicity of application, ability to use state-of-the-art thermometers, and ability to localize the heat dose.

References

Barranco, S. C.; Romsdahl, M. M.; and Humphrey, R. M. The radiation response of human malignant melanoma cells grown *in vitro*. *Cancer Res.* 31:830–833, 1971.

Ben-Hur, E.; Bronk, B. V.; and Elkind, M. M. Thermally enhanced radiosensitivity of cultured Chinese hamster cells. *Nature* 238:209–211, 1972.

Ben-Hur, E.; Elkind, M. M.; and Bronk, B. V. Thermally enhanced radioresponse of cultured Chinese hamster cells: inhibition of repair of sublethal damage and enhancement of lethal damage. *Radiat. Res.* 58:38–51, 1974.

Cetas, T. C. Thermometry in strong electromagnetic fields. In *The physical basis of electromagnetic internation with biological systems*, eds. L. S. Taylor and A. Y. Cheung. College Park, Md.: University of Maryland, 1978, pp. 261–270.

Cetas, T. C., and Connor, W. G. Thermometry considerations in localized hyperthermia. *Med. Phys.* 5:79–91, 1978.

Cetas, T. C.; Connor, W. G.; and Manning, M. R. Monitoring of tissue temperature during hyperthermia therapy. *Ann. NY Acad. Sci.* 335:281–297, 1980.

Dewey, W. C. et al. Cell biology of hyperthermia and radiation. In *Radiation biology in cancer research*, eds. R. E. Meyn and H. R. Withers. New York: Raven Press, 1980, pp. 589–621.

Feder, B. H.; Syed, A. M. N.; and Neblett, D. Treatment of extensive carcinoma of the cervix with the "transperineal parametrial butterfly." *Int. J. Radiat. Oncol. Biol. Phys.* 4:735–742, 1978.

Gerner, E. W. et al. The potential of localized heating as an adjunct to radiation therapy. *Radiology* 116:433–439, 1975.

Guy, A. W. Analyses of electromagnetic fields induced in biological tissues by thermographic studies on equivalent phantom models. *IEEE Trans. MTT* 19:205–214, 1971.

Guy, A. W., and Lehmann, J. F. On the determination of an optimum microwave diathermy frequency for a direct contact applicator. *IEEE Trans. BME* 13:76–87, 1966.

Hartman, J. T., and Crile, G., Jr. Heat treatment of osteogenic sarcoma. *Clin. Orthop.* 61:269–276, 1968.

Kaplan, H. S. Experimental frontiers in radiation therapy of cancer. In *Radiation research*, eds. S. Okada et al. Tokyo: Toppan, 1979, pp. 2–14.

Manning, M. R. et al. Clinical hyperthermia: results of a phase I trial employing hyperthermia alone or in combination with external beam or interstitial radiotherapy. *Cancer* 49:205–216, 1982.

Miller, R. C. et al. Potentiation of radiation myelitis by hyperthermia treatment. *Br. J. Radiol.* 49:895–896, 1976.

Miller, R. C. et al. Prospects for hyperthermia in human cancer therapy. I. Hyperthermic effects in man and spontaneous animal tumors. *Radiology* 123:489–495, 1977.

Miller, R. C. et al. Effects of interstitial irradiation alone, or in combination with localized hyperthermia on the response of a mouse mammary tumor. *J. Radiat. Res.* 19:175–180, 1978.

Miller, R. C.; Connor, W. G.; and Boone, M. L. M. Hyperthermia as an anticancer modality. *Applied Radiology* 6:57–60, 1977.

Robinson, J. E.; Wizenberg, M. J.; and McCready, W. A. Radiation and hyperthermia response of normal tissue *in situ*. *Radiology* 113:195–198, 1974.

Suit, H. D. Statement of the problem pertaining to the effect of close fractionation and total treatment time on response of tissue to x-irradiation. In *Time and dose relationship biology as applied to radiotherapy*, eds. V. P. Bond, H. D. Suit, and V. Marceal. Upton, N.Y.: Brookhaven National Laboratory, 1969, pp. 7–10.

Tak, W. K. T. Interstitial therapy in gynecological cancer. *Gynecol. Oncol.* 6:429–437, 1978.

Thrall, D. E.; Gillette, E. L.; and Dewey, W. C. Effect of heat and ionizing radiation on normal and neoplastic tissue of the C3H mouse. *Radiat. Res.* 63:363–377, 1975.

Urtasun, R. et al. Radiation and high-dose metronidazole in supratentorial glioblastomas. *N. Engl. J. Med.* 294:1364–1367, 1976.

Warren, S. L. Preliminary study of the effect of artificial fever upon hopeless tumor cases. *Am. J. Roentgenol.* 33:75–87, 1935.

CHAPTER 22
Clinical Thermoradiotherapy

Haim I. Bicher
Taljit S. Sandhu
Fred W. Hetzel

Introduction
Rationale
Clinical Thermoradiotherapy Trials
Conclusion

INTRODUCTION

The failure of conventional radiotherapy to control a significant number of tumors without any extensive adverse effects on the surrounding normal tissue is attributed mainly to the presence of a viable hypoxic fraction of cells in these tumors. As far back as 1909, Schwarz showed that restriction of blood circulation, thereby limiting oxygen supply, rendered human skin radioresistant. In general, tumors are heterogeneous structures with some regions containing relatively hypoxic, closely packed cells remote from blood vessels (Thomlinson and Gray 1955). Even a very small fraction of hypoxic cells in the total tumor volume may result in failure of radiotherapy to control the tumor. There have been a number of different approaches to overcome the limitation to response imposed by hypoxic cells.

An early attempt to solve this problem was to give the total radiation dose in several fractions. During the course of fractionated treatments potentially resistant hypoxic cells generally become sensitive oxygenated cells, at least during the latter phases of a multifraction treatment schedule. It is possible, however, that some tumor cells do not reoxygenate completely and remain radioresistant, such that some relatively hypoxic cells still survive treatment and are responsible for tumor recurrence. Even a modest number of remaining cells constitute a serious obstacle to curative radiotherapy. Thus, a significant effort has been made to compensate for this limitation.

Another potential solution has been to use high linear energy transfer (LET) radiation instead of ordinary x- and gamma-rays, such that the tumor response is independent of the presence or absence of molecular oxygen. The ideal radiation, which will have an oxygen enhancement ratio (OER) of unity (such as 2MeV alpha particles), has been incapable of penetrating sufficient surface tissue thickness to be of any therapeutic use. Neutrons, pions, or high-energy heavy ions hold promise; however, these options tend to be prohibitively expensive, but limited trials are beginning to evaluate clinical efficacy.

Another approach has been to place the patient, before radiotherapy, in an environment of pure oxygen at a pressure of two to three atmospheres, in an attempt to increase the ratio of oxygenated tumor cells. While theoretically valid, this technique so far has failed to produce any significant improvement in overall tumor control in humans.

Another approach has been the concomitant use of electron-affinic drugs that preferentially and differentially sensitize hypoxic cells to radiation (Goldfeder, Brown, and Berger 1979). The toxicity of these drugs at the dose levels required in the clinical situation has limited widespread use. A large number of experimental clinical trials involving these radiation sensitizers are in progress at various centers around the world and eventually may prove useful in some situations.

In recent years, there has been an increasing interest in the use of hyperthermia, either alone or in combination with radiation, in the treatment of cancer. Although reports (Bicher 1980) on the use of hyperthermia date back as far as 1866 (Bush), only in the past two decades has progressive research in the use of hyperthermia as a possible, clinically important, antitumor treatment modality been stimulated by the results of provocative thermoradiobiological studies. These investigations suggest that hyperthermia may be a practical solution to reduce radiation treatment failures (see chapter 20).

RATIONALE

Recent extensive cellular radiobiological studies and preliminary human clinical studies show that hyperthermia combined with radiation has a synergistic tumor-cell killing effect. Both in vitro (Kim, Kim, and Hahn 1975b; Sapareto, Raaphorst, and Dewey 1979; Sapareto, Hopwood, and Dewey 1978; Thrall et al. 1976) and in vivo (Gillette 1976; Gillette and Ensley 1979; Giovanella et al. 1973; Robinson, Wizenberg, and McCready 1974; Thrall, Gillette, and Bauman 1973) investigations show that thermal enhancement depends not only on the sequence of the two treatments, but also on the cell cycle. This synergistic effect appears most pronounced for radioresistant S-phase cells (Gerweck, Gillette, and Dewey 1975; Sapareto, Hopwood, and Dewey 1978). From a clinical point of view, the most important observation has been that the hypoxic cells may be equally or more sensitive to hyperthermia than aerobic cells (Gerweck, Gillette, and Dewey 1974; Kim, Kim, and Hahn 1975a, 1975b; Robinson, Wizenberg, and McCready 1974; Schulman and Hall 1974). The rationale for using hyperthermia above 43°C in combination with ionizing radiation is that the viable hypoxic tumor cells, which are radioresistant, may be destroyed by hyperthermia at these temperatures. Another interesting and potentially exploitable phenomenon is that mild hyperthermia (40°C to 42°C) may introduce a significant change in tumor blood flow. It has been shown (Bicher 1978, 1980; Bicher et al. 1980) that hyperthermia at 40°C to 42°C, both in mouse tumor and in tumors in humans, increases the blood flow (see chapter 8). This may sensitize effectively otherwise radioresistant hypoxic cells within the tumor. Dramatic changes in cell survival during hyperthermia also have been observed when pH of cells is only slightly altered (Gerweck 1977). Although there is no in vitro data indicating a differential response or sensitization of tumor cells as opposed to normal tissue cells, the cellular environment in the two cases is expected to be different in vivo. For example, several studies (Eden, Haines, and Kahler 1955; Gullino et al. 1965; Meyer et al. 1948; Newsland and Swenson 1953) have indicated that the pH of fluid in human and rodent solid tumors is lower than that of normal tissues. This could result in differential tumor cell killing with normal tissue sparing. Other studies (Alfieri, Hahn, and Kim 1980; LeVeen et al. 1976) indicate that some tumors have a sluggish blood flow as compared to normal tissue. This suggests that under similar heating conditions tumors may not be able to dissipate heat as efficiently as normal tissues, resulting in differential tumor heating. These findings, which suggest a different microenvironment in tumors versus normal tissue with regard to pH, blood flow, and nutritional state, probably inspired earlier investigators (Chen and Heidelberger 1969; Giovanella et al. 1973; Lambert 1912; Schrek 1966) to conclude that tumor cells were killed selectively by hyperthermia.

Based on these in vitro and in vivo radiobiological findings, several groups of investigators (Arcangeli et al. 1980; Bicher, Sandhu, and Hetzel 1980; Hornback et al. 1977; Johnson et al. 1979; Kim, Hahn, and Benjamin 1979; Manning et al. 1982) are engaged in clinical studies employing the combination of hyperthermia and ionizing radiation.

CLINICAL THERMORADIOTHERAPY TRIALS

Clinical applications of combined hyperthermia and radiation treatment modalities date back to the earlier part of the century. The early reports, unfortunately, are anecdotal in nature. Temperatures to which tumors and surrounding normal tissues were raised are not evaluable. This, combined with the lack of controls for radiation alone, have made it difficult to draw any quantitative conclusions. Moreover, since the measurements were not made, it has been impossible to establish in some cases that tumor temperature was raised at all.

The earliest report of combining hyperthermia and ionizing radiation is that of Schmidt in 1909. He proposed the use of diathermy for localized heating of tissue to treat malignancies in combination with ionizing radiation. In 1937, Arons and Sokoloff used radiofrequency (RF) currents for hyperthermic treatments of intrathoracic and intraabdominal tumors. Again, it is hard to establish whether they were able to raise the tissue temperature to therapeutic range. Woeber (1955) used ultrasound with simultaneous ionizing radiation to treat 20 patients with cutaneous malignancies. He reported a 40% reduction in radiation dose required for tumor control accompanied by a reduction in normal tissue reaction. Crockett and associates (1967), following some experimental work on normal dog bladder, treated seven patients with incurable bladder carcinomas with a combination of local hyperthermia and regional radiotherapy. The radiation dose varied from 4500 to 5500 rad. These authors reported striking reduction in tumor size with no serious adverse effects and suggested that the tumor response to radiation was enhanced by hyperthermia. The authors, however, did not make any attempt to determine heat distribution within the bladder. More recently, Hall (1975) reported regression of bladder papillomatosis without any complications. He heated the bladder for three hours, for 5 to 14 sessions, by irrigation with water heated to 42°C to 45°C.

Hartman and Crile (1968) reported treatment of osteogenic sarcoma with microwave heating and various doses of radiation. Five children were treated with this combination. Two of five, one seven-year-old girl with osteogenic sarcoma in the right midtibia, and a 16-year-old boy with an osteoblastic and osteolytic lesion of the metaphysis of the distal part of the left radius, were alive five years later. The other three patients survived 6 to 17 months and died of metastatic disease. The two patients who had long-term survival also retained function of the limb.

Stehlin (1975) reported a 67% five-year survival for 32 patients he treated using heated blood for perfusion of the extremities. These hyperthermia treatments were followed by radiation several weeks later. It was found that tissue temperatures of 40°C using his technique were tolerated very well for several hours, but perfusion of extremities with blood heated to 46°C resulted in complications.

More recent studies (Arcangeli et al. 1980; Bicher, Sandhu, and Hetzel 1980; Hornback et al. 1977; Johnson et al. 1979; Manning et al. 1982; U et al. 1980; Kim, Hahn, and Benjamin 1979) involving the combination of hyperthermia and ionizing radiation have made a serious effort to measure and document hyperthermia treatments accurately. In most cases a comparison with a control treated by radiation alone was made. Kim, Hahn, and Benjamin (1979) have treated 50 patients with a variety of cutaneous tumors. Improved results were reported for both the radiosensitive (i.e., mycosis fungoides) and radioresistant tumors (i.e., melanoma) with combined hyperthermia and radiation treatment as compared to either of these modalities used alone. These authors reported complete disappearance of multiple recurrent melanoma nodules without unusual normal skin reactions; however, combination therapy did produce enhanced skin reactions in patients whose treated areas included a skin graft or heavily scarred skin from extensive surgery, suggesting that tissues with reduced blood flow have reduced tolerance to thermoradiotherapy. Table 22.1 summarizes their results by histology. In their study the authors reported 78% overall tumor control rate after combined therapy as compared with 26% after radiation alone. These investigators applied two heating methods. Some patients with tumors on extremities were heated by immersion in a heated waterbath. The remaining patients were treated using RF (27.12 MHz) inductive heating. It should be pointed out that there is considerable variation in the radiation dose and the hyperthermia treatment duration, as well as in the number of fractions. The radiation dose varied from 800 rad in two fractions for melanoma to 2400 rad in eight fractions for Kaposi's sarcoma. Similarly, hyperthermia (43.5°C) treatments varied from two fractions of 30 minutes for melanoma to five fractions

Table 22.1
Results of Combined Hyperthermia and Radiation Treatment by Tumor Histology

Histology	Number of Patients	Number of Patients with Complete Response	Median Follow-up (Months)
Mycosis fungoides	14	11	18
Lymphoma	5	4	12
Kaposi's sarcoma	7	4	18
Malignant melanoma	22	19	9
Chondrosarcoma	1	1	7
Others	5	3	6

SOURCE: Kim, Hahn, and Benjamin 1979.

of 60 minutes for mycosis fungoides. The hyperthermia treatments followed immediately the radiation treatments in all cases. These data do not suggest any particular treatment schedule for a particular tumor; however, the study does demonstrate the improved effectiveness of combined thermoradiotherapy as compared to hyperthermia or radiation alone.

Hornback and co-workers (1977) treated 72 patients with advanced cancer using combined thermoradiotherapy. Fifty-three percent of the patients treated with heat prior to radiation therapy experienced complete remission of symptoms. Among the group of patients treated with heat after radiotherapy, 92% showed complete remission. Again there was no set protocol, and the radiation dose varied from 50 to 600 rad per day and total doses from 3000 to 6500 rad. Heat treatments were given using 433.92 MHz microwaves, thought to produce 41°C to 42°C at 7- to 8-cm depths. Although the authors mention having attempted to measure tumor temperature during these treatments, there is no mention of the tumor temperature achieved in the patients.

Johnson and colleagues (1979) conducted a pilot study to evaluate normal skin and melanoma thermal enhancement ratios of hyperthermia combined with radiation. The response of normal skin to the treatment was measured by evaluating the degree of erythema according to a standard numerical scoring system. Tumor response was assessed by measuring tumor diameter. Although the study was not conclusive, it did bring to light some of the problems associated with obtaining useful clinical data. The study involved patients with multiple cutaneous melanoma metastases. At least three lesions were evaluated on each patient. The patients were divided into three groups and given one, three, or four fractions, with a minimum interval of 72 hours between each fraction. Radiation doses per fraction for different lesions in a given patient varied from 500 to 900 rad. In some cases, single fractions of 1000, 1100, 1200, or 1300 rad were used. In all patients, one lesion was heated immediately following radiation therapy, and the other lesions, treated with radiation alone, were used for comparison. Hyperthermia treatments were administered using 915-MHz direct contact microwave applicators (Sandhu, Kowal, and Johnson 1978). Hyperthermia treatments varied between one and two hours at 41.5°C to 42.0°C. Skin enhancement ratio (SER) and tumor enhancement ratio (TER) were evaluated only for a limited number of patients because of lack of follow-up data. SER values varied from 1.2 to 1.7, and TER values in most cases were 1.3. It was realized that the stringent tumor regrowth time period end point made it extremely difficult to obtain clinically significant data. The study demonstrated, however, that superficial tumors, up to 4 cm in diameter and 2 cm in depth, could be heated to an accuracy of 0.5°C either during or after application of 915-MHz microwaves.

Manning and associates (1982) reported a limited study combining heat and radiation. Of 40 patients treated with hyperthermia, four received hyperthermia treatments combined with interstitial radiation. Each patient had a minimum of three nodules. One nodule received a heat treatment of 43°C for 40 minutes using RF waves. Another nodule received radiation alone from two radium needles to a dose of 4000 rad in 100 hours. A third lesion had the same dose of radiation plus simultaneous heat of 43°C for 40 minutes using radium needles as the heating electrodes. The response rate for heat-radiation combination was 80% to 90%, compared with 50% response rate for heat alone and radiation alone groups. The authors suggested a beneficial therapeutic ratio and minimal side effects from the combined treatment.

Another limited study by Dr. U and colleagues (1980) treated groups of patients with radiotherapy alone, hyperthermia alone, and combined treatment. One group of patients received 200- to 600-rad fractions two to five times per week, to a total of 1800 to 4200 rad in 5 to 14 fractions. Another group of patients received hyperthermic treatments of 42°C to 44°C two to three times per week, to a maximum of 10 sessions in four weeks. Another group received combined thermoradiotherapy treatments with radiation fractions of 200 to 600 rad, two to five times a week, to a total of 2000 to 4800 rad in 6 to 20 fractions. Hyperthermia treatments were given using either 2450- or 915-MHz microwaves. Of the group of eight patients treated with radiation plus hyperthermia, six experienced complete regression of lesions within one month of therapy. None of the tumors treated with hyperthermia alone regressed completely. In the group treated with radiation only, 73% showed tumor regression. Significantly, no adverse side effects were observed in normal tissue from the combined treatments.

Another interesting study was reported by Arcangeli and co-workers (1980). Fifteen patients with multiple neck lymph node metastases from head and neck cancer were treated with either radiation alone or in combination with hyperthermia. A total of 33 neck metastases were treated, 13 with radiation alone and the remainder with combination therapy. Hyperthermia was induced by 500-MHz microwaves, using a noncontact applicator. These investigators employed an innovative dosage schedule, described as a *multiple daily fractionation* (MDF) scheme, consisting of consecutive doses of 200, 150, and 150 rad per day, with a four- to five-hour interval between fractions, five days per week, up to a total of 4000 to 7000 rad total dose. All the lesions were irradiated with the same total dose, regardless of whether they received hyperthermia. The results are summarized in table 22.2. The results were compared also with their historical series on treating the same type of lesion with conventional fractionations of 200 rad per day. The MDF schedule alone resulted in 46% complete response, which was enhanced to 85% complete response with 15% partial response when combined with hyperthermia. It should be pointed out that in this treatment schedule, when MDF was combined with hyperthermia, heat was applied immediately after the second daily fraction of MDF. Moreover, the authors did not observe any abnormal reactions in areas that received combined treatment.

Bicher, Sandhu, and Hetzel (1980) proposed a fractionation scheme combining hyperthermia and radiation designed with a curative intent. The protocol below summarizes eight hyperthermia treatments combined with radiation (1600 rad delivered in four fractions over two weeks).

1. Hyperthermia only for four treatments at 45°C for 90 minutes (skin 36°C or below); treat every 72 hours.
2. Rest one week. Evaluate response to heat alone.
3. Radiation therapy and hyperthermia combined for four treatments at 42°C for 90 minutes; heat follows radiation (20-minute interval). The fraction administered at each treatment is 400 rad.

This protocol was devised to take advantage of the known effects of hyperthermia, either alone or in combination with ionizing radiation. Skin cooling was implemented to prevent normal tissue damage from hyperthermia. The number of fractions and duration of each treatment was arbitrarily chosen, taking into account possible biological factors such as thermotolerance and vascular response and repair, as well as patient compliance and patient comfort. In addition, the total dose of radiation employed was sufficiently low to allow for treatment of previously irradiated areas. Localized hyperthermia was induced by microwave radiation employing direct contact applicators. Applicators were essentially square or rectangular cross-section waveguides. Two types of microwave applicators were employed in this study. One type was a square or rectangular cross-section waveguide completely loaded with a low loss dielectric material and excited in TE_{10} mode (Sandhu, Kowal, and Johnson 1978.) The other type of applicator employed was a partially filled rectangular waveguide excited in the TEM mode. The latter design resulted in better heat distribution than that achieved with the TE_{10} mode applicator. It also allowed for air cooling of the skin of the treated area. Thermometry was accomplished using one or two ultramicrothermocouples, depending on the size and location of the tumor. Although it should be emphasized that great care was required during the initial planning and set-up for a patient, it was feasible to obtain satisfactory thermometry for every treatment of every field, with minimal patient discomfort. The tumor size, location, and depth determined the applicator size and frequency (915 or 300 MHz) employed. Heat was applied initially to tumors in four fractions, of 1.5-hour durations, at intervals of 72 hours. The tumor temperature was monitored to maintain a temperature of 45°C ± 0.5°C, while the overlying, normal skin was simultaneously maintained at or below 36°C with air cooling. Following four fractions the patient rested for one week. Four additional fractions of hyperthermia then were given in combination with a radiation dose of 400 rad to the tumor, immediately followed (within 20 minutes) by hyperthermia to a temperature of 42.0°C ±

Table 22.2
Results of Treatments Using Multiple Daily Fractionation (MDF) Scheme

Treatment Type	Response at End of Treatment	No. (%)
MDF and Hyperthermia	Complete response	17/20 (85)
	Partial response	3/20 (15)
MDF only	Complete response	6/13 (46)
	Partial response	4/13 (31)
Conventional radiation therapy (Historical)	Complete response	14/46 (30)
	Partial response	19/46 (41)

SOURCE: Arcangeli et al. 1980.
NOTE: Complete response: total tumor regression.
Partial response: ≥ 50% tumor regression.

Table 22.3
Results of Fractionation Scheme Combining Hyperthermia and Radiation in 62 Patients (82 tumors)

	Number (%)
Tumor Results	
Complete response	49 (60)
Partial response	28 (34)
No response	5 (6)
Recurrences	
Local	5 (6)
Marginal	3 (4)
Complications	
Skin burns	2 (2)
Oral burns	2 (2)
Other	1 (1)

SOURCE: Bicher 1980.

0.5°C. Such fractions again were separated by 72 hours.

To date, 62 patients have been treated, with a total of 82 fields. Tables 22.3 and 22.4 summarize the results. A complete response was observed in 49 treatment fields and a partial response in 28. No response was observed in five patients, including one with squamous cell carcinoma, one with adenocarcinoma, and one with sarcoma. The only complications that were directly attributable to treatment were two skin burns that resulted from inadequate surface cooling and two tongue and pharynx burns; all subsequently healed. One patient with a history of epilepsy experienced a grand mal seizure during treatment of a neck tumor. Two marginal recurrences and five local recurrences have been noted. Table 22.4 summarizes the results of therapy by histology. It suggests that malignant lymphoma is most responsive to combination treatment. It should be noted that tumor regression following completion of treatment was often slow and required approximately two months before a total response could be evaluated. Another interesting observation made during these treatments was that the microwave power required to maintain a desired treatment temperature declined following the first or second treatment. These phenomena may be related to thermally induced physiologic changes within the tumor that may be involved in the ultimate destruction of the tumor. Further physiologic studies are warranted. Another interesting outcome of this study was suggested by the marginal recurrences. They may indicate that such combined modality therapy is effective in the treatment of microscopic disease. Longer follow-up and more patients are required, however, before a true evaluation of the clinical efficiency of this and other experimental protocols can be determined.

CONCLUSIONS

Thermoradiotherapy has only recently been tested in the clinical arena. There is mounting evidence from in vitro and in vivo cellular and animal investigations and

Table 22.4
Results of Fractionation Scheme Combining Hyperthermia and Radiation by Tumor Histology

Tumor Histology	No. of Tumors	Response	Follow-up (Months)
Malignant melanoma	17	8 complete 6 partial 3 no response	2 months–1 year
Malignant lymphoma	8	8 complete	2–7 months
Squamous cell carcinoma	19	7 complete 11 partial 1 no response	2–6 months
Adenocarcinoma	33	24 complete 8 partial 1 no response	2–7 months
Other (transitional cell, basal cell, glioma, sarcoma)	5	2 complete 3 partial	2–9 months
Total	82	49 complete 28 partial 5 no response	1 year

SOURCE: Bicher 1980.
NOTE: Complete response = No tumor at two-month follow-up and thereafter.
Partial response = Tumor decreased in size to half or less at two-month follow-up.

early human trials to suggest that such combined therapy may have potential clinical efficacy. Oxygenated cells seem to be most sensitive to ionizing radiation, while hypoxic cells may be more responsive to hyperthermia. Since any given tumor probably contains both cell populations, it seems logical to combine these therapies. While optimal sequencing and timing of such combined therapies has yet to be determined, initial results to date support further investigation of thermoradiotherapy.

References

Alfieri, A. A.; Hahn, E. W.; and Kim, J. H. The relationships between the time of fractionated and single doses of radiation and hyperthermia on the sensitization of an in vivo mouse tumor. *Cancer* 36:893–903, 1975.

Arcangeli, G. et al. Effectiveness of microwave hyperthermia combined with ionizing radiation: clinical results on neck node metastases. *Int. J. Radiat. Oncol. Biol. Phys.* 6:143–148, 1980.

Arons, I., and Sokoloff, B. Combined roentgenotherapy and ultra-short wave. *Am. J. Smy.* 36:533–543, 1937.

Bicher, H. I. Increase in brain tissue oxygen availability induced by localized microwave hyperthermia. New York: Plenum Press, 1978.

Bicher, H. I. Changes in tumor tissue oxygenation induced by microwave hyperthermia. *Ann. NY Acad. Sci.* 335:20, 1980.

Bicher, H. I. et al. Effects of hyperthermia on normal and tumor microenvironment. *Radiology* 137:511–513, 1980.

Bicher, H. I.; Sandhu, T. S.; and Hetzel, F. W. Hyperthermia and radiation in combination: a clinical fractionation regime. *Int. J. Radiat. Oncol. Biol. Phys.* 6:867–870, 1980.

Bush, W. Uber den einfluss welchen heftigere erysipeln zuweilen auf organisierte neubildungen amiben. *Verhandi. Naturh. Preuss. Rhein Westpahl.* 23:28–30, 1866.

Chen, T. T., and Heidelberger, C. Quantitative studies on the malignant transformation of mouse prostate cells by carcinogenic hydrocarbons in vitro. *Int. J. Cancer* 4:166–178, 1969.

Crockett, A. T. et al. Enhancement of regional bladder megavoltage irradiation in bladder cancer using local bladder hyperthermia. *J. Urol.* 97:1034–1037, 1967.

Eden, M.; Haines, B.; and Kahler, H. The pH of rat tumors measured in vivo. *J. Natl. Cancer Inst.* 16:541–556, 1955.

Gerweck, L. E. Modification of cell lethality at elevated temperatures. The pH effect. *Radiat. Res.* 70:224–235, 1977.

Gerweck, L. E.; Gillette, E. L.; and Dewey, W. C. Killing of Chinese hamster cells in vitro by heating under hypoxic or aerobic condition. *Eur. J. Cancer* 10:691–693, 1974.

Gerweck, L. E.; Gillette, E. L.; and Dewey, W. C. Effect of heat and radiation on synchronous Chinese hamster cells: killing and repair. *Radiat. Res.* 64:611–623, 1975.

Gillette, E. L. Hyperthermia and therapeutic ratios. In *Proceedings of the Second International Symposium on Cancer Therapy by Hyperthermia and Radiation*, eds. M. J. Wizenberg and J. E. Robinson. Chicago: American College of Radiology, 1976, pp. 128–133.

Gillette, E. L., and Ensley, B. A. Effect of heating order on radiation response of mouse tumor and skin. *Int. J. Radiat. Oncol. Biol. Phys.* 5:209–213, 1979.

Giovanella, B. C. et al. Selective lethal effect of supranormal temperatures on mouse sarcoma cells. *Cancer Res.* 33:2568–2578, 1973.

Goldfeder, A.; Brown, D. M.; and Berger, A. Enhancement of radioresponse of a mouse mammary carcinoma to combined treatments with hyperthermia and radiosensitizer misonidazole. *Cancer Res.* 39:2966–2970, 1979.

Gullino, P. M. et al. Modifications of the acid-base status of the internal milieu of tumors. *J. Natl. Cancer Inst.* 34:857–869, 1965.

Hall, R. Treatment of bladder tumors by irrigating with hot saline. In *Proceedings of the Second International Symposium on Cancer Therapy by Hyperthermia and Radiation*, eds. M. J. Wizenberg and J. E. Robinson. Chicago: American College of Radiology, 1976, pp. 302.

Hartman, J. T., and Crile, G., Jr. Heat treatment of osteogenic sarcoma. Report of 5 cases. *Clin. Orthop.* 61:269–276, 1968.

Hornback, N. B. et al. Preliminary clinical results of combined 433 MHz microwave therapy and radiation therapy on patients with advanced cancer. *Cancer* 40:2854–2863, 1977.

Johnson, R. J. R. et al. A pilot study to investigate the therapeutic ratio of 41.5–42°C hyperthermia and radiation. *Int. J. Radiat. Oncol. Biol. Phys.* 5:947–953, 1979.

Kim, J. H.; Hahn, E. W.; and Benjamin, F. J. Treatment of superficial cancers by combination hyperthermia and radiation therapy. *Clin. Bull.* 9:13–16, 1979.

Kim, S. H.; Kim, J. H.; and Hahn, E. W. The radiosensitization of hypoxic tumor cells by hyperthermia. *Radiology* 114:727–728, 1975a.

Kim, S. H.; Kim, J. H.; and Hahn, E. W. Enhanced killing of hypoxic tumor cells by hyperthermia. *Br. J. Radiol.* 48:872–874. 1975b.

Lambert, R. A. Demonstration of the greater susceptibility to heat of sarcoma cells as compared with actively proliferating connective tissue cells. *JAMA* 59:2147–2148, 1912.

LeVeen, H. H. et al. Tumor eradication by radiofrequency therapy. *JAMA* 235:2198–2200, 1976.

Manning, M. R. et al. Clinical hyperthermia: results of the phase I clinical trial combining localized hyperthermia alone or in combination with external beam or interstitial radiotherapy. *Cancer* 49:205–216, 1982.

Meyer, K. A. et al. pH studies of malignant tissues in human beings. *Cancer Res.* 8:513–518, 1948.

Newsland, J., and Swenson, K. E. Investigations of the pH of malignant tumors in mice and humans after administration of glucose. *Acta Obstet. Gynecol. Scand.* 32:359–367, 1953.

Robinson, J. E.; Wizenberg, M. J.; and McCready, W. A. Radiation and hyperthermia response of normal tissues in situ. *Radiology* 113:195–198, 1974.

Sapareto, S. A.; Hopwood, L. E.; and Dewey, W. C. Combined effects of x-irradiation and hyperthermia on CHO cells for various temperatures and orders of applications. *Radiat. Res.* 73:221–223, 1978.

Sapareto, S. A.; Raaphorst, P. G.; and Dewey, W. C. Cell killing and the sequencing of hyperthermia and radiation. *Int. J. Radiol. Oncol. Biol. Phys.* 5:343–347, 1979.

Sandhu, T. S.; Kowal, H. S.; and Johnson, R. J. R. Development of microwave hyperthermia applicators. *Int. J. Radiat. Oncol. Biol. Phys.* 4:515–519, 1978.

Schmidt, H. E. Zur rontgenbehandling tiefgelegener tumoren. *Fortschr. Geb. Rontgenstrahlen* 14:134–136, 1909.

Schrek, R. Sensitivity of normal and leukemic lymphocytes and leukemic myeloblasts to heat. *J. Natl. Cancer Inst.* 37:649–654, 1966.

Schulman, N., and Hall, E. J. Hyperthermia: its effect on proliferative and plateau phase cell cultures. *Radiology* 113:207–209, 1974.

Schwarz, G. Uber dosensensibilisierung gegen roentgens und radium strahlen. *Muench. Med. Wochenschr.* 56:1217–1221, 1909.

Stehlin, J. S., Jr. Regional hyperthermia by perfusion. In *Proceedings of the Second International Symposium on Cancer Therapy by Hyperthermia and Radiation*, eds. M. J. Wizenberg and J. E. Robinson. Chicago: American College of Radiology, 1976, pp. 266–271.

Thomlinson, R. H., and Gray, L. H. The histological structure of some human lung cancers and the possible implications for radiotherapy. *Br. J. Cancer* 9:539–549, 1955.

Thrall, D. E. et al. Responses of cells in-vitro and tissues in-vivo to hyperthermia and x-irradiation. In *Advances in radiation biology*, vol. VI, eds. J. T. Lett, H. Adler, and M. Felle. 1976, pp. 211–227.

Thrall, D. E.; Gillette, E. L.; and Bauman, C. L. Effect of heat on the C_3H mammary adenocarcinoma evaluated in terms of tumor growth. *Eur. J. Cancer* 9:871–875, 1973.

U, R. et al. Microwave-induced local hyperthermia in combination with radiotherapy of human malignant tumors. *Cancer* 45:638–646, 1980.

Woeber, K. Die bedentung des ultraschalls fur die dermatelogie und seine anwendung bei hauttumoren in kombination mit roentgenstrahlen. *Strahlentherapie* 98:169–184, 1955.

CHAPTER 23
Immunologic Aspects of Hyperthermia

John A. Dickson
Sudhir A. Shah

Introduction
Historical Background
Concepts and Principles
Materials and Methods
Analysis of Data
Overview
Perspective

INTRODUCTION

Although the current resurrection of enthusiasm for hyperthermia as a possible anticancer treatment modality has reached almost tidal wave proportions, the value of heat in the management of common solid tumors in humans is unproved. Predictably, the major thrust of investigations concern in vitro and animal studies, with the ever-present pitfall of extrapolating results, implicitly or explicitly, to the human situation. All too often it is forgotten that cancer is a disease of the host, and that, as succinctly put by Black (1972), "Around every tumor there is a patient." This state of affairs is not helped by the difficulties of studies in the intact animal, not to mention the problems of human clinical research. In addition, the often attractive and compelling nature of results obtained in experimental systems, allied to strong circumstantial evidence in humans, can lead to plausible concepts and entrenched ideas, especially in the enthusiast.

The situation is well illustrated by the subject of hyperthermia and the immune system. This topic includes not only the immunology of cancer, but also several major aspects of neoplasia and host response—metastasis, the reticuloendothelial (mononuclear phagocyte) system, and the inflammatory response. Tumor blood flow, currently receiving increasing attention by hyperthermia investigators, also is a crucial factor, since access of the components of the host response (cells, antibodies) to the tumor and egress of tumor products (including antigenic components) depend upon a patent vascular system. It is apparent that these subjects constitute various aspects of the host-tumor relationship, the essential background for study of cancer therapy.

If we assume that the immune system and metastasis are related, then we must accept at the outset that our discussion of hyperthermia and immunity (and our experimental approach) is limited by two major difficulties: the assessment of host response and the quantitation of metastasis. The task is rendered more formidable when we recall the dismal history of attempts at immunotherapy in cancer patients and Currie's view (1980) that currently the study of immunology is simply "phenomenology," observations that are interesting (and often conflicting) but may have little relevance to the host with cancer. Our dilemma is compounded by the numerous actions of heat at the subcellular, cellular, organ, and host level, and our knowledge of these is primitive.

In view of these various considerations, it may be wise to withhold counsel on the subject of hyperthermia and the immune system. There is, however, a long-standing and entrenched belief that heating of tumors and/or the host can lead to a beneficial host response that may involve the immune system. At a time when the multimodality approach to cancer therapy is gaining popularity and the major nonsurgical anticancer treatments, radiotherapy and cytotoxic drugs, are immunosuppressive, a therapy that is both tumor destructive and host-stimulating has many would-be acolytes waiting in eager anticipation. There also is some data indicating that heat may increase the growth rate and enhance the dissemination of tumors. This has become of more immediate concern with the present increasing use of hyperthermia in humans. This chapter is an attempt to define and to clarify the possible role of the host response in thermotherapy.

HISTORICAL BACKGROUND

Just as the association between elevated temperature, in the form of pyrexia, and tumor regression first became apparent in humans, so too was the notion of involvement of the immune system in the tumor response derived from clinical studies. From 1893 to 1936, W. B. Coley and other physicians achieved some remarkable cures of advanced cancers by injection of bacterial filtrates (Coley's toxins). Some patients became immune to the toxins, and inevitably the possibility arose that the body's defense mechanisms were operating against the tumor as well as against the injected toxins (Nauts, Fowler, and Bogatko 1953). Earlier workers with animal tumors also surmised that hyperthermic tumor destruction was not solely the result of the elevated temperature. In his classic 1927 paper on cure of the Flexner-Jobling and Jensen's tumors in rats by diathermy heating, Westermark raised the possibility of the tumor "having been damaged to an extent sufficient for the organism itself later to deal with it." More specifically, Johnson (1940) suggested that the greater heat susceptibility of the Jensen sarcoma compared to the Walker carcinoma was attributable to the immunity that the rats had to the Jensen tumor.

There also was interest in the first half of this century in the effects of heat on the immune system of the normal host. In 1938, Neymann reviewed the data accumulated during the heyday of the use of physically induced hyperpyrexia for the treatment of nonmalignant conditions such as syphilis, arthritis, and asthma. At body temperatures of 39.7°C to 42.0°C, maintained for 6 to 10 hours, there was a fairly constant stimulation of the hemopoietic system confined almost exclusively to the polymorphonuclear leukocytes; there was a concomitant destruction of lymphocytes. Doan, Hargreaves, and Kester (1937), using normal rabbits, confirmed the occurrence of postfebrile leukocytosis and lymphopenia and showed that the lymphopenia was due to a direct destructive effect of the total-body heating on the lymph nodes and bone marrow.

In a series of investigations encompassing both animals and humans and spanning almost 60 years, Strauss and colleagues not only bridged the gap between the pioneers in the field and current workers, but also postulated a mechanism to connect tumor heating and the assumed immunostimulation. They believed that following tumor heating, necrotic cancer cells were absorbed and acted as an antigen, increasing the immunity of the tumor-bearing host with subsequent destruction of distant metastases (Strauss 1969; Strauss et al. 1965). Their concept of immunostimulation by tumor breakdown products also has been advocated by other surgeons currently practicing hyperthermia. From results in the treatment of limb tumors by regional hyperthermic perfusion, Cavaliere, Moricca, and Caputo (1976), Moricca and co-workers (1977), and Stehlin (1976) and associates (1977) maintain that antigenic products liberated from heated cancers lead to enhancement of an antitumor immune response that prevents and/or destroys metastases. In the early 1970s, reports on several animal systems confirmed that effective heating of a tumor led to regression of tumor at other anatomic sites (Dickson and Muckle 1972; Dickson and Suzanger 1974; Goldenberg and Langner 1971; Muckle and Dickson 1971), and the spectrum of animal tumors exhibiting this phenomenon, the *abscopal response* (Goldenberg and Langner 1971), has since been extended (Dickson, Calderwood, and Jasiewicz 1977; Dickson et al., in press).

From detailed histologic studies on a large number of tumor strains in mice, Overgaard and Overgaard (1972, 1976) concluded that the microscopic appearance of tumor cells after heat treatment was specific, and differed from the usual necrotic and necrobiotic processes seen in tumors. Further links in the tumor heating-immunostimulation chain were provided by Mondovi and colleagues (1972), who described an increased antigenicity in Ehrlich ascites cells heated at 42.5°C in vitro, and by Jasiewicz and Dickson (1976), who reported that the destructive effect of heat (42°C) on rat mammary cancer cells in vitro was potentiated by cell-specific antiserum.

These various speculations, observations, and findings of different workers introduced the possibility that heat led to a special kind of cell death in tumors, with a favorable alteration in tumor antigenicity and stimulation of a host antitumor immune response that could destroy metastatic cells. This possibility, allied to the body of data indicating that the destructive effects of heat can be selective for cancer cells (Rossi-Fanelli et al. 1977; Streffer et al. 1978; Wizenberg and Robinson 1976), has generated much enthusiasm for research into hyperthermia as an anticancer agent. Although this background information was available in the early 1970s, definitive data on the role of the immune system in tumor response to heating has been slow to appear. This is due to the numerous difficulties outlined here, including the thorny problem of how to measure host immune competence.

CONCEPTS AND PRINCIPLES

Tumor Immunology

In 1957, Prehn and Main demonstrated for the first time the existence of a tumor-specific antigen. This result was achieved in transplantation studies using a methylcholanthrene-induced tumor in an isogeneic strain of mice. A new era of interest in cancer immunology began, since it became feasible to separate clearly immunity to tumors from the immunity to normal tissue transplantation that concerned the major histocompatibility antigens shared by almost all body cells. Although the use of inbred strains of animals has greatly increased our knowledge of immunogenetics and of the various components of the immune system (Richards 1980) and has enabled more defined experiments to be designed, there has been little useful advance in our understanding of immune reactivity in the cancer-bearing host. It has proved difficult to demonstrate tumor-specific (or more correctly, tumor-associated) antigens in human tumors (Moore 1978a). Even in syngeneic animal systems, the demonstration of a host response to a tumor in the form of tumor-specific antibody and sensitized lymphocytes has been overshadowed by the unanswerable logic of the statement by Order (1925) that the relentless growth of a tumor in the host reflects per se the ineffectiveness of the immune system.

In Vitro Assays of Host Immune Competence

Much of our current knowledge of cancer immunology, therefore, has been obtained from manipulations of tumors that expressed strong tumor-associated antigens in syngeneic animals, and studies on transplantation resistance and hyperimmune animals have featured prominently. Much effort was expended in trying to develop in vitro assays of cellular and humoral immunity to parallel this in vivo work, in the hope that such assays could be used to assess host resistance in vivo. Investigations on the immunology of human cancer have been based on the results obtained with animal tumors. Since the in vivo transplantation approach has little application in humans, attention has been concentrated on obtaining in vitro techniques that reflected host immunocompetence. A plethora of tests has appeared for examining the integrity of various components of the immunologic apparatus, from lymphocytes and leukocytes to humoral factors and circulating immune complexes (Cochran 1978a; Moore 1978a). Many of these tests are complex and technically difficult, as well as tedious to perform and to control reproducibly; some involve subjective interpretation. Few, if any, are applicable to a number of different tumor systems. Usually, a particular test has to be modified to the system in question and may prove unsuitable, even after exhaustive examination of dose-response relationships (e.g., to ascertain the optimal quantity of ^{3}H-thymidine for lymphocyte labeling or the dose of phytohemagglutinin to employ in lymphocyte transformation testing). In microcytotoxicity assays, the results often are equivocal or require statistical analysis to record an effect that is not biologically impressive, and may in fact be unrealistic—such as when tumor cell inhibition is brought about by lymphocytes at an effector/target cell ratio of 500 or even 100 to 1. The various tests measure different facets of the immune competence of the host; no one test provides an overall view. The meaning of most (if not all) of the assays is incompletely understood, and the relevance of the results to the in vivo situation, or even the postulated mechanisms whereby the assays work, remains unproved. Even results obtained with syngeneic animal systems under well controlled, optimal conditions are increasingly being subjected to more critical scrutiny in terms of their application to the tumor-bearing host (Cochran 1978a; Moore 1978a). The existence of different types of lymphocyte (the T cell, B cell, "null" cell) is now well recognized, and there are data suggesting that the T-cell population comprises several subgroups separable according to their rosetting characteristics (Cochran 1978a). This heterogeneity of lymphocyte populations adds yet another unknown factor to the in vitro assay system. Another important consideration is that the host-tumor relationship is a dynamic one that is continually changing during the evolution and progress of a cancer in the host. During this tumor progression and the simultaneous response (or lack of response) of the host defense systems, various assays may give dissimilar results at different times (Terry 1975).

There has been in recent years an almost unseemly desire of research workers in tumor immunology to pay homage to the lymphocyte. Yet it is known that T cells, even lymphocytes as a whole, are not the sole mediators of immune defense. As well as cellular and humoral (antibody) reactions attributable to T and B lymphocytes, macrophages and polymorphonuclear leukocytes partake in host defense (Cochran 1978a). Cooperation between these various cell types is brought about by a series of complex molecules collectively known as *lymphokines*, activation products produced by lymphocytes confronted with an antigen to which they have been sensitized. About 20 such substances have been described; they include interferon and enable the activated lymphocytes to replicate and to react with and control leukocytes and macrophages. Besides being cytostatic and cytocidal to cancer cells under some conditions via toxins, macrophages also produce a range of biological materials that include hydrolytic enzymes, prostaglandins, and complement components, as well as substances that modulate lymphocyte activity (Eccles 1978). Simultaneously, therefore, many different cell types, their products and interactions, not to mention the vital role of complement itself, comprise the host defense system (Cochran 1978a; Moore 1979). The system probably has considerable reserve capacity, as pointed out by Cochran (1978b), since severe distortions of normal values observed in vitro often are not associated with infection in the host.

It is apparent, therefore, that with in vitro tests, depending on the antigens involved and the type of responses elicited, different answers can be obtained using different tests, different target cell types, and different animal systems (Stutman 1975). It may be that we have become obsessed with the concept of tumor-associated antigen. Howard (1975) has pointed out that there may be no such thing as a tumor antigen. Tumor antigen really is a "bundle of things" that should be viewed more as complex multiple changes in the tumor cell surface rather than as a single sought after chemically defined entity.

The discipline of tumor immunology adopts as its central working hypothesis the concept that tumors are antigenic in the hosts in which they arise. This has led over the past 20 years to the development of a growth industry devoted to detecting tumor cells and to characterizing antigens that might form the basis of cancer therapy by manipulation of the immune system. As pointed out by Klein (1978), the concept that tumors possess unique antigenic molecules gained surprisingly rapid acceptance in view of the considerable volume of earlier negative studies. Most of the results of the current era of tumor immunology are derived from artificial systems with chemically or virally induced tumors selected for their high antigenicity and immunogenicity in inbred animals. Spontaneous tumors rarely demonstrate such convenient properties (Moore 1978b), and there is no reason why all tumors should necessarily be recognized by the immune system. In addition, and perhaps most important of all, humans are allogeneic hosts. Consequently, the present discipline of tumor immunology has been established on "somewhat shaky foundations" (Klein 1978). The increasingly skeptical climate regarding the significance of results obtained in tumor immunology in recent years is gradually focusing attention on ideas recognized and accepted by earlier workers: namely, that a variety of host responses act against cancer and that all of these are not immunologic. Although such nonimmunologic aspects of host defense are difficult to investigate, the importance of the mononuclear phagocyte system (macrophages) is arousing renewed and escalating interest. The earlier pathologists were well aware of the potential of this cell type in host defense mechanisms.

These various considerations should be borne in mind in designing or evaluating experiments relating to hyperthermia and the immune system.

In Vivo Assays of Host Immune Competence

The demonstration in some animal systems of tumor-associated antigens led to the hope that a host would react to a tumor by mechanisms similar to those operating in normal tissue allograft rejection, and that cell-mediated reactions accordingly would be more important than humoral immunity. In many of the in vitro immunologic tests involving confrontation between tumor cells and lymphocytes, the effector cells can be replaced by, or supplemented with, host serum. This approach has featured prominently in studies on blocking antibody, for example. Defining the antitumor specificity of antiserum has proved difficult, however. In addition, the role of antitumor antibody in vivo remains unclear. Consequently, for in vivo work as well as in vitro testing, more emphasis has been placed on assessing cell-mediated function than on the humoral component of the immune system. It is apparent that if even a modest number of the numerous phenomena detected in vitro (ranging from cell [lymphocyte]-mediated cytotoxicity and migration inhibition by leukocytes to complement-mediated and antibody-dependent cytotoxicity) operate in the host, then any attempt to assess host response to cancer must take serum factors into account. The current state of the art in relation to the numerous tests available and their technical difficulties, as well as problems of interpreting the results and the unknown relevance to the in vivo situation, make attempts at such an approach somewhat daunting.

Theoretically, in vivo assay procedures present the opportunity to avoid some of the problems of in vitro testing, and to assess the numerous factors comprising host response. A positive skin reactivity test, for example, requires that the various components of the cellular immune response are intact and function sequentially and in the correct order. Skin testing has been the most popular test of host cellular reactivity because of its technical simplicity and ready application in the clinic. The test measures host-delayed hypersensitivity to intradermally applied antigen, which may be tumor derived or foreign. The reaction is essentially similar to the classical tuberculin reaction for testing hypersensitivity to tubercle bacilli, and a positive response is indicated by development of a palpable area of erythematous and sometime vesiculated induration at the injection site after 24 to 48 hours. Positive responses to solubilized (KCl) extracts of tumor have been obtained in syngeneic animals rendered immune by tumor inoculation (Meltzer et al. 1972; Oettgen et al. 1968) or surgical removal of tumor (Churchill et al. 1968; Wang 1968), although few sequential studies have been reported for animals bearing progressively growing tumors. In humans, a proportion of patients with different types of cancer respond to both autologous and allogeneic tumor material (Cochran 1978b). Careful attention to the use of control antigen and to tissue extraction procedures are required for reliable interpretation of results (Cochran 1978b). Currently, too few sequential studies have been done to know the extent to which this type of response relates to status of disease or prognosis in cancer patients.

Foreign antigens most commonly employed are recall type allergens that test immunologic memory of previous host exposure to the substance in question. They comprise semipurified extracts of bacteria, fungi, and viruses of widespread occurrence and include *Mycobacterium tuberculosis*, mumps, pertussis, trichophyton, some of the mycoses such as *Candida*, histoplasmosis, and coccidioidomycosis and *Streptococci* (streptokinase-streptodornase). In selecting suitable antigens for this type of test, due attention must be paid to local practices regarding prophylaxis, since host response depends upon previous exposure to the allergen (Cochran 1978b). A considerable number of studies on recall antigen responses in cancer patients have now been reported. The general consensus from the results is that patients with early cancer have no major deficit of immunologic memory, but that a decline in this type of activity occurs in advanced disease (Cochran 1978b).

Host response to new antigens is very important if it is accepted that tumor cells express antigens not otherwise encountered in adult life. For such testing it is a prerequisite that the majority of the normal population should not have encountered the antigen. A variety of such chemical and biological substances has been employed for skin contact testing (Cochran 1978b), and the small organic molecule dinitrochlorobenzene (DNCB) has received most attention. The test is not open to the criticism that the agent may vary in purity and composition as can bacterial extracts, and a simple and convenient semiquantitative form of the assay enables host response to be graded. (Bolton 1975).

As it is a previously unexperienced antigen, the DNCB assay involves injection of a sensitizing dose followed 10 to 14 days later by challenge dose(s) (Bolton 1975). Almost 100% of all normal individuals react to DNCB within 48 hours of challenge (Pinsky 1975). The test has been used most in humans, and cancer patients often exhibit a decreased reactivity that becomes more marked with progress of the disease. The response pattern depends not only on the extent of the disease, but on the particular type of cancer present. In patients studied sequentially, those who failed to respond to DNCB (or tumor extracts) had, in general, a poor prognosis compared to patients who responded well (Bolton 1975; Fass, Herberman, and Ziegler 1970; Harris and Sinkovics 1976; Leventhal et al. 1972; Morales and Eidinger 1976). In cancer patients after surgery (Eilber and Morton 1970) or undergoing chemotherapy (Hersh et al. 1974), prognosis was found to be related to the level of general immunocompetence (as measured by foreign antigen skin testing) and also to the level of specific antitumor immunity—the more vigorous the state of competence, the better the prognosis. In a comparative study on a large series of patients with a variety of solid tumors, Pinsky (1975) found that DNCB response correlated well with prognosis after surgery, while reactivity to any one of a battery of recall antigens did not correlate well. As emphasized by Cochran (1978b), improvement in prognosis as determined by skin sensitivity testing is only relative, and the evaluation must be interpreted in association with other significant prognostic factors, such as tumor staging and special examinations like x-ray and scanning procedures.

Humoral Factors

Several serum components have received attention in the hope that their measurement might prove rewarding for monitoring host immunity. Although quantitation of serum immunoglobulins and complement components is technically well established, the proportion of IgG that specifically reacts with a tumor is unknown and probably is very small. No single abnormality of the complement system has been found to be characteristic of tumor-bearing hosts (Cochran 1978b), although the tenet that interaction between tumor-associated antigens and at least some of the antibodies produced should cause activation of the complement system, and consumption of complement components is not unreasonable. Intercurrent infection, impairment of liver function, and tumor necrosis may cause abnormalities of these parameters. This absence of any consistent abnormality and the susceptibility to nonspecific alteration has rendered IgG and complement system measurement of little value to date as immunologic monitors.

More recently, hopes have revived of detecting antigen-antibody interaction in the form of their product, antigen-antibody complexes. There is evidence that in both animal and human tumor-bearing hosts such circulating immune complexes exist, and that their quantity correlates with stage of disease (Cochran 1978b). The use of exogenous $C1_q$ component of complement to detect these complexes in serum may prove of value as a monitoring test (Baldwin, Byers, and Robins 1979; Baldwin and Robins 1980; Höffken et al. 1978).

Free Circulating Antibody
Serum antitumor antibody is best detected in animal tumor systems following tumor amputation, or when irradiated tumor cells have been inoculated into the host after tumor amputation. Circulating tumor-directed antibodies have been detected in cancer patients (especially in patients with melanoma, sarcomas, and Burkitt's lymphoma) by a number of workers (Cochran 1978b). In general, however, antitumor antibodies are difficult to detect in tumor-bearing animals, and defining their antitumor specificity is an equally arduous task. Besides the classic cytotoxic antibody that lyses tumor cells in the presence of complement, importance is presently attached to antibody that sensitizes tumor cells for cell killing by a variety of cell types, including macrophages, granulocytes, lymphocytes, and the so-called *K cells* (killer lymphocytes). This constitutes antibody dependent cell-mediated cytotoxicity. Other types of antibody are blocking antibody, which blocks the cytotoxicity of both T and B lymphocytes involved in tumor cell killing (first described by Hellstrom [Richards 1980]) and antiantibodies. These latter have been suggested by Lewis and Abtekman (1952) to account for the decline of detectable antitumor antibody in advanced disease (Cochran 1978b).

Available information on antibodies is derived in large measure from in vitro experiments, and, as with cell-mediated toxicity, the data is governed by the particular details of how the assay is performed. Since not all antibodies are cytotoxic, and it has been found in humans that antibodies of different specificities may characterize different stages of disease (Cochran 1978b), the uncertain value of measuring the titer of even a relatively specific antibody is apparent. The situation is complicated further by the known poor penetration of antibody into solid tumors in vivo (Witz and Ran 1970).

Antitumor antibodies have been most easily detected and usually have been present at highest titer, in the serum during early disease when tumor volume is small; advancing disease is accompanied by disappearance of such antibody. Human melanoma and sarcoma have been intensively studied immunologically, and are often regarded as the classic examples illustrating variation of antibody titer with stage of disease. Antibody usually can be detected in patients with early disease. The titer decreases progressively with advancing disease, but an increase may occur following reduction of host tumor burden by tumor amputation, for example (Leventhal et al. 1972). It has been held that in the course of melanoma, metastases do not occur as long as the cytolytic titer of antibody is high, but that dissemination coincides with decreasing titer of cytolytic antibodies (Sinkovics 1979a). Melanomas have multiple neoantigens, both within the cell and on its membrane, and these induce multiple antibody formation. The gamut of antibody-mediated immune reactions, from complement dependent cytolytic antibody to blocking antibody and antibody-mediated and cell-mediated toxicity, directed against melanoma cells in vitro has been recognized (Sinkovics 1979a). A clear distinction between tumor-associated melanoma antigen and fetal antigens has not been accomplished (Seibert et al. 1977), however. It is recognized that in melanoma, regression at one site may accompany tumor growth at other sites. So the immune relationship between host and tumor is intricate and complex and is unlikely to be reliably assessable solely on antibody measurements. With the other intensively studied human tumors, sarcoma and Burkitt's lymphoma, various cell-associated antigens and humoral responses have been described (Moore 1979). Since there is strong evidence favoring a viral etiology of both these cancers, the humoral responses may not be strictly to tumor antigens but to neoantigens resulting from interaction of a virus with the cells.

Nonspecific Antibody
Host general immunocompetence can be measured by the ability to form antibody to foreign (nontumor) protein. For this, host response to a previously encountered antigen (secondary immunization) or to a new antigen (primary immunization) can be assessed; tetanus toxoid and *Salmonella* antigen have proved

suitable for these respective approaches in humans. Several studies using these antigens were performed on cancer patients in the 1950s and 1960s and gave a somewhat conflicting picture (Makay 1975). Some workers found that cancer patients produced a similar antibody titer to the control group, and even ill and debilitated patients with advanced disease had a normal antibody production. A few groups reported some reduction in antibody titer in patients with active, especially advanced, disease; follow-up showed that patients with low antibody titer often had a poor prognosis. Normal antibody production occurred in patients who had had cancer in the past.

Tests of this type have found little application in recent years.

Summary

The immune mechanisms whereby tumor cells may be damaged or destroyed are theoretically diverse. Several different cell types and subpopulations and their products and several forms of antibody may constitute the immune response, which is coded by distinct immune response (Ir) genes separate from those controlling transplantation antigens. Current limited knowledge of effector mechanisms has evolved from numerous in vitro tests; their relevance to the in vivo situation is unproved. Although the immune system is no longer a simple antigen-antibody response to foreign protein, this concept retains a major role embodied by circulating immune complexes. Failure of antigen binding in this way leaves the antigen free to swamp the system and interfere (at least in vitro) with various antitumor activities. On the other hand, inactivating antitumor antibody in the form of antigen-antibody complexes may deprive the host of a defense against circulating tumor cells. It is self-evident that any investigation of tumor immunity must attempt to assess both cell-mediated and humoral arms of the immune response.

A central theme of studies on tumor immunity has been the great difficulty in proving the antitumor specificity of the responses. The integral role of the macrophage as mediator in these responses, and the considerable capability of the macrophage as both a nonspecific and specific tumor cell killer, emphasize further that host defense against cancer does not necessarily equate with immune response. Another contender as a participant in host defense is the inflammatory response, and there are indications that this response, and also the activity of the mononuclear phagocyte system, is important in tumor regression after hyperthermia.

At present, few investigators attempt to make a comprehensive assessment of the integrity of host defenses as defined above. This is chiefly because we do not know the significance or value of many of the assays, and the in vitro tests require considerable and meticulous attention to detail to surmount the technical problems and obtain reproducible results. Research groups therefore tend to concentrate on one type of technique, and this is self-defeating in the effort to obtain an overall appraisal of host defense. Since there are no immunologic guidelines, it remains for individual workers to establish what appear to be the most appropriate and informative approaches for evaluating and monitoring host response in a particular tumor system. In spite of these unsatisfactory foundations, it appears that adequate in vivo assays and manipulations are available to answer the question of whether the immune system is involved in tumor response to hyperthermia and, even more important, to define the extent to which host defenses are implicated in the treatment outcome.

Animal Tumor Systems as Experimental Models for Human Cancer

The usefulness and limitations of tumor transplantation in cancer research have been extensively reviewed and discussed by several authors (Liebelt and Liebelt 1967). More recently, Clifton (1979) has traced the historical development of rodents as experimental models from their use by the seventeenth century chemist-physiologists to their present commanding position in biological research and virtual monopoly as tools for in vivo cancer research. It has been calculated that since 1970 over 95% of all animal experiments performed annually employed mice or rats, rats being used half as frequently as mice (Cochran 1978a). Clifton has specified the major criteria for an ideal experimental model of cancer. The enumerated guidelines (Cochran 1978a) include:

1. The model species should be inexpensive, readily available, and easy to obtain.
2. The disease should occur frequently and spontaneously or be readily inducible.
3. There should be considerable uniformity among model individuals in development of the disease, spontaneous or induced, and/or in the response of the disease to experimental manipulations.
4. In causation, progression, response to manipulation, and other aspects, the disease in the model species should be identical to that in humans—an ideal "hard to predict and rarely achieved."
5. Genetic homogeneity is desirable or essential for studies on transplantation or immunology.

It is readily seen why rats and mice have achieved wide acceptance as models, admirably fulfilling criteria 1, 3, and 5. Most types of cancer do not spontaneously occur in rodents with high frequency, and the use of induced tumors has been strongly criticized by Hewitt (1978). Criterion 2 essentially cannot be fulfilled, therefore. Although our current ideas of tumor pathogenesis owe much to Fould's (1969) concept of tumor progression, which evolved from studies in mice, and similarities to the human disease of hormone responsiveness in mammary tumors have "established the rat as the current queen of experimental models in this area" (Cochran 1978a), criterion 4 is the weakest link in the chain. Mammary tumors in the rat are chemically induced, and there is no evidence of such a connection in human breast cancer. The majority of rodent tumors do not resemble human cancer in site, in doubling time, or in response to manipulaton. Most rodent tumors are transplanted to easily accessible sites (foot, leg, flank), and many show a remarkable response to cytotoxic drugs, with host cure even when the tumor burden is large. Few, if any, of the common solid tumors of man can be cured by drugs. Although "mice, rats and humans share more attributes than those in which they differ" (Cochran 1978a), and this may justify the use of rodents to study mechanisms of cell regulation, it remains that spontaneous tumors of the internal organs are very rare in rats and mice. Most of the animal tumors in current use exhibit uniformity of behavior, as required by criterion 3, for volume of material and ease of study. This seemingly favorable point may be self-defeating, however, since a striking characteristic of human cancer is the unique and unpredictable nature of histologically similar tumors with regard to rate of growth and rapidity and site of spread.

Criterion 4 encompasses the relation between tumor and host. Too little attention has been given to cancer as a disease of the host; until as late as the 1950s tumors still were considered to be autonomous of their host. In the early 1950s, Greene (1951) reported a series of results from meticulous studies with animals, clearly supporting the sometimes ignored truism that a tumor-bearing host differs biologically and biochemically from its normal counterpart. Greene adduced evidence indicating that a tumor conditions the host in preparation for its spread and metastasis (Greene 1951; Greene and Newton 1948), and in 1956 Greenstein emphasized the biochemical aspects of this relationship by stating that "the key to the cancer problem lies in the host-tumor relationship." There is a substantial body of data detailing the multiple changes that occur in an animal host bearing tumor (Begg 1958; Greenstein 1954; Liebelt and Liebelt 1967). This emphasizes the spurious nature of results obtained from intravenous injection of cancer cells into normal animals. It also indicates the artificial nature of all transplantation experiments, since normal hosts are confronted with a large and usually overwhelming number of cancer cells, and are subjected to sudden and dramatic changes in a relatively short period of time following tumor inoculation, as contrasted to the slower and more subtle alterations that occur during spontaneous tumorigenesis. In the reawakening of enthusiasm for cancer immunotherapy over the last few years, the host-tumor relationship has often been ignored, and so-called immunotherapy experiments have concerned means of inhibiting or modifying the "take" of tumor cells in normal hosts, rather than studies in animals with an increasing tumor burden of primary tumor and metastases, or hosts with minimal residual tumor following therapy. That is, the experiments have concerned *immunoprophylaxis* and not *immunotherapy*. Similarly, too much of the work on animals has concentrated on cure of the *tumor*, which is not equated with cure of the *host* when metastasis is involved.

There also is evidence that the host component of the tumor-host relationship differs in animals and humans in significant ways. Tumors of the mouse and rat generally are sensitive to heat (Dickson 1977); hamster tumors (with the exception of melanomas) are unaffected, as judged by transplantability, by in vitro temperatures as high as 45°C for one hour, while the Rous sarcoma of fowl is resistant to heating at 50°C (Gericke et al. 1971). The reports of early workers on hyperthermia (Rohdenburg and Prime 1921; Westermark 1927) mention that mice and rats rapidly succumb when body temperature is elevated. In detailed studies on rats, Dickson and Ellis (1974) found that a body temperature of just over 41.6°C represented the 50% lethal dose of heat, and few rats survived a temperature of 42°C for one hour; there was no difference between normal and tumor-bearing rats in susceptibility to heat. This susceptibility to raised temperature circumscribes the use of rodents even for experiments on local hyperthermia, since the blood flow through the heated tumor can easily be large enough in relation to the host's small blood volume to cause a rapid increase in body temperature (Dickson and Ellis 1974). It therefore can be difficult (depending on anatomic location) to heat even small tumors in rodents for curative periods of time without subjecting the host "soil" to temperatures above normal. Elevation of body temperature in rabbits also presents difficulties, restricting experiments on total-body heating. It is difficult to achieve stable thermal gradients in these animals (Dickson and Muckle 1972). Oscillations in temperature have been reported after administration of pyrogens to rabbits, cats, and dogs, while such fluctuations occur less frequently in humans during pyrogenic fevers (Hardy 1961). In addition, rabbits will not withstand hyperthermia (local or total body) for times in any measure approaching those that at present are required to induce regression of human tumors (Calderwood and Dickson 1980; Dickson and Muckle 1972). Vermel and Kuznetsova (1970) emphasized that, since the thermoregulatory mechanisms in animals are different and less well developed than in humans, the results of hyperthermia experiments on laboratory animals may give little indication of the response when a human is the host. A further point in relation to total-body heating is that during rapid changes of body temperature, the small thermal gradients normally operative between tissues and organs in the animal body can increase considerably (Dickson and Muckle 1972). Efficient total-body hyperthermia requires stabilization of thermal gradients, and the time necessary for this to occur will depend upon the thermoregulatory mechanisms of the host and the heating conditions. The fluctuations in deep temperature encountered in the rabbit during radiant heating may be a reflection of insufficient time (owing to host intolerance) to stabilize thermal gradients (Dickson and Muckle 1972).

Hewitt (1978) has recently provided data indicating that the host-tumor relationship obtained in most animal tumor systems is artificial, and that the current structure of tumor immunology may have been erected on a false foundation. Hewitt has performed an exhaustive review of the tumors used in cancer research and points out that the vast majority were either chemically or virally induced, and this includes the pioneering experiments of Prehn and Main (1957) that initially demonstrated the occurrence of tumor-specific antigen. The presence of antigens that have been induced by chemical or viral means has two serious and misleading consequences. From the immunologic point of view, the host response is artifactual, and second, such immune resistance may make a considerable contribution to tumor cure by a therapeutic agent, be it drugs or immunostimulation. In a series of 27 spontaneous tumors transplanted into inbred mice over a period of 20 years, Hewitt has never demonstrated a host immunologic reaction that had any restraining effect on the growth of primary implants or metastases disseminated from them (Hewitt, Blake, and Walder 1976). There is a substantial body of evidence from other laboratories supporting the conclusion that for spontaneous tumors immunogenicity is scarcely ever demonstrable (Moore 1978b). Because of their artifactual immunogenicity and long history of indiscriminate transfer from one laboratory to another, the familiar veteran allogeneic tumors, such as Walker carcinosarcoma in rats and Crocker tumor in mice, should be abandoned as models of naturally occurring cancers (Hewitt 1978). Hewitt also emphasizes that there are numerous subtle sources of artifactual tumor immunogenicity that affect even nominally spontaneous tumors in isogeneic animals. These include inconsistencies in substrains of host carrying the tumor in different laboratories, genetic drift, accidental virus infection, and maintenance of the tumor cells in tissue culture for in vivo - in vitro studies. These hazards increase the longer a tumor has been in use from its date of origin.

With these considerations in mind, Hewitt (1978) analyzed in detail 19 tumors of spontaneous origin currently being used in cancer research. He found that 13 of the tumors possessed features rendering them unsuitable as tumor models. Even more sobering is the further calculation that over 90% of the tumors currently used in experimental therapy research are of very questionable status as models of spontaneous autochthonous cancer. Hewitt (1978) has proposed that to be a valid model of human cancer, an animal tumor should be of spontaneous origin, have a transplantation history of less than 15 years, and be transplanted within the substrain of origin. As a test for tumor immunogenicity, Hewitt suggested this should be determined from a rise in TD_{50} (number of tumor cells required for 50% successful transplantation) after immunization of recipients with lethally irradiated homologous tumor cells.

There is no doubt that Hewitt's findings and views crystallize much of the unease and dissatisfaction with the present state of tumor immunology, and deserve serious and soul-searching consideration. As accepted by Hewitt (1978), most of the frankly allografted tumors, or those of otherwise unspecified genetic constitution, arose in familiar laboratory species before inbred animal strains were developed, and their continued use is due largely to a reluctance to switch ongoing research programs based upon an accumulated wealth of data to the use of animals and tumors with which workers are unfamiliar. The breeding program and safeguards required to fulfill Hewitt's criteria for a valid animal tumor model are formidable, and would best be attained in large institutions. They represent standards difficult to achieve for short-term projects by workers with limited animal facilities.

Hewitt's belief that only nonimmunogenic tumors are acceptable as models of naturally occurring cancer, while serving to emphasize the general unsuitability of chemically or virally induced animal tumors may be an extreme view. In extensive investigations on a large number of tumors in syngeneic rats, Baldwin, Embleton, and Pimm (1979) compared spontaneous tumors with carcinogen-induced tumors of similar histologic types. They concluded that antigenic differences between the two categories of tumor were quantitative rather than qualitative. The tenet that spontaneous tumors lack tumor rejection antigens probably is not correct, therefore, but may simply reflect the inadequacy of present technology to detect such antigens and/or host response to them. There is also a large body of data indicating that in humans a host response to cancer does occur, and that most human cancers are antigenic in the autochthonous host (Cochran 1978a; Currie 1980; Harris and Sinkovics 1976; Moore 1978a, 1979; Richards 1980; Sinkovics 1979a). The current dilemma has arisen because evidence for human tumor-associated antigens is still largely circumstantial or has been derived from in vitro studies of tumor immune reactions, and so cannot be firmly correlated with tumor rejection (Pimm and Baldwin 1978). Second, even transplanted tumor cells within highly inbred rodent strains may not be in the same environment as in the host of origin. The primary causal condition that induced cancer initially may have been an endocrine imbalance or some other host dysfunction not present in other animals of the strain (Clifton 1979). Established transplantable tumor lines, no matter how "pure" and even when less than 15 years from their origin, are still the products of generations of cell selection in passage, with the attendant and inevitable progression. The characteristics of transplantable tumors do not remain static but exhibit periodic changes of growth rate and other parameters indicative of an altering host-tumor relationship. Such transplantable tumors, therefore, cannot be regarded as primary tumors. Third, because neither animal tumors nor animal hosts can be equated with human cancers or human beings, it is unlikely that the use of nonimmunogenic tumors as favored by Hewitt will prove any more rewarding from the point of view of human tumor therapy, although such use may reorient our approach to the human situation.

Investigations concerning hyperthermia have used the gamut of mouse and rat tumor systems as an extension of the availability of particular systems to individual workers. Mice are cheaper and more easily housed than rats; there is a much wider range of available strains of mice, including many having a high incidence of various naturally occurring malignancies (Hewitt 1978). On the other hand, the much larger size of the rat has numerous technical advantages, as well as reducing markedly the unfavorable comparison between mouse and man of tumor weight: host weight ratio. With large tumors in the mouse, this ratio increases to the extent where the host-tumor relationship is altered by gross host depletion of nutrients, rendering interpretation of data on tumor growth rate and response of tumor and host to therapy unreliable (Hewitt 1976). Rodents have disadvantages as subjects for heating by both local and total-body methods as outlined earlier. Rabbits represent a host of intermediate size that can be subjected to both local and systemic heating. Inbred strains do not exist, however, and tumors of the rabbit are very few in number. The animal also has very limited tolerance for whole-body heating (Dickson 1977; Dickson and Muckle 1972). Dogs and pigs have been used to obtain data on the physiology of hyperthermia (Dickson, MacKenzie, and McLeod 1979; Hahn et al. 1980) and for heating studies on spontaneous tumors in dogs centralized to a particular unit for veterinary care (Gillette, in press; Hahn et al. 1980). The fur coat of dogs makes control of body temperature difficult in systemic hyperthermia studies; the best animal subject for such studies is the pig. Although the pig does not sweat in response to thermal stimuli, its skin otherwise resembles that of humans, and similarities in body size and function are great enough to allow useful comparison to be made between the two species (Mount 1968; Mount and Ingram 1971). The pig tolerates prolonged total-body heating, and its temperature responses are very similar to those of humans (Dickson, MacKenzie, and McLeod 1979). The major limitations of larger species for experimental work, in addition to those enumerated, are the cost of purchase and maintenance, the scarcity of inbred strains, the spontaneous incidence of cancer is often low, and techniques for tumor induction have not been developed. Because of these various considerations, it is apparent that rodent tumor systems will hold the stage in hyperthermia studies, as well as in cancer research generally, for some time to come.

Summary

Chemically and virally induced tumors in rats and mice are usually immunogenic and, in studies of therapeutic agents, the results obtained may represent the combined effects of direct action of the agent and the artificially induced host resistance. Because of their induced (or uncertain) origin and a long and checkered history, the veteran allograft tumors, such as the Walker-256, should be abandoned in favor of spontaneous or weakly antigenic tumors in inbred animals. It is hoped that such animals represent a more defined system in which to investigate the tumor-host relationship and the effects of therapy. Weakly antigenic tumors, rather than either extreme of strongly antigenic or nonantigenic cancers, may be more akin to the human situation. The contribution of the veteran tumors and other immunogenic cancers must not be underestimated, however. Only by their use have we evolved techniques for detecting antigenicity and developed an awareness of factors that may cloud the issues in our studies of the host-tumor relationship.

A few other considerations must be kept in mind. There are human solid tumors for which a chemical induction has been established—notably cancer of the lung and bladder. Although well-defined tumors in syngeneic animals are appealing to the experimentalist, it must not be forgotten that humans are allogeneic hosts. There are only a few animal tumors (spontaneous mammary cancer in rats, osteosarcoma in dogs, for example) that in their characteristics and behavior resemble solid tumors in humans. The concept of animal tumor systems as models of human cancer, although a convenient one, implies a similarity that does not necessarily exist. No matter how attractive the tumor model, animals are not people, and great caution must be exercised in extrapolating results obtained in animals to the human situation. Rather, animal models should be regarded as useful for providing indications and guidance for our approach to investigations in humans. In a less than ideal world we must cut our coat to suit the cloth, and endeavor to interpret available data. As will become evident, our evaluation of data on tumor hyperthermia and host response is not only in keeping with the tenor of this discussion, but illustrates how our concepts of host defense and the tumor-host relationship are expanding beyond the confines of immunology and tumor antigens, which have monopolized our thinking for too long.

Adequacy of Tumor Heating

Strauss believed that, following tumor heating, tumor breakdown products entered the bloodstream and stimulated a host antitumor immune response. In work on electrocoagulation of the Brown-Pearce carcinoma in rabbits and carcinoma of the rectum in humans, Strauss and co-workers found that destruction of the whole tumor mass was not essential for tumor regression after heating (Strauss 1969; Strauss et al. 1965). They further observed that, if the Brown-Pearce tumor was transplanted into normal rabbits seven days after electrocoagulation, no growth occurred, although the injected material contained viable appearing cancer cells. They postulated, therefore, that 100% tumor destruction was not necessary for tumor regression after heating, but they did not succeed in quantitating the response. Shah and Dickson (1978a), again with a rabbit tumor, the VX-2 carcinoma, found that following curative tumor hyperthermia at 47°C to 50°C for 30 minutes by radiofrequency current, the VX-2 still grew on transplantation into fresh hosts for up to 17 days post-heating. The 17 days corresponded to the period during which host immune competence was increasing. Host animals with disseminated disease were cured by the treatment, and the authors concluded that an immune response was involved in regression of both primary and secondary tumors. Other workers also have found that a heat treatment that is inadequate to kill all the cells in a tumor may result in sterilization if the tumor is left in situ. Marmor, Hahn, and Hahn (1977) reported that after curative heating of the EMT-6 sarcoma in mice at 43°C to 44°C, survival of cells from the treated tumor, as determined by plating efficiency, was 10% to 80%. These workers concluded that direct cell killing by the heat could not account for the tumor cures, and that delayed mechanisms, either intratumor or involving host response, must play a major part in tumor eradication. With S180 tumors on the feet of mice, Crile (1963) found that following curative heating the tumor still grew when transplanted to fresh hosts within four hours after heating. Crile believed the four-hour delay reflected the time required for an effective antitumor inflammatory reaction to develop. With a fibrosarcoma growing in the footpad of mice, Suit (1977) found that a 7-mm tumor was cured by local hyperthermia at 43.5°C for 78 minutes. The tumor cells required a four-hour heating in vitro at this temperature to ensure no takes on transplantation into fresh hosts. It was concluded that the observed destruction of the growing tumor was not achieved by direct thermal killing of all cells. The literature indicates that tumors generally are more difficult (require a longer time) to inactivate by heat in vitro than in vivo. The in vitro data in this respect require careful evaluation, however, since it has become apparent in recent years that several factors, including pH and the presence of serum, may influence the response of cells to heating; standardization of incubation conditions is therefore especially important (Dickson 1977).

Host response to tumors in the rabbit involves in large measure a nonspecific reaction to histocompatibility (transplantation) antigens, and the evidence for an immune response in the results of Strauss and colleagues is largely circumstantial. Even when the tumor-lethal temperature is known from in vitro studies, as in the case of the VX-2 carcinoma (Dickson and Shah 1977), elective partial destruction of a tumor presents hazards as well as technical problems.

The difficulties of heating a tumor uniformly to a given temperature are now well recognized. A range of 0.5°C to 1.5°C may be recorded between three temperature sensors placed in a 1-ml heated tumor, and differences of 3°C or more can occur between multiple sensors in larger tumors (10 to 15 ml), animal (Dickson and Shah 1977, 1979) or human (Dickson and Shah 1979; Moricca et al. 1977). The precise temperature gradients will depend upon tumor type, size, site, and blood flow, as well as on the method of heating and temperatures achieved (Dickson and Shah 1979). Because of this nonuniformity of heating, the most feasible approach is to heat the tumor until the lowest-reading sensor registers the desired (the tumor-lethal, if possible) temperature, and then maintain this status quo for the desired heating period; several workers use this modus operandi (Dubois 1949; Dickson 1977; Dickson and Shah 1977). Water-bath heating usually is accepted as giving the best approximation to uniform tumor heating (Dickson 1977; Dickson and Muckle 1972; Robinson 1978). In a direct comparison of different heating techniques with the same tumor in rabbits, Dickson and Shah (1977) found that surrounding the tumor-bearing limb in a watercuff gave much smaller intratumor temperature gradients than heating by electrocoagulation, heat pipe, or radiofrequency current. Water-bath or cuff heating is limited by the danger of skin damage from the surrounding hot water required for adequate caloric input to overcome the heat loss via the circulation. The method has little application in human patients.

Even by the use of a minimum target for the lethal temperature in tumors, elective partial heating of the mass implies a decreasing temperature gradient across the unheated tumor. This represents a potential hazard. There is evidence from animal tumor studies of a temperature-stimulatory zone from 37°C to 41°C that precedes the tumor lethal temperature, usually regarded as beginning at 42°C. At temperatures in this transition zone, metabolism and growth rate of cancer (and normal) cells may be stimulated, as expected, on kinetic grounds (Dickson 1976, 1977; Dickson and Ellis 1974; Dickson and Muckle 1972; Dickson and Suzangar 1974; Robinson 1978). Enhanced tumor metastasis has occurred in some animal systems under these conditions (Dickson 1976, 1977; Dickson and Ellis 1974, 1976).

There is data also from studies on the treatment of tumors with radiation or chemotherapy that the biology of cells not destroyed by therapy becomes altered in a way that endangers the host. Sublethal tumor irradiation, both experimental and clinical, has been observed to promote the development of local and distant metastases (for review, see Fisher and Fisher 1964). Systemic cytotoxic drugs also have been reported to produce such an effect. Kondo and Moore (1961) demonstrated that experimental lung metastases were increased following administration of nitrogen mustard or actinomycin D, and Moore has emphasized repeatedly that chemotherapeutic agents in improper dosage can augment tumor growth (Fisher and Fisher 1964). Fisher and Fisher (1967) also reported an increased incidence of metastases in animals following local controlling doses of radiation. Tumor cells surviving the effects of radiation exhibited features of anaplasia when viewed by light and electron microscopy. The tumors also had augmented growth rates when transplanted to fresh hosts. Similar findings were noted by these workers following treatment of the Walker tumor in rats by isolated limb perfusion with nitrogen mustard at 35°C to 40°C. There was inhibition of tumor growth without increase in survival time, and the drug-perfused animals died with extensive metastases (Yates and Fisher 1962). When tumors were removed from the rats after treatment and inoculated into fresh animals, there was a significantly increased incidence of lung metastases compared to rats injected with control tumor. In addition, many of the hosts inoculated with drug-perfused tumor cells developed metastases in the lung while failing to grow tumors at the primary implantation site in the leg (Fisher and Fisher 1964). The results suggested that the major factor in the aggressive behavior of the tumors after treatment was an alteration in the biological properties of the cancer cells rather than host alterations.

Hyperthermia, like radiotherapy and cytotoxic drugs (Steel 1977), has a preferential damaging effect on replicating tumor cells, which is followed by recruitment of nondividing cells into the cell cycle (Calderwood and Dickson 1980; Dickson and Calderwood 1976). The timing of recovery of the tumor population has been used to plan fractionated therapy regimens (Dickson and Muckle 1972; Muckle and Dickson 1971; Steel 1977).

In their extensive work on regional hyperthermic perfusion for tumors of the limb in humans Cavaliere and co-workers (1967), Moricca and colleagues (1977), and Stehlin (1976) and associates (1977) found local tumor recurrence to be a major problem with all tumor types. Cavaliere solved the problem by amputation of the limb a short time after perfusion (Moricca et al. 1977), and Stehlin and co-workers improved their results by an aggressive program of postoperative irradiation and delayed local excision of the tumor (Stehlin 1976; Stehlin et al. 1977). From detailed pathologic studies on the heated cancers, Cavaliere and colleagues (1967) reported that only those patients who had total gross and histologic destruction of the tumor sustained lasting and complete remissions without recurrence. Although in some cases the histologic effects were immediate while in others they were delayed, in every case in which viable-looking tumor cells remained after heating, the histologist was able to forecast tumor recurrence. Similarly, in the animal tumors subjected to detailed histologic study after hyperthermia, all sections examined have shown virtually 100% tumor cell destruction in mouse (Overgaard and Overgaard 1972, 1976), rat (Copeland 1970; Dickson and Suzangar 1974), and rabbit (Dickson 1977; Muckle and Dickson 1971) tumors that have been cured.

Summary

Although the concept of subtotal destruction of a tumor, leaving the immune system and/or other host defense mechanisms such as the inflammatory response to deal with the remainder, is an intriguing one, our understanding of such processes is so primitive that any elective approach to the partial heating of human tumors can only be regarded as a recipe for disaster. Evidence from both animal and human studies indicates that cancer cells that are not destroyed by therapy may become altered biologically in a way that is hazardous to the host in terms of accelerated growth rate and/or enhanced metastasis. Although sublethal heating of tumors or regions of tumors undoubtedly occurs, owing to our current inadequate knowledge of tumor-lethal temperatures and technically unsatisfactory methods of hyperthermia, our goal *must* be total tumor destruction.

MATERIALS AND METHODS

It is now being increasingly recognized by experimentalists and clinicians alike that cancer is a systemic disease (Fisher and Fisher 1976; Mathé et al. 1980), and that the pathogenesis of a given neoplasm is governed by a balance operating between the growing tumor and the reaction of the host (Dickson and Suzangar 1976). The host-tumor relationship encompasses an interdependence of a primary tumor and its metastases (Dickson 1977; Dickson and Suzangar 1976) so that removal or other treatment of a primary tumor results in systemic alterations in the host and kinetic changes of metastases. In our experimental work, we have endeavored to take account of these considerations by using only tumors that metastasize. Metastasis is not nearly so conspicuous a feature of cancer in experimental animals as it is in humans. (Foulds 1969; Stewart et al. 1959), but to obtain results that may be meaningful for our approach to human cancer therapy, the models should replicate as closely as possible the disease in humans. Prior to establishing the experimental protocol, the tumor system must be carefully standardized in terms of host animals used (strain, sex, weight), number of viable tumor cells inoculated, site of inoculation, growth characteristics of the tumor, and pattern of metastasis (Dickson 1977). Spontaneous regression that characterizes some of the classic transplantable tumors (e.g., Jensen's sarcoma and Flexner-Jobling carcinoma) and the alterations in growth rate and incidence of metastasis that can occur with transplantable tumors must be remembered (Lewis and Aptekman 1962). These various deliberations must be borne in mind when evaluating data obtained in animal tumor systems.

In our own work, several animal tumor systems have been employed to ascertain the general applicability of the findings and at least to minimize the pitfalls of extrapolating results in terms of possible relevance to the human situation. Different hosts (mice, rats, rabbits), various sites (foot, leg, flank), and methods for primary inoculation (IM or SC) have been exploited, using standardized numbers of cells or weights of tumor.

Animal Models

Rabbits

The *VX-2 carcinoma* arose spontaneously as an anaplastic derivative of the Shope virus papilloma in a domestic rabbit 40 years ago (Kidd and Rous 1940). For five years after its origin, the blood of all rabbits bearing the VX-2 contained specific antibody to the papilloma virus. The antibody can no longer be detected in the blood of tumor-bearing rabbits, however, and it is not known whether the virus has disappeared entirely from the VX-2 carcinoma cells or still persists in some nonantigenic form (Rous, Kidd, and Smith 1952). The tumor is grown in the thigh muscles of New Zealand white rabbits (2.0 to 2.5 kg) from an inoculum of 1×10^6 cells. These have a viability in excess of 90% and are obtained by trypsinization of the donor VX-2 (Muckle and Dickson 1971). The tumor has a well characterized growth curve and pathogenesis, killing the host in 70 ± 6 days by metastasis to the regional and paraaortic lymph glands and lungs (Dickson and Muckle 1972; Muckle and Dickson 1971). Two distinct VX-2 strains exist from the point of view of heat sensitivity. One strain (London strain) is curable at 42.5°C; the other (Oxford strain) is resistant at temperatures below 47°C and thus resembles many types of human solid tumor (Dickson and Shah 1977; Dickson et al. 1977). The two strains are similar in histology and biological behavior. A further advantage of the rabbit model is that the animal tolerates a body temperature of 42°C for 90 minutes before becoming distressed (Dickson and Muckle 1972); the VX-2 system can therefore be used for both local and total-body heating experiments.

Rats

Three chemically induced rat tumors have been used routinely: the Yoshida sarcoma, MC-7 sarcoma and D-23 carcinoma. The *Yoshida tumor* had a multifocal origin in the genital omentum, pelvic peritoneum, and liver, following administration of O-aminoazotoluol to an albino rat and painting of the skin with potassium arsenite. It is classified as an undifferentiated tumor (Stewart et al. 1959). *Sarcoma MC-7* was the result of SC injection of 3-methylcholanthrene into inbred female Wistar rats in 1973 (Baldwin and Pimm 1973). The *D-23 tumor* is a syngeneic hepatocellular carcinoma that was induced 15 years ago by feeding young male rats a low protein diet containing 4-dimethylamino-azobenzene (Baldwin and Barker 1967).

These three tumors represent a spectrum of heat sensitivity: the Yoshida is heat sensitive at 42°C, the MC-7 can be cured at 43°C, while the D-23 requires 45°C for destruction. The Yoshida sarcoma is dark red in color, and this renders metastases easily visible. The D-23 tumor exists conveniently in both solid and ascitic forms. As the MC-7 and D-23 tumors are transplanted in opposite sexes of the same inbred strain of rats (WAB/Not) this has considerable economic advantages. The Yoshida grows in animals of both sexes.

The rat tumors are maintained by serial passage in the leg muscles (Yoshida) or flank (MC-7 and D-23). For experimental work, the tumors were inoculated as 100 mg (0.1 ml) of finely diced tumor slices (homogenate) SC into dorsum of the foot or flank, or IM into the thigh muscles. All rats weighed 200 to 250 g at time of transplantation. The D-23 ascites tumor was maintained by serial IP inoculation of 10^5 cells.

Mice

The *Ehrlich ascites tumor* (tetraploid strain) was passaged by injection of 10^7 cells IP into 30 to 35-g male outbred Swiss mice.

The mouse and rat ascites tumors were used only for in vitro heating experiments and for immunization procedures. Since rodents tolerate systemic heating very poorly, these tumors could not be heated in the host.

Immunology Procedures

Our current approaches for assessing host immunocompetence were selected and developed from extensive investigations on the rabbit VX-2 model. This system was chosen because the rabbit ear is suitably large for skin testing, the animal tolerates repeated blood sampling, and the procedure is technically easy. The VX-2 tumor is readily dissociated to yield single tumor cells for in vitro testing.

Rabbits

In Vitro Assays

Of the various separation techniques available, the carbonyl iron method gave the highest yield (70%) of lymphocytes from rabbit peripheral blood (Shah and Dickson 1974). The lymphocyte effector cells were then reacted at different effector:target cell ratios against VX-2 cells and against control fibroblasts from normal rabbit skin (the tissue of origin of the tumor) or kidney. For the interaction the commonly reported techniques for detecting cell-mediated immunity to tumor associated antigens were studied. These are based on the ability of the lymphocytes to prevent colony formation by the tumor cells (colony inhibition), detachment of the lymphocyte damaged tumor cells from the surface of the culture dish (microcytotoxicity assays), or release of isotopes such as ^{125}I-iododeoxyuridine or ^{51}Cr from the damaged tumor cells into the supernatant in the presence of lymphocytes in suspension. These assays were found technically unsatisfactory. Only 50% of VX-2 tumors formed colonies in vitro, and the isotope labeling was unsuitable. An inadequate number of ^{125}IUdR counts was taken up by the VX-2 cells, and with ^{51}Cr labeling there was a high spontaneous release of the isotope into the medium. These are recognized shortcomings of the various in vitro tests for cytotoxicity or cytostasis. Using ^{3}H-uridine-labled target cells and a lymphocyte:tumor cell ratio of 10:1, an inhibitory effect was obtained in the presence of 10% rabbit serum; however, there was often overlap in the range of effect with lymphocytes plus serum from cured rabbits versus lymphocytes plus serum from untreated tumor bearers (Dickson and Shah 1978).

Similarly, lymphocyte stimulation tests, using phytohemagglutinin (PHA) or a 3M KCl extract of VX-2 at different concentrations as mitogen, proved unrewarding. Uptake of labeled precursors (thymidine, uridine, and leucine) by the lymphocytes was low, and normal lymphocyte response to PHA varied widely. Although uptake was increased three to five times in the presence of PHA, this was a small increase compared to human lymphocytes, which can increase their uptake of labeled precursors 75-fold in the presence of PHA (Mavligit et al. 1974). No significant alteration occurred in the reactivity of host lymphocytes with advancement of disease or following cure of the animals. It is of note that most reported lymphocyte stimulation studies with PHA have concerned human lymphocytes, which are cultured relatively easily. Lymphocyte stimulation has not been extensively examined in experimental tumor systems.

Several established in vitro tests for detecting humoral immunity also were examined in detail. These included Ouchterlony diffusion techniques, immunofluorescence assay to detect antibody binding to the VX-2 cell surface, direct and indirect Coombs' tests using radioiodination to detect binding of gamma globulins from the host serum to the VX-2 cells or radioiodinated antirabbit gamma-globulin binding, as well as direct assays for damage or lysis of tumor cells by host serum in the presence of complement. The various microcytotoxicity tests detailed above were studied also with host serum or serum plus host lymphocytes in various amounts added to VX-2 tumor cells in microcytotoxicity wells.

These in vitro tests were very time-consuming, often yielded equivocal and inconsistent results when tumor and control (kidney, liver, skin) cells were used, and proved generally too inconvenient for routine use. In addition, the results did not relate antibody levels to clinical stage of the disease.

In vitro testing, therefore, proved to be of little value in assessing host immunocompetence in the rabbit, and was abandoned in favor of in vivo functional assays.

In Vivo Assays

Cell-Mediated Responses

Delayed hypersensitivity reactions performed on the ear were found very satisfactory for immunologic studies in the rabbit. For *specific cell-mediated immunity* assessment, a 3M KCl extract of VX-2 tumor (75 µg protein) in 0.1 ml of 0.9% NaCl was injected intradermally. The skin reaction was measured at maximal size 24 hours after injection as an area of skin swelling and erythema. This was quantitated by squaring the radius of the reaction (Shah and Dickson 1978a). Meltzer and colleagues (1971) reported that the KCl extraction procedure results in a recovery of 15% to 40% of the antigenic activity present on live diethylnitrosamine-induced guinea pig hepatoma cells, as measured by cutaneous hypersensitivity reactions. The 3M KCl extract of the VX-2 carcinomas used in the present work contained tumor-associated antigens. The preparation was biologically active, as demonstrated by positive skin tests in tumor-bearing animals following curative hyperthermia. Normal rabbit liver and kidney extracts, used as controls, failed to produce a positive reaction in such rabbits with regressing tumors.

Nonspecific cell-mediated immunity was monitored by DNCB hypersensitivity testing. The rabbits were sensitized with the DNCB at 20 mg/kg body weight and challenged 17 days later at a concentration of 1 mg/kg body weight. The sensitizing dose (in 0.1 ml acetone) was spread with a glass rod within an area of approximately 10 sq cm of shaved skin on the back, about 10 cm below the neck of the rabbit. A hair dryer then was used to evaporate the acetone. The challenge dose of DNCB was applied to the ear, after ear thickness was measured at six different places along its length with a micrometer. One-half the challenge dose was then applied in 0.1 ml acetone to each side of the ear. A positive reaction became apparent after six to eight hours as a gradually increasing erythema and induration of the ear, with maximal response at 24 hours after challenge. One hundred percent of the rabbits responded to this standard regimen.

Both the 3M KCl extract and DNCB sensitization tests were used repeatedly (at 75 µg protein and 1 mg/kg, respectively) on each experimental animal to monitor the response with time after tumor inoculation and heating. Again, it is of note that most of the work on skin testing with DNCB and tumor extracts reported in the literature relates to cancer patients. Less information is available on animal tumor systems and there are almost no sequential studies on animals with progressively growing tumors.

Humoral Immune Response

Antitumor antibody in the serum of rabbits was readily and reproducibly measured using passive hemagglutination of sensitized sheep red blood cells. The method of Onkelinx and co-workers (1969) was used, the 3M KCl VX-2 extract being bound to the red cells in the presence of glutaraldehyde. The host serum was then reacted against the sensitized red cells in V-shaped multiwell plates to determine the highest precipitating titer. Sensitization of the red cells, their storage (they were used within two weeks), and end-point determination for the assay were all performed at 4°C for optimal results.

Nonspecific humoral response in the rabbits was determined by response to foreign tissue protein in the form of bovine serum albumin (BSA). This was dissolved in 0.9% NaCl and mixed with an equal volume of Freund's adjuvant to give a concentration of 150 mg albumin/ml. Rabbits were given this mixture in the right hind leg muscle at a dose of 50 mg albumin/kg body weight. The secondary (anamnestic) response to BSA was obtained by repeating the above injection 28 days after the primary injection.

The animals were bled from the ear vein, and anti-BSA titers in the sera samples were determined by passive hemagglutination testing. Albumin was coupled to sheep RBCs in the presence of glutaraldehyde, the assay was otherwise similar to that for antitumor antibody.

Rats

For measuring immune competence in rats we have chosen the tests found to be satisfactory from our experience with rabbits. Namely, KCl tumor extract and DNCB skin testing for cellular immunity and passive hemagglutination of sensitized sheep RBCs for antitumor antibody and anti-BSA antibody. In the MC-7 sarcoma system, however, we have been unable to detect antitumor antibody in the serum of tumor-bearing hosts, even after surgical removal of the primary tumor. Various methods have been tried without success, including passive hemagglutination of RBCs sensitized with KCl extract of tumor. This emphasizes once again the generally unsatisfactory nature and unpredictability of presently available tests for host immune competence. Each tumor system is different and requires considerable time and effort to determine which techniques may be of value; the appropriate methods then require further effort to standardize and optimize them.

For the KCl tumor extract application, the same tumor protein dose as for the rabbit was used. The injection was given in 0.1 ml 0.9% NaCl SC into the dorsum of the right foot (the MC-7 tumor was implanted on the left foot). Increase in foot pad thickness was measured at maximum size 24 hours later as percentage change, taking the response to the first extract challenge as 100%. Foot thickness was measured at three different places using a micrometer, and allowance was made for change in thickness caused by an injection of 0.1 ml of 0.9% NaCl.

For DNCB testing, the same procedure and concentrations as for the rabbit were found satisfactory. Again, response to the DNCB was measured as change in ear thickness 24 hours after the challenge dose (one-half of the challenge dose was applied to each side of the ear). As for the rabbit, response to the first DNCB challenge (applied 14 days after the sensitizing dose) was taken as 100% value. Ear thickness was measured at three places, allowance being made for thickness of the normal ear.

The anamnestic response was monitored by injecting BSA (final concentration 5 mg/ml in equal volumes of 0.9% NaCl and Freund's complete adjuvant) IM into the right hind leg muscle at 5 mg/kg body weight. The secondary response was obtained by repeating this injection into the left hind leg muscle 21 days after the primary injection. The rats were bled from the jugular vein, and the BSA antibody titers determined by sheep RBC hemagglutination.

Table 23.1
Primary Tumors, Heat Sensitivity, and Sites of Distant Implanted Tumor or Metastases that Regressed Following Primary Tumor Heating

Host	Tumor Type	Heat Sensitivity	Metastases	Investigator
Hamster	Human adenocarcinoma of colon (xenogeneic)	39°C/30 min × 7	Contralateral cheek pouch	Goldenberg and Langner 1971
Rat*	Yoshida sarcoma (allogeneic)	42°C/60 min	Contralateral foot	Dickson, Calderwood, and Jasiewicz 1977
	Yoshida sarcoma (allogeneic)	42°C/60 min	Regional and distant lymph nodes	Dickson and Ellis 1976
	D-23 carcinoma (syngeneic)	45°C/60 min	Regional nodes	Dickson 1978
	MC-7 sarcoma (syngeneic)	43°C/120 min	Regional and distant nodes, lungs	Dickson and Shah, in press
Rabbit†	Brown-Pearce carcinoma (allogeneic)	70°C	Abdomen	Strauss 1969
	VX-2 carcinoma, London strain (allogeneic)	42.5°C/60 min × 3	Regional and distant lymph nodes, lungs	Muckle and Dickson 1971
	VX-2 carcinoma, Oxford strain (allogeneic)	47°C/30 min	Regional and distant lymph nodes, lungs	Shah and Dickson 1978a

*The rat tumors grew subcutaneously on the dorsum of the hind foot and were heated at 1.0 to 1.5 ml volume.
†The VX-2 rabbit tumors grew intramuscularly in the hind leg, and were heated at 15 to 20 ml; no precise data on tumor volume or degree of heating were given for the Brown-Pearce carcinoma.

ANALYSIS OF DATA

Animal Tumor Systems

The Abscopal Response

In the present revival of enthusiasm for the use of hyperthermia in the treatment of cancer, a finding of great interest has been the regression of tumor at other anatomic sites following curative heating of the primary tumor. This was first reported by Strauss (1969) for the Brown-Pearce carcinoma in the rabbit. Strauss found that not only did heating the primary tumor in the testis lead to regression of metastases in the abdomen, but the phenomenon worked in reverse—heating the abdominal mass induced disappearance of the testicular tumor. Goldenberg and Langner (1971) reported that after heating of a human colon carcinoma implanted in the hamster cheek pouch, a similar tumor in the contralateral cheek pouch regressed. These authors referred to this occurrence specifically as the abscopal antitumor action of hyperthermia. Mole (1953) coined the word *abscopal* in 1953 from the prefix *ab*, meaning "position away from" and *scopos*, meaning a "mark or target for shooting at." The response has been observed directly in rats as well as in rabbits and hamsters bearing measurable tumor transplanted at two or more sites. Others have reported the disappearance of established metastases in regional and distant lymph nodes and lungs after successful heating of the primary tumor in both allogeneic and syngeneic animal systems (table 23.1). These results were striking, and indicated that a systemic host reaction of some kind was involved in tumor response to heating.

In 1972 Dickson and Muckle made a direct comparison between local tumor heating and total body heating, using the IM VX-2 carcinoma in the rabbit. In both approaches, the tumor was maintained at 42.8°C for a total duration of three hours. After local hyper-

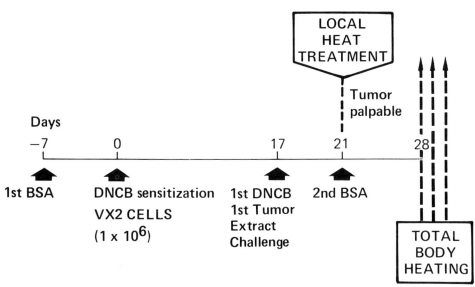

Figure 23.1. Experimental protocol for monitoring immunocompetence in VX-2 tumor-bearing rabbits treated by hyperthermia. Rabbits were given the first BSA injection (50 mg/kg) in the right hind leg muscle on day 7. On day 0 the animals were sensitized to DNCB (20 mg/kg), and VX-2 tumor cells (1×10^6) were injected into the left hind leg muscle. The first challenge with DNCB (1 mg/kg) and tumor extract (75 μg protein) was given on day 17. On day 21 albumin injection was repeated as above, and the IM tumor (volume 15 to 20 ml) was treated by local RF heating (47°C to 50°C for 30 minutes). Some animals were further treated by total-body hyperthermia (42°C for one hour) on days 28, 29, and 30. Because the VX-2 tumor was not heat sensitive at 42°C (Oxford strain; the London strain was lost in 1972), and the host did not tolerate body temperatures in excess of 42°C (Dickson and Muckle 1972), the effects of local and total-body heating could not be directly compared. The effect of total-body hyperthermia on the immune response generated after curative local hyperthermia was therefore examined. (Shah and Dickson 1978a. Reprinted by permission.)

thermia, the tumor regressed three times more rapidly than after total body heating (as measured from the regression equations for log tumor volume), and 50% of the treated animals were alive two years after heating. Only 1 of 14 rabbits (7%) survived for one year after systemic therapy; all control animals died with metastatic disease at approximately 70 days after tumor inoculation. The authors postulated that effective local tumor heating led to stimulation of an antitumor immune response that destroyed metastases, while total-body hyperthermia abrogated this response by a direct suppression of the immune system (Dickson and Muckle 1972).

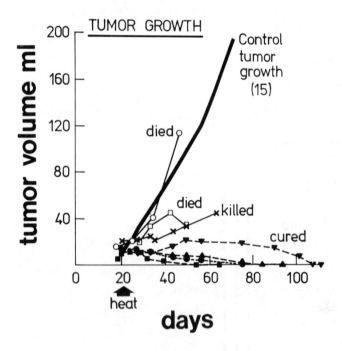

Figure 23.2. Changes in VX-2 tumor volume following curative RF heating. Twenty-one days after tumor cell inoculation, the tumor (15 to 20 ml) was treated by a single RF heat treatment at 47°C to 50°C for 30 minutes. Tumor volumes for the growth curve were obtained from caliper measurements. The number in parentheses for the control tumor growth curve denotes the number of tumors (animals) left untreated. One rabbit repeatedly chewed the flesh off its left foot after heating and had to be sacrificed at 64 days. (Shah and Dickson 1978a. Reprinted by permission.)

Is the Immune System Involved in Tumor Response to Hyperthermia?

Rabbits

Experimental Design

The schedule used in monitoring the response of the immune system in tumor-bearing rabbits to hyperthermia is detailed in figure 23.1. The timing of the various injections, of challenge with tumor extract, and of local and total-body heating was in relation to tumor cell inoculation (day 0).

Local Tumor Heating

Figure 23.2 shows the effect of local RF heating on primary tumor volume in 7 of the 13 treated rabbits. From an initial transplant of 1×10^6 VX-2 cells into the left hind leg muscles of the rabbit, the tumor became palpable by three weeks and increased in volume exponentially until about the sixty-fifth day after cell inoculation. Untreated rabbits died at 72 ± 7 (SD) days with metastases in the regional, iliac, and para-aortic lymph nodes and lungs. Heating was applied on day 21 (tumor volume, 15 to 20 ml) at the beginning of the logarithmic phase of tumor growth. Intratumor temperature was maintained at 47°C to 50°C for 30 minutes. Nine of 13 rabbits treated were cured; the tumor regressed completely within 60 to 80 days after

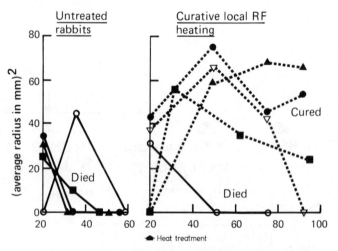

Figure 23.3. Effect of curative RF heating on the delayed cutaneous hypersensitivity reaction to tumor protein. Because of considerable quantitative differences in the response of the rabbits to the various immunologic tests, the results could not be expressed in a meaningful manner as means ± SD. Each rabbit acted as its own control, therefore, and each set of symbols in the figures refers to sequential results in an individual animal. (Shah and Dickson 1978a. Reprinted by permission.)

heating. Rabbits that were cured by heat treatment gained weight with time, whereas those that failed to respond to treatment did not. All animals that died had metastases in lungs and lymph nodes at autopsy. The nine cured rabbits were alive, with no signs of tumor, three years after treatment and were immune to challenge IM with up to 30×10^6 VX-2 cells. Heating the normal muscle of the right leg in tumor-bearing rabbits at 47°C to 50°C for 30 minutes did not affect the primary tumor or survival time of the host.

Effect of Local Hyperthermia on the Immune Response

Cellular Response: Skin Tests. Figure 23.3 illustrates the response of the VX-2 tumor-bearing rabbits to skin tests with 3M KCl extract of VX-2 carcinoma. In the untreated rabbits, the response changed from positive to negative or remained negative as the primary tumor increased in volume and the animals died. Following curative RF heating, tumor regression was accompanied by a marked increase in the skin response to challenge with tumor extract. In the unsuccessfully treated rabbits the response decreased with time as in untreated rabbits. Three of the four cured rabbits illustrated continued to react strongly against the tumor extract after the primary tumor had completely regressed. The response to a 3M KCl extract of normal rabbit liver (75 μg protein), tested on the same ear of the hosts as the VX-2 extract, was always negative.

Figure 23.4 shows the effect of curative local RF heating on the DNCB response in VX-2 tumor-bearing rabbits. In untreated rabbits the response to each of the DNCB challenges subsequent to the first at 17 days (fig. 23.1) decreased as the primary tumor volume increased, and the animals died. In successfully treated rabbits the DNCB response increased as the tumor regressed and the animals were cured. The animals that failed to respond to RF heating showed a decreasing response to DNCB challenge, comparable to that of untreated animals, as the primary tumor volume continued to increase with time.

Humoral Response: Antitumor Antibody. Figure 23.5 compares the antitumor antibody levels in untreated and in heat-treated tumor-bearing rabbits. In the untreated rabbits the antibody titers did not increase above 1:10 during the entire period of tumor growth. Following RF heating the antitumor antibody titers increased in about 80% of the rabbits. Titers continued to increase in animals that responded to the treatment and were cured, and a plateau level for antibody titers in the region of 1:1000 was achieved in some cases. The increase in antitumor antibody titers was delayed in rabbits that showed a slower than usual regression of the primary tumor following heat treat-

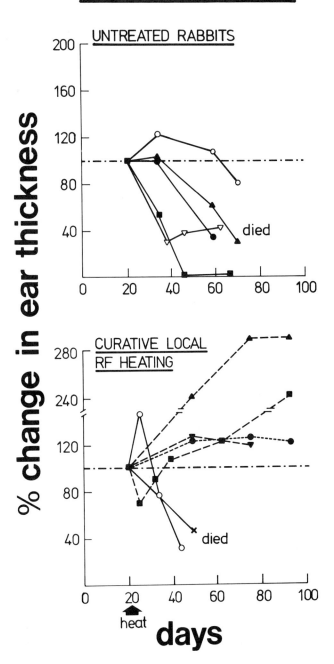

Figure 23.4. Effect of curative local RF heating on the DNCB response in VX-2 tumor-bearing rabbits. The response to the first DNCB challenge was taken as 100% value. Tumor-bearing rabbits were left untreated, or the primary tumor was treated by local RF heating. Intratumor temperature was maintained at 47°C to 50°C for 30 minutes. The response to subsequent DNCB challenges at 10 to 15 day intervals was compared with the first DNCB response, and the percentage of change obtained is plotted against the time in days after VX-2 tumor cell inoculation in the rabbits. (Shah and Dickson 1978a. Reprinted by permission.)

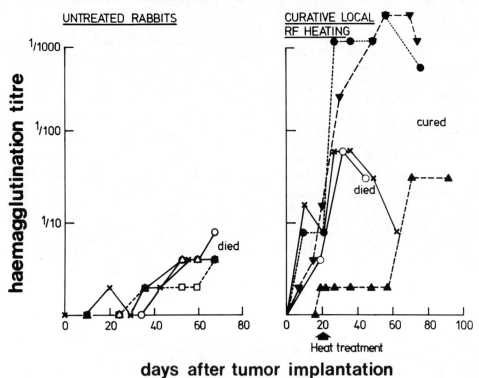

Figure 23.5. Effect of curative RF heating on antitumor antibody levels in VX-2 tumor-bearing rabbits. (Shah and Dickson 1978a. Reprinted by permission.)

ment. In rabbits that failed to respond to the treatment, the antibody levels began to decrease within about 10 days after the heating.

Anti-BSA Antibody
Figure 23.6 depicts the antibody response to BSA in untreated and in RF-treated rabbits. In tumor-bearing rabbits as in normal rabbits (Shah and Dickson 1978b), the antibody response to the second injection of BSA was always greater than the response to the first injection. The ability to respond to BSA was expressed as a ratio, the ratio of antibody titer present at different times after the second injection to the maximum titer developed after the first injection in each animal (secondary response/primary response). Two weeks after the second BSA injection, the BSA antibody levels in untreated rabbits achieved a ratio of up to 100:1, but this decreased to less than 10:1 as the tumor progressed, and the animals died. The secondary/primary antibody ratio to BSA in untreated tumor-bearing rabbits usually achieved a lower peak than in normal rabbits (100:1 versus 1000:1), and in the normal rabbits the high antibody levels were maintained with time. Following heat treatment, the secondary response to BSA in the cured rabbits increased markedly after the first four weeks as the tumor regressed. The pattern of response in rabbits that failed to respond to treatment was similar to that in untreated rabbits; that is, after an initial increase, the ratio decreased with progress of the disease until the animals died.

Local Tumor Heating Followed by Total-Body Hyperthermia
Figure 23.7 details the changes in tumor volume observed in six of the eight rabbits that were treated by local RF heating followed seven days later by total-body hyperthermia. The protocol for immunologic monitoring was as depicted in figure 23.1, and local RF heating was performed at 21 days as for the animals illustrated in figure 23.2. On days 28, 29, and 30, these tumor-bearing rabbits were further treated by total-body hyperthermia (fig 23.1). In five of eight animals treated in this manner, tumor growth was restrained temporarily, followed by return to an exponential rate of growth and death of the rabbits within 30 days of starting treatment. In a sixth rabbit, tumor growth remained in abeyance for almost 80 days after heating; there was then a rapid increase in tumor volume and the animal was sacrificed because of its cachectic state. The primary tumor in two rabbits regressed completely; however, one of these rabbits died with metastases in the iliac lymph node and lungs. At four weeks after curative RF heating alone (fig. 23.2, day 50), only one

BSA RESPONSE IN TUMOR BEARING RABBITS

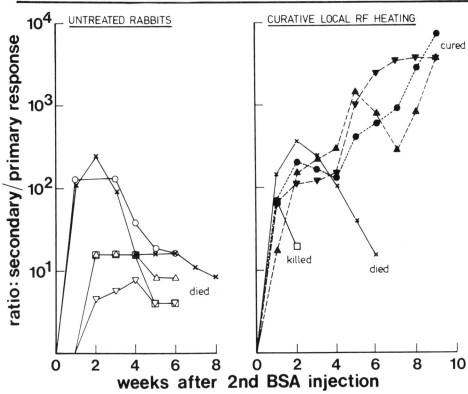

Figure 23.6. Effect of curative hyperthermia on the anamnestic response to BSA in VX-2 tumor-bearing rabbits. The BSA injections were given as denoted in figure 23.1. Anti-BSA antibody titers in serum from untreated or RF-treated rabbits were determined by sheep RBC hemagglutination. The results are expressed as the ratio of anti-BSA antibody titer present at weekly intervals after the second BSA injection to the maximum (plateau) titer developed after the first injection. (Shah and Dickson 1978a. Reprinted by permission.)

of ten rabbits had a primary tumor with a volume greater than 40 ml, whereas five of the eight rabbits treated by local heating followed by total-body hyperthermia had tumors 40 ml or greater at 50 days.

Effect on the Immune Response of Local plus Total-Body Hyperthermia

The markedly decreased survival rate of rabbits treated by local RF heating plus total-body hyperthermia was accompanied by abrogation of the stimulation of cell-mediated and humoral immune response that followed local RF heating. This is illustrated in figure 23.7 for the response to DNCB. Following the local RF heating, the response to DNCB increased, as already described. The response then decreased sharply following the application of total-body heating. The animals died with a negative skin reaction to tumor protein and low antitumor and anti-BSA titers, the results being comparable to the response in untreated tumor-bearing rabbits. In the one rabbit that was cured after this heating procedure, the DNCB response increased and remained elevated, as after curative local heating. The cured rabbit also gave a positive skin response to VX-2 tumor extract, and the antitumor and anti-BSA antibody titers in the serum were similar to those that occurred in animals cured by RF heating alone (figs. 23.3, 23.5 and 23.6).

In the rabbit VX-2 system, therefore, there was a rapid increase in "specific" anti-VX-2 cell-mediated and humoral immunity, and also in nonspecific immunity, following local heating of the primary tumor. The primary tumor still grew on transplantation of the in situ heated material until 17 days after heating; the VX-2 therefore contained viable tumor cells with replicative potential that had not been destroyed by heat. If the immune response was inhibited by total-body heating after the local heating, the primary tumor did not regress but continued to grow, and killed the animal from metastases. It was therefore concluded that regression of the primary tumor following heating, as well as disappearance of metastases, involved an immune response (figure 23.8).

It is important to realize at this point that although the anti-VX-2 immunity has been depicted as "specific," it was merely tumor associated. No evidence was

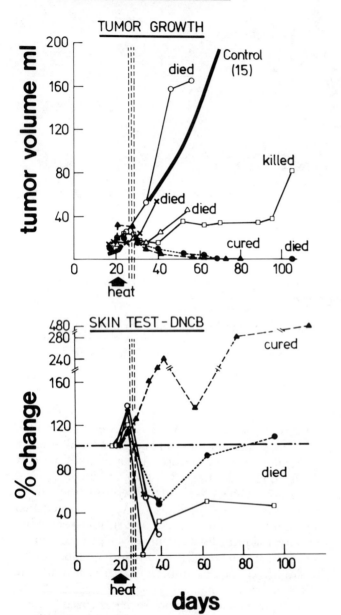

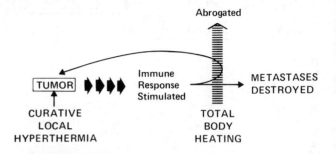

Figure 23.7. Effect of curative RF heating followed by total-body hyperthermia on tumor growth and DNCB response in rabbits. Rabbits were treated by RF heating on day 21 as described in figure 23.2, and the animals were further treated by total-body hyperthermia seven days later (*vertical broken lines*). Body temperature of the rabbits was elevated to 42°C for one hour on three consecutive days at 24-hour intervals, beginning on day 28 after VX-2 cell inoculation (fig. 23.1). (Shah and Dickson 1978a. Reprinted by permission.)

Figure 23.8. Schematic illustration of the beneficial effect on metastases and primary tumor of the immune response generated by curative local heating of the VX-2 tumor, and abrogation of the response by total-body hyperthermia (3 × 1-hr heatings at 42°C).

obtained that the VX-2 expressed on the cells a tumor-specific antigen. All results of microcytotoxicity testing, Ouchterlony gel diffusion and Coombs' tests suggested that the antigen detected on the VX-2 cells was a histocompatibility antigen differing only quantitatively from that in normal cells. This, of course, was in keeping with the VX-2 being an allogeneic tumor. It is also important, however, to note that the immunocompetence of the untreated VX-2 tumor-bearing rabbits and of the rabbits after hyperthermia, as assessed by skin tests with DNCB and tumor extract, antitumor antibody levels, and the anamnestic response to BSA, correlated well with survival of the animals. With increasing host tumor burden, the immunologic responses became progressively less; after curative heating, the change in skin reactivity to antigen and the increasing titer of antibodies was marked enough to be used as an index of successful treatment.

Strauss (1969; Strauss et al. 1965) treated a series of 276 rabbits bearing bilaterally implanted Brown-Pearce carcinomas. Electrocoagulation of the tumor in one testis was followed by regression of the contralateral testicular tumor and of abdominal secondary tumors if these were present. Conversely, heating of the metastatic tumor in the abdomen led to disappearance of tumors in the testes. Following electrocoagulation of the tumor, an inconsistent rise in antitumor antibody titers occurred. The titers increased to about 1:100 from two to three weeks after treatment and remained at this level for the following three to four weeks while the tumor was being resorbed. The antibody titers then decreased to zero during the next three to four months, even though all the cured rabbits were immune to challenge with Brown-Pearce carcinoma. The sera (0.2 ml undiluted) from immune rabbits caused lysis of Brown-Pearce tumor cells growing in petri dishes within three days (Strauss 1969; Strauss et al. 1965).

Strauss and colleagues also observed that, if the Brown-Pearce tumor was transplanted seven days after electrocoagulation (seven day in situ treated tumor), no growth occurred in the testis of normal rabbits although the injected material contained viable-looking cancer cells; the animals subsequently resisted repeated challenge with fresh carcinoma. The seven day in situ material, when injected into tumor-bearing rabbits, cured the animals, which also became immune to challenge with Brown-Pearce carcinoma. When sera from these immune rabbits were inoculated into normal rabbits, the normal animals rejected tumor challenges. Furthermore, when the sera from immune rabbits were inoculated into tumor-bearing rabbits, the tumors regressed and the animals were cured.

The results as described by Strauss are almost totally bereft of details. Tumor size, the time of treatment, the tumor temperature achieved during electrocoagulation, and the methodology and controls used for immune studies are not reported. The Brown-Pearce tumor has several controversial features (Stewart et al. 1959). Spontaneous regression of the tumor is a frequent occurrence, and apparent recovery of the host may take place even after extensive metastases have developed. It also has been reported that the metastatic growths in rabbits appeared within four to six weeks during cold weather, while regression of the primary without metastasis was the rule during hot weather. It therefore is difficult to evaluate the role of electrocoagulation and the part (if any) played by the immune system in the cure of tumor-bearing rabbits as reported by Strauss.

Rats

Figures 23.9 to 23.11 present our results for sequential monitoring of immunocompetence in rats bearing the syngeneic MC-7 sarcoma after curative local hyperthermia (43°C for 120 minutes), which resulted in a cure rate of 72% in treated animals. In untreated rats, there was a progressive decrease in the ability of the host to react to tumor extract. After curative hyperthermia, there was a significant increase in the skin response to tumor extract (fig. 23.9). This augmented T-cell mediated reaction was paralleled by a similar, but less marked, recovery of the response to DNCB after treatment (fig. 23.10). In the MC-7 system, antitumor antibody was not detected using passive hemagglutination, immunodiffusion, or immunoelectrophoresis techniques. Figure 23.11, however, shows that after heating the levels of antibody to BSA were maintained or increased, in contrast to the generally

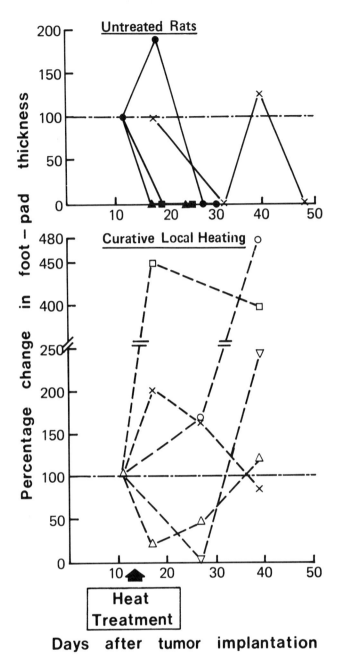

Figure 23.9. Effect of curative heating on delayed cutaneous hypersensitivity reaction to tumor protein in host rats. Response to the first extract challenge was taken as 100%. (Each symbol in figs. 23.9–23.11 represents the same rat tested sequentially.)

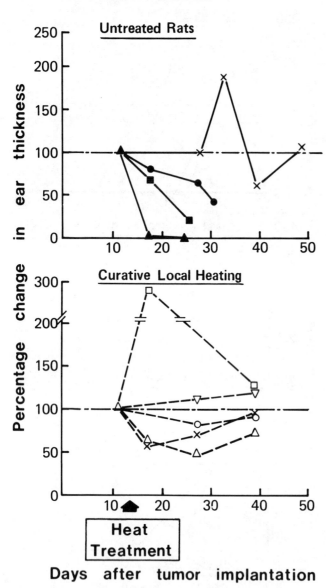

Figure 23.10. Effect of curative local heating on DNCB response in MC-7 tumor-bearing rats. Response to the initial challenge at 14 days after the sensitizing dose was taken as 100% value.

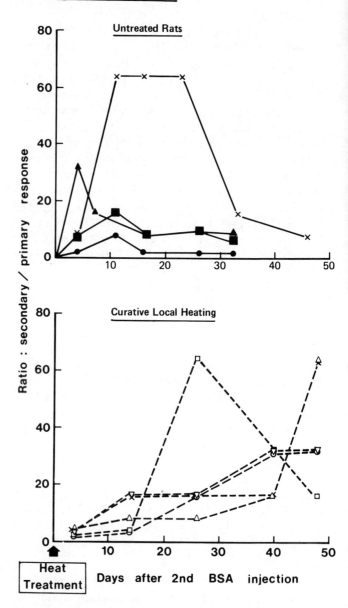

Figure 23.11. Effect of curative heating on anamnestic response to BSA in MC-7 tumor-bearing rats. The response was induced by a second IM injection of BSA given 21 days after the primary injection and 30 minutes before heating the foot tumors. Results are expressed as a ratio, as in figure 23.6.

low levels in untreated rats. These changes in cellular and humoral immunity were not detected in tumor-bearing animals not cured by heating. Cured rats were immune to large doses of tumor cells, 50% of the hosts rejecting a challenge of 2.5×10^6 MC-7 cells into the foot; 100% of normal rats succumbed to this dose of cells.

In outbred rats, Szmigielski and Janiak (1978) found that curative microwave heating (43°C for 60 minutes) of the allogeneic Guérin epithelioma caused stimulation of both tumor-directed and nonspecific cellular reactions and a nonspecific humoral immune response. After heating there was a significantly increased reactivity of host spleen cells to PHA and to mitomycin-inhibited tumor cells, as well as increased cytotoxicity (as measured by ^{51}Cr release assay) of both lymphocytes and macrophages to cultured Guérin cells. Augmented levels of anti-BSA antibody and lysozyme were detected in the host serum. The authors believed cell-mediated immunity to be the important effector arm in the response to heat. No details of the control target cells used for the in vitro tests or of the lymphocyte to target cell ratio were provided, however, so the tumor-directed specificity of the increased cell activity was not established.

Effect of Immunosuppression on Tumor Response to Curative Heating

Since host cure after effective tumor heating was accompanied by an immune response, and the treated animals remained immune to tumor challenge, the next step was to determine whether these events could occur in the absence of a functioning immune system. This was investigated by heating tumors in immunosuppressed hosts, and also challenging cured animals with tumor cells after similar suppression. The results are detailed in table 23.2.

In the MC-7-bearing rats, growth rate of the primary tumors was merely retarded, and the host cure rate reduced from 72% to zero. Host immunocompetence failed to increase following tumor heating. Similarly, when rats bearing the Yoshida sarcoma were immunosuppressed after tumor heating, the cure rate was reduced from 100% to 7%.

When heat-cured immune animals were immunosuppressed, all hosts succumbed to tumor challenge. Growth of the challenge inoculum to a tumor was accompanied by decreased specific and nonspecific T- and B-cell responses compared to cured nonsuppressed animals that rejected the challenge.

In these immunosuppression experiments, the host white blood cell count decreased to 10% to 20% of normal and took two to three weeks to regain control values. Of several protocols examined, including radiation or cortisone alone, as well as various x-ray doses, the rabbits tolerated the immunosuppression poorly; only 2 rabbits survived to be challenged.

Hahn, Alfieri, and Kim (in press) recently re-

Table 23.2
Effect of Immunosuppression on Host Cure and on Tumor Challenge after Curative Local Hyperthermia

Tumor Type	Heat Treatment	Animal Group	Immunosuppression	Effect on Immune Response	No. Animals Rejecting Tumor/Total	% Host Cure
MC-7 sarcoma	43°C/120 min	Tumor-bearing rats	—	Increased skin response to DNCB and tumor extract, increased anti-BSA antibody	15/21	72
	43°C/120 min	Tumor-bearing rats	Total-body irradiation and cortisone acetate	Decreased skin responses, anti-BSA antibody titer failed to rise, WBC and lymphocyte counts in blood decreased to 10%–20% normal	0/10	0
	43°C/120 min	Cured rats challenged with 2.5×10^6 MC-7 cells	—	Skin response to DNCB and tumor extract as well as anti-BSA antibody levels maintained	7/14	50
	43°C/120 min	Cured rats challenged with 2.5×10^6 MC-7 cells	Total-body irradiation and cortisone acetate	Decreased skin response to DNCB and tumor extract, decreased anti-BSA antibody	0/10	0
Yoshida sarcoma	42°C/60 min	Tumor-bearing rats	—	Not examined	40/40	100
	42°C/60 min	Tumor-bearing rats	Total-body irradiation and cortisone acetate	Decreased WBC and lymphocyte counts in blood (< 20% of normal)	1/14	7
VX-2 carcinoma	47°C–50°C/30 min	Cured rabbits challenged with 1×10^6 VX-2 cells	—	Skin response to DNCB and tumor extract as well as anti-BSA antibody levels maintained at titers > 1:10,000	20/20	100
	47°C–50°C/30 min	Cured rabbits challenged with 1×10^6 VX-2 cells	Total-body irradiation and cortisone acetate	Decreased WBC and lymphocyte counts (10%–20% of normal), decreased skin response to DNCB and tumor extract, decreased anti-BSA antibody to titers of 1:2000 and 1:4000	0/2	0

NOTE: Rats were immunosuppressed by total-body irradiation (150 rad × 3 at 48-hour intervals) plus simultaneous cortisone acetate (60 mg/kg body wt SC × 4 at 48-hour intervals). For tumor-bearing animals this regimen was started within two hours of tumor heating. Rabbits were given the same protocol except that the x-ray dose was increased to 200 rad per fraction.

ported briefly that the successful local control of a seven-day intradermally growing fibrosarcoma in intact mice (44.5°C for 20 minutes) was abrogated in immunosuppressed mice (500-rad whole-body irradiation) and in immunodeficient "nude" mice. Local tumor control was regained in mice reconstituted with 2.5×10^7 or more splenic T cells, but not with immune sera replacement.

Manipulation of the Mononuclear Phagocyte (Reticuloendothelial) System

In 1972, Dickson and Muckle called attention to the intense macrophage activity that accompanied regression of the VX-2 carcinoma in the rabbit after hyperthermia. In recent years there has been increasing interest and investigation of the role of the macrophage in tumor immunity (Eccles 1978; James, McBride, and Stuart 1978). Some workers (Hanna et al. 1972) have specifically classed the cells of the mononuclear phagocyte system (mononuclears, histiocytes, macrophages) as a major component of the immune system. The question therefore arose as to what effect abrogation or stimulation of this important cell system had on tumor response to hyperthermia.

Rabbits proved unsuitable for this work; they rapidly succumbed to the large doses of silica required (1 mg/g body weight) to suppress the macrophages.

In rats, when the macrophage toxin silica was injected after curative hyperthermia of the MC-7 sarcoma, the cure rate of the rats was reduced from 72% to 43% (table 23.3). The clearance rate of colloidal carbon from the blood was used to assess macrophage activity, and even the uptake of this particulate substance by the macrophages significantly reduced the host cure rate after heating. The unheated MC-7 had a spontaneous cure rate of 11%; this was reduced to zero by injection of silica (table 23.3). After silica treatment, all the cured rats (43%) succumbed to challenge with 2.5×10^6 MC-7 cells; 50% of the cured rats not given silica consistently rejected this challenge dose (table 23.5).

In nude mice and in semisyngeneic hybrids, Hahn, Alfieri, and Kim (in press) reported that inhibition of macrophage activity by chronic silica treatment abrogated the curative effect of 44.5°C for 20 minutes on a chemically-induced fibrosarcoma.

Table 23.4 shows that suboptimal heating of the MC-7 sarcoma (43°C for 1.5 hours) gave a cure rate of 31%. This increased to 65% when C. parvum was given IV (but not IP) prior to the hyperthermia. C. parvum by itself had no effect on the spontaneous cure rate of this tumor; nor did it improve the number of cured rats rejecting tumor after heat treatment (table 23.5).

Also in rats, Szmigielski and Janiak (1978) injected C. parvum into the Guérin tumor and found a synergistic effect with heat (43°C for 60 minutes), resulting in augmented tumor regression and an animal cure rate of 75% versus 41%; there was also increased macrophage activity against the tumor cells in vitro.

In mice, Bourdon and Halpern (1976) immunized syngeneic hosts with Moloney YC-8 lymphoma cells heated at 46°C to 52°C in vitro. Peritoneal macro-

Table 23.3
Influence of Macrophage Impairment on Regression of MC-7 Sarcoma and Host Cure by Hyperthermia

Animal Group	Heat Treatment	No. Rats Cured/Total	% Host Cure
Control	—	3/27	11
Colloidal carbon (IV)	—	1/10	10
Silica (IV and IP)	—	0/10	0
Control	43°C/2 hr	15/21	72
Colloidal carbon (IV)	43°C/2 hr	7/15	47
Silica (IV and IP)	43°C/2 hr	9/21	43

NOTE: Colloidal carbon in the form of commercial drawing ink was given at a dose of 8 mg/100 g body weight. Silica was used at 1 mg/g body weight, one-half the dose being introduced IP and the rest IV, starting one day before heating, with two additional doses being given at 10-day intervals. This regimen maintained macrophage activity (determined as phagocytic index, the clearance rate of colloidal carbon from the blood) at approximately 50% of normal. Heat treatment was by water-bath immersion, intratumor temperature being continuously monitored by a tissue-implantable thermocouple in each tumor.

Table 23.4
Influence of Macrophage Stimulation on Regression of MC-7 Sarcoma and Host Cure by Hyperthermia

Animal Group	Heat Treatment	No. Rats Cured/Total	% Host Cure
Control	—	3/27	11
C. parvum (IP)	—	1/8	12
C. parvum (IV)	—	1/12	8
Control	43°C/1.5 hr	5/16	31
C. parvum (IP)	43°C/1.5 hr	6/16	37
C. parvum (IV)	43°C/1.5 hr	11/17	65

NOTE: C. parvum (Burroughs Wellcome, Research Triangle Park, North Carolina, 7 mg dry weight organisms/ml) was given IP (1400 µg × 3) on each of the three days prior to tumor heating, and IV as a single injection (700 µg) on one of the three days before heating.

Table 23.5
Effect of MC-7 Tumor Challenge (2.5×10^6 cells) in Cured Rats

Animal Group	No. Rats with Tumor/Total	% Rejection
Control	6/6	0
Heat alone	7/14	50
Silica + heat	8/8	0
C. parvum (IV) + heat	4/8	50

NOTE: Protocol for silica or *C. parvum* plus heat was as in tables 23.1 and 23.2. The challenge dose of tumor cells was injected SC into the dorsum of the left hind foot.

phages from the immunized mice, when mixed with tumor cells in a ratio of 300:1 and given IP to normal mice, afforded 100% protection at 100 days. Control animals given macrophages from nonimmune mice along with tumor cells, or tumor cells alone, died in 25 to 30 days. With spleen lymphocytes from immunized mice, the survival rate was reduced to 30% at 100 days. More recently, Urano and colleagues (1978) have reported that the heating time at 43.5°C to control 50% (TCD_{50}) of treated methylcholanthrene-induced fibrosarcomas in mice could be reduced from 92 minutes to 25 minutes with adjuvant *C. parvum*. The effect was observed only when the adjuvant was given before the hyperthermia, and was less pronounced with a spontaneous (i.e., less immunogenic) fibrosarcoma (TCD_{50} was 141 minutes with heat versus 83 minutes with heat plus *C. parvum*) (Urano et al. 1979).

It therefore is apparent that in animal tumor systems (both allogeneic and syngeneic) there is now definitive evidence that the regression of both primary tumor and metastases after curative local heating, as well as the immune status of the host that subsequently develops, involve an augmented state of host immunocompetence. If the immune response is artificially suppressed or does not occur following heating, then the tumor continues to grow and the host dies from advanced disease. There is little evidence that the host reaction is a specific tumor-directed one, and this becomes more apparent the more "pure" (syngeneic) the system immunologically. Rather, the findings suggest that a nonspecific host response, comprising both T and B lymphocytes and involving a major macrophage component, is evoked after effective hyperthermia.

Humans

Local Hyperthermia

From his work on animals, Strauss and associates postulated that in humans also tumor heating by electrocoagulation led to stimulation of the host immune system by absorption of necrotic tumor products (Strauss 1969; Strauss et al. 1965). Although they treated a large number of patients with various types of cancer, these investigators reported no data for immunologic testing on them, and their concept of immunostimulation was totally bereft of details as far as their patients were concerned. Crile and Turnbull (1972) found, as did Strauss and co-workers, that a striking feature of patients treated by electrocoagulation for rectal cancer was that none of them died from carcinomatosis of the abdominal cavity; local recurrence and metastasis were less common than in patients with comparable disease treated by surgery. It could be argued that these results were attributable to nonintervention on the part of the surgeon, preventing spread of the cancer cells by handling at surgery, rather than to any involvement of the immune system.

Findings analogous to the abscopal response in animals have been observed following the treatment of human limb cancers by regional hyperthermic perfusion. In cases of multiple localization of tumors, destruction of the limb tumor resulted in disappearance of the others (Cavaliere, Moricca, and Caputo 1976; Moricca et al. 1977; Stehlin 1976), and heat treatment of primary cancers has strikingly reduced the incidence of metastases in tumors notoriously prone to rapid spread to distant sites, such as sarcoma and melanoma (Moricca et al. 1977; Stehlin et al. 1977). Experience with recurrent disease has led Cavaliere, Moricca, and Caputo (1976) and Stehlin (1976; Stehlin et al. 1977) to have recourse to delayed local excision or delayed amputation at 4 to 12 weeks after perfusion to cope with residual melanoma or sarcoma; both surgeons believe that during this time interval, antigenic products liberated from the heated tumor lead to enhancement of an antitumor immune response that destroys metastases.

Stehlin and co-workers (1977) obtained regression of melanomas of the limb by regional perfusion with melphalan at 38.8°C to 40°C. Following heating, the cytotoxic activity of plasma plus lymphocytes from the patient against his own melanoma cells cultured in vitro increased, to a peak level of 80% to 85% inhibition within two weeks; this was caused largely by an increased activity of the plasma. Activity then decreased to base-line value (10% inhibition) by 10 weeks after heating. No details of the preparation or purity of the lymphocyte populations used, no control data for target cells other than melanoma cells, no definition of tumor cell "inhibition," or of what percentage of treated patients exhibited the response are reported; no data for patients given hyperthermic perfusion alone are provided. Hence, compelling as are the clinical results (addition of heat to the melphalan perfusion has increased the five-year survival rate in recurrent melanoma from 22% to 77%), the claim (Stehlin et al. 1977) that the augmented activity of the plasma and lymphocytes represents stimulation of an immune response that destroys or inhibits latent or overt distant metastases remains unproved. Although other clinicians subscribe to the view that breakdown products from heated tumors may act as an immunostimulant, apart from the data of Stehlin and associates, the evidence for stimulation of a host response (immune or otherwise) in humans after tumor heating is entirely circumstantial.

Total-Body Heating

Cancer Biotherapy

Total-body hyperthermia had its origins in the early years of this century when W. B. Coley and numerous other physicians induced pyrexia by injecting mixed extracts of *Streptococcus pyogenes* and *Bacillus prodigiosus* and obtained some remarkable cures in patients with advanced cancers. Some patients became immune to the toxins and Coley surmised that an immune reaction might also be taking place against the host tumor. It is reasonable to believe that Coley's results were caused by stimulation of the immune system with a nonspecific immunotherapy and a sustained pyrexia of 104°F to 105°F (40°C to 40.6°C). The bacterial products caused an acute inflammatory reaction when injected into tumors, and there is evidence that such products can have a direct destructive effect on cancer cells, possibly even rendering them immunogenic (see Dickson 1977; Nauts 1976 for a review of effects of Coley's toxins and discussion in earlier chapters). Clinical experience with the toxins consistently emphasized the importance of inducing an adequate and prolonged systemic pyrexia to achieve maximal antitumor effects. Consequently, in the recent renewal of enthusiasm for hyperthermia, the participation of the other bacterial factors in toxin therapy has been ignored, although the temperatures induced by toxins were per se inadequate to damage tumors (Dickson 1974, 1977, 1978).

The importance of this oversight is further emphasized when it is realized that the field of cancer biotherapy—the use of microbial products to destroy tumor cells—includes a substantial body of results achieved in the *absence* of pyrexia. From a series of investigations on tumor regression in animals brought about by a variety of microorganisms, including *E. coli*, *Salmonella*, *Trypanosoma* species, and *Streptococci*, Klyuyeva and Roskin (1963) concluded that the pyrexia developing during infection by these organisms was not a necessary factor for inhibiting tumor growth. Using IM injections of a lysate of *Trypanosoma cruzi* in humans, these workers obtained regressions of histologically verified cancer of the lip, gastrointestinal tract, and breast that were comparable to the regressions achieved with Coley's toxins. Pyrexia was only an occasional feature following the thousands of *Trypanosoma* injections administered. Detailed serial histologic studies revealed the regressing tumors to be the site of inflammatory infiltrates and intense macrophage reaction with the presence of plasma cells. These changes were interpreted as representing the humoro-cellular responses of the body's defense mechanisms.

A major effect of Coley's toxins was upon tumor vasculature, leading to tumor hemorrhage and necrosis, while the *Trypanosoma* extracts had a direct cytostatic effect on the cancer cells. Both preparations altered the relations obtaining between the host and the malignant cells in a complex manner. Manipulation of this relationship to the benefit of the host did not always occur and was governed in an undefined manner by the type and potency of the preparation, dosage schedule, tumor type and extent, and unknown host factors. The part played by fever in the response is unproved; its occurrence with the most effective toxin preparations was a convenient clinical marker of potency, but may have been an irrelevant, if helpful, side effect.

The era of cancer biotherapy has provided a legacy both encouraging and frustrating. Since it is the approach that has produced most impressive results for advanced malignancy in the past, a very substantial investment in time and finances has been devoted to unraveling its mysteries. Although many different microbial products have been studied, no preparation has emerged to rival the mixed bacterial toxin of Coley, with its synergism of effect between the extracts of *S. pyogenes* and *B. prodigiosus*. Histologic studies suggested strongly that the filtrates had a direct, highly destructive effect on malignant cells in animals; a number of significant findings have confirmed that certain types of concurrent acute infection (especially streptococcal) may have a markedly beneficial effect on host resistance to cancer in animals; at the experimental and clinical levels, there are data indicating that preliminary administration of bacterial toxins may protect the host animal or patient's normal tissues against radiation damage, while potentiating the response of tumor to irradiation. Full references to these aspects are found in a recent review (Dickson 1977).

The mode of action of the toxins can be surmised from such data to have involved in large measure a nonspecific stimulation of host defenses. This notion is supported by the achievement of the best results when the toxins were given for a prolonged period of time—months or years rather than weeks; cessation of the injections often resulted in recurrence of the disease. In spite of the large number of patients treated by Coley and numerous other physicians in the heyday of toxin therapy (1891–1936), few guidelines emerged for patient management. The toxins defied biological standardization, their effects were capricious and inconsistent, and their purification in the 1940s proved self-defeating, yielding a polysaccharide that proved highly toxic when tested clinically (Dickson 1977). Although striking cures of advanced histologically verified cancer were obtained with toxins, it must be remembered also that in the majority of patients tumor regression was incomplete and the remissions were of limited duration.

There is also a substantial literature attesting to the lethal effects of the protozoal lysate from *Trypanosoma cruzi* on a large number of spontaneous and induced tumors in animals and in human cancer (Klyuyeva and Roskin 1963). A striking feature of the use of this endotoxin is the way in which experience with it paralleled closely that of physicians in a geographically different region of the world with Coley's toxins (the toxins were popularized in the United States; the *Trypanosoma* extract was used chiefly in Russia). Different *Trypanosoma* preparations varied in activity, even when the organisms were grown in equal numbers on identical culture media, and the biological activity was refractory to standardization; the best route, dose, and frequency of the extract were never established; the preparation had to be administered continuously over a prolonged period of time—interruption or irregular dosage was associated with decreased effectiveness. The lysates did not induce fever in the host, however, and Klyuyeva and Roskin were explicit in their belief that the extracts acted via the macrophages of the reticuloendothelial system as well as having a cancerostatic effect, progressing to a cancerolytic effect, on the tumor cells.

There is, therefore, considerable circumstantial evidence favoring activity of nonspecific host defenses (macrophages) as a major component of the response to biotherapy. Data from immunology studies indicate that such activated macrophages are capable of killing target (tumor) cells nonspecifically (Smith and Landy 1975). The pyrexia induced by Coley's toxins may have acted as a potentiator of these events. In spite of the large body of data generated by the era of biotherapy, the use of an ill-defined preparation like Coley's toxin represents an uncharted sea; patient compliance to the repeated pyrexia is poor, and the logistics of administering the preparation regularly on a long-term basis are formidable. For these reasons, therefore, Coley's work can only be regarded as heroics that will almost certainly not be repeated on a wide scale. Current approaches favor investigation of more defined macrophage stimulants, *C. parvum* for example, and the induction of high temperature by physical means.

Physically Induced Total-Body Hyperthermia

A number of workers have detailed myriad changes in components of the immune system after total-body heating in cancer patients at 40°C to 42°C (table 23.6).

Table 23.6
Whole-Body Hyperthermia and Immune Response in Humans

Tumor Type	Body Temperature	Response	Investigator(s)
Sarcoma, carcinoma and melanoma	Hyperpyrexia with bacterial toxins, 40°C–40.6°C/fever for 50–100 hr	Acute inflammatory reaction when toxins were injected into tumors	Coley and numerous other workers. Dickson and Ellis 1976; Nauts, Fowler, and Bogatko 1953
Melanoma, adenocarcinoma of stomach	41.7°C/4 hr	WBC count increased, no change in T- or B-cell count, significant decrease in C3 complement and cytotoxic ability of antibody and lymphocytes	De Horatius et al. 1977
Hodgkins disease	40°C/2 hr × 4	Lymphocyte response to PHA decreased significantly	Fabricius et al. 1979
Gastrointestinal cancers	41.8°C/2–4 hr	Short-lived general immunosuppression—PHA response, C3 complement levels, T-cell count, and antibody dependent lymphocyte cytotoxicity all decreased initially but subsequently recovered in 3–4 days	Gee et al. 1978
Advanced cancers	42°C/6 hr	WBC, bactericidal activity of neutrophils, T-cells, and PHA response increased; platelets decreased	Parks et al. 1978

These include a short-lived polymorphonuclear leukocytosis and lymphopenia (Gee et al. 1978; Parks et al. 1978). Such changes are characteristic of a stress response, and were well recognized at the height of the use of physically induced hyperpyrexia for the treatment of nonmalignant conditions such as syphilis and arthritis (Cohen and Warren 1935; Neymann 1938; Steel 1977). A decrease in C3 complement activity for three to four days after heating has been observed (De Horatius et al. 1977; Gee et al. 1978), but gross defects in this component are required to compromise host defense (Hyslop 1975). No general pattern of response is apparent from studies on serum immunoglobulin levels or the PHA sensitivity of lymphocytes from heated patients (Fabricius et al. 1978, 1979; Gee et al. 1978). The reported decrease in the circulating population of T lymphocytes (and the general lymphopenia) may be caused by physical damage to lymphoid tissue, as in the case of laboratory animals, or by stress (plasma cortisol levels in man increase two- to threefold during total body heating (Parks et al. 1978; McKenzie et al. 1976). On the other hand, it may simply reflect sequestration of these cells in organs such as spleen and liver under the altered physiologic conditions of high temperature. A further point is that total-body hyperthermia usually involves anesthetizing the patient, and there is some evidence that anesthetic agents may depress T-lymphocyte-mediated function, including response to mitogens (Bruce and Kahan 1975).

No data are available on tumor-directed immunocompetence in patients treated by whole-body heating. The lack of results and the nonspecific nature of the changes in general immunocompetence make it impossible to assess whether host immune status in humans is adversely affected by this form of heating in a manner similar to that described for some animal systems.

Clinical Fever versus Physical Heating
The importance of the bacterial factor in the patient response to Coley's toxin is emphasized by the comparative lack of success achieved by physically induced whole-body hyperthermia. Although higher temperatures in the region of 42°C can be achieved by physically applied heat, and an overall response rate of 47% for advanced cancer of various types has been reported, the beneficial effects have been of short duration, even with multiple heatings (Pettigrew et al. 1974). The physiologic response to externally applied heat differs radically from that obtaining in clinical fever. With exogenous physical hyperthermia, the heating overrides the body thermostat, and is controlled by the physician. In clinical fever, the body thermostat continues to operate and is merely set at a higher level than normal (Bligh 1973); the temperature rarely exceeds 41°C (Dubois 1949). It is usually assumed that such endogenous elevation of temperature has a beneficial effect on the immune system, since infectious diseases are accompanied by leukocytosis, and the pyrexia is often followed by development of a high antibody titer against the causative agent. It is apparent, therefore, that exogenous heating, no matter how efficient it may be, can not be equated with clinical fever.

Tumor Heating and the Paradox of Enhanced Metastasis

It was pointed out in an earlier section (adequacy of tumor heating) that a disturbing body of data exists indicating that tumor cells that survive therapy (radiation, drugs, perfusion) may be altered biologically and possibly may pose a threat to the host, both from augmented local growth and from metastasis. It probably is not even necessary that such surviving cells should be altered biologically, since cells populating both animal and human neoplasms are known to be heterogeneous with respect to their growth rate, metabolic characteristics, immunogenicity, and susceptibility to radiation and cytotoxic drugs (Fidler and Hart 1978; Poste and Fidler 1980). There is recent data that malignant tumors also contain subpopulations of cells with differing metastatic capabilities, as first demonstrated by Fidler and his group with the B16 melanoma, and subsequently confirmed by others for several other murine tumors (Fidler and Hart 1978; Fidler and Kripke 1980; Poste and Fidler 1980). Although experimental studies have yielded what appear to be contradictory results (the different tumor types, etiologies, and animal strains may be at least partly responsible), there is direct and indirect evidence that throughout the metastatic process, malignant cells are subject to attack by host defense systems, and that tumor immunogenicity determines to some extent the nature and degree of the influence of the immune system on the metastatic process. In view of these considerations (differing propensity of tumor subpopulations to metastasize, role of immunity in metastasis), it is not unexpected that a treatment that evokes a host immune response against both primary tumor and metastases should be associated with enhancement of metastasis in some circumstances. And indeed, in animals, enhanced tumor dissemination has been reported following both local and total-body hyperthermia.

Animal Systems

Local Hyperthermia

Dickson and Ellis (1974) found rapid and increased spread of the Yoshida sarcoma occurred via direct, lymphatic, and vascular routes after heating the tumor (10-ml volume) in the leg muscles of rats at 42°C for 60 minutes. The augmented spread involved not only predicted sites, such as heart and retroperitoneal tissues, but also unexpected organs. The incidence of metastatic tumor in the liver after heating, for example, was 12%; liver involvement was not found in any of 91 control rats. Because the heated tumors showed no restraint of growth, no microscopic evidence of destruction, and only a temporary decrease of metabolism, the authors concluded that the enhanced dissemination probably was due to inadequate heating. An additional factor may have been a more favorable soil for tumor growth created by the rise in body temperature to over 41.5°C by water-bath heating of these large tumors. Smaller (1.5 ml) Yoshida tumors are 100% curable by 42°C for 60 minutes, and in further studies with 1.5-ml foot tumors heated at 40°C for 60 minutes, Dickson and Ellis (1976) confirmed that inadequate heating of this tumor was accompanied by enhanced dissemination both to usual and rare sites (e.g., testis) in the *absence* of a rise in body temperature. Heating at 40°C led to a 50% increase in both respiration and anerobic glycolysis of the tumors, and the authors concluded that increased metabolism may have played a role in the augmented tumor spread. The metabolism and growth rate of several types of malignant cells are increased at 40°C to 41°C in vitro and in vivo, and this has given rise to the postulate of a stimulatory temperature range above normal body temperature and preceding the tumor lethal temperature zone (Dickson and Ellis 1976; Dickson 1977).

More recently, other workers have reported augmented tumor cell dissemination after local heating. Walker and colleagues (1978) found that in the mouse C_3H mammary system (flank tumor) there was a significantly increased metastasis rate after treatment of the tumor at 44°C for 60 minutes compared to radiotherapy at an equivalent cure rate (64%) or surgical removal (98% cure rate). The metastasis rate of 13% was reduced to zero by rapid heating to target temperature as opposed to slower heat input over 10 minutes. The metastasis rate was roughly related inversely to the cure rate of the heated tumor, a higher cure rate yielding fewer lung metastases; a 100% cure rate was not achieved. The augmented dissemination may therefore have been due to inadequate tumor heating. With the KHT fibrosarcoma in C_3H mice, Rappaport (in press) reported that heating (42°C to 44°C) of the foot before or after tumor inoculation enhanced lymphatic metastasis to regional nodes and in some instances to the lungs. With the SCK mammary carcinoma in similar host mice, only 44°C before, but not after, tumor inoculation enhanced metastasis, indicating participation of both host and tumor cells in the process. The SCK mammary tumors and the KHT fibrosarcoma are regarded as nonimmunogenic in their syngeneic host. Again in C_3H mice, Rice, Urano, and Suit (in press) found that after three 60-minute fractions at 41.5°C there was no significant increase in the number of lung colonies from a chemically-induced immunogenic fibrosarcoma, or of a spontaneous weakly immunogenic fibrosarcoma. In the work of Rice, Urano, and Suit, inadequate tumor heating was also operative, since 100% tumor cure was not achieved. With the KHT flank fibrosarcoma in mice heated inadequately at 43°C for 30 minutes (0% cures) Marmor, Hahn, and Hahn (1977) found no increase in the severity of macroscopic lung metastases compared to control mice. Similarly, Kim and Hahn (1979) reported that mild (40.5°C) or moderate (42.5°C) heating of the Dunn osteogenic sarcoma in mice did not increase metastatic spread of the tumor to the lung; tumor lethal temperature, cure rates, or number of metastases were not given by these workers.

In these various reports on augmented tumor dissemination after local tumor hyperthermia, there have been no concomitant immunology studies, so the role (if any) of the host defense systems remains purely speculative. Although there is data suggesting that the host response after tumor heating is related to tumor immunogenicity, the major component in the increased aggressive behavior of the tumors may well be heat-induced changes in the biochemistry or biology of the cancer cells.

Total-Body Hyperthermia

In the earlier experiments of Dickson and Ellis (1974) on augmented tumor dissemination, whole-body hyperthermia (> 41.5°C for 30 minutes) resulted from water-bath heating of large primary tumors. Schechter, Stowe, and Moroson (1978), on the other hand, obtained significant reduction in retroperitoneal metastases from mammary carcinomas in rats maintained at a body temperature of 40.7°C for 60 minutes on two occasions. This accompanied two water-bath heatings of the primary leg tumors (with resultant transient growth retardation) at 42.3°C for 90 minutes. Other workers have reported enhanced tumor dissemination after deliberately induced systemic hyperthermia. With body temperatures of

40.5°C to 41.9°C maintained for 10 to 20 minutes, Yerushalmi (1976) found an increase in the number of lung metastases from the Lewis tumor implanted IM in the leg of mice. This result was not clear-cut, however; the comparative data between heated and control animals was not subjected to statistical analysis, and the effect was less convincing for heating performed on established tumors (6 or 10 days postinoculation) rather than one day after inoculation, when considerable numbers of viable tumor cells would be expected in the circulation. Schaeffer (1976) injected suspensions of Dunn osteogenic sarcoma cells IV into mice and counted the resulting lung tumor colonies 30 days later. Whole-body immersion of the mice in a 42.5°C water bath for 15 minutes on day one after inoculation resulted in a statistically significant increase in colony count. The results may indicate a hazardous effect of whole-body heating on occult metastases, but the tumor model may have little relevance to a host bearing a primary tumor. Further caution in the interpretation of the results of Yerushalmi and of Schaeffer is indicated by the work of Brett and Schloerb (1962). These workers found that the take of the Walker-256 carcinoma in rats was prevented, delayed, or increased by whole-body heating, depending on various time-temperature schedules initiated five to six hours after tumor inoculation.

In recent experiments with mouse fibrosarcoma systems, Rice, Urano, and Suit (in press) found that with a weakly immunogenic tumor whole-body hyperthermia (41.5°C for 60 minutes × 3) led to a significantly increased incidence of lung metastases (79% vs 39% in controls). This did not occur when the tumors were 6 mm in diameter as opposed to 10 mm. In a chemically induced strongly immunogenic tumor, the same protocol of heating did not alter the occurrence of lung colonies. Kapp and Lord (in press) and Kapp, Lord, and Morrow (in press) have studied the effects of fractionated systemic hyperthermia in dogs with spontaneous osteosarcoma. A core temperature of 42°C for 60 minutes was induced by water-bath immersion and followed by 600-rad to the tumor; seven such treatments were given at five-day intervals. The animals were assessed by sequential chest films and bone scans. Six of seven dogs developed early disseminated metastases to the skeleton and other uncommon sites such as muscle, heart, and kidneys—a profound change from previously recorded predominance of lung metastases in untreated or locally treated dogs. These results are of special importance since dogs with spontaneous tumors probably are good animal models for human neoplasia. Canine osteosarcoma closely resembles the human disease in its histologic appearance and biological behavior.

Again, as with local tumor heating, the bulk of data implicating the immune system in the metastatic behavior of tumors after total-body heating is circumstantial. There is, however, a modicum of more definitive findings linking systemic hyperthermia with effects on host lymphoid tissues and immunocompetence.

In a direct comparison of local versus total-body heating using the VX-2 carcinoma in the rabbit, Dickson (1977) and Dickson and Muckle (1972) maintained the tumor at 42.6°C for total of three hours in each case. The rabbits had metastases in the lymph nodes and lungs at time of heating. Total-body heating was much less effective than local tumor heating; the respective cure rates for the two approaches were 7% and 50%. The authors postulated that suppression of the immune system at high body temperature might be responsible for the discrepancy in results, but inadequate heating or even stimulation of the metastases at the systemic temperatures of 40.5°C to 40.8°C achieved in the whole-body heating was also a possibility. As detailed earlier, evidence for this postulate has been provided by the finding of strongly positive in vivo specific and nonspecific cellular and humoral immune responses after local heating of the VX-2 carcinoma, and the abrogation of these reactions by total-body hyperthermia at 42°C (fig. 23.8) (Dickson and Shah 1978; Shah and Dickson 1978a). Many years ago, Doan, Hargreaves, and Kester (1937) described degeneration and fragmentation of lymphocytes in the lymph nodes and bone marrow of rabbits heated above 41°C for over 50 minutes by total-body hyperthermia. These destructive effects were in direct proportion to the height and duration of the temperature. More recently, Shah and Dickson (1978b) subjected normal rabbits to local (thigh muscle) heating at 42°C for one hour on three occasions by watercuff or 47°C to 50°C for 30 minutes by RF heating; rabbits also were treated by total-body hyperthermia at 42°C for one hour on three successive days. The host cellular response to DNCB was not affected by any of these procedures. The anamnestic response to BSA, however, was significantly reduced for three weeks from the second week following hyperthermia, and the depression was independent of the degree and method of heating. Shah and Dickson concluded that, in contrast to radiation, the B-lymphocyte population appeared to be more susceptible to heating than the T cells. The authors did point out, however, that in view of the importance of T cell-B cell cooperation for lymphocyte function and the possible participation of macrophages in this reaction, there may be an indirect component to the damaging effect of heat on B-lymphocytes, as well as a direct destructive action. Williams and Galt (1977) reported similar destructive histologic changes in the lymphoid tissues of rats maintained at 41°C for two to three hours, and also a temporary decrease in ability of lymphocytes from the heated animals to transform with PHA. Also in rats, Fletcher, Cetas, and Wilson (in press) found lymphocyte necrosis in the spleen and liver, and also necrosis of Kupffer cells in the liver, following whole-body heating at 41.5°C for 30 minutes. The liver temperature ran consistently 0.5°C above body temperature. In mice, whole-body microwave-induced hyperthermia has been reported by Roswkowski and Szmigielski (in press) to result in pronounced immunosuppression. Expression of contact hypersensitivity to oxazolone was inhibited, and beneficial effect of preimmunization of animals with killed tumor cells on subsequent challenge with live cells was completely abrogated by the heating. At the same time, cell proliferation in the regional draining lymph nodes was impaired as measured by uptake of ^{125}IUdR.

In animals, therefore, inadequate heating was apparent or must be suspected in a number of the reported cases of enhanced metastasis following local tumor hyperthermia. Although the metabolism and growth rate of several tumors increase at temperatures between 37°C and 41°C, only in the case of the Yoshida tumor in rats has data been proffered connecting this increase with augmented tumor dissemination following local heating. As a possible mechanism, alteration in tumor cell properties, such as mobility or adhesiveness and/or increased "showering" of cancer cells into the circulation induced by the altered local blood flow and tissue homeostasis, has been suggested (Dickson and Ellis 1974). Von Essen and Kaplan similarly attributed the increased frequency of metastasis following tumor irradiation (Kaal 1953; Kaplan and Murphy 1948; Von Essen and Kaplan 1952; Yamamoto 1936) to a transient disturbance of the host-tumor relationship, probably facilitating entry of tumor cells into blood vessels (Von Essen and Kaplan 1952). In most of these experiments the applied radiation dose was inadequate for cure of the primary tumor. It is difficult from the reported data and unsatisfactory experimental conditions in some of this earlier work to be sure what percentage of the tumors regressed, but the radiation was seemingly adequate for primary tumor cure in some cases. In considering the adequacy of any treatment modality to a primary tumor, especially in animal systems, the possible contribution of the host immune system, related to the immunogenicity of the tumor, must be borne in mind. More recent investigations, employing other experimental animal systems, have found a variety of responses to local tumor irradiation, including increased, decreased, and unchanged rates of metastasis compared to untreated or surgically treated animals. These different results have been attributed to the dissimilar tumor sizes employed, to the influence of the recognized growth-supporting capability of radiation-inactivated cells (Révész effect) and to the widely varying tumor-host relationships operating in the animal tumor systems (Von Essen and Stjernswärd 1978). Although the issue remains controversial at the experimental level, the overall conclusion supports the original proposal of Von Essen and Kaplan. Namely, it seems most likely the effect of radiation is principally focused at the level of the local tumor and tumor bed. It appears unlikely that systemic factors play a significant role, although generalized immunosuppression is recognized to occur following local irradiation (Berenbaum 1975; Von Essen and Kaplan 1952). In the clinical situation, the issue has been raised repeatedly, but as in the case of local hyperthermia (see below), there is no hard data indicating that inadequate treatment can accelerate the growth and spread of tumors. In view of the available animal data, however, it would be rash to ignore its possibility in the case of human tumors.

Following total-body hyperthermia, failure to control both local and metastatic disease and also enhanced tumor dissemination have been reported; host immunosuppression owing to a deleterious physical effect on the lymphoid system seems the direct cause of these effects. Total-body heating in laboratory animals is difficult, however, because of poor heat tolerance (Dickson 1977), and the contribution of inadequate tumor heating to the results at the degree (temperature × time) of heating used in these experiments must remain an open question. The analogy with radiotherapy is once more apparent. Total-body irradiation profoundly depresses immune responses in animals, owing to extensive destruction of small lymphocytes (Berenbaum 1975; Stjernswärd 1977). Again, such immunosuppression has been associated with enhanced tumor metastasis in experimental systems (Stjernswärd 1977). Such occurrence in humans must remain speculative, since the approach is not widely practiced at present, and its use generally has been confined to patients with already widespread metastatic disease, as in non-Hodgkins lymphoma.

Humans

The older texts on medical diathermy (Bierman 1942; Kovacs 1949), and also some more recent works (Scott 1957), list malignant disease as an absolute contraindication to heat therapy "because it is liable to cause spreading of the process and also increases the possibility of metastases. This is particularly the case with some of the faster-growing tumors such as sarcomata (Scott 1957). Supporting evidence for this belief is almost impossible to find for human cancer.

As described earlier, a number of workers have detailed changes in various components of the immune system after total-body heating in cancer patients at 40°C to 42°C. No overall or significant pattern of response has emerged from these studies, and no deleterious effects on tumor spread are documented. Since many of these patients had advanced disease, and metastasis cannot be quantified, it is apparent that clear-cut results are not to be expected in this context. Although no data are available on specific tumor-directed immunocompetence in patients treated by whole-body hyperthermia, there are currently no contraindications to this approach in humans.

Generation of the Host Response

The Concept of Altered Tumor Cell Immunogenicity

Several workers have examined the means whereby an immune response might be rallied after tumor heating, with a view to producing host immunity by injection of in vitro heated cancer cells (table 23.7).

Supporting Evidence

In Vitro Heating

Mondovi and co-workers (1972) immunized Swiss mice with Ehrlich ascites tumor cells preexposed in vitro to 42.5°C for three hours. More than 50% of the immunized mice were alive 35 days after challenge with 107 viable tumor cells; the mean survival time of controls was 19 days. Mice could not be immunized against this challenge dose by inoculation of in vitro irradiated cells (10,000 rad). The authors suggested that the results were caused by heat-induced changes in the cell membrane leading to increased immunogenicity of the tumor cells. In support of this postulate, Jasiewicz and Dickson (1976) found that after in vitro heating at 42°C for one hour, there was increased binding of cell-specific antiserum to rat breast cancer cells, and the destructive effect of heat on the cultures was increased fourfold. Similarly, Pantazatos, Tompkins, and Tick (1978) reported markedly increased damage in cells of a human colon tumor cell line incubated at 42°C to 44.5°C in the presence of specific antiserum. The authors concluded that heat modified the cell membrane, facilitating antibody-lytic reactions against surface antigens. Castillo and Goldsmith (1973) furnished evidence that 30% of syngeneic golden Syrian hamsters could be

Table 23.7
Immunization of Normal Animals with Heated Tumor Cells

Tumor	Animal	Heat Treatment	Effect	Investigator(s)
Brown-Pearce carcinoma	Rabbit	Electrocoagulation in vivo ≈ 70°C	+	Strauss 1969; Strauss et al. 1965
VX-2 carcinoma		47°C/30 min in vitro or in vivo	−	Shah and Dickson 1978a
Ehrlich ascites carcinoma	Mouse	42.5°C/180 min in vitro	+ −	Mondovi et al. 1972 Dickson, Jasiewicz, and Simpson, in press
Fibrosarcoma		43.5°C/240 min in vitro	+	Suit, Sedlacek, and Wiggins 1977
D-23 hepatoma	Rat	> 41°C/30 min in vitro*	−	Dennick, Price, and Baldwin 1979
D-23 hepatoma		43°C or 45°C/60 min in vitro	−	This chapter
D-23 ascites hepatoma		45°C/60 min in vitro	−	Dickson, Jasiewicz, and Simpson, in press
MC-7 sarcoma		43°C or 45°C/60 min in vitro	−	This chapter
Yoshida sarcoma		43°C or 45°C/60 min in vitro	−	This chapter

NOTE: + = Heat-treated cancer cells conferred immunity; − = no significant effect.
*These cells were rendered immunogenic by irradiation; subsequent heating (> 41°C/30 min) destroyed the immunogenicity.

Table 23.8
Tumor Immunogenicity and Thermal Sensitivity of Animal Tumors In Vivo

Host	Tumor	Site/Route	Volume (ml)	No. Temp. Sensors	Mode of Heating	Heat Sensitivity (%)	Investigator(s)	Immunogenic	Investigator(s)
Rabbit	VX-2 carcinoma	Leg (IM)	15–20	4 IT	RF	47°C/30 min (75)	Dickson and Shah 1977; Dickson et al. 1977	No	Dixon, Palmer, and Hodes 1960; Shah and Dickson 1978a
Rat	D-23 carcinoma	Foot (SC)	1.0–1.5	1–2 IT	RF	45°C/60 min (50)	Dickson 1978	Slightly	Baldwin and Barker 1967
	MC-7 sarcoma	Foot (SC)	1.0–1.5	1–2 IT	WB	42°C/180 min (100)	Jackson and Dickson 1979	Yes	Baldwin and Pimm 1973
	MC-7 sarcoma	Foot (SC)	1.0–2.5	1–2 IT	RF	45°C/15 min (100)	Dickson, Calderwood, and Jasiewicz 1977	Yes	Baldwin and Pimm 1973
	Yoshida sarcoma	Foot (SC)	1.0–1.5	1–2 IT	WB	42°C/60 min (100)	Dickson and Ellis 1976; Dickson and Suzangar 1974	Yes	Fox and Gregory 1972
Mouse	Mammary adenocarcinoma	Lower abdomen (SC)	0.1	1 SF	MW	43°C/45 min × 2 (100)	Mendecki, Friedenthal, and Botstein 1976	Yes	Baldwin and Embleton 1975
	EMT-6 sarcoma	Flank (ID)	0.1	1 IT	RF	44°C/20 min (100)	Marmor, Hahn, and Hahn 1977	Yes	Rockwell and Kallman 1973
	KHJJ carcinoma	Flank (ID)	0.1	1 IT	RF	44°C/20 min (50)	Marmor, Hahn, and Hahn 1977	Slightly	Rockwell and Kallman 1972
	SV-40 fibrosarcoma	Leg (SC)	0.1	1 IT	MW	44°C/28 min (0)	Yerushalmi 1975	No	Yerushalmi 1975
	KHT fibrosarcoma	Flank (ID)	0.1	1 IT	RF	43°C/30 min × 3 (0)	Marmor 1979	No	Rappaport, in press

NOTES: The figures in parentheses are the percentage cure rates obtained with the degree of heating indicated. The D-23 and MC-7 tumors are syngeneic in Wistar rats; the VX-2 and Yoshida are allogeneic tumors; the mouse tumors are syngeneic in C3H mice (mammary adenocarcinoma, KHT fibrosarcoma) or BALB/C mice (EMT-6 and KHJJ); and the SV-40 fibrosarcoma was maintained in F1 hybrid mice from the inbred BALB/C and C57BL strains. IM = intramuscular; SC = subcutaneous; ID = intradermal; IT = intratumor; SF = surface (skin); RF = radiofrequency (13.56 MHz); WB = waterbath; MW = microwave (2450 MHz).

rendered immune to a methylcholanthrene-induced sarcoma by pretreating the animals with extracts of the tumor heated in vitro at 43°C for 20 minutes. The part played by heat in this study is not clear, however, since unheated tumor extract was not tested.

In Vivo Data

Further suggestive evidence that tumor immunogenicity may influence heat sensitivity comes from recent data on animal tumors. Table 23.8 details the rabbit and rat tumors studied in this laboratory for thermal sensitivity, and available data from other centers on solid tumors in mice are given also. It can be seen that with immunogenic tumors (rat, mice) a 100% cure rate has been obtained by heating at a relatively low temperature and/or short exposure time, that is, a low degree of heating (temperature × time). With slightly immunogenic or nonimmunogenic tumors, a 100% cure rate has not been achieved, and tumor regression and host cure have required a considerably higher degree of heating. The findings suggest that immunogenic tumors are more heat sensitive than nonimmunogenic tumors (Dickson 1978). This connection was discussed almost 40 years ago when Johnson (1940) suggested that Jensen's sarcoma was more sensitive to heat than the Walker tumor because it could induce immunity in rats. More recently, other workers (Dickson, Calderwood, and Jasiewicz 1977; Dickson and Shah 1979; Marmor, Hahn, and Hahn 1977), whose results are presented in table 23.8, have also postulated that immune recognition may be involved in tumor eradication after hyperthermia, and that such recognition may be related to thermal sensitivity.

Recent data of Kim and Hahn (1979) are of relevance to the results in table 23.8 These workers found that with an immunogenic fibrosarcoma in mice, heat sensitivity was related to transplantation site of the tumor. At two different temperatures, the fibrosarcoma was markedly less susceptible to damage the deeper its site, sensitivity being greatest with intradermal tumors (table 23.9). Homogeneous tumor heating is notoriously difficult to achieve, and a cell inoculum will form a tumor at different rates in these various sites. In the absence of tumor volume data and intratumor temperature measurements, the results of Kim and Hahn must be viewed with caution. As emphasized by these workers, there is no doubt that site of implantation is an important variable with respect to the lethal effects of heat, and must be borne in mind when evaluating results from different laboratories.

It is noteworthy that the two tumor types in humans that have been most successfully treated by hyperthermia (using regional hyperthermic per-

Table 23.9
Effect of Transplantation Site on Tumor Control after Hyperthermia

Site of Transplantation	Percentage Tumor Control	
	42.5°C/60 min	43.5°C/30 min
ID	60	90
SC	30	70
IM	10	25

Source: Kim and Hahn 1979.
Note: Cells (0.3×10^6) of a methylcholanthrene-induced fibrosarcoma were inoculated into the thigh region of BALB/c mice. Heat treatment was by water-bath immersion on day 10 after transplantation (tumor volume not specified).

fusion) are melanoma and sarcomas (Dickson 1977). These are the classic types of human solid tumor in which a host immune response against the tumor has been demonstrated. Immune reactivity (level of antibody titer and in some cases, lymphocyte-mediated cytotoxicity toward the tumor cells) often shows a close correlation with clinical tumor burden, reactivity being apparent in early disease, decreasing with advance of the disease and reappearing after successful treatment (Harris and Sinkovics 1976; Sinkovics et al. 1974). It is also relevant to observe that there is a long history of clinical work suggesting that sarcomas are more heat sensitive than carcinomas. This was reported by many of the clinicians who induced hyperpyrexia in cancer patients by inoculation of bacterial filtrates in the early years of this century (Cavaliere et al. 1967; Nauts 1976; Nauts, Fowler, and Bogatko 1953). Within the last 10 years, similar findings have been documented following the use of externally applied heat to tumors, not only by local means (Moricca et al. 1977; Stehlin et al. 1977), but also by total-body hyperthermia in patients with advanced disease (Pettigrew and Ludgate 1977). These numerous clinical reports are in keeping with the results for animal tumors (detailed in table 23.8), which denote the more immunogenic sarcomas to be more heat sensitive than the less immunogenic carcinomas.

Mondovi and colleagues (1972) were the first workers who offered a plausible explanation for the stimulation of a host immune response following tumor heating and at the same time provided definitive findings as evidence. Much of the data in the literature, both on animal and human tumors, was in keeping with the concept of heat-induced immunogenic changes in tumors. The implications of extrapolating the findings and approach to other tumors were obvious: if a tumor as rapidly growing and aggressive as the Ehrlich could be altered in vitro by heat so that it induced immunity in the host to a dose of viable cells as large as 10⁷, then the potential for hyperthermia as a therapeutic modality in cancer would be increased enormously it seemed. Hence, it was essential that the widely quoted work of Mondovi and associates should be independently verified.

Opposing Evidence

In Vitro Heating
Dickson, Jasiewicz, and Simpson (1982) have recently investigated the Ehrlich tumor system in mice to define the conditions (degree of heating, relative importance of viable nonreplicating vs nonviable cells) promoting the immune response reported by Mondovi and co-workers. Using Ehrlich cells of the same ploidy, mice of the same strain, sex, and weight, and also similar incubation conditions to these workers, Dickson, Jasiewicz, and Simpson (in press) were unable to confirm the results of Mondovi and his group. Cells were heated at 42.5°C for up to six hours, at 45°C or at 60°C to produce populations of different percentage viability. These then were separated into viable and nonviable cells by ficoll-hypaque density centrifugation; viable cells were treated with mitomycin C to produce nonmultiplying metabolizing populations. None of these various separated populations of different viability after various degrees of heating produced a significant increase in survival of mice challenged by viable Ehrlich cells in doses as low as 10³ cells. On the other hand, mice inoculated with in vitro irradiated cells (15,000 rad) were alive 200 days after challenge with 10³ Ehrlich cells; control animals succumbed to this challenge in 15 ($\pm$1) days. These investigators also examined a syngeneic ascites tumor in rats, the D-23 hepatocellular carcinoma; this tumor is not heat sensitive below 45°C. Again, populations were heated in vitro for up to three hours and then separated into nonviable and viable cells, subpopulations of which were treated with mitomycin C; the metabolic state of each population was quantified by Warburg manometry, as for the Ehrlich cells. Rats were given 10⁷ nonreplicating cells IP at 10-day intervals and challenged 10 days later with 10⁵ live cells (TCD50 = 10⁴ cells). All rats died within the same time as controls (25 days). Following three inoculations of x-ray treated (15,000 rad, 240 KV) D-23 cells (10⁷), all rats survived challenge with 10⁵ viable cells 10 days later and were still alive 200 days after challenge. Similarly, Suit, Sedlacek, and Wiggins (1977) found that, after irradiation (10,000 rad) in vitro, methylcholanthrene-induced fibrosarcoma cells conferred better transplantation immunity in mice than in vitro heated (43.5°C for four hours) cells.

The results of Dickson, Jasiewicz, and Simpson (in press) and those of Suit, Sedlacek, and Wiggins (1977) are more in keeping with the general finding that under a considerable range of conditions and for a variety of tumors, radiation has proved to be the most effective agent for modifying the antigenicity of tumor cells to induce protective immunity in animal hosts (Prager and Baechtel 1973). Dennick, Price, and Baldwin (1979) reported that the ability of irradiated D-23 hepatoma cells to protect host rats against tumor challenge was destroyed by heating the irradiated cells at temperatures above 41°C; this effect was accompanied by significant inhibition of labeled thymidine, uridine, and leucine uptake into the cells. The authors postulated that interference with the metabolic integrity of the cell by heat treatment may have led to disruption of cytoskeletal elements and/or membrane fluidity in such a way as to alter specific spatial antigen configurations at the cell surface.

Attempts to immunize normal rabbits with in vitro heated VX-2 carcinoma, using various injection protocols, have been unsuccessful (Shah and Dickson 1978a). Similarly, we failed to induce immunity in rats by repeated injections of in vitro heated tumor slices; the cancers have included the syngeneic D-23 and MC-7 tumor and the allogeneic Yoshida sarcoma.

In Vivo Heating
Suit, Jasiewicz, and Simpson (1977) reported that after in vivo heat destruction (43.5°C for two hours) of a fibrosarcoma, the immunity induced in mice was no better than that following curative irradiation (5000 rad) or leg amputation. The results detailed earlier indicate that in rabbits and in rats there is an increase in host immunocompetence following curative tumor heating. There is little evidence, however, that this response is evoked by an altered immunogenicity of the heated cells; although tumor directed, the response is not tumor specific, and we have failed to detect any new antigens in the tumor cells after heating (within the limitations of the methodology used).

Strauss (1969) and colleagues (1965) observed that if the Brown-Pearce tumor was transplanted seven days after electrocoagulation, no growth occurred in normal rabbits, although the injected material contained viable-looking cancer cells; the animals subsequently resisted repeated challenge with fresh carcinoma. Slices removed from VX-2 tumors up to 17 days post-heating and inoculated into rabbits killed the hosts within 10 weeks, but the tumors regressed in the treated host (Shah and Dickson 1978a). In mice also, Marmor, Hahn, and Hahn (1977) found that after curative heating of the EMT-6 sarcoma, a large number of tumor cells remained capable of replication as determined by plating efficiency. These findings imply that delayed mechanisms play a significant part in tumor eradication in animals (Marmor, Hahn, and Hahn 1977; Shah and Dickson 1978a), and are in keeping with evidence that viable actively metabolizing cells induce a more effective and longer-lasting immunity than dead cells or acellular tumor antigen preparations (Davidson 1977; Dennick, Price, and Baldwin 1979; Everson and Cole 1966; Wagner 1973; Wagner, Rollinghoff, and Nossal 1973). The data also recall the earlier results of Foley (1953) and of Lewis and Aptekman (1952). These workers immunized mice and rats to tumors by occluding the tumor blood supply with a ligature or tourniquet. If the constriction was applied too tightly, this prevented resorption of the strangulated tumor and inhibited the production of resistance in the host to subsequent tumor implantation. Further experiments of this type in the early 1950s (reviewed by Pelner 1958) led to the conclusion that the optimum conditions for inducing immunity in rats were (1) that the tumor must actually grow in the host, (2) that it must die in the host, and (3) that it must degenerate under conditions that permit relatively slow absorption of degeneration products. Absorption of the products of dead and dying cells may also have played a part in the apparent immunity that followed tumor heating in the clinical studies of Strauss, Cavaliere, and Stehlin, since total tumor lysis took place over a delayed period (Moricca et al. 1977), and viable-looking cells were sometimes found in cancers removed years after treatment (Strauss 1969). In the experiments of Foley (1953) and of Lewis and Aptekman (1957), complete isolation of the tumor from the host did not induce immunity. Similarly, removal of tumors from humans immediately after heating has not led to immunity, but to rapid tumor recurrence (Cavaliere, Moricca, and Caputo 1976; Moricca et al. 1977). These findings concur with the conclusion that the beneficial effects following tumor destruction depend upon resorption of the tumor or its products (Pelner 1958).

Mechanism of Action of Heat

It appears from the various findings reviewed that generation of a host response following tumor hyperthermia occurs chiefly, if not exclusively, with tumors heated in vivo. The question arises as to the mechanism whereby the heated cells induce this response.

A multitude of different cell types, including lymphocytes, macrophages, and polymorphonuclear leukocytes, as well as their interactions, are now believed to operate in antitumor mechanisms in the host (Harris and Sinkovics 1976). Failure of immunologic control of cancer in the host is currently believed to be due in large part to antigen shed from the tumor cell surface and circulating as free antigen or as antigen complexed with antitumor antibody (Cochran 1978b; Currie 1980; Harris and Sinkovics 1976; Moore 1979). This antigenic "smokescreen" (Currie 1980) provides a protective cover for the tumor and at the same time inactivates the host defenses. It is apparent that in such a complex situation, the mechanism of action of heat is unlikely to be a straightforward one. The concept of tumor necrosis after heat leading to "antigenic products which are absorbed in the system and result in an immune response" (Strauss et al. 1965), or stimulation of antitumor immunity by antigenic material liberated when the tumor is destroyed in loco (Stehlin et al. 1977), or that the presence of the treated tumor "acts as an antigenic source to immunize against recurrences and metastases" (Moricca et al. 1977) are postulates that are difficult to envisage as helpful, or even as operating, in a host whose system already is swamped by excess tumor antigen.

In this connection, a recent report of Wallace and colleagues (1981) is of great interest. These workers treated a series of 49 patients with renal carcinoma by renal artery embolization followed by nephrectomy four to seven days later; all patients had metastases primarily to lung and/or bone. A response (50% or greater reduction in all metastatic lesions or stabilization of disease for at least 12 months) was noted in 18 (36%) of the patients. In seven patients, response was complete with disappearance of all metastases. Median survival in the group was 14 months, in contrast to six months for a control group of patients treated by nephrectomy alone. In the patients subjected to kidney infarction by embolization, immunologic reactivity improved after the procedure, as assessed by an increase in delayed hypersensitivity to a battery of recall antigens. Since spontaneous remission of metastases following nephrectomy alone occurs in less than 1% of cases (Everson and Cole 1966), Chuang and Wallace (in press) believe that after infarction the ischemic renal tumor may act as an "attenuated antigen" to stimulate an immune response. It is envisaged that infarction releases into the circulation a sudden shower of tumor antigen that abruptly alters the relative concentrations of tumor antigen and antibody. This, it is maintained, would affect the amount and solubility of circulating immune complexes, regardless of whether the patient had been in antigen or antibody excess. Such an alteration in quantity and efficiency of serum blocking factors could result in a heightened tumor-directed cell mediated immune response against metastatic foci after the tumor burden has been reduced by removal of the primary (Wallace et al. 1971). The postulate of alteration in circulating immune complexes is amenable to test by measurement of such complexes using, for example, the $C1_q$ binding assay. Wallace and colleagues state that they have made serial $C1_q$ measurements on a few patients before and after infarction and subsequent nephrectomy, and that "a consistent pattern of change has been noted." No data are provided for evaluation, however.

The concept of tumor debulking as a means of freeing or liberating the captive immune system has also been discussed by Ablin and Guinan (1979) in relation to the disappearance of metastases following cryosurgery of the prostate. These workers believe that the increased host immunocompetence after cryosurgery may be attributable not to immunopotentiation but to antigen depletion. Destruction of the tumor is viewed as removing the primary source of tumor antigen, thereby permitting the previously overwhelmed host to respond. Again, measurement of antigen-antibody complexes in the serum may help to answer this question. Cryosurgery has not produced consistent alterations in cell-mediated or humoral immunity in the case of the prostate, however, and with the rectum, antitumor antibody decreased after cooling the tumor (Langer et al. 1979). Nevertheless, antitumor antibody may not be a relevant indicator of host immunocompetence, even if the major component of host resistance *is* immune.

A fascinating aspect of the work of Wallace and associates is that embolization of the renal tumor by gelfoam and stainless steel coils resulted in an inflammatory reaction, with edema of the kidney and tumor. Fever occurred as part of a postembolization syndrome, which lasted for one to five days. Similar features, local inflammation and fever, were features of the response to Coley's toxins when the filtrates were given by intratumoral inoculation.

A complicating and unknown factor in the results of Wallace and co-workers (1981) is that, as well as embolization and nephrectomy, the patients received hormonal therapy in the form of methoxyprogesterone acetate (Depo-Provera™, 400 mg IM twice weekly). Administration of this progestogen has resulted in response rates of up to 17% in metastatic renal carcinoma, although in most recent series the response rate has been less than 5%.

We have examined recently the concept of breakdown products that may arise following tumor heating using three approaches: in vitro heating of cells that were then left to die slowly with production and leakage of degradation products, in vivo heating of tumors and collection of extracellular fluid for analysis, and inhibiting egress of blood from tumors after curative hyperthermia.

Tumor Cell Degradation Products In Vitro

The findings already discussed that regressing tumors may contain viable cancer cells and that viable metabolizing cells may be important in inducing immunity suggested that immunostimulation following heating may involve a characteristic pattern of lysis and resorption in a population of dead, dying, and viable tumor cells. Ehrlich ascites cell populations of different percentage viability were obtained by heat treatment at 42.5°C, 45°C, or 60°C, and then rendered nonreplicating with mitomycin C. These populations were left in vitro to die slowly over three days at 37°C, and then injected into mice on an immunization protocol. No survival benefit occurred in such mice (Dickson, Jasiewicz, and Simpson, in press). It appeared, therefore, that (in vitro, at least) no delayed heat-induced antigenic modifications occurred in the cancer cells, nor was there release of immunogenic products by dead or dying cells.

Analysis of Tumor Extracellular Fluid

In vivo, we have examined the extracellular fluid (ECF) of tumors for changes in composition following heating. For this, flank tumors were grown in rats around subcutaneously implanted millipore membrane diffusion chambers (figs. 23.12, 23.13) after the method of Gullino, Clark, and Grantham (1964). The ECF that accumulated in the chamber was collected via a Silastic™ tube draining onto the back of the animal at the neck (figs. 23.12, 23.13). The soluble proteins shed from the plasma membrane or leaking from the cells

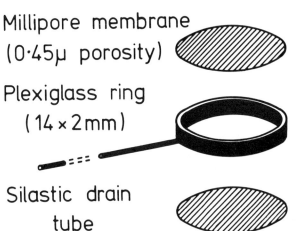

Figure 23.12. Exploded view of double membrane chamber for SC implantation. The porosity of the millipore membrane allows fluid but not cells to enter the chamber. The volume of the chamber is approximately 0.5 ml.

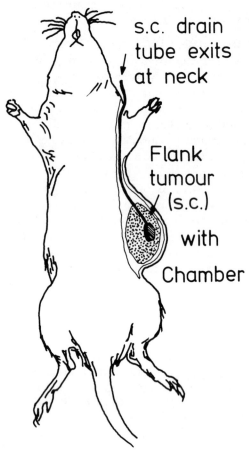

Figure 23.13. Chamber implanted in flank of rat with Silastic sampling tube placed subcutaneously to exit at back of neck. The rat is free-running, and is only restrained for chamber sampling. Initially, small pieces of tumor are placed around the chamber, which then becomes embedded in the growing tumor. Between sampling, the drain is closed by a small bung. The chamber can be drained at the rate of approximately 0.1 ml/24 hr.

were labeled with ^{125}I by the iodogen method of Salacinski and colleagues (1979) and then analyzed by electrophoresis on polyacrylamide gels. The profiles of labeled polypeptides obtained by this method show a shift from a predominantly high molecular weight pattern in untreated D-23 and MC-7 tumors to one in which the major component(s) after curative tumor heating is (are) in the low molecular weight region (fig. 23.14). This change was not seen after curative irradiation of the tumors. The profiles in figure 23.14 are also different from those obtained with normal rat serum before or after heating. It is of note that the molecular weight of the major polypeptide detected after heating is in the region of 15×10^3 daltons, well below the value of 50 to 60×10^3 daltons reported by Baldwin, Harris, and Price (1973) for the molecular weight of the D-23 and MC-7 tumor-associated antigens.

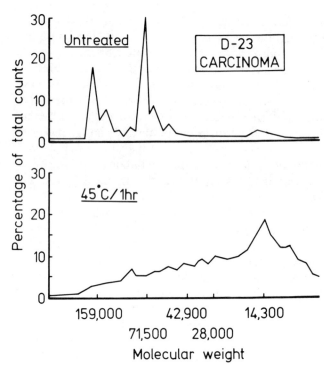

Figure 23.14. Electrophoretic patterns of iodinated proteins from D-23 tumor interstitial fluid following treatment. Each pattern represents 1 μl of interstitial (extracellular) fluid labeled with ^{125}I (Salacinski et al. 1979) before electrophoresis on phosphate buffered 8.5% polyacrylamide gel at 5 mA for 16 hours. The origin is on the *left* with the bromophenol blue front marker on the *right*. Mobilities of the molecular weight markers (B.D.H.) used to calibrate the gels are shown on the *abscissa*. Similar patterns were obtained after electrophoresis of iodinated samples of interstitial fluid from MC-7 tumors, before and after heating. (Dickson et al. 1980. Reprinted by permission.)

Inhibition of Tumor Blood Flow

In the MC-7 and D-23 tumors, tumor blood flow was inhibited by glucose loading of the host (Dickson et al., in press). Normal tissue blood flow remained unaffected or was increased. At a blood glucose level of 50 mM/l (1000 mg %) there was a rapid and complete inhibition of tumor blood flow (fig. 23.15). That this inhibition was 100% was evidenced by failure of exchange of substances such as water, chloride, glucose, and organic acids between the tumor and host, and by tumor angiograms obtained before and after glucose loading (Calderwood and Dickson 1980).

When tumor blood flow was inhibited for 12 hours by hyperglycemia and the rat tumors subjected to curative heating (table 23.1), tumors regressed as in normoglycemic hosts, but the animals died from metastases.

Uniqueness of Tumor Response to Hyperthermia

The abscopal response has formed a focal point in the

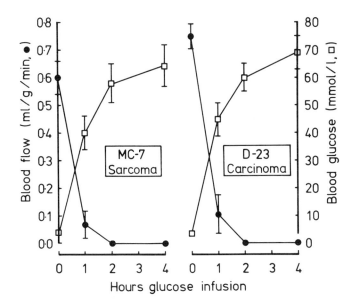

Figure 23.15. Inhibition of blood flow in the MC-7 sarcoma and D-23 carcinoma at elevated blood glucose levels. Hyperglycemia was achieved by infusion of 2% glucose into the femoral vein. Blood flow was measured by fractional distribution of ^{86}Rb in the tissues (Dickson and Calderwood 1980) 40 seconds after injection of 100 μCi isotope in 0.1 ml of 0.9% NaCl into the femoral vein. Blood glucose and blood flow values are means (± 1 SD) derived from five to seven animals. (Dickson et al., in press. Reprinted by permission.)

debate on the generation of a host response, immune or otherwise, following tumor heating. We must now address the question of whether such a response is unique to tumor treatment by hyperthermia.

Spontaneous regression of metastases following surgical removal of primary tumors is a recognized phenomenon in some animal tumor systems (Dickson 1977), and also has occurred occasionally in humans, especially with tumors of the breast and kidney. It is usually assumed that in such cases the host response has at least an immune component, although no supporting data is available. The significantly increased regression rate of lung and bone metastases following renal artery embolization and delayed nephrectomy for renal carcinoma (Chuang and Wallace, in press; Wallace et al. 1981) has been discussed earlier. Surgical removal of a primary tumor is an accepted method of inducing high-titer antibody in animal hosts (Harris and Sinkovics 1976), and Graham and Graham (1955) detected circulating antibodies in a number of patients only after excision of the primary cancer. The significance of such findings in relation to host immunocompetence is open to doubt, however, especially in view of recent work by Dennick, Price, and Baldwin (1979), showing that the presence of tumor-specific antiserum may not be related to the ability of a host animal to reject tumor. Regression of distant, untreated metastases has occurred occasionally in cases of melanoma, treated by local immunotherapy, such as BCG. In these cases, both circulating antibodies and cellular immunity appear to be responsible for tumor destruction (Morton 1972). Remission of metastatic disease in the lungs, bones or lymph nodes has also been described by several authors following cryosurgery (−15 to −20°C for two minutes or more) for carcinoma of the prostate (Ablin and Guinan 1979) and rectum (Langer et al. 1979) in humans. In the case of prostate, this abscopal response was accompanied by the presence in the serum of complement-mediated cytotoxic antibodies to the gland. The serum α_2 and γ globulins also were elevated in most of these patients, and the authors emphasized that an increase in general rather than specific immunocompetence might be the important factor in host response (Ablin and Guinan 1979).

The Inflammatory Response

References to participation of the inflammatory response in tumor regression can be traced through the history of cancer research, just as mention of the abominable snowman permeates the folklore of the Himalayas. The early monographs on cancer therapy as practiced over a century ago reveal that these early physicians induced severe inflammation at the site of the neoplasm by a variety of irritating or destructive agents such as poultices, vesicants, and caustics. At the same time, they stimulated the organism as a whole by the injection of bacterial or plant products at sites remote from the tumor (Tanchou 1844). William Coley, therefore, was by no means the first to indulge in stimulation of host defenses, nor to encourage local tissue reaction to a tumor.

In 1922, MacCarty (1922) first called attention to the favorable prognostic effect of chronic inflammatory infiltration in carcinoma of the breast, stomach, and rectum. MacCarty regarded the infiltrate "as a possible part of the defense mechanism." The beneficial effect of a host inflammatory reaction in tumor rejection has also been claimed in various animal model systems. The reports go back to the early part of the century also (1911–1912), but these initial observations on animal and human tumors lay unappreciated for more than three decades. The findings have been confirmed more recently for both spontaneous and transplanted tumors in outbred and inbred animals (Berman 1975).

In a long series of investigations that began in 1953, Black and his colleagues (1975) have investigated the relationship between lymphoid cell infiltration in human tumors (especially breast cancer) and the presence of histiocytes (macrophages) distending the sinusoids of regional lymph nodes (sinus histiocytosis). With in situ developing breast carcinoma, Black (1975) found that 70% of cases of noninvasive cancer had lymphoid cell infiltration of the breast in the region of the lesion and a concomitant sinus histiocytosis. These changes were interpreted as indicating immunogenicity in the developing cancer and a more favorable five-year survival rate. Other authors generally have confirmed Black's findings in regard to the prognostic significance of sinus histiocytosis, but only for breast cancer; it appears that axillary lymph nodes are particularly prone to this type of reaction (Berman 1975).

Black thus brought together a local inflammatory reaction (or local immune reaction, since lymphocytes are mediators of both) and a regional response of the reticuloendothelial system. Black termed this a response of the *lymphoreticuloendothelial system*, and the finding served to emphasize that the biological behavior of breast cancer did not depend solely upon properties of the cancer cells but reflected tumor-host interactions. Evidence is accumulating that cells of the classic inflammatory response (polymorphonuclear leukocytes, macrophages) play an important part in the immune response (McCluskey and Cohen 1974; Weiss 1972); the role of lymphocytes in the immune response is, of course, now well accepted. The inflammatory response can be initiated by immune reactions and by toxins. Most cytotoxic drugs depress not only the immune response but also inflammatory reactions. A sharp distinction between the immune response and the inflammatory response is frequently impossible to delineate, and in this perspective, inflammation can be regarded as a part of the entire immune response (Barland 1973). Cells of the reticuloendothelial system (mononuclears, macrophages, histiocytes) are now classed as a major component of the immune system (Hanna et al. 1972).

Although it is now well recognized that a more favorable prognosis usually is associated with lymphocytic infiltration in three histologic types of human tumor (medullary carcinoma of breast, testicular seminoma, and lymphocyte/histiocyte predominant forms of Hodgkins disease), the significance of inflammatory infiltrate in relation to other types of solid tumor continues to be debated in the literature. The picture is far from clear regarding the role of chronic inflammatory cells in these various tumor types. The cells possibly may have afferent or efferent (cytotoxic) immunologic function, both or neither, perhaps depending on the particular tumor-host system involved. Berman (1975) has recently reviewed the subject, with full references. An important point, emphasized by Berman, is that even if heavy chronic inflammatory infiltrate in an established tumor always indicated a host response, it still would represent only an effort to catch up to a neoplastic process, which in most cases was already getting out of hand.

Since heat represents an injurious physical agent, it is not surprising that the inflammatory reaction should be involved in the response that follows tumor heating. An association between hyperthermia and inflammation has been recognized for many years (table 23.10). When easily accessible tumors, such as those of the limbs, are heated, the cardinal signs of the inflammatory response—redness and swelling, with heat, pain, and disturbance of function—are easily observed. The inflammatory response is also a defense mechanism against infection, and hence it occurred when Coley's toxins were inoculated directly into tumors in the early days of the toxin era. Some workers in cancer thermotherapy have regarded inflammation as more than a natural accompaniment of heating. Crile (1962, 1963) believed that regression of S180 in mice after heating at 44°C for 30 to 45 minutes was due to the intense post-heating inflammatory response that developed in the tumor-bearing foot. The tumor grew well on transplantation into fresh mice for up to four hours after heating, which Crile regarded as the time required for an effective inflammatory response to develop. This response could be minimized by a short preheat (15 to 30 minutes at 44°C) of the tumor-bearing foot on the day before hyperthermia, but the cure rate of the S180 also was reduced significantly when the previously curative dose of heat was given. This result could have been due to the tumor being rendered thermotolerant by the preheating, but Crile

Table 23.10
Hyperthermia and Inflammatory Response

Investigator(s)	Response
Coley (Nauts 1976; Nauts, Fowler, and Bogatko 1953)	Inflammatory/Immune response
Crile 1962, 1963; Crile and Turnbull 1972	Postheating inflammatory reaction
Cavaliere et al. 1967; Cavaliere, Moricca, and Caputo 1976	Inflammatory reaction of the perfused limb (14 days)
Stehlin et al. 1977	Tissue reaction following perfusion (10 days)
Strauss 1969	Electrocoagulation
	Viable cancer cells enmeshed in fibrous tissue

NOTE: Coley (and subsequently his daughter: Nauts) injected patients or their tumors with bacteria or bacterial extracts, the classic stimulus for both the inflammatory and the immune responses. In the work of Cavaliere and co-workers (1980) and of Stehlin and associates (1977), reaction of the normal limb tissues following regional hyperthermic perfusion (redness, swelling, pain, loss of function) persisted for 10 to 14 days after heating.

(1963) showed further that the curative effect of heat was greatly augmented by injection of serotonin (one of the chemical mediators of inflammation) into S180 or S91 tumors.

From the clinical experience with hyperthermic limb perfusion, Cavaliere, Moricca, and Caputo (1976) have repeatedly stated their belief that the destructive action of heat on tumors requires an inflammatory reaction in the perfused limb. The postoperative swelling and venous stasis of the limb represented a beneficial response on the part of the normal tissues, and should not be avoided, but considered as one of the necessary components of hyperthermic treatment. The swelling and stasis could be minimized by perfusing the limb with a 1.6% solution of sodium bicarbonate after hyperthermia. There was an increased incidence of tumor recurrence in such cases, however (Moricca et al, 1977). A biochemical feature of inflammation is the acid pH that develops at the site (Menkin 1956). Although the evidence is purely circumstantial, it is tempting to speculate that Moricca and colleagues, by using postheating sodium bicarbonate, may have been neutralizing an important chemical participant in the inflammatory response. Stehlin and co-workers (1977) do not specifically discuss an inflammatory reaction in their patients, but they do refer to a "tissue reaction following perfusion," which probably is of a similar nature to that emphasized by Cavaliere. The work of Strauss (1965) indicates that the benefits of inflammation may extend beyond the acute reaction. In his treatise on electrocoagulation, Strauss describes cases in which viable-looking cancer cells were observed enmeshed in constricting connective tissue years after carcinoma of the rectum had been successfully treated.

There is, therefore, considerable circumstantial evidence implicating components of the inflammatory reaction in the response of tumors to heat. Infiltration of tumors after heating by the classic cells of the acute inflammatory response (polymorphonuclear leukocytes) is not a characteristic feature of the response to hyperthermia (Dickson 1977; Muckle and Dickson 1971; Overgaard and Overgaard 1972). This may favor Crile's postulate (1962, 1963) that it is the chemical mediators of inflammation that are involved.

OVERVIEW

The current state of the art in relation to tumor hyperthermia and host response provides instructive guidelines in relation to the value of in vitro experiments and the relevance of animal studies, as well as our philosophy concerning tumor immunology.

With due attention to appropriate controls (there is often disagreement as to what experimental conditions fulfill this requirement) and purified system components (absorbed antisera, "correct" lymphocyte to target cell ratio, etc.), results have been obtained indicating that interaction between tumor cells and lymphocytes or specific antiserum may be potentiated at elevated temperature (42°C region). There is only very limited (and questionable) data that following tumor heating in vivo, a *specific* antitumor immune response is evoked. Similarly, there is little convincing evidence that such a response (even a *non*specific one) can be generated by inoculation of in vitro heated cancer cells, tumor slices, or products of their breakdown in vitro.

From the results of work in the 1950s on tumor ligation, a host response of ill-defined nature can be assumed to occur following tumor breakdown, and more recent results with rabbit and rat tumor models indicate that the host response has a tumor-directed but nonspecific immune component. The response can be regarded as *autoimmunotherapy*, which is very effective in destroying the tumor and its metastases and rendering the host immune to further tumor inoculation. There is evidence that a major component of the response involves macrophage participation. The response is evoked by what we have simply termed "tumor breakdown products" (Dickson and Muckle 1972) entering the bloodstream. These products are tumor-associated in that if the heated tumor is isolated from the host by inhibiting access or egress of blood, metastases do not regress (Dickson et al., in press) (presumably because macrophages and/or lymphocytes or other white cells cannot gain access to the lysing tumor to be "armed," and/or breakdown products cannot leave the tumor); the response does not occur after heating normal host tissues, such as muscle (Shah and Dickson 1978a).

There are some suggestive data that generation of the host response may be governed by the way in which tumor cells die and are resorbed, and that the necrosis following tumor heating may have some specificity in this regard. Detailed studies on mouse mammary carcinomas have led Overgaard and Overgaard (1972) to conclude that the histologic appearance after heat treatment is specific and differs from the usual necrotic and necrobiotic processes seen in tumor tissue. Dickson and Suzangar (1974) found that the histologic appearance and metabolism of regressing Yoshida tumors in rats differed following treatment of the tumor by heat or drugs. The primary VX-2 tumor in the rabbit leg can be destroyed by radiotherapy, but tumor regression is not accompanied by disappearance of metastases (Muckle and Dickson 1973). Similarly, in the clinical situation, osteogenic sarcoma can be totally destroyed by megavoltage radiation (TD 8000 to 10,000 rad), but the patient still dies from metastases (Hoeffken 1976). On the other hand, Moricca (1977) and Stehlin (1977) and their associates have been impressed by the disappearance of metastases following hyperthermic perfusion for melanomas and sarcomas of the limbs. These results, in animals and in humans, all concern tumors that are recognized as being immunogenic. The evidence that heat sensitivity of tumors is related to tumor immunogenicity has been presented earlier. They also apply to hyperthermic temperatures of 42°C to 43°C (in the case of Stehlin et al. to 40°C in association with cytotoxic drugs). There is evidence that the majority of the common solid tumors of humans do not exhibit marked immunogenicity in the autochthonous host, and that higher temperatures in the 45°C to 50°C range will be required to produce tumor destruction (Dickson 1978; Dickson and Calderwood 1980; Dickson and Shah 1977; Storm, this volume, chapter 15; Storm et al. 1979, in press). Table 23.11 shows that a host response (defined or on the basis of circumstantial evidence) has been reported after tumor heating at temperatures ranging from 40°C to 70°C. Any hypothesis that heat produces characteristic or specific tumor cell damage and/or absorption of breakdown products must account for the effect encompassing such a wide range. It is apparent that the host response following tumor hyperthermia is not unique. The primary requisite for tumor regression and host cure is destruction of the tumor, and heat may be regarded simply as another and convenient means of producing tumor destruction, similar to ischemia or hypothermia. There are now substantial grounds for believing that immunogenicity of the type and degree possessed by the great majority of tumors used experimentally (and including the tumor ligation work of the 1950s and more recent hyperthermia results) is a laboratory artifact largely restricted to neoplasms induced by artificial means (Hewitt 1978; Moore 1978b). Even under the ideal controlled conditions afforded by these highly immunogenic tumors, there is little evidence that the host response to tumor destruction is tumor-specific. It is not unexpected, therefore, that evidence for the development of an immune response or augmentation of host resistance in humans following hyperthermia is less substantial than in animal tumor systems.

The belief that a proportion of human tumors may be capable of evoking immune responses is based upon in vitro studies, which are being subjected to increasingly critical scrutiny regarding their relevance to in vivo tumor rejection. Although there are clues from clinical medicine that immunologic reactions to human tumors may occur—such as cases of spontaneous regression of cancers, long periods of dormancy between removal of primary tumors and the development of metastases, and a higher prevalence of cancer found at postmortem examination than in clinical practice (Cochran 1978b; Currie 1980)—the presence of a growing tumor in itself implies a failure of the body's defenses. There is no unequivocal evidence that immunologic mechanisms can control the growth of established tumors in humans. Likewise, a review of the long history of immunotherapy in the treatment of human cancer has emphasized the singular lack of success with this approach (Currie 1972). There is, however, no doubt regarding the great lethal potential of macrophages and of the white blood cells (leuko-

Table 23.11
Stimulation of Antitumor Immune Response

Temp.	Subject	Investigator(s)
40°C	Humans	Coley (Nauts 1976; Nauts, Fowler, and Bogatko 1953), Stehlin et al. 1977
42°C	Rats	Dickson and Suzangar 1974
43°C	Humans	Cavaliere et al. 1967
43°C	Mice	Mendecki, Friedenthal, and Botstein 1976
44°C	Mice	Marmor, Hahn, and Hahn 1977
45°C	Rats	Dickson, Calderwood, and Jasiewicz 1977
47°C	Rabbits	Shah and Dickson 1978a
70°C	Humans	Strauss et al. 1969, 1965; Crile and Turnbull 1972

NOTE: Involvement of antitumor immune response (circumstantial or defined) in tumor regression after different elevated temperatures. Coley's patients had clinical fever and Stehlin and co-workers used regional hyperthermic perfusion at 40°C with chemotherapy. There is no evidence that 40°C per se is deleterious to tumors. The other temperatures were achieved by local heating of tumors.

cytes as well as lymphocytes) for tumor cells. The results from the era of cancer biotherapy are consonant with the view that a host reaction against cancer can be mobilized after tumor destruction. Such a response, although only tumor-directed, can be highly effective in humans and also in animal tumor systems.

In addition, it is generally accepted that the immune system is capable of dealing with only a small number of tumor cells, the minimal residual burden of the host (Biano 1975; Currie 1980; Harris and Sinkovics 1976; Kaiser and Reif 1975). In the tumor-bearing rabbits and rats studied in this laboratory, postmortem examination on numerous animals at the time of curative heating of the primary tumor revealed a considerable amount of tumor in the regional and paraaortic lymph nodes and in the lungs. As pointed out by Kaiser and Reif (1975), our concept of minimal residual mass may be too restricted, and the maximal number of tumor cells that can be eliminated may depend upon the host and its state of health, the tumor type, and numerous other nonimmunologic factors such as anatomic distribution of the tumor and normal tissue environment of tumor foci. It has been demonstrated, for example, that different mouse tumors possess widely different ranges for the maximal number of tumor cells that can be eliminated by the host. Nevertheless, regression of such amounts of residual tumor would be more in keeping with the participation of activated (armed) macrophages in association with other cell types than with the effects of tumor-specific killer T-cell activity. It is also of note in this context that the formation of autoantibodies has been reported following tumor necrosis; the presence of autoantibodies to normal tissues, such as muscle, in cancer patients is also well recognized (Cochran 1978b).

It may well be, therefore, that we have allowed ourselves to become too obsessed with the importance of tumor-associated antigens and the concept of a specific host response against them. As pointed out by Cochran (1978b), "Tumor associated antigens may merely be interesting epiphenomena and the immune responses which they evoke may have no significant capacity to cause cytostatic or cytotoxic effects." The concept of a less specific host response generated by tumor necrosis is not only in keeping with the data, both experimental and clinical, but enables a number of inconsistencies (e.g., regression of large tumor burden, requirement for long-term stimulation in biotherapy, difficulty in detecting tumor-specific reactions both in vivo and in vitro) to be viewed in a fresh light, if not to be explained.

The logical outcome of such considerations is the use of more refined immunostimulants, such as BCG or *C. parvum*, in association with hyperthermia. Although there is now a stronger rationale for employing such reticuloendothelial stimulation, experience in the use of such adjuvants to date is very reminiscent of that with Coley's toxins. Different preparations of BCG or *C. parvum* vary widely in their physical, chemical, and cultural characteristics, in their capacity to affect macrophages and lymphocytes in vitro, and in their capability to induce tumor regression in animal models. We still know very little regarding the optimal dose, route of administration, and timing of such adjuvants in animals or in humans. Although some of the problems encountered with BCG and *C. parvum* may be overcome by using other reticuloendothelial stimulants, a whole host of such materials is available. Even if good animal tumor models with characteristics (e.g., growth rate, metastatic properties, cell cycle characteristics, antigenicity) akin to human cancer were available, a "massive task of evaluation" would be needed to confirm or refute the utility of these substances against tumors (Cochran 1978b).

It is of note that stimulants such as BCG and *C. parvum* induce a chronic inflammatory response (granuloma) at the site of inoculation, and this may in part account for their effect when given by intralesional injection (Dickson 1977). Adjuvants (such as complete Freund's adjuvant) used to induce a powerful host antibody response in addition provoke a vigorous acute inflammatory response at the site of injection (Waynforth 1980). This attracts white blood cells and macrophages into the area, thus ensuring maximum contact between the antigen and host defenses. A similar mechanism may operate in the case of tumors treated by local hyperthermia. An inflammatory response undoubtedly occurs in this form of heating, and the work of Cavaliere, Moricca, and Caputo (1976) and Moricca and colleagues (1977) on regional perfusion would appear to confirm its value. There are a number of chemical mediators of the inflammatory response (Movat 1979), including serotonin, and the value of inducing an inflammatory response in the region of a tumor by such agents as a concomitant to hyperthermia deserves exploration. It is also of more than passing interest that the commonly employed immunostimulants (BCG, *C. parvum*) and the currently popular interferon all have as a side-effect (usually regarded as an undesirable effect) the induction of clinical fever (39°C to 40°C for four to eight hours). The local inflammatory response and the pyrexia are again reminiscent of the action of Coley's toxins.

William Coley may indeed have shown us the way to evoke an effective host response against cancer, but charting a safe and reliable course to this objective remains a formidable task. We must be prepared to accept the possibility that, as the workers of Coley's era found, the response may be effectively stimulated in only a percentage of patients, and at this time we do not even have monitoring assays to shorten the task by determining which patients will or will not respond.

PERSPECTIVE

Twenty-five years ago, Greenstein (1956) stated, "The key to the cancer problem lies in the host-tumor relationship." The pathogenesis of a given neoplasm is governed by the relationship operating between the tumor and the reaction of the host. Since the characteristics of a growing tumor are continually changing, the tumor-host relationship is a dynamic one, and there is considerable evidence that a primary tumor may influence the number and rate of growth of its metastases. This topic has been reviewed by Dickson (1977), and the consequences arising for the host from disruption of the myriad facets comprising host-tumor interaction have also been discussed recently (Tarin 1976). The outcome of cancer therapy also depends on the relationship obtained between tumor and host (Dickson and Suzangar 1976) (fig. 23.16), and in our studies on hyperthermia from this laboratory we have repeatedly emphasized the great importance of this concept (Dickson 1976, 1977; Dickson and Suzangar 1976). Heat therapy of tumors cannot be regarded simply as the physical application of heat to a mass of malignant cells, but must be viewed as a perturbation of the dynamic and continually changing host-tumor-metastasis relationship (Dickson 1976, 1977; Dickson and Calderwood 1980; Dickson and Suzangar 1976). Systemic host response and local normal tissue reaction may include not only the inflammatory response but also local immune defenses. Little is known about the latter, but support for organ specificity of host defenses comes from evidence that different types of lymphocytes are present in different lymphoid as well as in nonlymphoid organs (Reif 1975). The concept that each organ may possess its own immune defenses that act in concert with systemic immunity to maintain homeostasis is attractive, if presently undefined. A vitally important factor in the response of tumors to hyperthermia, including participation of the immune system, is blood flow through the heated region (Calderwood and Dickson 1980; Dickson and Calderwood 1980; Dickson et al., in press). Blood flow constitutes an integral part of the host-tumor association and the current state of the art apropos hyperthermia and tumor blood flow is discussed in chapters 8 and 9.

The recent work of Fidler and Hart (1978), Fidler and Kripke (1980), and Poste and Fidler (1980) has added a further dimension to the complexity of the host-tumor relationship. These workers showed that the B16 melanoma contained subpopulations of cells with differing metastatic capabilities, and similar heterogeneity in metastatic potential has now been demonstrated for several other murine tumors. The heterogeneous nature of tumors with regard to antigenicity and immunogenicity is well recognized, and it is also known that the antigenicity of a primary tumor and its metastases may differ. In view of this heterogeneity, it is not unexpected that results on the role of the immune response in metastasis have fallen into three groups, as specified by Fidler and Kripke (1980). In some tumor systems, depression of immune reactivity has led to an increase in both spontaneous and experimental metastasis, while in other tumor systems such depression decreased or prevented metastasis; in a third group of animal systems, tumor growth and metastasis were not influenced by manipulation of host immune reactivity. The authors concluded that although tumor immunogenicity influences the relationship between host immunity and tumor dissemination, the role of the immune system in experimental metastasis varies for different tumors, and no generalizations regarding the role of host immunity can be obtained from a single tumor system that will predict the outcome in another tumor system.

It is of note, however, that the findings of heterogeneity in animal tumors discussed above are paralleled by similar conclusions from clinical studies. In discussing metastasis in humans, Sugarbaker (1979) concluded that no unifying hypothesis was apparent to

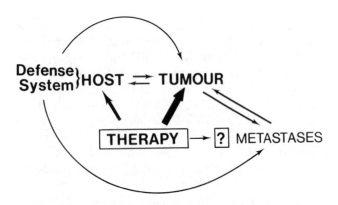

Figure 23.16. Host-tumor-therapy triad. Dynamic interrelationships operate between the host, primary tumor, and metastases. Therapy (surgery, radiation, cytotoxic drugs, heat) directed at the tumor affects the host also and may influence metastases directly, or via the tumor/host response, perturbing these relationships. Host defenses comprise the immune system, the mononuclear phagocyte (reticuloendothelial) system, and the inflammatory response (including fever), and can be envisaged as functioning at systemic and/or local (organ) level against tumor and metastases.

encompass the various features of metastatic disease, and that for the individual patient with cancer, a precise assessment of metastatic potential cannot be made. Tarin (1976) previously emphasized the same conclusion: in the human patient the pattern of metastasis can only be surmised from statistical evaluations. Since the eventual outcome of the metastatic process depends upon properties of both tumor and host and their dynamic interrelationship, it therefore is not surprising that difficulties may arise in applying conclusions derived from experimental systems to naturally occurring human neoplasms. This reiterates the view of earlier pathologists (Willis 1973) and clinicians (Strauss 1969) that host susceptibility and resistance to cancer is an individual-specific attribute.

There is little scientific data to support the concept of a specific antitumor immune response being generated following tumor heating in humans. Animal experiments indicate that a relatively high antigenicity is usually necessary for a tumor to elicit an immune response in the host (Baldwin and Robins 1976; Harris and Sinkovics 1976; Reif 1975); even when the *relative* antigenic strength is high, the *absolute* antigenic strength of tumors is low compared to other tissue (histocompatibility) antigens (Harris and Sinkovics 1976; Order 1925; Reif 1975). Tumor antigens detected on most human neoplasms differ in terms of their specificity from those associated with animal tumors induced either by oncogenic viruses or chemical carcinogens (Baldwin and Robins 1976). Although the results with the VX-2 demonstrate that an effective host response may occur after hyperthermic destruction of a nonimmunogenic tumor, the situation may be less well-defined in humans. The question then arises of whether the relatively nonimmunogenic nature of human tumors precludes generation following heating of an antitumor response as effective as that in the animal tumor systems studied to date, or whether we have, as believed by Hewitt (1978), become too obsessed with the belief that the host-tumor imbalance can be redressed by recruitment of host defenses. In this context it is of note that after many years of experience with electrocoagulation of animal and human tumors, Strauss, Saphir, and Appel (1956) concluded that the results in humans were "not as perfect as they are in the animal" because the immunity developing after hyperthermic tumor necrosis was only "relative" in humans, compared to "absolute" immunity in animals (i.e., the cured animal was immune to further tumor inoculation).

It is now established that heat (42°C to 50°C) can have a destructive effect on at least some types of human cancer. In both animal and human tumor studies, the bulk of evidence is in favor of a nonspecific host response partaking in the tumor regression that has followed hyperthermia or hyperpyrexia. If we accept these two tenets, are they of any benefit in our approach to cancer treatment? The most direct approach to mobilizing nonspecific host defenses is stimulation of host macrophages. There is now an impressive body of data attesting to the tumoricidal activity of macrophages under various conditions (Eccles 1978; James, McBride, and Stuart 1978). The macrophage-tumor-host relationship is, however, like the immune response and cancer, multifaceted, and of extreme complexity (Eccles 1978), and at present there are few guidelines as to how we should stimulate and control macrophages.

Sarcoma and melanoma may constitute a group of tumors that justify consideration distinct from the more common solid tumors of humans, since they are associated with evidence of an antitumor immune response in the host and they have proved more heat sensitive to date. The susceptibility of melanomas and sarcomas of the limb to heat may reflect a favorable local host reaction (inflammatory response) to this specialized form of heating (perfusion) at a particular anatomic site. There is growing support for a more conservative surgical approach to the treatment of osteogenic sarcoma, with emphasis on local radical en bloc resection of the lesion as an alternative to amputation (Marcove 1976; Salzer and Knahr 1976). At the same time, there is increasing interest in the role of host factors in the development and progression of sarcomas. The immune status is receiving considerable attention as a possible sensor of host resistance in such patients, although the results of immunologic testing in different series remain conflicting (Kotz, Rella, and Salzer 1976; Pritchard, Ivans, and Ritts 1976). The correlation between the appearance of pulmonary metastases and removal of the primary osteogenic sarcoma by amputation (Ferguson 1940) has also focused attention on the importance of intricate relationships between the host and the tumor and its metastases in humans (reviewed in Dickson 1977). The possible potential of hyperthermic destruction of sarcomas and melanomas to influence favorably the host capability to destroy metastases could have far-reaching consequences for therapy. It may be that such autoimmunotherapy could be effectively supported by the use of *C. parvum* or interferon, which has produced a significant increase in survival and disease-free interval in osteosarcoma in a recent Swedish trial (Sinkovics 1979b). In several ongoing clinical trials, intensive adjuvant chemotherapy following tumor resection appears to have achieved a significant advance in the control of metastases development from osteosarcoma (Sinkovics 1979b; Sutow 1975), and an aggressive approach of pulmonary wedge resection for overt metastases (Ivans and Pritchard 1976; Sinkovics 1979b) is becoming more accepted.

It would appear, therefore, that with osteosarcoma sufficient data are becoming available to institute a multidisciplinary approach to therapy and to justify consideration of a graded procedure—for example, hyperthermia or hyperthermia plus radiotherapy followed by resection four weeks later with chemotherapy as a backup, and pulmonary wedge resection if overt metastases appear. Ideally, the time scheduling for such combination therapy should be determined by immunologic monitoring in each individual patient. The development of sensitive immunologic monitoring tests that correlate with the clinical course of the disease is therefore an urgent requirement for our understanding of tumor and host response to therapy including heat, and for the design of therapy regimens.

With regard to the more common solid tumors in humans, the need at present is for more and well-defined data on tumor sensitivity to heat and for means of monitoring host response. The development of methodology for heating tumors in the clinic is as yet in its infancy, not to mention our lack of knowledge regarding protocols that may be appropriate for the destruction of human cancers. Most of the work to date in animals and humans has concerned superficial or accessible tumors. We have little information concerning the heat sensitivity of internal tumors or host response following such heating. There are few suitable animal models to use for guidance in this respect, although the use of spontaneous tumors in dogs looks promising. Current indications are that heat will achieve a place in the therapy of human cancer, most probably as an adjuvant to other modalities. That place cannot be predicted from animal work or cell survival curves (Dickson 1979b). There is no evidence that local tumor heating or total-body hyperthermia provoke tumor dissemination in humans, as may occur with animal tumor systems. More widespread effort must therefore be directed to the clinical situation.

Analysis of the clinical situation as it now stands seems to indicate clearly that for some malignancies (e.g., Hodgkins disease, testicular tumors, and some cases of breast and bowel cancer), curative results may be obtained while apparently disregarding features of potential immunity. For other neoplasms (bronchogenic carcinoma, glioblastoma), however, current therapy has little curative to offer. In only a few tumor types (melanoma, sarcoma) have we any evidence that a host response to the tumor may occur; even in such cases the significance of the reaction in terms of tumor rejection by the host is uncertain, and to what extent the response can be manipulated to benefit the host is unknown. Acquisition of more precise knowledge regarding host defense mechanisms (immune or otherwise) may significantly alter our therapeutic approaches to malignant disease. We must not delude ourselves, however. It may transpire that the beneficial value of hyperthermia in the treatment of human cancer does not include the immune system. If the response to heat involves and/or is rendered more effective by participation of nonspecific defenses, as current data seem to indicate, then their definition and manipulation are formidable tasks.

References

Ablin, R. J., and Guinan, P. D. Immunologic phenomena induced by cryosurgery. In *Immunodiagnosis and immunotherapy of malignant tumors: relevance to surgery*, ed. H. D. Flad. Heidelberg: Springer-Verlag, 1979, pp. 282–304.

Baldwin, R. W., and Barker, C. R. Tumor specific antigenicity of aminoazo dye induced rat hepatomas. *Int. J. Cancer* 2:355–364, 1967.

Baldwin, R. W.; Byers, V. S.; and Robins, R. A. Immune complexes in cancer. In *Tumour markers: impact and prospects*, eds. E. Boelsma and Ph. Rümke. New York: Elsevier North-Holland, 1979, pp. 151–165.

Baldwin, R. W., and Embleton, M. J. Immunology of experimental mammary cancer. In *Host defence against breast cancer*, ed. B. A. Stoll. London: Wm. Heinemann, 1975, pp. 78–94.

Baldwin, R. W.; Embleton, M. J.; and Pimm, M. V. Immune responses to carcinogen-induced and spontaneous rat tumors. In *Tumor-associated antigens and their specific immune response*, eds. F. Spreafico and R. Arnon. New York and London: Academic Press, 1979, pp. 21–30.

Baldwin, R. W.; Harris, J. R.; and Price, M. R. Fractionation of plasma membrane-associated tumor-specific antigen from an aminoazo dye-induced rat hepatoma. *Int. J. Cancer* 11:385–397, 1973.

Baldwin, R. W., and Pimm, M. V. BCG immunotherapy of a rat sarcoma. *Br. J. Cancer* 28:281–287, 1973.

Baldwin, R. W., and Robins, R. A. Host immune responses to tumors. In *Scientific foundations of oncology*, eds. T. Symington and R. L. Carter. London: Wm. Heinemann, 1976, pp. 514–520.

Baldwin, R. W., and Robins, R. A. Circulating immune complexes in cancer. In *Cancer markers*, ed. S. Sell. Clifton, N.J.: Humana Press, 1980, pp. 507–531.

Barland, P. The use of chemotherapeutic agents in the treatment of non-neoplastic diseases. In *Cancer chemotherapy*, 2nd edition, ed. E. S. Greenwald. London: Henry Kimpton, 1973, pp. 406–440.

Begg, R. W. Tumor host relations. *Adv. Cancer Res.* 5:1–54, 1958.

Berenbaum, M. C. Effects of cytotoxic drugs and ionizing radiation on immune responses. In *Host defence in breast cancer*, ed. B. A. Stoll. London: Wm. Heinemann 1975, pp. 147–171.

Berman, L. D. Immune parameters in the host response to neoplasia: morphological considerations. In *Immunity and cancer in man*, ed. A. E. Reif. New York: Marcel Dekker, 1975, pp. 103–117.

Biano, G. Clinical aspects of cancer immunotherapy. In *Immunity and cancer in man*, ed. A. E. Reif. New York: Marcel Dekker, 1975, pp. 47–79.

Bierman, W. *The medical applications of the short wave current.* Baltimore: Williams & Wilkins, 1942, p. 303.

Black, M. M. General discussion of interaction of humoral and cellular mechanisms in tumor immunity. *Natl. Cancer Inst. Monogr.* 35:276, 1972.

Black, M. M. Cell mediated response in human mammary cancer. In *Host defence in breast cancer*, ed. B. A. Stoll. London: Wm. Heinemann, 1975, pp. 48–77.

Bligh, J. *Temperature regulation in mammals and other vertebrates.* North Holland Research Monographs, Frontiers of Biology, vol. 30. Amsterdam: North Holland, 1973.

Bolton, P. M. DNCB sensitivity in cancer patients. A review based on sequential testing in 430 patients. *Clin. Oncol.* 1:59–69, 1975.

Bourdon, G., and Halpern, B. Antitumor immunity induced by the administration of homologous and isologous heated tumor cells. *Compt. Rend. Acad. Sci. D. (Paris)* 282:1571–1576, 1976.

Brett, D. E., and Schloerb, P. R. Intermittent hyperthermia on Walker 256 carcinoma. *Arch. Surg.* 85:154–157, 1962.

Bruce, D. L., and Kahan, B. D. Effects of anesthesia and surgery on immunity. In *Immunologic aspects of anesthetic and surgical practice*, eds. A. Mathieu and B. D. Kahan. New York and London: Grune & Stratton, 1975, pp. 171–181.

Calderwood, S. K., and Dickson, J. A. Influence of tumor volume and cell kinetics on the response of the solid Yoshida sarcoma to hyperthermia (42°C). *Br. J. Cancer* 42:22–32, 1980.

Calderwood S. K., and Dickson, J. A. Effect of hyperglycemia on blood flow, pH, and response to hyperthermia (42°C) of the Yoshida sarcoma in the rat. *Cancer Res.* 40:4728–4733, 1980.

Castillo, J., and Goldsmith, H. S. Immunological competence of heat-treated extracts of tumor or lymph nodes. *Arch. Surg.* 106:322–327, 1973.

Cavaliere, R. et al. Selective heat sensitivity of cancer cells. *Cancer* 20:1351–1381, 1967.

Cavaliere, R. et al. Heat transfer problems during local perfusion in cancer treatment. *Ann. NY Acad. Sci.* 335:311–326, 1980.

Cavaliere, R.; Moricca, G.; and Caputo, A. Regional hyperthermia by perfusion. In *Proceedings of the First International Symposium on Cancer Therapy by Hyperthermia and Radiation*, eds. M. J. Wizenberg and J. E. Robinson. Chicago: American College of Radiology, 1976, pp. 251–265.

Chuang, V. P., and Wallace, S. Arterial infusion and occlusion in cancer patients. *Semin. Roentgenol.*, in press.

Churchill, W. H. et al. Detection of antigens of a new diethylnitrosamine-induced transplantable hepatoma by delayed hypersensitivity. *J. Natl. Cancer Inst.* 41:13–22, 1968.

Clifton, K. H. Animal models of breast cancer. In *Endocrinology of cancer*, vol. 1, ed. D. P. Rose. Boca Raton, Fla.: CRC Press, 1979, pp. 1–20.

Cochran, A. J. *In vitro* testing of the immune response. In *Immunological aspects of cancer*, ed. J. E. Castro. Lancaster, England: MTP Press, 1978a, pp. 219–266.

Cochran, A. J. *Man, cancer and immunity.* London and New York: Academic Press, 1978b.

Cohen, P., and Warren, S. L. A study of the leucocytosis produced in man by artificial fever. *J. Clin. Invest.* 14:423–429, 1935.

Copeland, E. S. Effect of selective tumor heating on the localization of ^{131}I fibrinogen in the Walker carcinoma 256. I. Heating by immersion in warm water. *Acta Radiol.* 9:205–224, 1970.

Crile, G., Jr. Selective destruction of cancers after exposure to heat. *Ann. Surg.* 156:404–407, 1962.

Crile, G., Jr., The effects of heat and radiation on cancers implanted on the feet of mice. *Cancer Res.* 23:372–380, 1963.

Crile, G., Jr., and Turnbull, R. B. The role of electrocoagulation in the treatment of carcinoma of the rectum. *Surg. Gynecol. Obstet.* 135:391–396, 1972.

Currie, G. A. Eighty years of immunotherapy: a review of immunological methods used for the treatment of human cancer. *Br. J. Cancer* 26:141–153, 1972.

Currie, J. A. *Cancer and the immune response*, 2nd edition. London: Arnold, 1980.

Davidson, W. F. Cellular requirements for the induction of cytotoxic T cells *in vitro*. *Immunol. Rev.* 35:263–310, 1977.

De Horatius, R. J. et al. Immunologic function in humans before and after hyperthermia and chemotherapy for disseminated malignancy. *J. Natl. Cancer Inst.* 58:905–909, 1977.

Dennick, R. G.; Price, M. R.; and Baldwin, R. W. Modification of the immunogenicity and antigenicity of rat hepatoma cells. II. Mild heat treatment. *Br. J. Cancer* 39:630–635, 1979.

Dickson, J. A. Hyperthermia in the treatment of cancer. *Cancer Chemother. Rep.* 58:294–296, 1974.

Dickson, J. A. Hazards and potentiators of hyperthermia. In *Proceedings of the First International Symposium on Cancer Therapy by Hyperthermia and Radiation*, eds. M. J. Wizenberg and J. E. Robinson. Chicago: American College of Radiology, 1976, pp. 134–150.

Dickson, J. A. The effects of hyperthermia in animal tumor systems. *Recent Results Cancer Res.* 59:43–111, 1977.

Dickson, J. A. The sensitivity of human cancer to hyperthermia. In *Clinical prospects for hypoxic cell sensitisers and hyperthermia*, eds. W. L. Caldwell and R. E. Durand. Madison, Wis.: University of Wisconsin Press, 1978, pp. 174–193.

Dickson, J. A. Destruction of solid tumors by heating with radiofrequency energy. *Radio Science* 14 (Suppl. 6S):285–295, 1979a.

Dickson, J. A. Hyperthermia in the treatment of cancer. *Lancet* 1:202–205, 1979b.

Dickson, J. A. et al. Tumor eradication in the rabbit by radiofrequency heating. *Cancer Res.* 37:2162–2169, 1977.

Dickson, J. A. et al. Immune regression of metastases following hyperthermic treatment of primary tumors. *Proceedings EORTC Metastasis Conference*, eds. J. E. Castro and K. Hellmann. The Hague: Martinus Nijhoff, 1980, pp. 260–265.

Dickson, J. A., and Calderwood, S. K. *In vivo* hyperthermia of Yoshida tumor induces entry of nonproliferating cells into cycle. *Nature* 263:772–773, 1976.

Dickson, J. A., and Calderwood, S. K. Temperature range and selective sensitivity of tumors to hyperthermia: a critical review. *Ann. NY Acad. Sci.* 335:180–205, 1980.

Dickson, J. A.; Calderwood, S. K.; and Jasiewicz, M. L. Radiofrequency heating of tumors in rodents. *Eur. J. Cancer* 13:753–763, 1977.

Dickson, J. A., and Ellis, H. A. Stimulation of tumour cell dissemination by raised temperature (42°C) in rats with transplanted Yoshida tumours. *Nature* 248:354, 1974.

Dickson, J. A., and Ellis, H. A. The influence of tumor volume and the degree of heating on the response of the solid Yoshida sarcoma to hyperthermia (40–42°C). *Cancer Res.* 36:1188–1195, 1976.

Dickson, J. A.; Jasiewicz, M. L.; and Simpson, A. C. Immunogenicity of ascites tumor cells following *in vitro* hyperthermia. In *Proceedings of the Third International Symposium on Cancer Therapy by Hyperthermia, Drugs and Radiation*, eds. L. Dethlefsen and W. C. Dewey. Bethesda, Md.: *J. Natl. Cancer Inst.* (special journal supplement), in press.

Dickson, J. A.; MacKenzie, A.; and McLeod, K. Temperature gradients in pigs during whole-body hyperthermia at 42°C. *J. Appl. Physiol.* 47:712–717, 1979.

Dickson, J. A., and Muckle, D. S. Total body hyperthermia versus primary tumour hyperthermia in the treatment of the rabbit VX2 carcinoma. *Cancer Res.* 32:1916–1923, 1972.

Dickson, J. A., and Shah, S. A. Technology for the hyperthermic treatment of large solid tumors at 50°C. *Clin. Oncol.* 3:301–318, 1977.

Dickson, J. A., and Shah, S. A. Stimulation of an antitumor immune response in VX2 bearing rabbits by curative hyperthermia. In *Proceedings of the Second International Symposium on Cancer Therapy by Hyperthermia and Radiation*, eds. C. Streffer et al. Baltimore and Munich: Urban & Schwarzenberg, 1978, pp. 294–296.

Dickson, J. A., and Shah, S. A. Immunologic phenomena induced by hyperthermia in normal and tumor-bearing hosts and their relevance for cancer therapy. In *Immunodiagnosis and immunotherapy of malignant tumors*, eds. H. D. Flad, C. Herfarth, and M. Betzler. Berlin and New York: Springer-Verlag, 1979, pp. 266–280.

Dickson, J. A., and Shah, S. A. Hyperthermia, the immune response and tumor metastasis. In *Proceedings of the Third International Symposium on Cancer Therapy by Hyperthermia, Drugs and Radiation*, eds. L. Dethlefsen and W. C. Dewey. Bethesda, Md.: *J. Natl. Cancer Inst.* (special journal supplement), in press.

Dickson, J. A., and Suzangar, M. *In vitro-in vivo* studies on the susceptibility of the solid Yoshida sarcoma to drugs and hyperthermia (42°C). *Cancer Res.* 34:1263–1274, 1974.

Dickson, J. A., and Suzangar, M. The *in vitro* response of human tumors to cytotoxic drugs and hyperthermia (42°C) and its relevance to clinical oncology. In *Organ culture in biomedical research*, eds. M. Balls and M. A. Monnickendam. Cambridge, England: Cambridge University Press, 1976, pp. 417–461.

Dixon, F.; Palmer, C. G.; and Hodes, M. E. Transplantation of the V-2 carcinoma by suspensions containing known numbers of tumor cells. *Transplantation Bulletin* 7:91–93, 1960.

Doan, C. A.; Hargreaves, M. M.; and Kester, L. Differential reaction of bone marrow, connective tissue and lymph nodes to hyperpyrexia. In *The First International Conference on Fever Therapy*, New York: Paul B. Hoeber, 1937, pp. 40–42.

Dubois, E. F. Why are fever temperatures over 106°F rare? *Am. J. Med. Sci.* 217:361–368, 1949.

Eccles, S. A. Macrophages and cancer. In *Immunological aspects of cancer*, ed. J. E. Castro. Lancaster, England: MTP Press, 1978, pp. 123–154.

Eilber, F. R., and Morton, D. L. Impaired immunologic reactivity and recurrence following cancer surgery. *Cancer* 25:362–367, 1970.

Everson, T. C., and Cole, W. H. *Spontaneous regression of cancer*. Philadelphia: W. B. Saunders, 1966.

Fabricius, H.-A. et al. Changes in cellular immunological functions of healthy adults induced by a one-hour 40° hyperthermia. In *Proceedings of the Second International Symposium on Cancer Therapy by Hyperthermia and Radiation*, eds. C. Streffer et al. Baltimore and Munich: Urban & Schwarzenberg, 1978, pp. 309–311.

Fabricius, H.-A., et al. Discussion. In *Immunodiagnosis and immunotherapy of malignant tumors, relevance to surgery*, eds. H.-D. Flad, C. Herfarth, and M. Betzler. Berlin and New York: Springer-Verlag, 1979, p. 281.

Fass, L.; Herberman, R. B.; and Ziegler, J. Delayed cutaneous hypersensitivity reactions to autologous extracts of Burkitt lymphoma cells. *N. Engl. J. Med.* 282:776–780, 1970.

Ferguson, A. B. Treatment of osteogenic sarcoma. *J. Bone Joint Surg.* 22:92–96, 1940.

Fidler, I. J., and Hart, I. R. Host immunity in experimental metastasis. In *Immunological aspects of cancer*, ed. J. E. Castro. Lancaster, England: MTP Press, 1978, pp. 183–204.

Fidler, I. J., and Kripke, M. L. Tumor cell antigenicity, host immunity and cancer metastasis. *Cancer Immunology and Immunotherapy* 7:201–205, 1980.

Fisher, B., and Fisher, E. R. Effect of experimental perfusion upon biologic and pathologic behavior of tumors. *Surgery* 56:651–662, 1964.

Fisher, B., and Fisher, E. R. Metastases of cancer cells. In *Methods in cancer research*, vol. 1, ed. H. Busch. New York and London: Academic Press, 1967, pp. 243–286.

Fisher, B., and Fisher, E. R. Metastasis revisited. In *Fundamental aspects of metastasis*, ed. L. Weiss. Amsterdam and Oxford: North Holland, 1976, pp. 427–435.

Fletcher, A. M.; Cetas, T. C.; and Wilson, S. D. Liver temperatures in rats subjected to whole body hyperthermia, evidence for congestion and lymphocyte damage. In *Proceedings of the Third International Symposium on Cancer Therapy by Hyperthermia, Drugs and Radiation*, eds. L. Dethlefsen and W. C. Dewey. Bethesda, Md.: *J. Natl. Cancer Inst.* (special journal supplement), in press.

Foley, E. J. Antigenic properties of methylcholanthrene-induced tumors in mice of the strain of origin. *Cancer Res.* 13:835–840, 1953.

Foulds, L. *Neoplastic development*, vol. 1. London and New York: Academic Press, 1969.

Fox, B. W., and Gregory, C. J. A study of the immunosuppressive activity of Methylene Dimethane Sulphonate (MDMS) in relation to its effectiveness as an antitumor agent. *Br. J. Cancer* 26:84–89, 1972.

Gee, A. P. et al. Effects of whole body hyperthermia therapy on the general immunocompetence of the advanced cancer patient. In *Proceedings of the Second International Symposium on Cancer Therapy by Hyperthermia and Radiation*, eds. C. Streffer et al. Baltimore and Munich: Urban & Schwarzenberg, 1978, pp. 312–315.

Gericke, D. et al. *In vitro* thermosensibility of experimental tumours in small animals. *Naturwissenschaften* 3:155, 1971.

Gillette, E. Hyperthermia effects in animals with spontaneous tumors. In *Proceedings of the Third International Symposium on Cancer Therapy by Hyperthermia, Drugs and Radiation*, eds. L. Dethlefsen and W. C. Dewey. Bethesda, Md.: *J. Natl. Cancer Inst.* (special journal supplement), in press.

Goldenberg, D. M., and Langner, M. Direct and abscopal antitumor action of local hyperthermia. *Z. Naturforsch.* 26b:359–361, 1971.

Graham, J. B., and Graham, R. M. Antibodies elicited by cancer in patients. *Cancer* 8:409–414, 1955.

Greene, H. S. N. A conception of tumor autonomy based on transplantation studies: a review. *Cancer Res.* 11:899–903, 1951.

Greene, H. S. N., and Newton, B. L. Evolution of cancer of the uterine fundus in the rabbit. *Cancer* 1:82–99, 1948.

Greenstein, J. P. *Biochemistry of cancer*, 2nd edition. New York: Academic Press, 1954.

Greenstein, J. P. Some biochemical characteristics of morphologically separable cancers. *Cancer Res.* 16:641–653, 1956.

Gullino, P. M.; Clark, S. H.; and Grantham, F. H. The interstitial fluid of solid tumors. *Cancer Res.* 24:780–797, 1964.

Hahn, G. M. et al. Some heat transfer problems associated with heating by ultrasound, microwaves or radiotherapy. In *Thermal characteristics of tumors: applications in detection and treatment*, eds. R. K. Jain and P. M. Gullino. *Ann. NY Acad. Sci.* 335:327–346, 1980.

Hahn, E. W.; Alfieri, A. A.; and Kim, J. H. Tumor control with heat alone: evidence for cell-mediated interactions (abstr.). In *Proceedings of the Third International Symposium on Cancer Therapy by Hyperthermia, Drugs and Radiation*, eds. L. A. Dethlefsen and W. C. Dewey. Bethesda, Md.: *J. Natl. Cancer Inst.* (special journal supplement), in press.

Hanna, M. G. et al. Histopathology of *Mycobacterium bovis* (BCG)-mediated tumor regression. *Natl. Cancer Inst. Monogr.* 35:345–357, 1972.

Hardy, J. D. Physiology of temperature regulation. *Physiol. Rev.* 41:521–606, 1961.

Harris, J. E., and Sinkovics, J. G. *The immunology of malignant disease*. St. Louis: C. V. Mosby, 1976.

Hersh, E. M. et al. Primary and secondary immune response in the evaluation of immunocompetence and prognosis in cancer patients. *Recent Results Cancer Res.* 47:25–36, 1974.

Hewitt, H. B. Projecting from animal experiments to clinical cancer. In *Fundamental aspects of metastasis*, ed. L. Weiss. Amsterdam and London: North Holland, 1976, pp. 343–357.

Hewitt, H. B. The choice of animal tumors for experimental studies of cancer therapy. *Adv. Cancer Res.* 27:149–200, 1978.

Hewitt, H. B.; Blake, E. R.; and Walder, A. S. A critique of the evidence for active host defence against cancer, based on personal studies of 27 murine tumours of spontaneous origin. *Br. J. Cancer* 33:241–259, 1976.

Hoeffken, W. Malignant bone tumors. *Recent Results Cancer Res.* 54:262–270, 1976.

Höffken, K. et al. Circulating immune complexes in rats bearing chemically induced tumors. I. Sequential determination during growth of tumors at various body sites. *Int. J. Cancer* 21:496–504, 1978.

Howard, J. C. *In vitro* assays for tumor immunity. In *Immunobiology of the host tumor relationship*, eds. R. T. Smith and M. Landy. New York and London: Academic Press, 1975, pp. 71–124.

Hyslop, N. E. Host defence and the altered host. In *Immunologic aspects of anesthetic and surgical practice*, eds. A. Mathieu and B. D. Kahan. New York: Grune & Stratton, 1975, pp. 131–169.

Ivans, J. C., and Pritchard, D. J. Management of osteosarcoma at the Mayo Clinic. *Recent Results Cancer Res.* 54:221–230, 1976.

Jackson, D. J., and Dickson, J. A. Combination hyperthermia (42°C) and hyperglycemia in the treatment of the MC_7 sarcoma. *Br. J. Cancer* 40:306, 1979.

James, K.; McBride, B.; and Stuart, A. The macrophage and cancer. In *Proceedings of the European Reticuloendothelial Society*. Edinburgh, Scotland: Econoprint, 1978.

Jasiewicz, M. L., and Dickson, J. A. Potentiation of the destructive effect of heat (42°C) on synchronized cancer cells in culture by cell specific antiserum. *J. Thermal Biol.* 1:221–225, 1976.

Johnson, H. J. The action of short radio waves on tissues. III. A comparison of the thermal sensitivities of transplantable tumors *in vivo* and *in vitro*. *Am. J. Cancer* 38:533–550, 1940.

Kaal, S. Metastatic frequency of spontaneous mammary carcinoma in mice following biopsy and following local roentgen irradiation. *Cancer Res.* 13:744–747, 1953.

Kaiser, C. W., and Reif, A. E. Immunological monitoring and adjuvant immunotherapy of selected cancer patients. In *Immunity and cancer in man*, ed. A. E. Reif. New York: Marcel Dekker, 1975, pp. 19–46.

Kaplan, H. S., and Murphy, E. D. The effect of local irradiation on the biological behaviour of a transplantable mouse carcinoma. I. Increased frequency of pulmonary metastasis. *J. Natl. Cancer Inst.* 9:407–413, 1948.

Kapp, D. S., and Lord, P. F. Fractionated systemic hyperthermia and radiation on spontaneous canine osteosarcoma. In *Proceedings of the Third International Symposium on Cancer Therapy by Hyperthermia, Drugs and Radiation*, eds. L. A. Dethlefsen and W. C. Dewey. Bethesda, Md.: *J. Natl. Cancer Inst.* (special journal supplement), in press.

Kapp, D. S.; Lord, P. F.; and Morrow, D. Alteration in sites of metastases of spontaneous canine osteosarcoma after fractionated systemic hyperthermia and local irradiation. In *Proceedings of the Third International Symposium on Cancer Therapy by Hyperthermia, Drugs and Radiation*, eds. L. A. Dethlefsen and W. C. Dewey. Bethesda, Md.: *J. Natl. Cancer Inst.* (special journal supplement), in press.

Kidd, J. G., and Rous, P. A transplantable rabbit carcinoma originating in a virus-induced papilloma and containing the virus in masked or altered form. *J. Exp. Med.* 71:813–838, 1940.

Kim, J. H., and Hahn, E. W. Clinical and biological studies of localized hyperthermia. *Cancer Res.* 39:2258–2261, 1979.

Klein, G. Foreword to *Man, cancer and immunity*, by A. J. Cochran. London and New York: Academic Press, 1978.

Klyuyeva, N. G., and Roskin, G. I. *Biotherapy of malignant tumors*. Translated from the Russian by J. J. Oliver. London and New York: Pergamon Press, 1963.

Kondo, T., and Moore, G. E. Production of metastases by treatment with carcinostatic agents. I. Effects of carcinostatic agents. I. Effects of carcinostatic agents on the host. *Cancer Res.*:1396–1401, 1961.

Kotz, R.; Rella, W.; and Salzer, M. The immune status in patients with bone and soft tissue sarcomas. *Recent Results Cancer Res.* 54:197–205, 1976.

Kovacs, R. *Electrotherapy and light therapy*. London: Henry Kimpton, 1949, p. 226.

Langer, S. et al. Clinical and immunologic results of cryosurgery in patients with rectal cancer. In *Immunodiagnosis and immunotherapy of malignant tumors: relevance to surgery*, ed. H. D. Flad. New York and Heidelberg: Springer-Verlag, 1979, pp. 305–310.

Leventhal, B. G. et al. Immune reactivity of leukemic patients to autologous blast. *Cancer Res.* 32:1820–1825, 1972.

Lewis, M. R., and Aptekman, P. M. Atrophy of tumours caused by strangulation and accompanied by development of tumour immunity in rats. *Cancer* 5:411–416, 1952.

Liebelt, A. G., and Liebelt, R. A. Transplantation of tumors. In *Methods in cancer research*, vol. 1, ed. H. Busch. New York and London: Academic Press, 1967, pp. 143–242.

MacCarty, W. C. Factors which influence longevity in cancer. *Ann. Surg.* 76:9–12, 1922.

Makay, W. D. Circulating antibodies in human cancer. In *Host defence in breast cancer*, ed. B. A. Stoll. London: Wm. Heinemann, 1975, pp. 36–47.

Marcove, R. L. The treatment of malignant bone tumors by conservative surgery. *Recent Results Cancer Res.* 54:218–220, 1976.

Marmor, J. B. Interactions of hyperthermia and chemotherapy in animals. *Cancer Res.* 39:2269–2276, 1979.

Marmor, J. B.; Hahn, N.; and Hahn, G. M. Tumor cure and cell survival after localized radiofrequency heating. *Cancer Res.* 37:879–883, 1977.

Mathé, G. et al. Active immunotherapy of cancer for minimal residual disease: new trends and new materials. *Prog. Exp. Tumor Res.* 25:242–274, 1980.

Mavligit, G. M. et al. Cell mediated immunity to human solid tumors: *in vitro* detection by lymphocyte blastogenic responses to cell-associated and solubilized tumor antigens. *Recent Results Cancer Res.* 47:84–96, 1974.

McCluskey, R. T., and Cohen, S., eds. *Mechanisms of cell mediated immunity*. New York: John Wiley, 1974.

McKenzie, A. et al. Total body hyperthermia: techniques and patient management. In *Proceedings of the First International Symposium on Cancer Therapy by Hyperthermia and Radiation*, eds. M. J. Wizenberg and J. E. Robinson. Chicago: American College of Radiology, 1976, pp. 272–281.

Meltzer, M. S. et al. Tumor specific antigen solubilized by hypertonic potassium chloride. *J. Natl. Cancer Inst.* 47:703–709, 1971.

Meltzer, M. S. et al. Cell mediated tumor immunity measured *in vitro* and *in vivo* with soluble tumor-specific antigens. *J. Natl. Cancer Inst.* 49:727–734, 1972.

Mendecki, J.; Friedenthal, E.; and Botstein, C. Effects of microwave-induced local hyperthermia on mammary carcinoma in C3H mice. *Cancer Res.* 36:2113–2114, 1976.

Menkin, V. *Biochemical mechanisms in inflammation*, 2nd edition. Springfield, Ill.: Charles C Thomas, 1956.

Mole, R. H. Whole body irradiation—radiobiology or medicine? *Br. J. Radiol.* 26:234–240, 1953.

Mondovi, B. et al. Increased immunogenicity of Ehrlich ascites cells after heat treatment. *Cancer* 30:885–888, 1972.

Moore, M. Human tumor-associated antigens: methods of *in vitro* detection. In *Immunological aspects of cancer*, ed. J. E. Castro. Lancaster, England: MTP Press, 1978a, pp. 51–87.

Moore, M. Antigens of experimentally induced neoplasms: a conspectus. In *Immunological aspects of cancer*, ed. J. E. Castro. Lancaster, England: MTP Press, 1978b, pp. 15–50.

Moore, M. Tumour immunology. In *Medical immunology*, ed. W. J. Irvine. Edinburgh, Scotland: Teviot Medical Publications, 1979, pp. 425–464.

Morales, A., and Eidinger, D. Immune reactivity in renal cancer: a sequential study. *J. Urol.* 115:510–513, 1976.

Moricca, G. et al. Hyperthermic treatment of tumors: experimental and clinical observations. *Recent Results Cancer Res.* 59:112–152, 1977.

Morton, D. L. Immunotherapy of human melanomas and sarcomas. *Natl. Cancer Inst. Monogr.* 35:375–380, 1972.

Mount, L. E. *The climatic physiology of the pig.* London: Arnold, 1968, pp. 131–144.

Mount, L. E., and Ingram, D. L. *The pig as a laboratory animal.* New York and London: Academic Press, 1971.

Movat, H. Z. The acute inflammatory reaction. In *Inflammation, immunity and hypersensitivity—cellular and molecular mechanisms*, ed. H. Z. Movat. London and New York: Harper & Row, 1979, pp. 1–163.

Muckle, D. S., and Dickson, J. A. The selective inhibitory effect of hyperthermia on the metabolism and growth of malignant cells. *Br. J. Cancer* 25:771–778, 1971.

Muckle, D. S., and Dickson, J. A. Hyperthermia (42°C) as an adjuvant to radiotherapy and chemotherapy in the treatment of the allogeneic VX2 carcinoma in the rabbit. *Br. J. Cancer* 27:307–315, 1973.

Nauts, H. C. Pyrogen therapy of cancer. A historical overview and current activities. In *Proceedings of the First International Symposium on Cancer Therapy by Hyperthermia and Radiation*, eds. M. J. Wizenberg and J. E. Robinson. Chicago: American College of Radiology, 1976, pp. 239–250.

Nauts, H. C.; Fowler, G. A.; and Bogatko, F. H. A review of the influence of bacterial infections and bacterial products (Coley's toxin) on malignant tumors in man. *Acta Med. Scand.* 145 (Suppl. 276):1–103, 1953.

Neymann, C. A. *Artificial fever produced by physical means: its development and application.* London: Bailliere, Tindall & Cox, 1938.

Oettgen, H. F. et al. Delayed hypersensitivity and transplantation immunogenicity elicited by soluble antigens of chemically induced tumors in inbred guinea-pigs. *Nature* 220:295–297, 1968.

Onkelinx, E. et al. Glutaraldehyde as a coupling reagent in passive hemagglutination. *Immunology* 16:35–43, 1969.

Order, S. E. Immune parameters in the radiation therapy of cancer. In *Immunity and cancer in man*, ed. A. E. Reif. New York: Marcel Dekker, 1925, pp 91–102.

Overgaard, K., and Overgaard, J. Investigations on the possibility of a thermic tumor therapy. I. Short-wave treatment of a transplanted isologous mouse mammary carcinoma. *Eur. J. Cancer* 8:65–78, 1972.

Overgaard, K., and Overgaard, J. Pathology of heat damage. Studies on the histopathology in tumor tissue exposed *"in vivo"* to hyperthermia and combined hyperthermia and roentgen irradiation. In *Proceedings of the First International Symposium on Cancer Therapy by Hyperthermia and Radiation*, eds. M. J. Wizenberg and J. E. Robinson. Chicago: American College of Radiology, 1976, pp. 115–127.

Pantazatos, P.; Tompkins, W.; and Tick, N. Hyperthermia sensitization of a human colon tumor cell line (HCT-8R) to lysis by antibodies and complement. *Proc. Am. Assoc. Cancer Res.* 19:96, 1978.

Parks, L. C. et al. Anticancer hyperthermia. Metabolic, physiologic and immunologic responses of man (abstr.). *Fed. Proc.* 37:930, 1978.

Pelner, L. Host-tumor antagonism. XI. Recapitulation of various aspects: practical application of this knowledge to the care of patients. *J. Am. Geriatr. Soc.* 6:539–553, 1958.

Pettigrew, R. T. et al. Clinical effects of whole body hyperthermia in advanced malignancy. *Br. Med. J.* 4:679–683, 1974.

Pettigrew, R. T., and Ludgate, C. Whole body hyperthermia. A systemic treatment for disseminated cancer. *Recent Results Cancer Res.* 59:153–170, 1977.

Pinsky, C. M. *In vitro* assays for tumor immunity. In *Immunobiology of the host-tumor relationship*, eds. R. T. Smith and M. Landy. New York and London: Academic Press, 1975, pp. 71–124.

Pimm, M. V., and Baldwin, R. W. Immunology and immunotherapy of experimental and clinical metastases. In *Secondary spread of cancer*, ed. R. W. Baldwin. London and New York: Academic Press, 1978, pp 163–209.

Poste, G., and Fidler, I. J. The pathogenesis of cancer metastasis. *Nature* 283:139–146, 1980.

Prager, M. D., and Baechtel, F. S. Methods for modification of cancer cells to enhance their antigenicity. *Methods Cancer Res.* 9:339–400, 1973.

Prehn, R. T., and Main, J. M. Immunity to methylcholanthrene-induced sarcoma. *J. Natl. Cancer Inst.* 18:769–773, 1957.

Pritchard, D. J.; Ivans, J. C.; and Ritts, R. E. Immunologic aspects of human sarcomas. *Recent Results Cancer Res.* 54:185–196, 1976.

Rappaport, D. S. Hyperthermia and host resistance to lymphatic metastases. In *Proceedings of the Third International Symposium on Cancer Therapy by Hyperthermia, Drugs and Radiation*, eds. L. A. Dethlefsen and W. C. Dewey. Bethesda, Md.: *J. Natl. Cancer Inst.* (special journal supplement), in press.

Reif, A. E. Immune defenses against initiation of tumors. In *Immunity and cancer in man*, ed. A. E. Reif. New York: Marcel Dekker, 1975, pp. 1–18.

Rice, L.; Urano, M.; and Suit, H. D. Metastasis frequency with hyperthermia in spontaneous and in induced murine tumor systems. In *Proceedings of the Third Interna-*

tional Symposium on Cancer Therapy by Hyperthermia, Drugs and Radiation, eds. L. A. Dethlefsen and W. C. Dewey. Bethesda, Md.: *J. Natl. Cancer Inst.* (special journal supplement), in press.

Richards, V. Cancer immunology—an overview. *Prog. Exp. Tumor Res.* 25:1–60, 1980.

Robinson, J. E. Thermometry and waterbath heating. *Br. J. Radiol.* 51:933, 1978.

Rockwell, S. C., and Kallman, R. F. Cellular radiosensitivity and tumor radiation response in the EMT6 tumor cell system. *Radiat. Res.* 53:281–294, 1973.

Rockwell, S. C.; Kallman, R. F.; and Fajardo, L. F. Characteristics of a serially transplanted mouse mammary tumor and its tissue culture adapted derivative. *J. Natl. Cancer Inst.* 49:735–749, 1972.

Rohdenburg, G. L., and Prime, F. Effect of combined radiation and heat on neoplasms. *Arch. Surg.* 2:116, 1921.

Rossi-Fanelli, A. et al. *Selective heat sensitivity of cancer cells.* Berlin: Springer-Verlag, 1977, pp. 1–189.

Roszkowski, W., and Szmigielski, S. Immunological consequences of whole body microwave hyperthermia. In *Proceedings of the Third International Symposium on Cancer Therapy by Hyperthermia, Drugs and Radiation*, eds. L. A. Dethlefsen and W. C. Dewey. Bethesda, Md.: *J. Natl.Cancer Inst.* (special journal supplement), in press.

Rous, P.; Kidd, J. G.; and Smith, W. E. Experiments on the cause of the rabbit carcinomas derived from virus-induced papillomas. *J. Exp. Med.* 96:159–174, 1952.

Salacinski, P. et al. A new simple method which allows theoretical incorporation of radioiodine into proteins and peptides without damage. *J. Endocrinol.* 81:131–136, 1979.

Salzer, M., and Knahr, K. Resection of malignant bone tumors. *Recent Results Cancer Res.* 54:239–256, 1976.

Schaeffer, J. Treatment of metastatic osteogenic sarcoma in mice by whole body heating. In *Proceedings of the First International Symposium on Cancer Therapy by Hyperthermia and Radiation*, eds. M. J. Wizenberg and J. E. Robinson. Chicago: American College of Radiology, 1976, pp. 156–157.

Schechter, M.; Stowe, S. M.; and Moroson, H. Effects of hyperthermia on primary and metastatic growth and host immune response in rats. *Cancer Res.* 38:498–502, 1978.

Scott, B. O. *The principles and practice of diathermy.* London: Wm. Heinemann, 1957, p. 125.

Seibert, E. et al. Membrane associated antigens of human malignant melanoma. III. Specificity of human sera reacting with cultured melanoma cells. *Int. J. Cancer* 19:172–178, 1977.

Shah, S. A., and Dickson, J. A. Lymphocyte separation from blood. *Nature* 249:168–169, 1974.

Shah, S. A., and Dickson, J. A. Effect of hyperthermia on the immunocompetence of VX2 tumor bearing rabbits. *Cancer Res.* 38:3523–3531, 1978a.

Shah, S. A., and Dickson, J. A. Effect of hyperthermia on the immune response of normal rabbits. *Cancer Res.* 38:3518–3522, 1978b.

Sinkovics, J. G. *Medical oncology. An advanced course.* New York and Basel: Marcel Dekker, 1979a, pp. 156–158.

Sinkovics, J. G. *Medical oncology, an advanced course.* New York and Basel: Marcel Dekker, 1979b, pp. 131–133.

Sinkovics, J. G. et al. Intensification of immune reactions of patients to cultured sarcoma cells: attempt at monitored immunotherapy. *Semin. Oncol.* 1:351–365, 1974.

Simon, J. F. Effects of hyperpyrexia on the human blood count, blood chemistry and urine. *J. Lab. Clin. Med.* 21:400–405, 1935.

Smith, R. T., and Landy, M. *Immunobiology of the tumor-host relationship.* New York and London: Academic Press, 1975, pp. 125–199.

Steel G. G. *Growth kinetics of tumors.* Oxford: Clarendon Press, 1977, pp. 268–306.

Stehlin, J. S. Regional hyperthermia and chemotherapy by perfusion. In *Proceedings of the First International Symposium on Cancer Therapy by Hyperthermia and Radiation*, eds. M. J. Wizenberg and J. E. Robinson. Chicago: American College of Radiology, 1976, pp. 266–271.

Stehlin, J. S. et al. Hyperthermic perfusion of extremities for melanoma and soft tissue sarcoma. *Recent Results Cancer Res.* 59:171–185, 1977.

Stewart, H. L. et al. *Transplantable and transmissible tumors of animals.* Washington, D.C.: Armed Forces Institute of Pathology, 1959.

Stjernswärd, J. Radiotherapy, host immunity and cancer spread. In *Secondary spread in breast cancer. New aspects of breast cancer*, vol. 3, ed. B. A. Stoll. London: Wm. Heinemann, 1977, pp. 139–167.

Storm, F. K. et al. Normal tissue and solid tumor effects of hyperthermia in animal models and clinical trials. *Cancer Res.* 39:2245–2251, 1979.

Storm, F. K. et al. Clinical radiofrequency hyperthermia: a review. In *Proceedings of the Third International Symposium on Cancer Therapy by Hyperthermia, Drugs and Radiation*, eds. L. A. Dethlefsen and W. C. Dewey. Bethesda, Md. *J. Natl. Cancer Inst.* (special journal supplement), in press.

Strauss, A. A. *Immunologic resistance to carcinoma produced by electrocoagulation. Based on fifty-seven years of experimental and clinical results.* Springfield, Ill.: Charles C Thomas, 1969.

Strauss, A. A. et al. Immunologic resistance to carcinoma produced by electrocoagulation. *Surg. Gynecol. Obstet.* 121:989–996, 1965.

Strauss, A. A.; Saphir, O.; and Appel, M. The development of an absolute immunity in experimental animals and a relative immunity in human beings due to a necrosis of malignant tumors. *Swiss Med. J.* 86(suppl. 20):606–610, 1956.

Streffer, C. et al., eds. *Proceedings of the Second International Symposium on Cancer Therapy by Hyperthermia and Radiation.* Baltimore and Munich: Urban & Schwarzenberg, 1978.

Stutman, O. *In vitro* assays for tumor immunity. In *Immunobiology of the host tumor relationship*, eds. R. T. Smith and M. Landy. New York and London:Academic Press, 1975, pp. 71–124.

Sugarbaker, E. V. Some characteristics of metastasis in man. *Am. J. Pathol.* 97:623–632, 1979.

Suit, H. D. Hyperthermic effects on animal tissues. *Radiology* 123:483–487, 1977.

Suit, H. D.; Sedlacek, R. S.; and Wiggins, S. Immunogenicity of heated tumor cells. *Cancer Res.* 37:3836–3839, 1977.

Sutow, W. W. *Cancer chemotherapy—fundamental concepts and recent advances.*Chicago: Year Book, 1975, p. 203.

Szmigielski, S., and Janiak, M. Reaction of cell-mediated immunity to local hyperthermia of tumors and its potentiation by immunostimulation. In *Proceedings of the Second International Symposium on Cancer Therapy by Hyperthermia and Radiation*, eds. C. Streffer et al. Baltimore and Munich: Urban & Schwarzenberg, 1978, pp. 80–88.

Tanchou, S. *Récherches sur le traitement médical des tumeurs cancéreuses du sein. Ouvrage pratique basé sur trois cents observations (extraits d'un grand nombre d'auteurs).* Paris: Germer Bailliere, 1844.

Tarin, D. Cellular interactions in neoplasia. In *Fundamental aspects of metastasis*, ed. L. Weiss. New York: American Elsevier, 1976, pp. 151–187.

Terry, W. D. *In vitro* assays for tumor immunity. In *Immunobiology of the host tumor relationship*, eds. R. T. Smith and M. Landy. New York and London: Academic Press, 1975, pp. 69–124.

Urano, M. et al. Enhancement by *Corynebacterium parvum* of the normal and tumor tissue response to hyperthermia. *Cancer Res.* 38:862–864, 1978.

Urano, M. et al. Enhancement of the thermal response of animal tumors by *Corynebacterium parvum*. *Cancer Res.* 39:3454–3457, 1979.

Vermel, E. M., and Kuznetsova, L. B. Hyperthermia in the treatment of malignant diseases. *Probl. Oncol. USSR* (Engl. transl.) 16:96–102, 1970.

Von Essen, C. F., and Kaplan, H. S. Further studies on metastasis of a transplantable mouse mammary carcinoma after roentgen irradiation. *J. Natl. Cancer Inst.* 12:883–892, 1952.

Von Essen, C. F., and Stjernswärd, J. Radiotherapy and metastases. In *Secondary spread of cancer*, ed. R. W. Baldwin. London and New York: Academic Press, 1978, pp. 73–99.

Wagner, H. Cell mediated immune response *in vitro*. IV. Metabolic studies on cellular immunogenicity. *Eur. J. Immunol.* 3:84–90, 1973.

Wagner, H.; Rollinghoff, M.; and Nossal, G. J. V. T-cell mediated immune responses induced *in vitro*. A probe for allograft and tumor immunity. *Transplant. Rev.* 17:3–11, 1973.

Walker, A. et al. Promotion of metastasis of C_3H mouse mammary carcinoma by local hyperthermia. *Br. J. Cancer* 38:561–563, 1978.

Wallace S. et al. Embolization of renal carcinoma—experience with 100 patients. *Radiology* 138:563–570, 1981.

Wang, M. Delayed hypersensitivity to extracts from primary sarcomata in the autochthonous host. *Int. J. Cancer* 3:483–490, 1968.

Waynforth, H. B. *Experimental and surgical technique in the rat.* New York: Academic Press, 1980, p. 223.

Weiss, L. *The cells and tissues of the immune system.* Englewood Cliffs, N.J.: Prentice Hall, 1972.

Westermark, N. The effect of heat upon rat tumors. *Scand. Arch. Physiol.* 52:257–322, 1927.

Williams, A. E., and Galt, J. M. Studies on the effect of whole body hyperthermia on immunological function in rats (abstr.). In *Proceedings of the Second International Symposium on the Treatment of Cancer by Hyperthermia and Radiation*, eds. C. Streffer et al. Baltimore and Munich: Urban & Schwarzenberg, 1977, p. 67.

Willis, R. A. *The spread of tumors in the human body*, 3rd edition. London: Butterworth, 1973, p. 151.

Witz, I., and Ran, M. Solid tumor response to antitumor antibody. In *Immunity and tolerance in oncogenesis*, ed. L. Severi. Perugia, Italy: University of Perugia, 1970, pp. 345–354.

Wizenberg, M. J., and Robinson, J. E., eds. *Proceedings of the First International Symposium on Cancer Therapy by Hyperthermia and Radiation.* Chicago: American College of Radiology, 1976.

Yamamoto, T. Experimental study of effect of x-rays on metastasis of malignant tumor especially in bones. Jpn. J. Obstet. Gynecol. 19:559–569, 1936.

Yates, A., and Fisher, B. Experimental studies in regional perfusion. *Arch. Surg.* 85:827–833, 1962.

Yerushalmi, A. Cure of a solid tumor by simultaneous administration of microwaves and x-ray irradiation. *Radiat. Res.* 64:602–610, 1975.

Yerushalmi, A. Influence on metastatic spread of whole body or local tumor hyperthermia. *Eur. J. Cancer* 12:455–463, 1976.

CHAPTER 24
Bioeffects of Microwave and Radiofrequency Radiation

Stephen F. Cleary

Introduction
Human Exposure Effects
Effects on Experimental Animals
  Mortality
  Neuroendocrine Alterations
  Hematologic and Immunologic Effects
  Experimental Microwave Cataractogenesis
  Teratogenesis
  Chromosomal and Mutagenic Effects
Conclusions

INTRODUCTION

There are ample precedents involving the misuse or misapplication of physical or chemical agents in the treatment of disease to justify concern for the damage potential of microwave or radiofrequency (RF) radiation as used for the induction of tissue hyperthermia. The experience of early workers in the field of radiology, for example, drew attention not only to potential damage to patients from overexposure to ionizing radiation, but to the hazards to the clinicians and technicians who administered the treatments and who experienced a significant increase in morbidity and mortality resulting from ionizing radiation exposure. Awareness of the problem led to the development and implementation of dose-limiting procedures that have been successful in reducing the risk of radiation damage both to radiologic workers and to patients. Of concern in the context of microwave- or RF-induced hyperthermia are possible exposure effects and consequent methods of reducing or eliminating such deleterious effects.

The analogy of microwave and RF bioeffects to the effects of ionizing radiation may be pursued somewhat further since in either case the development of a basic understanding of bioeffects, and subsequently safe exposure practices, depends to a great extent upon dosimetric information. The dosimetry of ionizing radiation, owing to a great expenditure of resources during the past few decades, has provided data that have been used in the development of interaction mechanisms that have been used in turn to relate a radiation dose to biological effects. The dosimetry of microwave or RF radiation presently lags far behind dosimetry of ionizing radiation. The state of advancement in microwave and RF dosimetry is related in part to a lack of concern and support, but to a great extent it is due to the inherent physical complexities involved in the interaction of microwave and RF fields with biological systems and with sensing elements, as described in chapter 13. Limitations imposed by dosimetry as well as by the type and extent of data on the biological effects of microwave and RF radiation have prevented the development of adequate models of interaction mechanisms and dose-response relationships in most instances. Thus, although a significant variety of biological alterations have been attributed to microwave exposure, the usefulness of such data in the establishment of safe exposure levels is limited.

In spite of limitations in our understanding of the biological effects of microwave and RF exposure, the clinical application of such methods demands a summary of current information and potential problems associated with their use. It should be noted that there is some suggestion that certain low-level microwave exposure effects may be beneficial rather than harmful. This chapter is devoted to a phenomenologic description of the present status of research on the biological effects of microwave and RF radiation, with primary emphasis on in vivo and in vitro effects in mammalian systems. The majority of the current research effort in this field is directed toward studies of the effects of low-intensity microwave and RF fields, prompted by concern for potential deleterious effects on human populations exposed to a variety of sources ranging from microwave ovens to citizens-band radios. Such effects would be presumed to occur as a result of exposure to field intensities that do not result in perceptible increases in body temperature, so-called nonthermal effects. Although such effects are of concern to anyone potentially exposed to microwave or RF radiation, since no morbidity or mortality has been associated with such exposure to date, and since clinical applications involve high-intensity thermogenic fields, exposure effects that have been generally associated with field intensities that result in tissue heating are emphasized in this chapter. For the majority of reported effects in humans, most of which have resulted from occupational exposure, it is generally impossible, because of a lack of exposure data, to differentiate thermogenic exposure effects from nonthermogenic effects. It is assumed, however, that adherence to recommended safe exposure levels has in most instances limited exposures to intensities of less than 10 mW/cm^2. Consequently, human exposure effects usually would not be considered to be thermally induced, with certain exceptions such as cataract induction.

HUMAN EXPOSURE EFFECTS

The known or suspected effects of microwave and RF radiation on humans are derived primarily from studies of workers occupationally exposed in the course of testing, repairing, or using microwave (principally radar) and RF generating equipment. Broad-based psychological and physiologic surveys, as well as studies of specific end points, have been conducted. In many cases, the data are not amenable to statistical analysis because of inadequate control procedures which, when coupled with the lack of accurate dosimetric information, introduces a high degree of uncertainty regarding dose-response relationships for human exposure. The difficulties encountered in exposure-dose estimation have resulted in the categorization of exposure either in terms of exposure groups such as "high," "medium," and "low," or in attempts to relate exposure effects to the duration of employment.

Baranski and Edelwejn (1975) divided their study group of workers involved in the fabrication, repair, and use of microwave equipment into a low-exposure group (average field intensities of tens of microwatts/cm^2), a moderate-exposure group (average intensities of hundreds of microwatts/cm^2 to approximately 1 mW/cm^2), and a high-exposure group of individuals intermittently exposed to average field intensities in the range of 1 to 10 mW/cm^2. Each group was then subdivided into five subgroups based upon duration of employment. Subjective complaints, such as headaches and sweating, were reported frequently, and flat electroencephalographic (EEG) recordings were encountered in individuals with long durations of exposure. No definite conclusions were drawn from these results because of the complexity of the environmental and occupational factors and the lack of appropriate controls.

Gorden (1970) reported the results of a long-term study of workers who were exposed to a variety of continuous wave and pulse-modulated microwave fields. Disturbances in cardiac rhythm (principally bradycardia) and hypotonia were encountered in individuals exposed to centimeter-wavelength radiation at power densities of from 0.1 to 10 mW/cm^2. The same symptoms were encountered in individuals exposed for a number of years to centimeter and millimeter radiation at time-averaged intensities of 10 to 100 μW/cm^2, but in this instance the symptoms were detected less frequently and remitted after the affected workers were removed from the microwave-exposure environment. The author concluded that chronic occupational microwave exposure may lead to neurocirculatory disturbances, which occur in three stages, depending on duration and extent of exposure. The initial stage of this syndrome consists of lability of cardiac rhythm and blood pressure, both effects being reversible in the sense of complete remission upon cessation of occupational exposure. The second stage of the neurocirculatory syndrome includes an accentuation of the first stage symptoms in addition to EEG changes, hyperactivity of the thyroid, and the induction of an *asthenic* state characterized by headaches, irritability, increased excitability, fatigue, and cardiac pain. The third stage consists of more pronounced manifestations of the above symptoms together with changes in electrocardiogram (ECG). The induction of asthenia in workers occupationally exposed to microwave or RF radiation has been described by other East European investigators (Mielczarek 1971; Petrov 1972), who refer to exposure-related increases in the incidence of general irritability, headaches, weakness, sleeplessness, decreased libido, and chest pains as typical responses to chronic exposure to low-intensity microwave radiation. There have been instances, however, where such effects have not been detected (Baranski and Czerski 1966, 1976a, 1976b).

Baranski and Czerski (1976a, 1976b) reported the results of an epidemiologic study of Polish microwave workers who were categorized according to average exposure intensities. Three groups were defined: a low-power-density group consisting of individuals exposed to microwave intensities of tens of microwatts per square centimeter, a group exposed to moderate power densities in the range of hundreds of microwatts per square centimeter up to about 1 mW/cm^2, and a group exposed to time-averaged intensities of from 1 to 10 mW/cm^2. The authors noted that difficulties were encountered in obtaining a control group consisting of individuals with similar distributions of age and environmental and socioeconomic factors. No microwave exposure-related effects were found in the group exposed to the lowest microwave intensities for an average duration of 10 years. The incidence of asthenic symptoms, such as headache and fatigue, was 45% in the group exposed to the highest radiation intensities as compared to 32% in the group exposed to moderate levels of microwave radiation. Thirty percent of the workers exposed to the lowest intensities reported that they experienced these effects during their initial year of occupational microwave exposure. The symptoms appeared to remit for about two years and then recurred after three to five years of exposure, suggesting an adaptive response. The statistical significance of these data is not discussed. Additional physiologic studies detected blood pressure alteration in the high-exposure-intensity group, but no direct correlation was found between heart rate and exposure, nor were ECG changes detected. Hematologic studies revealed lymphocytosis in 10.5% of the individuals exposed to the highest intensities; these subjects also had monocytosis with a total white cell count in excess of 10,000/mm^3. Prolonged exposure to moderate or high intensity microwave fields (as defined by the authors) resulted in a decrease in the average number and amplitude of alpha waves in the EEG and increased sensory stimuli thresholds, the statistical significance of which was not indicated.

The incidence of functional disturbances (such as headache, fatigue, emotional instability, sleep and memory disturbances, hand tremors, and excessive sweating), cardiocirculatory disturbances (including ECG abnormalities), and circulatory disorders were investigated in a group of 841 radar workers (Czerski, Siekierzynski, and Gidynski 1974; Siekierzynski et al. 1974). The group was subdivided into 507 persons potentially exposed to mean time averaged intensities in the range of 0.2 to 6 mW/cm^2 and 334 workers potentially exposed to maximum microwave intensities of 0.2 mW/cm^2. No microwave-intensity-dependent differences or employment-duration-dependent differences were detected, but the incidence of functional disorders increased with increasing age of the workers in both groups. The authors could not draw any conclusions regarding the causal relationship between occupational microwave exposure and the induction of functional disturbances. The results of this study provide no indication that occupational microwave exposure at levels of up to 6 mW/cm^2 results in cumulative alterations in the end-points investigated. Contradictory results have been reported, however, where cumulative irreversible effects have been detected.

Medvedev (1973) investigated the relationship between occupational microwave exposure and cardiovascular disease in 80 workers who had been potentially exposed during the period 1948 to 1967; these workers had not had any known exposure for four to seven years prior to the study. A control group of 80 men was selected on the basis of similar age distribution to the exposed group. Exposure was estimated to have been to microwave intensities of up to 10 mW/cm^2 for seven hours per day during the duration of employment. Abnormal ECGs were recorded from 13 exposed individuals, as compared to four abnormal recordings from the control group. Ischemic heart disease, hypertension, and abnormal blood lipid indices were more prevalent in the exposed group. The authors concluded that chronic occupational microwave exposure facilitated the development of cardiovascular disorders, especially in individuals with a predisposition to such conditions.

The effects of exposure to 3 to 30 GHz microwave radiation on 162 workers were reported by Klimkova-Deutschova (1974). Seven control groups consisting of workers from other industries, such as radio and television transmission and metal and plastic welding, were used as controls. Characteristic symptoms of microwave exposure included headache, fatigue, increased excitability, various other indications of autonomic nervous system and cerebellar disturbances, and EEG synchronization and increases in slow rhythms of high amplitude. There was no significant correlation of the incidence of these symptoms with age. Klimkova-Deutschova concluded that the symptoms were characteristic responses to microwave exposure.

Although the reported effects of occupational microwave exposure are not consistent, certain generalizations emerge. Central nervous system (CNS) disturbances appear to occur in progressive stages, depending on the duration of employment or the age of the workers. The first stage of CNS alterations is characterized by a neurasthenic syndrome involving circulatory and digestive disturbances and subjective symptoms such as fatigue, sleeplessness, and headache. Prolonged exposure reportedly leads to similar disturbances, but also to some degree of damage to motor systems that may progress to an eventual state of encephalopathy in some instances. EEG changes typically involve an increase in slow rhythms of high amplitude that, together with other clinical and biochemical changes, suggest alterations localized in the mesodiencephalic brain region. Klimkova-Deutschova (1974) suggested that such effects are related to nonuniform absorption of pulse-modulated or continuous-wave microwave radiation in the brain, resulting in thermally mediated functional changes. It is difficult to assess the validity of such mechanisms because of limitations on existing knowledge of thermophysiology of the brain. It should be noted, however, that in the brain (and other tissues) the nature of the temporal and spatial heating patterns from microwave absorption present the organism with unique challenges relative to other physical agents, such that normal adaptive mechanisms may not be able to compensate adequately for such stress.

Difficulties in microwave and RF internal dosimetry and field characterization make it difficult to establish relationships between CNS effects and exposure intensity, but if it is assumed that the effects are related to brain heating, alterations would not be anticipated at incident power densities of less than approximately 1 mW/cm^2. Effects such as reversible neurasthenia, attributed to field-induced alterations in synaptic transmission and enzymatic activities (Klimkova-Deutschova 1974), and altered EEGs (Baranski and Edelwejn 1975) have been reported to occur as a consequence of occupational exposure to pulse-modulated microwaves at time-averaged intensities of less than 1 mW/cm^2, suggesting that unknown factors (other than brain heating) may be involved in CNS alterations in humans.

The reports of the CNS effects of occupational microwave and/or RF exposure have all emanated from the USSR, Poland, or Czechslovakia and in general contain limited information regarding exposure parameters. Exposure to pulse-modulated fields of potentially high instantaneous intensities, but of relatively low time-averaged intensities, appears to produce greater degrees of CNS alterations than continuous wave fields of equivalent power density. The dearth of exposure data and apparent deficiencies in statistical methodology and translations into English make it difficult to interpret and assess the significance of the reported CNS and other effects of occupational microwave exposure.

There have been no comparable epidemiologic studies conducted in the United States or other western nations. As an alternative, animal experimentation has been employed in this country (as well as in the Communist bloc) to provide data relative to the relationship of microwave and RF exposure to alterations in neural, CNS, behavioral, and other physiologic end points. Data from such studies have been of value in increasing our understanding of these effects, but differences in the distribution of internally absorbed energy, exposure conditions, and the generally short-term or acute timing of animal studies limit their applicability to the assessment of the effects of chronic or long-term human exposure to microwave or RF fields.

Although there is little quantitative information regarding occupational microwave and RF exposure levels, it may be assumed that the voluntary exposure standard of 10 mW/cm^2 that has been in effect in the United States and other western nations and the 10 μW/cm^2 standard of the Communist bloc has served to limit exposure to personnel to less than these maximal levels. Hence, the majority of the effects reported to result from occupational exposure may be considered to be low-intensity exposure effects. There have appeared, however, a few reports of what may be presumed to be effects of accidental exposure of humans to high-intensity microwave fields.

McLaughlin (1957) reported an incident involving a fatal exposure of a 42-year-old worker who was accidentally exposed within 3.1 m of the antenna of a radar transmitter of unspecified frequency and intensity. A sensation of abdominal heating was experienced by the worker almost immediately, forcing him to move out of the radar beam. Acute abdominal pain and vomiting commenced within 30 minutes after exposure, and the patient was in mild shock when examined one hour after exposure. The patient exhibited auricular fibrillation, and his blood pressure was 90/30 mmHg, with a radial pulse rate of 72. Upon admission to a hospital, the patient's leukocyte count was 15,700/cu mm, his abdomen was greatly distended and showed the general appearance of acute peritonitis. Six hours following the onset of pain an operation was performed and the patient's appendix was removed. Symptoms of bowel obstruction were noted several days later, and subsequent evisceration of the gut through the abdominal incision required a second operation, which revealed an oval perforation of the bowel. The patient died within hours after the second operation. McLaughlin concluded that the "cooked" and hemorrhagic gross appearance of the bowel and the results of post mortem pathology were consistent with a local absorption of heat as a consequence of whole-body microwave exposure. McLaughlin also noted that the observed hemorrhagic infarcts of the spleen of the patient were similar to those seen by him in two other patients exposed to microwave radiation.

Further indications of the possible effects of exposure to presumably high intensity (viz., > 10 mW/cm^2) microwave radiation have been reported. McLaughlin (1962) described a syndrome consisting of increased capillary fragility, inadequate clot retraction, and abnormal bleeding that occurred in 115 workers engaged in the manufacture of microwave equipment. The author described four representative cases from this group of workers exposed from one to three hours per day for six months to three years at distances of from 0.3 to 15.2 m from radar antennas. The first patient was a 39-year-old man who developed a large area of ecchymosis on his hand following a machine shop accident. The patient recovered after being treated with ice packs, bed rest, and whole-blood transfusion. A 27-year-old male worker was examined for complaints of localized tenderness and pain and general malaise, resulting in hospitalization. His blood pressure was 90/60, and he was diagnosed as suffering from temporary adrenal insufficiency and a generalized stress syndrome, which was successfully treated with cortisone. A 28-year-old female worker complained of malaise and dermal red spots on her arm that were similar in appearance to those detected in the area of pain, as in the previous patient. After fainting at work, the patient was hospitalized and was tentatively diagnosed as suffering from a subarachnoid hemorrhage. Her blood chemistries appeared normal with the exception of the fibrin-clot volume. Following a period of apparent recovery, the patient died two weeks after release from the hospital. Autopsy revealed evidence of recent bleeding at the site of the previous hemorrhage. The final case involved incidents of persistent bleeding in a 26-year-old woman who developed ecchymoses from her knee to toes subsequent to leg trauma. A large hematoma was evacuated from the site of injury and the wound healed.

The effects of exposure of a microwave oven repairman were reported by Rose and colleagues (1969). Exposure of a 40-year-old man occurred for a period exceeding five years, typically involved fields of 2.45-GHz microwaves at intensities of from 10 to 22 mW/cm^2 for a minimum of four minutes per exposure, with the frequency of exposure varying from zero to multiple exposures throughout the working day. Symptoms included loss of visual acuity, impotence attributed to nodules in his penis, low sperm count, and episodic skin eruptions on the abdomen and thigh, apparently caused by vasculitis. The authors suggested that the microwave exposure was a possible etiology for the skin lesions, since the anatomic location and the hemorrhagic appearance of the lesions were consistent with this history; however, similar abnormalities were not detected in seven co-workers exposed to varying intensities of microwave radiation.

Doury, Boisselier, and Bernard (1970) reported the apparent effects of exposure of a radar repairman to 1.3- to 3-GHz microwaves for a period of three years. Symptoms included tachycardia, multiple venous thromboses, endocrine dysfunction, and weight loss. The patient was treated with anticoagulants and hormones and was removed from the work environment, with subsequent improvement of symptoms.

Prominent among the effects of acute, presumably high intensity, microwave exposure of humans is cataract induction, an irreversible effect generally attributed to thermal damage of lens tissue. Hirsch and Parker (1952) reported a case of cataract induction in a 32-year-old microwave worker who had worked for a period of approximately one year with pulse-modulated microwave equipment that emitted 1.67- to 3.33-GHz radiation at an average power output of 100 W. The worker was exposed consistently to 5 mW/cm^2 microwave radiation from a horn antenna, and during the course of his work he often looked directly into the antenna with his face in close proximity to the source, where the field intensity was calculated to be on the order of 100 mW/cm^2. Following three days of exposure to these fields for a total of two hours per day, visual loss was experienced, and conjunctivitis was noted. Lens examination revealed a slight roughening of both lenses and moderate degrees of nuclear opacification. The authors concluded that precautionary measures were needed in situations involving microwave exposure, but owing to the circumstantial nature of the evidence, they did not claim that the visual difficulties were definitely attributable to microwave exposure.

In 1959, another case of cataract induction was described by Shimkovich and Shilyaev and involved a 22-year-old microwave worker exposed four to five times to microwaves in the frequency range of 300 MHz to 3 GHz for two to four minutes at power intensities estimated to be at least 300 mW/cm^2. The worker described heating of his hands during exposure, after which he reported ocular disturbances including tearing, pain, and photophobia. There was a marked decrease in visual acuity, and bilateral lens opacities were detected. The patient was treated with iontophoresis and vitamins, and no further lens changes were observed. The lens opacification was diagnosed as "occupational (ultra-high frequency) cataracts of both eyes."

Seven years of occupational exposure to a variety of microwave and RF sources was reported to result in the development of bilateral cataracts in a 51-year-old worker. The cataracts were described by Kurz and Einaugler (1968) as dense, centrally located opacities below the posterior lens capsule.

Three other cases of cataract induction by occupational microwave exposure were reported by Zaret in 1964. In each case, cataracts developed in the worker's eye that was close to a radiating tube or waveguide, and the intensities were estimated to be greater than 350 mW/cm^2. The exposures, which were for durations of several minutes per day, occurred from several times per day to several times per week for periods of weeks to months. Zaret also described cases of microwave cataract induction in a 51-year-old woman exposed to microwave oven leakage (1974), in a 53-year-old radar repairman (1975), and in nine operational aviation technicians (Zaret and Snyder 1977).

The potential importance of the exposure conditions in the induction of ocular damage from microwaves is indicated by the failure of therapeutic microwave exposure to reveal cataract induction. Raue (1963) used 2.5-GHz microwaves for the treatment of postoperative or traumatic retinal edema and other ocular diseases without apparent ocular damage. The results of microwave diathermic treatment of 75 cases involving various ophthalmic diseases were reported by Clark (1952). Therapy consisted of three treatments per week for three to six weeks, with some treatments prolonged for up to nine months. No lens damage was detected from such treatments.

Epidemiologic studies by Cleary and Pasternack (1966) and others (Majewska 1968; Kheifets 1970; Cleary, Pasternack, and Beebe 1965; Appleton 1973; Odlund 1973; Siekierzynski et al. 1974b; Aurell and Tengroth 1973; Zydecki 1974; Shaklett, Tredici, and Epstein 1975; Hathaway et al. 1977) have failed to detect a statistically significant increased risk of cataractogenesis as a result of occupational microwave exposure, although in some instances it has been found that such exposure appears to increase the number of non-vision-impairing lens defects.

Another source of information relative to the effects in humans of intense absorption of RF and microwave radiation is derived from published reports of therapeutic applications of such fields, which date back to an 1893 report by D'Arsonval of the use of electromagnetic fields for electrotherapy. Carpenter and Page (1930) found that an electric field strength of 40 V/cm was more effective in inducing fevers in humans when applied at a frequency of 10 MHz than at higher frequencies. In 1931, Bell and Ferguson exposed five volunteers at 1.3 m from the antenna of a 55 MHz radio generator at output power levels of from 10 to 18.5 kW for the purpose of determining the maximum tolerable intensity. The subjects held a metal rod in one hand during exposure to serve as an antenna to maximally couple to the RF field. The effects of exposure in the order of occurrence were a warm sensation and discomfort in the hand holding the rod, followed by warming sensations in other parts of the body, cramps in the hand, sweating, and finally, headache, drowsiness, and fatigue. The authors suggested a psychogenic origin of the responses to the field and the similarity to the induced symptoms to heatstroke.

Bierman, Horowitz, and Levenson (1935) used 10 MHz radiation to treat a variety of diseases in a series of 24 patients in whose bodies temperature rises of approximately 3.6°C were induced. The response consisted of a transient leukopenia followed by leukocytosis, a pattern noted first in the neutrophils, then in the monocytes, and finally in the lymphocyte cell fraction. These results were interpreted as indicating that the bone marrow was the first immunologic tissue to be affected by exposure to 10-MHz radiation. Bierman, Horowitz, and Levenson reported that repeated RF exposures appeared to induce decreasing stimulatory responses, suggesting an adaptive reaction. Jung (1935) also reported alterations in leukocyte count as a result of hyperpyrexia induced by diathermy.

Counterindications for the application of RF diathermy for the induction of hyperthermia were noted by Leavy (1935). The excessive heat stress reportedly could not be tolerated by patients with certain cardiac disorders, pulmonary diseases, functional neuroses, or of advanced age. Beneficial results were reported in the treatment of articular diseases of the spine, arthritis, angina pectoris, synovitis, peripheral vascular diseases, and bone fractures.

The relationship of RF radiation frequency to tissue heating was investigated by Coulter and Carter (1936), who measured the skin, subcutaneous, and muscle temperatures in the thighs of six volunteers exposed to 12.5, 16.67, 25, or 50 MHz diathermy fields. As the wavelength of the radiation decreased, the skin temperature decreased, whereas the deep muscle temperature increased for equivalent applied intensities. RF-radiation-induced core temperature increases have also inhibited spermatogenesis in two volunteers whose core temperatures were increased to 40.5°C and 41°C during a 45-minute exposure period.

Sperm counts remained normal for 18 days postexposure, after which they decreased to a minimum between 44 and 50 days after exposure. It was concluded that the alterations were thermally induced and the latency was due to the cyclical nature of spermatogenesis.

Horvath, Miller, and Hutt (1948) analyzed the use of 2.45-GHz microwaves for selective tissue heating. Maximum temperature increases of 10.7°C induced in subcutaneous tissue and 4°C in muscle were not associated with detectable vasodilation or increased rectal temperature. Although there were detectable temperature increases at depths of up to 6 cm, the volunteers reported only a pleasant warming sensation in the skin. These findings were in agreement with those of Osborne and Frederick (1948), who used microwaves to induce temperatures of up to 40.1°C at tissue depths of 5.1 cm.

In summary, data relative to the effects of microwave and RF on humans are not sufficiently quantitative or extensive enough to derive dose-response relationships nor to adequately assess late exposure effects. Occupational exposures at time-averaged intensities of less than approximately 10 mW/cm^2 appear to induce functional alterations in the CNS of exposed individuals. Such alterations, which are to a great extent subjective in nature, are described as neurasthenia, although some objective symptoms such as altered EEG and ECG patterns have been reported, as well as other apparently reversible physiologic changes of undetermined significance with respect to morbidity or mortality. Accidental exposures to microwave fields at unspecified, but presumably high intensities (i.e., well in excess of 10 mW/cm^2), can lead to irreversible tissue damage such as cataracts, internal hemorrhages, and burns. The therapeutic use of RF and microwaves in diathermy has revealed that such exposures can result in complex internal patterns of tissue heating, leading to a number of potentially beneficial as well as adverse effects such as alterations in the hematopoietic, reproductive, and central nervous systems. Late effects on morbidity or mortality associated with microwave or RF exposure of humans in occupational or therapeutic settings have not been reported, but it appears that few if any studies of such effects have been undertaken.

Limitations on the available human data on the effects of microwave and RF radiation have directed attention to data derived from the use of experimental animals. Such data have increased knowledge of the effects of such exposure, but two problems have emerged in attempting to extrapolate such data to assess exposure effects in humans. First, there is very limited data on late exposure effects and the effects of chronic exposure of experimental animals. Second, the physical interactions of microwave and RF with mammalian (and nonmammalian) species depend in a complex manner on the size, shape, orientation, and configuration of the body, leading to wavelength-dependent variations in coupling to the electromagnetic field and internal patterns of energy absorption. Coupled with well-known interspecies variations in physiologic responses, the extrapolation of data from animal experiments to humans is difficult at best. With these considerations in mind, the remainder of this chapter reviews the results of animal experimentation, specifically those data most pertinent to the effects of thermogenic microwave or RF fields, with some mention of low-intensity field effects, in order to provide data on comparative responses of experimental animals and humans.

EFFECTS ON EXPERIMENTAL ANIMALS

Inherent difficulties in dosimetry and in field characterization and other experimental artifacts, such as field perturbation and coupling by measuring electrodes, have impeded progress in experimental investigations of the biological effects of microwave and RF radiation. These complications most probably are among the causes of inconsistencies in the reports of such studies. Advances in experimental methods have, however, reduced the extent of such problems. Standardized methods of dosimetry and field characterization recently have been developed, leading to a greater degree of concordance in the reported data. Despite the limitations inherent in quantitative extrapolation of experimental data to human exposure effects, qualitative comparisons provide indications of possible deleterious effects in humans.

Mortality

Mortality has been investigated in experimental animals acutely exposed to microwave radiation primarily as a function of field intensity and secondarily as a function of radiation wavelength. As indicated previously, there are few data on life span reduction or late mortality in experimental animals as a result of either fractionated or chronic microwave or RF exposure. The relation between the size, shape, and orientation of the animal and the wavelength of the radiation determines the amount of whole-body radiation absorption and, consequently, mortality would be expected to be related to wavelength. Since other factors, such as the modes of internal energy absorption, also would be expected to be involved in the induction of death from tissue damage, a simple relationship of mortality to wavelength would not be anticipated—and indeed, has not been demonstrated by mortality data. Empirical, as well as theoretical, studies have indicated that the approximate radiation frequency ranges for maximum absorption are 100 to 400 MHz for monkeys, 100 to 300 MHz for dogs, 300 to 700 MHz for rabbits, 500 to 800 MHz for guinea pigs, 0.5 to 1 GHz for rats, and 1 to 3 GHz for mice (Durney et al. 1978; Gandhi, Hunt, and D'Andrea 1977).

The results of acute lethality experiments in various species of experimental animals and under different exposure conditions are summarized in table 24.1

No clear pattern is discernible from these data, owing at least in part to the differences in exposure conditions and the variety of species used in these studies. Comparison of the dose lethal to 50% of the animals (LD_{50}) in a given experiment involving exposures at intensities that differ by at least one order of magnitude (Polson et al. 1974; Addington et al. 1961; Schrot and Hawkins 1974) suggests intensity-time relationships characteristic of dose reciprocity. Inter-experiment comparison of exposure doses for comparable end points, such as LD_{50}, suggests a common mechanism for microwave-induced acute lethality that is not strongly dependent upon species or radiation wavelength or intensity within the ranges of the independent variables employed in these studies.

In studies in which rectal temperature was measured (Susskind 1958; Baranski and Czerski 1976a, 1976b; Deichmann, Bernal, and Keplinger 1959; Searle, Imig, and Dahlen 1959), lethality was associated with an increase of approximately 6°C, although Jacobson and Susskind (1958) reported no lethality in animals whose rectal temperatures were elevated 4.5°C as a result of microwave exposure. These results and post mortem pathologic findings, such as blood vessel damage leading to edema and hemorrhage (Polson et al. 1974), support the hypothesis that death was due to excessive heat stress.

The effect of cholinergic and anticholinergic drugs on the survival of rats exposed to 2.4-GHz microwaves at an intensity of 150 mW/cm^2 was investigated by Koldaev (1970). Survival was found to increase when cholinergic drugs, such as pilocarpine, were administered immediately after microwave exposure, whereas anticholinergic drugs produced the opposite effect. He concluded that high-intensity microwave radiation preferentially damaged the parasympathetic components of the autonomic nervous system. Susskind (1958) reported that pretreatment of mice with chlorpromazine (6 mg/kg) produced an average decrease in body temperature of 10°C, which significantly increased the survival time during exposure. Lethality occurred, however, when the exposure was prolonged until the radiation induced a critical temperature elevation. Other factors reported to affect microwave-induced lethality were oxygen tension and environmental temperature. Koldaev (1970)

Table 24.1
Microwave-Induced Acute Lethality

Species	Frequency (GHz)	Intensity (mW/cm^2)	Exposure Duration (min)	Exposure Dose (J/cm^2)	Comments	Reference
Rat	24.0	250	17.4–47	261–705	Duration of survival dependent upon ambient temperature	Deichmann, Bernal, and Keplinger 1959
Rat	0.95	178–278	4.9–3.1	52–52	LD_{50}	Polson et al. 1974
Rat	2.45	342–3920	3.4–0.2	70–45	LD_{50}	Polson et al. 1974
Rat	4.54	762–12,376	3.2–0.16	146–119	LD_{50}	Polson et al. 1974
Rat	7.44	629–5720	3.1–0.3	117–96	LD_{50}	Polson et al. 1974
Rat	2.4	150	40	360	100% lethality	
Rat	3.0	400–2400	4–0.5	96–72	LD_{50}	Schrot and Hawkins 1974
Dog	2.45	800	268 min mean time to death	12,864	Partial-body exposure of head during anesthesia	Searle, Imig, and Dahlen 1959
Dog	0.2	38–330	60–7	137–138	Lethality related to orientation and rectal temperature rise	Addington et al. 1961
Mouse	10.0	5	180	54	LD_{50}	Baranski and Czerski 1976b
Mouse	9.1	117–438	10–2	70–79	LD_{50}	Susskind 1958
Mouse	10.0	120–440	12–2	86–53	LD_{50}	Jacobson and Susskind 1958
Mouse	9.27 (PW)	14–58	23	19–80	No lethality	Prausnitz and Susskind 1962
Mouse	9.27 (PW)	68–380	18–2	73–46	LD_{50}	
Mouse	2.4	57–67	15–16	55–60	47%–52% lethality	Koldaev 1976

found that exposure of rats to a 40% oxygen atmosphere for 10 minutes prior to exposure increased the survival time of rats exposed to 150 mW/cm², 2.4-GHz microwave radiation by 37%. Irradiation following exposure to low oxygen tension reduced survival time by 29% to 57%. Muroff and Samaras (1969) used cool air circulation to successfully counteract the lethal effects of exposure of rats to 2.45-GHz microwaves at an intensity of 100 mW/cm².

The results of acute lethality experiments involving exposure of experimental animals to high-intensity thermogenic microwave radiation suggest hyperpyrexia as the cause of death. The extent to which nonuniform energy absorption affects such lethality cannot be determined since there is insufficient data on the wavelength dependency of this end point in a given species. On general principles, it would be anticipated that microwave energy absorption would not produce the same effects as exposure to other heat-inducing agents that result in equivalent whole-body temperature increases as determined, for example, by colonic temperature measurements. Detailed information on the maximal tissue temperature elevation within a body exposed to microwave radiation is necessary to assess the potential for localized damage.

Neuroendocrine Alterations

Data on the effects of microwave radiation on humans are not adequate to determine the sensitivity of the neuroendocrine system. Afanas'yev (1968) and Szady and co-workers (1976) found no effects of occupational exposure on the serum levels of 17-ketosteroids or 17-corticosteroids. Increased adrenal secretion of tetrahydrocortisone and cortisone was attributed by Dumkin and Korenevskaya (1974) to occupational exposure. Neuroendocrine alterations in microwave and RF workers also were reported by Fofanov (1969a, 1969b); Gembitskiy, Kolesnik, and Malyshev (1969); D'yachenko (1970); and Kleyner and associates (1975), but no consistent response pattern is evident from these studies.

The results of animal experimentation help to explain the inconsistent and, in some instances, apparently contradictory results of human studies of neuroendocrine effects of microwave and RF exposure. Animal experiments have revealed that the response of the neuroendocrine system depends on the exposure duration, intensity, wavelength, and other exposure conditions. Demokidova (1974a) exposed rats to 14.9-MHz RF radiation at a field strength of 70 V/m for one hour per day for one to four months and to a 69.7-MHz field at 4 to 49 V/m for one hour per day for 1.5 months and found an increase in pituitary and adrenal weight and a decrease in thyroid weight. Exposure of the same species to a dose of 0.153 mW/cm², 3-GHz microwave radiation during a one- to two-hour period caused a decrease in adrenal and pituitary weight and an increase in thyroid weight (Demokidova 1974b).

Rats exposed to an output power of 600 W in the near-field of a 57.2-MHz RF generator for 5, 15, or 30 minutes had significantly reduced ascorbic acid levels in the adrenal gland, an effect also induced by exposure to infrared radiation and attributed by the authors to the effects of heat stress (Adler and Magora 1964). Novitski'i and colleagues (1977) reported power-density-dependent increases in corticotropin-releasing factor activity in the hypothalamus, pituitary adrenocorticotropic hormone activity, and in the concentration of 11-oxycorticosteroids and the Na^+ to K^+ ratio in the blood of rats exposed to 2.38-GHz microwaves for 30 minutes. Repeated 30-minute per day exposures for 10, 20, or 30 days at an intensity of 1 mW/cm² led to a progressive decrease in the response variables after a maximum level had been attained after 10 days of exposure, suggesting an adaptive response.

Gunn, Gould, and Anderson (1961) investigated the effects of acute scrotal exposure of young male rats to 24-GHz microwaves for five minutes at an intensity of 250 mW/cm². Zinc uptake in the dorsolateral prostate gland was reduced to 55% of control values 13 days after exposure but returned to normal by the 29th day postexposure. Exposures of 10 and 15 minutes decreased uptake to 45% and 30% of controls, respectively, and mean prostate weight was reduced to 80% of control values. Hormone studies (gonadotropin or testosterone) indicated that microwave exposure initially affected luteinizing hormone output by the pituitary. The trophic hormone response of the testicular interstitial tissue was prevented by prolonged microwave exposure, suggesting the possibility of irreversible tissue damage. This comparison of microwave effects on the testes with effects of infrared and heat exposure revealed that comparable effects were induced by microwaves at a lower testicular temperature, thus suggesting a microwave-specific effect not directly related to heating.

The effects of 3-GHz microwaves on thyroid function in adult rabbits was investigated by Baranski, Ostrowski, and Stodolnik-Baranska (1972). Following exposure for three hours per day for four months to an intensity of 5 mW/cm², there was a 50% increase in the incorporation of radioiodine into the thyroid, and the concentration of serum protein-bound iodine increased 117%, indicative of increased thyroid secretory activity. Electron microscopic examination of the thyroid showed an increase in the number of cytosomes and enlargement of Golgi apparatus and endoplasmic reticulum, suggesting that the increased activity was due to the stimulation of cellular activity and not to hyperplasia.

Demokidova (1974a) exposed rats to 69.7-MHz RF radiation at field strengths of 5, 12, and 48 V/m for one hour per day for 1.5 months. Irradiation led to an increase in body weight, a decrease in the thickness of the tibial cartilage and in the activity of alkaline phosphatase, and increased adrenal and pituitary weights. One, two, or four months of exposure to 14.88 MHz at 70 V/m resulted in increased body weight and decreased thyroid and adrenal weight, which was attributed to thyroid inhibition.

In contrast to the results obtained from long-term RF exposure, single exposures to 3-GHz microwaves caused a decrease in pituitary and adrenal weight and increased thyroid weight (Demokidova 1974b). In this study, the rats were exposed to 0.15 mW/cm^2 for one hour or irradiated for three variable intervals within a two-hour period at power densities of 0.06, 0.24, or 0.32 mW/cm^2. Altered water excretion and urine electrolyte concentrations were detected 0.5, 1, 2, and 4 months postexposure, but all dependent variables returned to normal after five months, except urinary potassium concentrations and pituitary weight.

The effect of exposure duration on the function of thyroid, adrenocortical, and adrenomedullary cells of rats was investigated by Parker (1973). Exposure to 2.45-GHz microwaves at intensities of 10, 15, 20, and 25 mW/cm^2 caused increases in rectal temperatures of 0.4°C, 0.5°C, 1.0°C, and 1.7°C, respectively, following 16 hours of exposure. At 10 mW/cm^2, an increase of 26% was noted in the activity of phenylethanolamine-N-methyl-transferase activity, and there was a 35% decrease in epinephrine concentration, with no alteration in plasma corticosterone levels or adrenal weight. No differences were detected in these variables following a four-hour exposure. Exposure for 16 hours did not affect thyroid function, but a 60-hour exposure at 15 mW/cm^2 decreased serum protein-bound iodine by 28%, serum thyroxine by 53%, and induced a 67% decrease in the concentration of radioiodine in the thyroid. Parker concluded that microwave exposure induced endocrine changes mediated by the sympathetic nervous system at or below the level of the hypothalamus, and that nonuniform internal microwave absorption may have accounted for the alterations.

The adrenocortical response of rats to acute exposure to 2.45-GHz microwaves was investigated by Lotz and Michaelson (1978). A positive correlation was noted between mean corticosterone concentration in the plasma and exposure-induced increases in colonic temperature. Power density thresholds were found to depend upon exposure duration, with decreasing thresholds for increasing durations. Stimulation of the adrenal axis was dependent on adrenocorticotropin secretion by the pituitary, as demonstrated by the occurrence of normal levels of corticosterone after microwave exposure in hypophysectomized rats or in rats pretreated with dexamethasone.

Low-intensity microwave radiation also has been reported to alter the endocrine system of rats (Novitski'i et al. 1977). The threshold power density for the stimulation of the hypophyseal-hypothalamic-adrenal axis was 0.01 mW/cm^2 for a 30-minute exposure to 2.38-GHz microwaves. The response to such exposure included increased secretion of corticotropin-releasing factor by the hypothalamus and of ACTH by the pituitary, 11-oxycorticosteroids in the plasma, and altered Na^+ to K^+ ratio in the blood. There was a power density-dependent increase in these variables in the range of 0.01 to 1 mW/cm^2, after which a constant response level was reached, up to a power density of 75 mW/cm^2. Daily 30-minute exposure at 1 mW/cm^2 resulted in maximum increase in these dependent variables after 10 days of exposure, followed by a progressive decrease in all variables after 20 and 30 days of exposure, at which time the values were below control levels.

The results of studies of the effects of microwave and RF radiation suggest that the mammalian neuroendocrine system is sensitive to radiation-induced stress, the response depending in an apparently complex manner, on the intensity and duration of the exposure. Responses to repeated exposures suggest an adaptive response characteristic of endocrine system function.

Hematologic and Immunologic Effects

The hematologic response of experimental animals to microwave and RF exposure has been the subject of investigation since the earliest reports of hematologic alterations in humans, dating back to 1940 to 1950 and the advent of high-powered radar devices. The majority of studies on the effects of microwave exposure on the immunologic system of experimental animals have been conducted recently in response to reports of findings of East European epidemiologic investigations (Baranski and Czerski 1966; Volkova and Fukalova 1974).

Deichmann, Miale, and Landeen reported in 1964 that 24-GHz pulse-modulated microwaves at 20 mW/cm^2 caused leukocytosis, lymphocytosis, and neutrophilia in rats after a seven-hour exposure.

Normal values were reattained one week post-exposure. A three-hour exposure at 10 mW/cm^2 produced similar changes that returned to baseline values two days after exposure. Kitsovskaya (1964) irradiated rats with 3-GHz microwaves at intensities of 10, 40, or 100 mW/cm^2 for various exposure durations. Erythropenia, leukopenia, lymphopenia, and granulocytosis, which persisted for months after exposure, were induced in rats exposed at 40 and 100 mW/cm^2, with no alterations at 10 mW/cm^2. The differences in the persistence of the effects reported by these investigators may be attributed to differences in exposure conditions. Other investigators have reported that the qualitative and quantitative hematologic microwave responses depend on the radiation wavelength and intensity and exposure duration (Michaelson, Thompson, and El Tamami 1964).

In 1962, Prausnitz and Susskind exposed mice once a day for 59 weeks to 9.27-GHz pulsed microwaves at a time-averaged power density of 100 mW/cm^2. The 4.5-minute exposure caused an average colonic temperature increase of 3.5°C. Testicular damage was detected, as well as myeloid leukemia or monocytic leukosis, in 21 of 60 irradiates (35%) that died during the experiment, as compared to 4 of 40 (10%) unirradiated control Swiss mice. This report is unique in its finding that high intensity microwave exposure may cause an increased incidence of cancer in mammals. The lack of information on the relationship of microwave or RF exposure to cancer induction may be because this end point has not been investigated, as suggested by the absence in the literature of reports of such studies.

The effects of low-intensity microwaves (3.5 mW/cm^2, 3 GHz, pulse modulated or continuous wave) on rabbits and guinea pigs have been reported by Baranski (1971b, 1972). A three-month exposure for three hours per day resulted in lymphocytosis, abnormalities in nuclear structure, and mitosis in the erythroblastic cell series in the bone marrow and in lymphoid cells in lymph nodes and spleen. Changes in cell number in peripheral blood were found to be correlated with altered cellularity of the lymph nodes and spleen.

Rotkovska and Vacek (1975) compared the effects of thermogenic microwave exposure of mice (2.45 GHz, 100 mW/cm^2, five-minute exposure) to exposure to an ambient temperature of 43°C for five minutes; both treatments resulted in a 2°C elevation of rectal temperature. Microwave exposure caused a reduction in total cell volume of the spleen and bone marrow and an increase in the number of hematopoietic stem cells in these organs as measured by colony-forming assay. There was a decrease in the incorporation of ^{59}Fe in spleen 24 hours after microwave exposure, whereas exposure to the elevated ambient temperature decreased the colony-forming units in bone marrow and spleen and increased the percentage of ^{59}Fe incorporation. The results of this study indicate a marked difference in the kinetic response of the hematopoietic system to two types of heat stress, which the authors suggest may be due to a direct, specific cellular effect of microwave exposure.

In 1977 Rotkovska and Vacek reported on the effects of exposure of mice to 300 to 750 rad of ionizing radiation, followed at various time intervals by exposure to 2.45 GHz for five minutes at an intensity of 100 mW/cm^2. The combined exposure increased the rate of recovery of hematopoietic tissue, increased myelopoiesis and erythropoiesis, and led to an increased survival compared to mice exposed to ionizing radiation alone. The effect of microwave exposure was attributed to an effect upon the mechanisms that activate the stem-cell pool, either by enhancing repair of sublethal x-ray damage or by increasing the stem cell proliferative capacity. Acceleration of hematopoietic recovery in dogs exposed simultaneously to ionizing radiation and microwaves was also reported by Michaelson and associates (1963). Lappenbush and co-workers (1973) noted that a 30-minute exposure of Chinese hamsters to 60 mW/cm^2, 2.45-GHz microwaves five minutes after exposure to 725 to 950 rad of ionizing radiation increased the LD$_{50(30)}$ compared to x-rays alone or pretreatment with microwaves. The radioprotective effect was associated with a delay in the decrease of circulating leukocytes, a reduction in the period of low cell density, and a complete replenishment of white blood cells within 30 days of the combined exposure. Hamsters exposed to microwaves before x-ray exposure demonstrated more severe leukocyte changes than hamsters treated with x-rays only; this effect was due to a more rapid decrease in leukocytes, resulting in leukopenia. The results of these studies have obvious implications with respect to combined ionizing radiation and microwave-induced hyperthermia for the treatment of neoplasms.

The mammalian immunologic system has also been demonstrated to be sensitive to microwave exposure. Czerski (1975), for example, noted an increase in lymphoblastoid transformations in rabbit lymphocytes cultured for seven days following sampling from the peripheral blood of rabbits exposed for two hours per day, six days per week for six months to 2.95-GHz pulse-modulated microwaves at an intensity of 5 mW/cm^2. Maximum responses were detected after one or two months of exposure, after which the rate of transformation returned to the baseline, only to increase again one month after termination of exposure. He also found that exposure of mice for six hours per day to the same microwave source at an intensity of 0.5 mW/cm^2 for 6 or 12 weeks increased the number of lymphoblasts in the lymph nodes. Miro, Toubiere, and Pfirster (1974), after exposing mice for 145 hours to 3.1-GHz pulsed microwaves at an incident intensity of 2 mW/cm^2, found increased numbers of lymphoblastic cells in the spleen and lymphoid tissues. Similar increases in lymphoblastoid transformed cells were observed in Chinese hamsters exposed to 2.45-GHz microwaves for 15 minutes per day for five days at 4 mW/cm^2. A maximum rate of lymphoblastoid transformation was seen in cultures from hamsters exposed to 30 mW/cm^2, an exposure that resulted in a 0.9°C increase in rectal temperature (Huang et al. 1977). Phytohemagglutinin-stimulated mitosis was decreased in cells obtained from hamsters exposed to 5, 15, 30, or 45 mW/cm^2. The incidence of chromosomal aberrations was not detectably altered by microwave exposure at these intensities. The exposure effects were reversible, with all values returning to normal within 5 to 10 days. Similar effects have been reported by Prince and colleagues (1972), who exposed rhesus monkeys to thermogenic 10.5-, 19.3-, and 26.6-MHz RF radiation.

Additional evidence of the sensitivity of the mammalian immunologic system to microwave exposure is provided also by the work of Czerski (1975), who exposed mice to 2.95-GHz microwaves pulse-modulated with a time-averaged intensity of 0.5 mW/cm^2 for six weeks. Alterations included increases in the number of antibody-producing cells and increased serum antibody titers following immunization with sheep red blood cells. Increased responses were not, however, detected in mice similarly exposed for 12 weeks, suggesting an adaptive response. Immunologic effects of microwave exposure of experimental animals have been reported also by Wiktor-Jedrzejczak with Ahmend and Czerki (1977a, 1977b) and with Sell (1977). They noted that microwave radiation did not stimulate lymphoid cell proliferation per se, but appeared to serve as a polyclonal B-cell activator, resulting in the early maturation of noncommitted B cells. A significant decrease in the primary immune response to sheep red blood cells, a thymus-dependent antigen, in mice immunized prior to their first exposure to microwaves suggested that microwave exposure caused a nonspecific cell maturation before they were activated by the sheep red blood cell antigen, thus increasing the proportion of unresponsive cells. It was suggested that the inability of Smialowicz and co-workers (1979) to reproduce these effects may have been due to differences in exposure systems as well as to the different strains of mice used in these experiments. Thermogenic levels of exposure to 26-MHz RF radiation involving a 2°C to 3°C rise in rectal temperature have been shown to induce changes in splenic lymphocyte populations, which have been interpreted as a stress response (Liburdy 1979).

Alterations in the development of immune responses of rats have been reported as a consequence of exposure to 424-MHz and 2.45-GHz radiation (Smialowicz et al. 1979; Smialowicz, Kinn, and Elder 1979), and 3-GHz exposures have been found to alter the phagocytic activity of rabbit leukocytes (Szmigielski, Jaljaszewicz, and Wiranowska 1975).

In vitro studies of the effects of microwave and RF exposure on lymphocytes (Baranski and Czerski 1976a, 1976b; Hamrick and Fox 1977; Holm and Schneider 1970; Stodolnik-Baranska 1967), lymphoblasts (Lin and Peterson 1977), macrophages (Mayers and Habershaw 1973), and granulocytes (Szmigielski, Jeljaszewicz, and Wiranowska 1975) suggest the involvement of radiation-induced heating in the reported cellular alterations.

The use of microwave or RF radiation alone or in combination with ionizing radiation for the treatment of cancer in humans and experimental animals has provided data on the effects of radiation-induced hyperthermia on the immune system. Shah and Dickson (1978a, 1978b) induced temperatures of 47°C to 50°C in VX-2 carcinomas in rabbits with local application of 13.56-MHz RF radiation. In addition to tumor regression, cell-mediated immunity was induced as indicated by skin reactivity to dinitrochlorobenzene and tumor extract, and a 100-fold increase in serum levels of antitumor antibody and increased response to bovine serum albumin antigen were noted. In contrast to the effects of localized tumor heating, they reported only temporary control of tumor growth, followed by a return to an exponential increase in tumor volume and death in rabbits treated with whole-body hyperthermia. Whole-body hyperthermia did not result in the enhancement of cellular and humoral immune responsiveness induced by localized RF hyperthermia.

Similar results were reported by Szmigielski and associates (1978). They used 2.45-GHz microwaves to locally heat Guérin epithelioma in Wistar rats to 43°C. Stimulation of the immune response was evidenced by antibody response to bovine serum albumin, reactivity of spleen lymphocytes to the mitogen PHA, and increased serum lysozyme levels. These investigators also reported tumor-specific effects that included increased cytotoxicity of peritoneal macrophages and spleen cells to cultured tumor cells. Marmor, Hahn, and Hahn (1977) obtained similar results by exposing

EMT-6 mouse tumors to 13.56-MHz radiowaves. The tumor cure rate was found to depend upon induced temperature elevation and exposure duration with a five-minute elevation to 44°C, resulting in a 50% cure rate. They suggested that cell killing could not account for the tumoricidal effect and, consequently, that the hyperthermia may stimulate a tumor-directed immune response, a hypothesis also supported by Szmigielski and colleagues.

Negative effects of hyperthermia on tumor dissemination and on the immune response also have been reported. Local hyperthermia induced by hot water bath immersion of solid Yoshida sarcomas in the feet of rats enhanced tumor dissemination when the treatment was not sufficient to cause complete tumor destruction (Dickson and Ellis 1976). Metastasis of C3H mouse mammary carcinoma was promoted by local heating in a study reported by Walker and associates (1978), and a suppression of B-lymphocyte-mediated immune response was reported by Shah and Dickson (1978a, 1978b) following the application of localized hyperthermia, regardless of the degree and method of heating of the rabbit thigh muscles.

In summary, mammalian hematologic and immunologic systems are sensitive to microwave and RF fields over a wide range of intensities and wavelengths. Generally, data from in vivo and in vitro studies support the hypothesis that the responses are related to the induction of thermal stress, but alterations have been reported under conditions of low-intensity exposure with no evidence of whole-body heating. Lacking further information on the mechanisms of interaction of microwave and RF fields, it may be suggested that the qualitative and quantitative responses are related to the specific patterns of internal energy absorption that result in nonuniform heating and that pose unique physiologic challenges to mammalian systems. In view of the reported sensitivity of the mammalian central nervous system and neuroendocrine system to microwave and RF exposure, and differences in in vitro and in vivo responses, it may be further suggested that there are indirect as well as direct exposure effects on the mammalian hematologic and immunologic systems. Such complexities in the interaction of such fields help to explain the positive as well as negative results of radiation-induced hyperthermia in the treatment of cancer. It is obvious that more information is needed to permit the evaluation and optimization of treatment modalities involving microwave- and RF- induced hyperthermia.

Experimental Microwave Cataractogenesis

Recognition of cataracts as the most apparent irreversible effect in humans exposed to high intensity microwave fields has led to a number of experimental and theoretical studies. The results of these studies, which have been reviewed in detail by Cleary (1980), typically have involved acute partial-body (e.g., ocular region) exposure of rabbits. In the case of theoretical studies, computer techniques have been applied to calculations of microwave-induced intraocular temperature rises (Taflove and Brodwin 1975; Al-Badwaiky and Youssef 1957; Emery et al. 1975).

Cataract induction in experimental animals, principally New Zealand white rabbits, has been described by a number of investigators who have employed various microwave conditions (Guy et al. 1975; Appleton, Hirsch, and Brown 1975; Carpenter, Ferri, and Hagan 1974). The majority of the cataract studies have involved the use of near-zone fields that limited exposure to the animal's head or eye. Acute exposure to 2.45-GHz microwaves at intensities in the range of 100 to 300 mW/cm^2 results in detectable lens opacification 24 to 48 hours postexposure, a significantly shorter latency than for cataract induction from exposure to ionizing radiation, thus suggesting differences in the mechanisms of cataract induction from these types of radiation. Theoretical and experimental evidence suggests that microwave cataractogenesis is due to thermal damage of lens fibers. Whole-body hypothermia was used in dogs (Baillie 1970) and rabbits (Kramer et al. 1975) to limit intraocular temperatures during exposure to microwaves to less than 43°C, thus preventing cataract induction. Studies in which ocular heating by microwaves was compared to other types of heating have indicated, however, that the induction of cataracts by microwaves depends on factors other than heating, such as heating rates and thermal gradients (Carpenter, Hagan, and Donovan 1977). Variations in the sensitivity of different species of animals to cataract induction by microwaves further indicates a marked dependency on the anatomic configuration of the animal's head (Kramer et al. 1976) and the wavelength of the radiation (Hagen and Carpenter 1975).

Time-intensity thresholds for cataract induction in rabbits have been determined at 2.45, 5.4, 5.5, 8.2 and 10 GHz (Birenbaum et al. 1969; Carpenter, Biddle, and Van Ummersen 1960; Carpenter and Van Ummersen 1968; Guy et al. 1975; Williams et al. 1955). In these acute exposure studies the minimum exposure time required to induce lens opacification that manifested itself within a few days after exposure was determined as a function of microwave intensity. Figure 24.1 summarizes time and power density cataract thresholds in rabbits irradiated in the near-zone of 2.45-GHz sources as determined by various investigators (Carpenter and Van Ummersen 1968; Guy et al.

1975; Williams et al. 1955). Variations in microwave field measurement techniques preclude quantitative comparisons of these data, but in instances where similar dosimetric methods were used there was agreement in the data. The hyperbolic form of the curves is indicative of dose reciprocity, which implies that in the rabbit, for single acute exposures, lens opacification appears to be a threshold phenomenon related to the total absorbed energy and thus indirectly suggestive of a thermal exposure effect. In the experiments summarized in figure 24.1, cataracts were not produced at intensities of 100 mW/cm^2 or less for exposure times of up to 100 minutes.

Microwave cataractogenesis following multiple exposures to intensities that were below threshold for single exposures was investigated by Carpenter, Ferri, and Hagan (1974). Lens opacification was detected in one of eleven rabbits exposed 20 to 24 times for one hour to an 80 mW/cm^2, 2.45-GHz field, and four opacities were detected in ten rabbits exposed at 100 mW/cm^2. Eighty percent of the rabbits exposed to 120 mW/cm^2 for one hour developed lenticular opacities, even though a single one-hour exposure at that intensity produced no opacities. Subsequent revaluation of dosimetry later led Carpenter (1977) to revise the microwave intensity to a value of 180 mW/cm^2 rather than 120 mW/cm^2 for the opacification consequent to the 4- or 4.5-hour exposures (Carpenter, Hagan, and Donovan 1977).

In general, the data obtained from experimental studies of microwave cataractogenesis support the hypothesis that this effect is due to acute thermal damage of lens tissue. Anatomic differences and differences in experimental conditions limit the application of cataract thresholds for rabbits to quantitative predictions of cataract thresholds in humans, but it appears reasonable to assume that in the absence of contradictory data, microwave intensities that do not induce temperature elevations in excess of 1°C would not induce cataracts. Additional data is required, however, on the effects of repeated exposure to subcataract threshold intensities on the human lens.

Teratogenesis

The teratogenic potential of microwave and RF radiation in humans is unknown. Imrie (1971) noted that one of three women treated with shortwave diathermy during early pregnancy aborted. Rubin and Erdman (1959) reported that one of four women exposed to 2.45-GHz diathermy for the treatment of chronic pelvic inflammatory disease aborted on the sixty-seventh day of pregnancy. The paucity of controlled data and awareness that large numbers of women of childbearing age are routinely occupationally exposed to RF and microwave fields has served as the impetus for animal studies (principally mice and rats) of microwave-and RF-induced teratogenesis.

In 1974, Rugh and colleagues studied the effects of exposure of over 200 litters of CF-1 mice to 2.45-GHz microwaves at doses of 12.5 to 33.5 J/g. Exposures were for a five-minute duration at specific absorption rates of 112 mW/g. A linear dose-dependent increase in specific malformations and resorptions was noted, with exencephaly being the most common abnormality. In other studies, CF-1 mice were exposed to 2.45-GHz microwaves at specific absorption rates that were varied from 78.8 to 136.2 mW/g (Rugh 1976; Rugh, Ho, and McManaway 1976). Induced fetal abnormalities were most frequent when exposure occurred on day 8 or 10 of gestation. Using ionizing radiation as a comparison teratogen, the authors concluded that 2.45 GHz was a teratogen under the conditions of their experiment. Rugh and McManaway (1977) subsequently reported a 68% incidence of

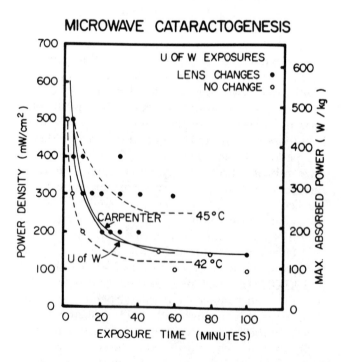

Figure 24.1. Experimental 2.45 GHz near-field microwave intensity-time thresholds for induction of lens opacities in the rabbit following single acute exposures. The *dashed lines* are the computed retrolental temperature-time-intensity relationships for acute exposures. (Emery et al. 1975. Reprinted by permission.)

resorbed, dead, or abnormal fetuses following exposure to 107.4 mW/g on day eight of gestation, compared to 28% incidence following similar exposure on day ten and 20% incidence as a result of exposure on days three or four. Two primary peaks of teratogenic sensitivity were noted, one on day four corresponding to the time of implantation of the embryo in the uterine horn, and the second peak on day eight, the beginning of organogenesis.

Berman, Kinn, and Carter (1978) recently exposed over 300 litters of CD-1 mice to 3.4 mW/cm^2, 13.6 mW/cm^2, or 20 mW/cm^2, 2.45-GHz microwaves for 100 minutes per day for the duration of gestation (days 1 to 18). No significant elevations in rectal temperature were detected as a treatment effect. The mean weight of the fetuses exposed at 20 mW/cm^2 was significantly decreased, and when summed for all treatment groups there was a statistically significant increase in the incidence of cranioschisis. These results suggest that long-term exposure to relatively low intensity (i.e., nonthermogenic) microwave fields may result in teratogenesis in mice.

Teratogenesis in rats exposed to RF and microwave radiation, generally at thermogenic intensities, has been reported, indicating that the effects may not be species specific (Brent, Franklin, and Wallace 1971; Chernovetz, Justesen, and Oke 1977; Chernovetz, Justesen, and Levinson 1979; Dietzel 1975).

The most frequently reported effect of fetal exposure of mice or rats to microwave or RF radiation is a reduction in body weight, although the persistence of this effect after birth has not been determined. Other commonly encountered teratogenic effects of microwave exposure include exencephaly, cranioschisis, and fetal resorption. Although such effects generally have been induced by acute exposures to relatively high-intensity fields, implicating a thermal damage mechanism, insufficient data on both the effects of long-term, low-intensity radiation and on effects in other mammalian species do not permit the conclusion that microwave- and RF-induced teratogenesis are caused solely by field-induced heating.

Chromosomal and Mutagenic Effects

Chromosomal and mutagenic effects of microwave and RF radiation have been described in a limited number of biological systems in a narrow range of exposure conditions. Mutagenic effects have been reported in nonmammalian species exposed to RF fields, whereas most chromosomal studies have been conducted at microwave frequencies.

Leach (1976) exposed Chinese hamsters for 15 minutes to 2.45-GHz microwaves at an intensity of 200 mW/cm^2. A decrease in mitotic index and increased chromosomal "stickiness" (between chromosomes and sister chromatids) were detected five hours postexposure. The chromosomes of cells examined seven days postexposure were less sticky than immediate postexposure samples, but there was still a significantly higher incidence of structural aberrations, including breaks and rearrangements. This suggested that the induction of chromosomal aberrations by microwave radiation may involve multiple mechanisms of interaction.

The effects of RF exposure on chromosomes were reported in 1963 by Mickey, who exposed cells of meristem of garlic root tips to conductive RF fields at frequencies of 5 to 40 MHz. Exposure for periods of from five minutes to eight hours at 0.25 to 6 kV/cm induced chromosomal stickiness leading to clumping, pyknosis, and bridge formation during anaphase and telophase, as well as chromosomal fragmentation with loss of chromatin and the formation of micronuclei and cell death. Chen, Samuel, and Hoopingarner (1974) exposed human amnion cells and Chinese hamster cells in vitro to 2.45-GHz microwaves at intensities of from 20 to 85 mW/cm^2 for 4 to 20 minutes, with maximum temperature elevations of up to 43°C. Chromosomal aberrations resulted from microwave exposure under a variety of exposure conditions, but thermal control studies using conventional heating led the authors to conclude that microwave-induced aberrations were not due to heat alone.

Yao and Jiles (1970) exposed kangaroo rat bone marrow and choroid cells in vitro to 2.45-GHz microwaves for 10 to 30 minutes at 200 mW/cm^2. They reported chromatid breaks, isochromatid breaks, dicentrics and rings, and chromatid exchanges. Thirty-minute exposure at 200 mW/cm^2 caused cell death. A specific type of chromosomal aberration referred to as despiralization, not known to result from other physical agents such as ionizing radiation, was noted to be an effect of microwave exposure.

The effects of low-intensity RF fields on chromosomes have been investigated also. Mittler (1976), for example, detected no increase in chromosomal aberrations in *Drosophila melanogaster* following 12-hour exposures to a 0.625 V/cm, 146-MHz field or to a 6 V/cm, 29-MHz field. Holm and Schneider (1970) found no significant effects on cell growth, DNA synthesis, or mitosis in human lymphocytes exposed in vitro for 1 to 84 hours to low-intensity 27-MHz RF radiation. It was noted, however, that cultures irradiated for 72 hours or more had seven times more chromosome breaks than unirradiated control cells.

A specific type of chromosomal aberration, referred to as "bandedness" by George (1978), was induced in germinated broad bean seedlings (*Vicia faba*) following 6- to 24-hour exposures to a 4.5 V/cm, 1.6-MHz field generated by a Tesla coil. The bandedness, which apparently was due to chromosome fragmentation, was detected at all stages of mitosis and appeared to be caused by the dispersal of chromatin material from specific regions of the chromosomes, thereby producing gaps.

Chromosomal aberrations, including chromatid breaks, dicentrics, and exchanges, have been induced by the in vitro exposure of human lymphocytes to 3-GHz microwaves at intensities of from 7 to 14 mW/cm^2 (Stodolnik-Baranska 1966). The maximum cell temperature during exposure was reportedly 38°C. In vivo exposures of guinea pigs and rabbits to 3.5 to 7 mW/cm^2 microwave radiation for three hours per day for either two to three weeks or three to four months induced nuclear structural alterations in erythroblasts and lymphoid cells, including fragmentation, micronuclei formation, nonuniform chromatin staining, and chromosome bridges between daughter nuclei. Baranski (1971a, 1971b) noted that alterations were induced in erythroblasts and lymphoid cells only, and not in granulocytic precursor cells, suggesting cell-specific effects of microwave exposure.

Mutagenic studies of RF and microwave fields have produced varied results. Mittler (1976) detected no effects of a 29-MHz field at 6 V/cm or a 146 MHz, 0.63 V/cm field on the incidence of sex-linked lethal mutations in *D. melanogaster* following 12 hours of exposure. Mickey (1963) reported the induction of sex-linked dominant and recessive lethals in *Drosophila* following exposures of from five minutes to one hour to 3- to 40-MHz fields pulse modulated at repetition rates of from 500 to 1000 Hz, pulse durations of 30 to 50 μsec, at field strengths of 0.8 to 3 kV/cm.

The mutagenic effects of pulse-modulated X-band microwaves on *Escherichia coli* were reported by Dutta and co-workers (1978). Bacteria were exposed to 1, 10, or 20 mW/cm^2 fields for periods of 1, 5, 10, or 15 hours at frequencies of 8.6, 8.8, and 9.0 GHz. No alterations in repair indices, the mutagenic end point investigated, were detected at 1 mW/cm^2, but more than 10 hours of exposure at 10 or 20 mW/cm^2 resulted in significant alterations. They concluded that temperature elevations to 39°C or less did not induce reparable genetic damage. Alterations induced by exposure to 20 mW/cm^2 were attributed to thermal and nonthermal mechanisms, whereas effects of 10 mW/cm^2 fields were not due to thermal mechanisms. Exposure of pollen of *Antirrhinum majus* for 4, 12, or 44 hours to a 1.5 V/m, 200-MHz field resulted in an increase in embryonic lethality and in mutations for characters of seedlings and young plants, results that were consistent with those obtained from exposure of pollen or plants of *Oenothera hookeri* under similar conditions (Harte 1973a, 1973b, 1975).

In summary, microwave and RF radiations at both thermogenic and nonthermogenic intensities have been demonstrated to induce chromosomal aberrations and mutations. Although chromosomal effects cannot be adequately classified because of limited data, high-intensity thermogenic fields appear to result in the induction of chromosome stickiness and breakage. The possibility that microwave and/or RF exposure may induce specific alterations in chromosomes is suggested by reports of banding and partial despiralization, and there is some evidence of cell-specific chromosomal sensitivities to such exposure. Thermal damage mechanisms appear to be involved in high-intensity field exposures, but conventional heating of equivalent magnitudes does not appear to cause the same type or extent of chromosomal damage as microwave or RF exposure. Limited data on the mutagenic effects of microwave and RF exposure also suggest the possible involvement of nonthermal mechanisms for the alteration of genetic material. The possibility of genetic damage, as well as the correlation between chromosomal aberrations and mutagenesis and carcinogenesis, emphasized the need for additional data on the chromosomal and mutagenic effects of microwave and RF exposure.

CONCLUSIONS

Exposure of humans and other mammals to microwave and RF radiation has been associated with a wide variety of psychological and physiologic alterations, ranging from subtle behavioral changes with exposure at low intensities to death from exposure to thermogenic, high-intensity fields. In addition to a dependency on field intensity and exposure duration, such effects appear to be dependent upon wavelength, which determines both the extent of coupling of the body to the field and the complex nonuniform distribution of internally absorbed energy. Such factors may provide an explanation for the difference in the biological response to thermogenic microwave and RF fields as compared to heating by other modalities.

The therapeutic application of microwave or RF fields for tissue hyperthermia must involve detailed knowledge of the internally induced temperature distributions in order to minimize the risk of damage to healthy tissue and to maximize absorption at the treatment site. The development of nonperturbing tissue-implantable temperature or field-strength probes will, in conjunction with theoretical methods for the estimation of absorbed energy distributions, provide information needed for the optimal therapeutic application of RF and microwave radiations.

Human and animal data suggest microwave- and RF- specific effects (especially in the case of exposure to low-intensity fields) that presently are not well understood. Prominent among such effects are functional alterations in the central nervous system and behavioral changes; these are reversible effects that are not explainable in terms of thermal stress reactions. Hematologic, immunologic, and neuroendocrine alterations also are demonstrated responses of mammalian systems to microwave and RF exposure, and although the available data is inconclusive, there are again indications of specific sensitivities to such fields. The mammalian immunologic response has been reported to be either enhanced or suppressed, depending upon the exposure intensity and duration and the wavelength of the radiation, a finding of potential significance in therapeutic applications for cancer therapy.

In addition to the difficulties with nonuniform absorption encountered in the interpretation of the biological effects of microwave and RF exposure, the mechanisms of interaction of such fields with biological systems are not adequately understood. Investigations of microwave and RF effects on calcium binding to neuronal membranes (Adey 1980) and the mammalian blood brain barrier (Justesen 1980) implicate biomembranes as the primary organizational structure involved in the interaction of such fields. Membrane-field interactions could explain the apparent cell-specific effects of microwave and RF exposure. Advances in our understanding of such interactions should prove of great value in the optimization of therapeutic applications of microwave and RF radiation.

References

Addington, C. H. et al. Biological effects of microwave energy at 200 mc. In *Proceedings of the Fourth Annual Tri-Service Conference on the Biological Effects of Microwave Radiation*, vol. 1, ed. M. F. Peyton. New York: Plenum, 1961, pp. 177–186.

Adey, W. R. Frequency and power windowing in tissue interactions with weak electromagnetic fields. *Proc. IEEE* 68:119–125, 1980.

Adler, E., and Magora, A. Experiments on the relations between shortwave therapy and the ascorbic acid content of the adrenal glands in rats. *Arch. Phys. Ther.* 16:223–230, 1964 (Ger).

Afanas'yev, B. G. The functional condition of the adrenal cortex in ship specialists who are subjected to the action of a superhigh frequency field. *Voen. Med. Zh.* 1:73–74, 1968 (Rus).

Al-Badwaiky, K. A., and Youssef, A. A. Biological thermal effect of microwave radiation on human eyes. In *Proc. URSI/USNC Annu. Meeting*, Boulder, Colorado, 1975, pp. 61–78.

Appleton, B. *Results of clinical surveys for microwave ocular effects*. Rockville, Md.: DHEW publication (FDA) 73-8031, 1973.

Appleton, B.; Hirsch, S. E.; and Brown, P. V. K. Investigation of single-exposure microwave ocular effects of 3000 MHz. *Ann. NY Acad. Sci.* 247:125–135, 1975.

Aurell, E., and Tengroth, B. Lenticular and retinal changes secondary to microwave exposure. *Acta Ophthalmol.* 51:764–771, 1973.

Baillie, H. D. Thermal and non-thermal cataractogenesis by microwaves. *Biological Effects and Health Implications of Microwave Radiation*, ed. S. F. Cleary. Rockville, Md.: DHEW publication BRH/DBE 70-2:59, 65, 1970.

Baranski, S. Effect of chronic microwave irradiation on the blood forming system of guinea pigs and rabbits. *Aerospace Med.* 42:1196–1199, 1971a.

Baranski, S. Influence of microwaves on white blood cells response. *Acta Physiol. Pol.* 22:898, 1971b (Pol).

Baranski, S. Effect of microwaves on the reactions of the white blood cell systems. *Acta Physiol. Pol.* 23:685–695, 1972 (Pol).

Baranski, S., and Czerski, P. Investigation of the behavior of the corpuscular blood constituents in persons exposed to microwaves. *Lek. Wojskowy* 4:903–909, 1966 (Pol).

Baranski, S., and Czerski, P. *Biological effects of microwaves*. Pennsylvania: Dowden, Hutchinson, Ross, 1976a.

Baranski, S., and Czerski, P. Biological effects of microwaves. Experimental data. In *Biological effects of microwaves*. Pennsylvania: Dowden, Hutchinson, Ross, 1976b, p. 133.

Baranski, S., and Edelwejn, Z. Experimental morphologic and electroencephalographic studies of microwave effects on the nervous system. *Ann. NY Acad. Sci.* 247:109–116, 1975.

Baranski, S.; Ostrowski, K.; and Stodolnik-Baranska, W. Effect of microwaves on the morphology and function of the thyroid gland. *Acta Physiol. Pol.* 23:997–1009, 1972 (Pol).

Bell, W. H., and Ferguson, D. Effects of super-high-frequency radio current on health of men exposed under service conditions. *Arch. Phys. Ther. X-Ray Radium* 12:477–490, 1931.

Berman, E.; Kinn, J. B.; and Carter, H. G. Observations of mouse fetuses after irradiation with 2.45 GHz microwaves. *Health Phys.* 35:791–801, 1978.

Bierman, W.; Horowitz, E. A.; and Levenson, C. L. Fever therapy in pelvic conditions. Results of experimental and clinical studies. *Arch. Phys. Ther. X-Ray Radium* 16:520–525, 1935.

Birenbaum, L. et al. Effect of microwaves on the rabbit eye. *J. Microwave Power* 4:232–243, 1969.

Brent, R. L.; Franklin, J. G.; and Wallace, J. P. The interruption of pregnancy using microwave radiation (abstr.). *Teratology* 4:484A, 1971.

Carpenter, C. M., and Page, A. B. The production of fever in man by short radio waves. *Science* 71:450–452, 1930.

Carpenter, R. L. Microwave radiation. In *Handbook of physiology*, ed. D. H. K. Lee. Bethesda, Md.: American Physiological Society, 1977.

Carpenter, R. L.; Biddle, D. K.; and Van Ummersen, C. A. Biological effects of microwave radiation with particular reference to the eye. In *Proc. 3rd Int. Conf. Med. Electron.*, London, England: 1960, p. 401.

Carpenter, R. L.; Ferri, E. S.; and Hagan, G. L. Assessing microwaves as a hazard to the eye—progress and problems. *Proc. Int. Symp. Biologic Effects, Health Hazards Microwave Radiation.* Warsaw: Polish Medical Publishers, 1974, pp. 178–185.

Carpenter, R. L.; Hagan, G. J.; and Donovan, G. L. Are microwave cataracts thermally caused? In *Symp. of Biological Effects and Measurement of RF and Microwaves.* Rockville, Md.: HEW publication (FDA) 77-8026, 1977, pp. 352–379.

Carpenter, R. L., and Van Ummersen, C. A. The action of microwave power on the eye. *J. Microwave Power* 3:3–19, 1968.

Chen, K. M.; Samuel, A.; and Hoopingarner, R. Chromosomal aberrations of living cells induced by microwave radiation. *Environ. Letters* 6:37–46, 1974.

Chernovetz, M. E.; Justesen, D. R.; and Levinson, D. M. Acceleration and declaration of fetal growth of rats by 2450 MHz microwave radiation, ed. S. Stuchly. Edmonton, Alberta, Canada: IMPI, 1979, pp. 175–193.

Chernovetz, M. E.; Justesen, D. R.; and Oke, A. F. A teratological study of the rat: microwave and infrared radiations compared. *Radio Science* 12:191–197, 1977.

Clark, W. B. Microwave diathermy in ophthalmology—clinical evaluation. *Trans. Am. Acad. Ophthalmol. Otolaryngol.* 56:600–607, 1952.

Cleary, S. F. Microwave cataractogenesis. *Proc. IEEE* 68:49–55, 1980.

Cleary, S. F., and Pasternack, B. S. Lenticular changes in microwave workers. *Arch. Environ. Health* 12:23–29, 1966.

Cleary, S. F.; Pasternack, B. S.; and Beebe, G. W. Cataract incidence in radar workers. *Arch. Environ. Health* 11:179–182, 1965.

Coulter, J. S., and Carter, H. A. Heating of human tissues by short wave diathermy. *JAMA* 106:2063–2066, 1936.

Czerski, P. Microwave effects on the bloodforming system with particular reference to the lymphocyte. *Ann. NY Acad. Sci.* 247:232–242, 1975.

Czerski, P.; Siekierzynski, M.; and Gidynski, A. Health surveillance of personnel occupationally exposed to microwaves. I. Theoretical considerations and practical aspects. *Aerospace Med.* 45:1137–1142, 1974.

D'Arsonval, A. The generation of high-frequency and high-intensity currents and their physiological effects. *CR Soc. Biol.* 45:122–124, 1893 (Fre).

Deichmann, W. B.; Bernal, E.; and Keplinger, M. Effects of environmental temperature and air volume exchange on survival of rats exposed to microwave radiation of 24,000 megacycles. *Ind. Med. Surg.* 28: 535–538, 1959.

Deichmann, W. B.; Miale, J.; and Landeen, K. Effect of microwave radiation on the hemopoietic system of the rat. *Toxicol. Appl. Pharmacol.* 6:71–74, 1964.

Demokidova, N. K. Certain data on the biological effects of continuous and intermittent microwave radiation. In *Biological effects of radiofrequency electromagnetic fields*, ed. Z. V. Gordon. Arlington, Va.: U. S. Joint Publications Research Service, JPRS 63321, 1974a, pp. 113–119.

Demokidova, N. K. The effects of radiowaves on the growth of animals. In *Biological effects of radiofrequency electromagnetic fields*, ed. Z. V. Gordon. Arlington, Va.: U. S. Joint Publications Research Service, JPRS 63321, 1974b, pp. 237–242.

Dickson, J. A., and Ellis, H. A. The influence of tumor volume and the degree of heating on the response of the solid Yoshida sarcoma to hyperthermia (40–42°C). *Cancer Res.* 36:1188–1195, 1976.

Dietzel, F. Effects of electromagnetic radiation on implantation and intrauterine development of the rat. *Ann. NY Acad. of Sci.*: 367–376, 1975.

Doury, P.; Boisselier, P.; and Bernard, J. G. Pathologic effects of UHF electromagnetic radiation from Aeriens radar on humans—a case report. *Sem. Hop. Paris* 46:2681–2683, 1970 (Fre).

Dumkin, V. N., and Korenevskaya, S. P. Glucocorticoid function of the adrenals in radiowave sickness. In *Biological effects of radiofrequency electromagnetic fields*, ed. Z. V. Gordon. Arlington, Va.: U. S. Joint Publications Research Service, JPRS 63321, 1974, pp. 72–74.

Durney, C. H. et al. *Radiofrequency radiation dosimetry handbook*, 2nd edition, SAM-TR-78-22. Brooks Air Force Base, Tex.: U. S. Dept. of the Air Force, Air Force Systems Command, Aerospace Medical Division, School of Aerospace Medicine, 1978, p. 141.

Dutta, S. K. et al. Effects of chronic non-thermal exposures of pulsed microwaves on a repair-deficient mutant of *Escherichia coli. Mutat. Res.* 53:91–92, 1978.

D'yachenko, N. A. Change in thyroid function after chronic exposure to microwave radiation. In *Effects of microwave irradiation*, Arlington, Va.: U.S. Joint Publications Research Service, JPRS 51238, 1970, pp. 6–8.

Emery, A. F. et al. Microwave induced temperature rises in rabbit eyes in cataract research. *J. Heat Transfer* 97:123–128, 1975.

Fofanov, P. N. *Hemodynamic changes in individuals working under microwave irradiation*. Arlington, Va.: U.S. Joint Publications Research Service, JPRS 48481, 1969a.

Fofanov, P. N. Clinical picture of CW SHF-UHF electromagnetic radiation on man. *Sov. Med.* 31:107–111, 1969b (Rus).

Gandhi, O. P.; Hunt, E. L.; and D'Andrea, J. A. Deposition of electromagnetic energy in animals and in models of man with and without grounding and reflector effects. *Radio Science* 12:39–47, 1977.

Gembitskiy, Y. V.; Kolesnik, F. A.; and Malyshev, V. M. Changes in the blood system during chronic exposure to a superhigh-frequency field. *Voen. Med. Zh.* 5:21–23, 1969 (Rus).

George, K. Chromosome band inducing effect of high frequency electromagnetic field. *Indian J. Exp. Biol.* 16:390–393, 1978.

Gorden, Z. *Biological effects of microwaves in occupational hygiene*. Proc. Izdatelstro Medicina, (NASA TTF633, 1970.

Gunn, S. A.; Gould, T. C.; and Anderson, W. A. D. The effect of microwave radiation on morphology and function of rat testes. *Lab Invest.* 10:301–314, 1961.

Guy, A. W. et al. Effect of 2450 MHz radiation on the rabbit eye. *IEEE Trans. MTT* 23:492–498, 1975.

Hagan, G. J., and Carpenter, R. L. Relative cataractogenic potencies of two microwave frequencies. Paper read at the Proc. URSI/USNC Annu. Meeting, 1975, Boulder, Colorado.

Hamrick, P. E., and Fox, S. S. Rat lymphocytes in cell culture exposed to 2450 MHz (CW) microwave radiation. *J. Microwave Power* 12:125–132, 1977.

Harte, C. Mutationen bei *Oenothera hookeri* nach dauerein wirkung von meterwellen wahrend einer vegtationsperiode. *Theor. Appl. Genet.* 43:6–12, 1973a.

Harte, C. Genetische mosaike in der M_2 bei *Oenothera* nach behandlung mit meterwellen. *Theor. Appl. Genet.* 43:54–58, 1973b.

Harte, C. Mutagenesis by radiowaves in *Antirrhinum majus. L. Mutat. Res.* 29:71–75, 1975.

Hathaway, J. A. et al. Ocular medical surveillance on microwave and laser workers. *J. Occup. Med.* 19:683–688, 1977.

Hirsch, F. G., and Parker, J. T. Bilateral lenticular opacities occurring in a technician operating a microwave generator. *AMA Arch. Ind. Hyg. Occup. Med.* 6:512–517, 1952.

Holm, D. A., and Schneider, L. K. The effects of nonthermal radiofrequency radiation on human lymphocytes *in vivo. Experientia* 26:992–994, 1970.

Horvath, S. M.; Miller, R. N.; and Hutt, B. K. Heating of human tissues by microwave radiation. *Am. J. Med. Sci.* 216:430–436, 1948.

Huang, A. T. et al. The effect of microwave radiation (2450 MHz) on the morphology and chromosomes of lymphocytes. *Radio Science* (suppl.) 12:173–177, 1977.

Imrie, A. H. Pelvic short wave diathermy given inadvertently in early pregnancy. *J. Obstet. Gynecol.* 78:91–92, 1971.

Jacobson, B. S., and Susskind, C. Review of the work conducted at University of California—effects of microwave irradiation on internal temperature and viability in mice. In *Proceedings of the Second Tri-Service Conference on Biological Effects of Microwave Energy*, ARDC-TR-58-54, eds. E. G. Pattishall and F. W. Banghart. Griffiss Air Force Base, N.Y., USDAF, ARDC, Rome Air Development Center, 1958, pp. 234–41.

Jung, R. W. Immunologic studies in hyperpyrexia. *Arch. Phys. Ther. X-Ray Radium* 16:397–404, 1935.

Justesen, D. R. Microwave irradiation and the blood-brain barrier. *Proc. IEEE* 68:60–66, 1980.

Kheifets, N. S. Biomicroscopic characteristics of crystalline lenses in persons exposed to the effects of electromagnetic fields of ultrahigh frequency. *Vestn. Oftalmol.* 6:70–72, 1970 (Rus).

Kitsovskaya, I. A. The effect of centimeter waves of different intensities on the blood and hemopoietic organs of white rats. *Gig. Tr. Prof. Zabol.* 8:14, 1964.

___, A. I. et al. *Clinical aspects of the effect of metric range electromagnetic fields*, JPRS 66434, Arlington, Va.: U.S. Joint Publications Research Service, 1975.

Klimkova-Deutschova, E. Neurologic findings in persons exposed to microwaves. In *Biologic effects and health hazards of microwave radiation*. Proceedings of an International Symposium. Warsaw: Polish Medical Publishers, 1974, pp. 268–272.

Koldaev, V. M. Effect of an ultrahigh-frequency electromagnetic field on rats combined with changes in intensity of oxidative processes. *Bull. Exp. Biol. Med. (USSR)* 70:1294–1295, 1970.

Koldaev, V. M. Effect of cholinergic agents on survival of mice after acute microwave irradiation. *Bull. Exp. Biol. Med. (USSR)* 81:325–326, 1976.

Kramer, P. O. et al. The ocular effects of microwaves on hypothermic rabbits: a study of microwave cataractogenic mechanisms. *Ann. NY Acad. Sci.* 247:155–165, 1975.

Kramer, P.O. et al. Acute microwave irradiation and cataract formation in rabbits and monkeys. *J. Microwave Power* 11:135–136, 1976.

Kurz, G. H., and Einaugler, R. B. Cataract secondary to microwave radiation. *Am. J. Ophthalmol.* 66:866–869, 1968.

Lappenbush, W. L. et al. Effect of 2450 MHz microwaves on the radiation response of x-irradiated Chinese hamsters. *Radiat. Res.* 54:294–303, 1973.

Leach, W. M. On the induction of chromosomal aberrations by 2450 MHz microwave radiation. *J. Cell Biol.* 70:387a, 1976.

Leavy, I. M. Physical therapy in chronic diseases—with special reference to peripheral vascular disease and ulcerations. *Arch. Phys. Ther. X-Ray Radium* 16:145–149, 1935.

Liburdy, R. P. Radiofrequency radiation alters the immune system: modulation of T- and B-lymphocyte levels and cell-mediated immunocompetence. *Radiat. Res.* 77:34–46, 1979.

Lin, J. C., and Peterson, W. D. Cytological effects of 2450 MHz CW microwave radiation. *J. Bioeng.* 1:471–478, 1977.

Lotz, W. G., and Michaelson, S. M. Temperature and corticosterone relationships in microwave-exposed rats. *J. Appl. Physiol.* 44:438–445, 1978.

Majewska, K. Investigations on the effect of microwaves on the eye. *Pol. Med. J.* 7:989–994, 1968.

Marmor, J. B.; Hahn, N.; and Hahn, G. M. Tumor cure and cell survival after localized radiofrequency heating. *Cancer Res.* 37:879–883, 1977.

Mayers, C. P., and Habershaw, J. A. Depression of phagocytosis: a nonthermal effect of microwave radiation as a potential hazard to health. *Int. J. Radiat. Biol.* 24:449–461, 1973.

McLaughlin, J. T. Tissue destruction and death from microwave radiation (radar). *California Medicine* 86:336–339, 1957.

McLaughlin, J. T. Health hazards from microwave radiation. *West. J. Med.* 3:126–132, 1962.

Medvedev, V. P. Cardiovascular diseases in persons with a history of exposure to the effect of an electromagnetic of extra-high frequency. *Gig. Tr. Prof. Zabol.* 17:6–9, 1973.

Michaelson, S. M. et al. The influence of microwaves on ionizing radiation exposure. *Aerospace Med.* 34:111, 1963.

Michaelson, S. M.; Thompson, R. A. E.; and El Tamami, M. Y. The hematologic effects of microwave exposure. *Aerospace Med.* 35:824–829, 1964.

Mickey, G. M. Electromagnetism and its effect on the organism. *NY State J. Med.* 63:1935–1942, 1963.

Mielczarek, H. The neurological syndrome of the so-called microwave illness. *Lek. Wojsk.* 47:442, 1971.

Miro, L.; Toubiere, R.; and Pfirster, A. Effects of microwaves on the cell metabolism of the reticulohistocytic system. In *Biological effects and health hazards of microwave radiation*. Proceedings of an International Symposium. Warsaw: Polish Medical Publishers, 1974, pp. 89–97.

Mittler, S. Failure of 2- and 10-meter radio waves to induce genetic damage in *Drosophila melanogaster*. *Environ. Res.* 11:326–330, 1976.

Muroff, L. R., and Samaras, G. Prolongation of life in a microwave field by means of an environmental chamber. In *Radiation bio-effects—summary report*. BRH/DBE 70-1. Rockville, Md.: DHEW, PHS, Consumer Protection and Environmental Health Service, Environmental Control Administration, Bureau of Radiological Health, 1970, pp. 59–61.

Novitski'i, A. A. et al. Functional state of the hypothalamus-hypophysis-adrenal cortex system as a criterion in setting standards for superhigh frequency electromagnetic radiation. *Voen. Med. Zh.* 10:53–56, 1977 (Rus).

Odlund, L. T. Radio-frequency energy. A hazard to workers? *Ind. Med. Surg.* 92:23–26, 1973.

Osborne, S. L., and Frederick, J. N. Microwave radiations—heating of human and animal tissues by means of high frequency current with wavelength of twelve centimeters (the microtherm). *JAMA* 137:1036–1040, 1948.

Parker, L. N. Thyroid suppression and adrenomedullary activation by low-intensity microwave radiation. *Am. J. Physiol.* 224:1388–1390, 1973.

Petrov, I. R., ed. Influence of microwave radiation in the organism of man and animals. Leningrad: Medicina, (NTIS Order No. NASA TT-F-708) 1972.

Polson, P. et al. *Mortality in rats exposed to CW microwave radiation at 0.95, 2.45, 4.54, and 7.44 GHz.* Menlo Park, Calif.: Stanford Research Institute, 1974.

Prausnitz, S., and Susskind, C. Effects of chronic microwave irradiation on mice. *IRE Trans. Bio. Med. Electron.* 9:104–108, 1962.

Prince, J. E. et al. Cytologic aspect of radiofrequency radiation in the monkey. *Aerospace Med.* 43:759–761, 1972.

Raue, H. The use of microwaves in ophthalmology. *J. Klin. Monatsbl. Augenheilkd.* 142:563–567, 1963 (Ger).

Rose, V. E. et al. Evaluation and control of exposures in repairing microwave ovens. *Am. Ind. Hyg. Assoc. J.* 30:137–142, 1969.

Rotkovska, D., and Vacek, A. The effect of electromagnetic radiation on the hematopoietic stem cells of mice. *Ann. NY Acad. Sci.* 247:243–250, 1975.

Rotkovska, D., and Vacek, A. Modification of repair of x-irradiation damage of hemopoietic system of mice by microwaves. *J. Microwave Power* 12:119–123, 1977.

Rubin, A., and W. J. Erdman. Microwave exposure of the human female pelvis during early pregnancy and prior to conception. *Am. J. Phys. Med.* 38:219–220, 1959.

Rugh, R. The relation of sex, age, and weight of mice to microwave radiation sensitivity. *J. Microwave Power* 11:172–132, 1976.

Rugh, R. et al. Are microwaves teratogenic? In *Biological effects and health hazards of microwave radiation.* Proceedings of an International Symposium. Warsaw: Polish Medical Publishers, 1974, pp. 98–107.

Rugh, R.; Ho, H. S.; and McManaway, M. The relation of dose rate of microwave radiation to the time of death and total absorbed dose in the mouse. *J. Microwave Power* 11:279–281, 1976.

Rugh, R., and McManaway, M. Comparison of ionizing and microwave radiations with respect to their effects on the rodent embryo and fetus (abstr.). *Proc. of the 16th Annual Meeting of the Congenital Anomalies Research Association of Japan, 1976.*

Schrot, J., and Hawkins, T. D. Lethal effects of 3000 MHz radiation on the rat. *Radiat. Res.* 59:504–512, 1974.

Searle, G. W.; Imig, C. J.; and Dahlen, R. W. Studies with 2450 MC-CW exposures to the heads of dogs. In *Proceedings of the Third Annual Tri-Service Conference on Biological Effects of Microwave Radiating Equipment.* RADC-TR-59-140. Girffiss Air Force Base, N.Y.: USDAF, ARDC, Rome Air Development Center, 1959, pp. 54–61.

Shacklett, D. E.; Tredici, T. J.; and Epstein, D. L. Evaluation of possible microwave-induced lens changes in the United States Air Force. *Aviat. Space Environ. Med.* 46:1403–1406, 1975.

Shah, S. A., and Dickson, J. A. Effect of hyperthermia on the immune response of normal rabbits. *Cancer Res.* 38:3518–3522, 1978a.

Shah, S. A., and Dickson, J. A. Effect of hyperthermia on the immuno-competence of VX2 tumor-bearing rabbits. *Cancer Res.* 38:3523–3531, 1978b.

Shimkovich, I. S., and Shilyaev, V. G. Cataracts of both eyes which developed as a result of repeated short exposure to an electromagnetic field of high density. *Vestn. Oftalmol.* 72:12–15, 1959 (Rus).

Siekierzynski, M. et al. Health surveillance of personnel occupationally exposed to microwaves. II. Functional disturbances. *Aerospace Med.* 45:1143–1145, 1974a.

Siekierzynski, M. et al. Health surveillance of personnel occupationally exposed to microwaves. III. Lens translucency. *Aerospace Med.* 45:1143–1145, 1974b.

Smialowicz, R. J. et al. Evaluation of lymphocyte function in mice exposed to 2450 MHz CW microwaves. In *Proceedings of a Symposium on Electromagnetic Fields in Biological Systems*, ed. S. S. Stuckly. Edmonton, Canada: Int. Microwave Power Inst., 1979, pp. 122–152.

Smialowicz, R. J. et al. Exposure of rats to 425 MHz CW microwave radiation: effects on lymphocytes. *J. Microwave Power*, in press.

Smialowicz, R. J.; Kinn, J. B.; and Elder, J. A. Perinatal exposure of rats to 2450 MHz CW microwave radiation: effects on lymphocytes. *Radio Science* 14:(suppl.) 1979.

Stodolnik-Baranska, W. The influence of vibrations and microwaves on cells and chromosomes. Doctoral thesis, Academia Medyczna, Warsaw, 1966.

Stodolnik-Baranska, W. Lymphoblastoid transformations of lymphocytes *in vitro* after microwave irradiation. *Nature* 214:102–103, 1967.

Susskind, C. *Biological effects of microwave radiation—annual scientific report (1957–58)*, RADC-TR-59-298. Griffiss Air Force Base, N.Y.: USDAF, ARDC, Rome Air Development Center, 1958.

Szady, J. et al. Effects of microwaves on the twenty-four-hour rhythm, and twenty-four-hour urinary excretion of seventeen hydroxycorticoids and seventeen ketosteroids. *J. Microwave Power* 11:139–140, 1976.

...ski, S. et al. Local microwave hyperthermia (43°C) ... stimulation of the macrophage and T-lymphocyte systems in the treatment of Guerin epithelioma in rats. *Z. Krelosforsch* 91:35–48, 1978.

Szmigielski, S.; Jeljaszewicz, J.; and Wiranowska, M. Acute staphylococcal infections in rabbits irradiated with 3 GHz microwaves. *Ann. NY Acad. Sci.* 247:305–311, 1975.

Taflove, A., and Brodwin, M. E. Computation of the electromagnetic fields and induced temperatures within a model of the microwave-irradiated human eye. *IEEE Trans. MTT* 23:888–896, 1975.

Volkova, A. P., and Fukalova, P. P. Changes in certain protective reactions of an organism under the influence of SW in experimental and industrial conditions. In *Biological effects of radiofrequency electromagnetic fields*, ed. Z. V. Gordon. JPRS 63321, Arlington, Va.: U. S. Joint Publication Research Service, 1974, pp. 168–174.

Walker, A. et al. Promotion of metastasis of C3H mouse mammary carcinoma by local hyperthermia. *Br. J. Cancer* 38:561–563, 1978.

Wiktor-Jedrezejczak, W.; Ahmend, A.; and Czerski, P. Immune response of mice to 2450 MHz microwave radiation: overview of immunology and empirical studies of lymphoid splenic cells. *Radio Science* 12(suppl.):209–219, 1977a.

Wiktor-Jedrzejczak, W.; Ahmend, A.; and Czerski, P. Increase in the frequency of FC receptor (FCR) bearing cells in the mouse spleen following a single exposure of mice to 2450 MHz microwaves. *Biomedicine* 27:250–252, 1977b.

Wiktor-Jedrzejczak, W.; Ahmend, A.; and Sell, K. Microwaves induce an increase in the frequency of complement receptor-bearing lymphoid spleen cells in mice. *J. Immunol.* 118:1499–1502, 1977.

Williams, D. B. et al. Biologic effects of microwave radiation: time and power thresholds for the production of lens opacities by 12.3 cm microwaves. USAF School Aviation Medicine, Randolph Air Force Base, Tex. Rep. No. 55-94, 1955.

Yao, K. T. S., and Jiles, M. M. Effects of 2450 MHz microwave radiation on cultivated rat kangaroo cells. In *Biological effects and health implications of microwave radiation*, ed. S. Cleary. Rockville, Md.: USDHEW, Report BRH/DBE 70-2 (PB 193 858), 1970.

Zaret, M. M. *An experimental study of the cataractogenic effects of microwave radiation.* RADC-TDR-64-273. Griffiss Air Force Base, N.Y.: USDAF, AFSC, RTD, Rome Air Development Center, Display Techniques Branch, 1964.

Zaret, M. M. Cataracts following use of microwave oven. *NY State J. Med.* 74:2032–2048, 1974.

Zaret, M. M. Blindness, deafness, and vestibular dysfunction in a microwave worker. *Eye, Ear, Nose and Throat Monthly* 54:49–52, 1975.

Zaret, M. M., and Snyder, W. Z. Cataracts and avionic radiations. *Br. J. Ophthalmol.* 61:380–384, 1977.

Zydecki, S. Assessment of lens translucency in juveniles, microwave workers and age-matched groups. In *Biologic effects and health hazards of microwave radiation.* Proceedings of an International Symposium. Warsaw: Polish Medical Publishers, 1974, pp. 306–308.

Index

Abscopal response, 488, 503–504, 526–527
 in humans, 513
Absorption coefficient, 334–335
Acidic environment, 174–178
Acidity, effects of extracellular, 178–180
Acidosis, 93
 effect on extracellular pH, 87–88
Acids and bases, weak, 80–81
Actinomycin D
 and increased metastasis, 498
 and thermal tolerance, 157, 159
Adenocarcinoma, 473–475
 of the colon, 323–324
 of the intestine, 434
 of the lung, 409–410
 in mouse, 374
 radiofrequency therapy for, 322–324
Adriamycin, 373, 391, 410, 428–429
 cardiotoxic effects of, 429
 combined with heat, 404
 and thermal tolerance, 157, 159
Aerobic glycolysis. See Glycolysis
Airway obstruction, 411
Alkalosis, effect on extracellular pH, 87–88
Aminoglycosides, 429
Amphotericin B, 454
Anal carcinoma, microwave therapy for, 322
Anamnestic response, 501–502
 in animals, 518
Anemia, 429
Anesthesia
 in experimental animals, 82
 management, 419–420
 nitrous oxide, 423–424
Angiogenesis, 188
 in tumor, 77
Angle of incidence, 34
Animal tumor investigations, 319–322
Animal tumor models, 493–497, 499–502
 data analysis for, 503–513
Animal tumors, experimental methodology for, 81–91
Anisotonic media, and thermal tolerance, 148
Anorexia, 428, 429–430
Antiantibodies, 492
Antibodies
 antitumor, 505–506
 blocking, 492
 nonspecific, 492–493
 tumor specific, 492
Antibody titer, 492
Anti-BSA antibody, 506, 507–509
Anticholinergic drugs, effect on animals exposed to radiation, 552–553
Antigen
 attenuated, 524
 foreign, 491
 tumor-associated, 489, 490, 495
Antigenic smokescreen, 524
Antigenic strength of tumor, 533
Antimony electrode, 210
Antineoplastic effects, 438–441
Antizyme, 155
Aperture, choice of, 337
Applicator cones, for ultrasonic hyperthermia, 345
Applicators, microwave, 288–292
 circularly polarized, 291
 direct contact type, 289–292, 299–301
 effective frequencies of, 297
 phase array contact, 292, 301
 radiating type of, 288–289, 298–299
 SAR with, 294–301
 slab-loaded rectangular waveguide, 291–292
Applicators, shortwave, 287–288
 advantages of, 299
 SAR with, 293
Arrhenius analysis, 47–52, 372
Arrhenius equation, 47–52
 concepts and principles, 48–49
 Eyring extension of, 48
Arrhenius plots, 64
Arterial perfusion, 65
Arteriovenous (AV) shunt, 412–418
 technique, 403
Ascites, 411
Asthenia, in microwave workers, 546–547
Autoimmunotherapy, 529

BA-1112 rhabdomyosarcoma of rats, vascular response to heat of, 196
Bacillus prodigiosus, 514
Bacterial luminescence, 47
Bacterial toxins. See Mixed bacterial toxins
Bandedness, 559–560
Barbiturate metabolism, 420
BCNU, 427–428, 436. See also Nitrosoureas
 and thermal tolerance, 157
Beam steering, 338–340
Bernard-Horner syndrome, 385
BHTE. See Bio-heat transfer equation
Biliary obstruction, 411
Bio-heat transfer, 9–38
Bio-heat transfer equation (BHTE), 10, 68, 258
 boundary conditions in, 16
 heat sources in, 16
 improvements in, 11
 limitations of, 11
 tissue geometry in, 16
Biosynthesis, inhibition of, 371
Black's maxim, 65
Blanket technique, 30
Bleomycin, 373, 425
Blood, as heat transfer mechanism, 377
Blood flow
 and alteration of tumor, 110–113
 distribution in tumors, 14–15
 equation for determination during heating, 199
 equations for relationship to tumor weight, 191–192
 inhibition by hyperglycemia, 110, 113, 526
 inhibition in tumors after hyperthermia, 100–104, 198–199
 intratumor variations in, 100–101
 measurement of, 88–91
 measurement in tumors of, 190-192
 in normal tissue vs tumor tissue, 191–192
 pattern after heating, 103
 reduced in tumors, 180–181
 role in combined treatments, 201–202
 significance during hyperthermia, 187–202
 and tissue temperature during heating, 199–200
 in tumor, 480
 through tumor capillaries, 189–190
 in tumor vs normal tissue, 191–192, 218–220
 and tumor size, 191–192
Blood flow rate, 12–15
 contributions to heat transfer, 11
 and effect of heat on tumor, 20
 and effective thermal conductivity, 12
 equation for tumor, 13
 exponential decay in, 14
 role in hyperthermia, 66–68
 techniques for measurement of, 14–15
 and tumor weight, 14, 15

Blood flow regulation, in response to heat, 4
Blood pressure, and tumor blood flow, 114
Blood transfusion, 420–421
Blood velocity vector, 11
Blood volume, 193–194
Bone involvement, 411–412
Bone marrow, and radiofrequency radiation, 550
Bone metastases, radiofrequency therapy for, 323–324
Boston University Medical Center, 316
Bovine serum albumin (BSA), 501–502, 506
Bowel obstruction, 411
Breaking point, 212
Breast carcinoma, microwave therapy for, 322
Breast neoplasms, therapy, for, 409–410, 419
Bronchogenic carcinoma, combination treatment for, 436
Bronchogenic neoplasms, combination therapy for, 425–429
Burns, from plane wave ultrasound, 353
Busch, W., 401
Busulfan lung, 428

Calcium gluconate, precautions for administration of, 423
Cancer biotherapy, 514–515
Canine tumors, response to ultrasound, 358–361
Capacitive component, 470–471
Capacitive electrodes, 316–317
Capacitor electrodes, 287–288
Capillary development in tumors, 189–190
Carbonyl iron test, 500
Carcinoma. See also specific types
  anal, 322
  of the bladder, 481
  of the breast, 322
  bronchogenic, 409–410, 436
  Brown-Pearce, 497, 503, 508–509
  of the colon, 403–404
  D-23, 81, 500
  Flexner-Jobling, 319–320, 370
  gastric, 403–404
  HB mammary, 320
  KHJJ, 320, 321
  of the lung, 323–324
  medullary of breast, 528
  metastatic to bone, 323, 324
  nasopharyngeal, 322
  oat cell, 410, 428
  ovarian, 434
  rectal, 513
  squamous cell, 473–475
  VX-2, 69–70, 81, 321, 497, 499–501, 505–509, 518–519
  Walker-256, 194, 320
Carcinomatous meningitis, 425
Carcinosarcoma, 473–475
Cardiopulmonary stress, 410
Cardiovascular disorders, in microwave workers, 547
Cardiovascular stress, during whole-body hyperthermia, 402–403
Cataracts
  in animals exposed to radiation, 557–558
  in microwave workers, 549–550
Cat brain, in study of tumor microcirculation, 210
Cauterization, 407
Cavitation, 334
  collapse, 341–342
Cell birth rate, 76
Cell counts, 59
Cell culture chemosensitivity, 373
Cell cycle
  blockade of progression of, 85–86
  effects of heat on, 448–449, 452
  effects of radiation on, 448–449
  influence on hyperthermic response, 459
  thermal sensitivity of stages of, 372
Cell cycle phases, 70–71
  and sensitivity to heat, 5–6
Cell cycle progression, blockade of, 71, 95–100, 109
Cell cycle time, 70–71, 74
  measurements of, 74
Cell death
  after combined heat and radiation, 375
  in hypoxic tumors, 375
  and inhibition of cell repair, 374
Cell growth state, and thermal tolerance, 148–149
Cell killing, 142, 143
  delayed, 2, 200–201
  detection of, 59–60
  kinetics of, 448
  mechanisms of hyperthermic, 152–155, 180–181
  and oxygen concentration, 148
  by protein damage, 153
  and synergism of heat and radiation, 455–458
  thermodynamic analysis of, 146
  time-temperature relationship of, 68–70
  in vitro, 401
  in vivo, 99–100
Cell killing effect, 71–72
Cell loss
  equation for determination of, 76
Cell loss factor, 71–72, 75–76
Cell-mediated responses, 501
Cell membrane
  changes from hyperthermia, 153
  permeability of, 153
Cell population kinetics, 70–77, 92
  mechanisms in Yoshida sarcoma, 93–94
  and multiple-dose heating, 109–110
  use of data on, 100
Cell proliferation, effects of treatment on, 77
Cell repair processes, 100, 410
Cell reproduction, after hyperthermic insult, 453
Cell sensitivity, 372, 448
Cell structure, heat-induced changes in, 372, 453–455
Cell survival response, in transformed cells, 151
Cell synchrony, 109
Cellular inactivation rate, 448
Cellular metabolism, during hyperthermia, 143
Cellular response
  in animals, 518
  to heat with radiation, 448–449
Cell viability, criterion for, 59
Central cold spot, 306–307, 310, 311
Central leak, in lens design, 361
Central nervous system
  biochemical support of, 435
  damage from treatment, 436
  disorders in microwave workers, 548
  dysfunction after systemic hyperthermia, 403
  treatment tolerance of, 434
Chamber systems, 208
Chemotherapeutic drugs
  interactions with heat, 157
  effect on tissues in combined therapy, 202
Chemotherapy, with hyperthermia, 424–429
Chinese hamster ovary (CHO) cells, cellular inactivation rate of, 448
Chinese hamster V-79 fibroblasts, reaction to hyperthermia of, 372
CH mouse mammary adenocarcinoma, in study of tumor microenvironment, 209
Cholesterol, and cell membrane permeability, 153–155
Cholinergic drugs, effect on animals exposed to radiation, 552–553
Chondrosarcoma, radiofrequency

therapy for, 322–324
Chromel-constantan thermocouple, 352
Chromosomal aberrations, 453–454, 461–462
  and radiation, 156
Chromosomal effects of radiation, 559–560
Chromosomal proteins, and cell death, 372
Cisplatin, with hyperthermia, 429
Clinical cancer trials, 322–329
Clinical fever vs physical heating, 516
Clinical treatment, significant factors in, 374
Clinical trials with hyperthermia, 438–441
Clonogenic cells, measurement of, 77
Coagulation necrosis, 334, 360
Coley, W. B., 63, 401, 514
Coley's toxins, 109, 401, 488, 514–515, 525
  effect on tumor vasculature, 514
Colon carcinoma, response to hyperthermia alone, 403–404
Colony counts, 59–60
Colony inhibition, 500
Combination therapy, 279–281, 373–375, 425
  cardiotoxic effects of, 429
  for hypoxic tumors, 375
  with radiation, 435–436
  timing sequence of, 404
  with ultrasound, 365
Computer
  interpretation, 348
  system for ultrasound hyperthermia, 347–348
Congestive heart failure, and whole-body hyperthermia, 409
Cooled capacitance electrodes, 317
Cooling, 108
Coombs' test, 501, 508
Copper-constantan microthermocouple, 211
Core temperature, measurement in animals, 83
Corynebacterium parvum, 512
Cost control, 418
Creatinine, in systemic hyperthermia, 403
Crushed limb syndrome, 92
Cryosurgery, 525
Cuff heating, 498
Cyclophosphamide, with heat, 404
Cystitis, 428
Cytokinetic studies, in experimental animals, 83–86
Cytoplasmic reaction, in tumor cells, 169–171
Cytoreduction in situ, 360

Cytotoxic agents, effects of, 71–72
Cytotoxicity, 159. *See also* Cell killing
Cytoxan™, 425

Dacron shunt, 412–418, 419
D'Arsonval, A., 550
Data Logger Fluke™, 383
Decubitus ulcer, 32
Degassing, before ultrasonic hyperthermia, 346
Delayed hypersensitivity reaction, 491
Dewey, William C., 152–153
Diarrhea, after systemic hyperthermia, 403
Diathermy, definition of, 1
Dielectric cavities, 283
Dielectric heating, 25. *See also* Capacitive heating; Radio-frequency heating
Dielectric properties of biological tissues, 281–283
Dihydroxybutylaldehyde, 3
Dinitrochlorobenzene (DNCB), 491, 501–502, 505, 507–509
Direct contact applicators
  advantages of, 301
  SAR measurement with, 299–301
Direct draining method, 190
Distributed parameter approach, 10–16
Distributed parameter models, 10–16
DL-glyceraldehyde, 3
DMO method, 80–81
  for intracellular pH measurement, 86–87
DNA
  biosynthesis of, 372
  in the cell cycle, 77
  in hyperthermic cell killing, 153
  radiation damage to, 461–462
  synthesis after heating, 453–454
Dose quality, influence on hyperthermic response, 459
Dose-rate influence on hyperthermic response, 459
Dose-response relationship, 69–70
Dose-time thermal effect, 2
Dosimetry, 472–473
  problems in hyperthermia, 5
  technique, 470
Doss, J. D., 469
Doubling time, 70
Drug metabolism, 424
Drugs
  and heat sensitivity of cells, 454–455
  and increased metastases, 498
Drug synergism, 424–429
Drug therapy, 425–429

gastrointestinal effects of, 428
Drug tolerance, heat-induced, 157
DS carcinosarcoma of rats, vascular response to heat of, 197
D-23 carcinoma, derivation of, 500
Dye exclusion test, 59
Dyes, blood-borne in tumors, 14

Edema, after heating, 193
Ehrlich ascites cells, 488, 520–523
  derivation of, 500
  and inhibition of tumor growth, 375–376
Electrical safety, during ultrasonic hyperthermia, 346, 350–352
Electrocautery, 316
Electrocoagulation, 32–33, 508–509
  in clinical cancer trials, 322
  in humans, 513
  and rectal carcinoma, 322
Electrode calibration, 210
Electrofulguration, 316
  in clinical cancer trials, 322
Electrolyte
  imbalance, 429
  support, 420–421
Electromagnetic energy
  advantages of, 338
  equation for, 284–285
Electromagnetic fields
  effects on tissues, 281
  for electrotherapy, 550
Electromagnetic hyperthermia, 1, 4–5, 279–301
  dosimetry in, 5
  mechanisms of, 4–5
  methods of, 286–292
  quantitation of, 283–284
Electromagnetic interference, 225–226
  in microwave coupling, 234
  in radiofrequency coupling, 234
Electromagnetic radiation, 334–340
Electromagnetic waves
  biological effects of, 5
  penetration characteristics of, 5
Embolization, 524–525
Embolization, tumor
  technique for, 114
  and tumor blood flow, 114
EMT-6 tumor, changes after hyperthermia, 372
Endocrine changes, after radiation exposure, 553–554
Endothelial proliferation factor (EPF), 188
Endotoxins, 402
Energy requirements for hyperthermia, 340

Environmental factors, and hyperthermic response, 178–180
Enzymes, thermal inactivation of, 461–462
Enzyme substrate reaction, 47
Ependymoblastoma of C57BL mice, vascular response to heat of, 197
Epithelioma, Guérin, 512
Erysipelas, 1
Erythema, after heating, 192–193
Escherichia coli, 514
   thermal sensitivity of, 454
Esophageal neoplasms, therapy for, 410
Ethanol, 2
Experimental models, guidelines for, 493
Exponential horn insonator, 345, 359
Extracellular fluid, of tumors, 525–526
Extracellular pH, 78, 147, 178–180, 181
   after heating, 104
   measurement of, 79
Extracorporeal circuit (ECC), 412–418
   complications with, 417–418
Extracorporeal circulation, 377. See also Perfusion hyperthermia
   of heated blood, 3
Extracorporeal heat exchanger, 403
Extracorporeal induction
   procedure for, 419
   technique for, 412–418
   treatment duration of, 433
Eyring extension. See Arrhenius equation

Faraday cage, 23
Federal Communications Commission (FCC), 23, 25
Fed plateau cultures, 148–149
Feedback controls, in thermal regulation, 19–21
Feedforward controls, in thermal regulation, 19–21
Ferromagnetic seeds, 272–274
Fever, after systemic hyperthermia, 403
Fiberoptic temperature probe, 226–227. See also *specific types*
Fibrosarcoma, 512–513
   thermal sensitivity of, 370–372
Fibrotic scar formation, 171–173
Filipin, 2
Flash [H]-TdR labeling, 97
Flash labeling index, 83–84
Flexner-Jobling carcinoma
   early tests with, 319–320
   thermal sensitivity of, 370–372
Flow microfluorometry, 77

Fluid replacement, 420–421
Fluorescent-type temperature sensor, 227
Focused beams, 337
Focusing lenses, for ultrasonic hyperthermia, 345
Fractional distribution, of Rb, 88–90
Fractionated heating and radiation, cellular response to, 461–462
Fractionation, 150, 372, 433, 461–462, 482–484
   and thermal tolerance, 155–157
Fraction labeled mitoses (FLM) curves, 74
Frazier, Howard O., 434, 435
Frequencies, electromagnetic wave, 281–282
Freund's adjuvant, 502
Fructose diphosphate (FDP), 435

Galactose, and blood flow inhibition, 110
Gallium melting-point standard, 224
Gastric carcinoma, response to systemic hyperthermia alone, 403–404
Gelfoam particles, for tumor embolization, 114
Generator, low frequency RF, 469
Genetic heat resistance, 143–145
Gibbs' free energy, 48
Glass pH microelectrode, 210–211
Glioblastoma multiforme, 467
Glucose. See also Hyperglycemia
   and blood flow inhibition, 110–113
   concentration in tumor, 106–107
   consumption, 15, 21
   content in interstitial fluid, 451–452
   effect on lactate efflux from tumor, 116–118
   loading, 526
   use in tumors, 116–118
Glucose administration, effect on tumor physiology, 114–116
Glycolysis of tumors, 78
Golgi apparatus, 167
Granulocytes, 492
Green's function, 11
Growth fraction (GF), 70–71, 74–75, 151
   equation for, 75
Growth rate, 72–73. See also Tumor growth
Guérin epithelioma, 510, 512

Hazards of treatment, 441, 443
HB mammary carcinoma, 320
   tests with, 320–322

Heat boxes, 401
Heat dissipation, 334
Heat dose, effects on thermal gradients, 310
Heat equation, 258
Heat generation rates, 336–337
Heat-induced cell death, thermodynamic analysis of, 48–52
Heating, inadequate, 91–92
Heating cabinet, for systemic hyperthermia, 3
Heating patterns, 293–301
   difficulties in evaluation of, 24–25
Heating and radiation, cellular response to, 462–463
Heat reservoir, 316
Heat resistance, 174–178
   in mutant cells, 144–145
Heat sensitivity. See also Sensitization
   in malignant cells, 150–151
   phenomenon of increasing, 163
Heat sources, 16, 18
Heat transfer, 29–36, 336–337. See also Thermal conductivity and diffusivity
   countercurrent, in tissues, 11
   in hot-wax bath, 32
   major problems in analyzing, 10
   mechanisms of, 10–11, 16–21
   modeling, 340–341
   in radiofrequency heating, 36
Hemagglutination titer, 501
Hematologic changes, in microwave workers, 547
Hematologic effects of radiation, 554–557
Heparinization, 421
Hepatic toxicity, in systemic hyperthermia, 403
Hepatoma
   thermal sensitivity of Morris-5123, 370
   thermal sensitivity of Novikoff, 370–372
Heterogeneous heating, 92–93
Hewitt, H. B., 495
High-resistance lead wires, 226
Hinke electrode, 210–211
Histamine, 125
Histiocytes. See Mononuclear phagocyte system
Histopathologic effects, of hyperthermia, 163–182, 211–218
Histopathologic examination, for heat-induced necrosis, 2
Histopathologic reaction to hyperthermia
   early, 164–171
   late, 171–173
Hodgkin's disease, 528
Hormonal therapy, 525

Host response, review of, 529–532
Host tolerance, 4
Host-tumor relationship, 65, 519
Host-tumor-therapy triad, 532
Hot-air incubator, 28–30
Hot-water bath, 28–30, 149
Hot-wax bath (Pettigrew technique), 31–32
[H]-TdR. *See also* Thymidine, tritiated
   flash labeling with, 75
   labeled mitosis technique, 74, 77
   repeated labeling, with, 75
Human cancer cells, 370
Humans
   immune response in, 513–516
   ultrasonic hyperthermia in, 361–365
Humoral response, 491–492, 501–502
Hydrogen detector, 211
Hydrogen ion concentration, 79
   response of cells to, 450–452
Hyperalimentation, 429
Hyperbaric oxygen chambers, 467
Hyperglycemia, 119–121
   and blood flow inhibition, 110–113
   effect on extracellular pH, 87–88
   and tumor blood flow, 116–118
   and tumor cure rate, 118–119
Hypernephroma
   radiofrequency therapy for, 323–324
   therapy for, 409–410
Hyperthermia. *See also specific techniques*
   biophysical data required in, 236
   cardiotoxic effects of, 429
   categories of, 29
   and cell killing in vivo, 99–100
   clinical implications of, 181–182
   combined with chemotherapy, 424–429
   combined with ionizing radiation, 373–375, 435–436
   control systems, 235
   dosage, 433–435
   dose schedule, 158–159
   dosimetry, 5, 470–473
   early history of, 1
   effect on cell cycle, 498
   effects on cell kinetics, 76–77
   effects on drug action, 424–429
   effects on experimental tumors, 370–372
   effects on local blood flow, 212
   effects on normal tissues, 164, 333–334
   effects on pH in vivo, 104–109
   effects on solid tumors, 164–178
   effects on tumors, 333–334
   energy requirements for, 340
   extreme, and tumor changes, 217–218
   hazards of, 433–434, 441, 443
   immunologic effects of, 437–438
   inadequacy of, 66
   instrumentation, 233–236
   interaction with drugs, 157–159
   interaction with radiation, 155–157, 159
   intermediate, and tumor changes, 216
   local, 29
   main methods of, 108
   mechanism of tumor response to, 524–527
   mild, and tumor changes, 215–216
   molecular effects of, 453–455
   patient management during, 418–430
   patient preparation for, 419
   physiologic responses to, 207–220
   potentiators of, 109–126
   precautions for use of, 410
   pretreatment evaluation for, 410–412
   pulse, 435
   with radiation, 447–463
   radiowave, 315–330
   rationale for use of, 157–159
   reactions to, 328–329
   regional, 29, 369–395
   response to, 164–178
   targets of, 371–372
   techniques, 28–36, 233–236 (*see also specific techniques*)
   temperature range in regional, 371
   temperature restriction, 66
   therapeutic advantage of, 207
   treatment duration of, 433–435
   treatment schedules for, 5–6, 316–330 passim
   with ultrasound, 5, 333–365
   vascular changes from, 192–198
   whole-body, 29 (*see also* Total-body hyperthermia)
Hyperthermic antiblastic perfusion, 384, 387
Hyperthermic cell destruction
   mechanisms of, 180–181
   in vivo, 180–182
Hyperthermic effect, potentiation of, 372–375
Hypocalcemia, 422–423
Hypokalemia, 420
Hypomagnesemia, 403
Hypophosphatemia, 403, 420, 422
Hypothermia, 557
Hypovolemia, 420–421
Hypoxia, 93, 468
   effects of chronic, 180
   and thermal sensitivity in vivo, 124
   and thermal tolerance, 148
   of tumor, 106–107, 449–452
   tumor cell, 467
Hypoxic cell radiosensitizers, 467

IEEE Transactions Microwave Theory and Techniques, 280
Immune competence
   monitoring protocol, 506
   in rabbits, 500–502
   in rats, 502, 509–511
   in vitro assays for, 489–490, 500–501
   in vivo assays for, 490–493, 501–502
Immune complexes, 525
Immune response, 4, 32, 65
   to heat, 375–377
   in humans, 513–516
   to hyperthermia, 487–535, 505–509
Immunization, 492
Immunofluorescence assay, 501
Immunogenicity, altered tumor, 520–524
Immunologic effects of hyperthermia, 437–438
Immunologic effects of radiation, 554–557
Immunologic monitoring, 206
Immunologic monitors, 491–492
Immunology procedures, 500–502
Immunoprophylaxis, 494
Immunostimulation, effects of heat and, 4
Immunosuppression, 503–504
   effects of heat and, 4
   effects on tumor response to heat, 511–512
Implantation of a heat source, 32–33
Indirect cell killing, *See* Cell killing, delayed
Induced thermal resistance, 146
   See also Thermal tolerance
Induction coils, 287–288
Inductive heating mode, 25–27. *See also* Radiofrequency heating
Indwelling electrodes, for radiofrequency coupling, 234
Infectious diseases, and hyperthermic treatment, 164
Inflammatory response, 124–125, 527–529
Insonation, 345–346
Insonation head, 345
Insonator, positioning of, 346–347
Intensity modulation, 338–340

Intensity scans, 348
Interofermetric technique, 299–231
Interferon, 281, 490
Interstitial hyperthermia, 316
   and phantoms, 270, 272–274
Interstitial thermoradiotherapy. See Thermoradiotherapy, interstitial
Intracellular pH, 403
   calculation of, 87, 88
   determination of, 86–88
   during incubation period, 372
   measurement of, 79–81
Intratumor pH, after heating, 201
Invar rod mechanism, 229–230
Irreversible destruction, 47–48
Isotope labeling, 500
Isotopes
   in blood flow measurement, 88–91
   in determination of intracellular pH, 88

Jehovah's Witnesses, 443–444

Kaposi's sarcoma, radiofrequency therapy for, 323–324
K cells, 492
Ketone bodies, 435
KHJJ carcinoma, tests with, 320–322
Krebs-Mehrschritt Therapie, 120

Labeled mitosis technique, 74
Labeled precursors, 500
Labeling, 74–77. See also specific types
   radioisotope, 76
Labeling index, 94–100
Laboratory assessment, during hyperthermia, 424–425
Lactic acid, production in tumors, 15, 78
Leimyosarcoma, 472
Leukocytosis, 516
Light-emitting diode (LED), in fiber-optic probes, 226–227
Limb salvage protocol, 389
Linear energy transfer (LET) radiation, 459, 467, 479
Local hyperthermia, 233–235, 504–509
   clinical methods of, 316
   damage from, 164
   and enhanced metastasis, 517
   followed by total-body hyperthermia, 506–509
   and immune response, 505–509
   for large tumors, 324
   physical aspects of, 279–301, 305–312
   vs total-body hyperthermia, 518–519
Local-regional hyperthermia, 377–395

Log phase cultures, 148–149
Longmire operation, 411
Los Alamos Scientific Laboratory, 469
L-phenylalanine mustard, 425
Luciferase, 47–48
Lumped parameter models, 17–21, 29
   compartmental approach, 17–18
   whole-body, 17–18
Lung carcinoma
   radiofrequency therapy for, 323–324
   response to systemic hyperthermia alone, 403–404
Lymphadenectomy, and perfusion hyperthermia, 384, 386
Lymphocyte heterogenicity, 489
Lymphocytes, 492, 516
   and duration of heat, 518
   killer, 492
   reaction to heat of, 376
   stimulation of, 500–501
Lymphoid cell infiltration of tumor, 527–529
Lymphokines, 490
Lymphoma
   Burkitt's, 492
   Moloney YC-8, 512–513
   non-Hodgkin's, 519
Lymphoma cutis, radiofrequency therapy for, 323–324
Lymphopenia, 516
Lymphoreticuloendothelial system, 528
Lysosomal activity, role in cell killing of, 180–181
Lysosomal enzyme activation, 120–121
Lysosomal reaction, in tumor cells, 170–171, 179
Lysosome function, after hyperthermia, 372
Lysosome labilization, 119–121

MacCarty, W. C., 527
Macro, in computer systems, 348
Macrophages, 64, 490, 492, 512–515, 527. See also Mononuclear phagocyte system
Magnetic induction heating
   and phantom use, 270–274
   power disposition equation in, 270
Magnetic-loop induction hyperthermia, 5, 305–318
   applicator, 36, 305–306
   local, 305–312
Magnetrode™, 317–318
   and animal tissue phantoms, 307–308
   for deep subcutaneous tumors, 323

   design, 305–306
   heat distribution by, 307–311
   for internal tumors, 323
   magnetic field distribution of, 306–307
Mammary adenocarcinoma of mice, vascular response to heat of, 194–195
Massachusetts Institute of Technology, 361
Mathematical models of thermal systems, 9–38
Maxwell-Boltzmann equation, 48
MC-7 sarcoma, derivation of, 500
M. D. Anderson Hospital, 435
Mechanism of action of heat, 524–527
Melanoma, 467, 473–475, 481
   antibody titer in human, 492
   B16, 516
   combination therapy for, 425
   in humans, 513–514
   and immune response, 533–534
   microwave therapy for, 322
   perfusion therapy for, 386–387, 391, 392
   radiofrequency therapy for, 322–324
   recurrence after regional perfusion, 69–70
   response in humans to heat and radiation by, 375
   response to systemic hyperthermia alone by, 403–404
   therapy for, 429
Melanomatosis, 419
Melphalan, 3, 384, 513
Membrane permeability modulators, 124
Membrane proteins, 153, 155
Metabolic acidosis, 425
Metabolic heat generation, 15
Metabolic stress, 421–424
Metabolism, cell
   heat-induced changes in, 453–455
Metal casings of radium needles, 470
Metaphase arrest technique, with vincristine, 85–86
Metastasis
   enhanced in animals, 516–519
   enhanced in humans, 519–520
   and immune response in humans, 513–516
   increased by drugs, 498
   increased after heating, 376
   regression in animals, 507–513
   spontaneous regression following surgery, 527
Metastatic carcinoma, microwave therapy for, 322
Microcalomel electrode, 210

Microcirculation, changes after heat, 219
Microcirculation of tumor, and thermal response, 107–109
Microcytotoxicity tests, 500–501, 508
Microelectrodes, 79–80
 alternative, 210–211
Microenvironment, tumor, 77–78
 and potentiation of hyperthermia 110–125
 and thermal response, 107–109, 449–452
Microflow, 211
Micropore chambers, 79
Microprocessor-controlled feedback-loop, 402
Microsensor amplifier, 210
Microwave applicators. See Applicators, microwave
Microwave coupling, 234
 hazards of, 234
Microwave diathermy
 apparatus, 32
 standard, 261
Microwave energy absorption, measurement of, 5
Microwave heating, and phantom use, 263–267
Microwave phase array, 5
Microwave radiation, 4–5, 23–25
 animal and clinical studies with, 315–330
 bioeffects of, 545–561
 and cataracts in animals, 557–558
 chromosomal effects of, 559–560
 in clinical cancer trials, 322
 difficulties in evaluation of, 24–25
 effects on animals of, 551–560
 effects on humans of, 546–551
 FCC limitations and, 23
 hazards of occupational exposure to, 546–551
 hematologic effects of, 554–557
 immunologic effects of, 554–557
 mortality in animals exposed to, 551–553
 mutagenic effects of, 559–560
 and neuroendocrine alterations, 553–554
 for ocular diseases, 550
 properties in biological tissue, 281–283
 temperature distribution with, 35–36
 and teratogenesis, 558–559
 thermal energy absorbed during, 21
Microwave syringe, 316
Microwave waveguides, 316
Miniature glass electrodes, 79, 86
Misonidazole, 373

MIT Clinical Research Center, 361
Mitochondria, reaction to hyperthermia of, 372
Mitomycin C, 523
Mitosis, after hyperthermia, 166
Mitotic arrest techniques, 95–100
Mixed bacterial toxins (MBT), 29, 33, 63–64, 514–515. See also Coley's toxins
Mode-mode coupling hypothesis, 236
Molten wax, in systemic hyperthermia, 3
Monode, 287–288
Mononuclear cells. See Mononuclear phagocyte system
Mononuclear phagocyte (reticuloendothelial) system, 64, 490, 512–513
Morphological effects of hyperthermia, 164–178
Morphology of tumor microcirculation, 207–220
Mortality in extracorporeal induction, 439
Multiple-beam superposition, 335–336
Multiple daily fractionation scheme, 482–484
Multiple-dose heating, and cell kinetics, 109–110
Multiple-dose treatment, factors in, 72
Murine tumors, and ultrasonic hyperthermia, 357–358
Mutagenic effects of radiation, 559–560
Myasthenia gravis, 444
Mycosis fungoides, 481, 482
 radiofrequency therapy for, 323–324

Narcotic administration, 420
Nasopharyngeal carcinoma, microwave therapy for, 322
National Cancer Institute (NCI), 402, 432
National Council on Radiation Protection and Measurements, 284
Nausea, after systemic hyperthermia, 403
Necrosis, and tumor growth, 192
Needle electrodes, 470–473
Needle probes, 433
Neoplastic cells, thermal sensitivity of, 55–61
Neuroblastoma, treated by systemic hyperthermia alone, 403–404
Neurocirculatory syndrome, from occupational microwave exposure, 546–551
Neuroendocrine alterations, from radiation exposure, 553–554

Neutropenia, 429
Nitrogen mustard, and increased metastasis, 498
Nitrosoureas, with heat, 404
Noninvasive circumferential electrode, 5
Noninvasiveness, rationale for, 334–344
Noninvasive ultrasonic tomography, 231
Nonperturbable implantable transducers, 235
Nonperturbing temperature probes, 224, 226–227
Nonproliferating cells, 70–71
 and thermal tolerance, 148
Nonspecific cell-mediated immunity, 501
Normal cells, in hyperthermia, 181
Normal tissues
 effects of hyperthermia on, 164
 perfusion of, 108
Nosocomial infections, 418
Nucleic acid, biosynthesis in tumors after hyperthermia, 371
Nutrient deprivation, 468
 and thermal tolerance, 148

Occupational hazards, for workers exposed to microwave radiation, 546–551
Ohmic heating, 25. See also Radio frequency heating
Optical rotation sensor, 227
Osteosarcoma
 limb salvage protocol for, 389
 treated by perfusion, 387–391
Ouchterlony diffusion techniques, 501, 508
Outpatient follow-up, 430
Oxygenation, and effect on treatment, 76
Oxygen concentration, and cell killing, 148
Oxygen consumption, 15, 21
Oxygen deprivation, 459–461
Oxygen electrode, 210
Oxygen enhancement ratio (OER), 374, 479
Oxygen tension, 424
Oxygen ultra-microelectrodes, 210

Pain, 318, 319
 after perfusion, 384–386
 and tissue temperature, 286
 during ultrasonic hyperthermia, 363, 364, 365
Pancake coil. See Induction coil
Pancreatic neoplasms, therapy for, 409–410
Papillomatosis, bladder, 481

Paraffin, immersion in, 408
Paraffin in systemic hyperthermia, 402
Patient management during hyperthermia, 418–430
Peclet number, 12
Pennes, H. H., 11
Parameters of the bio-heat transfer equation of, 11–16
Pentobarbitone anesthesia, 99
Percentage labeled mitoses (PLM) curves, 74
Perfusion hyperthermia, 377–395
  of animal internal organs, 379–380
  of animal limbs, 378–379
  clinical applications of, 382–386
  clinical results of, 386–393
  criteria for, 382–383
  of extremities, 481
  with heated blood, 32
  methods of, 383–386
  survival rate following, 386–393
Pericardial effusions, management of, 410–411
Peripheral neuropathy, after systemic hyperthermia, 403
Pettigrew, Robert, 402
Pettigrew technique, 31–32
pH, 147–148, 468. See also Extracellular pH; Intracellular pH
  electrode, 211
  and hyperthermic lethality, 450–451
  and hyperthermic response, 121–124
  at inflammation site, 529
  influence on hyperthermic response, 459–461
  low, 64, 65, 78, 147–148
  measurement in vivo, 78–81, 86–88
  and physiologic response to heat, 214–215
  reduced, 178–180
  in tumor microenvironment, 78
pH, tumor, 450–452
  after heating, 198–199
  and hyperthermia, 104–106
  and tumor blood flow, 114–118
  and tumor cure rate, 118–119
Phantom carrier design, 307
Phantoms, 257–277, 470–473
  advantages of, 261
  animal tissue, and magnetrode, 307–308
  in applicator development, 261–263
  applications of, 263
  dead animal, in magnetrode, 308
  dual chamber, 260–261
  dynamic, 274–276
  and ferromagnetic implants, 272–274
  in interstitial heating, 270
  in magnetic induction heating, 270–274
  materials used in, 258–261
  in microwave heating, 263–267
  planar, 259–260
  in power deposition studies, 258
  in radiofrequency current heating, 267–270
  in SAR measurement, 292–293
  static, 258–274
  techniques in use of, 259–261
  theoretical considerations of dynamic, 274–275
  thermal patterns in animal tissue, 308
  in thermographic studies, 261–263
  use in ultrasound, 353
Phased array beamed steering, in ultrasonic coupling, 235
Phased array contact applicator, 292
  SAR measurement with, 301
Phospholipids, 153–155
Physical models. See Phantoms
Physical properties of human tissues and electromagnetic heating, 285–286,
Physiologic response to hyperthermia, 207–220
Physiology of tumor microcirculation, 207–220
Phytohemagglutinin, 500
Plane wave fields
  intensity of, 334–335
  temperature distribution in, 334–335
Plane wave ultrasound, 353–354
Plasma membrane
  and cell killing, 454
  as hyperthermia target, 371
Plateau phase cells, 148–149
  heat sensitivity of, 180
Plating, cell, 60
pO, 93
  after heating, 104
Polarization phenomena, 281
Polarograms, 210–211
Polyamine biosynthetic enzymes, 153
Polyamines, 147
  intracellular levels of, 153
  and killing of thermotolerant cells, 153–155
  as modulators of permeability, 153
  role in cell survival, 153
  synergistic effects of, 372–375
Polyenic antibiotics, 372

Polymorphonuclear leukocytes (PMN), 437
Pondville Hospital, 361
Portal circulation, heating of, 434
Posthyperthermia care, 429–430
Potentiators of hypothermia, 109–126
Power deposition
  patterns, 258, 261
  preferential heating and, 272
Power requirements, determination of, 337
Probes. See also specific types
  for temperature measurement, 35
Progression to regression phenomenon, 430
Proliferating cells, 70–71, 73
Protein biosynthesis, inhibition of, 373
Proteins
  biosynthesis in tumors after hyperthermia, 371
  reaction to hyperthermia, 372
  synthesis inhibition, 146
Pulmonary fibrosis, 428
Pulmonary function, adequacy of, 410–411
Pulse hyperthermia, 435
Pyrex® glass pipette, 210–211
Pyrexia, 514–515
Pyrogenic agents, 29, 33, 63

Qualitative selectivity, 58
Quantitative selectivity, 58
Quasi-plateau, 210

Radiation, 29. See also specific types
  bioeffects of, 545–561
  and cataracts in animals, 557–558
  chromosomal effects of, 559–560
  combined with hyperthermia, 435–436
  damage and cell repair processes, 374
  dosimetry, 545
  effects when combined with heat, 447–463
  fractionated dosage of, 479
  hematologic effects of, 554–557
  immunologic effects of, 554–557
  and influence of dose rate, 459
  infrared, 31
  low-dose-rate, 468
  mutagenic effects of, 559–560
  synergistic effect with hyperthermia, 201
  and teratogenesis, 558–559
  visible, 31
Radiation pressure gauge, for ultrasonic hyperthermia, 346

Radiation-type applicators, SAR measurement with, 298–299
Radioactive micospheres, 191
  in study of perfusion rates, 14
Radioactive tracers, for measurement of blood flow, 190
Radiofrequency coupling, 233–234
Radiofrequency currents
  and phantom use, 267–270
  thermal energy absorbed with, 21
Radiofrequency radiation, 25–27, 83, 545–561
  bioeffects of, 545–561
  and cataracts in animals, 557–558
  chromosomal effects of, 559–560
  in clinical cancer trials, 322–329
  counterindications for therapy, 550
  effects on animals, 551–560
  effects on humans, 546–551
  FCC regulations on, 25
  heat transfer with, 36
  hematologic effects of, 554–557
  immunologic effects of, 554–557
  mortality in animals treated with, 551–553
  mutagenic effects of, 559–560
  and neuroendocrine alterations, 553–554
  techniques for, 25
  temperature distribution with, 36
  and teratogenesis, 558–559
Radiofrequency waves, 4–5
Radioistotope labeling. *See* Labeling
Radioisotopes, 88–91
Radioprotective effect, 555
Radiosensitivity, 455–458
  of cells, 448
Radiosensitization
  mechanism of thermal, 461–462
  and thermotolerance, 155–157
Radiosensitizing drugs, 373
Radio-stimulating effect, 374
Radiotherapy, interstitial
  engineering considerations in, 469–472
Radium needles, 470
Rate of heating, and thermal tolerance, 148–149
Rate theory, 47
Raytheon Magnetron™, 211
Reaction kinetics, 342
Recurrence, tumor, 165
  pattern and mechanism of, 173–178
Reference electrode, 211
Reflection, of ultrasonic energy, 343–344
Refraction
  of acoustic waves, 343–344
  of electromagnetic waves, 343

Regional hyperthermia, 65–66
  clinical methods of, 316–318
Regional perfusion hyperthermia, 369–395
  in humans, 513–514
  of limbs, 2, 377–379, 383–395
Regrowth analysis, 77
Relative biological effectiveness (RBE), 374
Renal function, adequacy of, 411
Repeated [H]-TdR labeling, 84–85, 95–100
Repopulation, 99–100
Residual current, 210
Resistance sensor, 225
Resistant subpopulations of cells, 109–110
Resistive component, 470
Respiratory insufficiency, 442–443
Response
  anamnestic, 501–502
  cell-mediated, 501
  cellular, 505
  humoral, 501–502, 505–506
Reticuloendothelial system, 528
Reversible reaction, 48
Révész effect, 519
Rhabdomyosarcoma, treated by perfusion, 392
RNA
  biosynthesis in tumors after hyperthermia, 371
  synthesis inhibition, 171
Rodents as experimental models, 493–497
Rubidium, in blood flow measurement, 88–91

Safety of treatment, 58–59
Salmonella, 514
SAR. *See* Specific absorption rate
Sarcoma, 473–475. *See also specific types*
  canine osteogenic, 518
  EMT-6, 320–322, 497
  and immune response, 492–497, 533–534
  Jensen's, 370–372
  Kaposi's, 481
  MC-7, 81, 500, 512
  microwave therapy for, 322
  mouse mammary, 151
  180, thermal sensitivity of, 370
  osteogenic, 322, 481
  recurrence after regional perfusion, 69–70
  response to systemic hyperthermia alone, 403–404
  Rous fowl, 320
  soft tissue, treated by perfusion, 391–392

  therapy for, 428
  thermal sensitivity of, 370–372
  Yoshida, 81, 500, 517
Scatter, of ultrasonic energy, 343–344
Schwarz, G., 479
SCK tumor, in study of tumor microcirculation, 209
Seizures, after systemic hyperthermia, 403
Selective thermal sensitivity, 2–3
Selective tumor heating
  clinical methods of, 316–318
  damage from, 393
Semiconductor temperature sensor, 227
Seminoma, testicular, 528
Sensitization, 107, 118
Sensitizers, 479–480. *See also specific types*
  cytotoxic agents, 467
Sensitizing effect, of heat and radiation, 455–458
Sepsis, 411, 438
Serotonin, 4, 64, 125
Shortwave applicators. *See* Applicators, shortwave
Shortwave-induced hyperthermia, 2–3
Showering of cancer cells, 519
Silica, 512
Sinus histiocytosis, 528
Skin induction technique. *See* Systemic hyperthermia
Skin testing, 491
Smeared focus lenses, 341
Sodium oxamate, 3
Solid tumors, morphology of, 164–178
Sonic propagation speed, variations in, 228–229
Space-suit technique, 30
Specific absorption rate (SAR), 21, 258
  equations for, 284
  measurement of, 292–293
Specific cell-mediated immunity, 501
Spermatogenesis
  inhibition of, 164
  and radiofrequency radiation exposure, 550–551
Split-graded heat dose, 142–143, 148
  and radiation, 156
Squamous cell carcinoma, in study of tumor microcirculation, 209
Squamous cell carcinoma of the head and neck, radiofrequency therapy for, 323–324
Squamous cell carcinoma of Syrian hamsters, vascular response to heat of, 196–197

Staff requirements, 418
Step-down heating (SDH), 50, 64
Stepper motors, 346
Steroids, in posthyperthermia period, 430
Stop effect, 382, 389
Strauss, A. A., 488, 503, 508–509, 513, 529
Streptococcus pyogenes, 514
Stress, and tumor growth, 99
Sublethal tumor irradiation, 498
Superficial tumors
   microwave therapy for, 322
   response to heat of, 324–325
SupraHT, 434
Surface heating, 28–29
Surgery, role in treatment of, 441–442
Surgical procedures, healing after, 412
Survival curve analysis, 143
Survival rates, 439
Sympathectomy, and perfusion hyperthermia, 384–386
Synergism, 402
   between heat and chemotherapeutic drugs, 372–375
   between heat and radiation, 372–375, 455–462
Systemic factors, in thermal response, 125–126
Systemic hyperthermia, 3, 65–66, 149, 401–404, 407–444. *See also* Extracorporeal induction; Total-body hyperthermia; Whole-body hyperthermia
   background of, 401–402
   combined with regional heating, 435
   extracorporeal induction technique for, 412–418
   hepatic response to, 404
   history of, 407–408
   methodology for, 402–403

Target volume, 337–338
Temperature
   of blood, 431–432
   control in treatment area, 418
   effective range of, 431, 433–435
   elevated range of, 371
   maximum, with ultrasound, 342
   measurement of, 431–433
   monitoring, 82–83, 432
   pattern, 258
   stimulatory zone, 498
   in tumor vs normal tissue, 200
   uniformity, 338–340
Temperature dependence, of heat-induced cell killing, 49–52
Temperature distribution, 333–334
   equations for transient, 17–18
   in hot-wax bath, 32
   during hyperthermia, 28–36
   limitations on monitoring of, 35
   during normothermia, 32–33
   in perfused tumors, 32–33
   in plane wave fields, 334–335
   in radiofrequency coupling, 234
   in radiofrequency heating, 36
   in tumors treated by ultrasound, 352–365
Temperature-duration history, 333–334
Temperature field, identification in tissue, 10
Temperature intensity, and multiple beams, 335–336
Temperature measurement
   devices for, 5
   during ultrasonic hyperthermia, 346
Temperature probes, 83, 292. *See also specific types*
   calibration of, 224
   nonperturbing, 224
Temperature range, for tumor destruction, 69
Temperature regulation device (TRD), 412–418
Temperature-sensitive mutants, 141
Teratogenesis, and radiation exposure, 558–559
Therapeutic gain factor, 374
Thermal conduction, determination of, 10
Thermal conductivity
   concept of effective, 11
   and diffusivity, 9–38
   measurement by invasive probe, 12
   measurement by noninvasive probe, 12
Thermal convection, determination of, 10
Thermal death time, 66–70, 326–328
   defined, 68
Thermal dialysis, 434
Thermal dilution technique, for measurement of blood flow rate, 14
Thermal distribution
   in live dog, 309–311
   in live vs dead dog, 311
Thermal effects
   on normal extremities, 318–319
   on normal skin and subcutaneous tissue, 318
   on normal viscera, 319
Thermal enhancement ratio (TER), 374
   of hyperthermia and radiation combined, 482
Thermal history, 47–48
Thermal properties of human tissues, 284–286
Thermal properties of tumors, 11–14
Thermal regulation, 19–21
Thermal resistance, 64
   of normal tissues, 101–104
Thermal responses
   Arrhenius analysis of, 47–52
   by tumor size, 324
Thermal sensitivity
   of cell cycle stages, 372
   concepts and principles of, 66–81
   concepts and principles of in vitro, 58–60
   environmental factors for in vivo, 64
   of experimental tumors, 370–382
   history of in vitro investigations of, 55–58
   history of in vivo investigations of, 63–66
   of neoplastic cells in vitro, 55–61
   of neoplastic tissues in vivo, 63–129
   of noncancerous tissues, 329
   perspectives on, 126–128
   research methodology for in vitro, 60
Thermal systems, mathematical models of, 9–38
Thermal tolerance, 2, 51, 141–159
   and cell growth state, 148–149
   and cell transformation, 150–151
   characteristics of, 144–145
   decreased, 2
   defined, 142–146
   development of, 147, 151
   duration of, 144
   effect on radiosensitization of, 155–157
   environmental conditions for, 147–148
   factors affecting, 146–151
   and fractionation, 155–157
   vs genetic heat resistance, 143–145
   increase in, 66
   and initial thermal dose, 147
   and interaction of heat and drugs, 157
   nonheritability of, 141
   parameters of development of, 150
   and rate of heating, 149–150
   survival responses and, 142–143
   temperatures required for, 146–147
   and tissue differentiation, 150
   in vivo vs in vitro, 150

Thermal toxicity, 328–329
Thermistor, 5, 83, 225–226, 472
   catheter, 432–433
Thermochemotherapy, 3–4
Thermocouple, 5, 83, 225, 231, 472
   chromel-constantan, 346
Thermodynamic analysis
   in vitro, 49–51
   in vivo, 51–52
Thermodynamic formulas, historical background of, 47–48
Thermographic camera, 258–260, 470, 472
Thermography, 223–231
Thermojunction, location of, 343–344
Thermometry, 82–83, 223–231, 235, 431–433
   accuracy of, 223–224
   electromagnetic interference in, 225–226
   instrument design, 231
   with RF generators, 469
   resolution, 223–225
   spatial resolution in, 224–225
   system specifications for, 223
   temporal resolution of, 225
   in ultrasound, 352–353
Thermoradiotherapy, 3. *See also* Hyperthermia; Radiation; Synergism
   cellular kinetics of, 447–463
   clinical, 479–487
   clinical trials of, 480–484
   molecular effects of, 453–455
   molecular kinetics of, 447–463
   rationale for clinical, 480
Thermoradiotherapy, interstitial, 272–274, 467–476
   clinical results of, 473–475
   complications from, 474
   dosimetry, 472–473
   rationale for, 468–469
Thermosensitivity. *See* Thermal sensitivity
Thermotolerance. *See* Thermal tolerance
Thermotolerant cells, mechanisms for killing, 151–155
Thiotepa, 373
Third International Symposium on Cancer Therapy by Hyperthermia, Drugs and Radiation, 281
Thomas electrode, 210–211
Thomas shunt, 418–419
Threshold temperature, for vascular changes, 193
Thrombocytopenia, 427, 429
Thymidine labeling, 77
Thymidine, tritiated, in DNA after hyperthermia, 371, 373

Thymoma, 444
Thyroid function, after radiation exposure, 553–554
Thyroid neoplasm, radiofrequency therapy for, 323–324
Time-temperature relationship, 68–70
Tissue oxygen tension (TpO), 210, 212
Tissue temperature
   and blood flow during heating, 199–200
   characterization of, 229–231
Total-body hyperthermia, 66. *See also* Whole-body hyperthermia
   and enhanced metastasis, 517–519
   in humans, 514–516
   and immune suppression, 504
   and inadequate heating, 92
Transducer-driving electronics, 346
Transducers
   multiple, 342–343
   specifications for, 345
Transition state concept, 48
Translation and rotation subsystem, 346–347
Transplantation
   of animal tumors, 81–82
   guidelines for research on, 493–496
   and immunity, 523
   site and tumor control, 522
Treatment area design, 418
Trypanosoma cruzi, 514
Trypanosoma extracts, effect on cancer cells of, 514
Trypsinization, 60
Tumor
   age distribution of cells in, 73–74
   breakdown products of, 92, 529
   control, 72–73
   cure, 72–73
   doubling time, 70, 72–73
   fatal lysis of, 404
   growth analysis of, 72–77
   growth rate, 72–77, 93–100
   gynecologic, 467
   as heat reservoir, 102
   hemisection of, 98–99
   histology posttreatment, 436–437
   immunology, 489–493
   infarction, 524–525
   ligature, 524
   local control of, 467–476
   measurement of response of, 77
   measurement of thermal properties of, 10–11
   and metastatic growth, 65
   microenvironment of, 64

   nutrient levels and hyperthermia, 106–107
   pathology after treatment, 436–437
   pH and hyperthermia, 104–106
   pH measurement in vivo, 78–81
   physiologic response of, 21
   recurrence, 498–499
   regression, 514–515
   regrowth time, 73
   response to combination therapy, 436
   response in humans to ultrasound, 363–365
   response to hyperthermia, ✓ 320–329, 526–527
   response to ultrasound, 357–362
   response to whole-body hyperthermia, 403–404
   selective heating of, 200
   size and thermal response, 324, 411, 430
   subpopulations of, 532
   temperature in, 32–33, 108
   temperature profile in, 27–28
   temperature during ultrasonic hyperthermia, 364
   thermal sensitivity of, 449–452
   used in research, 495
Tumor, human
   in study of tumor microcirculation, 209
   vascular response to heat of, 197–198
Tumor acidity, effect on vascular occlusion of, 199
Tumor angiogenesis factor (TAF), 77, 188
Tumor blood flow, 65, 532
   and tumor volume, 92–93
Tumor cell
   altered immunogenicity of, 520–524
   degradation products of, 525
   kinetics and thermal sensitivity, 93–94
Tumor cure rate, after hyperglycemia and hyperthermia, 118–119
Tumor extract, 3M KCl, 500–502, 505
Tumor heating, adequate, 494–495, 497–499
Tumor heating capacity, in clinical trials, 324
Tumor histology, in clinical trials, 324
Tumor-host relationship, 494
Tumor interstitial fluid (TIF), 78
   and pH measurement, 79
Tumor lethal temperature, 498

Tumor membranes, as hyperthermia targets, 371
Tumor microcirculation
  analysis of data on, 212–218
  and hyperthermia, 207–222
  research methodology of, 209–212
Tumor microenvironment, and thermal sensitivity, 100–126
Tumor-necrotizing fraction, 109
Tumor transplantation. See Transplantation
Tumor vasculature
  architecture of, 189
  central type of, 189
  development and characteristics of, 188–190
  morphologic classification of, 189
  patterns of, 189
  response to heat of, 404
Tumor volume, 72–73
  effect of hyperthermia on, 94–100
  equation for, 72
  and thermal response, 91–93

Ultrasonic computed tomography, 227–231
Ultrasonic coupling, 234–235
Ultrasonic energy absorption, measurement of, 5
Ultrasonic hyperthermia, 333–365
  in combination therapy, 365
  fractionation regimens for, 150
  objectives of treatment in humans, 361–362
  patient eligibility for, 362
  procedures in humans for, 362–363
  reactions in humans, 362
  responses in humans, 363–365
  results in dogs, 360–361
  technique for, 348–351
  treatment plan in humans, 362
  of tumors in humans, 361–365
Ultrasonic parameters of tissues, 231
Ultrasonic temperature mapping, 227–231, 235

Ultrasound, 5, 22–23. See also Ultrasonic hyperthermia
  advantages of, 338
  advantages of steered, focused, 356
  applicability of, 344
  constraints on use of, 22
  focused, 339–340
  instrumentation for, 344–348
  maximum temperature with, 342
  rationale for use of, 333–354
  safety features of, 351–352
  steered, focused, 354–357
  temperature distribution in, 33–34
  thermal energy absorbed from, 21–23
  thermometry, 352–353
  and tissue interfaces, 343
Ultrastructural changes, in tumor after hyperthermia, 167–171
Ultratherm 608™, 381
University of Mississippi Medical Center, 408
Unknown primary, radiofreqeuncy therapy for, 322–324
Urine phosphate level, after AV shunt, 403

Vascular changes by hyperthermia
  in animal tumors, 194–197
  in normal tissues, 192–194
Vascular damage from heating
  mechanisms of, 198–199
  and tumor cell survival, 200–201
Vascular permeability, 193
Vascular physiology, 190–192
Vasoconstrictor drugs, and tumor blood flow, 114
Vasodilator drugs, and tumor blood flow, 114
Vincristine arrest, 76, 85–86
Visceral tumors, response to heat of, 325
Voltage-controlled amplifier, 351
Volume heating, 28–29
Von Ardenne, M., 117, 119–120
VX-2 carcinoma, derivation of, 499

Walker-256 carcinoma, early tests with, 320
Walker-256 carcinoma of SD rats, vascular response to heat of, 194
Warren, Stafford, 401
Warm-water bath, 2, 82, 224, 481, 498
  immersion technique, 82–83
Warm-water circulating suit, 3, 402
Wavelength, choice of, 337–338
Wax. See Paraffin
Weak acids and bases, 80–81
Westermark, Nils, 316, 488
Whole-body hyperthermia, 29–30, 159, 233. See also Total-body hyperthermia
  contraindications for, 409
  dosage and effectiveness of, 433–435
  effects compared, 408
  effects of perfusion on dog central nervous system, 381
  eligibility for treatment by, 408–410
  and immune response in humans, 515–516
  methods of, 408
  perfusion in animals, 380–382
  perfusion in humans, 393–393
  perfusion technique for, 380–382
  with radiation, 404
  safety of, 408–409
  system criteria, 408
  tumor responses with, 403–404

Xenon (Xe) clearance technique, in blood flow measurement, 90–91
X-Y-Z translator tables, 346

Yale-New Haven Hospital, 361
Yoshida sarcoma, derivation of, 500